THE ELECTRONIC COMMUNICATIONS PROBLEM SOLVER®

REGISTERED TRADEMARK

**Staff of Research and Education Association,
Dr. M. Fogiel, Director**

Research and Education Association
505 Eighth Avenue
New York, N. Y. 10018

THE ELECTRONIC COMMUNICATIONS PROBLEM SOLVER®

Copyright © 1984 by Research and Education Association. All rights reserved. No part of this book may be reproduced in any form without permission of the publisher.

Printed in the United States of America

Library of Congress Catalog Card Number 84-61814

International Standard Book Number 0-87891-558-3

PROBLEM SOLVER is a registered trademark of
Research and Education Association, New York, N.Y. 10018

WHAT THIS BOOK IS FOR

Students have generally found electronic communications a difficult subject to understand and learn. Despite the publication of hundreds of textbooks in this field, each one intended to provide an improvement over previous textbooks, students continue to remain perplexed as a result of the numerous conditions that must often be remembered and correlated in solving a problem. Various possible interpretations of terms used in electronic communications have also contributed to much of the difficulties experienced by students.

In a study of the problem, REA found the following basic reasons underlying students' difficulties with electronic communications taught in schools:

(a) No systematic rules of analysis have been developed which students may follow in a step-by-step manner to solve the usual problems encountered. This results from the fact that the numerous different conditions and principles which may be involved in a problem, lead to many possible different methods of solution. To prescribe a set of rules to be followed for each of the possible variations, would involve an enormous number of rules and steps to be searched through by students, and this task would perhaps be more burdensome than solving the problem directly with some accompanying trial and error to find the correct solution route.

(b) Textbooks currently available will usually explain a given principle in a few pages written by a professional who has an insight in the subject matter that is not shared by students. The explanations are often written in an abstract manner which leaves the students confused as to the application of the principle. The explanations given are not sufficiently detailed and extensive to make the student aware of the wide range of applications and different aspects of the principle being studied. The numerous possible variations of principles and their applications are usually not discussed, and it is left for the students to discover these for themselves while doing exercises. Accordingly, the average student is expected to rediscover that which has been long known and practiced, but not published or explained extensively.

(c) The examples usually following the explanation of a topic are too few in number and too simple to enable the student to obtain a thorough grasp of the principles involved. The explanations do not provide sufficient basis to enable a student to solve problems that may be subsequently assigned for homework or given on examinations.

The examples are presented in abbreviated form which leaves out much material between steps, and requires that students derive the omitted material themselves. As a result, students find the examples difficult to understand--contrary to the purpose of the examples.

Examples are, furthermore, often worded in a confusing manner. They do not state the problem and then present the solution. Instead, they pass through a general discussion, never revealing what is to be solved for.

Examples, also, do not always include diagrams/graphs, wherever appropriate, and students do not obtain the training to draw diagrams or graphs to simplify and organize their thinking.

(d) Students can learn the subject only by doing the exercises themselves and reviewing them in class, to obtain experience in applying the principles with their different ramifications.

In doing the exercises by themselves, students find that they are required to devote considerably more time to electronic communications than to other subjects of comparable credits, because they are uncertain with regard to the selection and application of the theorems and principles involved. It is also often necessary for students to discover those "tricks" not revealed in their texts (or review books), that make it possible to solve problems easily. Students must usually resort to methods of trial-and-error to discover these "tricks", and as a result they find that they may sometimes spend several hours to solve a single problem.

(e) When reviewing the exercises in classrooms, instructors usually request students to take turns in writing solutions on the boards and explaining them to the class. Students often find it difficult to explain in a manner that holds the interest of the class, and enables the remaining students to follow the material written on the boards. The remaining students seated in the class are, furthermore, too occupied with copying the material from the boards, to listen to the oral explanations and concentrate on the methods of solution.

This book is intended to aid students in electronic communications in overcoming the difficulties described, by supplying detailed illustrations of the solution methods which are usually not apparent to students. The solution methods are illustrated by problems selected from those that are most often assigned for class work and given on examinations. The problems are arranged in order of complexity to enable students to learn and understand a particular topic by reviewing the problems in sequence. The problems are illustrated with detailed step-by-step explanations, to save the students the large amount of time that is often needed to fill in the gaps that are usually found between steps of illustrations in textbooks or review/outline books.

The staff of REA considers electronic communications a subject that is best learned by allowing students to view the methods of analysis and solution techniques themselves. This approach to learning the subject matter is similar to that practiced in various scientific laboratories, particularly in the medical fields.

In using this book, students may review and study the illustrated problems at their own pace; they are not limited to the time allowed for explaining problems on the board in class.

When students want to look up a particular type of problem and solution, they can readily locate it in the book by referring to the index which has been extensively prepared. It is also possible to locate a particular type of problem by glancing at just the material within the boxed portions. To facilitate rapid scanning of the problems, each problem has a heavy border around it. Furthermore, each problem is identified with a number immediately above the problem at the right-hand margin.

To obtain maximum benefit from the book, students should familiarize themselves with the section, "How To Use This Book," located in the front pages.

To meet the objectives of this book, staff members of REA have selected problems usually encountered in assignments and examinations, and have solved each problem meticulously to illustrate the steps which are difficult for students to comprehend. Special gratitude is expressed to them for their efforts in this area, as well as to the numerous contributors who devoted brief periods of time to this work.

Gratitude is also expressed to the many persons involved in the difficult task of typing the manuscript with its endless changes, and to the REA art staff who prepared the numerous detailed illustrations together with the layout and physical features of the book.

The difficult task of coordinating the efforts of all persons was carried out by Carl Fuchs. His conscientious work deserves much appreciation. He also trained and supervised art and production personnel in the preparation of the book for printing.

Finally, special thanks are due to Helen Kaufmann for her unique talents to render those difficult border-line decisions and constructive suggestions related to the design and organization of the book.

Max Fogiel, Ph.D.
Program Director

HOW TO USE THIS BOOK

This book can be an invaluable aid to students in electronic communications as a supplement to their textbooks. The book is subdivided into 19 chapters, each dealing with a separate topic. The subject matter is developed beginning with RL and RC circuits, Fourier series and transforms, Laplace transforms, spectral analysis and extending through frequency response in linear systems, random variables and processes, amplitude, frequency, and pulse modulation systems. Also included are problems in signal noise considerations, transmission lines and antennas. An extensive number of applications have been included, since these appear to be more troublesome to students.

TO LEARN AND UNDERSTAND A TOPIC THOROUGHLY

1. Refer to your class text and read the section pertaining to the topic. You should become acquainted with the principles discussed there. These principles, however, may not be clear to you at that time.

2. Then locate the topic you are looking for by referring to the "Table of Contents" in front of this book, "The Electronic Communications Problem Solver."

3. Turn to the page where the topic begins and review the problems under each topic, in the order given. For each topic, the problems are arranged in order of complexity, from the simplest to the more difficult. Some problems may appear similar to others, but each problem has been selected to illustrate a different point or solution method.

To learn and understand a topic thoroughly and retain its contents, it will be generally necessary for students to review the problems several times. Repeated review is essential in order to gain experience in recognizing the principles that should be applied, and in selecting the best solution technique.

TO FIND A PARTICULAR PROBLEM

To locate one or more problems related to a particular subject matter, refer to the index. In using the index, be certain to note that the numbers given there refer to problem numbers, not page numbers. This arrangement of the index is intended to facilitate finding a problem more rapidily, since two or more problems may appear on a page.

If a particular type of problem cannot be found readily, it is recommended that the student refer to the "Table of Contents" in the front pages, and then turn to the chapter which is applicable to the problem being sought. By scanning or glancing at the material that is boxed, it will generally be possible to find problems related to the one being sought, without consuming considerable time. After the problems have been located, the solutions can be reviewed and studied in detail. For this purpose of locating problems rapidly, students should acquaint themselves with the organization of the book as found in the "Table of Contents".

In preparing for an exam, locate the topics to be covered on the exam in the "Table of Contents," and then review the problems under those topics several times. This should equip the student with what might be needed for the exam.

CONTENTS

Chapter No.   **Page No.**

1 NATURAL RESPONSE OF RL & RC CIRCUITS   1

    RL Circuits (Natural Response)   1
    RC Circuits (Natural Response)   13

2 FORCED RESPONSE OF RL & RC CIRCUITS   30

    Unit Step Function   30
    RL Circuits (Forced Response)   35
    RC Circuits (Forced Response)   49

3 THE FOURIER SERIES   74

    Trigonometric Fourier Series   74
    Complex Fourier Series   89
    Convergence of Fourier Series   102
    Symmetry Conditions   110
    Useful Functions of Fourier Series   112

4 THE FOURIER TRANSFORM 134

Definition of Fourier Transform 134
Properties of Fourier Transform 143
Applications of Fourier Transform 158

5 THE LAPLACE TRANSFORM 193

Definition of Laplace Transform 193
Properties of Laplace Transform 206
Simple Functions 214
Inverse Laplace Transform 238
Applications of Laplace Transform 252
Periodic Functions 264
Initial and Final Value Theorems 275
Convolution 276

6 SPECTRAL ANALYSIS 293

Fourier Series Representation 293
Exponential Fourier Series 298
Response of a Linear System 308
Normalized Power 311
Power Content of Signals 322
Fourier Transform Representation 327
Parseval's Theorem 333
Bandlimiting of Waveforms 344
Autocorrelation 349

7 FREQUENCY RESPONSE IN LINEAR SYSTEMS 368

Transfer Function of a System 368
Frequency Domain Analysis 378
Steady State Response 400

8 RANDOM VARIABLES AND PROCESSES 424

Probability 424

Probability Density Function of Random
Variables   434
Cumulative Distribution Function   439
Joint Distribution and Density Function   454
Average Value, Mean and Variance   467
Random Processes   495

9 AMPLITUDE MODULATION SYSTEMS (AM)   499

Amplitude Modulation   499
AM Demodulators   510
Balanced Modulators   515
Amplitude Modulation Receiver   518
Single Sideband Modulation   523
Square Law Modulation   530

10 FREQUENCY MODULATION SYSTEMS (FM)   539

Angle Modulation   539
Frequency Modulation   542
Phase Modulation   542
Frequency Demodulator   572
Bessel's Equation   577

11 PULSE - MODULATION SYSTEMS (PM)   589

The Sampling Theorem   589
Pulse-Amplitude Modulation   603
Natural and Flat-Topped Sampling   608
Signal Recovery - Holding and Crosstalk   615
Methods of Generating Pulse-Time and Pulse-Duration
Modulation Signals   620

12 PULSE - CODE MODULATION (PCM)   625

Quantization of Signals   625
The PCM System   635
Delta Modulation   647

Binary Communications - ON-OFF keying, PSK, DPSK and FSK   654

13 MATHEMATICAL REPRESENTATION OF NOISE   662

Frequency Domain Representation and Spectral Characteristics of Noise   662
Filtering   685
Noise Bandwidth   696
Shot Noise   703

14 NOISE IN MODULATION SYSTEMS   711

Noise in AM   711
Noise in FM   725
Noise in PCM and DM   742

15 COMMUNICATION SYSTEMS AND NOISE CALCULATIONS   758

Tuned Circuits, Amplifier, and Oscillator in Communication Systems   758
Thermal Noise   777
Noise Power and Bandwidth Calculations   790
Noise Figure and Noise Temperature Calculations   793

16 DATA TRANSMISSION   822

17 INFORMATION THEORY AND CODING   855

Information Theory and Entropy   855
Signal Transmission: Channel Representation and Channel Capacity   870
Coding and Error-Detection (Syndrome)   879

18 ANTENNAS 895

Fundamental Laws of Radiation 895
Antenna Characteristics 899
Directional Characteristics of Simple Antennas 912
Power Relations 921

19 TRANSMISSION LINES 929

A Reflected Pulse 929
Characteristic Impedance and Line Input Impedance 945
Transmission Line Distributed Parameters 962
Reflection and Transmission Coefficients and Voltage Standing Wave Ratio 966
Smith Chart 987
Matching 996

INDEX 1030

CHAPTER 1

NATURAL RESPONSE OF RL & RC CIRCUITS

RL CIRCUITS (NATURAL RESPONSE)

● **PROBLEM 1-1**

For the circuit shown in Fig. 1, find i and v as functions of time for $t > 0$.

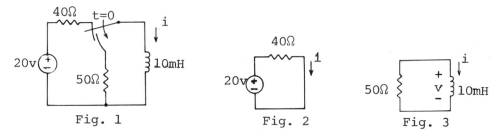

Fig. 1 Fig. 2 Fig. 3

Solution: First, find the current through the inductor i, just before the switch is thrown. Fig. 2 shows the circuit at $t = 0^-$. In Fig. 2, $i = \frac{20v}{40\Omega} = 0.5A$. Fig. 3 shows the circuit in Fig. 1 at $t = 0^+$.
In Fig. 3, in order for the voltage drops to sum to zero around loop, the voltage v must be $-(50\Omega)(0.5A) = -25V$. The time constant is found to be
$$\frac{L}{R} = \frac{10mH}{50\Omega} = \frac{1}{5000} .$$

Write the response,
$$i(t) = 0.5e^{-5000t} A ; t > 0$$
$$v(t) = -25e^{-5000t} V ; t > 0 .$$

● **PROBLEM 1-2**

Each circuit shown in Fig. 1 has been in the condition shown for an extremely long time. Determine $i(0)$ in each circuit.

1

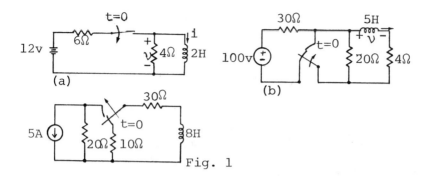

Fig. 1

Solution: Since the current through an inductor cannot change instantaneously, the current through each inductor in Fig. 1 is the same for $t = 0^-$ as for $t = 0$. (a) An inductor behaves as a short circuit in the D.C. steady state condition.

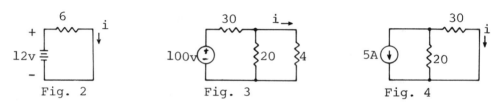

Redrawing the circuit in Fig. 1(a) gives the D.C. steady state circuit of Fig. 2

$$i = \frac{12V}{6\Omega} = 2A$$

$$i(0) = 2A$$

b) The D.C. steady state circuit is shown in Fig. 3.

Using the voltage division concept, obtain the voltage across the 4-Ω resistor, so that the current i is found.

$$i = \frac{V}{R} = \frac{100(20||4)}{30 + (20||4)} \Big/ 4$$

$$i = \frac{100(20(4)/(20+4))}{[30 + (20(4)/(20+4))]4} = \frac{100(3.33)}{[33.33]4} = 2.5A$$

$$i(0) = 2.5A.$$

c) Again the D.C. steady state circuit is shown in Fig. 4. Using the current division concept yields

$$i = -\frac{5(20)}{20 + 30} = -2A$$

$$i(0) = -2A.$$

● **PROBLEM 1-3**

At the instant just after the switches are thrown in the circuits of Fig. 1, find v.

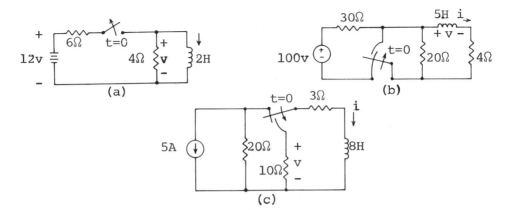

Solution: In order to find the voltages indicated just after the switches are thrown, find the currents in the inductors at $t = 0^-$. Since the current in an inductor cannot change instantaneously and remains the same before, during, and after the switch is thrown, we can find v in each case by simple loop methods. For each of the three parts of the problem Fig. 2 shows the circuit just before the switch is thrown in D.C. steady state condition. Fig. 3 shows the circuit just after the switch is thrown.

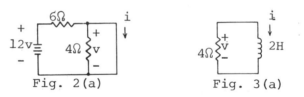

a) In Fig. 2(a), i is found to be $12v/6\Omega = 2A$. In Fig. 3(a), if $i = 2A$, then $v = iR = -2(4) = -8V$.

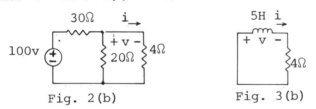

b) In Fig. 2(b) by KCL find the voltage at node n_1 and i.

$$0 = \frac{v-100}{30} + \frac{v}{20} + \frac{v}{4}$$

$$v = \frac{100}{30(\frac{1}{30} + \frac{1}{20} + \frac{1}{4})} = \frac{100}{10} = 10V$$

$$i = \frac{v}{4} = \frac{10}{4} = 2.5A$$

We can find the voltage across the 4-Ω resistor in Fig. 3(b). $v_{4-\Omega} = 4(2.5) = 10V$ and since the voltage drops around the loop must sum to zero, v must be $-10V$.

c) The current i in Fig. 2(c) by current division is

$$i = -\frac{5(20)}{20 + 30} = -2A$$

3

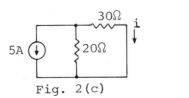
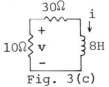

Fig. 2(c)   Fig. 3(c)

The voltage v in Fig. 3(c) becomes $v = iR = -(-2)(10) = 20V$.

• **PROBLEM 1-4**

After $t = 0$, each of the circuits in Fig. 1 is source-free. Find expressions for i and v in each case for $t > 0$.

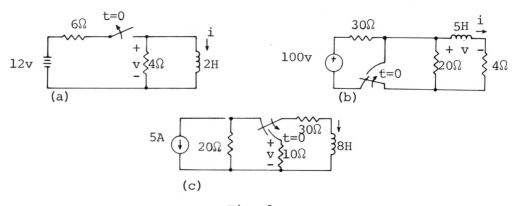

Fig. 1

Solution: The response of each circuit for $t > 0$ will be exponential in the form

$$i(t) = i(0^+)e^{-\frac{t}{\tau}}$$
$$v(t) = v(0^+)e^{-\frac{t}{\tau}}$$

where $\tau = L/R$ for each circuit we must find $i(0^+)$, $v(0^+)$ and R/L. Use the values of $i(0^+)$ and $v(0^+)$ found in the preceeding problems.

a) $\tau = L/R = 2/4 = 0.5s$
$i(t) = 2e^{-2t}$ A; $t > 0$
$v(t) = -8e^{-2t}$ V; $t > 0$

Fig. 2

b) $\tau = L/R = 5/4 = 1.25$
$i(t) = 2.5e^{-0.8t}$ A; $t > 0$
$v(t) = 10e^{-0.8t}$ A; $t > 0$

Fig. 3

c)
$$\tau = \frac{1}{R} = \frac{8}{30+10} = 0.2s$$

$$i(t) = -2e^{-5t} A; \quad t > 0$$

$$v(t) = 20e^{-5t} V; \quad t > 0$$

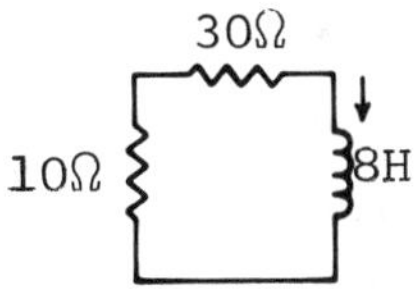

Fig. 4

• **PROBLEM 1-5**

The two switches in the circuit of Fig. 1 are thrown simultaneously at $t = 0$. (a) Find $i_1(0^+)$, $i_2(0^+)$. (b) Find L_{eq} and τ. (c) Write $i(t)$ for $t > 0$. (d) To find $i_1(t)$ or $i_2(t)$, it is necessary to include a possible constant current present in the inductive loop. This may be done by finding the voltage across the 6-Ω resistor, which is the voltage across each coil, and integrating the inductor voltage to find the current; the known initial value is used as the constant of integration. Find $i_1(t)$ and $i_2(t)$. (e) Show that the sum of the energies remaining in the two coils as $t \to \infty$ plus that dissipated since $t = 0$ in the resistor is equal to the sum of the inductor energies at $t = 0$.

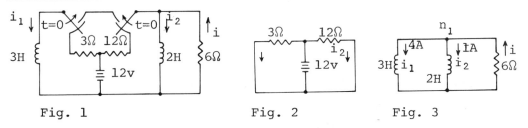

Fig. 1   Fig. 2   Fig. 3

Solution: (a) Just before the two switches are thrown, the currents i_1 and i_2 can be found from Fig. 2.

$$i_1 = \frac{12v}{3\Omega} = 4A$$

$$i_2 = \frac{12v}{12\Omega} = 1A$$

Since these two currents are inductor currents, they remain at the same value at $t = 0^+$. The circuit shown in Fig. 3 is for $t = 0^+$. An equation for node n, gives $i = 4A + 1A = 5A$ so for $t = 0^+$ we have $i_1 = 4A$, $i_2 = 1A$ and $i = 5A$.

(b) Fig. 3 shows that L_{eq} is the parallel combination of the two inductors

$$L_{eq} = \frac{3 \times 2}{3+2} = \frac{6}{5} = 1.2H$$

and the time constant is

$$\frac{L_{eq}}{R} = \frac{1.2}{6} = 0.2s$$

(c) From the information obtained in parts (a) and (b) we can write an equation for $i(t)$; $t > 0$; $i(t) = 5e^{-5t}$ A; $t > 0$

(d) Note that the voltage across the 6-Ω resistor can be written, $(6)(5)e^{-5t}V$. Paying attention to the current directions shown in Fig.3, observe that the voltage across each of the inductors is $-30e^{-5t}V$. By using the relation,

$$i(t) = \frac{1}{L}\int_{t_0}^{t} v(\tau)d\tau + i(t_0), \qquad (1)$$

we can find i_1 and i_2 for $t > 0$. Substituting into eq. (1) for i_1, yields

$$i_1(t) = \frac{1}{3}\int_{0}^{t} -30e^{-5\tau}d\tau + 4$$

$$i_1(t) = \frac{-30}{3(-5)} e^{-5\tau}\Big|_{0}^{t} + 4$$

$$i_1(t) = 2(e^{-5t} - 1) + 4$$

$$i_1(t) = 2 + 2e^{-5t} A. \qquad (2)$$

Substituting into eq. (1) for i_2, yields

$$i_2(t) = \frac{1}{2}\int_{0}^{t} -30e^{-5\tau}d\tau + 1$$

$$i_2(t) = \frac{-30}{2(-5)} e^{-5\tau}\Big|_{0}^{t} + 1$$

$$i_2(t) = 3(e^{-5t} - 1) + 1$$

$$i_2(t) = -2 + 3e^{-5t} A. \qquad (3)$$

Note that $i_1 + i_2 = 5e^{-5t} = i$ and that $i_1(0^+)$ and $i_2(0^+)$ still equal 4A and 1A respectively.

(e) The energy stored in the inductors at $t = 0$ is $\frac{1}{2}L[i(0)]^2$. The energy stored at $t \to \infty$ is $\frac{1}{2}L[i(\infty)]^2$ and the energy dissipated in the resistor since $t = 0$ is $\int_{0}^{\infty} P_R\, dt$. We are asked to prove

$$\frac{1}{2}L_1[i_1(\infty)]^2 + \frac{1}{2}L_2[i_2(\infty)]^2 + \int_{0}^{\infty} P_R\, dt$$

$$= \frac{1}{2}L_1[i_1(0)]^2 + \frac{1}{2}L_2[i_2(0)]^2 \qquad (4)$$

Evaluate:

$$i_1(0), i_2(0), i_1(\infty), \text{ and } i_2(\infty).$$

Eqs. (2) and (3) yields, $i_1(0) = 4A$, $i_2(0) = 1A$, $i_1(\infty) = 2A$, $i_2(\infty) = -2A$, and $P_R(t) = (30e^{-5t})(5e^{-5t}) = 150e^{-10t}$. Substituting into eq. (4),

$$\frac{1}{2}(3)(2)^2 + \frac{1}{2}(2)(-2)^2 + \int_{0}^{\infty} 150e^{-10t}\, dt$$

$$= \frac{1}{2}(3)(4)^2 + \frac{1}{2}(2)(1)^2$$

$$6 + 4 + \frac{150}{-10}e^{-10t}\Big|_{0}^{\infty} = 24 + 1$$

$$6 + 4 + \frac{150}{-10}[0-1]_0 = 24 + 1$$

$$[6 + 4 + 15 = 24 + 1] \text{ J}$$

$$25 = 25$$

• **PROBLEM 1-6**

A 30-mH inductor is in series with a 400-Ω resistor. If the energy stored in the coil at $t = 0$ is 0.96 μJ, find the magnitude of the current at (a) $t = 0$; (b) $t = 100$ μs; (c) $t = 300$ μs.

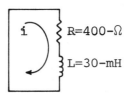

Solution: (a) Find the initial current $(i(0))$ by making use of the energy relationship for an inductor

$$E = \tfrac{1}{2} Li^2.$$

Since we are given E and asked to find i,

$$i = \sqrt{\frac{2E}{L}}$$

$$i = \sqrt{\frac{2(0.96 \times 10^{-6})}{0.03}}$$

$$i = \sqrt{\frac{1.92 \times 10^{-6}}{3 \times 10^{-2}}} = \sqrt{64 \times 10^{-6}}$$

$$i = 8 \times 10^{-3} = 8 \text{mA}$$

(b) After $t = 0$ the current through the inductor is governed by the response of the series RL circuit.

$$i(t) = I_0 e^{-Rt/L}.$$

To find i at 100 μs,

$$i(100 \text{μs}) = (.008)\exp\left[\frac{-400(100 \times 10^{-6})}{0.03}\right]$$

$$i(100 \text{μs}) = (.008)(2.64) = 2.11 \text{mA}.$$

(c) To find i at 300 μs,

$$i(300 \text{ μs}) = (.008) \exp\left[\frac{-400(300 \times 10^{-6})}{0.03}\right]$$

$$i(300 \text{ μs}) = (.008)e^{-4} = (.008)(.018) = 0.15 \text{mA}.$$

• PROBLEM 1-7

Solve the equation
$$L \frac{dI}{dt} + RI = E_0 \qquad (a)$$
for the case in which an initial current I_0 is flowing and a constant emf E_0 is impressed on the circuit at time $t = 0$.

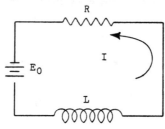

Solution: We consider only $t \geq 0$ as being of interest. We solve (a) by separation of variables:

$$L \frac{dI}{dt} = E_0 - RI,$$

$$\frac{dI}{E_0 - RI} = \frac{dt}{L}.$$

Integrating,
$$\int \frac{dI}{E_0 - RI} = \int \frac{dt}{L} + C, \qquad (c)$$

where C is the constant of integration. We use a change of variable:

$$p = E_0 - RI,$$

$$dp = -R\, dI.$$

Substituting this into (b) yields

$$-\frac{1}{R} \int \frac{dp}{p} = \frac{t}{L} + C,$$

$$-\frac{1}{R} \ln p = \frac{t}{L} + C,$$

$$\ln p = -\frac{R}{L} t - RC.$$

Replacing p by $E_0 - RI$ and exponentiating:

$$E_0 - RI = e^{-Rt/L - RC}$$

$$I = \frac{E_0}{R} - e^{-Rt/L - RC}. \qquad (c)$$

Since $I = I_0$ at $t = 0$, (c) becomes

$$I_0 = \frac{E_0}{R} - e^{-RC},$$

$$e^{-RC} = \frac{E_0}{R} - I_0.$$

Substituting this into (c) gives
$$I = \frac{E_0}{R} - e^{-Rt/L}\left(\frac{E_0}{R} - I_0\right) \quad (d)$$
as the final solution.

The E_0/R term in (d) is called the steady-state part, and the $\left(I_0 - \frac{E_0}{R}\right)e^{-Rt/L}$ term is called the transient part (which goes to zero as $t \to \infty$). The solution may thus be written as

$$I = \left\{\begin{array}{c}\text{STEADY-STATE}\\\text{PART}\end{array}\right\} + \left\{\begin{array}{c}\text{TRANSIENT}\\\text{PART}\end{array}\right\}.$$

● **PROBLEM 1-8**

The RL-circuit:

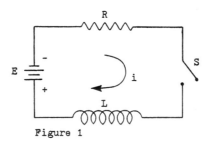

Figure 1

The above circuit has a constant impressed voltage E, a resistor of resistance R, and a coil of impedance L. Find the current $i = i(t)$ flowing in the circuit.

Solution: The differential equation governing this circuit is derived from setting the sum of the voltage drops around the circuit equal to the impressed voltage. Thus,

$$L\frac{di}{dt} + Ri = E, \quad (a)$$

where $L\frac{di}{dt}$ is the voltage drop across the coil, and Ri is the voltage drop across the resistor. We may solve equation (a) by the use of an integrating factor. Divide (a) by L:

$$\frac{di}{dt} + \frac{R}{L}i = \frac{E}{L},$$

and let $\varphi(t)$ be the function such that

$$\varphi\frac{di}{dt} + \frac{R}{L}\varphi i = \frac{d}{dt}(\varphi i). \quad (b)$$

Expanding the right-hand side of (b):

$$\varphi\frac{di}{dt} + \frac{R}{L}\varphi i = \varphi\frac{di}{dt} + i\frac{d\varphi}{dt}.$$

Subtract $\varphi\, di/dt$:

$$\frac{R}{L}\varphi i = i\frac{d\varphi}{dt},$$

and divide by i to obtain

$$\frac{d\varphi}{dt} = \frac{R}{L}\varphi.$$

This equation may be solved for φ by separation of variables:

$$\frac{d\varphi}{\varphi} = \frac{R}{L}dt.$$

Integrating both sides,

$$\int \frac{d\varphi}{\varphi} = \frac{R}{L}\int dt,$$

or

$$\ln \varphi = \frac{R}{L}t.$$

Exponentiating both sides,

$$\varphi = e^{\frac{R}{L}t}.$$

This function φ is the integrating factor which will allow us to write (a) as

$$e^{\frac{R}{L}t}\frac{di}{dt} + e^{\frac{R}{L}t}\frac{R}{L}i = \frac{E}{L}e^{\frac{R}{L}t},$$

or, using (b),

$$\frac{d}{dt}\left(e^{\frac{R}{L}t}i\right) = \frac{E}{L}e^{\frac{R}{L}t}.$$

We integrate both sides to obtain

$$\int \frac{d}{dt}\left(e^{\frac{R}{L}t}i\right)dt = \frac{E}{L}\int e^{\frac{R}{L}t}dt + C, \quad (c)$$

where C is the constant of integration. Evaluating (c) we have

$$e^{\frac{R}{L}t}i = \frac{E}{L}\left(\frac{L}{R}\right)e^{\frac{R}{L}t} + C.$$

Multiplication by $e^{-\frac{R}{L}t}$ gives

$$i = \frac{E}{R} + Ce^{-\frac{R}{L}t}. \quad (d)$$

To evaluate the constant of integration, we assume that the switch S is closed at time $t = 0$, and the initial current is thus $i(0) = 0$. Using this information and (d) we have

$$i(0) = \frac{E}{R} + Ce^{0},$$

$$0 = \frac{E}{R} + C,$$

$$C = \frac{-E}{R}.$$

Our final solution is therefore,

$$i(t) = \frac{E}{R} - \frac{E}{R} e^{-\frac{R}{L} t},$$

$$i(t) = \frac{E}{R}\left(1 - e^{-\frac{R}{L} t}\right). \tag{e}$$

We see that

$$\lim_{t \to \infty} i(t) = \frac{E}{R} - \frac{E}{R} \lim_{t \to \infty} e^{-\frac{R}{L} t}$$

$$= \frac{E}{R} - \frac{E}{R}(0)$$

$$= \frac{E}{R}.$$

Therefore, $i(t)$ approaches the steady-state value E/R exponentially. The graph of $i(t)$ is shown in Figure 2.

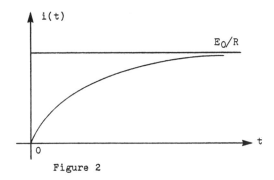

Figure 2

● **PROBLEM 1-9**

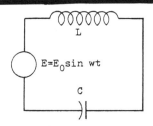

In the above circuit, the inductor has impedance L, the capacitor has capacitance C, and the impressed emf is $E = E_0 \sin wt$. At $t = 0$, current flow is zero, and the initial charge on the capacitor is q_0. Find the charge on the capacitor $q = q(t)$ for all $t > 0$.

Solution: The differential equation for the circuit is obtained by setting the sum of the voltage drops around the circuit equal to the impressed emf. Thus,

$$L \frac{di}{dt} + \frac{q}{C} = E \tag{a}$$

where $L\, di/dt$ is the drop on the inductor and q/C is the drop on the capacitor. Since $i = dq/dt$, $di/dt = d^2q/dt^2$. Substituting this into (a) we have

$$L\frac{d^2q}{dt^2} + \frac{q}{C} = E_0 \sin \omega t . \qquad (b)$$

Equation (b) is inhomogeneous; its solution will be the sum of the homogeneous solution and a particular solution.

The homogeneous equation is
$$L\frac{d^2q}{dt^2} + \frac{q}{C} = 0$$
or
$$\frac{d^2q}{dt^2} + k^2 q = 0 \qquad (c)$$
where
$$k^2 = \frac{1}{LC} .$$

By inspection or analogy to the harmonic oscillator equation we see that the solution of (c) is
$$q = A \cos kt + B \sin kt \qquad (d)$$
where A and B are constants. To get a particular solution of (b) we use the method of undetermined coefficients. Let
$$q_p = U \sin \omega t + V \cos \omega t$$
where U and V are the undetermined coefficients and q_0 is the particular solution. Thus,
$$\frac{d^2 q_p}{dt^2} = -\omega^2 q_p$$
and (b) becomes
$$L(-\omega^2 q_p) + \frac{q_p}{C} = E_0 \sin \omega t .$$
Thus,
$$\left(\frac{U}{C} - \omega^2 UL\right) \sin \omega t + \left(\frac{V}{C} - \omega^2 VL\right) \cos \omega t = E_0 \sin \omega t . \qquad (e)$$

Equating coefficients of sine and cosine on either side of (e) yields
$$\frac{U}{C} - \omega^2 LU = E_0 \qquad (f)$$
and
$$\frac{V}{C} - \omega^2 LV = 0 . \qquad (g)$$

From (f)
$$U = \frac{CE_0}{1-\omega^2 LC}$$
and from (g)
$$V = 0 \qquad (\text{if } 1/C - \omega^2 L \ne 0).$$
Thus,
$$q_p = \frac{CE_0}{1-\omega^2 LC} \sin \omega t . \qquad (h)$$

The general solution is the sum of (d) and (h):
$$q(t) = A \cos kt + B \sin kt + \frac{CE_0}{1-\omega^2 LC} \sin \omega t . \qquad (j)$$

We now evaluate A and B from the initial conditions:
$$q(0) = A \cos(0) + B \sin(0) + \frac{CE_0}{1-\omega^2 LC} \sin(0),$$
$$q(0) = A(1) + B(0) + \frac{CE_0}{1-\omega^2 LC} (0)$$
$$q_0 = A . \qquad (k)$$

Next we use $i(0) = 0$:

$$i(t) = \frac{dq}{dt} = -kA \sin kt + kB \cos kt + \frac{\omega CE_0 \cos \omega t}{1-\omega^2 LC}.$$

Thus,
$$i(0) = -kA(0) + kB(1) + \frac{\omega CE_0(1)}{1-\omega^2 LC}$$

$$0 = kB + \frac{\omega CE_0}{1-\omega^2 LC}$$

$$B = -\frac{\omega CE_0}{(1-\omega^2 LC)k}. \qquad (\ell)$$

Our final answer is

$$q(t) = q_0 \cos kt + \frac{CE_0}{1-\omega^2 LC}\left(\sin \omega t - \frac{\omega}{k} \sin kt\right)$$

RC CIRCUITS (NATURAL RESPONSE)

● **PROBLEM** 1-10

After having been closed for a long time, the switch in the network of Fig. 1 is opened at $t = 0$. Find $v_c(t)$ for $t > 0$.

Fig. 1 Fig. 2

Solution: The voltage across the capacitor v_c at steady state $(t = 0^-)$ can be found in Fig. 2. to be 20V. After the switch is opened, the capacitor is no longer in an RC loop. Therefore the voltage v_c remains at 20V for $t > 0$.

● **PROBLEM** 1-11

In the circuit of Fig. 1, $v(0^+) = 10$. Solve for $i(t)$ directly, without finding $v(t)$.

Fig. 1

Solution: Directly write a relation for i_0

$$i_0 = \frac{v_0}{R_{eq}} e^{-t/R_{eq}C}$$

where R_{eq} is the equivalent resistance across the capacitor.

$$R_{eq} = \frac{10(10)}{20} + \frac{20(5)}{25} = 9\,\Omega$$

$$i_0 = \frac{10}{9} e^{-t/(9)(1/3)} = \frac{10}{9} e^{-t/3} \text{ A }.$$

Write a relation for i in terms of i_0. By using the concept of current division, we have

$$i_{ma} = \frac{i_0(10)}{20}$$

$$i_{mb} = \frac{i_0(10)}{20}$$

$$i = \frac{i_{ma} \cdot 5}{5+20} - \frac{i_{mb} \cdot 20}{5+20}$$

$$i = i_0 \frac{1}{2}\left(\frac{5}{25} - \frac{20}{25}\right) = i_0 \frac{1}{2}\left(-\frac{15}{25}\right)$$

$$i = i_0\left(-\frac{15}{50}\right) = -\frac{3}{10} i_0$$

thus

$$i = -\frac{3}{10} \cdot \frac{10}{9} e^{-t/3} \text{ A} = -\frac{1}{3} e^{-t/3} \text{ A} ; \; t > 0$$

• **PROBLEM 1-12**

In the circuit of Fig. 1, $v_C(0^+) = 10$. Find v_{ab} for all $t \geq 0^+$.

Solution: Write a relationship for v_{ab} in terms of v_C. By using voltage division concepts, note that,

$$v_{mn} = \frac{v_C[(4+8)||(3+1)]}{9 + [(4+8)||(3+1)]}$$

and

$$v_{an} = \frac{v_{mn}(8)}{4+8}$$

$$v_{bn} = \frac{v_{mn}(1)}{3+1}$$

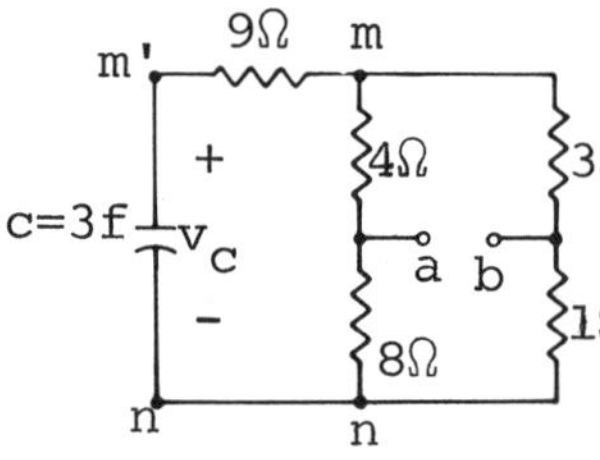

Fig. 1

Finally

$$v_{ab} = v_{an} - v_{bn} = v_{mn}\left(\frac{8}{12}\right) - v_{mn}\left(\frac{1}{4}\right)$$

$$v_{ab} = \frac{v_C[(4+8)||(3+1)]}{9 + [(4+8)||(3+1)]} \left(\frac{8}{12} - \frac{1}{4}\right)$$

$$v_{ab} = \frac{v_C(3)}{12}\left(\frac{1}{2.4}\right) = v_C \frac{3}{28.8} = \frac{v_C}{9.6} \quad.$$

Since we have a source free RC circuit, the response will be in the form
$$v_C = V_0 e^{-t/RC}.$$
For this circuit, R is the equivalent resistance across the capacitor.
$$R = 9 + [(4+8)||(3+1)] = 12\Omega$$
$$\tau = RC = 12(3) = 36s$$
and since
$$V_0 = 10V \text{ and } v_{ab} = \frac{v_C}{9.6}$$
we have,
$$v_{ab} = \frac{10e^{-t/36}}{9.6} = \frac{25}{24} e^{-t/36} V; \quad t \geq 0^+.$$

• **PROBLEM 1-13**

After being closed for a long time, the switch in the circuit of Fig. 1 is opened at $t = 0$. Find $i(t)$ for $t > 0$.

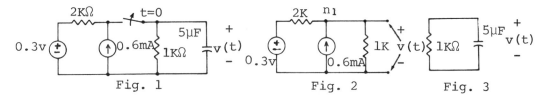

Fig. 1   Fig. 2   Fig. 3

Solution: Fig. 2 shows the circuit in Fig. 1 at $t = 0^-$. Find $v(0^-)$ by writing a KCL equation for node n_1.

$$0.6(10^{-3}) = \left(\frac{v-0.3}{2} + \frac{v}{1}\right)\frac{1}{10^3}$$

$$0.6(10^{-3}) + \frac{0.3}{2}(10^{-3}) = \left(\frac{v}{2} + \frac{v}{1}\right)10^{-3}$$

$$\frac{3}{2} v = 0.75$$

$$v(0^-) = \frac{2(.75)}{3} = 0.5V.$$

We know that $v(0^-) = v(0^+)$ since the voltage across the capacitor cannot change instantaneously. At $t = 0^+$ we have the RC circuit shown in Fig. 3. The time constant is $RC = (1K)(5\mu F) = 1/200$. Write $v(t)$ as:
$$v(t) = 0.5e^{-200t} V; \quad t > 0.$$

• **PROBLEM 1-14**

In the circuit of Fig. 1 the switch is open for $0 < t < t_1$ and closed for $t > t_1$. If $v_{ab}(0^+) = V_0$, formulate $v_{ab}(t)$ and $i(t)$ for all $t \geq 0^+$.

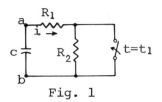

Fig. 1

Solution: We have the solution for the interval $t = 0$ to $t = t_1$:

$$v_{ab} = V_0 \, e^{-t/(R_1+R_2)C} \qquad (1)$$

$$i = \frac{V_0}{R_1+R_2} \, e^{-t/(R_1+R_2)C} \qquad (2)$$

At $t = t_1$, the switch shorts the resistor R_2. This changes the time constant for the circuit and the current i. But since all currents are finite (no impulse currents), the voltage v_{ab} just before t_1 and just after t_1 must be equal. Using eqs. (1) and (2) write the relation for voltage v_{ab} and current i at t_1^-.

$$v_{ab}(t_1^-) = V_0 \, e^{-t_1/(R_1+R_2)C} \, ; \qquad (3)$$

$$i(t_1^-) = \frac{V_0}{R_1+R_2} \, e^{-t_1/(R_1+R_2)C} \, ; \qquad (4)$$

By calling the new initial voltage $v_{ab}(t_1^-)$, V_{t_1}, we have the new response for $t \geq t_1$

$$v_{ab}(t) = V_{t_1} \, e^{-\frac{(t-t_1)}{R_1 C}} \qquad (5)$$

Since the current at t_1 is discontinuous $i(t_1^-) \neq i(t_1^+)$

$$i(t_1^+) = \frac{V_0}{R_1} \, e^{-t_1/(R_1+R_2)C} = \frac{v_{ab}(t_1^-)}{R_1} \, .$$

So that, for $t \geq t_1$

$$i(t) = \frac{V_{t_1}}{R_1} \, e^{-(t-t_1)/R_1 C}$$

• **PROBLEM 1-15**

The independent source in the circuit of Fig. 1 is 140V for $t < 0$ and 0 for $t > 0$. Find $i(t)$ and $v_0(t)$ for $t > 0$.

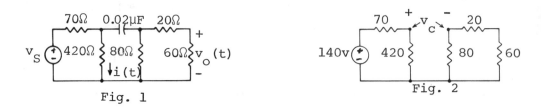

Fig. 1   Fig. 2

Solution: In order to find the current $i(t)$ and voltage $v_0(t)$, first find the voltage across the capacitor. Since the voltage across a capacitor cannot change instantaneously, we can find it at $t = 0^-$. Fig. 2 shows the circuit in Fig. 1 at $t = 0^-$.

Note that the voltage v_C is the voltage across the 420-Ω resistor. By voltage division,

$$v_C = \frac{140(420)}{420+70} = 120V.$$

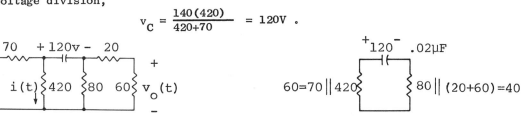

Fig. 3

Fig. 4

The circuit in Fig. 3 is at $t = 0^+$. In order to find $i(t)$ and $v_0(t)$, further reduce the above circuit. In Fig. 4, find the voltage drop across each of the resistors.

$$v_{60-\Omega}\text{ resistor} = \frac{120(60)}{100} = 72V$$

$$v_{80-\Omega}\text{ resistor} = \frac{120(40)}{100} = 48V.$$

Now, one can find

$$i(0^+) = \frac{72V}{420\Omega} = .1714A$$

$$v_0(0^+) = \frac{48}{80}(60) = 36V.$$

From Fig. 4 the time constant is found to be

$$\tau = RC = (60 + 40)(0.02\mu F) = \frac{1}{500,000} \text{ s}.$$

We can now write

$$i(t) = .1714e^{-500,000t} A \; ; \; t > 0$$

$$v_0(t) = 36e^{-500,000t} V \; ; \; t > 0.$$

● **PROBLEM 1-16**

With reference to the circuit shown in Fig. 1, let $v(0) = 9V$. Find $i(t)$ for $t > 0$.

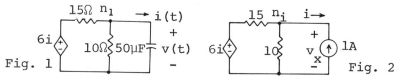

Fig. 1

Fig. 2

Solution: By writing a KCL equation at node n_1, $i(0)$ can be found.

$$0 = \frac{v - 6i}{15} + \frac{v}{10} + i$$

since $v(0) = 9V$, and we must find $i(0)$ we have,

$$0 = \frac{9 - 6i(0)}{15} + \frac{9}{10} + i$$

$$-\frac{6}{15} i(0) + i(0) = -\left(\frac{9}{10} + \frac{9}{15}\right)$$

$$0.6 i(0) = -1.5$$

$$i(0) = \frac{-1.5}{0.6} = -2.5 A .$$

Find R_{eq} across the terminals of the capacitor. By connecting a 1-A source in place of the capacitor as shown in Fig. 2, we find

$$R_{eq} = \frac{v_x}{1} .$$

Writing a KCL equation at node n_1 yields

$$1 = \frac{v_x - 6i}{15} + \frac{v_x}{10}$$

since $i = -1A$,

$$1 = \frac{v_x + 6}{15} + \frac{v_x}{10}$$

$$\left(\frac{1}{15} + \frac{1}{10}\right) v_x = 1 - \frac{6}{15}$$

$$R_{eq} = \frac{v_x}{1} = \frac{1 - 6/15}{\left(\frac{1}{15} + \frac{1}{10}\right)} = 3.6 \Omega .$$

The time constant can now be found

$$\tau = R_{eq} C = 3.6 (50 \mu F) = 1.8 \times 10^{-4} s .$$

$i(t)$ becomes

$$i(t) = -2.5 e^{-t/1.8 \times 10^{-4}} A ; t > 0 .$$

● **PROBLEM 1-17**

For each of the circuits shown in Fig. 1, determine the transient response.

Solution: In order to find the transient response for $t > 0$, for each circuit first find v at $t > 0^-$ and then i at $t = 0^+$. We know that the voltage across a capacitor cannot change instantaneously so

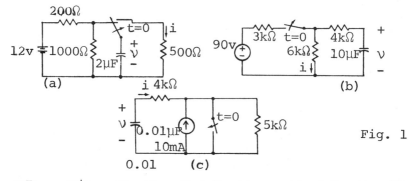

Fig. 1

that $v(0^-) = v(0^+)$. Finally, find the time constant for the RC cir-

cuit (1/RC) and write the response in the form:
$$v(t) = v(0^+)e^{-t/RC} \; ; \quad t > 0$$
$$i(t) = i(0^+)e^{-t/RC} \; ; \quad t > 0$$

For each part of the problem, Fig. 2 shows the circuit at $t = (0^-)$ and Fig. 3 shows the circuit at $t = (0^+)$. At $t = (0^-)$, the circuit is in D.C. steady state so that the capacitor is an open circuit.

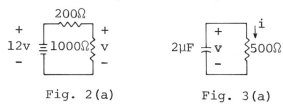

Fig. 2(a)   Fig. 3(a)

(a) In Fig. 2(a) by voltage division, we have
$$v = \frac{12(1000)}{200 + 1000} = 10V \; .$$

In Fig. 3(a), the current i becomes $i = \frac{v}{R} = \frac{10}{500} = 20mA$. The time constant is $RC = 500(2\mu F) = 500(2 \times 10^{-6}) = .001s$. The responses are:
$$v(t) = 10e^{-1000t} V \; ; \quad t > 0$$
$$i(t) = 20e^{-1000t} mA \; ; \quad t > 0$$

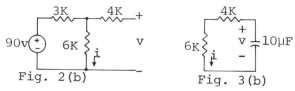

Fig. 2(b)   Fig. 3(b)

(b) In Fig. 2(b), find v by voltage division
$$v = \frac{90(6K)}{(3+6)K} = 60V \; .$$

In Fig. 3(b) the voltage across the 6K-Ω resistor can be found by voltage division.
$$v_{6-\Omega} = \frac{60(6K)}{6K + 4K}$$
$$i = \frac{v_{6-\Omega}}{6K} = \frac{60}{6K + 4K} = 6mA \; .$$

The time constant is $RC = (6K + 4K)(10 \mu F) = .1s$. Write the response
$$v(t) = 60e^{-10t} V \; ; \quad t > 0$$
$$i(t) = 6e^{-10t} mA \; ; \quad t > 0 \; .$$

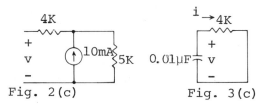

Fig. 2(c)   Fig. 3(c)

(c) In Fig. 2(c) the voltage v is the voltage across the current source. Since for a D.C. current a capacitor acts as an open circuit, the current through the 4KΩ resistor, $i_{4K\Omega} = 0$. In order for the voltage around the loop to be zero, the voltage v must be (10mA)(5K) = 50V. In Fig. 3(c) the current i must be $v/R = 50/4K = 12.5$ mA in order for the voltage drops around the loop to sum to zero. The time constant is RC = $(4K)(.01 \mu F) = 4 \times 10^{-5}$. Write the response

$$v(t) = 50e^{-25,000t} V \; ; \; t > 0$$
$$i(t) = 12.5e^{-25,000t} mA \; ; \; t > 0 \; .$$

• **PROBLEM** 1-18

For the circuit shown in Fig. 1, find $v_1(t)$ and $v_2(t)$ for $t > 0$

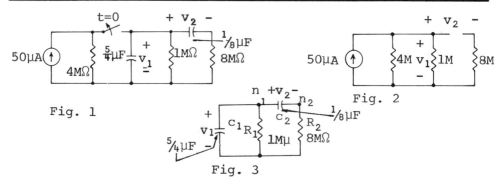

Fig. 1

Fig. 2

Fig. 3

Solution: First, find the initial voltages for the two capacitors. Do this by finding $v_1(0^-)$ and $v_2(0^-)$. Fig. 2 shows the circuit at $t = 0^-$. By current division

$$v_1 = \frac{50 \mu A (1M)}{1M + 4M} (4M) = 40V \; .$$

Since the 8M-Ω resistor is not part of the "live" circuit voltage, $v_2 = 40V$. The initial conditions are,

$$v_1(0) = v_2(0) = 40V \; .$$

When the switch is opened at $t = 0$, we have the circuit in Fig. 3. Notice that in Fig. 3 we cannot combine capacitors or resistors. Solve for v_1 and v_2 by differential equations. Calling the voltage at node n_1, v_1 and that at node n_2, $v_1 - v_2$, write the KCL equations for these two nodes.

$$n_1: \; 0 = C_1 \frac{dv_1}{dt} + \frac{v_1}{R_1} + C_2 \frac{d}{dt}[v_1 - (v_1 - v_2)]$$

$$n_2: \; 0 = C_2 \frac{d}{dt}[(v_1 - v_2) - v_1] + \frac{v_1 - v_2}{R_2}$$

Simplifying the above equations yields

$$C_1 \frac{dv_1}{dt} + C_2 \frac{dv_2}{dt} + \frac{v_1}{R_1} = 0 \quad (1)$$

$$-C_2 \frac{dv_2}{dt} + \frac{v_1}{R_2} - \frac{v_2}{R_2} = 0 \quad (2)$$

Knowing that the solution to the above set of differential equations must be in the form,

Let,
$$C_1 e^{s_1 t} + C_2 e^{s_2 t}.$$

$$v_1 = A e^{s_1 t} + B e^{s_1 t} \tag{3}$$

and
$$v_2 = C e^{s_1 t} + D e^{s_1 t}. \tag{4}$$

Substituting eqs. (3) and (4) and the capacitor and resistor values into equations (1) and (2) yields

$$\frac{5}{4}(As_1 e^{s_1 t} + Bs_2 e^{s_2 t}) + \frac{1}{8}(Cs_1 e^{s_1 t} + Ds_2 e^{s_2 t}) + (A e^{s_1 t} + B e^{s_2 t}) = 0$$

$$-\frac{1}{8}(Cs_1 e^{s_1 t} + Ds_2 e^{s_2 t}) + \frac{1}{8}(A e^{s_1 t} + B e^{s_2 t}) - \frac{1}{8}(C e^{s_1 t} + D e^{s_2 t}) = 0.$$

Rearranging the terms yields

$$(\frac{5}{4} As_1 + \frac{1}{8} Cs_1 + A)e^{s_1 t} + (\frac{5}{4} Bs_2 + \frac{1}{8} Ds_2 + B)e^{s_2 t} = 0 \tag{5}$$

$$(-\frac{1}{8} Cs_1 - \frac{1}{8} C + \frac{1}{8} A)e^{s_1 t} + (-\frac{1}{8} Ds_2 - \frac{1}{8} D + \frac{1}{8} B)e^{s_2 t} = 0 \tag{6}$$

In order for equations (5) and (6) to be zero, each term in parenthesis must be zero. This enables us to write four equations:

$$(\frac{5}{4} s_1 + 1)A + \frac{1}{8} Cs_1 = 0 \tag{7}$$

$$(\frac{5}{4} s_2 + 1)B + \frac{1}{8} Ds_2 = 0 \tag{8}$$

$$-\frac{1}{8}[(s_1 + 1)C - A] = 0 \tag{9}$$

$$-\frac{1}{8}[(s_2 + 1)D - B] = 0 \tag{10}$$

Dividing equation (7) by $(5/4\, s_1 + 1)$, and equation (9) by $(-1/8)$, and adding them yields

$$\frac{1/8\, s_1 C}{(5/4\, s_1 + 1)} + (s_1 + 1)C = 0 \tag{11}$$

eliminating C, we obtain

$$\frac{1/8\, s_1}{(5/4\, s_1 + 1)} = (-s_1 - 1).$$

Simplifying

$$\frac{1}{8} s_1 = (\frac{5}{4} s_1 + 1)(-s_1 - 1)$$

$$\frac{1}{8} s_1 = -\frac{5}{4} s_1^2 - [(1 + \frac{5}{4})s_1] - 1$$

$$0 = -\frac{5}{4} s_1^2 - [(1 + \frac{5}{4} + \frac{1}{8})s_1] - 1$$

Multiplying both sides of the equation by -1 and combining fractions, we have

$$0 = \frac{10}{8}s_1^2 + \frac{19}{8}s_1 + \frac{8}{8}$$

$$0 = 10s_1^2 + 19s_1 + 8 \tag{12}$$

Solving the quadratic equation (12) we obtain two values for s_1.

$$s_1 = \frac{-b \pm \sqrt{b^2 - 4ac}}{2a}$$

$$s_1 = \frac{-19 \pm \sqrt{19^2 - 4(10)(8)}}{20}$$

$$s_1 = \frac{-19 \pm \sqrt{41}}{20} = \frac{-19 \pm 6.4}{20}$$

$$s_1 = \frac{-19 + 6.4}{20} = 0.63$$

$$s_1 = \frac{-19 - 6.4}{20} = -1.27$$

If we were to solve equations (8) and (10) for s_2, the same results would be obtained since the same equation as eq. (12) would result when we eliminate the variables C and D. Since

$$s_1 = -0.63, -1.27$$

$$s_2 = -0.63, -1.27$$

and

$$v_1 = Ae^{s_1 t} + Be^{s_2 t} \tag{13}$$

$$v_2 = Ce^{s_1 t} + De^{s_2 t} \tag{14}$$

We can choose s_1 and s_2 to be any of the above values as long as they are different. $s_1 = -0.63$ and $s_2 = -1.27$ were chosen. At $t = 0$ we found $v_1 = v_2 = 40V$, eqs. (13) and (14) become

$$40 = A + B \tag{15}$$
$$40 = C + D \tag{16}$$

since $e^0 = 1$. Taking eqs. (7)-(10) and substituting s_1 and s_2, yields

$$.213A - .079C = 0 \tag{17}$$
$$-.588B - .159D = 0 \tag{18}$$
$$.370C - 1.000A = 0 \tag{19}$$
$$-.270D - 1.000B = 0 \tag{20}$$

Solving eqs. (15) and (16) for A and C respectively and substituting into eq. (19) produces,

$$B - .37D = 25.2 \tag{21}$$

Solving eqs. (15) and (16) for B and D respectively, and substituting into eq. (20), yields

$$A + .27C = 50.8 \tag{22}$$

By using determinants, take eq. (21) and (18) and solve for B and D.

$$B = \frac{\begin{vmatrix} 0 & -.159 \\ 25.2 & -.370 \end{vmatrix}}{\begin{vmatrix} -.588 & -.159 \\ 1 & -.360 \end{vmatrix}} = \frac{-(-4.00)}{.218 - (-159)} = 10.6$$

$$D = \frac{\begin{vmatrix} -.588 & 0 \\ 1 & 25.2 \end{vmatrix}}{.377} = \frac{-14.82}{.377} = -39.3$$

By using determinants, take eqs. (22) and (17) and solve for A and C.

$$A = \frac{\begin{vmatrix} 0 & -.079 \\ 50.8 & +.27 \end{vmatrix}}{\begin{vmatrix} .213 & -.079 \\ 1 & +.27 \end{vmatrix}} = \frac{-(-4.01)}{(.058) - (-.079)} = 29.4$$

$$C = \frac{\begin{vmatrix} .213 & 0 \\ 1 & 50.8 \end{vmatrix}}{0.137} = \frac{10.82}{0.137} = 79.3$$

The final results are obtained by substituting A, B, C, D, s_1 and s_2 into eqs. (3) and (4), yielding

$$v_1(t) = 29.4 e^{-.630t} + 10.6 e^{-1.270t} \quad ; \quad t > 0$$
$$v_2(t) = 79.3 e^{-.630t} - 39.3 e^{-1.270t} \quad ; \quad t > 0$$

• **PROBLEM** 1-19

An RC-circuit has an impressed emf of 400 cos 2t volts, a resistance of 100 ohms, and a capacitance of 10^{-2} farads. Initially there is no charge on the capacitor. Find the current $i = i(t)$ flowing in the circuit.

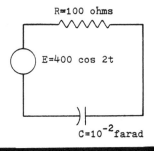

Solution: The differential equation governing this circuit is obtained by setting the voltage drops around the circuit equal to the impressed emf. We have

$$Ri + \frac{q}{C} = E, \tag{a}$$

where Ri is the voltage drop across the resistor, q/C the drop across the capacitor, and E the impressed voltage (q represents the charge on the capacitor). Since $i = dq/dt$ we may rewrite (a), with substituted values for R, C and E, as

$$100 \frac{dq}{dt} + \frac{q}{10^{-2}} = 400 \cos 2t .$$

Dividing by 100 we obtain

$$\frac{dq}{dt} + q = 4 \cos 2t , \quad (b)$$

which is the differential equation to be solved. When this is solved we will have found $q = q(t)$. We may then obtain an expression for $i(t)$ since $i(t) = d/dt \, q(t)$.

We solve (b) by use of an integrating factor, $\varphi(t)$. We want $\varphi(t)$ such that

$$\varphi \frac{dq}{dt} + \varphi q = \frac{d}{dt}(\varphi q) . \quad (c)$$

Expanding the right side,

$$\varphi \frac{dq}{dt} + \varphi q = \varphi \frac{dq}{dt} + q \frac{d\varphi}{dt} .$$

Subtracting $\varphi \frac{dq}{dt}$, and dividing by q, we obtain

$$\varphi = \frac{d\varphi}{dt} .$$

The function which is its own derivative is

$$\varphi = e^t .$$

Thus we multiply (b) by φ:

$$e^t \frac{dq}{dt} + e^t q = 4e^t \cos 2t ,$$

and substitute from (c) to obtain

$$\frac{d}{dt}(e^t q) = 4e^t \cos 2t .$$

Integrating,

$$\int \frac{d}{dt}(e^t q) dt = 4 \int e^t \cos 2t \, dt + C$$

where C is the integration constant.

$$e^t q = 4 \int e^t \cos 2t \, dt + C . \quad (d)$$

It is necessary to evaluate the integral

$$I = \int e^t \cos 2t \, dt \quad (e)$$

in equation (d). This will be done by integrating twice by parts. Recall that the formula for integrating by parts is

$$\int u \, dv = uv - \int v \, du . \quad (f)$$

We let

$$u = e^t , \quad dv = \cos 2t \, dt ,$$
$$du = e^t \, dt, \quad v = \tfrac{1}{2} \sin 2t ,$$

and $I = \int u\, dv$. Substituting in formula (f):

$$I = (e^t)(\tfrac{1}{2} \sin 2t) - \int (\tfrac{1}{2} \sin 2t) e^t\, dt,$$

$$I = \tfrac{1}{2} e^t \sin 2t - \tfrac{1}{2} \int e^t \sin 2t\, dt. \qquad (g)$$

We now have the problem of evaluating the integral in (g):

$$J = \int e^t \sin 2t\, dt,$$

which we again do by parts. Let

$$u = e^t, \quad dv = \sin 2t\, dt,$$

$$du = e^t dt, \quad v = -\tfrac{1}{2} \cos 2t.$$

Then $J = \int u\, dv$, and from formula (f),

$$J = (e^t)(-\tfrac{1}{2} \cos 2t) - \int (-\tfrac{1}{2} \cos 2t) e^t dt,$$

$$J = -\tfrac{1}{2} e^t \cos 2t + \tfrac{1}{2} \int e^t \cos 2t\, dt.$$

Since $\int e^t \cos 2t\, dt = I$, we can write

$$J = -\tfrac{1}{2} e^t \cos 2t + \tfrac{1}{2} I.$$

Substituting this value of J into equation (g):

$$I = \tfrac{1}{2} e^t \sin 2t - \tfrac{1}{2}\{-\tfrac{1}{2} e^t \cos 2t + \tfrac{1}{2} I\},$$

$$I = \tfrac{1}{2} e^t \sin 2t + \tfrac{1}{4} e^t \cos 2t - \tfrac{1}{4} I,$$

$$\tfrac{5}{4} I = \tfrac{1}{2} e^t (\sin 2t + \tfrac{1}{2} \cos 2t),$$

$$I = 2/5\, e^t (\sin 2t + \tfrac{1}{2} \cos 2t).$$

This value for I may now be substituted into (d):

$$e^t q = 4(2/5\, e^t (\sin 2t + \tfrac{1}{2} \cos 2t)) + C,$$

or

$$q = 8/5 \sin 2t + 4/5 \cos 2t + Ce^{-t}. \qquad (h)$$

We must now evaluate the constant C in (h) by use of the initial condition $q(0) = 0$.

$$q(0) = 8/5 \sin(0) + 4/5 \cos(0) + Ce^0,$$

$$0 = 8/5\,(0) + 4/5\,(1) + C(1),$$

$$C = -4/5.$$

Finally,

$$q(t) = 8/5 \sin 2t + 4/5 \cos 2t - 4/5\, e^{-t}. \qquad (k)$$

Our last step is to find $i(t)$ by differentiating (k):

$$i(t) = 16/5 \cos 2t - 8/5 \sin 2t + 4/5\, e^{-t}.$$

The term $4/5\, e^{-t} \to 0$ as $t \to \infty$ and is called the transient part of the solution; the term $16/5 \cos 2t - 8/5 \sin 2t$ is called the steady state part.

● PROBLEM 1-20

An RC-circuit, has resistance R and capacitance C. The impressed emf is $E = E_0 \sin \omega t$. If no initial current is flowing at $t = 0$, find the current $i = i(t)$ for all $t > 0$.

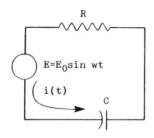

Solution: The differential equation for this circuit comes from Kirchhoff's Law, which states that the sum of the voltage drops around the circuit equals the impressed emf. Thus,

$$Ri + \frac{q}{C} = E, \tag{a}$$

where Ri is the voltage drop across the resistor and q/C is the drop across the capacitor. The charge q on the capacitor is related to the current by

$$i = \frac{dq}{dt}. \tag{b}$$

Hence, our differential equation for q is

$$R \frac{dq}{dt} + \frac{q}{C} = E_0 \sin \omega t. \tag{c}$$

To solve this linear first-order equation we use the method of undetermined coefficients and assume a solution of the form

$$q = A \cos \omega t + B \sin \omega t, \tag{d}$$

where A and B are constants yet to be determined. Thus,

$$\frac{dq}{dt} = -\omega A \sin \omega t + \omega B \cos \omega t. \tag{e}$$

Substituting (d) and (e) in (c) we obtain

$$R(-\omega A \sin \omega t + \omega B \cos \omega t) + \frac{1}{C}(A \cos \omega t + B \sin \omega t) = E_0 \sin \omega t,$$

or

$$\left(\frac{B}{C} - \omega RA\right) \sin \omega t + \left(\frac{A}{C} + \omega RB\right) \cos \omega t = E_0 \sin \omega t. \tag{f}$$

Since $\sin \omega t$ and $\cos \omega t$ are linearly independent, we equate the coefficients of $\sin \omega t$ and $\cos \omega t$ on either side of equation (f):

$$\left(\frac{B}{C} - \omega RA\right) = E_0, \tag{g}$$

$$\left(\frac{A}{C} + \omega RB\right) = 0. \tag{h}$$

From (h), $A = -\omega RCB$ which we substitute into (g) to obtain

$$\frac{B}{C} - \omega R(-\omega RCB) = E_0,$$

$$B\left(\frac{1}{C} + \omega^2 R^2 C\right) = E_0 \,,$$

$$B = \frac{E_0 C^2}{1+\omega^2 R^2 C^2} \,. \tag{i}$$

Therefore,

$$A = -\omega RC\left(\frac{E_0 C^2}{1+\omega^2 R^2 C^2}\right),$$

$$A = \frac{-\omega RC^3 E_0}{1+\omega^2 R^2 C^2} \,. \tag{j}$$

Substituting (i) and (j) into (d), our solution is

$$q(t) = \frac{E_0 C^2}{1+\omega^2 R^2 C^2} \{\sin \omega t - \omega RC \cos \omega t\} \,.$$

To find $i(t)$ we differentiate $q(t)$ according to (b):

$$i(t) = \frac{\omega E_0 C^2}{1+\omega^2 R^2 C^2} \{\cos \omega t + \omega RC \sin \omega t\} \,. \tag{k}$$

Equation (k) is a particular solution of (c) since it does not contain an arbitrary constant. To obtain the required arbitrary constant we must add to (k) the solution of the homogeneous equation

$$R \frac{dq}{dt} + \frac{q}{C} = 0 \,.$$

Separating variables, we have

$$\frac{dq}{q} = -\frac{1}{RC} dt \,.$$

Integrating,

$$\ln q = -\frac{1}{RC} t + K_1 \,,$$

where K_1 is a constant of integration. Exponentiating,

$$q = K_2 e^{-1/RC\, t} \,,$$

where

$$K_2 = e^{K_1} \,.$$

Thus,

$$i = \frac{dq}{dt} = -\frac{K_2}{RC} e^{-t/RC} \,, \tag{ℓ}$$

Adding (ℓ) to (k):

$$i(t) = \frac{\omega E_0 C^2}{1+\omega^2 R^2 C^2} \{\cos \omega t + \omega RC \sin \omega t\} - \frac{K_2}{RC} e^{-t/RC} \,.$$

Using the initial condition $i(0) = 0$,

$$i(0) = \frac{\omega E_0 C^2}{1+\omega^2 R^2 C^2} \{(1) + \omega RC(0)\} - \frac{K_2}{RC} (1) \,,$$

$$0 = \frac{\omega E_0 C^2}{1+\omega^2 R^2 C^2} - \frac{K_2}{RC} \,.$$

Therefore,

$$K_2 = \left(\frac{\omega E_0 C^2}{1+\omega^2 R^2 C^2}\right) \,.$$

Our final general solution is therefore,

$$i(t) = \frac{\omega E_0 C^2}{1+\omega^2 R^2 C^2}\{\cos \omega t + \omega RC \sin \omega t - e^{-t/RC}\} .$$

● **PROBLEM 1-21**

The RC-circuit

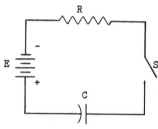

Figure 1

The above circuit has a constant impressed voltage E, a resistance R, a capacitance C and switch S. Find the current $i = i(t)$ flowing in the circuit.

<u>Solution</u>: The differential equation governing the system is obtained by setting the voltage drops around the circuit equal to the impressed voltage. Thus,

$$Ri + \frac{1}{C} q = E , \qquad (a)$$

where Ri is the voltage drop across the resistor and q/C is the voltage drop on the capacitor. The charge, q, is defined by $i = dq/dt$. We differentiate (a) in order to eliminate q:

$$R \frac{di}{dt} + \frac{1}{C} \frac{dq}{dt} = 0 ,$$

$$R \frac{di}{dt} + \frac{1}{C} \cdot i = 0 ,$$

$$\frac{di}{dt} + \frac{1}{RC} i = 0 . \qquad (b)$$

We may solve (b) by separation of variables:

$$\frac{di}{dt} = - \frac{1}{RC} i$$

$$\frac{di}{i} = - \frac{1}{RC} dt .$$

Integrating,

$$\int \frac{di}{i} = - \frac{1}{RC} \int dt + C_1 ,$$

where C_1 is the constant of integration.

$$\ln i = - \frac{1}{RC} t + C_1 .$$

Exponentiating,

$$i = e^{-t/RC + C_1} ,$$

$$i = e^{C_1} e^{-t/RC}.$$

Let $e^{C_1} = C_2$. Then,
$$i = C_2 e^{-t/RC}. \tag{c}$$

We may solve for C_2 by using the initial condition that when the switch is closed at $t = 0$ the capacitor has no charge on it. Thus, the voltage drop across it, q/C, is zero. The entire voltage E is applied to R and Ohm's law states that $i(0) = E/R$. Thus, in equation (c) we have
$$i(0) = C_2 e^0,$$
$$\frac{E}{R} = C_2.$$

Our general solution is therefore,
$$i(t) = \frac{E}{R} e^{-t/RC}. \tag{d}$$

The graph of (d) is Fig. 2.

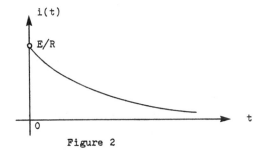

Figure 2

CHAPTER 2

FORCED RESPONSE OF RL & RC CIRCUITS

UNIT STEP FUNCTION

● **PROBLEM** 2-1

Develop an analytical expression using step function notation for the signal g(t) shown in Figure 1.

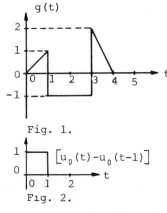

Fig. 1.

Fig. 2.

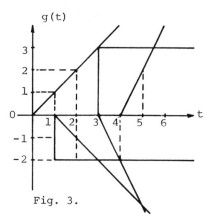

Fig. 3.

Solution: As the function appears to consist of three distinct regions, it is convenient to separate g(t) into three components.

$$g(t) = g_1(t) + g_2(t) + g_3(t), \quad (1)$$

where $g_1(t) = g(t)$ in the interval $(0 \le t < 1)$; $g_2(t) = g(t)$ in the interval $(1 \le t < 3)$; $g_3(t) = g(t)$ in the interval $(3 \le t < 4)$.

Begin by combining step functions to form a new function, one which is equal to unity in the interval of interest and equal to zero elsewhere. For example, in the first interval $(0 \le t < 1)$ we sum

$$U_0(t) = \begin{cases} 0, & t < 0 \\ +1, & t \ge 0 \end{cases} \qquad (2)$$

with

$$-U_0(t-1) = \begin{cases} 0, & t < 1 \\ -1, & t \ge 1 \end{cases} \qquad (3)$$

The result is the function

$$[U_0(t) - U_0(t-1)] = \begin{cases} +1, & (0 \le t < 1) \\ 0, & \text{elsewhere} \end{cases}, \qquad (4)$$

which has the form shown in Figure 2

The product of this function with t gives $g_1(t)$, a function that linearly increases from $g_1(0) = 0$ to $g_1(1) = 1$ and is zero elsewhere:

$$g_1(t) = t[U_0(t) - U_0(t-1)]. \qquad (5)$$

Now, combine step functions to form a function which is equal to one within the second interval, $(1 \le t < 3)$, and is zero elsewhere. This new function is

$$[U_0(t-1) - U_0(t-3)] = \begin{cases} +1, & (1 \le t < 3) \\ 0, & \text{elsewhere} \end{cases}. \qquad (6)$$

The function $g_2(t)$ is equal to -1 in this interval and, therefore,

$$g_2(t) = -[U_0(t-1) = U_0(t-3)]. \qquad (7)$$

For the third interval, we again form a function which is unity between $t = 3$ and $t = 4$ and is zero elsewhere:

$$[U_0(t-3) - U_0(t-4)] = \begin{cases} +1, & (3 \le t < 4) \\ 0, & \text{elsewhere} \end{cases}. \qquad (8)$$

The function $g_3(t)$ falls linearly from $g_3(3) = 2$ to $g_3(4) = 0$, and we must multiply (8) by a function that exhibits this behavior. The function $(4-t)$ goes from one to zero in this interval and, therefore, $2(4-t)$, or $-2(t-4)$, satisfies our requirements. The function $g_3(t)$ is then

$$g_3(t) = -2(t-4)[U_0(t-3) - U_0(t-4)]. \qquad (9)$$

Hence, the complete expression for $g(t)$ is, from (1),

$$g(t) = t[U_0(t) - U_0(t-1)] - [U_0(t-1) - U_0(t-3)]$$
$$-2(t-4)[U_0(t-3) - U_0(t-4)].$$

A second approach to this problem would be to break the function up into a series of step and ramp functions as shown in fig. 3.

The reasoning behind this construction is as follows: For 0 < t < 1 a unit ramp function is required, so draw a line through the origin, with slope = 1, and extend it out infinitely. Next consider the interval 1 < t < 3. At t=1 we need to level off the function and then pull it down to -1, where it will remain until t=3. This can be accomplished by adding to the unit ramp function another ramp function with slope = -1. Note that this will level off g(t) at the value 1 (since the slopes cancel each other, the function neither rises nor falls, but remains fixed at the value g(t) = 1). What is now needed is a step function of magnitude -2 to pull g(t) down to the value -1. Next consider what happens at t=3. The function is pulled up to a value g(t) = 2. Since, for 1 < t < 3, it has been "sitting" at g(t) = -1, a step function of magnitude 3 acting at t=3 will pull g(t) up to the value 2. Next a ramp function of slope -2 is needed, again applied at t = 3. For t ≥ 4 we need g(t) = 0. Note that the function at this point is at t = 0 (at t = 3, g(t) = 2 and the ramp function of slope -2 has pulled it down to g(t) = 0 by t = 4). However, it is necessary to cancel out the ramp function so that g(t) remains at 0. This is easily accomplished with a ramp function of slope +2 starging at t = 4.

Now it is easy to read off g(t), which is simply the sum of all the ramp and step functions, from fig. 3:

$$g(t) = tU(t) - (t-1)U(t-1) - 2U(t-1) + 3U(t-3) -$$
$$- 2(t-3)U(t-3) + 2(t-4)U(t-4)$$

Just as a check, note what happens for given intervals:
t < 1: g(t) = t

1 < t < 3: g(t) = t - (t-1) - 2; g(t) = -1

3 < t < 4: g(t) = t - (t-1) - 2 + 3 - 2(t-3); g(t) = -2t + 8

t > 4: g(t) = t-(t-1) - 2 + 3 - 2(t-3) + 2(t-4); g(t)=0

Also note that this solution is exactly the same as the previous one.

● **PROBLEM 2-2**

Develop an analytical expression for the waveform shown.

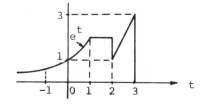

Solution: As seen from the figure, the given curve can be considered to consist of the following three functions:

$$f_1(t) \equiv e^t u(-t) \quad \text{for } t < 0 \qquad (1)$$

$$f_1(t) = e^t[u(t) - u(t-1)] \quad \text{for } 1 > t > 0 \qquad (2).$$

In this equation, $[u(t) - u(t-1)]$ represents a rectangularly shaped function of unit magnitude which "turns on" at $t = 0$ and "turns off" at $t = 1$.

$$f_2(t) \equiv e[u(t-1) - u(t-2)] \quad \text{for } 1 < t < 2 \qquad (3).$$

In this equation, the function $f_2(t) = e$ "turns on" at $t = 1$ and "turns off" at $t = 2$. The last straight line portion of the curve shown is represented by

$$f_3(t) \equiv (2t-3)[u(t-2)-u(t-3)] \quad \text{for } 2 < t < 3. \qquad (4)$$

In this equation, $[u(t-2)-u(t-3)]$ represents a rectangular pulse with a magnitude of 1 which turns on at $t = 2$ and turns off at $t = 3$. Therefore, the total response is

$$f(t) = f_1(t) + f_2(t) + f_3(t) = e^t[u(-t) + u(t) - u(t-1)]$$
$$+ e[u(t-1) + u(t-2)] + (2t-3)[u(t-2) - u(t-3)] \qquad (5)$$

The function $[u(-t) + u(t) - u(t-1)]$ represents a unit step function which was "turned on" at $t = -\infty$ and "turned off" at $t = 1$. The above function is the same as $u(t+1)$ if t is replaced by $-t$.

The resulting function, $u(1-t)$, is a unit step function of unity magnitude as long as $1-t \geq 0$, which means the entire range $-\infty \leq t \leq +1$. So,

$$u(1-t) = [u(-t) + u(t) - u(t-1)] \qquad (6)$$

Then,

$$f(t) = e^t u(1-t) + e[u(t-1) + u(t-2)] + (2t-3)[u(t-2)$$
$$-u(t-3)] \qquad (7)$$

● PROBLEM 2-3

Evaluate (i) $t\delta(t - 1)$; (ii) $t\delta^{(1)}(t - 1)$.

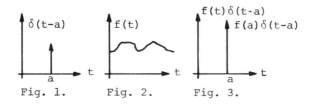

Fig. 1.   Fig. 2.   Fig. 3.

Solution: In this problem δ is an impulse function. An impulse function is defined as a function that has infinite value at one point and zero value at all other points. Expressed mathematically:

$$\delta(t - a) = 0 \text{ for } t \neq a$$
$$= \infty \text{ for } t = a$$

Fig. 1 is the graphical expression of an impulse function.

(i) From Figs 2 and 3 it is seen that

$$f(t) \delta(t - a) = f(a) \delta(t - a)$$

In this problem $f(t) = t$ and $a = 1$; therefore, $f(1) = 1$. So

$$t\delta(t - 1) = 1 \cdot \delta(t - 1) = \delta(t - 1)$$

(ii) The result can be found by using the sampling property which states that

$$f(t) \delta^{(1)}(t-a) = f(a) \delta^{(1)}(t - a) - f^{(1)}(a) \delta(t - a)$$

where $\delta^{(1)}(t-a) = \frac{d}{dt}[\delta(t - a)]$

and $f^{(1)}(a) = \frac{d}{dt}[f(a)]$

In this case $f(t) = t$, $a = 1$ and $f^{(1)}(a) = \left.\frac{d\,t}{dt}\right|_{t=a} = 1$

Therefore,

$$t \delta^{(1)}(t - 1) = 1 \cdot \delta^{(1)}(t - 1) - 1 \cdot \delta(t - 1)$$
$$= \delta^{(1)}(t - 1) - \delta(t - 1).$$

• **PROBLEM 2-4**

Write an expression for the wave forms shown in fig. 1 using unit step function notation.

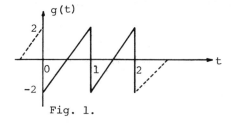

Fig. 1.

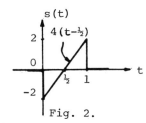

Fig. 2.

Solution: The waveform g(t) is periodic with period T = 1s. By writing an expression for the waveform shown in fig. 2 and then shifting it by T left and right and summing we can obtain g(t).

$$s(t) = 4(t - \tfrac{1}{2})[u(t) - u(t-1)] \tag{1}$$

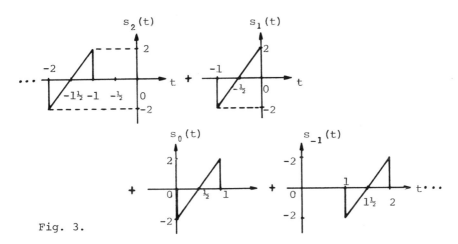

Fig. 3.

Hence, g(t) can be formed by summing the shifted waveforms shown in fig. 3.

$$g(t) = \sum_{n=-\infty}^{\infty} s_n(t + n) ; \tag{2}$$

$$= \ldots s_2(t) + s_1(t) + s_0(t) + s_{-1}(t) + s_{-2}(t) \ldots$$

Substituting (1) into (2) and replacing t by t + n gives

$$g(t) = \sum_{n=-\infty}^{\infty} 4(t + n - \tfrac{1}{2})[u(t + n) - u(t + n - 1)].$$

RL CIRCUITS (FORCED RESPONSE)

● **PROBLEM 2-5**

Draw the waveform for the current and voltage across inductor L for the circuit of Fig. 1, if the switch is in position 1 at t=0.

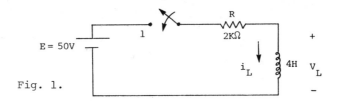

Fig. 1.

Solution: Time constant of the circuit, $\tau = \dfrac{L}{R} = \dfrac{4h}{2000\Omega}$

$$= 2 \times 10^{-3} \text{ sec} = 2 \text{ msec.}$$

At $t = 0^+$, the current (i_L) in the inductor will be zero because i_L cannot change instantaneously. The current in the circuit will be maximum at steady state.

Hence, at steady state,

$$I_{max} = \text{Maximum current} = \dfrac{E}{R} = \dfrac{50V}{2000\Omega} = 25 \times 10^{-3} \text{ amp}$$

$$= 25 \text{ ma.}$$

Now, $i_L(t)$ = Instantaneous current in the inductor

$$= I_{max}(1-e^{-t/\tau})$$

$$= 25 \times 10^{-3}(1-e^{-t/2\times 10^{-3}}) \text{ A}$$

$v_L(t)$ = Instantaneous voltage across the inductor

$$= E\, e^{-t/\tau}$$

$$= 50\, e^{-t/2\times 10^{-3}} \text{ V}$$

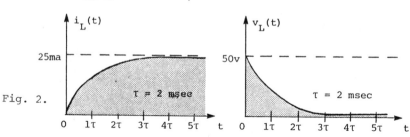

Fig. 2.

The waveforms for $i_L(t)$ and $v_L(t)$ are sketched in Fig. 2.

• **PROBLEM 2-6**

An inductor of 500 mh is in series with a resistor of 5 ohms, as shown in Fig. 1. Draw the waveforms for the voltage across L and the current through L. Assume that the initial current in the inductor is of magnitude 0.3 amp.

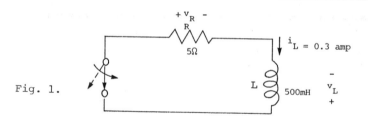

Fig. 1.

Solution: The voltages across the resistor and inductor are equal

i.e. at $t = 0^+$

$$v_L = v_R = iR = (0.3)(5) = 1.5 \text{ volts}$$

where v_L is the same value E in expressions:

$$v_L(t) = Ee^{-t/\tau}$$

$$i_L(t) = \left(\frac{E}{R}\right) e^{-t/\tau}$$

$$v_R(t) = Ee^{-t/\tau}$$

The time constant of the circuit τ is given by

$$\tau = L/R = \frac{500 \times 10^{-3} h}{5 \Omega} = 100 \times 10^{-3} \text{ sec} = 100 \text{ m sec.}$$

The inductor voltage, v_L, is

$$v_L(t) = Ee^{-t/\tau} = 1.5 \, e^{-t/100 \times 10^{-3}} \text{ volts}$$

The inductor current, i_L, is

$$i_L(t) = \frac{E}{R} e^{-t/\tau} = \frac{1.5}{5} e^{-t/100 \times 10^{-3}} = 0.3 \, e^{-t/100 \times 10^{-3}} \text{ volt}$$

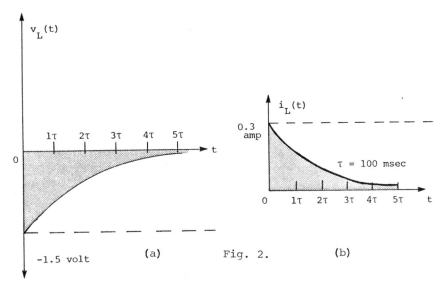

Fig. 2.

The waveforms for $i_L(t)$ and $v_L(t)$ are sketched in Fig. 2.

● **PROBLEM 2-7**

Given the circuit shown in Fig. 1:

(a) Derive the relationship for the instantaneous current in the inductor when the switch is at position 1 at t = 0. Calculate the value of the current i_L at t = 6 msec.

(b) Repeat part (a) for v_L and v_{R_1}.

(c) Find a relationship for the voltages across the inductor (v_L) and the resistor (v_{R_1}) and the instantaneous current in the inductor (i_L), if the switch is moved instantaneously to position 2 at t = 24 msec.

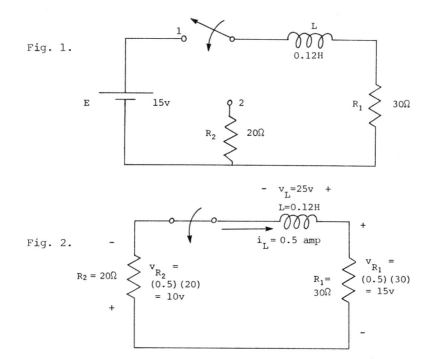

Solution: (a) The time constant of the circuit, τ, is

$$\tau = \frac{L}{R} = \frac{0.12}{30} = 4 \times 10^{-3} \text{ sec} = 4 \text{ msec.}$$

I_m = Maximum current in the circuit

$$= \frac{E}{R_1} = \frac{15}{30} = 0.5 \text{ amp}$$

$i_L(t)$ = Instantaneous current in the inductor

$$= I_m(1-e^{-t/\tau}) = 0.5(1-e^{-t/4\times 10^{-3}}) \text{ A}$$

At t = 6 msec:

$$i_L(t) = 0.5\left[1-\exp\left(-\frac{6\times 10^{-3}}{4\times 10^{-3}}\right)\right]$$

$$= 0.5\left[1-e^{-1.5}\right]$$

$$= 0.5\,(1-0.223) = (0.5)(0.777)$$

$$= 389 \text{ mA}$$

(b) $v_L(t)$ = Instantaneous voltage across the inductor

$$= Ee^{-t/\tau} = 15e^{-t/4\times 10^{-3}} \text{ V}$$

At t = 6 msec:

$$v_L(6 \text{ msec}) = 15\exp\left(-\frac{6\times 10^{-3}}{4\times 10^{-3}}\right) = (15)(0.223)$$

$$= 3.35 \text{ V}$$

For v_{R_1}:

$$v_{R_1}(t) = E - v_L(t) = E - Ee^{-t/\tau} = E(1-e^{-t/\tau})$$

$$= 15(1-e^{-t/4\times 10^{-3}}) \text{ V}$$

Hence, at t = 6 msec,

$$v_{R_1}(6 \text{ msec}) = 15\left[1-\exp\left(-\frac{6\times 10^{-3}}{4\times 10^{-3}}\right)\right]$$

$$= 15(1-0.223) - (15)(0.777)$$

$$= 11.65 \text{ volts}$$

(c) For $t > 5\tau$, it can be assumed that v_L, i_L and v_R have reached their final (steady state) values.

Hence, for t = 24 msec, which is greater than 5τ, we assume,

$$i_L = 0.5 \text{ amp}, \quad v_L = 0 \text{ V}, \text{ and } v_R = 15 \text{ V}$$

The equivalent circuit of Fig. 1, at the instant switch is changed from position 1 to 2, is shown in Fig. 2.

At this instant, the numerical value of the voltage across the inductor should be equal to the sum of the numerical values of the voltages across R_1 and R_2 respectively.

Therefore, voltage across the inductor = 25 volts.

The new time constant, τ, is

$$\tau = \frac{L}{R_1 + R_2} = \frac{0.12 \text{ H}}{50 \, \Omega} = 2.4 \text{ msec}.$$

Let
$$t' = t - 24 \text{ msec},$$

therefore,
$$v_L(t) = 25e^{-t'/2.4 \times 10^{-3}} \text{ V}$$

$$i_L(t) = 0.5e^{-t'/2.4 \times 10^{-3}} \text{ A}$$

and
$$v_R(t) = 15e^{-t'/2.4 \times 10^{-3}} \text{ V}$$

● **PROBLEM 2-8**

For the circuit shown in Fig. 1, find $i_L(t)$.

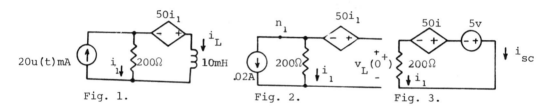

Fig. 1.   Fig. 2.   Fig. 3.

<u>Solution:</u>   Note that $i_L(0) = 0$ and that the open circuit voltage v_{oc} across the inductor is $v_L(0^+)$. This voltage also gives the Thevenin equivalent circuit that will enable us to find R_{eq} across the inductor.

Writing the KCL equation for node n_1 in Fig. 2, gives $i_1 = 0.02$ A

$$v_L(0^+) = 200(.02) + 50(.02) = 5 \text{ V}.$$

If we open circuit the 2-A current source and place a 5-V source in series with the 50i-V dependent source, we have the Thevenin equivalent circuit. Short circuiting the output yields, i_{sc} and

$$\frac{v_{oc}}{i_{sc}} = \frac{v_L(0^+)}{i_{sc}} = R_{eq} .$$

A KVL equation around the loop in Fig. 3 gives

$$50 i_1 + 5V = i_{sc}(200).$$

Since $i_1 = -i_{sc}$,

$$5 = i_{sc}(200 + 50)$$

$$i_{sc} = \frac{5}{250}$$

$$R_{eq} = 5\left(\frac{250}{5}\right) = 250 \, \Omega.$$

Write the relation for the voltage across the inductor.

$$v_L(t) = 5 \, e^{-\frac{250}{.01}t} = 5 \, e^{-25,000t}$$

using, $i_L(t) = i_0 + \frac{1}{L}\int_{t_0}^{t} v_L(T) \, dT$

give $i_L(t) = 0 + \frac{1}{.01}\int_{0}^{t} 5 \, e^{-25,000T} \, dT$

$$i_L(t) = \frac{5}{.01} \frac{1}{(-25,000)} \left[e^{-25,000T} \right]_0^t$$

$$i_L(t) = -.02 \, (e^{-25,000t} - 1) \, u(t) \, A.$$

$$i_L(t) = .02 \, (1 - e^{-25,000t}) \, u(t) \, A.$$

● **PROBLEM 2-9**

For the circuit shown in Fig. 1, find: (a) i_L; (b) v_L.

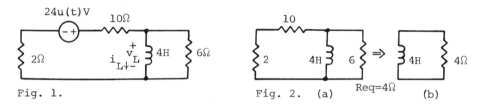

Fig. 1.   Fig. 2.   (a)   Req=4Ω   (b)

Solution: We can determine the complete response $i_L(t)$ by finding the sum of the natural and forced responses.

In the circuit of Fig. 1, the natural response would be the response which occurs when the 24-V source is "off" or short-circuited.

The natural response for the circuit in Fig. 2(b) is,

therefore,

$$i_n = A e^{-t}$$

The forced response is the response that would occur at t = ∞. Fig. 3 shows the circuit for t = ∞, with the inductor short circuited.

$$i_f = i_L(\infty) = \frac{24 \text{ V}}{12 \text{ }\Omega} = 2A$$

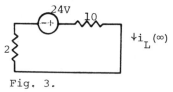

The complete response becomes

$$i_L(t) = i_n + i_f = Ae^{-t} + 2,$$

Fig. 3.

since the current through an inductor cannot change instantaneously and $i(0^-) = 0$. Thus,

$$i(0^+) = 0$$

$$i_L(0^+) = A + 2 = 0$$

giving, A = -2

$$i_L(t) = 2(1 - e^{-t}) u(t) \text{ A}.$$

Find the voltage v_L in the same way.

$$v_L = v_n + v_f$$

We already know that $v_f = v_L(\infty) = 0$ so that

$$v_L = v_n + 0 = V_0 e^{-t}.$$

V_0 is the initial voltage across the inductor. We can find this voltage by observing that at $t = 0^+$, $i_L = 0$ so that the voltage across the 6-Ω resistor is equal to $v_L(0^+)$. Since all the current in the loop flows through the 6-Ω resistor

$$v_L(0^+) = \frac{24}{6 + 2 + 10} (6) = 8V$$

thus $v_L(t) = 8 e^{-t} u(t) \text{ V}.$

● **PROBLEM 2-10**

For the RL circuit calculate the current $i_L(t)$. The initial condition is $i_L(0^-) = -1$ A.

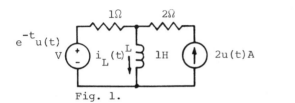
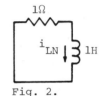

Fig. 1.   Fig. 2.

<u>Solution</u>: The complete response is the sum of the natural and forced responses. The natural response i_{LN} can be calculated from the network of figure 2, in which both sources have been set to zero value (Note that the 2Ω resistor, being in series with the current source, has no effect on i_L).

Thus, $i_{LN}(t) = Ae^{-\frac{R}{L}t} = Ae^{-t}$ for $t \geq 0$.

The forced response can be obtained by superposition, using the differential equation approach. The forced response due to the current source is obtained from figure 3. The forced response is a dc current, so the inductor is a short circuit.

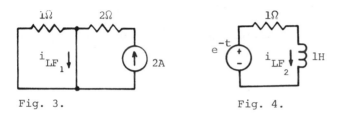

Fig. 3.   Fig. 4.

The forced component due to the current source is $i_{LF_1} = 2A$. Figure 4 is the equivalent source for calculation of the forced response due to the exponential voltage source.

Now $\frac{di_{LF_2}}{dt} + i_{LF_2} = e^{-t}$ has the characteristic equation

$$s + 1 = 0$$

which has the root $s = -1$. Since this is the complex frequency associated with e^{-t}, we must solve the differential equation directly. First, multiply by an integrating factor ϕ (to be determined):

$$\phi \frac{di_{LF_2}}{dt} + \phi i_{LF_2} = \phi e^{-t}$$

Imposing the condition that $\frac{d\phi}{dt} = \phi$ leads to $\phi(t) = e^t$.

Hence,

$$\frac{d}{dt}(e^t i_{LF_2}) = e^t \cdot e^{-t} = 1$$

so

$$e^t i_{LF_2}(t) = t + c$$

where c is a constant to be determined. Thus,

$$i_{LF_2}(t) = te^{-t} + ce^{-t}$$

Now since the forced response is merely <u>any one</u> solution of the nonhomogeneous equation, we can pick $c = 0$. Therefore,

$$i_L(t) = i_{F_1} + i_{F_2} + Ae^{-t} ; \quad t \geq 0$$

or

$$i_L(t) = 2 + te^{-t} + Ae^{-t} ; \quad t \geq 0$$

In order to evaluate A, we can use the initial condition, $i_L(0+) = i_L(0-) = -1 = 2 + A$. Thus $A = -3$, and

$$i_L(t) = (2 + te^{-t} - 3e^{-t}) U_{-1}(t)$$

● **PROBLEM** 2-11

A current source of $0.2\,u(t)$ A, a 100-Ω resistor, and a 0.4-H inductor are in parallel. Find the magnitude of the inductor current (a) as $t \to \infty$; (b) at $t = 0^+$; (c) at $t = 4$ ms.

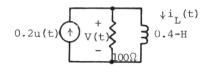

Fig. 1.

<u>Solution</u>: (a) The circuit described above is shown in Fig. 1 below.

We know that as $t \to \infty$, the inductor becomes a short circuit, so that the full current, 0.2A, from the source must flow through it: $i_L(\infty) = 12$A.

(b) At $t = 0^-$ the current source is "off", the entire circuit is "dead". At $t = 0$ the current source is turned on. The current through the inductor cannot change instantaneously for a finite voltage. Thus, $i_L(0^+) = 0$.

(c) Using the relation

$$i_L(t) = i_0 + \frac{1}{L} \int_{t_0}^{t} v\, dT,$$

find the current through the inductor given the initial current. We know that $i_0 = 0$, and $v(t)$ must be in the form $V_0 e^{-\frac{R}{L}t}$. We can find V_0 because it is known that at $t = 0^+$ no current was flowing through the inductor, thus the current through the 100-Ω resistor must have been 0.2A. Therefore, $V_0 = 100(0.2) = 20$V. We can now solve the integral

$$i_L(t) = \frac{1}{0.4} \int_0^t 20 \, e^{-\frac{100}{.4}T} \, dT$$

$$i_L(t) = \frac{20}{0.4\,(-250)} \left[e^{-250T} \right]_0^t$$

$$i_L(t) = -0.2 \left(e^{-250t} - 1 \right)$$

$$i_L(t) = 0.2 \left(1 - e^{-250t} \right) \tag{1}$$

$$i_L(.004) = 0.2 \left(1 - e^{-250(.004)} \right) = 0.2(1 - 0.368)$$

$$i_L(.004) = 0.126 \text{ A.}$$

Equation (1) can be written in a more general form:

$$i_L(t) = \frac{V_0}{R} \left(1 - e^{-\frac{R}{L}t} \right) u(t).$$

• **PROBLEM 2-12**

Consider applying a unit ramp voltage source to a series RL circuit as shown in fig. 1. Compute the voltages $v_R(t)$ and $v_L(t)$ with zero initial condition for (a) L = 0.1H; (b) L = 1H; (c) L = 10H.

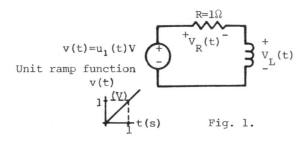

Fig. 1.

Solution: The solution to this problem requires solving a first order linear differential equation. The voltages of interest in the circuit are expressed as,

$$v(t) = t \quad \text{(unit ramp function)}$$

$$V_R = iR$$

$$V_L = L\frac{di}{dt}$$

where i is the current in the loop. Using KVL we write
$$v(t) = V_R + V_L$$
hence, the differential equation is found to be
$$t = iR + L \frac{di}{dt} \quad . \tag{1}$$
First we solve the homogeneous differential equation
$$iR + L \frac{di}{dt} = 0.$$
The equation can be written as
$$-\frac{R}{L} dt = \frac{di}{i}$$
Integrating both sides of the equation,
$$-\int \frac{R}{L} dt = \int \frac{di}{i}$$

$$K_1 t = \ln i \quad \text{Where } K_1 \text{ is the constant of integration.}$$
and taking the exponential of both sides gives the homogeneous solution,
$$e^{-\frac{R}{L}t + K_1} = e^{\ln i} \quad \text{Which gives } e^{K_1} \cdot e^{-\frac{R}{L}t} = i$$

$$K_0 e^{-\frac{R}{L}t} = i_n(t) \quad \text{Where } K_0 = e^{K_1} \tag{2}$$

The particular solution is guessed to be of the form
$$i_p(t) = K_1 t + K_2 . \tag{3}$$
Substituting $i_p(t)$ into the differential equation [Eq(1)] gives
$$t = R[K_1 t + K_2] + L[K_1]$$
$$t = RK_1 t + RK_2 + LK_1$$
Equating coefficients of like terms yields
$$RK_1 = 1; \quad K_1 = \frac{1}{R}$$
and $RK_2 + LK_1 = 0$

Hence, $K_2 = -\frac{L}{R^2}$

The total solution is the sum of the particular and homogeneous solutions, hence
$$i_T(t) = i_p(t) + i_h(t)$$
$$i_T(t) = \frac{1}{R} t - \frac{L}{R^2} + K_0 e^{-\frac{R}{L}t}$$

Find the constant K_0 by applying the initial condition $i_T(0) = 0$.

$$i_T(0) = 0 = -\frac{L}{R^2} + K_0$$

$$K_0 = \frac{L}{R^2}$$

Then, the total current and the voltages v_R and v_L are given by

$$i(t) = \frac{1}{R}t + \frac{L}{R^2}(e^{-\frac{R}{L}t} - 1)$$

$$v_R(t) = i(t)R = t + \frac{L}{R}(e^{-\frac{R}{L}t} - 1)$$

$$v_L(t) = L\frac{di(t)}{dt} = \frac{L}{R}(1 - e^{-\frac{R}{L}t})$$

Finally, since $R = 1 \Omega$, we have

$$i(t) = t + L(e^{-t/L} - 1)$$

$$v_R(t) = t + L(e^{-t/L} - 1)$$

$$v_L(t) = L(1 - e^{-t/L})$$

(a) When $L = 0.1H$

$$V_R = t + \frac{1}{10}(e^{-10t} - 1)V; \quad V_L = \frac{1}{10}(1 - e^{-10t})V$$

(b) When $L = 1H$

$$V_R = t + (e^{-t} - 1)V; \quad V_L = (1 - e^{-t})V$$

(c) When $L = 10H$

$$V_R = t + 10(e^{-\frac{t}{10}} - 1)V; \quad V_L = 10(1 - e^{-\frac{t}{10}})V$$

• **PROBLEM 2-13**

Use the impulse-train response to evaluate the steady-state component in the RL network shown. Assume that the input is $u_T(t) = \sin t$.

<u>Solution</u>: The impulse response ($h(t)$) in an RL circuit is given by the current i:

$$i(t) = h(t) = e^{At}bu_0(t) \quad \text{where } A = -\frac{R}{L} \text{ and } b = \frac{1}{L}.$$

Therefore, $h(t) = \frac{1}{L}e^{-(R/L)t}u_0(t)$. \hfill (1)

This result implies that if a voltage impulse function is applied to the RL series network, a current with exponential decay will flow in the network in the direction shown.

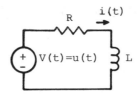

And so, when the voltage u(t) is the driving function and the current i(t) through the inductor is the desired response, the impulse response is h(t) (eq (1)).

When the impulse response is in the form $e^{-at}u_0(t)$, the impulse-train response is given as,

$$\frac{e^{-at}}{1 - e^{-at}} \quad . \tag{2}$$

Substituting eq. (1) for e^{-at} in eq. (2) the impulse-train response becomes

$$h_T(t) = \frac{(1/L)e^{-(R/L)t}}{1 - e^{-(R/L)T}} \quad 0 < t < T. \tag{3}$$

Use h_T to find the steady-state response ("finite-convolution") integral:

$$y_T(t) = \int_b^{b+T} u_T(\tau) h_T(t-\tau) d\tau \tag{4}$$

or $\quad y_T(t) = \int_b^{b+T} U_T(t-\tau) h_T(\tau) d\tau \tag{5}$

where constant b is arbitrary.

Substituting equation (3) into equation (5) where $T = 2\pi$ and b is chosen to be 0, yields,

$$y_T(t) = \int_0^{2\pi} \left[\frac{1/L}{1 - e^{-(R/L)2\pi}} \right] e^{-(R/L)\tau} \sin(t-\tau) d\tau$$

evaluation of this integral yields,

$$i_T(t) = \frac{1/L}{1 + (R/L)^2} \left[\frac{R}{L} \sin t - \cos t \right]$$

and

$$i_T(t) = \frac{1}{\sqrt{R^2 + L^2}} \sin(t-\theta)$$

where $\tan \theta = (L/R)$.

NOTE: For a sinusoidal input, the steady state response is also sinusoid of the same frequency but with a phase shift and different magnitude.

RC CIRCUITS (FORCED RESPONSE)

• **PROBLEM 2-14**

For the network shown in Fig. 1, the switch is closed after v_c has reached its final magnitude of 40 volts. Obtain the mathematical relationships and plot the curves for $v_c(t)$, $i_c(t)$ and $v_R(t)$.

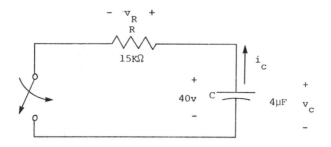

Solution: The time constant of the circuit,

$$\tau = RC = (15 \times 10^3 \Omega)(4 \times 10^{-6} F)$$

$$= 60 \text{ msec.}$$

Hence, $v_c(t)$, the voltage across the capacitor is equal to

$$Ee^{-t/\tau} = 40e^{-t/60 \times 10^{-3}} \text{ V}$$

$i_c(t)$ = current through the capacitor = $\frac{E}{R} e^{-t/\tau}$

$$= 2.67 \times 10^{-3} e^{-t/60 \times 10^{-3}} \text{ A}$$

and $v_R(t)$ = The voltage across the resistor

$$= Ee^{-t/\tau} = 40 e^{-t/60 \times 10^{-3}} \text{ V}$$

The curves for $v_c(t)$, $i_c(t)$ and $v_R(t)$ are plotted in Fig. 2.

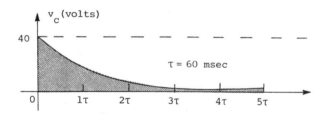

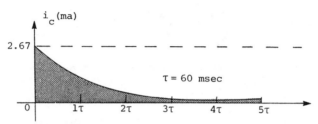

Fig. 2.

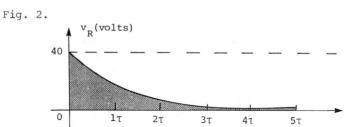

• **PROBLEM 2-15**

For the circuit in Fig. 1, the switch is closed at t = 0. Derive the relationships and sketch the curves for v_c, i_c and v_R.

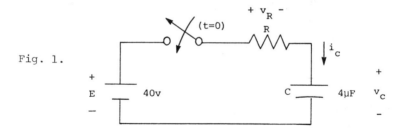

Fig. 1.

Solution: The time constant τ of the circuit is

$$= RC = (15 \times 10^3 \Omega)(4 \times 10^{-6} F) = 60 \times 10^{-3} \text{ sec}$$

$$= 60 \text{ m sec.}$$

The voltage, v_c, across the capacitor is

$$v_c = E(1 - e^{-t/\tau}) = 40(1 - e^{-t/60 \times 10^{-3}}) V$$

Hence,

$$i_c = \frac{E}{R}(e^{-t/\tau}) = \frac{40 \text{ V}}{15 \times 10^3 \Omega} e^{-t/60 \times 10^{-3}}$$

$$= 2.67 \cdot 10^{-3} e^{-t/60 \times 10^{-3}} A$$

The voltage, v_R, across the resistor is

$$v_R = E e^{-t/\tau} = 40 \, e^{-t/60 \times 10^{-3}} V$$

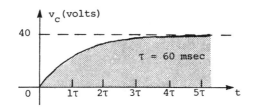

Fig. 2.

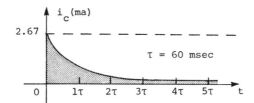

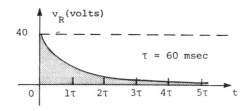

The curves for v_C, i_C, and v_R are sketched in Fig. 2.

● **PROBLEM 2-16**

(a) For the given network shown in Fig. A, if the switch is placed at position 1 at t=0, derive an expression for the instantaneous voltage across the capacitor (v_C). Compute the value of v_C at t = 10 m sec.

(b) Repeat part (a) for i_C (the current through the capacitor) and v_{R1} (the voltage across the resistor R_1).

(c) Assuming, the capacitor's leakage resistance to be infinite, derive relationships for the voltages v_C, v_{R1} and the current i_C when the switch is placed at position 2 at t = 30 m sec.

(d) If the position of the switch is changed from 2 to 3 at t = 48 m sec, derive relationships and obtain values for the voltages v_C, $v_{R1} + v_{R2}$, the current i_C, also at t = 100 m sec.

(e) For the cases (a) through (d), sketch the waveforms with the same time axis.

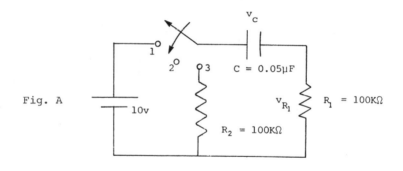

Fig. A

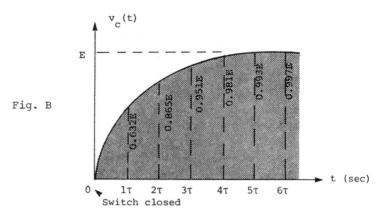

Fig. B

Solution: When the switch is in position 1,

τ = Time constant of the RC circuit

$= R_1 C = (100K)(0.05 \text{ }\mu F) = (100 \times 10^3 \Omega)(0.05 \times 10^{-6} F)$

$= 5 \times 10^{-3}$ sec = 5 ms

and $v_c(t) = E(1 - e^{-t/\tau})$.

Hence, $v_c(t) = 10 \ (1 - e^{-t/5 \times 10^{-3}})$ V

When $t = 10$ m sec,

$$v_c(t) = 10\left[1 - e^{-\left(\frac{10 \times 10^{-3}}{5 \times 10^{-3}}\right)}\right]$$

$= 10\left[1 - e^{-2}\right]$

$= 10(1 - 0.135)$

$= 8.65$ volts

Also note:

$t = 10$ m sec $= 2(5$ m sec$) = 2\tau$

Then, from Fig. B, when $t = 2\tau$, $v_c(t) = 0.865E$

$$= (0.865)(10) = 8.65 \text{ volts.}$$

which yields the same value.

(b) The voltage across resistor $R_1 = v_{R_1} = E - v_c$

Hence,
$$v_{R_1}(t) = E - v_c(t)$$
$$= E - E(1-e^{-t/\tau})$$
$$= E(e^{-t/\tau})$$
$$= 10e^{-t/\tau}$$
$$= 10e^{-t/5\times 10^{-3}}$$

Now,
$$i_c(t) = i_{R_1} = \frac{v_{R_1}(t)}{R_1} = \frac{10e^{-t/\tau}}{100K}$$
$$= \frac{10}{100\times 10^3} e^{-t/\tau}$$
$$= 10^{-4} e^{-t/5\times 10^{-3}} \text{ A}$$

When $t = 10$ m sec

$$i_c(t) = 10^{-4} e^{-\left(\frac{10\times 10^{-3}}{5\times 10^{-3}}\right)}$$
$$= 10^{-4} e^{-2}$$
$$= (10^{-4})(0.135)$$
$$= 1.35 \times 10^{-6} \text{ amp} = 13.5 \text{ }\mu\text{amp}$$

and
$$v_{R_1}(t) = 10e^{-t/\tau} = 10e^{-t/5\times 10^{-3}}$$
$$= 10e^{-\left(\frac{10\times 10^{-3}}{5\times 10^{-3}}\right)}$$
$$= 10e^{-2}$$
$$= 10(0.135)$$
$$= 1.35 \text{ volts}$$

(c) At $t = 30$ m sec, the position of the switch is changed from 1 to 2.

Since $t = 30$ m sec $> 5\tau = 5 \times 5 = 25$ m sec, we can assume that the capacitor has reached the final value of voltage (10 volts) just before changing the position of the switch from 1 to 2. Note: the voltage across the capacitor cannot change instantaneously and there is no path for the current to flow (i.e., open circuit). When the switch is in position 2, the value of the voltage across the capacitor will stay at the same value just before t = 30 m sec and, the current through the capacitor will be zero. Since the current in the resistor is zero (open-circuit), the voltage drop across the resistor is also zero.

i.e. $\quad v_c(t) = E = 10$ volts

$$v_R(t) = 0$$

$$i_c(t) = 0$$

(d) When the switch is placed into position 3 at t = 48 m sec, the capacitor c and the resistors R_1 and R_2 are in series and there is no source of energy. The voltage across the capacitor just before t = 48 m sec, is equal to 10 volts because there was no discharge of voltage across the capacitor from t = 30 m sec to t = 48 m sec. That is the reason why the voltage across the capacitor will discharge from 10 volts to zero volts with a time constant equal to τ', where τ' is given by,

$$\tau' = R'C \qquad \text{where } R' = R_1 + R_2$$

$$= (100 \text{ K} + 100 \text{ K})(0.05 \text{ }\mu\text{F})$$

$$= (200 \times 10^3 \Omega)(0.05 \times 10^{-6})$$

$$= 10 \times 10^{-3} \text{ sec}$$

$$= 10 \text{ m sec}$$

Hence for t > 48 m sec,

$$v_c(t) = 10 \, e^{-t'/\tau'}$$

where t' = t - 48 m sec.

t' accounts for the time delay of $v_c(t)$ by 48 m sec.

Therefore, $\quad v_c(t) = 10 e^{-(t-48 \times 10^{-3})/10 \times 10^{-3}}$ V

$$v_{R_1} + v_{R_2} = i_c(t)[R_1 + R_2]$$

where $\quad i_c(t)$ = current through the capacitor

= current in R_1 and R_2

$$= \frac{(v_{R_1} + v_{R_2})}{R_1 + R_2}$$

Now $v_{R_1} + v_{R_2}$ = Sum of the voltages across R_1 and R_2
= $-v_c$ [From Kirchoff's current law]
= $-10e^{-(t-48 \times 10^{-3})/10 \times 10^{-3}}$ V

$$i_c(t) = -\frac{10}{(R_1 + R_2)} e^{-(t-48 \times 10^{-3})/10 \times 10^{-3}}$$

$$= -\frac{10}{200 \times 10^3} e^{-(t-48 \times 10^{-3})/10 \times 10^{-3}}$$

$$= 5 \times 10^{-5} e^{-(t-48 \times 10^{-3})/10 \times 10^{-3}} \text{ A}$$

When t = 100 msec,

$$v_c(t) = 10e^{-(100 \times 10^{-3} - 48 \times 10^{-3})/10 \times 10^{-3}} = v_c(100 \text{ msec})$$

$$= 10e^{-5.2} = 10(0.006) = 0.06 \text{ volt}$$

$v_R(100 \text{ msec}) = v_{R_1}(100 \text{ msec}) + v_{R_2}(100 \text{ msec})$
= $-v_c(100 \text{ msec})$ = -0.06 volt

$$i_c(100 \text{ msec}) = 5 \times 10^{-5} e^{-(100 \times 10^{-3} - 48 \times 10^{-3})/10 \times 10^{-3}}$$

$$= 5 \times 10^{-5} e^{-5.2}$$

$$= 5 \times 10^{-5} (0.006)$$

$$= 3 \times 10^{-7} \text{ amp}$$

$$= 0.3 \text{ μamp}$$

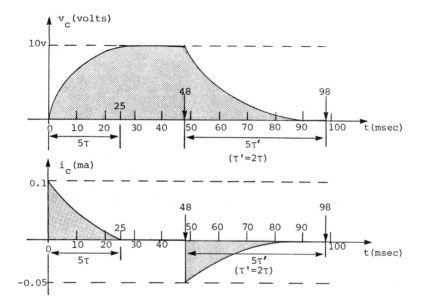

Fig. C

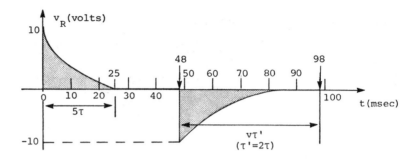

(e) The plots are shown in Fig. C. Note that the graphs corresponding to part (d) are delayed by 48 msec.

• **PROBLEM 2-17**

(a) Given the network shown in Fig. 1, derive a mathematical relationship for $v_c(t)$, when the switch is placed into position 1 at $t = 0$.

(b) Repeat part (a) if the position of the switch is changed from 1 to 2 at $t = 15$ msec.

(c) Plot $v_c(t)$ for both the above cases on the same time axis.

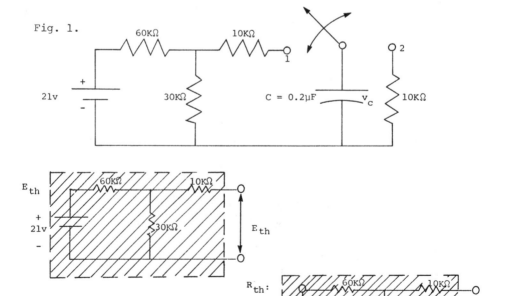

Solution: (a) When the switch is in position 1, the network to the left of the capacitor can be replaced by using Thevenin's equivalent.

Fig. 2 shows the equivalent diagrams for computing E_{th} (the open-circuit voltage across the capacitor) and R_{th} (the output resistance looking into the left of the terminals across the capacitor). Note that in order to compute R_{th}, the source (d.c. voltage) is to be replaced with a short circuit.

From Fig. 2,

$$E_{th} = \left(\frac{30K}{30K + 60K}\right)21 \text{ volts} = \frac{30}{90} \times 21 = 7 \text{ volts}$$

$$R_{th} = 10K + \frac{(30K)(60K)}{(30K + 60K)} = 10K + 20K = 30K\Omega$$

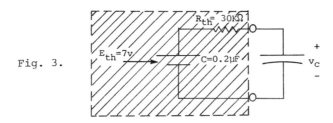

Fig. 3.

Fig. 3 shows the resultant equivalent circuit when the part of the network to the left of C is replaced with its Thevenin equivalent

From Fig. (3),

The instantaneous voltage across the capacitor,

$$v_c(t) = E_{th}(1-e^{-t/\tau}) \text{ V}$$

where τ = Time constant of the network shown in Fig. 3

$$\tau = R_{th}C = (30 \times 10^3 \Omega)(0.2 \times 10^{-6} \text{ F})$$

$$= 6 \times 10^{-3} \text{ sec} = 6 \text{ msec.}$$

Hence, $v_c(t) = 7(1-e^{-t/6 \times 10^{-3}}) \text{ V}$

(b) The voltage across a capacitor cannot change instantaneously. Therefore, the voltage across the capacitor just before t = 15 msec and just after t = 15 msec will be the same.

Therefore, $v_c(t = 15 \text{ msec}^+) = v_c(t = 15 \text{ msec}^-)$.

$v_c(t = 15 \text{ msec}^-)$ can be computed from the

equation for $v_c(t)$ given in part (a) as,

$$v_c(t = 15 \text{ msec}^-) = v_c(t = 15 \text{ msec}^+) = E$$

$$= 7\left[1-e^{-\left(\frac{15 \times 10^{-3}}{6 \times 10^{-3}}\right)}\right]$$

$$= 7\left[1-e^{-\frac{5}{2}}\right]$$

$$= 7(0.918)$$

$$= 6.426 \text{ volts}$$

When the switch is changed from position 1 to position 2, the capacitor will have acquired a charge of 6.426 volts. Therefore, when the switch is in position 2, since there is no source of energy, the voltage across the capacitor (6.426 volts) will discharge with a time constant τ', where τ' is the new time constant given by

$$\tau' = RC \quad \text{where } R = 10K\Omega$$

$$= (10 \times 10^3 \Omega)(0.2 \times 10^{-6} \text{ F})$$

$$= 2 \times 10^{-3} \text{ sec}$$

$$= 2 \text{ msec}$$

Hence, when the switch is in position 2, the instantaneous voltage across the capacitor is

$$v_c(t) = Ee^{-t'/\tau'} \text{ V}$$

where $\quad E = v_c(t = 15 \text{ msec}^-) = v_c(t = 15 \text{ msec}^+)$

$$= 6.426 \text{ volts}$$

and $\quad t' = (t-15)$ msec.

Resulting in $\quad v_c(t) = 6.426 \, e^{-t'/2 \times 10^{-3}}$ V

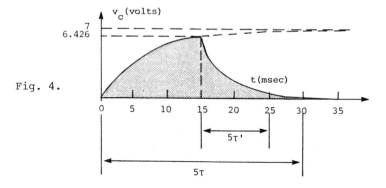

Fig. 4.

(c) $v_c(t)$ is plotted in Fig. 4. Note that the plot for $v_c(t)$ corresponding to the case (b) is delayed by $t = 15$ msec.

• **PROBLEM 2-18**

If the switch is placed to a closed position at $t = 0$ in the circuit of Fig. A, derive an expression for the voltage $v_c(t)$.

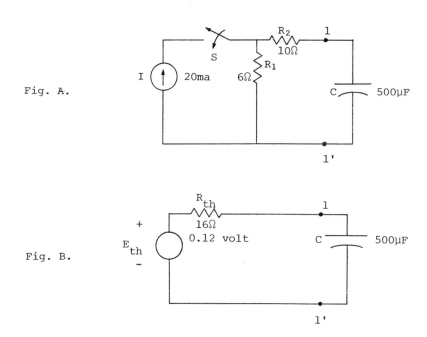

Fig. A.

Fig. B.

Solution: When the switch is closed at $t = 0$, the part of the network to the left of the terminals across the capacitor (1, 1') can be replaced by using Thevenin equivalent as shown in Fig. B.

From Fig. A,

E_{th} = open-circuit voltage across the terminals (1, 1')

$= IR_1 = (20 \times 10^{-3})(6) = 0.12$ volt

R_{th} = Output resistance across the terminals (1, 1') looking into the left

$= R_1 + R_2$
$= 6\Omega + 10\Omega = 16\Omega$

Note that R_{th} can be found by open circuiting the current source.

From Fig. B, the time constant of the circuit is

$$\tau = R_{th}C = (16\Omega)(500\ \mu F)$$

$$= (16 \times 500 \times 10^{-6}) = 8 \times 10^{-3}\ \text{sec}$$

$$= 8\ \text{msec}$$

Hence,
$$v_c(t) = E_{th}(1-e^{-t/\tau})$$

$$= 0.12\ (1-e^{-t/8 \times 10^{-3}})\ V$$

● **PROBLEM 2-19**

Derive a relationship for the instantaneous voltage $v_c(t)$ across the capacitor shown in Fig. 1, if the switch is in the closed position. Assume that the initial voltage of the capacitor is equal to 80 volts.

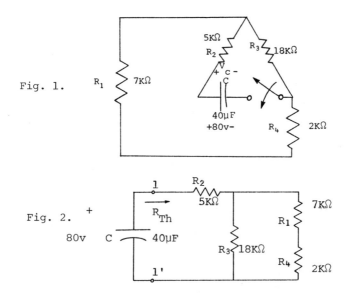

Solution: When the switch is closed, the initial voltage across the capacitor will slowly discharge to zero volts at a time constant τ which is found below. The equivalent circuit of Fig. 1 is shown in Fig. 2.

$$\tau = \text{Time constant of the circuit}$$

$$= R_{th}C$$

where R_{th} is the output resistance across the terminals (1, 1') looking into the right of the circuit of Fig. 2.

R_{th} can be obtained by taking the equivalent resistance of resistors R_1, R_4 and R_3.

Hence,
$$R_{th} = 5K + \{18K \parallel (7K + 2K)\}$$
$$= 5K + \{18K \parallel 9K\}$$
$$= 5K + \frac{(18K)(9K)}{(18K + 9K)}$$
$$= 5K + \frac{(18K)(9K)}{(27K)}$$
$$= 5K + 6K$$
$$= 11 \text{ K}\Omega$$

Therefore, $\tau = R_{th}C$
$$= (11 \times 10^3)(40 \times 10^{-6}) = 0.44 \text{ sec}$$

$v_c(t)$ = Instantaneous voltage across the capacitor
$$= v_0 (e^{-t/\tau}) \text{ V} \tag{1}$$

where v_0 = Initial voltage across the capacitor
$$= 80 \text{ volts}$$

Substituting in values for τ and v_0 into Eq. 1,
$$v_c(t) = 80e^{-t/0.44} \text{ V}$$

• **PROBLEM 2-20**

For the circuit shown in Fig. 1, find $v_c(t)$ and $i_c(t)$ if i_s = (a) $25u(t)$ mA; (b) $10 + 15u(t)$ mA.

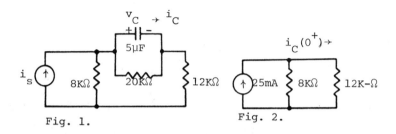

Fig. 1. Fig. 2.

Solution: (a) Note that $v_c(0^+) = 0$, this means that at $t = 0^+$ the capacitor short-circuits the 20k-Ω resistor, Fig. 2 shows the circuit used to find $i_c(0^+)$.
 Using current division,

$$i_c(0^+) = \frac{25\,(8)}{20} = 10 \text{ mA}.$$

The voltage across the capacitor can be found using the relation,

$$v_c(t) = V_0 + \frac{1}{C}\int_{t_0}^{t} i(T)\,dT$$

$$v_c(t) = 0 + \frac{1}{5\mu F}\int_{0}^{t} 10 \times 10^{-3}\, e^{-\frac{T}{R_{eq}C}}\,dT$$

R_{eq} can be found by open circuiting the current source.

$$R_{eq} = ((8+12)||20)k = 10k\,\Omega$$

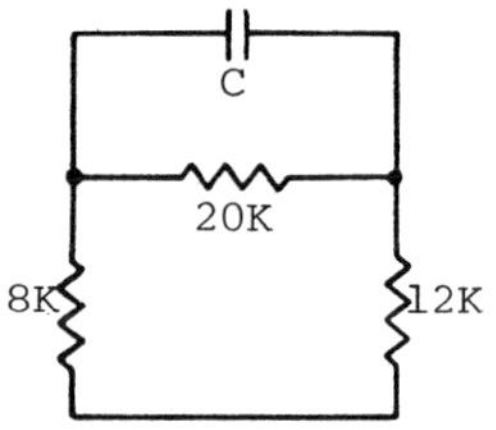

$$R_{eq}C = 5\mu F\,(10k) = .05$$

$$v_c(t) = \frac{10 \times 10^{-3}}{5 \times 10^{-6}} \left(-\frac{1}{20}\right) \left[e^{-20T} \right]_0^t$$

Fig. 3.

$$v_c(t) = 100\,(e^{-20t} - 1)\,u(t)\,V$$

$$i_c(t) = 10\,e^{-20t}\,u(t)\,mA.$$

(b) Fig. 4 shows a plot of i_s for this problem.

Knowing that $v_c(0^+) = v_c(0^-)$, find $v_c(0^+)$ for the circuit when it is in steady state condition with $i_s = 10$ mA. Fig. 5 shows that the voltage across the 20-kΩ resistor is

$$v_c(0^-) = v_c(0^+).$$

By current division,

$$v_c(0^+) = v_c(0^-) = \frac{10\,(8)}{40}\,(20) = 40\text{ V}.$$

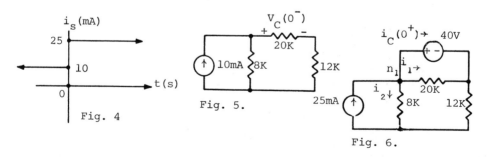

Fig. 4

Fig. 5.

Fig. 6.

One can now find $i_c(0^+)$ by replacing the capacitor with a 40-V source in the circuit for $t = 0^+$ shown in Fig. 6.

By summing the currents at node n_1 we have

$$25 \text{ mA} = i_1 + i_2 + i_c(0^+)$$

$$i_1 = \frac{40v}{20k} = 2 \text{ mA}.$$

Find i_2 by using the superposition theorem on the circuit in Fig. 6.

$$i_a = \frac{25 \ (12)}{20} = 15 \text{ mA}$$

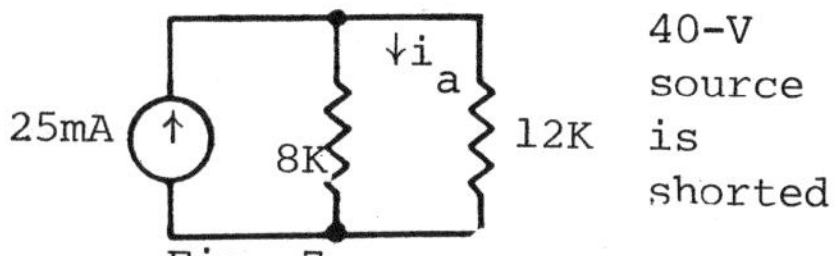

Fig. 7. 25mA source, 8K, 12K, 40-V source is shorted, $\downarrow i_a$

$$i_b = \frac{40v}{20k} = 2 \text{ mA}$$

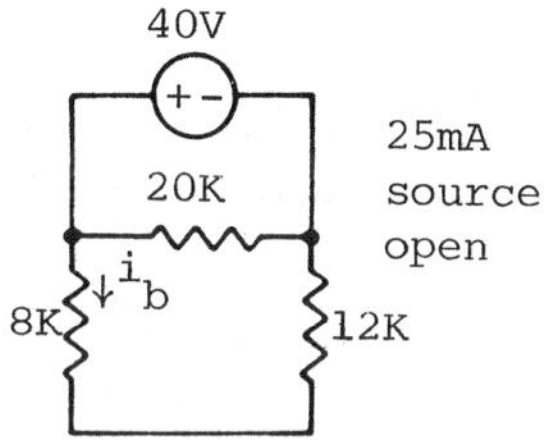

Fig. 8. 40V, 20K, 8K, 12K, 25mA source open, $\downarrow i_b$

$$i_2 = i_a + i_b = (15 + 2) \text{ mA} = 17 \text{ mA}$$

$$i_c(0^+) = (25 - 2 - 17) \text{ mA} = 6 \text{ mA}$$

Knowing $i_c(0^+) = 6$ mA and $v_c(0^+) = 40$ V makes it possible to find $i_c(t)$ and $v_c(t)$.

$$i_c(t) = 6 \ e^{-20t} \ u(t) \text{ mA}$$

$$v_c(t) = v_0 + \frac{1}{C} \int_{t_0}^{t} i(T) \ dT$$

$$v_c(t) = 40 + \frac{1}{5\mu F} \int_{0}^{t} 6 \ e^{-20T} \text{ mA } dT$$

$$v_c(t) = 40 + \frac{6 \times 10^{-3}}{5 \times 10^{-6}} \left(-\frac{1}{20}\right) \left[e^{-20T}\right]_0^t$$

$$v_c(t) = 40 - 60 \ (e^{-20t} - 1) \ u(t) \text{ V}.$$

$$v_c(t) = 40 + 60(1 - e^{-20t}) \ u(t) \text{ V}.$$

● PROBLEM 2-21

For the network of fig. 1

(a) Choose i(t) as the unknown of the circuit and write a first-order differential equation in i(t).

(b) Give the homogeneous solution to the equation obtained in (a).

(c) Give the particular solution to the equation obtained in (a).

(d) Write down the total solution of the equation obtained in (a) and determine the unknown constant using the initial condition.

(e) Sketch the plot of i(t) versus t.

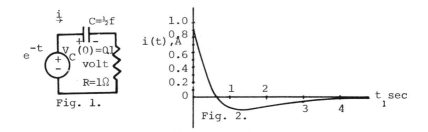

Fig. 1.

Fig. 2.

Solution: (a) In general, for a series RC circuit,

$$\frac{1}{C}\int i(t)\,dt + Ri(t) = v(t).$$

For this problem we have

$$2\int i\,dt + i = e^{-t}.$$

A differential equation is obtained by differentiating the above equation with respect to t. Thus

$$2i + \frac{di}{dt} = -e^{-t}.$$

(b) The homogeneous equation is

$$2i + \frac{di}{dt} = 0.$$

Assume a solution of the form $i = e^{mt}$.

Substituting this into the differential equation gives the characteristic equation,

$$2 + m = 0$$

whose solution is m = 2. Hence

$i_h = Ke^{-2t}$ A.

(c) The particular solution is found by assuming a current of the same form as the driving function. $i = ae^{-t}$. Substituting this into the differential equation gives

$$2ae^{-t} - ae^{-t} = -e^{-t}.$$

Solving for a

$$2a - a = -1$$

or

$$a = -1.$$

The particular solution is

$$i_p = -e^{-t} \text{ A.}$$

(d) The total solution is

$$i = i_h + i_p = Ke^{-2t} - e^{-t}.$$

At t = 0 the total voltage across R is $e^{-0} - 0.1 = 0.9$ volts. The current through R is 0.9/1 = 0.9 amps. Evaluating i for t = 0 gives

$$i(0) = K - 1 = 0.9 \text{ A.}$$

Solving for K we get K = 1.9.

The final solution is

$$i(t) = 1.9e^{-2t} - e^{-t} \text{A.}$$

(e) A plot of i(t) is shown in fig. 2.

• **PROBLEM 2-22**

Find the unit impulse response of the network of fig. 1. Assume zero initial conditions and $v(t) = \delta(t)$.

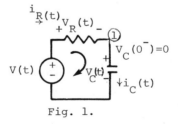

Fig. 1.

Solution: Writing Kirchoff's voltage law around the loop gives

$$v_R(t) + v_C(t) = v(t) \tag{1}$$

Using voltage current relationship of the capacitor and Kirchoff's current law at node 1 we get;

$$i_R(t) = i_C(t) = C \frac{dv_C}{dt} \qquad (2)$$

Using $v_R(t) = i_R(t)R$ and combining eqs (1) and (2) yields the differential equation defining the response $v_C(t)$:

$$RC \frac{dv_C(t)}{dt} + v_C(t) = \delta(t) \qquad v_C(0^-) = 0$$

to get the homogeneous solution of this equation we set the equation equal to zero and use definitions given below.

NOTE: The diff. input function is given by

$$a_1 \frac{dx(t)}{dt} + a_0 x(t) = 0$$

then the homogeneous solution is

$$x_h(t) = k e^{-(a_0/a_1)t}$$

The homogeneous solution is therefore;

$$[v_C(t)]_h = k e^{-t/RC}$$

The particular solution to a differential equation is any solution which satisfies the equation:

$$[v_C(t)]_p = \frac{1}{RC} e^{-t/RC} u_0(t)$$

The total solution is the addition of homogeneous and particular solutions.

$$v_C(t) = [v_C(t)]_h + [v_C(t)]_p$$

Since the initial conditions are zero the homogeneous solution, which describes the response of the circuit before the application of any forcing function, will be zero and the total solution is

$$v_C(t) = [v_C(t)]_p = \frac{1}{RC} e^{-t/RC} u_0(t).$$

● **PROBLEM 2-23**

In the circuit of fig. 1 the switch has been closed for a long time and is opened at $t = t_0$. Find the voltage $V_C(t)$ for $t \geq t_0$. How does the phase angle ϕ affect voltage $V_C(t)$?

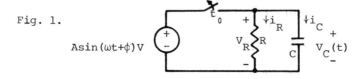

Fig. 1.

Solution: We write the first-order homogeneous differential equation by removing the voltage source and writing KVL and KCL equations for the remaining RC circuit.

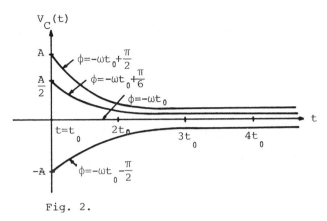

Fig. 2.

The KVL equation is

$V_C(t) = V_R(t)$, and the KCL equation is

$i_R(t) = -i_C(t) \quad t \geq t_0$.

Since $i_C = C\dfrac{dV_C}{dt}$ and $i_R(t) = \dfrac{V_R(t)}{R}$, substitute into the KCL equation above to obtain

$$\dfrac{V_R(t)}{R} + C\dfrac{dV_C}{dt} = 0.$$

Substitute $V_C(t)$ for $V_R(t)$ from the KVL equation, obtaining the differential equation.

$$RC\dfrac{dV_C(t)}{dt} + V_C(t) = 0 \quad t \geq t_0$$

Recognize that the solution of this equation is of the form

$$V_C(t) = ke^{-t/RC} \quad t \geq t_0 \quad (1)$$

where k must be determined from the initial conditions.

Find the initial condition $V_C(t_0^+)$ by noting that $V_C(t_0^-) = V_C(t_0^+)$, since the voltage across a capacitor cannot change instantaneously. Noting that $V_C(t_0^-)$ is the source voltage just before the switch is opened,

$$V_C(t_0^-) = V_C(t_0^+) = A \sin(\omega t_0 + \phi), \quad (2)$$

gives an equation for k. From the solution of the homogeneous differential equation Eq (1) at $t = t_0$ we have

$$V_C(t_0) = ke^{-t_0/RC} = A \sin(\omega t_0 + \phi)$$

solving for k gives

$$k = \frac{A \sin(\omega t_0 + \phi)}{e^{-t/RC}} = A \sin(\omega t_0 + \phi) e^{t/RC}$$

Substituting this equation for k into our solution Eq (1) gives

$$V_C(t) = A \sin(\omega t_0 + \phi) e^{t_0/RC} e^{-t/RC}$$

$$V_C(t) = A \sin(\omega t_0 + \phi) e^{-(t-t_0)/RC}.$$

We note that the phase angle ϕ affects the amplitude B of $V_C(t)$ where $B = A \sin(\omega t_0 + \phi)$.

We note that $B = A$ when $\omega t_0 + \phi = \frac{\pi}{2}, \frac{-3\pi}{2}, \frac{5\pi}{2}, \frac{-7\pi}{2} \ldots$.
(see fig. 2)

● **PROBLEM 2-24**

Find the transient and steady-state current if the waveforms shown in fig. 1 are applied to

(i) series RC network ($R = C = 1$).

(ii) series RL network ($R = L = 1$).

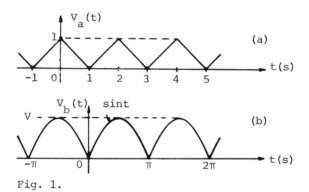

Fig. 1.

Solution: In each part of this problem, the procedure will be to solve a first order linear differential equation for the particular integral solution, and then, choosing the appropriate constraints, solve for the integration constants. The transient solution is found second because its integration constant is determined from the steady-state solution.
(i) The differential equation for this circuit is found from the voltage equation,

$$V(t) = Ri + \frac{1}{C}\int i \, dt.$$

Differentiating, we get

$$\frac{dV(t)}{dt} = R\frac{di}{dt} + \frac{i}{C}i = \frac{di}{dt} + i \qquad (1)$$

where we have used R = C = 1.

Waveform (a):
There is no equation for a triangular wave so we divide the wave into the two sections which we can then treat separately. $0 \le t \le 1$: The voltage is $V_a(t) = 1 - t$.

Substitute this into equation (1) to get

$$-1 = \frac{di_1}{dt} + i_1$$

The solution is found using the integration factor $e^{\int dt} = e^t$. Thus

$$-1\, e^t = e^t \frac{di_1}{dt} + e^t i_1 = \frac{d}{dt}(i_1 e^t).$$

Integrating,

$$-e^t + k_1 = i e^t$$

or

$$i_1 = k_1 e^{-t} - 1. \qquad (2)$$

$1 \le t \le 2$: The voltage is $V_a(t) = t - 1$.

Substituting $V_a(t)$ into equation (1) we get

$$1 = \frac{di_2}{dt} + i_2.$$

Solving for i_2 as above gives

$$i_2 = k_2 e^{-t} + 1. \qquad (3)$$

Next we solve for k_1 and k_2 by applying constraints which tie the two solutions together. The first is

$$i_1(0) = i_2(2)$$

which is necessary for periodicity.

Placing equations (2) and (3) into this constraint gives

$$k_1 - 1 = 1 + k_2 e^{-2}$$

or
$$k_2 = (k_1 - 2)\, e^2 \qquad (4)$$

The second constraint is

$$i_1(1) = i_2(1).$$

Again using equations (2) and (3) gives

$$-1 + k_1 e^{-1} = 1 + k_2 e^{-1}$$

or $\quad k_2 = k_1 - 2e.$ (5)

Equating equations (4) and (5), and solving for k_1, yields

$$k_1 = 1.462 \qquad (6)$$

Putting this into equation (5) we find

$$k_2 = 1.462 - 2e. \qquad (7)$$

The final step is to place the values of k_1 and k_2 into their respective equations, (2) and (3). The results are

$$i_1(t) = 1.462\, e^{-t} - 1. \qquad 0 \le t \le 1$$

and $\quad i_2(t) = (1.462 - 2e)\, e^{-t} + 1$

$$\qquad\qquad = 1.462 e^{-t} - 2e^{-(t-1)} + 1 \qquad 0 \le t \le 2,$$

for the steady-state solution.

The transient solution is found from the homogeneous part of equation (1) which is

$$\frac{di}{dt} + i = 0$$

The characteristic equation is $m + 1 = 0$ so the solution is

$$i_{tr} = A e^{-t}.$$

The total current for $0 \le t \le 1$ is the sum of the transient and steady-state solutions,

$$i(t) = A e^{-t} + 1.462 e^{-t} - 1 \qquad 0 \le t \le 1.$$

The voltage is

$$V(t) = i + \frac{di}{dt} = 0.462 + A - t$$

which is found by substituting $i(t)$ into the equation. At $t = 0$, the applied voltage is 1. Hence

$$V(0) = 1 = 0.462 + A$$

and $A = 0.538$.

The complete transient solution is

$$i_{tr}(t) = 0.538 e^{-t} \qquad t \ge 0$$

or $\quad i_{tr}(t) = 0.538 e^{-t} u(t).$

Waveform (b):

The solution for i proceeds the same as for waveform (a), except in this case it is easier since we do not have to divide the wave into two segments. Again use equation (1).

$$\frac{dV_b(t)}{dt} = \frac{di(t)}{dt} + i(t).$$

Setting $V_b(t) = \sin t$ we get

$$\cos t = \frac{di}{dt} + i. \qquad 0 \le t \le \pi$$

Multiplying by the integration factor and solving,

$$e^t \cos t = e^t \frac{di}{dt} + e^t i = \frac{d}{dt}(e^t i)$$

$$\tfrac{1}{2} e^t (\cos t + \sin t) + k_3 = i e^t$$

or $\quad i(t) = \tfrac{1}{2}(\cos t + \sin t) + k_3 e^{-t}$

To solve for k_3 use $i(0) = i(\pi)$. Then

$$\tfrac{1}{2} + k_3 = -\tfrac{1}{2} + k_3 e^{-\pi}$$

or $\quad k_3 = -1.045.$

The steady-state solution is

$$i(t) = \tfrac{1}{2}(\cos t + \sin t) - 1.045 e^{-t}$$

or $\quad i(t) = 0.707 \sin(t + 45°) - 1.045 e^{-t} \qquad 0 \le t \le \pi$

The form of the transient solution is the same as the triangular wave case,

$$i_{tr}(t) = B e^{-t}.$$

The total solution is

$$i(t) = B e^{-t} + 0.707 \sin(t + 45°) - 1.045 e^{-t}.$$

At $t = 0$, the applied voltage is zero so the initial current is zero. Hence

$$i(0) = 0 = B + 0.5 - 1.045.$$

Thus $B = 0.545$.

The transient solution is

$$i_{tr}(t) = 0.545 e^{-t} u(t).$$

(ii) RL network.

The differential equation for this circuit is

$$V(t) = Ri(t) + L \frac{di(t)}{dt} = i + \frac{di}{dt}. \qquad (8)$$

Waveform (a):

Time interval $0 \le t \le 1$:

The applied voltage is $V(t) = 1 - t$. Then equation (8) gives

$$1 - t = i_1 + \frac{di_1}{dt}$$

whose solution is

$$i_1(t) = 2 - t + k_4 e^{-t}. \qquad (9)$$

Time interval $1 \le t \le 2$:

The applied voltage is $V(t) = t - 1$. Using this equation (8) becomes

$$t - 1 = i_2 + \frac{di_2}{dt}$$

whose solution is

$$i_2(t) = t - 2 + k_5 e^{-t}. \qquad (10)$$

Two constraints used to solve for k_4 and k_5 are

$$i_1(0) = i_2(2)$$
$$i_1(1) = i_2(1).$$

Applying these constraints to equations (9) and (10) gives

$$k_5 = (2 + k_4)e^2$$

and $\quad k_5 = 2e + k_4.$

Solving for the constants, we get

$$k_4 = -1.462$$
$$k_5 = 2e - 1.462.$$

Placing the constants into equations (9) and (10) gives the final steady-state solution,

$$i_1(t) = 2 - t - 1.462\, e^{-t} \qquad 0 \le t \le 1$$
$$i_2(t) = t - 2 - 1.462\, e^{-t} + 2e^{-(t-1)} \qquad 1 \le t \le 2$$

To find the transient solution use the homogeneous form of equation (8),

$$i + \frac{di}{dt} = 0$$

which has a solution of the form ce^{-t}. The total current in the region $0 \le t \le 1$ is

$$i(t) = ce^{-t} + 2 - t - 1.462e^{-t}.$$

The inductor prohibits current from flowing at t = 0, so

$$i(0) = 0 = C + 2 - 1.462$$

and $C = -0.538$.

The transient solution is

$$i_{tr}(t) = -0.538e^{-t} u(t).$$

Waveform (b):

In the time interval $0 \le t \le \pi$ the differential equation with applied voltage sin t is

$$\sin t = i + \frac{di}{dt}.$$

Solving with the integration factor e^t,

$$e^t \sin t = \frac{d}{dt}(e^t i)$$

so $i(t) = \frac{1}{2}(\sin t - \cos t) + k_6 e^{-t}$

Determine k_6 from the condition $i(0) = i(\pi)$. Then

$$-\frac{1}{2} + k_6 = \frac{1}{2} + k_6 e^{-\pi}$$

and $k_6 = 1.045$

The steady-state solution is

$$i(t) = \frac{1}{2}(\sin t - \cos t) + 1.045\, e^{-t}$$

or

$$i(t) = 0.707 \sin(t - 45°) + 1.045 e^{-t} \quad 0 \le t \le \pi$$

The form of the transient solution is

$$i_{tr}(t) = De^{-t}$$

and the total current is

$$i(t) = De^{-t} + 0.707 \sin(t - 45°) + 1.045\, e^{-t}.$$

At t = 0, the current is zero so

$$i(0) = 0 = D - 0.5 + 1.045$$

and

$$D = -0.545.$$

The transient solution is

$$i_{tr}(t) = -0.545 e^{-t} u(t)$$

CHAPTER 3

THE FOURIER SERIES

TRIGONOMETRIC FOURIER SERIES

● **PROBLEM 3-1**

Consider the infinite trigonometric series

$$\frac{a_0}{2} + \sum_{n=1}^{\infty} (a_n \cos nx + b_n \sin nx)$$

and assume that it converges uniformly for all $x \in (-\pi, \pi)$. It can then be considered as a function f of x with period 2π, i.e.

$$f(x) = \frac{a_0}{2} + \sum_{n=1}^{\infty} a_n \cos nx + b_n \sin nx. \qquad (1)$$

Determine the values of a_n, b_n in terms of $f(x)$.

Solution: It is this computation which leads to the definition of the Fourier Series of a given function $f(x)$. First multiply both sides of (1) by $\cos mx$ where m is a positive integer which we will vary later. This yields

$$f(x) \cos mx = \frac{a_0}{2} \cos mx + \sum_{n=1}^{\infty} a_n \cos nx \cos mx$$

$$+ \sum_{n=1}^{\infty} b_n \sin nx \cos mx. \qquad (2)$$

The next step is to integrate both sides of equation (2) from $-\pi$ to π. In order to integrate the two series on the right term by term these two series would have to be uniformly convergent, but since this exercise is intended only to motivate a definition, we will simply assume that term-wise integration is valid. Thus, (2) becomes

$$\int_{-\pi}^{\pi} f(x)\cos mx\,dx = \frac{a_0}{2}\int_{-\pi}^{\pi}\cos mx\,dx + \sum_{n=1}^{\infty}\left(a_n\int_{-\pi}^{\pi}\cos nx\cos mx\,dx\right)$$

$$+ \sum_{n=1}^{\infty}\left(b_n\int_{-\infty}^{\infty}\sin nx\cos mx\,dx\right) \quad (3)$$

This rather formidable expression yields useful information if one recalls the trigonometric identities

$$\sin nx \sin mx = \frac{1}{2}\cos(n-m)x - \frac{1}{2}\cos(n+m)x \quad (4)$$

$$\cos nx \cos mx = \frac{1}{2}\cos(n+m)x + \frac{1}{2}\cos(n-m)x \quad (5)$$

$$\sin nx \cos mx = \frac{1}{2}\sin(n+m)x + \frac{1}{2}\sin(n-m)x. \quad (6)$$

Using these three identities, the following equations may be verified by carrying out the integrations:

$$\int_{-\pi}^{\pi}\sin nx\cos mx\,dx = 0 \quad \text{(for all } n,m>0) \quad (7)$$

$$\int_{-\pi}^{\pi}\cos nx\cos mx\,dx = \begin{cases} 0 & \text{(if } n \neq m) \\ \pi & \text{(if } n = m) \end{cases} \quad (8)$$

$$\int_{-\pi}^{\pi}\sin nx\sin mx\,dx = \begin{cases} 0 & \text{(if } n \neq m) \\ \pi & \text{(if } n = m) \end{cases} \quad (9)$$

For instance, using the identity (4) in the integral of equation (9) yields

$$\int_{-\pi}^{\pi}\sin nx\sin mx\,dx = \frac{1}{2}\int_{-\pi}^{\pi}\cos(n-m)x\,dx - \frac{1}{2}\int_{-\pi}^{\pi}\cos(n+m)x\,dx. \quad (10)$$

If $n \neq m$, then

$$\int_{-\pi}^{\pi}\cos(n-m)x\,dx = \left.\frac{\sin(n-m)x}{n-m}\right|_{-\pi}^{\pi} = 0$$

and if $n = m$, then

$$\int_{-\pi}^{\pi} \cos(n-n)x\,dx = \int_{-\pi}^{\pi} dx = 2\pi.$$

Also,

$$\int_{-\pi}^{\pi} \cos(n+m)x\,dx = \left.\frac{\sin(n+m)x}{n+m}\right|_{-\pi}^{\pi} = 0 \quad \text{(for all } n,m>0\text{)}.$$

Using these results in (10) yields the result quoted in (9) and the other formulas are established in a similar fashion. These formulas are called the orthogonality properties of sin and cos.

Returning to the series in (3), it is seen that all terms in the second sum are zero (by equation (7)) and that for any m, only one term in the first sum is nonzero by equation (8). That is, for $m>0$,

$$\int_{-\pi}^{\pi} \cos mx\,dx = 0$$

so that (3) gives

$$\int_{-\pi}^{\pi} f(x)\cos mx\,dx = a_m \pi \quad (m>0). \tag{11}$$

The coefficients b_n are treated similarly, that is the expansion (1) is multiplied by sin mx and integrated. Again the orthogonality properties (7)-(9) are employed to yield

$$\int_{-\pi}^{\pi} f(x)\sin mx\,dx = b_m \pi. \tag{12}$$

Finally, to obtain a_o, simply integrate the expansion (1) as it stands from $-\pi$ to π. This results in

$$\int_{-\pi}^{\pi} f(x)\,dx = a_o \pi. \tag{13}$$

The results of equations (11), (12), (13) may be summarized as

$$a_n = \frac{1}{\pi}\int_{-\pi}^{\pi} f(x)\cos nx\,dx \quad (n \geq 0) \tag{14}$$

$$b_n = \frac{1}{\pi}\int_{-\pi}^{\pi} f(x)\sin nx\,dx \quad (n>0). \tag{15}$$

Thus, it has been proved that if a function f is representable by a uniformly convergent trigonometric series then that series must have the coefficients of equations (14) and (15).

• **PROBLEM 3-2**

Find the Fourier sine series of $f(x) = x^2$ over the interval $(0,1)$.

Solution: The Fourier sine series of a function defined on an interval $(0,c)$ is given by

$$f(x) \sim_s \sum_{n=1}^{\infty} b_n \sin\left(\frac{n\pi x}{c}\right) \tag{1}$$

where

$$b_n = \frac{2}{c} \int_0^c f(x) \sin\left(\frac{n\pi x}{c}\right) dx \tag{2}$$

We now turn to the given problem. Here $f(x) = x^2$, and $c = 1$. Thus we obtain the Fourier series

$$x^2 \sim_s \sum_{n=1}^{\infty} b_n \sin n\pi x \tag{3}$$

where

$$b_n = 2 \int_0^1 x^2 \sin n\pi x \, dx.$$

We must evaluate the b_n. Using integration by parts, we obtain

$$b_n = 2 \int_0^1 x \sin n\pi x \, dx$$

$$= 2 \left\{ \left[\frac{-x^2}{n\pi} \cos n\pi x \right]_0^1 + \frac{2}{n\pi} \int_0^1 x \cos n\pi x \, dx \right\}$$

$$= 2 \left\{ \frac{-(-1)^n}{n\pi} + \frac{2}{n^2 \pi^2} x \sin n\pi x \Big|_0^1 \right.$$

$$\left. - \int_0^1 \frac{2}{n^2 \pi^2} \sin n\pi x \, dx \right\}$$

$$= 2 \left\{ \frac{(-1)^{n+1}}{n\pi} + \frac{2}{n^3 \pi^3} [(-1)^n - 1] \right\}. \tag{4}$$

Substituting (4) for b_n in (3), the required Fourier sine series over $0 < x < 1$ is

$$x^2 \underset{s}{\sim} 2 \sum_{n=1}^{\infty} \left[\frac{(-1)^{n+1}}{n\pi} - \frac{2\{1-(-1)^n\}}{n^3\pi^3} \right] \sin n\pi x.$$

● **PROBLEM 3-3**

Find the Fourier sine series for the function defined by

$f(x) = 0 \qquad 0 \leq x < \pi/2$

$f(x) = 1 \qquad \pi/2 < x \leq \pi.$

Solution: The Fourier sine series of a function defined on $0 \leq x \leq L$

$$f(x) \underset{s}{\sim} \sum_{n=1}^{\infty} b_n \sin \frac{n\pi x}{L} \qquad (n = 1, 2, \ldots) \qquad (1)$$

where

$$b_n = \frac{2}{L} \int_0^L \sin nx \, dx. \qquad (2)$$

Recall that the Fourier sine series is equivalent to finding the Fourier trigonometric series of an odd function, i.e., a function such that $f(-x) = -f(x)$.

In the given problem, since $f(x) = 0$ for $0 \leq x \leq \pi/2$ we need find its series development only for the interval $\pi/2 < x \leq \pi$. Taking $L = \pi$,

$$b_n = \frac{2}{\pi} \int_{\pi/2}^{\pi} \sin nx \, dx = -\frac{2}{n\pi} \left(\cos nx \Big|_{\pi/2}^{\pi} \right)$$

$$= \frac{2}{n\pi} \left[\cos \frac{n\pi}{2} - \cos n\pi \right] = \frac{2}{n\pi} \left[\cos \frac{n\pi}{2} + (-1)^{n+1} \right]. \qquad (3)$$

Substituting (3) into (1),

$$f(x) \underset{s}{\sim} \frac{2}{\pi} \left[\frac{\sin x}{1} - \frac{2\sin 2x}{2} + \frac{\sin 3x}{3} + \frac{\sin 5x}{5} \right.$$

$$\left. - \frac{2\sin 6x}{6} + \ldots \right].$$

• **PROBLEM 3-4**

Let
$$f(x) = \begin{cases} \pi, & -\pi \leq x < 0 \\ x, & 0 \leq x \leq \pi \end{cases}$$

have the trigonometric Fourier series relative to the orthonormal system $\{\cos n\pi x, \sin n\pi x;\ n = 0,1,2,\ldots\}$

$$f(x) \sim \frac{3\pi}{4} + \sum_{n=1}^{\infty}\left[\frac{(-1)^n - 1}{\pi n^2} \cos nx - \frac{1}{n}\sin nx\right]. \tag{a}$$

What is the trigonometric Fourier series of the function

$$g(x) = \begin{cases} \pi, & -\pi \leq x < 0 \\ \frac{\pi}{2}, & x = 0 \\ x, & 0 < x \leq \pi \end{cases}.$$

Solution: Examining $g(x)$, we see that it is identical with $f(x)$ in the interval $-\pi \leq x \leq \pi$ except at the point $x = 0$. At $x = 0$, $f(0) = 0$ while $g(0) = \pi/2$. Therefore, we can reasonably assume that $g(x)$ will have a trigonometric Fourier series closely similar to (a).

Now, it is a theorem of calculus that if $g(x)$ is identical with $f(x)$ over a given interval except possibly for a finite number of points, then

$$\int_a^b f(x)dx = \int_a^b g(x)dx.$$

An intuitive justification for this is that single points have zero "area". Therefore the presence or absence of such points does not affect the value of the definite integral of a function.

By the above reasoning,

$$\frac{1}{\pi}\int_{-\pi}^{\pi} g(x) \cos \frac{n\pi x}{\pi} dx = \frac{1}{\pi}\int_{-\pi}^{\pi} f(x) \cos \frac{n\pi x}{\pi} dx$$

$$\frac{1}{\pi}\int_{-\pi}^{\pi} g(x) \sin \frac{nx}{\pi} dx = \frac{1}{\pi}\int_{-\pi}^{\pi} f(x) \sin \frac{nx}{\pi} dx.$$

That is, $g(x)$ has the same Fourier coefficients as $f(x)$. It follows that $g(x)$ also has the same trigonometric Fourier series as $f(x)$,

i.e.,
$$g(x) \sim \frac{3\pi}{4} + \sum_{n=1}^{\infty}\left[\frac{(-1)^n - 1}{\pi n^2} \cos nx - \frac{1}{n}\sin nx\right], \tag{a}$$

on the interval $-\pi \leq x \leq \pi$.

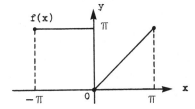
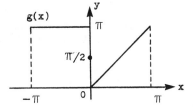

Comparing the graphs of f and g, we see that they are the same except at the point $x = 0$. Which of these functions does the series converge to? Although not proved here, it is true that Fourier series at a point of discontinuity converge to the midpoint (or average) of the right and left side values so $g(x)$ is the precise limit of the series.

• **PROBLEM 3-5**

Consider the function

$$f(x) = \begin{cases} \pi, & -\pi \leq x < 0 \\ x, & 0 \leq x \leq \pi \end{cases}$$

and defined for all other x by the periodicity condition $f(x+2\pi) = f(x)$ for all x. Does the trigonometric Fourier series of f converge for all values of x?

Solution: The solution to this problem requires three main details after finding the trigonometric Fourier series; 1) what it means for a function to be periodic, 2) under what conditions the Fourier series of a function converges and 3) what is the value to which the series converges for a given value of x?

We begin by finding the trigonometric Fourier series of $f(x)$. Assume that $f(x)$ has the series expansion

$$f(x) \sim \frac{1}{2} a_0 + \sum_{n=1}^{\infty} \left(a_n \cos \frac{n\pi x}{L} + b_n \sin \frac{n\pi x}{L} \right)$$

where

$$a_n = \frac{1}{L} \int_{-L}^{L} f(x) \cos \frac{n\pi x}{L} dx,$$

$$b_n = \frac{1}{L} \int_{-L}^{L} f(x) \sin \frac{n\pi x}{L} dx.$$

We find a_n and b_n for the given function

$$a_0 = \frac{1}{\pi} \int_{-\pi}^{0} f(x) \cos(0)x \, dx = \frac{1}{\pi} \int_{-\pi}^{0} \pi \, dx + \frac{1}{\pi} \int_{0}^{\pi} x \, dx$$

$$= 3\pi/2 \; .$$

$$a_n = \frac{1}{\pi} \int_{-\pi}^{\pi} f(x) \cos \frac{n\pi x}{\pi} dx \qquad (n = 1,2,3,\ldots)$$

$$= \frac{1}{\pi} \int_{-\pi}^{0} \pi \cos nx \, dx + \frac{1}{\pi} \int_{0}^{\pi} x \cos nx \, dx$$

$$= \left[\frac{\sin nx}{n} \right]_{-\pi}^{0} + \frac{1}{\pi} \left[\frac{x \sin nx}{n} + \frac{\cos nx}{n^2} \right]_{0}^{\pi}$$

$$= \frac{1}{\pi} \left[\frac{\cos n\pi - 1}{n^2} \right] = \frac{(-1)^n - 1}{n^2 \pi} = \begin{cases} -2/n^2 \pi & n \text{ odd} \\ 0 & n \text{ even.} \end{cases}$$

$$b_n = \frac{1}{\pi} \int_{-\pi}^{\pi} f(x) \sin \frac{n\pi x}{\pi} dx = \frac{1}{\pi} \int_{-\pi}^{0} \pi \sin nx \, dx + \frac{1}{\pi} \int_{0}^{\pi} x \sin nx \, dx$$

$$= \left[-\frac{\cos nx}{n}\right]_{-\pi}^{0} + \frac{1}{\pi}\left[-x\frac{\cos nx}{n} + \frac{\sin nx}{n^2}\right]_{0}^{\pi}$$

$$= -\frac{1}{n} + \frac{(-1)^n}{n} + \frac{(-1)^{n+1}}{n} = \frac{-1 + (-1)^n + (-1)^{n+1}}{n}$$

$$= -1/n \ .$$

Thus
$$f(x) \sim \frac{3\pi}{4} + \sum_{n=1}^{\infty}\left(\frac{[(-1)^n - 1]}{\pi n^2}\cos nx - \frac{1}{n}\sin nx\right)$$

$$= \frac{3\pi}{4} + \left(\frac{2}{\pi}\cos x + \sin x\right) - \frac{1}{2}\sin 2x$$

$$- \left(\frac{2}{9\pi}\cos 3x + \frac{1}{3}\sin 3x\right) - \frac{1}{4}\sin 4x - \ldots \qquad (a)$$

We now discuss the periodicity and convergence of (a). A function is said to be periodic with period P if, for $P > 0$, $F(x + P) = F(x)$ for all x in the domain of f. For example, the function $\sin n\pi x/L$ and $\cos n\pi x/L$ are periodic of period $2L/n$. They are also periodic of period $2L$ since F is also periodic of period nP.

Now, if a function of period L can be expanded into a trigonometric Fourier series over the interval $-L \leq x \leq L$ and the series converges for all x in this interval then the Fourier series can represent the periodic function for all x, provided the function is piecewise smooth on the interval $-L \leq x \leq L$. Moreover, the trigonometric Fourier series of f converges at every point to
$$\frac{f(x+) + f(x-)}{2},$$
where $f(x+)$ and $f(x-)$ denote the right-hand and left-hand limits of f at x. If f is continuous at x, then $f(x+) = f(x-)$ and the trigonometric Fourier series of f at x converges to $f(x)$.

Apply the above reasoning to the given function. It is periodic with period 2π and is piecewise continuous on the interval $-\pi \leq x \leq \pi$. Thus the series (a) converges to
$$\frac{f(x+) + f(x-)}{2}$$
for all x. From the definition of f we see that it is continuous at all points except where $x = \pm 2n\pi$, $(n = 0,1,2,\ldots)$. Thus F converges to $f(x)$ at every x except where $x = \pm 2n\pi$, $(n = 0,1,2,\ldots)$. For example when $x = \pi$, $f(\pi) = \pi$. Thus substituting π for x in (a),

$$\frac{3\pi}{4} + \sum_{n=1}^{\infty} \frac{[(-1)^n - 1]\cos n\pi}{\pi n^2} - \frac{1}{n}\sin n\pi \ . \qquad (b)$$

For n even, (b) reduces to $3\pi/4$. For n odd, (b) is:

$$\frac{3\pi}{4} + \frac{2}{\pi}\sum_{n=1}^{\infty}\frac{1}{(2n-1)^2} = \frac{3\pi}{4} + \frac{2}{\pi}\left[\frac{1}{1^2} + \frac{1}{3^2} + \ldots\right].$$

Thus,
$$\pi = \frac{3\pi}{4} + \frac{2}{\pi}\left[\frac{1}{12} + \frac{1}{3^2} + \ldots\right].$$

Finally we consider the points of discontinuity, $x = \pm 2n\pi$ $(n = 0,1,2,\ldots)$. At each such point, $f(x+) = f(0+) = 0$, $f(x-) = f(0-) = \pi$. Hence,
$$\frac{f(x+) + f(x-)}{2} = \frac{\pi}{2} \ .$$

As an example, at $x = 0$, (a) is

$$\frac{3\pi}{4} + \sum_{n=1}^{\infty} \frac{(-1)^n - 1}{\pi n^2} = \frac{3\pi}{4} - \frac{2}{\pi} \sum_{n=1}^{\infty} \frac{1}{(2n-1)^2},$$

and this converges to $\pi/2$.

We see that the trigonometric Fourier series of f converges for all x, to the function $g(x)$ defined by

$$g(x) = \begin{cases} \pi, & -\pi \le x < 0 \\ \pi/2, & x = 0 \\ x, & 0 < x \le \pi \end{cases},$$

$g(x+2\pi) = g(x)$ for all x.

• **PROBLEM 3-6**

Express the function

$$f(x) = |x|, \quad -\pi \le x \le \pi,$$

as a trigonometric Fourier series on the interval $-\pi \le x \le \pi$.

<u>Solution</u>: If a function satisfies certain conditions (piecewise continuity, piecewise differentiability, etc.) it has, for its trigonometric Fourier series,

$$f(x) \sim \frac{1}{2} a_0 + \sum_{n=1}^{\infty} \left[a_n \cos \frac{n\pi x}{L} + b_n \sin \frac{n\pi x}{L} \right] \quad (a)$$

for $-L \le x \le L$, where

$$a_n = \frac{1}{L} \int_{-L}^{L} f(x) \cos \frac{n\pi x}{L} dx \quad (n = 0, 1, 2, 3, \ldots), \quad (b)$$

$$b_n = \frac{1}{L} \int_{-L}^{L} f(x) \sin \frac{n\pi x}{L} dx \quad (n = 1, 2, 3, \ldots). \quad (c)$$

In solving the given problem we make use of the notions of odd functions and even functions. A function $f(x)$, is said to be even if $f(-x) = f(x)$. Some examples of even functions are $y = \cos x$, $y = x^2$, and $y = |x|$. One characteristic of such functions is that they are symmetric with respect to the y-axis.

A function $f(x)$ is called odd if $f(-x) = -f(x)$. Examples of odd functions are $y = \sin x$, $y = x$, $y = x^3$. We note further that the product of two even functions is itself an even function while the product of an odd with an even function is an odd function. Finally, if $f(x)$ is even on the interval $-A \le x \le A$, then

$$\int_{-A}^{A} f(x) dx = 2 \int_{0}^{A} f(x) dx,$$

and if $f(x)$ is odd over the same interval, $\int_{-A}^{A} f(x) dx = 0$.

In the given problem, $f(x) = |x|$ is an even function. The evaluation of the Fourier coefficients (a_n, b_n as in (b) and (c)) is now shown to be considerably simplified. First, consider

$$a_n = \frac{1}{L} \int_{-L}^{L} |x| \cos \frac{n\pi x}{L} dx$$

for $-\pi \leq x \leq \pi$. Since $|x| = x$ for $0 \leq x \leq \pi$, this may be written as

$$a_n = \frac{2}{L} \int_0^L x \cos \frac{n\pi x}{L} dx .$$

Next, since $\sin x$ is odd, the coefficients

$$b_n = \frac{1}{L} \int_{-L}^{L} f(x) \sin \frac{n\pi x}{L} dx , \quad (n = 1,2,3,\ldots)$$

are all equal to zero. Hence we need consider only the a_n in the series (a).

Now, $$a_n = \frac{2}{\pi} \int_0^\pi x \cos \frac{n\pi x}{\pi} dx = \frac{2}{\pi} \int_0^\pi x \cos nx \, dx .$$

Integrating by parts,

$$a_n = \frac{2}{\pi} \left[\frac{\cos nx}{n^2} + \frac{x \sin nx}{n} \right]_0^\pi$$

$$= \frac{2}{\pi} \left[\frac{\cos n\pi - 1}{n^2} \right] = \frac{2}{\pi} \left[\frac{(-1)^n - 1}{n^2} \right]$$

(since $\cos n\pi = -1$ for odd n and 1 for even n).

$$\frac{2}{\pi} \left[\frac{(-1)^n - 1}{n^2} \right] = \begin{cases} \frac{-4}{\pi n^2} , & n \text{ odd} \\ 0 , & n \text{ even} \end{cases} \quad n = 1,2,3,\ldots .$$

The above integration and evaluation is not valid for $n = 0$. To find a_0,

$$a_0 = \frac{2}{\pi} \int_0^\pi f(x) \cos(0) \frac{\pi x}{\pi} dx = \frac{2}{\pi} \int_0^\pi f(x) dx = \frac{2}{\pi} \int_0^\pi x \, dx = \frac{x^2}{\pi} \Big]_0^\pi = \pi .$$

Thus, the required series is

$$\frac{\pi}{2} - \frac{4}{\pi} \sum_{\substack{n=1 \\ (n \text{ odd})}}^{\infty} \frac{\cos nx}{n^2} = \frac{\pi}{2} - \frac{4}{\pi} \sum_{n=1}^{\infty} \cos \frac{(2n-1)x}{(2n-1)^2} .$$

That is,

$$|x| \sim \frac{\pi}{2} - \frac{4}{\pi} \sum_{n=1}^{\infty} \frac{\cos(2n-1)x}{(2n-1)^2} , \quad -\pi \leq x \leq \pi .$$

● **PROBLEM 3-7**

Find the trigonometric Fourier series of the function f defined by

$$f(x) = \begin{cases} \pi , & -\pi \leq x < 0 \\ x & 0 \leq x \leq \pi \end{cases}$$

on the interval $-\pi \leq x \leq \pi$.

Solution: A function, $f(x)$ has a trigonometric Fourier series if

$$f(x) \sim \frac{1}{2} a_0 + \sum_{n=1}^{\infty} \left(a_n \cos \frac{n\pi x}{L} + b_n \sin \frac{n\pi x}{L} \right) , \quad (a)$$

for $-L \leq x \leq L$, where

$$a_n = \frac{1}{L} \int_{-L}^{L} f(x) \cos \frac{n\pi x}{L} \, dx \quad \text{and} \tag{b}$$

$$b_n = \frac{1}{L} \int_{-L}^{L} f(x) \sin \frac{n\pi x}{L} \, dx. \tag{c}$$

The trigonometric Fourier series are a special class of the more general Fourier series of a given function f. Using the concept of an orthonormal system the following definition is made: let $\{\varphi_n\}$, (n = 1,2,3,...) be an orthonormal system with respect to a weight function r(x) on $a \le x \le b$. That is,

$$\int_a^b \varphi_m(x) \varphi_n(x) r(x) dx = 0, \quad [\varphi_m(x), \varphi_n(x) \in \{\varphi_n\}, \, m \ne n].$$

$$\int_a^b [\varphi_m(x)]^2 r(x) dx = K \quad (K \text{ a constant} > 0).$$

Now, let f be a function such that, for each n = 1,2,3,..., the product $f\varphi_n r$ is integrable on $a \le x \le b$. Then the series

$$\sum_{n=1}^{\infty} c_n \varphi_n$$

where

$$c_n = \int_a^b f(x) \varphi_n(x) r(x) dx, \quad (n = 1, 2, \ldots)$$

is called the Fourier series of f relative to the system $\{\varphi_n\}$,

$$f(x) \sim \sum_{n=1}^{\infty} c_n \varphi_n(x), \quad a \le x \le b.$$

In the case of the trigonometric Fourier series, the orthonormal system, $\{\varphi_n\}$ consists of functions of the form $\sin n\pi x$, $\cos n\pi x$. An infinite series involving these two functions arises as a general solution to such problems as finding the displacement of a vibrating string, the flow of heat through a slab and the solution of Laplace's equation, $\nabla^2 v = 0$, in potential theory.

The main problem in determining the Fourier series of a function (after checking to see that it satisfies the mathematically theoretical requirements necessary to be expressible as a Fourier series) is finding the coefficients a_n and b_n. We now turn to the given problem and attempt to do this.

The function is neither even nor odd. Thus we cannot assume that one set of coefficients - either a_n or b_n - will vanish. We use the formulae (b) and (c) to find the coefficients in (a). Then, after substitution we make the resulting series as elegant as possible.

Thus, evaluating the b_n first

$$b_n = \frac{1}{L} \int_{-L}^{L} f(x) \sin \frac{n\pi x}{L} dx = \frac{1}{\pi} \int_{-\pi}^{\pi} f(x) \sin \frac{n\pi x}{\pi} dx$$

$$= \frac{1}{\pi} \left[\int_{-\pi}^{0} \pi \sin \frac{n\pi x}{\pi} dx + \int_{0}^{\pi} x \sin \frac{n\pi x}{\pi} dx \right]$$

(since $\int_a^d f(x) dx = \int_a^b f(x) dx + \int_c^d f(x) dx$)

$$= \frac{1}{\pi} \left[-\frac{\pi \cos nx}{n} \right]_{-\pi}^{0} + \frac{1}{\pi} \left[\frac{\sin nx}{n^2} - \frac{x \cos nx}{n} \right]_{0}^{\pi}$$

(where the second integral was evaluated using integration by parts)

$$= \frac{1}{\pi}\left(-\frac{\pi}{n}\right) = -\frac{1}{n}, \quad (n = 1,2,3,\ldots). \tag{d}$$

Next, we find an expression for the a_n. Here, we are faced with the complication that when $n = 0$, integrating

$$\frac{1}{L}\int_{-L}^{L} f(x) \cos\frac{n\pi x}{L} dx$$

will result in division by zero — an undefined operation. Thus we treat the case $n = 0$ first and then proceed to find the values of the remaining a_n.

$$a_0 = \frac{1}{\pi}\int_{-\pi}^{\pi} f(x) \cos\frac{(0)\pi x}{\pi} dx$$

$$= \frac{1}{\pi}\left[\int_{-\pi}^{0} \pi \, dx + \int_{0}^{\pi} x \, dx\right] = \frac{3\pi}{2} \tag{e}$$

$$a_n = \frac{1}{L}\int_{-L}^{L} f(x) \cos\frac{n\pi x}{L} dx = \frac{1}{\pi}\int_{-\pi}^{\pi} f(x) \cos nx \, dx$$

$$= \frac{1}{\pi}\left[\int_{-\pi}^{0} \pi \cos nx \, dx + \int_{0}^{\pi} x \cos nx \, dx\right]$$

$$= \frac{1}{\pi}\left\{\left[\frac{\pi \sin nx}{n}\right]_{-\pi}^{0} + \left[\frac{\cos nx}{n^2} + \frac{x \sin nx}{n}\right]_{0}^{\pi}\right\}$$

$$= \frac{1}{\pi}\left[\frac{\cos n\pi - 1}{n^2}\right] = \frac{(-1)^n - 1}{n\pi^2} = \begin{cases} -2/n\pi^2, & n \text{ odd} \\ 0, & n \text{ even.} \end{cases} \tag{f}$$

Substituting (d), (e) and (f) into (a):

$$f(x) \sim \frac{3\pi}{4} + \sum_{n=1}^{\infty}\left[\frac{(-1)^n - 1}{\pi n^2} \cos nx - \frac{1}{n} \sin nx\right]$$

$$= \frac{3\pi}{4} - \left(\frac{2}{\pi} \cos x + \sin x\right) - \frac{1}{2} \sin 2x$$

$$- \left(\frac{2}{9\pi} \cos 3x + \frac{1}{3} \sin 3x\right) - \frac{1}{4} \sin 4x - \ldots .$$

● **PROBLEM 3-8**

Find the trigonometric Fourier series of the function f defined by $f(x) = x$, $-4 \leq x \leq 4$, on the interval $-4 \leq x \leq 4$.

<u>Solution</u>: We first sketch a graph of the function over the required interval.

The trigonometric Fourier series of a function is given by

$$f(x) \sim \frac{1}{2} a_0 + \sum_{n=1}^{\infty}\left(a_n \cos\frac{n\pi x}{L} + b_n \sin\frac{n\pi x}{L}\right) \tag{a}$$

for $-L \leq x \leq L$, where

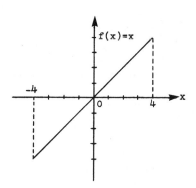

$$a_n = \frac{1}{L}\int_{-L}^{L} f(x) \cos \frac{n\pi x}{L} dx \quad \text{and}$$

$$b_n = \frac{1}{L}\int_{-S}^{L} f(x) \sin \frac{n\pi x}{L} dx .$$

We assume that these two integrals exist, for otherwise the problem could not be solved by this approach. The series (a) arises naturally as a general solution to the vibrating string problem. The mathematical analysis of this physical problem results in a partial differential equation with certain boundary conditions that the solution to the equation must satisfy.

We are concerned with evaluating the coefficients in (a). Notice that the given function $f(x) = x$ is an odd function, i.e., $f(-x) = -f(x)$. Thus, for example, if $x = 4$, $f(-4) = -4$ which is $-f(4)$. In such cases, it is necessary to find only one set of coefficients - the members of the other set will all equal zero.

We show that the last statement is true in the given problem.

$$a_n = \frac{1}{L}\int_{-L}^{L} f(x) \cos \frac{n\pi x}{L} dx = \frac{1}{4}\int_{-4}^{4} x \cos \frac{n\pi x}{4} dx .$$

Integrating the above by parts

$$\frac{1}{4}\int_{-4}^{4} x \cos \frac{n\pi x}{4} dx = \frac{1}{n\pi} x \sin \frac{n\pi x}{4} \Big]_{-4}^{4} + \frac{4}{n^2 \pi^2} \cos \frac{n\pi x}{4} \Big]_{-4}^{4} = 0 .$$

Then, evaluating the coefficients b_n,

$$b_n = \frac{1}{L}\int_{-L}^{L} f(x) \sin \frac{n\pi x}{L} dx = \frac{1}{4}\int_{-4}^{4} x \sin \frac{n\pi x}{4} dx .$$

Integrating by parts,

$$b_n = \frac{1}{4}\left[\frac{-4}{n\pi} x \cos \frac{n\pi x}{4}\right]_{-4}^{4} + \frac{1}{n\pi}\int_{-4}^{4} \cos \frac{n\pi x}{4} dx$$

$$= \frac{-8}{n\pi} \cos n\pi + 0 = \frac{-8}{n\pi}(-1)^n = \frac{8}{n\pi}(-1)^{n+1} .$$

Since both $f(x)$ and $\sin x$ are odd functions their product is an even function. Thus the b_n could also have been calculated by noting that

$$\frac{1}{L}\int_{-L}^{L} f(x) \sin \frac{n\pi x}{L} dx = \frac{2}{L}\int_{0}^{L} f(x) \sin \frac{n\pi x}{L} dx .$$

The series (a) is therefore

$$f(x) \sim \sum_{n=1}^{\infty} (-1)^{n+1} \frac{8}{n\pi} \sin \frac{n\pi x}{4}, \quad -4 \leq x \leq 4$$

or,
$$x \sim \frac{8}{\pi} \sum_{n=1}^{\infty} \frac{(-1)^{n+1}}{n} \sin \frac{n\pi x}{4}, \quad -4 \leq x \leq 4. \quad \text{(b)}$$

The expression (b) is the trigonometric Fourier series for the odd function, $f(x) = x$. Note that the a_n being zero reduced (a), the general trigonometric Fourier series to the form (b).

● **PROBLEM 3-9**

Find a trigonometric Fourier series for the function defined by

$$f(x) = 1 \qquad -\pi < x < 0 \qquad \text{1)}$$
$$f(x) = 2 \qquad 0 < x < \pi. \qquad \text{(1')}$$

Solution: We first sketch the graph of (1), (1'). Notice that $f(x)$ is not defined at the points $-\pi, 0, \pi$.

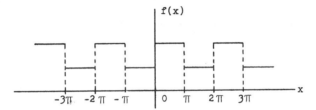

The trigonometric Fourier series of a function $f(x)$ defined for $-L \leq x \leq L$, that satisfies certain conditions (of continuity and differentiability) is

$$f(x) \sim \frac{1}{2} a_0 + \sum_{n=1}^{\infty} \left(a_n \cos \frac{n\pi x}{L} + b_n \sin \frac{n\pi x}{L} \right), \quad (1)$$

where
$$a_n = \frac{1}{L} \int_{-L}^{L} f(x) \cos \frac{n\pi x}{L} dx \quad (2)$$

and
$$b_n = \frac{1}{L} \int_{-L}^{L} f(x) \sin \frac{n\pi x}{L} dx. \quad (3)$$

Since the given function is composed of two intervals, and it is continuous over each of these intervals, we have, letting $L = \pi$,

$$a_0 = \frac{1}{2\pi} \int_{-\pi}^{0} dx + \frac{1}{2\pi} \int_{0}^{\pi} 2dx$$
$$= \frac{1}{2\pi} \left[x \Big|_{-\pi}^{0} \right] + \frac{1}{2\pi} \left[2x \Big|_{0}^{\pi} \right] = 3/2, \quad (4)$$

$$a_n = \frac{1}{\pi} \int_{-\pi}^{0} \cos nx \, dx + \frac{2}{\pi} \int_{0}^{\pi} \cos nx \, dx$$
$$= \frac{1}{n\pi} \left[\sin nx \Big|_{-\pi}^{0} \right] + \frac{2}{n\pi} \left[\sin nx \Big|_{0}^{\pi} \right] = 0, \quad n > 0, \quad (5)$$

87

$$b_n = \frac{1}{\pi} \int_{-\pi}^{0} \sin nx \, dx + \frac{2}{\pi} \int_{0}^{\pi} \sin nx \, dx$$

$$= -\frac{1}{n\pi}\left[\cos nx \Big|_{-\pi}^{0}\right] + \frac{2}{n\pi}\left[\cos nx \Big|_{0}^{\pi}\right] = \frac{1-(-1)^n}{n\pi} \quad (6)$$

Substituting (4), (5) and (6) into (1),

$$f(x) \sim \frac{3}{2} + \frac{2}{\pi}\left(\frac{\sin x}{1} + \frac{\sin 3x}{3} + \frac{\sin 5x}{5} + \ldots\right)$$

$$= \frac{3}{2} + \frac{2}{\pi} \sum_{k=0}^{\infty} \frac{\sin (2k+1)x}{(2k+1)} \quad (7)$$

when $x = -\pi, 0, \pi$, i.e., at the points of discontinuity, (7) has the value 3/2. Since $f(x)$ was originally undefined at these points we can assign the value 3/2 to them.

The series (7) through its periodicity properties extends the original domain of definition $(|x| < \pi)$ to all x.

• PROBLEM 3-10

Find the Fourier series of the function $g(x)$ defined as follows

$$\begin{aligned} g(x) &= x, & (-\pi < x < \pi) \\ g(\pi) &= g(-\pi) = 0, & \\ g(x+2\pi) &= g(x). & \end{aligned} \quad (1)$$

Solution: The function defined by (1) is periodic. We are asked to approximate $f(x)$ by the trigonometric series

$$s(t) = \frac{a_0}{2} + \sum_{k=1}^{\infty} \left(a_k \cos kwt + b_k \sin kwt\right)$$

which is the Fourier series expansion. It has been shown that if a function, $g(t)$ is continuous, is periodic and has a piecewise continuous derivative, then the Fourier series converges to $g(t)$. The constants a_k and b_k are determined by the formulae

$$a_k = \frac{w}{\pi} \int_{0}^{2\pi/w} g(t) \cos kwt \, dt \qquad (k = 0,1,\ldots,n)$$

$$b_k = \frac{w}{\pi} \int_{0}^{2\pi/w} g(t) \sin kwt \, dt \qquad (k = 1,2,\ldots,n).$$

In the given problem, $f(x)$ has period 2π. Thus the periodicity factor, w, is equal to one. The Fourier coefficients become

$$a_k = \frac{1}{\pi} \int_{0}^{2\pi} f(x) \cos kx \, dx$$

$$b_k = \frac{1}{\pi} \int_{0}^{2\pi} f(x) \sin kx \, dx.$$

But the domain of definition is $-\pi < x < \pi$. We note that $f(x)$, $\cos kx$ and $\sin kx$ are all of the same period. Thus we may integrate over any period interval, in particular over $(-\pi,\pi)$. Hence

$$a_k = \frac{1}{\pi} \int_{-\pi}^{\pi} x \cos kx \, dx \qquad (k = 0,1,\ldots,n) \qquad (2)$$

$$b_k = \frac{1}{\pi} \int_{-\pi}^{\pi} x \sin kx \, dx \qquad (k = 1,2,\ldots,n) \qquad (3)$$

The coefficients determined by (2) all reduce to zero. The integrals in (3) are evaluated using integration by parts:

$$b_k = \frac{1}{\pi} \left(x \frac{\cos kx}{-k} \Big|_{-\pi}^{\pi} + \int_{-\pi}^{\pi} \frac{\cos kx}{k} \, dx \right)$$

$$= -\frac{2}{k} \cos k\pi$$

$$= (-1)^{k+1} \frac{2}{k} \qquad \left(\cos k\pi = \begin{cases} -1, & k \text{ odd} \\ +1, & k \text{ even} \end{cases} \right)$$

The Fourier series for $f(x)$ is therefore

$$f(x) \sim 2 \sum_{n=1}^{\infty} (-1)^{n+1} \frac{\sin nx}{n} \qquad (4)$$

For $-\pi < x < \pi$, the series in (4) converges to x. For $x = \pm \pi$, the series is zero. Thus, the Fourier series (4) satisfies the conditions given by (1). Moreover, it is periodic with period 2π. (See Fig.)

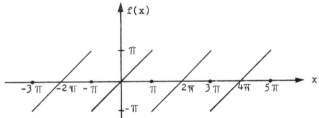

COMPLEX FOURIER SERIES

• **PROBLEM 3-11**

Find the exponential Fourier series of the periodic waveform from the graph shown below.

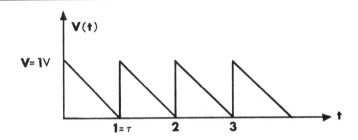

Solution: This is a typical example of the way Fourier

series are used to represent functions in electrical engineering. The function to be represented here is $v(t)$ = the voltage at a certain point in a circuit at time t, where v is periodic in time with period $\tau = 2\frac{T}{2} = 1$. The complex form of the Fourier series of such a function is given by

$$v(t) = \sum_{-\infty}^{\infty} c_n e^{i\left(\frac{n2\pi t}{\tau}\right)}, \quad c_n = \frac{1}{\tau}\int_a^{a+\tau} v(t) e^{-i\left(\frac{n2\pi t}{\tau}\right)} dt. \quad (1)$$

It is convenient to define $\omega = \frac{2\pi}{\tau}$, take the limits of integration from 0 to τ and let j stand for the number whose square is -1, (i.e. j = i). This is convenient in electrical applications since i is usually used for current. In this case (1) becomes

$$v(t) = \sum_{-\infty}^{\infty} c_n e^{jn\omega t}, \quad c_n = \frac{1}{\tau}\int_0^{\tau} v(t) e^{-jn\omega t} dt. \quad (2)$$

where

$$v(t) = 1 - t, \quad t \in (0,\tau)$$
$$v(t+\tau) = v(t) \quad \text{all } t \;.$$

First evaluate c_0 by

$$c_0 = \frac{1}{\tau}\int_0^{\tau} v(t) dt$$

(note that this may be interpreted as the average value of the voltage v over the interval $(0,\tau)$). This gives

$$c_0 = \frac{1}{\tau}\int_0^{\tau} (1-t) dt = \left.\left(\frac{1}{\tau} t - \frac{t}{2}\right)\right|_0^{\tau}$$

and evaluating yields

$$c_0 = \frac{1}{\tau}\left(\tau - \frac{\tau^2}{2}\right) = 1 - \frac{\tau}{2} = \frac{1}{2} \quad (3)$$

since $\tau = 1$ second.

Similarly, evaluate equation c_n, $n \neq 0$:

$$c_n = \frac{1}{\tau}\int_0^{\tau} (1-t) e^{-jn\omega t} dt,$$

$$c_n = \frac{1}{\tau}\int_0^{\tau} e^{-jn\omega t} dt - \frac{1}{\tau}\int_0^{\tau} t e^{-jn\omega t} dt$$

but

$$\int e^{ax} dx = \frac{e^{ax}}{a}$$

and, using integration by parts,

$$\int xe^{ax} dx = \frac{xe^{ax}}{a} - \frac{e^{ax}}{a^2}.$$

Using these results in the integrals above yields

$$c_n = \frac{1}{\tau} \left[\frac{e^{-jn\omega t}}{(-jn\omega)} - \frac{te^{-jn\omega t}}{(-jn\omega)} + \frac{e^{-in\omega t}}{(-jn\omega)^2} \right] \Bigg|_0^\tau ,$$

so that

$$c_n = \frac{1}{\tau} \left[\frac{e^{-jn\omega\tau} - e^0}{(-jn\omega)} - \frac{\tau e^{-jn\omega\tau} - 0}{(-jn\omega)} + \frac{e^{-jn\omega\tau} - e^0}{(-jn\omega)^2} \right]. \quad (4)$$

We would seek ways to simplify equation (4) before proceeding further. One way is recalling that

$$\omega = \frac{2\pi}{\tau} \quad \text{rad/s}.$$

Since $\tau = 1$ in this instance, from the graph.

$$\omega = 2\pi \quad \text{rad/s}. \quad (5)$$

Substituting equation (5) into equation (4),

$$c_n = \left[\frac{e^{-j2\pi n} - 1}{(-j2\pi n)} - \frac{e^{-j2\pi n}}{(-j2\pi n)} + \frac{e^{-j2\pi n} - 1}{(-j2\pi n)^2} \right]. \quad (6)$$

A further simplification may be made by evaluating the exponential quantity with the use of Euler's equation,

$$e^{jx} = \cos x + j \sin x, \quad (7)$$

$$e^{-j2\pi n} = \cos(-2\pi n) + j \sin(-2\pi n) = +1 \quad (8)$$

since the cosine or sine of $\pm 2\pi$ or any whole-number multiple thereof is +1 or zero, respectively. Now, substituting equation (8) into equation (6),

$$c_n = \left[\frac{1-1}{(-j2\pi n)} - \frac{1}{(-j2\pi n)} + \frac{1-1}{(-j2\pi n)^2} \right] \quad \text{and}$$

$$c_n = \frac{1}{j2\pi n}. \quad (9)$$

A further step in simplification can be made by recalling that

$$\frac{1}{j} = -j = 0 - j = e^{-2\frac{\pi}{2}} \quad (10)$$

which can be proven with equation (7). Substituting equation (10) into equation (9),

$$c_n = \frac{1}{2\pi n} e^{-j\frac{\pi}{2}} \quad (11)$$

Finally, substitute equations (3), (5) and (11) into equation (2):

$$v(t) = \frac{1}{2} + \sum_{n=-\infty}^{\infty} (\frac{1}{2\pi n}) e^{-j\frac{\pi}{2}} e^{j2\pi nt} .$$

This can be rewritten as

$$v(t) = \frac{1}{2} + \sum_{n=-\infty}^{\infty} (\frac{1}{2\pi n}) e^{j(2\pi nt - \frac{\pi}{2})} \text{ volts}$$

since $e^a e^b = e^{a+b}$.

• **PROBLEM 3-12**

Find the exponential Fourier series of the periodic waveform below.

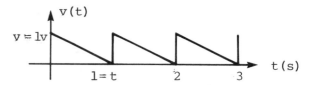

Solution: The Fourier series may be expressed in exponential form as follows:

$$v(t) = c_o + \sum_{n=-\infty}^{\infty} c_n e^{jn\omega t} \quad (1)$$

where c_o is the average value of $v(t)$, ω is the fundamental frequency in radians per second, and c_n is the coefficient of the nth harmonic. The values for c_n are determined from

$$c_n = \frac{1}{\tau} \int_0^\tau v(t) e^{-jn\omega t} dt \quad (n=0, \pm 1, \pm 2, \pm 3, \ldots). \quad (2)$$

First, evaluate the average value of the waveform.

By inspection, this can be seen to be 1/2. A more formal approach would be

$$c_o = \frac{1}{\tau} \int_0^\tau v(t)\,dt \tag{3}$$

where the graph shows

$$v(t) = 1 - t \qquad (t=0 \text{ to } t=\tau). \tag{4}$$

Therefore,

$$c_o = \frac{1}{\tau} \int_0^\tau (1-t)\,dt = \frac{1}{\tau}\left[t - \frac{t^2}{2}\right]_0^\tau \tag{5}$$

and evaluating yields

$$c_o = \frac{1}{\tau}\left[\tau - \frac{\tau^2}{2}\right] = 1 - \frac{\tau}{2} = \frac{1}{2} \tag{6}$$

since $\tau = 1$ second.

Similarly, evaluate equation (2):

$$c_n = \frac{1}{\tau}\int_0^\tau (1-t)e^{-jn\omega t}\,dt, \tag{7}$$

$$c_n = \frac{1}{\tau}\int_0^\tau e^{-jn\omega t}\,dt - \frac{1}{\tau}\int_0^\tau t e^{-jn\omega t}\,dt \tag{8}$$

by using integral tables, where we find

$$\int e^{ax}\,dx = \frac{e^{ax}}{a} \qquad \text{and} \tag{9}$$

$$\int x e^{ax}\,dx = \frac{xe^{ax}}{a} - \frac{e^{ax}}{a^2}. \tag{10}$$

Following the integral tables, integrate

$$c_n = \frac{1}{\tau}\left[\frac{e^{-jn\omega t}}{(-jn\omega)} - \frac{te^{-jn\omega t}}{(-jn\omega)} + \frac{e^{-jn\omega t}}{(-jn\omega)^2}\right]_0^\tau \tag{11}$$

and evaluate:

$$c_n = \frac{1}{\tau}\left[\frac{e^{-jn\omega\tau} - e^0}{(-jn\omega)} - \frac{\tau e^{-jn\omega\tau} - 0}{(-jn\omega)} + \frac{e^{-jn\omega\tau} - e^0}{(-jn\omega)^2}\right]. \tag{12}$$

We should seek ways to simplify equation (12) before proceeding further. One way is recalling that

$$\omega = \frac{2\pi}{\tau} \quad \text{rad/s.} \tag{13}$$

Since $\tau = 1$ in this instance, from the graph.

$$\omega = 2\pi \quad \text{rad/s.} \tag{14}$$

Substituting equation (14) into equation (12),

$$c_n = \left[\frac{e^{-j2\pi n}-1}{(-j2\pi n)} - \frac{e^{-j2\pi n}}{(-j2\pi n)} + \frac{e^{-j2\pi n}-1}{(-j2\pi n)^2} \right]. \tag{15}$$

A further simplification may be made by evaluating the exponential quantity with the use of Euler's equation,

$$e^{jx} = \cos x + j \sin x, \tag{16}$$

$$e^{-j2\pi n} = \cos(-2\pi n) + j \sin(-2\pi n) = +1 \tag{17}$$

since the cosine or sine of $+2\pi$ or any whole-number multiple thereof is +1 or zero, respectively. Now, substituting equation (17) into equation (15),

$$c_n = \left[\frac{1-1}{(-j2\pi n)} - \frac{1}{(-j2\pi n)} + \frac{1-1}{(-j2\pi n)^2} \right] \quad \text{and} \tag{18}$$

$$c_n = \frac{1}{j2\pi n}. \tag{19}$$

A further step in simplification can be made by recalling that

$$\frac{1}{j} = -j = 0 - j = e^{-2\frac{\pi}{2}} \tag{20}$$

which can be proven with equation (16). Substituting equation (20) into equation (19),

$$c_n = \frac{1}{2\pi n} e^{-j\frac{\pi}{2}} \tag{21}$$

Finally, substitute equations (6), (14) and (21) into equation (1):

$$v(t) = \frac{1}{2} + \sum_{n=-\infty}^{\infty} (\frac{1}{2\pi n}) e^{-j\frac{\pi}{2}} e^{j2\pi nt}. \tag{22}$$

This can be rewritten as

$$v(t) = \frac{1}{2} + \sum_{n=-\infty}^{\infty} (\frac{1}{2\pi n}) e^{j(2\pi nt - \frac{\pi}{2})} \quad \text{volts} \tag{23}$$

since $e^a e^b = e^{a+b}$. (24)

● **PROBLEM 3-13**

Find the Fourier sine series of $f(x) = x^2$ over the interval $0 < x < 1$.

Solution: Since $f(x) = x^2$ is an even function we are inclined to believe that a Fourier sine series representing $f(x)$ does not exist. But this would be true only if the interval of definition were symmmetrical with respect to the x-axis. In the given problem, $0 < x < 1$.

We shall first consider the general problem of determining the Fourier sine series of a non-symmetric continuous function with piecewise smooth derivative. Then we shall apply the results to the given problem.

Let f be defined on the interval $0 \le x \le L$ and assume an orthonormal system $\{\varphi_n\}$. We wish to find the Fourier series of f relative to the orthonormal system $\varphi_n(x) = \sqrt{2/L} \sin \frac{n\pi x}{L}$, $0 \le x \le L$, $(n = 1, 2, \ldots)$. (This system arises as the set of characteristic functions associated with the Sturm-Liouville problem

$$\frac{d^2 y}{dx^2} + \lambda y = 0$$

$y(0) = 0, y(\pi) = 0$.)

The desired series is of the form

$$\sum_{n=1}^{\infty} c_n \varphi_n \tag{1}$$

where
$$c_n = \int_a^b f(x) \varphi_n(x) r(x) dx = \int_a^b f(x) \sqrt{2/L} \sin \frac{n\pi x}{L} dx ,$$

$(n = 1, 2, 3, \ldots)$.
Substituting for c_n and φ_n in (1), the series is

$$\sum_{n=1}^{\infty} \sqrt{2/L} \int_a^b f(x) \sin \frac{n\pi x}{L} dx \left[\sqrt{2/L} \sin \frac{n\pi x}{L} \right]$$

$$= \sum_{n=1}^{\infty} b_n \sin \frac{n\pi x}{L} , \tag{2}$$

$$b_n = \frac{2}{L} \int_a^b f(x) \sin \frac{n\pi x}{L} dx . \tag{3}$$

We now turn to the given problem. Here $f(x) = x^2$, $[a,b] = (0,1)$ and $L = 1$. Thus we obtain the Fourier series

$$x^2 \sim \sum_{n=1}^{\infty} b_n \sin n\pi x \tag{4}$$

where $b_n = 2 \int_0^1 x^2 \sin n\pi x \, dx$.

We must evaluate the b_n. Using integration by parts, we obtain

$$b_n = 2 \int_0^1 x^2 \sin n\pi x \, dx$$

$$= 2\left\{ \left[\frac{-x^2}{n\pi} \cos n\pi x\right]_0^1 + \frac{2}{n\pi} \int_0^1 x \cos n\pi x \, dx \right\}$$

$$= 2\left\{ \frac{-(-1)^n}{n\pi} + \frac{2}{n^2\pi^2} x \sin n\pi x \Big|_0^1 - \int_0^1 \frac{2}{n^2\pi^2} \sin n\pi x \, dx \right\}$$

$$= 2\left\{ \frac{(-1)^{n+1}}{n\pi} + \frac{2}{n^3\pi^3}\left[(-1)^n - 1\right] \right\}. \tag{5}$$

Substituting (5) for b_n in (4), the required Fourier sine series over $0 < x < 1$ is

$$x^2 \sim 2 \sum_{n=1}^\infty \left[\frac{(-1)^{n+1}}{n\pi} - \frac{2\{1-(-1)^n\}}{n^3\pi^3}\right] \sin n\pi x .$$

● **PROBLEM 3-14**

Find the Fourier sine series for the function defined by

$f(x) = 0 \qquad 0 \le x < \pi/2$

$f(x) = 1 \qquad \pi/2 < x \le \pi$.

Solution: The Fourier sine series of a function defined on $0 \le x \le L$

$$f(x) \sim \sum_{n=1}^\infty b_n \sin \frac{n\pi x}{L} \qquad (n = 1, 2, \ldots) \tag{1}$$

where $b_n = \frac{2}{L} \int_0^L \sin nx \, dx.$ \qquad (2)

Note that the Fourier sine series is equivalent to finding the Fourier trigonometric series of an odd function, i.e., a function such that $f(-x) = -f(x)$.

In the given problem, since $f(x) = 0$ for $0 \le x < \pi/2$ we need find its series development only for the interval $\pi/2 < x \le \pi$. Taking $L = \pi$,

$$b_n = \frac{2}{\pi} \int_{\pi/2}^\pi \sin nx \, dx = -\frac{2}{n\pi}\left[\cos nx \Big|_{\pi/2}^\pi\right]$$

$$= \frac{2}{n\pi}\left(\cos \frac{n\pi}{2} - \cos n\pi\right) = \frac{2}{n\pi}\left(\cos \frac{n\pi}{2} + (-1)^{n+1}\right). \tag{3}$$

Substituting (3) into (1),

$$f(x) = \frac{2}{\pi}\left(\frac{\sin x}{1} - \frac{2\sin 2x}{2} + \frac{\sin 3x}{3} + \frac{\sin 5x}{5} - \frac{2\sin 6x}{6} + \ldots\right).$$

● **PROBLEM 3-15**

Find a cosine series which represents $f(x)$ in $0 \le x \le \pi$ if $f(x)$ is defined as

$$f(x) = 0 \qquad 0 \le x < \pi/2$$
$$f(x) = 1 \qquad \pi/2 < x \le \pi.$$

Solution: The Fourier cosine series for a function $f(x)$, defined over $-L \le x \le L$, that is piecewise smooth is given by the formula

$$f(x) \sim \frac{1}{2} a_0 + \sum_{n=1}^{\infty} a_n \cos \frac{n\pi x}{L} \qquad (1)$$

where

$$a_n = \frac{2}{L} \int_0^L f(x) \cos \frac{n\pi x}{L} dx. \qquad (2)$$

In the given problem $0 \le x \le \pi$. Thus we may take L as equal to π. Then the coefficients as given by (2) are

$$a_0 = \frac{1}{\pi}\int_{\pi/2}^{\pi} dx = 1/2, \qquad (3)$$

$$a_n = \frac{2}{\pi}\int_{\pi/2}^{\pi} \cos nx\, dx = \frac{2}{n\pi} \sin nx \Big|_{\pi/2}^{\pi}$$

$$= -\frac{2}{n\pi} \sin \frac{n\pi}{2}, \quad n > 0. \qquad (4)$$

Since $\sin \frac{n\pi}{2} = 1$ for $n = 1, 2, 3, \ldots$, we have, upon substituting (3) and (4) into (1),

$$f(x) = \frac{1}{2} - \frac{2}{\pi}\left(\frac{\cos x}{1} - \frac{\cos 3x}{3} + \frac{\cos 5x}{5} - \ldots\right)$$

$$= \frac{1}{2} - \frac{2}{\pi}\sum_{n=0}^{\infty} (-1)^n \frac{\cos(2n+1)}{2n+1}. \qquad (5)$$

We may use (5) to obtain an expression for $\pi/4$. Thus, letting $x = 0$ in (5), since $f(x) = 0$,

$$\frac{2}{\pi}\left(1 - \frac{1}{3} + \frac{1}{5} - \ldots\right) = \frac{1}{2}$$

or

$$\frac{\pi}{4} = 1 - \frac{1}{3} + \frac{1}{5} - \ldots = \sum_{n=0}^{\infty} \frac{(-1)^n}{(2n+1)}$$

• **PROBLEM 3-16**

Find the Fourier cosine series over the interval $0 < x < c$ for the function $f(x) = x$.

Solution: If $f(x)$ has a Fourier cosine series over $0 < x < c$, it is of the form

$$f(x) \sim \frac{1}{2} a_0 + \sum_{n=1}^{\infty} a_n \cos \frac{n\pi x}{c}. \tag{1}$$

where

$$a_n = \frac{2}{c} \int_0^c f(x) \cos \frac{n\pi x}{c} dx. \quad (n = 0, 1, \ldots) \tag{2}$$

In the given problem, $f(x) = x$; hence (2) may be rewritten

$$a_n = \frac{2}{c} \int_0^c x \cos \frac{n\pi x}{c} dx. \quad (n = 0, 1, 2, \ldots) \tag{3}$$

The problem at hand is the evaluation of the coefficients a_n. We first find a_n for $n \neq 0$. The integral in (3) may be evaluated using integration by parts. Thus,

$$a_n = \frac{2}{c} \left[\frac{c}{n\pi} x \sin \frac{n\pi x}{c} \Big|_0^c - \frac{c}{n\pi} \int_0^c \sin \frac{n\pi x}{c} dx \right], \quad n \neq 0$$

$$= \frac{2}{c} \left(\frac{c}{n\pi}\right)^2 \cos \frac{n\pi x}{c} \Big|_0^c = \frac{2c}{(n\pi)^2} \left[(-1)^n - 1\right]$$

$$= -\frac{2c}{(n\pi)^2} \left[1 - (-1)^n\right].$$

We now find the coefficient a_0 separately.

$$a_0 = \frac{2}{c} \int_0^c x \cos \frac{(0\pi x)}{c} dx = \frac{2}{c} \frac{x^2}{2} \Big|_0^c = c.$$

Thus the Fourier cosine series over the interval $0 < x < c$ for the function $f(x) = x$ is

$$f(x) \sim \frac{1}{2} c - \frac{2c}{\pi^2} \sum_{n=1}^{\infty} \frac{1-(-1)^n}{n^2} \cos \frac{n\pi x}{c}$$

or,

$$f(x) \sim \frac{1}{2} c - \frac{4c}{\pi^2} \sum_{k=0}^{\infty} \frac{\cos[(2k+1)\pi x/c]}{(2k+1)^2} \qquad (4)$$

since, when n is even, $1 - (-1)^n = 0$.

● **PROBLEM 3-17**

Determine the coefficient c_k in the complex Fourier series for the waveforms shown.

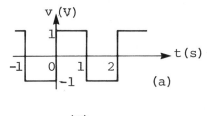

(a)

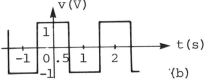

(b)

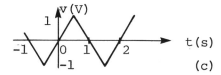

(c)

Solution: The general expression for the coefficient is

$$c_k = \frac{1}{\tau} \int_{\tau_1}^{\tau_1 + \tau} v(t) e^{-jk\omega_o t} dt$$

where τ is the period, $\omega_o = 2\pi/\tau$ is the fundamental angular frequency and τ_1 can be chosen to facilitate evaluation of c_k.

(a) $\tau = 2s$ and $\omega_o = \pi$ rad/s. τ_1 is chosen as $-1s$, so

99

that v(t) changes values only once over the interval of integration. Thus

$$c_k = \frac{1}{2}\int_{-1}^{1} v(t)e^{-jk\pi t}dt$$

$$= \frac{1}{2}\left[-\int_{-1}^{0} e^{-jk\pi t}dt + \int_{0}^{1} e^{-jk\pi t}dt\right].$$

Evaluation of the integrals gives

$$c_k = \frac{1}{2}\left[\frac{1}{jk\pi} e^{-jk\pi t}\Big|_{-1}^{0} - \frac{1}{jk\pi} e^{-jk\pi t}\Big|_{0}^{1}\right]$$

$$= \frac{1}{2jk\pi}[(1 - e^{jk\pi}) + (1 - e^{-jk\pi})].$$

This expression is rearranged to collect the exponential terms:

$$c_k = \frac{1}{jk\pi}[1 - (\frac{e^{jk\pi} + e^{-jk\pi}}{2})].$$

The exponential form of $\cos k\pi$ is recognized, so that

$$c_k = -j\frac{1}{k\pi}(1 - \cos k\pi).$$

(b) This is the same waveform as part (a), but advanced by $\frac{1}{2}$ s. The time τ, is thus chosen as $-\frac{1}{2}$ s, and

$$c_k = \frac{1}{2}\left[\int_{-\frac{1}{2}}^{-\frac{1}{2}} e^{-jk\pi t}dt - \int_{\frac{1}{2}}^{\frac{3}{2}} e^{-jk\pi t}dt\right]$$

$$c_k = \frac{-1}{2jk\pi}\left[e^{-jk\pi t}\Big|_{-\frac{1}{2}}^{\frac{1}{2}} - e^{-jk\pi t}\Big|_{\frac{1}{2}}^{\frac{3}{2}}\right]$$

$$= \frac{-1}{2jk\pi}\left[2e^{-jk\frac{\pi}{2}} - e^{jk\frac{\pi}{2}} - e^{-j\frac{3k\pi}{2}}\right].$$

To simplify this, we first note that

$$e^{-j\frac{3k\pi}{2}} \quad e^{j\frac{k\pi}{2}} = e^{-j2k\pi} = e^{j\frac{k\pi}{2}} \quad k = 0, \pm 1, \pm 2, \ldots$$

Therefore

$$c_k = \frac{-1}{jk\pi}\left[e^{-jk\frac{\pi}{2}} - e^{j\frac{k\pi}{2}}\right] = \frac{2}{k\pi}\left[\frac{e^{j\frac{k\pi}{2}} - e^{-j\frac{k\pi}{2}}}{2j}\right].$$

The exponential form of $\sin\left(\frac{k\pi}{2}\right)$ is recognized, so that

$$c_k = \frac{2}{k\pi} \sin\left(\frac{k\pi}{2}\right).$$

(c) The period τ, and hence ω_o, are unchanged from the values of the first two parts. In considering a choice of τ_1, it is noted that $v(t) = 2t$ for $-\frac{1}{2} \le t \le \frac{1}{2}$ and $v(t) = 2 - 2t$ for $\frac{1}{2} \le t \le \frac{3}{2}$. Thus $\tau_1 = -\frac{1}{2}$ is a reasonable choice, and

$$c_k = \frac{1}{2}\left[\int_{-\frac{1}{2}}^{\frac{1}{2}} 2t e^{-jk\pi t}dt + \int_{\frac{1}{2}}^{\frac{3}{2}} (2-2t) e^{-jk\pi t}dt\right].$$

This can be evaluated making use of the integral

$$\int xe^{ax}dx = \frac{e^{ax}}{a^2}(ax-1).$$

However, we will approach the evaluation of c_k in another way. Let $v_b(t)$ be the waveform of part (b) and $v_c(t)$, that of part (c). Now if $v_b(t)$ were doubled in amplitude, it would be exactly the time derivative of $v_c(t)$. That is,

$$2v_b(t) = \frac{d}{dt} v_c(t).$$

In this equation, $v_b(t)$ and $v_c(t)$ are replaced by their complex Fourier series representation, with the known values of c_k (for the $v_b(t)$ series) inserted:

$$2\sum_{k=-\infty}^{\infty} \left(\frac{2}{k\pi}\right)\sin\left(\frac{k\pi}{2}\right)e^{jk\pi t} = \frac{d}{dt}\sum_{k=-\infty}^{\infty} c_k e^{jk\pi t}.$$

The differentiation can be carried inside the summation to give

$$\sum_{k=-\infty}^{\infty} \left(\frac{4}{k\pi}\right)\sin\left(\frac{k\pi}{2}\right)e^{jk\pi t} = \sum_{k=-\infty}^{\infty} jk\pi c_k e^{jk\pi t}.$$

The coefficients of $e^{jk\pi t}$ are equated term for term:

$$\frac{4}{k\pi} \sin\left(\frac{k\pi}{2}\right) = jk\pi c_k$$

or $\quad c_k = -j\dfrac{4}{k^2\pi^2} \sin\left(\dfrac{k\pi}{2}\right).$

CONVERGENCE OF FOURIER SERIES

• **PROBLEM 3-18**

Define the following properties of a real valued function f of a real variable:
(a) The "limit from the right" of f at x_o
(b) The "limit from the left" of f at x_o
(c) f is piecewise continuous on (a,b).
(d) The right and left hand derivatives of f at x_o.

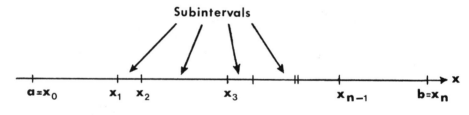

Fig. 1

Solution: The limit from the right of f(x) as x approaches x_o is denoted by

$$f(x_o+) = \lim_{x \to x_o+} f(x) \equiv \lim_{\substack{h \to 0 \\ h > 0}} f(x_o + h) = L. \qquad (1)$$

This definition may be stated in words as follows: the limit from the right of f(x) as x approaches x_o is L if for any positive number ε, there exists a positive number δ such that for all x satisfying $x_o < x < x_o + \delta$ one has

$$|f(x) - L| < \varepsilon.$$

That is, for x sufficiently close to x_o (and greater than x_o), f(x) is as close to L as desired. This definition is especially important when f is only defined for $x > x_o$ or when f is discontinuous at x_o.

102

(b) The limit from the left of f(x) as x approaches x_o is denoted by

$$f(x_o-) = \lim_{x \to x_o-} f(x) \equiv \lim_{\substack{h \to 0 \\ h > 0}} f(x_o - h) = L. \qquad (2)$$

This equation means that the required limit is L if for any positive ε there exists a positive δ such that for all x satisfying $x_o - \delta < x < x_o$ one has

$$|f(x) - L| < \varepsilon.$$

That is, for x sufficiently close to x_o (and less than x_o), f(x) is as close to L as desired.

(c) A function in a closed interval is called piecewise continuous if the interval can be split up into a finite number of subintervals such that in each subinterval (see Fig. 1) the two following conditions hold:

    i) f(x) is continuous

    ii) f(x) possesses (finite) limits at the left and right hand ends of each subinterval.

Condition (ii) means that $f(x_o-)$ and $f(x_o+)$ exist for all $i = 1, 2, \ldots, n$, and also that $f(a+)$ and $f(b-)$ exist.

(d) If $f(x_o+)$ exists at a point x_o, then f is said to have a right hand derivative at x_o if

$$f'_+(x_o) = \lim_{\substack{t \to 0 \\ t > 0}} \frac{f(x_o+t) - f(x_o+)}{t}$$

exists. Similarly, the left hand derivative of f at x_o is the limit

$$f'_-(x_o) = \lim_{\substack{t \to 0 \\ t < 0}} \frac{f(x_o+t) - f(x_o-)}{t}.$$

• **PROBLEM 3-19**

State the most general Uniform Convergence Theorem for Fourier series (i.e. the one with the weakest premises). Discuss its meaning.

Solution: Let f be continuous on an interval (a,b) with period 2c and suppose that f' is piecewise continuous on (a,b). Then the Uniform Convergence Theorem states that the Fourier series for f converges uniformly to f on (a,b).

Note that if the function is defined only on the interval (a,a+2c) for some real a, then the conclusion still holds at all points in the domain of definition of f and the Fourier series is the periodic extension of f on the rest of the real axis. In this case the series will converge uniformly in any interval (a+2nc, a+2(n+1)c) but not at the endpoints where f is not even defined.

The importance of uniform convergence arises from the manipulations that can be performed when it exists. It can be proved that if a series of functions $\Sigma(f_n)$ converges uniformly to f and all of the f_n are Riemann integrable and differentiable then

$$\int_a^b f\,dx = \sum_{n=1}^{\infty} \int_a^b f_n(x)\,dx \qquad (1)$$

and

$$f'(x) = \sum_{n=1}^{\infty} f'_n(x). \qquad (2)$$

That is, interchange of summation with integration and differentiation is valid on any interval where the series is uniformly convergent. This is useful in determining the Fourier series of f'(x) if the series for f(x) is known.

Finally, in most cases of physical application, especially those in electrical engineering, the functions dealt with are only piecewise continuous so that the theorem stated here is only valid on intervals of continuity and the Pointwise Convergence Theorem is of more practical use.

● **PROBLEM 3-20**

Define convergence in the mean and discuss its connection with ordinary convergence and with the concept of mean squared deviation (or variance) used in statistics.

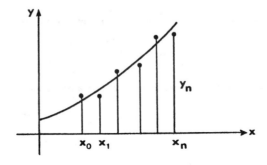

Solution: A sequence of functions $\{f_n(x)\}$ is said to converge in the mean to a function $f(x)$ on an interval (a,b) if

$$\lim_{n\to\infty} \int_a^b [f(x) - f_n(x)]^2 dx = 0. \tag{1}$$

Recall that ordinary convergence requires that, for

$$f: R \to R, \quad \lim_{n\to\infty} \|f_n(x) - f(x)\| = 0$$

where $\|g(x)\|$ = absolute value of $g(x)$ for real valued g and if g is a vector function,

$$\|g(x)\| = \|g_1(x), g_2(x), \ldots, g_m(x)\| =$$

$$= \sqrt{g_1^2 + g_2^2 + \ldots + g_m^2}. \tag{2}$$

The function on $g(x)$ defined in (2) is an example of a norm on the vector space consisting of all such $g(x)$ and the expression in equation (1) is simply another example of a norm on this vector space, i.e.

$$\int_a^b [g(x)]^2 dx = \|g(x)\|_2 \tag{3}$$

and $\|\cdot\|_2$ is called the mean square norm. Thus (1) may be written as

$$\lim_{n\to\infty} \|f(x) - f_n(x)\|_2 = 0 \tag{4}$$

and (4) is now the condition for $\{f_n(x)\}$ to converge in the mean to $f(x)$. In the case of Fourier series, f_n is replaced by the partial sum

$$g_n(x) = \frac{a_0}{2} + \sum_{k=1}^n a_k \cos kx + b_k \sin kx \tag{5}$$

and it is said that the Fourier series for $f(x)$ converges to f in the mean if

$$\lim_{n\to\infty} \int_a^b [f(x) - g_n(x)]^2 dx = 0. \tag{6}$$

Finally, the concept of variance is approached as follows. Suppose a set of n measurements, $y_1, y_2, \ldots, y_n$ are made to try to determine a value for the fixed quantity y. Then the best estimate of y is taken to be the mean of this set of measurements

$$\langle y \rangle \equiv \frac{1}{n} \sum_{i=1}^{n} y_i \tag{7}$$

and the precision of this estimate is usually described by the mean square deviation,

$$d^2 \equiv \frac{1}{n} \sum_{i=1}^{n} [y_i - \langle y \rangle]^2 \tag{8}$$

which is simply the average value of the quantities $[y_i - \langle y \rangle]^2$. Now suppose that there is a theoretical relationship in which the value of y depends on another measurable quantity, x (e.g. y might be the voltage across a resistor, R, and x might be the current through R in which case y = Rx). Then

$$y = y(x) \tag{9}$$

is the theoretical relationship and each measurement y_i is accompanied by a measurement of x, namely x_i. Then to test how well the data fits (9), one would form the mean square deviation

$$D = \frac{1}{n} \sum_{i=0}^{n} [y_i - y(x_i)]^2 \tag{10}$$

(see Figure). If the measurements y_i were taken at equal intervals Δx, then $\Delta x \cdot n = L$ where $L = x_n - x_o$ is the length of interval. In this case

$$D = \frac{1}{L} \sum_{i=0}^{n} [y_i - y(x_i)]^2 \Delta x$$

As n now increases the sum approaches an integral. It seems logical then to measure the extent of fit of two functions $y(x)$ and $\bar{y}(x)$ by the integral

$$D = \frac{1}{L} \int_a^b [y(x) - \bar{y}(x)]^2 dx \tag{11}$$

or if the functions are $f(x)$ and $g_n(x)$

$$D_n = \frac{1}{L} \int_a^b [f(x) - g_n(x)]^2 dx. \tag{12}$$

Thus from (6) we see that convergence in the mean of a Fourier series $\Sigma(g_n)$ to f is equivalent to

$$\lim_{n\to\infty} D_n = 0$$

where D_n is defined in (12).

• **PROBLEM 3-21**

A piecewise continuous function $f(x)$ is to be approximated in the interval $(-\pi,\pi)$ by a trigonometric polynomial of the form

$$g_n(x) = \frac{A_o}{2} + \sum_{k=1}^{n} A_k \cos kx + B_k \sin kx, \qquad (1)$$

where A_k, B_k, A_o are undetermined. Prove that the total square deviation

$$D_n = \int_{-\pi}^{\pi} [f(x) - g_n(x)]^2 dx \qquad (2)$$

is minimized by choosing A_o, A_k, B_k to be the Fourier coefficients of f, a_k, b_k, a_o.

Solution: Straightforward calculation leads to

$$D_n = \int_{-\pi}^{\pi} [f(x) - g_n(x)]^2 dx \qquad (3)$$

$$= \int_{-\pi}^{\pi} [f(x)]^2 dx - 2 \int_{-\pi}^{\pi} f(x) g_n(x) dx$$

$$+ \int_{-\pi}^{\pi} [g_n(x)]^2 dx. \qquad (4)$$

Now,

$$\int_{-\pi}^{\pi} f(x) g_n(x) dx = \frac{1}{2} A_o \int_{-\pi}^{\pi} f(x) dx + \sum_{k=1}^{n} A_k \int_{-\pi}^{\pi} f(x) \cos kx \, dx$$

$$+ \sum_{k=1}^{n} B_k \int_{-\pi}^{\pi} f(x) \sin kx \, dx. \qquad (5)$$

Also, the orthogonality relations for $\sin kx$ and $\cos kx$ give

$$\int_{-\pi}^{\pi} [g_n(x)]^2 dx = \int_{-\pi}^{\pi} \left[\frac{A_o^2}{4} + \sum_{k=1}^{n} A_k^2 \cos^2 kx \right.$$

$$\left. + B_k^2 \sin^2 kx \right] dx + \int_{-\pi}^{\pi} \left[\sum_{\substack{i,j \\ i \neq j}}^{n} A_i B_j \cos ix \sin jx \right] dx. \quad (6)$$

But

$$+ \int_{-\pi}^{\pi} \left[\sum_{\substack{i,j=1 \\ i \neq j}}^{n} A_i B_j \cos ix \sin jx \right] dx$$

$$= \int_{-\pi}^{\pi} [A_1 B_2 \cos x \sin 2x + A_1 B_3 \cos x \sin 3x + \ldots$$

$$+ A_1 B_n \cos x \sin nx + \ldots + A_n B_{n-1} \cos nx \sin (n-1)x] dx$$

Since $\int_a^b \cos mx \sin nx \, dx = 0$ ($m \neq n$, m,n integers)

the integral of the sum reduces to zero also. Hence,

$$\int_{-\pi}^{\pi} [g_n(x)]^2 dx = \int_{-\pi}^{\pi} \frac{A_o^2}{4} dx + \sum_{k=1}^{n} \left[A_k^2 \int_{-\pi}^{\pi} \cos^2 kx \, dx \right.$$

$$\left. + B_k^2 \int_{-\pi}^{\pi} \sin^2 kx \, dx \right].$$

But

$$\sum_{k=1}^{n} \left[A_k^2 \int_{-\pi}^{\pi} \cos^2 kx \, dx + B_k^2 \int_{-\pi}^{\pi} \sin^2 kx \, dx \right]$$

$$= \pi \Sigma A_k^2 + B_k^2. \quad \text{Thus} \quad \int_{-\pi}^{\pi} [g_n(x)]^2 dx$$

$$= \left\{ \pi \left[\frac{1}{2} A_o^2 + \sum_{k=1}^{n} (A_k^2 + B_k^2) \right] \right\}. \quad (7)$$

Thus, using (5) and (7) in (4):

$$D_n = \int_{-\pi}^{\pi} [f(x)]^2 dx + \left\{ \frac{A_o^2 \pi}{2} - A_o \int_{-\pi}^{\pi} f(x) dx \right\}$$

$$+ \sum_{k=1}^{n} \left\{ \pi A_k^2 - 2A_k \int_{-\pi}^{\pi} f(x) \cos kx\, dx \right\}$$

$$+ \sum_{k=1}^{n} \left\{ \pi B_k^2 - 2B_k \int_{-\pi}^{\pi} f(x) \sin kx\, dx \right\}. \qquad (8)$$

This whole expression is minimized if each $\{\cdot\}$ is minimized. For instance the quantity

$$\delta_k = \pi A_k^2 - 2A_k \int_{-\pi}^{\pi} f(x) \cos kx\, dx \qquad (k=1,2,\ldots,n)$$

has an extremum if

$$\frac{d\delta_k}{dA_k} = 2\pi A_k - 2 \int_{-\pi}^{\pi} f(x) \cos kx\, dx = 0$$

or

$$A_k = \frac{1}{\pi} \int_{-\pi}^{\pi} f(x) \cos kx\, dx = a_k \quad (k=1,\ldots,n) \qquad (9)$$

In this fashion we also obtain

$$B_k = \frac{1}{\pi} \int_{-\pi}^{\pi} f(x) \sin kx\, dx = b_k \quad (k=(1,\ldots,n) \qquad (10)$$

$$A_o = \frac{1}{\pi} \int_{-\pi}^{\pi} f(x) dx = a_o. \qquad (11)$$

Thus (9), (10) and (11) are statements of the fact that we have proven the Fourier coefficients of $f(x)$ to be those coefficients which minimize the total square deviation of equation (2).

SYMMETRY CONDITIONS

• **PROBLEM 3-22**

Assuming $f(x)$ is an even function and periodic with period L, prove that

$$a_n = \frac{4}{L} \int_0^{L/2} f(x) \cos n \frac{2\pi}{L} x \, dx$$

Solution: The Fourier expansion of $f(x)$ is

$$f(x) = a_0 + \sum_{n=1}^{\infty} a_n \cos n \frac{2\pi}{L} x + \sum_{n=1}^{\infty} b_n \sin n \frac{2\pi}{L} x$$

where

$$a_0 = \frac{1}{L} \int_{-L/2}^{+L/2} f(x) \, dx \tag{1}$$

$$a_n = \frac{2}{L} \int_{-L/2}^{+L/2} f(x) \cos n \frac{2\pi}{L} x \, dx \tag{2}$$

and

$$b_n = \frac{2}{L} \int_{-L/2}^{+L/2} f(x) \sin n \frac{2\pi}{L} x \, dx \tag{3}$$

Now,

$$a_n = \frac{2}{L} \int_{-L/2}^{+L/2} f(x) \cos n \frac{2\pi}{L} x \, dx,$$

can be written as

$$a_n = \frac{2}{L} \int_{-L/2}^{0} f(x) \cos n \frac{2\pi}{L} x \, dx$$

$$+ \frac{2}{L} \int_0^{L/2} f(x) \cos n \frac{2\pi}{L} x \, dx \tag{4}$$

Substituting $x = -y$ in the first integral of Eq. (4), we obtain

$$2/L \int_{-L/2}^{0} f(x) \cos n \frac{2\pi}{L} x \, dx$$

$$= -\frac{2}{L} \int_{+L/2}^{0} f(-y) \cos n \frac{2\pi}{L} y \, dy$$

$$= \frac{2}{L} \int_0^{L/2} f(y) \cos n \frac{2\pi}{L} y \, dy$$

$$= \frac{2}{L} \int_0^{L/2} f(x) \cos n \frac{2\pi}{L} x \, dx$$

Note for an even function $f(-y) = f(y)$ and since y is a dummy variable, y can be replaced by x.

Therefore, $a_n = \frac{2}{L} \int_0^{L/2} f(x) \cos n \frac{2\pi}{L} x \, dx$

$$+ \frac{2}{L} \int_0^{L/2} f(x) \cos n \frac{2\pi}{L} x \, dx$$

$$= \frac{4}{L} \int_0^{L/2} f(x) \cos n \frac{2\pi}{L} x \, dx$$

• **PROBLEM 3-23**

If a function is even, prove that the coefficient b_n (for $n = 1, 2, 3, \ldots$) corresponding to sine terms is zero.

SOLUTION: Method I. Assume that $f(x)$ is a periodic function with period L, so that $f(x+nL) = f(x)$ where n is an integer.

The Fourier expansion of $f(x)$ is given by

$$f(x) = a_0 + \sum_{n=1}^{\infty} a_n \cos \frac{n2\pi}{L} x + \sum_{n=1}^{\infty} b_n \sin \frac{n2\pi}{L} x$$

where a_0 = Average value of the function $f(x)$

$$= \frac{1}{L} \int_{-L/2}^{+L/2} f(x) \, dx$$

$$a_n = \frac{2}{L} \int_{-L/2}^{+L/2} f(x) \cos n \frac{2\pi}{L} x \, dx$$

and $b_n = \frac{2}{L} \int_{-L/2}^{+L/2} f(x) \sin n \frac{2\pi}{L} x \, dx$

Now, consider the expression for b_n. Since $f(x)$ is even and

$\sin n \frac{2\pi}{L} x$ is an odd function, the product of an even function and an odd function yields an odd function. If an odd function is integrated within the limits symmetric with respect to the origin, it yields a zero value. So in the expression for b_n, since the integrand is an odd function and the limits are symmetric with respect to the origin, the value of the integral is thus zero. Thus, the coefficient b_n is zero.

<u>Method II</u>: The Fourier expansion of $f(x)$ is

$$f(x) = a_0 + \sum_{n=1}^{\infty} \left(a_n \cos \frac{2n\pi x}{L} + b_n \sin \frac{2n\pi x}{L} \right)$$

Note that $f(x) = f(-x)$, provided $f(x)$ is even

Hence, $a_0 + \sum_{n=1}^{\infty} \left(a_n \cos \frac{2n\pi}{L} x + b_n \sin \frac{2n\pi}{L} x \right)$

$$= a_0 + \sum_{n=1}^{\infty} \left(a_n \cos \frac{2n\pi x}{L} - b_n \sin \frac{2n\pi}{L} x \right)$$

By comparison, $\sum_{n=1}^{\infty} b_n \sin \frac{2n\pi}{L} x = - \sum_{n=1}^{\infty} b_n \sin \frac{2n\pi}{L} x,$

resulting in $b_n = 0$.

Therefore, sine terms do not appear in the Fourier expansion of an even function.

USEFUL FUNCTIONS OF FOURIER SERIES

• **PROBLEM 3-24**

Write the Fourier series for the three voltage waveforms in Fig. 1.

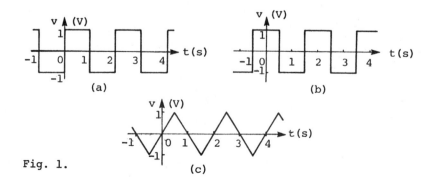

Fig. 1.

Solution: In general, the Fourier coefficients are given by

$$a_n = \frac{2}{T}\int_0^T f(t) \cos n\omega_o t \, dt \tag{1}$$

$$b_n = \frac{2}{T}\int_0^T f(t) \sin n\omega_o t \, dt \tag{2}$$

$$a_o = \frac{1}{T}\int_0^T f(t) \, dt \tag{3}$$

(a) This waveform is an odd function, which means that $a_o = 0$ and $a_n = 0$. The period, T, is 2 and the waveform over one period given by $v(t) = \begin{cases} 1; & 0 < t < T/2 \\ -1; & T/2 < 0 < T \end{cases}$.

Also, since the function is odd, $b_n = \frac{4}{T}\int_0^{T/2} f(t) \sin n\omega_o t \, dt$.

Then $b_n = \frac{4}{2}\int_0^1 \sin n\omega_o t \, dt = 2 \left[\frac{1}{n\omega_o}(-\cos n\omega_o t) \Big|_0^1 \right]$

Recall that $\omega_o = \frac{2\pi}{T} = \frac{2\pi}{2} = \pi$

$b_n = \frac{2}{n\pi}(-\cos n\pi + \cos 0)$

$b_n = \frac{2}{n\pi}(1 - \cos n\pi)$

$\cos n\pi = \begin{cases} 1 & n \text{ even} \\ -1 & n \text{ odd} \end{cases}$.

Then $b_n = \begin{cases} \frac{2}{n\pi}(0), & n \text{ even} \\ \frac{2}{n\pi}(2), & n \text{ odd} \end{cases}$

$b_n = \begin{cases} 0, & n \text{ even} \\ \frac{4}{n\pi}, & n \text{ odd} \end{cases}$

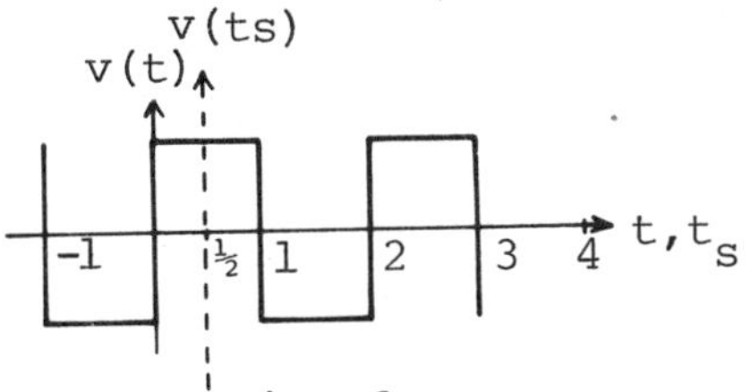

So the Fourier series is

Fig. 2.

$$v(t) = \frac{4}{\pi}\left(\sin \pi t + \frac{1}{3}\sin 3\pi t + \frac{1}{5}\sin 5\pi t + \ldots\right).$$

(b) For this problem, we can make use of the results

of the previous problem and use a very nice trick that eliminates the integration. (The integration in this problem is easy, but this method can be of great use when the integrations involved are more tedious.)

Use the waveform of part (a) and shift the v-axis so that it intersects the t-axis at $t = \frac{1}{2}$ s. (See Fig. 2.)
Then $t = t_s + \frac{1}{2}$. We know from part (a) that

$$v(t) = \frac{4}{\pi}(\sin \pi t + \frac{1}{3}\sin 3\pi t + \frac{1}{5}\sin 5\pi t + \ldots).$$

The required waveform for part (b) is $v(t_s)$. Therefore

$$v(t_s) = \frac{4}{\pi}(\sin \pi(t_s+\frac{1}{2}) + \frac{1}{3}\sin 3\pi(t_s+\frac{1}{2})$$

$$+ \frac{1}{5}\sin 5\pi(t_s+\frac{1}{2}) + \ldots)$$

$$= \frac{4}{\pi}(\sin(\pi t_s+\frac{\pi}{2}) + \frac{1}{3}\sin(3\pi t_s+\frac{3\pi}{2})$$

$$+ \frac{1}{5}\sin(5\pi t_s+\frac{5\pi}{2}) + \ldots).$$

Dropping the subscript s for simplicity, we obtain

$$v(t) = \frac{4}{\pi}(\cos \pi t - \frac{1}{3}\cos 3\pi t + \frac{1}{5}\cos 5\pi t + \ldots).$$

(c) Again, this waveform is odd so $a_n = a_o = 0$ and

$$b_n = \frac{4}{T}\int_0^{\frac{T}{2}} f(t)\sin n\omega_o t\, dt = \frac{4}{T}\int_{-\frac{T}{4}}^{\frac{T}{4}} f(t)\sin n\omega_o t\, dt$$

$$f(t) = v(t) = 2t \quad \text{for} \quad -\frac{T}{4} < t < \frac{T}{4}$$

$$b_n = \frac{4}{2}\int_{-\frac{1}{2}}^{\frac{1}{2}} 2t \sin n\omega_o t\, dt = 4\int_{-\frac{1}{2}}^{\frac{1}{2}} t \sin n\omega_o t\, dt.$$

Integrating by parts: $\quad u = t \quad du = dt$

$$dv = \sin n\omega_o t\, dt \qquad v = -\frac{1}{n\omega_o}\cos n\omega_o t.$$

Then

$$b_n = 4\left[\left.\frac{-t}{n\omega_o}\cos n\omega_o t\right|_{-\frac{1}{2}}^{\frac{1}{2}} + \frac{1}{n\omega_o}\int_{-\frac{1}{2}}^{\frac{1}{2}}\cos n\omega_o t\, dt\right];$$

$$\omega_o = \frac{2\pi}{T} = \pi$$

$$b_n = 4\left[\frac{-t}{n\pi}\left(\cos\left(\frac{n\pi}{2}\right) - \cos\left(\frac{-n\pi}{2}\right)\right) + \frac{1}{n^2\omega_o^2}(\sin n\omega_o t)\right]_{-\frac{1}{2}}^{\frac{1}{2}}$$

Recall that $\cos\theta = \cos(-\theta)$

$$b_n = 4\left[\frac{1}{n^2\pi^2}\left(\sin\frac{n\pi}{2} - \sin\left(\frac{-n\pi}{2}\right)\right)\right]$$

$$\sin(-\theta) = -\sin\theta.$$

So, $$b_n = \frac{8}{n^2\pi^2}\sin\frac{n\pi}{2}$$

$$\sin\frac{n\pi}{2} = \begin{cases} 0, & n \text{ even} \\ 1, & n=1, 5, 9..... \\ -1, & n=3, 7, 11..... \end{cases}$$

So,

$$b_n = \begin{cases} 0, & n \text{ even} \\ \frac{8}{n^2\pi^2}, & n=1, 5, 9..... \\ \frac{-8}{n^2\pi^2}, & n=3, 7, 11..... \end{cases}$$

and the Fourier series is

$$v(t) = \frac{8}{\pi^2}\left(\sin\pi t - \frac{1}{9}\sin 3\pi t + \frac{1}{25}\sin 5\pi t - ....\right).$$

• **PROBLEM 3-25**

Represent the triangular waveform shown in Fig. 1 as a Fourier series.

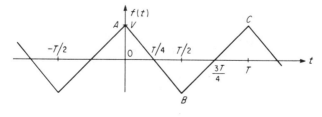

Fig. 1.

Solution: We choose the origin to make the function even; therefore, $b_n = 0$. Also the function has half-wave symmetry; therefore, $a_o = 0$. So we can write the Fourier series

$$f(t) = \sum_{n=1}^{\infty} a_n \cos n\omega t; \quad \omega = \frac{2\pi}{T}$$

where $a_n = \frac{2}{T}\int_0^T f(t) \cos n\omega t \, dt$.

We can choose the periodic interval $[t_o, t_o+T]$ rather than $[0,T]$. In this case it is convenient to choose $[-\frac{T}{2}, \frac{T}{2}]$ because of the form of $f(t)$. Hence we can write the alternate form of a_n as

$$a_n = \frac{2}{T}\int_{-\frac{T}{2}}^{\frac{T}{2}} f(t) \cos n\omega t \, dt.$$

Since both $f(t)$ and $\cos n\omega t$ are even functions, the integral must have the same value between $-\frac{T}{2}$ to 0 and between 0 and $\frac{T}{2}$. We then have

$$a_n = \frac{4}{T}\int_0^{\frac{T}{2}} f(t) \cos n\omega t \, dt.$$

Hence,

$$f(t) = -\frac{4V}{T}\left(t - \frac{T}{4}\right) \quad \text{where}$$

$-\frac{4V}{T}$ is the slope of the line for the interval $0 \leq t \leq \frac{T}{2}$. Therefore,

$$a_n = -\frac{16V}{T^2}\int_0^{\frac{T}{2}} \left(t - \frac{T}{4}\right) \cos n\omega t \, dt$$

$$a_n = -\frac{16V}{T^2}\int_0^{\frac{T}{2}} t \cos n\omega t \, dt + \frac{4V}{T}\int_0^{\frac{T}{2}} \cos n\omega t \, dt$$

$$a_n = -\frac{16V}{T^2}\left[\frac{1}{n^2\omega^2}\cos n\omega t + \frac{t}{n\omega}\sin n\omega t\right]_0^{\frac{T}{2}}$$

$$+ \frac{4V}{T}\left[\frac{\sin n\omega t}{n\omega}\right]_0^{\frac{T}{2}}; \quad \omega = \frac{2\pi}{T}$$

$$a_n = -\frac{16V}{T^2}\left[\frac{T^2}{n^2 4\pi^2}\{\cos n\pi - 1\} + \frac{T^2}{n4\pi}\{\sin n\pi\}\right]$$

$$+ \frac{2V}{\pi n}\{\sin n\pi\}.$$

Hence, $\sin n\pi = 0$ for all n and $\cos n\pi = 1$ for n = even, and -1 for n = odd, which yields

$$f(t) = \sum_{n \text{ odd}}^{\infty} \frac{8V}{n^2 \pi^2} \cos n\omega t$$

$$f(t) = \frac{8V}{\pi^2}[\cos \omega t + \frac{1}{9} \cos 3\omega t + \frac{1}{25} \cos 5\omega t + \ldots].$$

• **PROBLEM 3-26**

Find the Fourier series representation of a periodic function with period 2π, such that

$$f(t) = t^2 \quad -\pi \leq t \leq \pi.$$

Also compute the Fourier coefficients.

Solution: The Fourier series representation for a periodic function f(t) whose period is τ is

$$f(t) \sim a_o + \sum_{n=1}^{\infty}[a_n \cos(\frac{2\pi nt}{\tau}) + b_n \sin(\frac{2\pi nt}{\tau})] \tag{1}$$

where
$$a_o = \frac{1}{\tau}\int_{t_o}^{t_o+\tau} f(t)dt \tag{2}$$

$$a_n = \frac{2}{\tau}\int_{t_o}^{t_o+\tau} f(t)\cos(\frac{2\pi nt}{\tau})dt \tag{3}$$

$$b_n = \frac{2}{\tau}\int_{t_o}^{t_o+\tau} f(t)\sin(\frac{2\pi nt}{\tau})dt. \tag{4}$$

Using Eq. (2) with $\tau = 2\pi$, $f(t) = t^2$ yields

$$a_o = \frac{1}{2\pi}\int_{-\pi}^{\pi} t^2 dt. \tag{5}$$

Evaluate the integral:

$$a_o = \frac{1}{2\pi} \left.\frac{t^3}{3}\right|_{-\pi}^{\pi} = \frac{\pi^2}{3}. \tag{6}$$

Using Eq. (3) with $\tau = 2\pi$, $f(t) = t^2$ yields:

$$a_n = \frac{2}{2\pi} \int_{-\pi}^{\pi} t^2 \cos(nt)\,dt. \tag{7}$$

The expression can be simplified slightly by taking advantage of symmetry. The integral of an even function over a symmetric interval is twice the integral of the function over the upper half of the interval. Therefore,

$$a_n = \frac{2}{\pi} \int_0^{\pi} t^2 \cos(nt)\,dt. \tag{8}$$

The integral can be evaluated by a double application of integration by parts, summarized as follows:

$$\frac{2}{\pi} \int_0^{\pi} t^2 \cos(nt)\,dt = \left.\frac{2t^2 \sin(nt)}{\pi n}\right|_0^{\pi} - \frac{4}{\pi n} \int_0^{\pi} t\,\sin(nt)\,dt$$

$$= 0 - \frac{4}{\pi n} \left[\left.\frac{-t\cos(nt)}{n}\right|_0^{\pi} + \frac{1}{n} \int_0^{\pi} \cos(nt)\,dt \right]$$

$$= -\frac{4}{\pi n} \left(\frac{-\pi\cos(n\pi)}{n} + 0 \right)$$

$$= \frac{4\cos(n\pi)}{n^2}. \tag{9}$$

The expression $\cos(n\pi)$ takes the value +1 if n is even, -1 if n is odd; therefore, we can write $\cos(n\pi) = (-1)^n$. Eq. (9) is then

$$a_n = (-1)^n \frac{4}{n^2}. \tag{10}$$

The expression for b_n involves integrating an odd function over a symmetric interval; so by symmetry $b_n = 0$.

$$b_n = \frac{2}{2\pi} \int_{-\pi}^{\pi} t^2 \sin(nt)\,dt = 0 \tag{11}$$

The Fourier series is thus

$$f(t) \sim \frac{\pi^2}{3} + \sum_{n=1}^{\infty} [(-1)^n \frac{4}{n^2} \cos(nt)].$$

Note: We use the symbol $\sim$, "corresponds to," rather than =, "equals," because at discontinuities in $f(t)$ the series converges to the mean of $f(t^-)$ and $f(t^+)$. Thus if $f(t)$ is continuous, the Fourier series is equal to $f(t)$. If $f(t)$ has discontinuities, the series produces the same value at all values of t except those at which discontinuities occur, where it produces the average of $f(t^-)$ and $f(t^+)$.

● **PROBLEM 3-27**

Find the Fourier series of the half-wave rectified sinusoid as shown in Fig. 1.

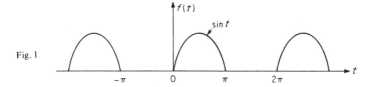

Fig. 1

Solution: The function $f(t)$ can be represented over one period as

$$f(t) = \begin{cases} \sin t & 0 \leq t < \pi \\ 0 & \pi \leq t < 2\pi \end{cases}$$

Since there is no half-wave symmetry we can expect a d.c. value. Hence,

$$a_o = \frac{1}{T} \int_o^T f(t) dt = \frac{1}{2\pi} \int_o^\pi \sin t\, dt + \frac{1}{2\pi} \int_\pi^{2\pi} 0\, dt$$

$$a_o = \frac{1}{2\pi} [-\cos t]_o^\pi = \frac{1}{2\pi} [1 + 1] = \frac{1}{\pi}.$$

However we cannot say that the function has odd or even symmetry. Therefore, we must determine a_n and b_n.

The Fourier representation of a periodic function is

$$a_o + \sum_{n=1}^{\infty} (a_n \cos n\omega_o t + b_n \sin n\omega_o t)$$

where

$$a_n = \frac{2}{T} \int_0^T f(t) \cos n\omega_o t \, dt$$

and

$$b_n = \frac{2}{T} \int_0^T f(t) \sin n\omega_o t \, dt.$$

First we find

$$b_n = \frac{2}{2\pi} \int_0^\pi \sin(t) \sin(nt) \, dt; \qquad \omega_o = 1$$

by the trigonometric function-product relationship

$$\sin t \sin nt = \tfrac{1}{2} \cos(t-nt) - \tfrac{1}{2} \cos(t+nt).$$

We obtain

$$b_n = \frac{1}{\pi} \int_0^\pi [\tfrac{1}{2} \cos t(1-n) - \tfrac{1}{2} \cos t(1+n)] \, dt$$

$$b_n = \frac{1}{\pi} [\tfrac{1}{2(1-n)} \sin t(1-n) - \tfrac{1}{2(1+n)} \sin t(1+n)]_0^\pi$$

$$b_n = \frac{1}{\pi} [\tfrac{1}{2(1-n)} \sin(\pi-n\pi) - \tfrac{1}{2(1+n)} \sin(\pi+n\pi) - 0 + 0]$$

$$b_n = \frac{1}{2\pi} (\tfrac{1}{1-n} \sin(\pi-n\pi) - \tfrac{1}{1+n} \sin(\pi+n\pi)).$$

When we substitute any positive value for n into the above expression we obtain $b_n = 0$; for n = 1 we obtain the undetermined form $\frac{0}{0}$ for the first term. Using L'Hospital's rule we can evaluate $\frac{1}{1-n} \sin(\pi-n\pi)$ for n = 1 as follows:

Let $\qquad g(n) = \dfrac{g_1(n)}{g_2(n)} = \dfrac{\sin(\pi-n\pi)}{1-n}$

where $g_1(n) = \sin(\pi-n\pi)$ and $g_2(n) = 1-n$.

By L'Hospital's rule we can evaluate

$$G(n) = \dfrac{g_1'(n)}{g_2'(n)} = \dfrac{-\pi \cos(\pi-n\pi)}{-1}.$$

If G(1) exists then g(1) = G(1). Hence.

$$g(1) = \pi \cos(\pi-\pi) = \pi \qquad \text{and}$$

$$b_1 = \frac{1}{2\pi}(\pi) = \frac{1}{2}.$$

The only sine component in the Fourier series is $\frac{1}{2}\sin t$. We proceed to find a_n:

$$a_n = \frac{1}{\pi}\int_0^\pi \sin(t)\cos(nt)\,dt$$

$$a_n = \frac{1}{\pi}\int_0^\pi [\frac{1}{2}\sin t(1+n) + \frac{1}{2}\sin t(1-n)]\,dt$$

$$a_n = -\frac{1}{2\pi}[\frac{1}{1+n}\cos t(1+n) + \frac{1}{1-n}\cos t(1-n)]_0^\pi$$

$$a_n = -\frac{1}{2\pi}[\frac{1}{1+n}\cos(\pi+n\pi) + \frac{1}{1-n}\cos(\pi-n\pi)$$

$$-(\frac{1}{1+n}+\frac{1}{1-n})].$$

When $n = 1$ we obtain

$$a_1 = -\frac{1}{2\pi}[\frac{1}{2} + \frac{1}{0} - (\frac{1}{2} + \frac{1}{0})]$$

where the two indeterminate terms are both equal to zero by L'Hospital's rule, giving $a_1 = 0$.

When $n = 2$ we obtain

$$a_2 = -\frac{1}{2\pi}[-\frac{1}{3} + 1 - \frac{1}{3} + 1] = -\frac{2}{3\pi}.$$

When $n = 3$ we again find

$$a_3 = -\frac{1}{2\pi}[\frac{1}{4} - \frac{1}{3} - \frac{1}{4} + \frac{1}{3}] = 0.$$

Continuing in this way we find that for all odd harmonics ($n = 1,3,5,\ldots$) a_n is zero and all even harmonics ($n = 2,4,6\ldots$)

$$a_n = -\frac{1}{2\pi}[-\frac{1}{1+n} - \frac{1}{1-n} - \frac{1}{1+n} - \frac{1}{1-n}]$$

$$a_n = -\frac{1}{2\pi}[-\frac{2}{1+n} - \frac{2}{1-n}] = \frac{1}{2\pi}[\frac{2}{1+n} + \frac{2}{1-n}]$$

$$a_n = \frac{1}{2\pi}[\frac{2(1-n)+2(1+n)}{n^2-1}] = \frac{2}{\pi}\frac{1}{1-n^2}.$$

Hence, we can express the half-wave rectified sine wave as

$$f(t) = \frac{1}{\pi} + \frac{1}{2} \sin t + \frac{2}{\pi} \sum_{n=2,4,6\ldots} \frac{1}{1-n^2} \cos nt.$$

• **PROBLEM 3-28**

Find the Fourier series for the sawtooth waveform of Fig. 1.

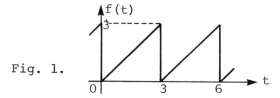

Fig. 1.

<u>Solution</u>: We note that this waveform repeats every 3 seconds. Thus the period $\tau = 3s$ and since $\omega_o = \frac{2\pi}{\tau}$, $\omega_o = \frac{2\pi}{3}$.

A periodic function can be represented as a Fourier series in the form

$$f(t) = a_o + \sum_{n=1}^{\infty} (a_n \cos n\omega_o t + b_n \sin n\omega_o t) \qquad (1)$$

where a_o is the average value of the function $f(t)$ and is defined as

$$a_o = \frac{1}{\tau} \int_0^{\tau} f(t)\,dt. \qquad (2)$$

$f(t)$ represents one period of the entire periodic function, thus $f(t) = t$; $0 < t < 3$.

Substituting these values yields:

$$a_o = \frac{1}{3} \int_0^3 t\,dt$$

$$a_o = \frac{1}{3} [\frac{t^2}{2}]_0^3$$

$$a_o = \frac{1}{3} [\frac{9}{2} - 0] = \frac{3}{2}.$$

The coefficients b_n and a_n are defined as

$$b_n = \frac{2}{\tau} \int_0^{\tau} f(t) \sin n\omega_o t\,dt \qquad (3)$$

$$a_n = \frac{2}{\tau} \int_0^\tau f(t) \cos n \omega_o t \, dt. \qquad (4)$$

Hence

$$b_n = \frac{2}{3} \int_0^3 t \sin n \left(\frac{2\pi}{3}\right) t \, dt$$

$$b_n = \frac{2}{3} \left[\frac{1}{\left(\frac{n2\pi}{3}\right)^2} \sin n \left(\frac{2\pi}{3}\right) t - \frac{t}{\frac{n2\pi}{3}} \cos n \left(\frac{2\pi}{3}\right) t \right]_0^3$$

$$b_n = \frac{2}{3} \left[\frac{1}{\left(\frac{n2\pi}{3}\right)^2} \sin n2\pi - \frac{9}{n2\pi} \cos n2\pi \right]$$

$$b_n = \frac{3}{n^2 2\pi^2} \sin n 2\pi - \frac{3}{n\pi} \cos n 2\pi.$$

The sine term is zero for all n since any multiple of 2π in the sine term is zero.

Hence

$$b_n = - \left(\frac{3}{\pi} \frac{1}{n}\right) \qquad n = 1, 2, 3, \ldots$$

since the cos term is 1 for any multiple of 2π.

$$a_n = \frac{2}{3} \int_0^3 t \cos n\omega_o t \, dt$$

$$a_n = \frac{2}{3} \left[\frac{1}{\left(\frac{n2\pi}{3}\right)^2} \cos \frac{n2\pi}{3} t + \frac{t}{\frac{n2\pi}{3}} \sin \frac{n2\pi}{3} t \right]_0^3$$

$$a_n = \frac{2}{3} \left[\frac{9}{n^2 4\pi^2} \cos n2\pi - \frac{9}{n^2 4\pi^2} \cos 0 \right]$$

But $\cos n2\pi = \cos 0$ for all n; therefore, $a_n = 0$ for all n.

The Fourier representation of this waveform is written

$$f(t) = \frac{3}{2} - \frac{3}{\pi} \sum_{n=1}^{\infty} \frac{1}{n} \sin n \frac{2\pi}{3} t$$

$$f(t) = \frac{3}{2} - \frac{3}{\pi} \left(\sin \frac{2\pi}{3} t + \frac{1}{2} \sin \frac{4\pi}{3} t + \frac{1}{3} \sin \frac{6\pi}{3} + \ldots \right).$$

● PROBLEM 3-29

Find the Fourier series for the sawtooth waveform shown in Fig. 1.

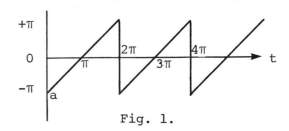

Fig. 1.

Solution: We note that the sawtooth waveform has odd-symmetry and its average value is zero; thus $a_0 = 0$, $a_n = 0$, and

$$b_n = \frac{2}{\tau} \int_0^\tau f(t) \sin n\omega_0 t \, dt.$$

Hence, we can express the sawtooth waveform as the Fourier series

$$\sum_{n=1}^{\infty} b_n \sin n\omega_0 t.$$

Note that the waveform is a straight-line variation, ranging from $-\pi$ to $+\pi$ over one complete cycle; thus $\tau = 2\pi$ and $\omega_0 = \frac{2\pi}{\tau} = 1$.

The straight-line variation that represents one period of the waveform can be written as

$$f(t) = t - \pi \quad ; \quad 0 < t < 2\pi.$$

Therefore

$$b_n = \frac{2}{\tau} \int_0^\tau f(t) \sin n\omega_0 t \, dt = \frac{1}{\pi} \int_0^{2\pi} (t-\pi) \sin nt \, dt$$

$$b_n = \frac{1}{\pi} \left[\int_0^{2\pi} t \sin nt \, dt - \int_0^{2\pi} \sin nt \, dt \right]$$

$$b_n = \left[\frac{1}{n^2 \pi} \sin nt - \frac{t}{n\pi} \cos nt + \frac{1}{n} \cos nt \right]_0^{2\pi}$$

$$b_n = -\frac{2}{n}\cos(2\pi n) + \frac{1}{n}\cos(2\pi n) - \frac{1}{n}$$

$$b_n = -\frac{2}{n} \quad \text{for } n = 1,2,3,4,\ldots$$

Hence, the Fourier series which represents the sawtooth wave is

$$\sum_{n=1}^{\infty} -\frac{2}{n}\sin nt = -2(\sin t + \frac{1}{2}\sin 2s + \frac{1}{3}\sin 3t + \ldots)$$

● **PROBLEM 3-30**

First obtain the Fourier series of the waveform below. Then find the sum of the first four terms of the series at t = 2.

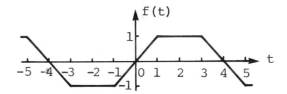

Solution: The general Fourier series expansion is

$$f(t) = a_o + \sum_{n=1}^{\infty}(a_n \cos(n\omega_o t) + b_n \sin(n\omega_o t))$$

where $\omega_o = 2\pi/\tau = 2\pi/8 = \pi/4$ rad/s.

Note that the waveform has odd symmetry (i.e. $f(t) = -f(-t)$; therefore a_n for all n, including a_o, are zero. The waveform also has half-wave symmetry (i.e. $f(t) = -f(t\pm\tau/2)$, so only the fundamental and odd harmonics are present. The b coefficients are then given by

$$b_n = \frac{4}{\tau}\int_o^{\tau/2} f(t)\sin(n\omega_o t)dt = \frac{1}{2}\int_o^{4} f(t)\sin(\frac{n\pi t}{4})dt$$
for odd n

$b_n = 0$ for even n

To perform this integration, f(t) is expressed by three different functions over the range $0 \le t \le 4$:

$f(t) = t \quad 0 \le t \le 1$

$f(t) = 1 \quad 1 \le t \le 3$

$f(t) = 4-t \quad 3 \le t \le 4.$

The integral is broken up into three corresponding parts:

$$b_n = \frac{1}{2}\int_0^1 t\sin(\frac{n\pi t}{4})dt + \frac{1}{2}\int_1^3 \sin(\frac{n\pi t}{4})dt$$

$$+ \frac{1}{2}\int_3^4 (4-t)\sin(\frac{n\pi t}{4})dt \text{ for n odd.}$$

Evaluation of the first and last integrals is facilitated by making use of the following relation, from integral tables:

$$\int x\sin x\, dx = \sin x - x\cos x$$

where x would be made equal to $n\pi t/4$.

The first of the three integrals becomes

$$\frac{1}{2}\int_0^1 t\sin(\frac{n\pi t}{4})dt = \frac{4^2}{2n^2\pi^2}\int_0^1 (\frac{n\pi t}{4})\sin(\frac{n\pi t}{4})(\frac{n\pi dt}{4})$$

$$= \frac{8}{n^2\pi^2}\left[\sin(\frac{n\pi t}{4}) - (\frac{n\pi t}{4})\cos(\frac{n\pi t}{4})\right]_{t=0}^1$$

$$= \frac{8}{n^2\pi^2}\left[\sin(\frac{n\pi}{4}) - (\frac{n\pi}{4})\cos(\frac{n\pi}{4})\right].$$

The second integral is

$$\frac{1}{2}\int_1^3 \sin(\frac{n\pi t}{4})dt = \frac{2}{n\pi}\int_1^3 \sin(\frac{n\pi t}{4})(\frac{n\pi dt}{4})$$

$$= -\frac{2}{n\pi}\cos(\frac{n\pi t}{4})\Big|_1^3 = \frac{2}{n\pi}[\cos(\frac{n\pi}{4}) - \cos(\frac{3n\pi}{4})].$$

The third integral is

$$\frac{1}{2}\int_3^4 (4-t)\sin(\frac{n\pi t}{4})dt = 2\int_3^4 \sin(\frac{n\pi t}{4})dt - \frac{1}{2}\int_3^4 t\sin(\frac{n\pi t}{4})dt$$

$$= \frac{-8}{n\pi}\cos(\frac{n\pi t}{4})\Big|_3^4 - \frac{8}{n^2\pi^2}\left[\sin(\frac{n\pi t}{4}) - (\frac{n\pi t}{4})\cos(\frac{n\pi t}{4})\right]_3^4$$

$$= \frac{8}{n\pi}[\cos(\frac{3n\pi}{4}) - \cos(n\pi)] + \frac{8}{n^2\pi^2}[n\pi\cos(n\pi)$$

$$+ \sin(\frac{3n\pi}{4}) - \frac{3n\pi}{4}\cos(\frac{3n\pi}{4})]$$

$$= \frac{2}{n\pi} \cos\left(\frac{3n\pi}{4}\right) + \frac{8}{n^2\pi^2} \sin\left(\frac{3n\pi}{4}\right).$$

The results of these three integrations are collected to obtain

$$b_n = \frac{8}{n^2\pi^2} \left[\sin\left(\frac{n\pi}{4}\right) + \sin\left(\frac{3n\pi}{4}\right)\right] \qquad n \text{ odd}.$$

When evaluated for n = 1, 3, 5 and 7, the coefficients are

$$b_1 = \frac{8}{\pi^2}\left[\frac{\sqrt{2}}{2} + \frac{\sqrt{2}}{2}\right] = \frac{8\sqrt{2}}{\pi^2}$$

$$b_3 = \frac{8}{9\pi^2}\left[\frac{\sqrt{2}}{2} + \frac{\sqrt{2}}{2}\right] = \frac{8\sqrt{2}}{9\pi^2}$$

$$b_5 = \frac{8}{25\pi^2}\left[-\frac{\sqrt{2}}{2} - \frac{\sqrt{2}}{2}\right] = -\frac{8\sqrt{2}}{25\pi^2}$$

$$b_7 = \frac{8}{49\pi^2}\left[-\frac{\sqrt{2}}{2} - \frac{\sqrt{2}}{2}\right] = \frac{8\sqrt{2}}{49\pi^2}$$

The first four terms of the Fourier series are therefore

$$f(t) \approx \frac{8\sqrt{2}}{\pi^2}\left[\sin(\omega_o t) + \frac{1}{9}\sin(3\omega_o t) - \frac{1}{25}\sin(5\omega_o t) - \frac{1}{49}\sin(5\omega_o t)\right]$$

$$\approx \frac{8\sqrt{2}}{\pi^2}\left[\sin\left(\frac{\pi t}{4}\right) + \frac{1}{9}\left(\frac{3\pi t}{4}\right) - \frac{1}{25}\sin\left(\frac{5\pi t}{4}\right) - \frac{1}{49}\sin\left(\frac{7\pi t}{4}\right)\right].$$

The value of f(t) at t = 2 sec is

$$f(t=2) \approx \frac{8\sqrt{2}}{\pi^2}\left[1 - \frac{1}{9} - \frac{1}{25} + \frac{1}{49}\right] = 0.9965.$$

• **PROBLEM 3-31**

Show geometrically that for each angle θ, the transformation

$$T_\theta: R^2 \to R^2,$$

defined by $T_\theta(x,y) = (x\cos\theta - y\sin\theta, x\sin\theta + y\cos\theta)$, is an orthogonal transformation.

Solution: Let V be an inner product space. A linear transformation $T: V \to V$ is said to be orthogonal if

$$\|Tv\| = \|v\|,$$

i.e., if it preserves the norms or magnitudes for each vector v in V. Thus, orthogonal transformations are those linear transformations which preserve distances.

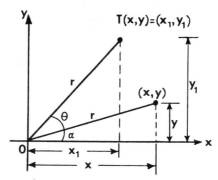

The matrix of the transformation with respect to the standard basis is

$$A_\theta = \begin{bmatrix} \cos\theta & -\sin\theta \\ \sin\theta & \cos\theta \end{bmatrix}.$$

From the figure, it is seen that $x = r\cos\alpha$, $y = r\sin\alpha$. Then,

$$(x_1, y_1) = T(x,y)$$

or

$$x_1 = r\cos(\alpha + \theta) = r\cos\alpha\cos\theta - r\sin\alpha\sin\theta$$

$$= x\cos\theta - y\sin\theta$$

$$y_1 = r\sin(\alpha + \theta) = r\cos\alpha\sin\theta + r\sin\alpha\cos\theta$$

$$= x\sin\theta + y\cos\theta.$$

Thus,

$$\begin{bmatrix} x_1 \\ y_1 \end{bmatrix} = \begin{bmatrix} \cos\theta & -\sin\theta \\ \sin\theta & \cos\theta \end{bmatrix} \begin{bmatrix} x \\ y \end{bmatrix} = T_\theta(x,y).$$

The figure describes a rotation through an angle θ, and from the figure we derive the given transformation. Hence, the given transformation is a rotation. A rotation preserves magnitudes; however, we can check explicitly that T_θ is orthogonal. Let

$$v = (x,y), \quad \|v\| = \sqrt{x^2 + y^2}.$$

Now,

$$T_\theta v = (x \cos \theta - y \sin \theta, x \sin \theta + y \cos \theta),$$

so

$$\|T_\theta v\| = \sqrt{(x \cos \theta - y \sin \theta)^2 + (x \sin \theta + y \cos \theta)^2}$$

$$= \sqrt{x^2 \cos^2 \theta + y^2 \sin^2 \theta - 2xy \cos \theta \sin \theta + x^2 \sin^2 \theta + y^2 \cos^2 \theta + 2xy \cos \theta \sin \theta}$$

$$= \sqrt{x^2(\cos^2 \theta + \sin^2 \theta) + y^2(\sin^2 \theta + \cos^2 \theta)}$$

$$= \sqrt{x^2 + y^2}$$

(since $\sin^2 \theta + \cos^2 \theta = 1$)

$$= \|v\|.$$

• **PROBLEM 3-32**

Given a square waveform f(t) as shown in Fig. 1. Represent this function by a set of Legendre functions.

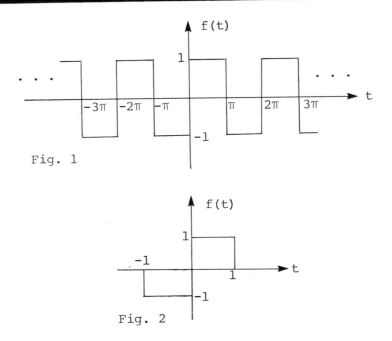

Fig. 1

Fig. 2

Solution: The generating function for Legendre Polynomials is given by the formula

$$P_n(t) = \frac{1}{2^n n!} D^n (t^2-1)^n \tag{1}$$

where
$$D_n = \frac{d^n}{dt^n}$$

Legendre Polynomials can also be generated by using the difference equation

$$n P_n(t) = (2n-1)t P_{n-1}(t) - (n-1)P_{n-2}(t) \tag{2}$$

The basis normalized functions corresponding to the Legendre Polynomials which form an orthogonal set on the interval $[-1,1]$ are given as follows:

$$\phi_0(t) = \frac{1}{\sqrt{2}}, \quad \phi_1(t) = t\sqrt{\frac{3}{2}}, \quad \phi_2(t) = \sqrt{\frac{5}{2}}\left(\frac{3}{2}t^2 - \frac{1}{2}\right),$$

$$\phi_3(t) = \sqrt{\frac{7}{2}}\left(\frac{5}{2}t^3 - \frac{3}{2}t\right), \quad \ldots, \quad \phi_n(t) = \left(\frac{2n+1}{2}\right)^{\frac{1}{2}} P_n(t) \tag{3}$$

The function $f(t)$ is to be redefined as shown in Fig.2 with a substitution $t' = t/\pi$, because the Legendre functions are orthonormal on $[-1, 1]$.

The formula for Fourier coefficients is

$$c_n = \int_{-1}^{+1} f(t) \phi_n(t) dt, \quad n = 0, 1, 2, \ldots \tag{4}$$

in which the functions $\phi_n(t)$ are given by Eq.(3). Now, the generalized Fourier coefficients are as follows:

$$c_0 = \int_{-1}^{+1} \frac{f(t)}{\sqrt{2}} dt = 0$$

$$c_1 = \int_{-1}^{+1} \sqrt{\frac{3}{2}} \, t \, f(t) \, dt = +\sqrt{\frac{3}{2}}$$

$$c_2 = \int_{-1}^{+1} \sqrt{\frac{5}{2}} \left(\frac{3}{2} t^2 - \frac{1}{2}\right) f(t) \cdot dt = 0$$

$$c_3 = \int_{-1}^{+1} \sqrt{\frac{7}{2}} \left(\frac{5}{2} t^3 - \frac{3}{2} t\right) f(t) dt = \sqrt{\frac{7}{16}} \tag{5}$$

Since f(t) is an odd function, c_n is equal to zero for even n.

Therefore, f(t) can be written as follows:

$$f(t) = \sum_{i=0}^{\infty} c_i \phi_i(t) = \frac{3}{2} t + \frac{7}{4\sqrt{2}} \left(\frac{5}{2} t^3 - \frac{3}{2} t \right) + \cdots \quad (6)$$

● **PROBLEM 3-33**

Given a periodic square waveform f(t) of period 2π as shown in Fig. 1, approximate f(t) based on one term, two terms, then three terms by the set of mutually orthogonal functions $\{\sin(n\omega_0 t)\}$ where n = 1, 2, 3,...

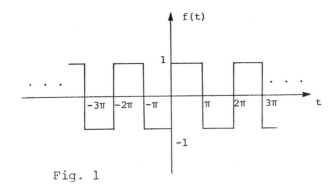

Fig. 1

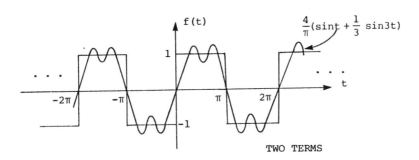

TWO TERMS

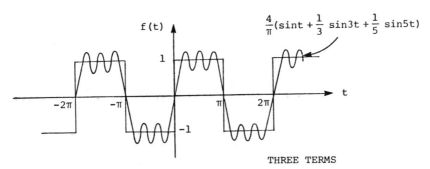

THREE TERMS

Fig. 2

Solution: The set of functions $\{\sin(n\omega_0 t)\}$ for $n = 1, 2, 3 \ldots$ is orthogonal on the interval $(t_0, t_0 + 2\pi/\omega_0)$ for any t_0. This can be demonstrated as shown below:

$$(\sin n\omega_0 t, \sin m\omega_0 t) = \int_{t_0}^{t_0 + 2\pi/\omega_0} \sin n\omega_0 t \sin m\omega_0 t \, dt$$

$$= \tfrac{1}{2} \int_{t_0}^{t_0 + 2\pi/\omega_0} \left[\cos(n-m)\omega_0 t - \cos(n+m)\omega_0 t \, dt\right]$$

$$= 0$$

Note: If a cosine function is integrated within the limits $(t_0, t_0 + 2\pi/\omega_0)$, it yields a zero value.

$$\text{Therefore, } (\sin n\omega_0 t, \sin m\omega_0 t) = \begin{cases} 0, & n \neq m \\ \pi, & n = m \end{cases}$$

The set $\{\sin nt\}$ where $n = 1, 2, \ldots$ is one possible basis set used in representing $f(t)$.

$f(t)$ can be approximated as shown below:

$$f(t) = c_1 \sin t + c_2 \sin 2t + \ldots + c_n \sin nt$$

where the coefficient c_n can be obtained by the relation

$$c_n = \frac{\int_0^{2\pi} f(t) \sin nt}{\int_0^{2\pi} \sin^2 nt \, dt}$$

Now, since $\int_0^{2\pi} \sin^2 nt \, dt = \tfrac{1}{2} \int_0^{2\pi} (1 + \cos 2nt) \, dt$

$$= \tfrac{1}{2} \int_0^{2\pi} dt = \tfrac{1}{2} \int_0^{2\pi} \cos 2nt \, dt$$

$$= \tfrac{1}{2}(2\pi) + \tfrac{1}{2}(0)$$

$$= \pi$$

$$c_n = \frac{1}{\pi} \int_0^{2\pi} f(t) \sin nt \, dt$$

$$= \frac{1}{\pi} \int_0^{\pi} f(t) \sin nt \, dt + \frac{1}{\pi} \int_{\pi}^{2\pi} f(t) \sin nt \, dt$$

$$= \frac{1}{\pi} \int_0^\pi \sin nt \, dt - \frac{1}{\pi} \int_\pi^{2\pi} \sin nt \, dt$$

$$= \begin{cases} \frac{4}{n\pi}, & n \text{ odd} \\ 0, & n \text{ even} \end{cases}$$

Next, for $n = 1$, one term approximation is used

Therefore,
$$c_1 = \frac{4}{\pi(1)} = \frac{4}{\pi}$$

and
$$f(t) \simeq \frac{4}{\pi} \sin t$$

For $n = 1, 2$ and 3, two terms approximation is used because two non-zero Fourier coefficients are required.

Hence,
$$c_1 = \frac{4}{\pi}$$

$$c_2 = 0 \quad (\text{for even } n)$$

$$c_3 = \frac{4}{3\pi}$$

So $f(t) \simeq \frac{4}{\pi} \sin t + \frac{4}{3\pi} \sin 3t$

$$\simeq \frac{4}{\pi} \left(\sin t + \frac{\sin 3t}{3} \right)$$

For $n = 1, 2, 3, 4$ and 5, a three term approximation is used because three non-zero Fourier coefficients are required

Hence,
$$c_1 = \frac{4}{\pi}$$

$$c_2 = c_4 = 0 \quad (\text{for even } n)$$

$$c_3 = \frac{4}{3\pi}$$

and
$$c_5 = \frac{4}{5\pi}$$

Therefore, $f(t) \simeq \frac{4}{\pi} \sin t + \frac{4}{3\pi} \sin 3t + \frac{4}{5\pi} \sin 5t$

$$= \frac{4}{\pi} \left(\sin t + \frac{\sin 3t}{3} + \frac{\sin 5t}{5} \right)$$

The approximations for the cases of two and three terms are shown in Fig. 2.

CHAPTER 4

THE FOURIER TRANSFORM

DEFINITION OF FOURIER TRANSFORM

● **PROBLEM 4-1**

Develop the definition of the Fourier transform of a function f(x) be extending the definition of the Fourier series of f to the case where the discrete spectrum of Fourier coefficients becomes a continuous spectrum.

<u>Solution</u>: Let f be representable by its Fourier series on (-L,L). Then f can be written as

$$f(x) = \sum_{n=-\infty}^{\infty} C_n e^{i\frac{n\pi x}{L}} \quad (-L < x < L) \tag{1}$$

where C_n are the complex Fourier coefficients of f given by

$$C_n = \frac{1}{2L} \int_{-L}^{L} f(x) e^{-i\frac{n\pi x}{L}} dx. \tag{2}$$

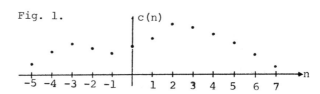

Fig. 1.

To emphasize the functional dependence of C_n on n, write $C(n) = C_n$. The function $C(n)$ is called the Fourier spectrum of f(x) and a typical example is plotted in Figure 1 (for convenience, f(x) is assumed real and even

so that the C(n) are real and the graph may be made).
Now make the substitutions

$$k = \frac{n\pi}{L}, \quad \left(\frac{L}{\pi}\right) C_n = C_L(k)$$

where k is called the wave number of the nth term (or kth term) in the Fourier series expansion of f. Then equations (1) and (2) may be written as

$$C_L(k) = \frac{1}{2\pi} \int_{-L}^{+L} f(x) e^{-ikx} \, dx,$$

$$f(x) = \sum_{\frac{Lk}{\pi} = -\infty}^{\infty} C_L(k) e^{ikx} \Delta k \tag{3}$$

since

$$\Delta k = \frac{\pi}{L} \Delta n = \frac{\pi}{L}.$$

With the change of scale from n to k the Fourier spectrum may be plotted versus wave number as in Figure 2.

Evidently as L approaches infinity, the wave number spectrum approaches a continuous spectrum, i.e., trigonometric functions of all wave numbers must be summed to represent f. Hence, as $L \to \infty$ (the function is no longer periodic) the sum in the second equation of (3) becomes an integral since $\Delta k \to 0$ and we write

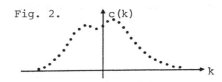

Fig. 2.

$$C(k) = \lim_{L \to \infty} C_L(k) = \frac{1}{2\pi} \int_{-\infty}^{\infty} f(x) e^{-ikx} dx \tag{4}$$

and

$$f(x) = \int_{-\infty}^{\infty} C(k) e^{ikx} \, dk. \tag{5}$$

With a slight change in notation, formulas (4) and (5) become the Fourier transformation. Thus, define a new function F by

$$F(k) = \sqrt{2\pi} \; C(-k)$$

so that the formulas now read

$$F(k) = \frac{1}{\sqrt{2\pi}} \int_{-\infty}^{\infty} f(x) e^{ikx} \, dx \tag{6}$$

$$f(x) = \frac{1}{\sqrt{2\pi}} \int_{-\infty}^{\infty} F(k) e^{-ikx} \, dx \tag{7}$$

$F(k)$ is known as the Fourier transform of the function $f(x)$ and, conversely, $f(x)$ is called the inverse Fourier transform of $F(k)$. To emphasize the interpretation of the Fourier transform as an operator operating on $f(x)$, we will write

$$F(k) = \Phi\{f(x)\}$$

where Φ denotes the operation on f described in equation (6). Finally, note that (6) is merely a defintion of $F(k)$, just as (2) is a definition of the coefficients C_n, so that there is no question about the validity of this formula. That is, if $f(x)$ is integrable and the integral in (6) converges then $F(k)$ exists. On the other hand there is a question about whether the original function $f(x)$ can be retrieved by the formula (7) since this formula was obtained by a limiting process on the Fourier series of f and there is a question as to whether this series represents f. The theory surrounding the convergence of the integral in (7) to $f(x)$ is very similar to that of the convergence of Fourier series and will be discussed later in this chapter.

• **PROBLEM 4-2**

Find the Fourier transform, $F(k) = \Phi\{f(x)\}$ of the Gaussian probability function

$$f(x) = Ne^{-\alpha x^2} \qquad (N, \alpha = \text{constant}). \tag{1}$$

Show directly that $f(x)$ is retrievable from the inverse transform. I.e., show that

$$f(x) = \frac{1}{\sqrt{2\pi}} \int_{-\infty}^{\infty} F(k) e^{-ikx} \, dx = \Phi^{-1}\{\Phi\{f(x)\}\} \, .$$

<u>Solution</u>: The defintion of the Fourier transform of $f(x)$ is

$$F(k) = \Phi\{f(x)\} = \frac{1}{\sqrt{2\pi}} \int_{-\infty}^{\infty} f(x) e^{ikx} \, dx \, . \tag{2}$$

Using the Gaussian function of (1) yields

$$F(k) = \frac{1}{\sqrt{2\pi}} \int_{-\infty}^{\infty} Ne^{-\alpha x^2} e^{ikx} \, dx = \frac{N}{\sqrt{2\pi}} \int_{-\infty}^{\infty} e^{(-\alpha x^2 + ikx)} \, dx. \quad (3)$$

It is convenient to complete the square in the integrand of (3) to get

$$-\alpha x^2 + ikx = -\left(x\sqrt{\alpha} - \frac{ik}{2\sqrt{\alpha}}\right)^2 - \frac{k^2}{4\alpha}. \quad (4)$$

Now make the change of variables

$$x\sqrt{\alpha} - \frac{ik}{2\sqrt{\alpha}} = u \text{ in (4) to obtain}$$

$$-\alpha x^2 + ikx = -u^2 - \frac{k^2}{4\alpha}, \quad dx = \frac{1}{\sqrt{\alpha}} du \quad (5)$$

and substitute (5) into (3) to find

$$F(k) = \frac{N}{\sqrt{2\pi\alpha}} e^{-k^2/4\alpha} \int_{-\infty}^{\infty} e^{-u^2} du = N \frac{1}{\sqrt{2\alpha}} e^{-k^2/4\alpha}. \quad (6)$$

Note that $F(k)$ is also a Gaussian probability function with a peak (the mean) at $x = 0$. Also note that if $f(x)$ is sharply peaked due to a large α, then $F(k)$ is broadened and vice versa (see Figure 1). This has important applications in Quantum physics where

$$|f(x)|^2$$

represents the probability of finding a (one-dimensional

Fig. 1.

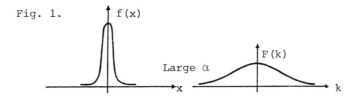

Large α

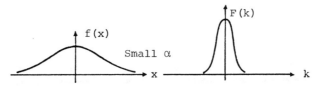

Small α

particle at point x and $|F(k)|^2$ represents the probability of finding this particle with momentum $p = \hbar k$ where $\hbar$ is a known constant. Thus, the better one is able to predict the location of the particle (narrow $f(x)$), the harder it is to predict its momentum (broad F(k)) and vice versa. This is the basic idea behind what is known as the Heisenberg's uncertainty principle and the fact that "narrow" functions have "broad" Fourier transforms and vice versa is a property of all Fourier transforms.

The inverse Fourier integral is given by

$$\Phi^{-1}\{F(k)\} = \frac{1}{\sqrt{2\pi}} \int_{-\infty}^{\infty} F(k) e^{-ikx} dx = \frac{1}{\sqrt{2\pi}} \frac{N}{\sqrt{2\alpha}}$$

$$\times \int_{-\infty}^{\infty} e^{-k^2/4\alpha} e^{-ikx} dk \qquad (7)$$

and it is desired to see whether this equals f(x) as one would expect from the theory. The integral in (7) is calculated in the same way as that in (3). In fact, as a short-cut to that calculation, set

$$\alpha^1 = \frac{1}{4\alpha} \quad \text{and} \quad x^1 = -x$$

to deduce

$$\frac{1}{\sqrt{2\pi}} \int_{-\infty}^{\infty} e^{-\alpha^1 k^2} e^{ix^1 k} dk = \frac{1}{\sqrt{2\alpha^1}} e^{-(x^1)^2/4\alpha^1}$$

$$= \sqrt{2\alpha}\, e^{-\alpha x^2}$$

so that from (7)

$$\Phi^{-1}\{F(k)\} = \frac{1}{\sqrt{2\pi}} \int_{-\infty}^{\infty} F(k) e^{-ikx} dk = \frac{N}{\sqrt{2\alpha}} \sqrt{2\alpha}\, e^{-\alpha x^2}$$

$$= N e^{-\alpha x^2}$$

or

$$\Phi^{-1}\{F(k)\} = f(x)$$

as expected.

• **PROBLEM 4-3**

Find the Fourier transforms, $F_i(k) = \Phi\{f_i(t)\}$, of the functions whose graphs are shown in the accompanying figure.

Solution: The Fourier transform of a function f(t) is defined by

$$F(k) = \frac{1}{\sqrt{2\pi}} \int_{-\infty}^{\infty} f(t) e^{-jkt} \, dt \qquad (1)$$

where $j = \sqrt{-1}$.

a) The "box" function graphed in Figure 1(a) may be written analytically as

$$f_a(t) = \begin{cases} 1 & |t| \leq d \\ 0 & |t| > d \end{cases} \qquad (d > 0).$$

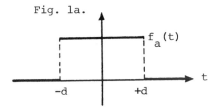

Fig. 1a.

Using (1), its Fourier transform is found to be

$$F_a(k) = \frac{1}{\sqrt{2\pi}} \int_{-d}^{d} e^{jkt} \, dt = \frac{1}{\sqrt{2\pi}} \frac{e^{jkd} - e^{-jkd}}{jk} . \qquad)2)$$

Using the formulas

$$e^{j\theta} = \cos\theta + j\sin\theta$$
$$e^{-j\theta} = \cos\theta - j\sin\theta$$

in (2) yields

$$F_a(k) = \sqrt{\frac{2}{\pi}} \frac{\sin dk}{k} \qquad (3)$$

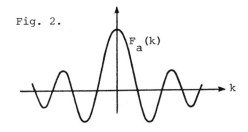

Fig. 2.

$F_a(k)$ is graphed in Figure 2. Note the property that as f_a becomes narrower (small d) F_a becomes "broader" and vice versa.

b) The function $f_b(t)$ of Figure 1 can be written in analytic form as

$$f_b(t) = \begin{cases} \dfrac{-V_0}{b}(t-b) & ; \quad 0 \le t \le b \\ 0 & ; \quad t < 0, \ t > b \end{cases}$$

Hence, from (1)

$$F_b(k) = \dfrac{-V_0}{b\sqrt{2\pi}} \int_0^b (t-b) e^{-jkt}\, dt$$

Fig. 1b.

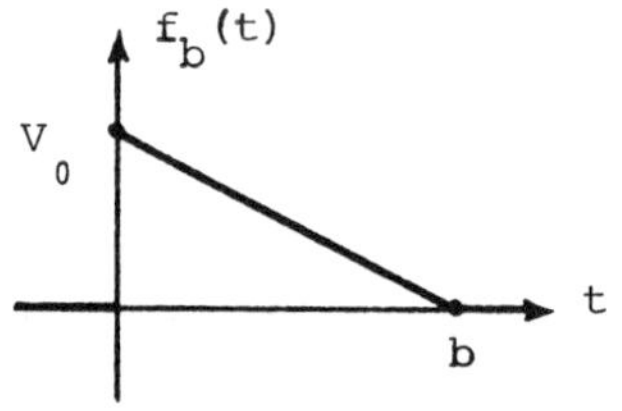

$f_b(t)$

V_0

b

t

$$= \dfrac{-V_0}{b\sqrt{2\pi}} \int_0^b t e^{-jkt}\, dt \quad \dfrac{V_0}{\sqrt{2\pi}} \int_0^b e^{-jkt}\, dt \ .$$

Using integration by parts on the first integral yields

$$F_b(k) = \left[\dfrac{-V_0}{b\sqrt{2\pi}} \left(\dfrac{t}{-jk} e^{-jkt} \right) \Big|_0^b - \dfrac{V_0}{b\sqrt{2\pi}} \int_0^b \dfrac{1}{jk} e^{-jkt}\, dt \right]$$

$$+ \left(\dfrac{V_0}{\sqrt{2\pi}} \dfrac{1}{-jk} e^{-jkt} \right) \Big|_0^b$$

$$= \dfrac{V_0}{b\sqrt{2\pi}} \left(\dfrac{t}{-jk} e^{-jkt} - \dfrac{1}{(jk)^2} e^{-jkt} \right) \Big|_0^b$$

$$+ \left[\dfrac{V_0}{\sqrt{2\pi}} \dfrac{1}{-jk} e^{-jkb} + \dfrac{1}{jk} \right]$$

$$\frac{V_0}{\sqrt{2\pi}} \frac{1}{jk} e^{-jkb} - \frac{V_0}{\sqrt{2\pi} \, bk^2} e^{-jkb} +$$

$$\frac{V_0}{\sqrt{2\pi} \, bk^2} - \frac{V_0}{\sqrt{2\pi} \, jk} e^{-jkb} + \frac{V_0}{\sqrt{2\pi} \, jk}$$

$$= \frac{V_0}{\sqrt{2\pi}} \left[\frac{-1}{bk^2} e^{-jkb} + \frac{1}{bk^2} + \frac{1}{jk} \right] .$$

c) For the functions in Fig. 1(c) use ω as the transfer parameter instead of k. Then

$$f_c(t) = \sin t; \qquad -\pi \leq t \leq \pi .$$

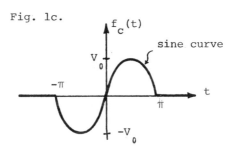

Fig. 1c.

Hence,

$$F_c(\omega) = V_0 \int_{-\pi}^{\pi} \sin t \, e^{-j\omega t} \, dt$$

$$= V_0 \int_{-\pi}^{\pi} \left(\frac{e^{jt} - e^{-jt}}{2j} \right) e^{-j\omega t} \, dt$$

$$= V_0 \int_{-\pi}^{\pi} \frac{e^{jt(1-\omega)} - e^{-jt(1+\omega)}}{2j} \, dt$$

$$= V_0 \left[\frac{e^{jt(1-\omega)}}{2j(j(1-\omega))} + \frac{e^{-jt(1+\omega)}}{2j(j(1+\omega))} \right] \Bigg|_{-\pi}^{\pi}$$

$$= V_0 \left[- \frac{e^{jt(1-\omega)}}{2(1-\omega)} - \frac{e^{-jt(1+\omega)}}{2(1+\omega)} \right] \Bigg|_{-\pi}^{\pi}$$

$$= V_0 - \left[\frac{e^{j\pi(1-\omega)}}{2(1-\omega)} + \frac{e^{-j\pi(1+\omega)}}{2(1+\omega)} - \frac{e^{-j\pi(1-\omega)}}{2(1-\omega)} \right.$$

$$\left. - \frac{e^{j\pi(1+\omega)}}{2(1+\omega)} \right]$$

$$= \frac{V_0 j}{(1-\omega)} \left[\frac{-e^{j\pi(1-\omega)} + e^{-j\pi(1-\omega)}}{2j} \right]$$

$$+ \frac{V_0 j}{(1+\omega)} \left[\frac{e^{j\pi(1+\omega)} - e^{-j\pi(1+\omega)}}{2j} \right]$$

$$= -\frac{jV_0}{1-\omega} \sin\pi(1-\omega) + \frac{jV_0}{1+\omega} \sin\pi(1+\omega)$$

This can be simplified by combining the two terms

$$F_c = \frac{-jV_0(1+\omega)\sin\pi(1-\omega) + jV_0(1-\omega)\sin\pi(1+\omega)}{(1-\omega)(1+\omega)}$$

$$= \frac{-jV_0(1+\omega)\sin\pi\omega - jV_0(1-\omega)\sin\pi\omega}{1-\omega^2}$$

or

$$F_c(\omega) = \frac{-j2V_0 \sin\pi\omega}{1-\omega^2}$$

• **PROBLEM 4-4**

Determine the Fourier transform of the waveform shown.

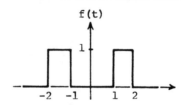

Solution: The definition of $F(j\omega)$ which in the Fourier transform of $f(t)$ is given by

$$F(j\omega) \equiv \int_{-\infty}^{\infty} f(t)e^{-j\omega t}dt. \qquad (1)$$

In the given function, $f(t) = 1$ in ranges of $t = -2$ to -1 and $+1$ to $+2$. In the rest of t, $f(t) = 0$. Therefore

$$F(j\omega) \equiv \int_{-\infty}^{-2} 0 \cdot e^{-j\omega t}dt + \int_{-2}^{-1} 1 \cdot e^{-j\omega t} dy + \int_{-1}^{+1} 0 \cdot e^{-j\omega t}dt$$

$$+ \int_{+1}^{+2} 1 \cdot e^{-j\omega t}dt + \int_{+2}^{+\infty} 0 \cdot e^{-j\omega t}dt$$

$$= 0 - \frac{1}{j\omega}(e^{+j\omega} - e^{+2j\omega}) + 0 - \frac{1}{j\omega}(e^{-2j\omega} - e^{-j\omega}) + 0$$

$$= -\frac{1}{j\omega}\{(e^{+j\omega} - e^{-j\omega}) - (-e^{2j\omega} - e^{-j\omega})\} \qquad (2)$$

Using Euler's identity

$$e^{j\theta} = \cos\theta + j\sin\theta, \qquad (3)$$

yields:

$$F(j\omega) = \frac{j}{\omega}\{(\cos\omega + j\sin\omega - \cos(-\omega) - j\sin(-\omega))$$

$$- (\cos 2\omega + j\sin 2\omega - \cos(-2\omega) - j\sin(-2\omega))\}$$

$$= \frac{j}{\omega}(2j\sin\omega - 2j\sin 2\omega)$$

$$= \frac{2}{\omega}(-\sin\omega + \sin 2\omega)$$

$$= \frac{2}{\omega}(\sin 2\omega - \sin\omega). \qquad (4)$$

PROPERTIES OF FOURIER TRANSFORM

● **PROBLEM 4-5**

State and prove the symmetry theorem.

Solution: *Symmetry theorem*: If $f(t)$ and $F(w)$ form a Fourier transform pair, then $F(t)$ and $2\pi f(-\omega)$ will also form a Fourier transform pair.

i.e.   If $f(t) \longleftrightarrow F(\omega)$

then, $F(t) \longleftrightarrow 2\pi f(-\omega)$

Proof: $F(\omega)$ is given by

$$F(\omega) = \int_{-\infty}^{\infty} f(t)\, e^{-j\omega t}\, dt$$

$$= \int_{-\infty}^{\infty} f(\lambda)\, e^{-j\omega \lambda}\, d\lambda \quad \text{(since t is a dummy variable)}$$

Replacing the variable 'ω' with 't'

$$F(t) = \int_{-\infty}^{+\infty} f(\lambda)\, e^{-jt\lambda}\, d\lambda$$

Let $u = -\lambda$

So,

$$F(t) = -\int_{u=+\infty}^{u=-\infty} f(-u)\, e^{jtu}\, du$$

$$= \int_{-\infty}^{+\infty} f(-u)\, e^{jtu}\, du$$

$$= \int_{-\infty}^{+\infty} f(-\omega)\, e^{j\omega t}\, d\omega \quad \text{(since u is a dummy variable)}$$

$$= \frac{1}{2\pi} \int_{-\infty}^{+\infty} 2\pi f(-\omega)\, e^{j\omega t}\, d\omega$$

Therefore, $F(t)$ is the inverse Fourier transform of $2\pi f(-\omega)$, which means $F(t)$ and $2\pi f(-\omega)$ form a Fourier transform pair.

● **PROBLEM 4-6**

Prove that $F[f(at)] = \frac{1}{|a|} F\left(\frac{\omega}{a}\right)$, i.e. The time scaling property where $F[f(t)] = F(\omega)$.

Solution:
$$F[f(at)] = \int_{-\infty}^{+\infty} f(at)\, e^{-j\omega t}\, dt$$

Substituting $u = at$

hence, $t = \dfrac{u}{a}$

$dt = \dfrac{1}{a}\, du$

(i) case: $a > 0$

$$F[f(at)] = \frac{1}{a}\int_{-\infty}^{+\infty} f(u)\, e^{-j\frac{\omega}{a}u}\, du$$

$$= \frac{1}{a}\int_{-\infty}^{+\infty} f(t)\, e^{-j\frac{\omega}{a}t}\, dt$$

$$= \frac{1}{a}\, F\!\left(\frac{\omega}{a}\right)$$

(ii) case: $a < 0$

$$F[f(at)] = \frac{1}{a}\int_{u=+\infty}^{u=-\infty} f(u)\, e^{-j\frac{\omega}{a}u}\, du$$

$$= -\frac{1}{a}\int_{u=-\infty}^{u=+\infty} f(u)\, e^{-j\frac{\omega}{a}u}\, du$$

$$= \frac{1}{-a}\int_{-\infty}^{+\infty} f(t)\, e^{-j\frac{\omega}{a}t}\, dt$$

$$= \frac{1}{-a}\, F\!\left(\frac{\omega}{a}\right)$$

So, $F[f(at)] = \begin{cases} \dfrac{1}{a}\, F\!\left(\dfrac{\omega}{a}\right), & a > 0 \\ \dfrac{1}{-a}\, F\!\left(\dfrac{\omega}{a}\right), & a < 0 \end{cases}$

Hence, $F[f(at)] = \dfrac{1}{|a|}\, F\!\left(\dfrac{\omega}{a}\right)$

● **PROBLEM 4-7**

Prove the following properties of Fourier transforms:

a) The Fourier transform of f(x) exists if f is absolutely integrable over $(-\infty, +\infty)$,

b) If f(x) is real valued then

$$F(-k) = F^*(k)$$

where $F^*(k)$ is the complex conjugate of $F(k)$.

Solution: By definition, if f is absolutely integrable over $(-\infty, +\infty)$ then

$$\int_{-\infty}^{\infty} |f(x)| \, dx \qquad (1)$$

exists. The Fourier transform of f is given by

$$F(k) = \frac{1}{\sqrt{2\pi}} \int_{-\infty}^{\infty} f(x) \, e^{ikx} \, dx . \qquad (2)$$

Now, recalling that $|e^{iy}| = 1$ for all y we have that

$$|f(x) \, e^{ikx}| = |f(x)|$$

so that

$$\frac{1}{\sqrt{2\pi}} \int_{-\infty}^{\infty} |f(x) \, e^{ikx}| \, dx = \frac{1}{\sqrt{2\pi}} \int_{-\infty}^{\infty} |f(x)| \, dx ,$$

and we know that this second integral exists. Thus,

$$f(x) \, e^{ikx}$$

is absolutely integrable over $(-\infty, \infty)$ and is therefore integrable over $(-\infty, \infty)$. Hence, $F(k)$ exists.

b) This proof is immediate. From (2)

$$F(-k) = \frac{1}{\sqrt{2\pi}} \int_{-\infty}^{\infty} f(x) \, e^{-ikx} \, dx \qquad (3)$$

and, recalling the identity $e^{iy} = \cos y + i \sin y$,

$$F^*(k) = \left(\frac{1}{\sqrt{2\pi}} \int_{-\infty}^{\infty} f(x) \, e^{ikx} \, dx\right)^*$$

$$= \left(\frac{1}{\sqrt{2\pi}} \int_{-\infty}^{\infty} f(x) \cos kx \, dx + i \frac{1}{\sqrt{2\pi}} \int_{-\infty}^{\infty} f(x) \sin kx \, dx\right)^*$$

$$= \frac{1}{\sqrt{2\pi}} \int_{-\infty}^{\infty} f(x) \cos kx \, dx - i \frac{1}{\sqrt{2\pi}} \int_{-\infty}^{\infty} f(x) \sin kx \, dx \qquad (f \text{ real})$$

$$= \frac{1}{\sqrt{2\pi}} \int_{-\infty}^{\infty} f(x) \cos(-kx) \, dx + i \frac{1}{\sqrt{2\pi}} \int_{-\infty}^{\infty} f(x) \sin(-kx) \, dx$$

$$= \frac{1}{\sqrt{2\pi}} \int_{-\infty}^{\infty} f(x) \, (\cos(-kx) + i \sin(-kx)) \, dx$$

$$= \frac{1}{\sqrt{2\pi}} \int_{-\infty}^{\infty} f(x) \, e^{-ikx} \, dx \, . \qquad (4)$$

In the third step the facts that $\cos(-y) = \cos y$ and $\sin(-y) = -\sin y$ were used. Equating (3) and (4) yields

$$F(-k) = F^*(k), \text{ for real } f(x) \, .$$

● **PROBLEM 4-8**

a) Prove the attenuation property of Fourier transforms:

$$\Phi\{f(x) \, e^{ax}\} = F(k-ai)$$

where

$$F(k) = \Phi\{f(x)\} \, .$$

b) Prove the shifting property of Fourier transforms:

$$\Phi\{f(x-a)\} = e^{ika} F(k) .$$

c) Prove the derivative properties of Fourier transforms:

$$\Phi\{f'(x)\} = -ik\Phi\{f(x)\}$$

$$\Phi\{f''(x)\} = -k^2\Phi\{f(x)\} .$$

Solution: a) The Fourier transform of $f(x)$ is defined by

$$F(k) = \Phi\{f(x)\} = \frac{1}{\sqrt{2\pi}} \int_{-\infty}^{\infty} f(x) e^{ikx} dx . \quad (1)$$

The Fourier transform of $g(x) = f(x) e^{ax}$ is then

$$G(k) = \Phi\{f(x) e^{ax}\} = \frac{1}{\sqrt{2\pi}} \int_{-\infty}^{\infty} f(x) e^{ax} e^{ikx} dx$$

$$= \frac{1}{\sqrt{2\pi}} \int_{-\infty}^{\infty} f(x) e^{(a+ik)x} dx$$

$$= \frac{1}{\sqrt{2\pi}} \int_{-\infty}^{\infty} f(x) e^{i(k-ia)x} dx . \quad (2)$$

Now make the change of variable $r = k - ia$ in (2) to give

$$\Phi\{f(x) e^{ax}\} = \frac{1}{\sqrt{2\pi}} \int_{-\infty}^{\infty} f(x) e^{irx} dx = F(r) = F(k-ia) .$$

b) Suppose $f(x)$ is shifted a length a to the right. Then the Fourier transform of this new function $f(x-a)$ is

$$\Phi\{f(x-a)\} = \frac{1}{\sqrt{2\pi}} \int_{-\infty}^{\infty} f(x-a) e^{ikx} dx \quad (3)$$

Now make the substitution $x' = x-a$. Then (3) becomes

$$\Phi\{f(x-a)\} = \frac{1}{\sqrt{2\pi}} \int_{-\infty}^{\infty} f(x') \, e^{ik(x'+a)} \, dx'$$

$$= e^{ika} \frac{1}{\sqrt{2\pi}} \int_{-\infty}^{\infty} f(x') \, e^{ikx'} \, dx'$$

$$= e^{ika} F(k) .$$

c) Suppose that $\Phi\{f'(x)\}$ exists. Then by definition (1),

$$\Phi\{f'(x)\} = \frac{1}{\sqrt{2\pi}} \int_{-\infty}^{\infty} f'(x) \, e^{ikx} \, dx . \qquad (4)$$

Integrating (4) by parts gives

$$\Phi\{f'(x)\} = \frac{1}{\sqrt{2\pi}} f(x) \, e^{ikx} \Big|_{-\infty}^{\infty} - \frac{ik}{\sqrt{2\pi}} \int_{-\infty}^{\infty} f(x) \, e^{ikx} \, dx . \qquad (5)$$

If the Fourier transform of $f(x)$ exists, this usually implies that $f(x) \to 0$ as $x \to \pm\infty$ (this is sometimes not the case, but then $f(x)$ can be treated as a distribution so that the derivative formula is then still valid). Thus, the first term in (5) is zero and

$$\Phi\{f'(x)\} = -ik \, \Phi\{f(x)\} . \qquad (6)$$

If

$$\Phi\{f''(x)\}$$

exists, it is given by

$$\Phi\{f''(x)\} = \frac{1}{\sqrt{2\pi}} \int_{-\infty}^{\infty} f''(x) \, e^{ikx} \, dx .$$

Again integrating by parts yields

$$\Phi\{f''(x)\} = \frac{1}{\sqrt{2\pi}} f'(x) \, e^{ikx} \Big|_{-\infty}^{\infty} - \frac{ik}{\sqrt{2\pi}} \int_{-\infty}^{\infty} f'(x) \, e^{ikx} \, dx . \qquad (7)$$

For reasons mentioned above, it is expected that $f'(x) \to 0$ as $x \to \pm\infty$ so that (7) gives

$$\Phi\{f''(x)\} = -ik\, \Phi\{f'(x)\}.$$

Using (6) gives

$$\Phi\{f''(x)\} = -k^2\, \Phi\{f(x)\}.$$

The obvious extension is made by using $f^{(n-1)}(x) = g(x)$ in the formula of (6) to give

$$\Phi\{g'(x)\} = -ik\, \Phi\{g(x)\}$$

or

$$\Phi\{f^{(n)}(x)\} = -ik\, \Phi\{f^{(n-1)}(x)\},$$

which yields upon iteration

$$\Phi\{f^{(n)}(x)\} = (-ik)^n\, \Phi\{f(x)\}.$$

• **PROBLEM 4-9**

a) Prove that if the functions $g(x)$ and $F(k)$ are absolutely integrable on $(-\infty, +\infty)$ and that the Fourier inversion integral for $f(x)$ is valid for all x except possibly at a countably infinite number of points, then

$$\int_{-\infty}^{\infty} F(k)\, G(-k)\, dk = \int_{-\infty}^{\infty} f(x)\, g(x)\, dx \qquad (1)$$

where

$$F(k) = \Phi\{f(x)\},\quad G(k) = \Phi\{g(x)\}.$$

This is known as the second Parseval theorem of Fourier transform theory.

b) From the above equation (1), prove the first Parseval theorem of Fourier transform theory,

$$\int_{-\infty}^{\infty} |F(k)|^2\, dk = \int_{-\infty}^{\infty} |f(x)|^2\, dx. \qquad (2)$$

Solution: a) The Fourier transform of a function $f(x)$ is defined by

$$F(k) = \frac{1}{\sqrt{2\pi}} \int_{-\infty}^{\infty} f(x) e^{ikx} dx \qquad (3)$$

so that by definition

$$G(-k) = \frac{1}{\sqrt{2\pi}} \int_{-\infty}^{\infty} g(x) e^{-ikx} dx . \qquad (4)$$

Therefore,

$$\int_{-\infty}^{\infty} F(k) G(-k) dk = \int_{-\infty}^{\infty} F(k) dk \int_{-\infty}^{\infty} \frac{1}{\sqrt{2\pi}} g(x) e^{-ikx} dx . \qquad (5)$$

Now $F(k)$ and $g(x)$ are absolutely convergent on $(-\infty, +\infty)$, that is, the integrals

$$\int_{-\infty}^{\infty} |F(k)| dk , \qquad \int_{-\infty}^{\infty} |g(x)| dx$$

are convergent, so that

$$\int_{-\infty}^{\infty} F(k) e^{-ikx} dx , \qquad \int_{-\infty}^{\infty} g(x) e^{-ikx} dx$$

are absolutely convergent (since

$$|F(k) e^{-ikx}| = |F(k)| |e^{-ikx}| = |F(k)|$$

and

$$|g(x) e^{-ikx}| = |g(x)|) .$$

Hence, the order of integration in (5) may be interchanged giving

$$\int_{-\infty}^{\infty} F(k) G(-k) dk = \int_{-\infty}^{\infty} g(x) dx \frac{1}{\sqrt{2\pi}} \int_{-\infty}^{\infty} F(k) e^{-ikx} dk . \qquad (6)$$

Since the Fourier inversion integral is valid,

$$\frac{1}{\sqrt{2\pi}} \int_{-\infty}^{\infty} F(k) e^{-ikx} \, dk = f(x) \tag{7}$$

and using this result in (6) gives the second Parseval theorem:

$$\int_{-\infty}^{\infty} F(k) G(-k) \, dk = \int_{-\infty}^{\infty} g(x) f(x) \, dx \quad . \tag{8}$$

The validity of (8) is insured even if the Fourier inversion integral for f(x) has a countably infinite number of discrepancies with f(x) since this will not affect the equality of the integrals

$$\int_{-\infty}^{\infty} g(x)(f(x)) \, dx \quad \text{and} \quad \int_{-\infty}^{\infty} g(x) \frac{1}{\sqrt{2\pi}} \int_{-\infty}^{\infty} F(k) e^{-ikx} \, dk \, dx \quad .$$

b) The first Parseval theorem is a corollary to the second Parseval theorem stated in equation (8) which follows by letting $f(x) = g(x)$ so that $F(k) = G(k)$ and recalling that (assuming $f = g$ real) $G(-k) = G^*(k)$ where $G^*(k)$ is the complex conjugate of $G(k)$ where $G^*(k)$ is the complex congugate of $G(k)$. Noting that

$$G(k) G^* = |G(k)|^2$$

and using these results in (8) gives

$$\int_{-\infty}^{\infty} G(k) G^*(k) \, dk = \int_{-\infty}^{\infty} [g(x)]^2 \, dx$$

or

$$\int_{-\infty}^{\infty} |G(k)|^2 \, dk = \int_{-\infty}^{\infty} |g(x)|^2 \, dx \quad .$$

• **PROBLEM 4-10**

a) State conditions under which the Fourier integral formula

$$f(x) = \frac{1}{2\pi} \int_{-\infty}^{\infty} e^{-ikx} \, dk \int_{-\infty}^{\infty} f(\xi) e^{ik\xi} \, d\xi \tag{1}$$

is valid. Discuss its validity for the examples in problems 2 and 3 of this chapter.

b) Recast (1) in real form assuming that f(x) is real.

Solution: The most widely used sufficient conditions for pointwise convergence are as follows:

If $f(x)$ is absolutely integrable and piecewise very smooth on $(-\infty,\infty)$, then the Fourier integral theorem is valid in the sense that

$$\frac{1}{2\pi}\int_{-\infty}^{\infty} e^{-ikx}\,dk \int_{-\infty}^{\infty} f(\xi)e^{ik\xi}\,d\xi = \frac{1}{2}[f(x+) + f(x-)]$$

where $f(x+)$ and $f(x-)$ are the right and left hand limits of f at x respectively. Recall that a function f is piecewise very smooth if its second derivative $f''(x)$ is piecewise continuous.

Each of the functions in problems 2 and 3 satsify these conditions so that the Fourier integral theorem is valid for each. For example,

$$f(x) = Ne^{-\alpha x^2}$$

is absolutely integrable on $(-\infty,\infty)$ since

$$\int_{-\infty}^{\infty} \left|Ne^{-\alpha x^2}\right| dx = \int_{-\infty}^{\infty} Ne^{-\alpha x^2}\,dx$$

which is a convergent integral. Also, since f has a continuous second derivative it is piecewise very smooth.

b) Write (1) as

$$f(x) = \frac{1}{\sqrt{2\pi}} \int_{-\infty}^{0} F(k)e^{-ikx}\,dk + \frac{1}{\sqrt{2\pi}} \int_{0}^{\infty} F(k)e^{-ikx}\,dk .$$

Make the change of variable $k' = -k$ in the first integral to obtain

$$f(x) = \frac{1}{\sqrt{2\pi}} \int_{\infty}^{0} F(-k')e^{ik'x}\,(-dk') + \frac{1}{\sqrt{2\pi}} \int_{0}^{\infty} F(k)e^{-ikx}\,dk$$

$$= \frac{1}{\sqrt{2\pi}} \int_0^\infty F(-k')e^{ik'x} \, dk' + \frac{1}{\sqrt{2\pi}} \int_0^\infty F(k)e^{-ikx} \, dk$$

$$= \frac{1}{\sqrt{2\pi}} \int_0^\infty [F^*(k)e^{ikx} + F(k)e^{-ikx}] \, dk \tag{2}$$

where the fact that $F(-k) = F^*(k)$ has been used in the last step. Now since the Fourier integral formula is assumed valid for f,

$$F(k)e^{-ikx} = \frac{1}{\sqrt{2\pi}} \int_{-\infty}^\infty f(\xi)e^{ik(\xi-x)} \, d\xi \tag{3}$$

and taking the complex conjugate of this formula,

$$F^*(k)e^{ikx} = \frac{1}{\sqrt{2\pi}} \int_{-\infty}^\infty f(\xi)e^{-ik(\xi-x)} \, dk \, . \tag{4}$$

Adding (3) to (4) and recalling that

$$\cos\theta = \frac{e^{i\theta} + e^{-i\theta}}{2}$$

gives

$$F(k)e^{-ikx} + F^*(k)e^{ikx} = \frac{1}{\sqrt{2\pi}} \int_{-\infty}^\infty f(\xi) 2\cos k(\xi-x) \, d\xi$$

which, upon substitution into (2), yields the real form of the Fourier integral formula

$$f(k) = \frac{1}{\pi} \int_0^\infty dk \int_{-\infty}^\infty f(\xi) \cos k\,(\xi-x) \, d\xi \quad .$$

• **PROBLEM 4-11**

Define the Dirac delta function, $\delta(x)$, and prove the sifting property of $\delta(x)$ for all functions $f(x)$ which are continuous at $x = 0$,

$$\int_{-\infty}^\infty \delta(x)f(x) \, dx = f(0) \, .$$

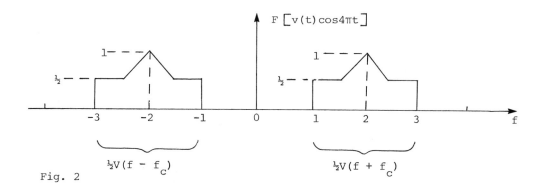

Fig. 2

The sketch for $F[v(t) \cos 4\pi t]$ is shown in Fig. 2. Note: $V(f-f_c)$ has the same shape as $V(f)$, but translated 2 units to the right. Similarly, $V(f+f_c)$ is a 2 unit left translation of $V(f)$.

● **PROBLEM 4-14**

(a) Determine the Fourier transform of the gate function shown in Fig. 1.

$f(t) = Ap_T(t)$, where $p_T(t)$ is a pulse of unit magnitude, width T, and centered at $t = 0$.

(b) Derive the Fourier transform of the function $g(t) = \frac{\sin at}{t}$ from the result obtained in part (a)

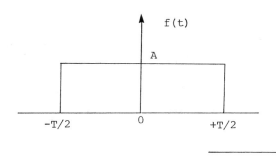

Fig. 1

Solution:

$$F[f(t)] = \int_{-\infty}^{+\infty} f(t) e^{-j\omega t} dt$$

$$= \int_{-T/2}^{+T/2} (A) e^{-j\omega t} dt$$

$$= A \left[\frac{e^{-j\omega t}}{-j\omega} \right]_{-T/2}^{+T/2}$$

$$= \frac{A}{j\omega} \left[2j \sin \omega \frac{T}{2} \right]$$

$$= \frac{AT}{\left(\frac{\omega T}{2}\right)} \sin \left(\frac{\omega T}{2}\right)$$

$$= AT \, \text{sinc} \left(\frac{\omega T}{2}\right)$$

where $\text{sinc}(\alpha) = \frac{\sin \alpha}{\alpha}$

(b) From symmetry theorem,

if $\quad f(t) \longleftrightarrow F(\omega)$,

then $\quad F(t) \longleftrightarrow 2\pi f(-\omega)$

Therefore, from part (a),

since

$$A(2a) \, \text{sinc}(at) \longleftrightarrow 2\pi \left[Ap_{2a}(-\omega) \right]$$

$$Ap_T(t) \longleftrightarrow AT \, \text{sinc} \left(\frac{\omega T}{2}\right)$$

and letting $\frac{T}{2} = a$,

Therefore, $\quad \text{sinc}(at) \longleftrightarrow \frac{2\pi A}{2Aa} p_{2a}(-\omega)$

$$\frac{\sin(at)}{at} \longleftrightarrow \frac{\pi}{a} p_{2a}(-\omega)$$

$$\frac{\sin(at)}{t} \longleftrightarrow \pi \, p_{2a}(-\omega)$$

Since $p_{2a}(\omega)$ is an even function

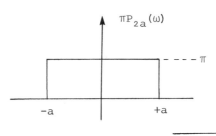

Fig. 2

Hence, the Fourier transform of the function $\frac{\sin(at)}{t}$ is a pulse centered at $\omega=0$ and is of magnitude π and width 2a. It is shown in Fig. 2.

● **PROBLEM 4-15**

Prove that the Fourier transform of the function

$$\int_{-\infty}^{t} f(\lambda)\, d\lambda$$

is given by

$$\pi F(\omega)\, \delta(\omega) + \frac{F(\omega)}{j\omega}$$

Solution: Let $g(t) = \int_{\infty}^{t} f(\lambda) d\lambda$, then

consider a function $h(t) = f(t) \times u(t)$,

where $u(t)$ is a unit step function.

Therefore, $h(t) = \int_{-\infty}^{+\infty} f(\lambda)\, u(t-\lambda)\, d\lambda$

Now, $u(t-\lambda) = 1$, if $t-\lambda > 0$

i.e. If $t > \lambda$ or $\lambda > t$

Hence,
$$h(t) = \int_{-\infty}^{t} f(\lambda)\, d\lambda$$

So,
$$g(t) = h(t) = f(t) * u(t) = \int_{-\infty}^{t} f(\lambda)\, d\lambda.$$

Now, $G(\omega)$ = Fourier transform of $g(t)$
$$= F(\omega)\, U(\omega)$$

where $U(\omega)$ = Fourier transform of a unit step function
$$= \pi \delta(\omega) + \frac{1}{j\omega}$$

Therefore, $G(\omega) = \pi \delta(\omega) F(\omega) + \frac{1}{j\omega} F(\omega)$.

Note: If $F(\omega) = 0$ at $\omega = 0$, $G(\omega) = \frac{1}{j\omega} F(\omega)$

Because $\delta(\omega) F(\omega) = \delta(\omega) F(0)$.

So, $G(\omega) = \frac{1}{j\omega} F(\omega)$, provided $F(0) = 0$, that means

$$F(\omega)\Big|_{\omega=0} = \int_{-\infty}^{+\infty} f(t)\, dt = \text{Area under the curve } f(t) = 0$$

● **PROBLEM 4-16**

> Prove that the convolution of a function $f(t)$ with delta function $\delta(t)$ yields the same function $f(t)$.

<u>Solution</u>: Consider a function $g(t) = f(t) * \delta(t)$

Taking the Fourier transform of $g(t)$,
$$G(\omega) = F(\omega)(1) = F(\omega)$$

The above step follows from the fact that the Fourier transform of $\delta(t)$ is a constant function of unit magnitude.

Hence,
$$f(t) = f(t) * \delta(t)$$

• **PROBLEM 4-17**

(a) Let $F(\omega) = R(\omega) + j\, X(\omega)$ be the Fourier transform of $f(t)$. Prove that $R(\omega)$ is an even function of 'ω' and $X(\omega)$ is an odd function of 'ω'. Also prove that, $X(\omega)$ is zero if $f(t)$ is an even function.

(b) Determine the time function $f(t)$, whose frequency domain is a band pass filter as shown in Fig. 1.

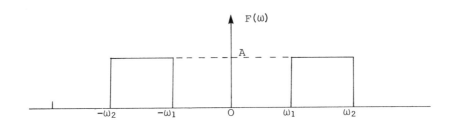

Fig. 1

Solution: (a) By definition,

$$F(\omega) = \int_{-\infty}^{+\infty} F(t)\, e^{-j\omega t}\, dt$$

$$= \int_{-\infty}^{+\infty} f(t)\,(\cos \omega t - j \sin \omega t)\, dt$$

$$= \int_{-\infty}^{+\infty} f(t) \cos \omega t\, dt + j \int_{-\infty}^{+\infty} -f(t) \sin \omega t\, dt$$

$$= R(\omega) + j\, X(\omega)$$

where
$$R(\omega) = \int_{-\infty}^{+\infty} f(t) \cos \omega t\, dt$$

and
$$X(\omega) = \int_{-\infty}^{+\infty} -f(t) \sin \omega t\, dt$$

Since $R(-\omega) = R(\omega)$ is an even function of 'ω'.

$$R(-\omega) = \int_{-\infty}^{+\infty} f(t) \cos(-\omega t)\, dt = \int_{-\infty}^{+\infty} f(t) \cos \omega t\, dt$$

Similarly, $X(-\omega) = \int_{-\infty}^{+\infty} -f(t) \sin(-\omega t) \, dt = \int_{-\infty}^{+\infty} f(t) \sin \omega t \, dt$

$$= -X(\omega)$$

Since $X(-\omega) = -X(\omega)$, $X(\omega)$ is an odd function of 'ω'.

If $f(t)$ is an even function, $f(t) \sin \omega t$ will be an odd function because the product of an even function and an odd function yields an odd function, and the odd function integrated with symmetric limits yields zero. So $X(\omega)$ is zero if $f(t)$ is an even function.

(b) By definition,

$$f(t) = \frac{1}{2\pi} \int_{-\infty}^{+\infty} F(\omega) \, e^{j\omega t} \, d\omega$$

$$= \frac{1}{2\pi} \left[\int_{-\omega_2}^{-\omega_1} A \, e^{j\omega t} \, d\omega + \int_{\omega_1}^{\omega_2} A \, e^{j\omega t} \, d\omega \right]$$

$$= \frac{A}{2\pi} \left[\left\{ \frac{e^{j\omega t}}{jt} \right\}_{-\omega_2}^{-\omega_1} + \left\{ \frac{e^{j\omega t}}{jt} \right\}_{\omega_1}^{\omega_2} \right]$$

$$= \frac{A}{2\pi jt} \left[e^{-j\omega_1 t} - e^{-j\omega_2 t} + e^{j\omega_2 t} - e^{j\omega_1 t} \right]$$

$$= \frac{A}{2\pi jt} \left[2j \sin(\omega_2 t) - 2j \sin(\omega_1 t) \right]$$

$$= \frac{A}{\pi t} \left[\sin(\omega_2 t) - \sin(\omega_1 t) \right]$$

$$= \frac{A}{\pi} \left[\frac{\omega_2 \sin(\omega_2 t)}{\omega_2 t} - \frac{\omega_1 \sin(\omega_1 t)}{\omega_1 t} \right]$$

$$= \frac{A}{\pi} \left[\omega_2 \, \text{sinc}(\omega_2 t) - \omega_1 \, \text{sinc}(\omega_1 t) \right]$$

• **PROBLEM 4-18**

(a) Find the Fourier transform of the function shown in Fig. 1.

(b) Determine the Fourier transform of the function $\dfrac{\sin^2 at}{t^2}$, using the solution obtained in part (a).

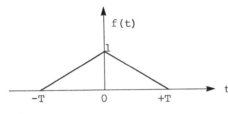

Fig. 1.

<u>Solution</u>: Let $f(t)$ be represented by $Q_{2T}(t)$, where $Q_{2T}(t)$ is an isosceles triangle of peak value 1, base 2T, and centered at $t = 0$.

<u>Method I</u>:

$$\text{For } -T < t < 0, \quad f(t) = 1 + \tfrac{1}{T} t$$

$$\text{For } 0 < t < t, \quad f(t) = 1 - \tfrac{1}{T} t$$

Hence, $f(t) = Q_{2T} = 1 - \dfrac{|t|}{T}$

$F(\omega)$ = Fourier transform of $f(t)$

$$= \int_{-\infty}^{+\infty} f(t)\, e^{-j\omega t}\, dt$$

$$= \int_{-T}^{0} (1 + \tfrac{1}{T} t)\, e^{-j\omega t}\, dt + \int_{0}^{T} (1 - \tfrac{t}{T})\, e^{-j\omega t}\, dt$$

$$= \int_{-T}^{T} e^{-j\omega t}\, dt + \tfrac{1}{T}\int_{-T}^{0} t e^{-j\omega t}\, dt - \tfrac{1}{T}\int_{0}^{T} t e^{-j\omega t}\, dt$$

So,

$$F(\omega) = I_1 + \tfrac{1}{T} I_2 - \tfrac{1}{T} I_3 \tag{1}$$

where
$$I_1 = \int_{-T}^{T} e^{-j\omega t} \, dt,$$

$$I_2 = \int_{-T}^{0} t e^{-j\omega t} \, dt,$$

and
$$I_3 = \int_{0}^{T} t e^{-j\omega t} \, dt.$$

Now,
$$I_1 = \int_{-T}^{T} e^{-j\omega t} \, dt = \int_{-T}^{T} (\cos \omega t - j \sin \omega t) \, dt$$

$$= \int_{-T}^{T} \cos \omega t \, dt = 2 \int_{0}^{T} \cos \omega t \, dt$$

(The above step follows from the fact that the integral of an odd function with symmetric limits is zero.)

So,
$$I_1 = 2 \left[\frac{\sin \omega t}{\omega} \right]_0^T = \frac{2}{\omega} \sin \omega T$$

$$= 2T \left(\frac{\sin \omega T}{\omega T} \right) = 2T \, \text{sinc} \, (\omega T)$$

$$I_2 = \int_{-T}^{0} t \, e^{-j\omega t} \, dt$$

where
$$\int t \, e^{-j\omega t} \, dt = (t) \left(\frac{e^{-j\omega t}}{-j\omega} \right) - (1) \left(\frac{e^{-j\omega t}}{-\omega^2} \right)$$

$$= \frac{j\omega t}{\omega^2} e^{-j\omega t} + \frac{1}{\omega^2} e^{-j\omega t}$$

$$= \frac{1}{\omega^2} \left(e^{-j\omega t} + j\omega t \, e^{-j\omega t} \right)$$

Hence,
$$I_2 = \int_{-T}^{0} t \, e^{-j\omega t} \, dt = \frac{1}{\omega^2} \left(e^{-j\omega t} + j\omega t \, e^{-j\omega t} \right)_{-T}^{0}$$

$$= \frac{1}{\omega^2} \left[1 - \left(e^{j\omega T} - j\omega T \, e^{j\omega T} \right) \right]$$

$$= \frac{1}{\omega^2}\left(1 - e^{j\omega T} + j\omega T\, e^{j\omega T}\right)$$

Similarly, $\quad I_3 = \frac{1}{\omega^2}\left(e^{-j\omega t} + j\omega t\, e^{-j\omega t}\right)\Big|_0^T$

$$= \frac{1}{\omega^2}\left[\left(e^{-j\omega T} + j\omega T\, e^{-j\omega T}\right) - (1)\right]$$

$$= \frac{1}{\omega^2}\left(e^{-j\omega T} + j\omega T\, e^{-j\omega T} - 1\right)$$

Substituting the values I_1, I_2, and I_3 into equation (1).

$$F(\omega) = 2T\left(\frac{\sin \omega T}{\omega T}\right) + \frac{1}{\omega^2 T}\left(1 - e^{j\omega T} + j\omega T\, e^{j\omega T} - e^{-j\omega T} - j\omega T\, e^{-j\omega T} + 1\right)$$

$$= 2T\left(\frac{\sin \omega T}{\omega T}\right) + \frac{1}{\omega^2 T}\left(2 - 2\cos \omega T + j\omega T\,(2j \sin \omega T)\right)$$

$$= 2T\left(\frac{\sin \omega T}{\omega T}\right) + \frac{1}{\omega^2 T}\left[2(1 - \cos \omega T) - 2\omega T \sin \omega T\right]$$

$$= \frac{2}{\omega^2 T}(1 - \cos \omega T) = \frac{2}{\omega^2 T}\left(2 \sin^2 \frac{\omega T}{2}\right) = \frac{4 \sin^2 \frac{\omega T}{2}}{\omega^2 T}$$

$$= \frac{4}{\omega^2 T} \frac{\sin^2 \frac{\omega T}{2}}{\left(\frac{\omega T}{2}\right)^2} \left(\frac{\omega T}{2}\right)^2$$

$$= \frac{4}{\omega^2 T} \frac{\omega^2 T^2}{4} \operatorname{sinc}^2\left(\frac{\omega T}{2}\right)$$

where $\quad \operatorname{sinc}\left(\frac{\omega T}{2}\right) = \frac{\sin\left(\frac{\omega T}{2}\right)}{\left(\frac{\omega T}{2}\right)}$

<u>Method II</u>. $Q_{2T}(t)$ can be written in terms of $P_T(t)$ as follows:

$$Q_{2T}(t) = \frac{1}{T}\int_{-T}^{t}\left[P_T\left(t + \frac{T}{2}\right) - P_T\left(t - \frac{T}{2}\right)\right]dt$$

$$= \frac{1}{T}\int_{-T}^{t} g(t)\, dt$$

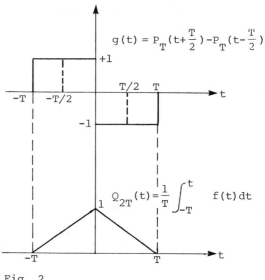

Fig. 2

where $g(t) = P_T(t + \frac{T}{2}) - P_T(t - \frac{T}{2})$ and $P_T(t-t_0)$ is a pulse of unit magnitude, width T, and centered at $t = t_0$.

$Q_{2T}(t)$ is constructed from $P_T(t)$ as shown in Fig. 2.

Since $g(t) = 0$ for $t < -t$,

$$f(t) = Q_{2T} = \frac{1}{T} \int_{-\infty}^{t} g(t)\, dt$$

Let $F(\omega)$ = Fourier transform of $f(t)$

$$F(\omega) = \frac{1}{T}\left[\pi\delta(\omega)\, G(0) + \frac{G(\omega)}{j\omega}\right]$$

Because $\delta(\omega)\, G(\omega) = \delta(\omega)\, G(0)$.

Now, since $F\left[P_T(t)\right] = T\, \text{sinc}\left(\frac{\omega T}{2}\right)$,

where $\text{sinc}(x) = \frac{\sin x}{x}$ from the time-shifting theorem, and:

$g(t) = P_T(t + \frac{T}{2}) - P_T(t - T/2)$

$G(\omega)$ = Fourier transform of $g(t)$

$$= \left[T\, \text{sinc}\left(\frac{\omega T}{2}\right)\right] e^{j\omega\frac{T}{2}} - \left[T\, \text{sinc}\left(\frac{\omega T}{2}\right)\right] e^{-j\omega\frac{T}{2}}$$

$$= T \text{ sinc}\left(\frac{\omega T}{2}\right)\left[e^{j\omega\frac{T}{2}} - e^{-j\omega\frac{T}{2}}\right]$$

$$= T \text{ sinc}\left(\frac{\omega T}{2}\right)\left[2j \sin \frac{\omega T}{2}\right]$$

$$= 2jT\left(\frac{\omega T}{2}\right) \text{ sinc}\left(\frac{\omega T}{2}\right)\left[\frac{\sin\left(\frac{\omega T}{2}\right)}{\left(\frac{\omega T}{2}\right)}\right]$$

$$= j\omega T^2 \text{ sinc}^2\left(\frac{\omega T}{2}\right)$$

Note: $G(0) = 0$

Hence, $\quad F(\omega) = \frac{1}{j\omega T} G(\omega)$

$$= T \text{ sinc}^2\left(\frac{\omega T}{2}\right)$$

$$= T \frac{\sin^2\left(\omega\frac{T}{2}\right)}{\left(\frac{\omega T}{2}\right)^2}$$

● PROBLEM 4-19

a) If a > 0 show that the Fourier transform of the function defined by

$$f(t) = e^{-at} \cos\omega_d t \quad t \geq 0$$
$$= 0 \quad t < 0$$

is $(a+j\omega)^2/[(a+j\omega)^2 + \omega_d^2]$.

Then find the total 1Ω energy associated with the function

$$f = e^{-t} \cos t \quad t \geq 0$$
$$= 0 \quad t < 0$$

by using:

b) time domain integration. That is, find the total energy by integrating

$$W = \int_0^\infty [f(t)]^2 \, dt.$$

c) frequency domain integration. That is, find the total energy by integrating

$$W = \frac{1}{2\pi} \int_{-\infty}^{\infty} |F(\omega)|^2 \, d\omega$$

where $F(\omega)$ is the Fourier transform of the function $f(t)$.

Solution: Using the relation

$$F(\omega) = \int_{-\infty}^{\infty} f(t) e^{-j\omega t} \, dt$$

we can find the Fourier transform of

$$f(t) = e^{-at} (\cos \omega_d t)$$

Hence,

$$F(\omega) = \int_0^\infty e^{-at} (\cos \omega_d t) e^{-j\omega t} \, dt$$

$$= \int_0^\infty e^{-at} \left(\frac{e^{j\omega_d t} + e^{-j\omega_d t}}{2} \right) e^{-j\omega t} \, dt$$

$$= \int_0^\infty \left[\frac{e^{(j\omega_d - j\omega - a)t}}{2} + \frac{e^{(-j\omega_d - j\omega - a)t}}{2} \right] dt$$

$$= \left[\frac{e^{(j\omega_d - j\omega - a)t}}{2(j\omega_d - j\omega - a)} + \frac{e^{-(j\omega_d + j\omega + a)t}}{2(-j\omega_d - j\omega - a)} \right]_0^\infty$$

$$= \left[\frac{-1}{2(j\omega_d - j\omega - a)} - \frac{1}{2(-j\omega_d - j\omega - a)} \right]$$

Hence,

$$F(\omega) = \left[\frac{a + j\omega}{(a + j\omega)^2 + \omega_d^2}\right] \quad .$$

b) In the time domain, the total energy is found by integrating

$$[f(t)]^2$$

as follows:

$$W = \int_{-\infty}^{\infty} [f(t)]^2 \, dt$$

$$W = \int_{0}^{\infty} \left[e^{-t}(\cos t)\right]^2 \, dt$$

$$= \int_{0}^{\infty} e^{-2t}\left(\frac{1}{2} + \frac{1}{2}\cos 2t\right) dt$$

$$= \int_{0}^{\infty} \left[\frac{e^{-2t}}{4} + \frac{e^{-2t}}{2}\left(\frac{e^{+j2t}}{2} + \frac{e^{-j2t}}{2}\right)\right] dt$$

$$= \left[-\frac{e^{-2t}}{4} + \frac{e^{-2t}e^{j2t}}{4(-2+j2)} - \frac{e^{-2t}e^{-j2t}}{4(2+j2)}\right]_{0}^{\infty}$$

$$W = \frac{8 - (-2 - j2) + (2 - j2)}{32} = \frac{12}{32} = \frac{3}{8} \quad .$$

c) In the frequency domain the total energy is found by integrating

$$W = \frac{1}{2\pi} \int |F(\omega)|^2 \, d\omega \quad .$$

We found in (a)

$$F(\omega) = \frac{a + j\omega}{(a+j\omega)^2 + \omega_d^2} \quad .$$

If $a = 1$ and $\omega_d = 1$ then

$$F(\omega) = \frac{1 + j\omega}{(1 + j\omega)^2 + 1}$$

and

$$|F(\omega)|^2 = \frac{1 + \omega^2}{(2-\omega^2)^2 + 4\omega^2} = \frac{1 + \omega^2}{4 + \omega^4}.$$

The energy is

$$W = \frac{1}{2\pi} \int_{-\infty}^{\infty} \frac{1 + \omega^2}{4 + \omega^4} d\omega = \frac{1}{\pi} \int_0^{\infty} \frac{1 + \omega^2}{4 + \omega^4} d\omega.$$

Now, by an integral formula,

$$\int_0^{\infty} \frac{x^2 + 1}{x^4 + 4} dx = \frac{3\pi}{8}.$$

Hence, we have

$$W = \frac{1}{\pi} \int_0^{\infty} \frac{1 + \omega^2}{4 + \omega^2} d\omega = \frac{1}{\pi} \cdot \frac{3\pi}{8}.$$

Thus

$$W = \frac{3}{8}$$

verifying Parseval's identity in this example.

• **PROBLEM 4-20**

Use Fourier transform methods to find the time-domain response of a network having a system function

$$j2\omega/(1 + 2j\omega),$$

if the input is

$$V(t) = \cos t$$

(For a sinusodial input cos t, the Fourier transform is

$$\pi[\delta(\omega+1) + \delta(\omega-1)]).$$

Solution: The time domain response for a particular input V(t) can be obtained by finding the product of the system function H(jω) and the Fourier transform of the input. The inverse Fourier transform of the resulting function is the time-domain response.

For a sinusoidal input cos t, the Fourier transform pair

$$\cos t \iff \pi[\delta(\omega+1) + \delta(\omega-1)]$$

allows us to find the response

$$f(t) = F^{-1}\left\{\frac{j2\omega}{1+2j}\pi(\delta(\omega+1) + \delta(\omega-1))\right\}$$

$$f(t) = F^{-1}\left\{\frac{j2\pi\omega\,\delta(\omega+1)}{1+2j\omega} + \frac{j2\pi\omega\,\delta(\omega-1)}{1+2j\omega}\right\}$$

$$f(t) = F^{-1}\left\{\frac{j\pi\omega\,\delta(\omega+1)}{\frac{1}{2}+j\omega} + \frac{j\pi\omega\,\delta(\omega-1)}{\frac{1}{2}+j\omega}\right\}$$

Using the sifting property of the unit impulse, we obtain:

$$f(t) = F^{-1}\left\{-\frac{j\pi\,\delta(\omega+1)}{\frac{1}{2}-j} + \frac{j\pi\,\delta(\omega-1)}{\frac{1}{2}+j}\right\}$$

$$f(t) = F^{-1}\left\{-\frac{j\pi\,\delta(\omega+1)(\frac{1}{2}+j)}{\frac{1}{4}+1} + \frac{j\pi\,\delta(\omega-1)(\frac{1}{2}-j)}{\frac{1}{4}+1}\right\}$$

$$f(t) = F^{-1}\left\{\frac{\pi\,\delta(\omega+1)}{\frac{5}{4}} - \frac{j\frac{1}{2}\pi\,\delta(\omega+1)}{\frac{5}{4}} + \frac{\pi\,\delta(\omega-1)}{\frac{5}{4}} + \frac{j\frac{1}{2}\pi\,\delta(\omega-1)}{\frac{5}{4}}\right\}$$

$$f(t) = F^{-1}\left\{\frac{4}{5}\pi(\delta(\omega+1) + \delta(\omega-1)) - \frac{2}{5}\pi(j\,\delta(\omega+1) - j\,\delta(\omega-1))\right\}$$

$$f(t) = \frac{4}{5}\cos t - \frac{2}{5}\sin t.$$

• **PROBLEM 4-21**

Given a function f(t) as shown in Fig. 1,

(a) Obtain the Fourier transform of f(t).

(b) Draw the amplitude spectrum of F(ω), where F(ω) is the Fourier transform of f(t).

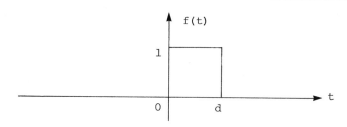

Fig. 1

Solution: f(t) can be represented as

$$f(t) = \begin{cases} 1, & \text{for } 0 < t < d \\ 0, & \text{elsewhere} \end{cases}$$

The Fourier transform for a function f(t) is defined as

$$F(\omega) = \int_{-\infty}^{+\infty} f(t) e^{-j\omega t} dt$$

Hence, for the given function f(t)

$$F(\omega) = \int_0^d (1) e^{-j\omega t} dt$$

$$= \left(\frac{e^{-j\omega t}}{-j\omega} \right)_0^d$$

$$= \frac{-1}{j\omega} \left(e^{-j\omega d} - 1 \right)$$

$$= \frac{1}{j\omega} \left(1 - e^{-j\omega d} \right)$$

$$= \frac{1}{\omega j} \left[1 - (\cos \omega d - j \sin \omega d) \right]$$

174

$$= \frac{-j}{\omega}\left[(1-\cos \omega d) + j \sin \omega d\right]$$

$$= \frac{1}{\omega}\left[\sin \omega d - j(1-\cos \omega d)\right]$$

The magnitude of $F(\omega)$ is

$$|F(\omega)| = \frac{1}{\omega}\sqrt{\sin^2 \omega d + (1-\cos \omega d)^2}$$

$$= \frac{1}{\omega}\sqrt{\sin^2 \omega d + \cos^2 \omega d + 1 - 2\cos \omega d}$$

$$= \frac{1}{\omega}\sqrt{2 - 2\cos \omega d}$$

$$= \frac{(\sqrt{2})\sqrt{1-\cos \omega d}}{\omega}$$

$$= \frac{1}{\omega}\sqrt{2(1-\cos \omega d)}$$

$$= \frac{1}{\omega}\sqrt{4\left[\frac{(1-\cos \omega d)}{2}\right]}$$

$$= \frac{2}{\omega}\sqrt{\sin^2 \frac{\omega d}{2}}$$

$$= 2\,\frac{\sin\left(\frac{\omega d}{2}\right)}{\omega}$$

$$= 2\left(\frac{d}{2}\right)\frac{\sin\left(\frac{\omega d}{2}\right)}{\left(\frac{\omega d}{2}\right)}$$

$$= d\,\frac{\sin\left(\frac{\omega d}{2}\right)}{\left(\frac{\omega d}{2}\right)}$$

$$= d\,\frac{\sin \beta}{\beta}$$

in which $\beta = \frac{\omega d}{2}$

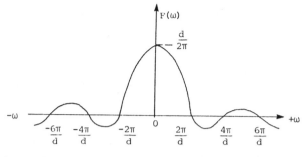

Fig. 2

Therefore, the amplitude spectrum of $F(\omega)$ is a sampling function of the form $K \frac{\sin X}{X} = K \, sa(X)$ with peak magnitude of d.

The amplitude spectrum of $F(\omega)$ is shown in Fig. 2.

• **PROBLEM 4-22**

Derive the Fourier transform of the function f(t) where f(t) is a cosine function of angular frequency ω_0 modulated by a rectangular pulse.

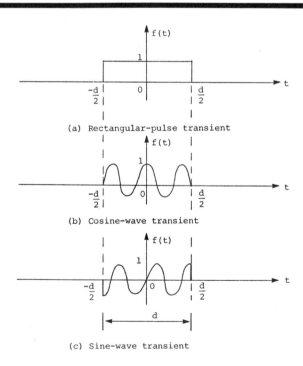

(a) Rectangular-pulse transient

(b) Cosine-wave transient

(c) Sine-wave transient

Fig. 1 Generation of sinusoidal transients

Solution: $f(t)$ can be represented as

$$f(t) = \begin{cases} \cos \omega_0 t, & -\frac{d}{2} < t < \frac{d}{2} \\ 0, & \text{elsewhere} \end{cases}$$

$$F(\omega) = \text{Fourier transform of } f(t)$$

$$= \int_{-\infty}^{+\infty} f(t) e^{-j\omega t} dt$$

$$= \int_{-d/2}^{+d/2} (\cos \omega_0 t)(e^{-j\omega t}) dt$$

$$= \frac{1}{2} \int_{-\frac{d}{2}}^{+\frac{d}{2}} (e^{j\omega_0 t} + e^{-j\omega_0 t})(e^{-j\omega t}) dt$$

$$= \frac{1}{2} \int_{-\frac{d}{2}}^{+\frac{d}{2}} \left[e^{j(\omega_0-\omega)t} + e^{-j(\omega+\omega_0)t} \right] dt$$

$$= \frac{1}{2} \int_{-\frac{d}{2}}^{+\frac{d}{2}} \left\{ \left[\cos(\omega_0-\omega)t + \cos(\omega+\omega_0)t \right] \right. $$

$$\left. + j \left[\sin(\omega_0-\omega)t - \sin(\omega+\omega_0)t \right] \right\} dt$$

Note: Integral of a sine function with symmetric limits yields a zero value.

Hence, $F(\omega) = \int_{0}^{d/2} \left[\cos(\omega_0-\omega)t + \cos(\omega+\omega_0)t \right] dt$

$$= \left[\frac{\sin(\omega_0-\omega)t}{(\omega_0-\omega)} + \frac{\sin(\omega+\omega_0)t}{(\omega+\omega_0)} \right]_0^{d/2}$$

$$= \left[\frac{\sin(\omega_0-\omega)\frac{d}{2}}{(\omega_0-\omega)} + \frac{\sin(\omega+\omega_0)\frac{d}{2}}{(\omega+\omega_0)} \right]$$

$$-\frac{d}{2}\left[\frac{\sin(\omega_0-\omega)\frac{d}{2}}{(\omega_0-\omega)\frac{d}{2}} + \frac{\sin(\omega+\omega_0)\frac{d}{2}}{(\omega+\omega_0)\frac{d}{2}}\right]$$

$$= \frac{d}{2}\left(\frac{\sin\alpha}{\alpha} + \frac{\sin\beta}{\beta}\right)$$

in which $\alpha = \dfrac{(\omega-\omega_0)d}{2}$ and $\beta = \dfrac{(\omega+\omega_0)d}{2}$

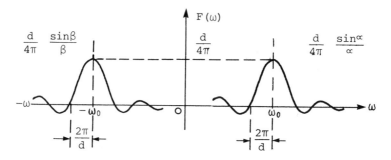

Fig. 2 Continous spectrum of a cosine-wave transient.

The amplitude spectrum of the given cosine-wave transient is shown in Fig. (2).

• **PROBLEM** 4-23

(a) Derive the Fourier transform of the function given in Fig. 1 and obtain its frequency response.

(b) Obtain the Fourier transform and the frequency response of a unit impulse function by making use of the solution in part (a).

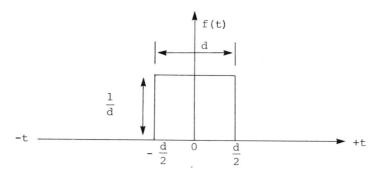

Fig. 1

Solution: (a) f(t) can be represented as

$$f(t) = \begin{cases} \frac{1}{d}, & |t| < \frac{d}{2} \\ 0, & \text{elsewhere} \end{cases}$$

Now, $F(\omega)$ = Fourier transform of $f(t)$

$$= \int_{-\infty}^{+\infty} f(t) e^{-j\omega t} dt$$

Since $f(t)$ is an even function

$$F(\omega) = \int_{-\infty}^{+\infty} f(t) \cos \omega t \, dt$$

$$= \int_{-\frac{d}{2}}^{+\frac{d}{2}} \left(\frac{1}{d}\right) \cos \omega t \, dt$$

$$= (2) \frac{1}{d} \int_0^{\frac{d}{2}} \cos \omega t \, dt$$

$$= \frac{2}{d} \left[\frac{\sin \omega t}{\omega}\right]_0^{d/2} = \frac{2}{d} \left(\frac{\sin \frac{\omega d}{2}}{\omega}\right)$$

$$= \frac{\sin \frac{\omega d}{2}}{\frac{\omega d}{2}} = \text{sinc}\left(\frac{\omega d}{2}\right)$$

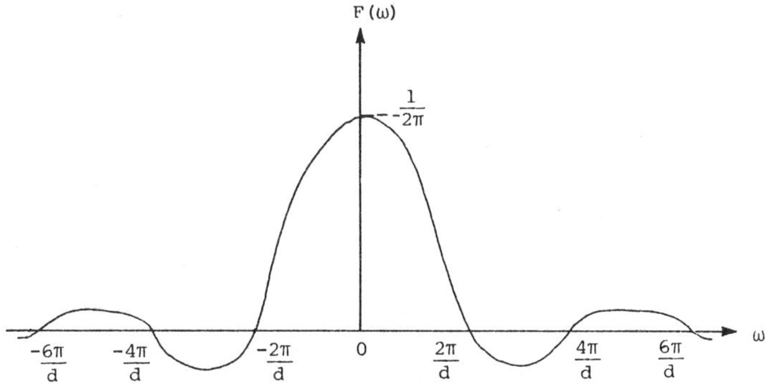

Fig. 2 Frequency distribution of f(t).

Hence, the frequency distribution of f(t) is a sampling function of the form $\frac{\sin x}{x}$, which is shown in Fig. 2.

If d is made smaller until it reaches zero as a limit, then f(t) becomes a pulse of infinite amplitude in an arbitrarily small time interval around t = 0. This is defined as a unit impulse, $\delta(t)$.

i.e., $$\delta(t) = \lim_{d \to 0} f(t)$$

The Fourier transform of $\delta(t) = \lim_{d \to 0} F(\omega)$

$$= \lim_{d \to 0} \frac{\sin\left(\frac{\omega d}{2}\right)}{\left(\frac{\omega d}{2}\right)}$$

By definition, $\lim_{x \to 0} \frac{\sin x}{x} = 1$

therefore, $F(\omega) = 1$

Hence, the frequency distribution of a unit impulse is a constant and equal to 1.

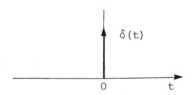

Fig. 3 Unit impulse function

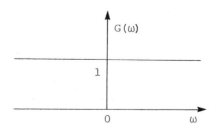

Fig. 4 Frequency distribution of Unit Impulse Function.

The unit impulse function is shown in Fig. 3 and its frequency distribution is shown in Fig. 4.

• **PROBLEM 4-24**

To the circuit shown in Fig. 1, a unit step voltage is applied at the input. Obtain the transient response, using the Fourier transform technique.

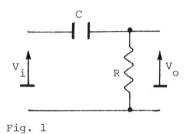

Fig. 1

<u>Solution</u>: The output in the frequency domain can be written as

$$V_0(\omega) = H(\omega) V_i(\omega) \quad (1)$$

where $H(\omega)$ = Transfer function of the network

$$= \frac{R}{R + \frac{1}{j\omega c}} = \frac{j\omega RC}{1 + j\omega RC} \quad (2)$$

$V_0(\omega)$ = Fourier transform of the output

$V_i(\omega)$ = Fourier transform of the input

= Fourier Transform of the unit step function

$$= \lim_{\alpha \to 0} \left[\text{Fourier transform of } e^{-\alpha t} u(t) \right]$$

where $u(t)$ is a unit step function.

Now, Fourier transform of $e^{-\alpha t} u(t)$
$$= \int_0^\infty e^{-\alpha t} e^{-j\omega t} \, dt$$

$$= \int_0^\infty e^{-(\alpha + j\omega)t} \, dt$$

$$= \left[\frac{e^{-(\alpha + j\omega)t}}{-(\alpha + j\omega)} \right]_0^\infty$$

$$= \frac{-1}{\alpha+j\omega} \quad (0-1)$$

$$= \frac{1}{\alpha+j\omega}$$

Therefore,
$$V_i(\omega) = \lim_{\alpha\to 0} \left(\frac{1}{\alpha+j\omega}\right)$$

$$= \lim_{\alpha\to 0} \frac{\alpha-j\omega}{\alpha^2+\omega^2}$$

$$= \lim_{\alpha\to 0} \frac{\alpha}{\alpha^2+\omega^2} - j \lim_{\alpha\to 0} \frac{\omega}{\alpha^2+\omega^2}$$

$$= \pi\delta(\omega) - \frac{j}{\omega}$$

$$= \pi\delta(\omega) + \frac{1}{j\omega} \quad (3)$$

Substituting Eqs. (2) and (3) into Eq. (1),

$$V_0(\omega) = \left(\frac{j\omega RC}{1+j\omega RC}\right)\left(\pi\delta(\omega) + \frac{1}{j\omega}\right)$$

$$= \frac{j\omega RC}{1+j\omega RC}\pi\delta(\omega) + \frac{RC}{1+j\omega RC}$$

Using the property of the delta function,
i.e., $\quad f(t)\delta(t) = f(0)\delta(t)$

$$V_0(\omega) = 0 + \frac{RC}{1+j\omega RC} = \frac{1}{\frac{1}{RC}+j\omega}$$

Hence, inverse Fourier transform of $V_0(\omega)$ is

$$V_0(t) = e^{-t/RC} u(t)$$

Therefore,
$$V_0(t) = \begin{cases} e^{-t/RC}, & t > 0 \\ 0, & \text{elsewhere} \end{cases}$$

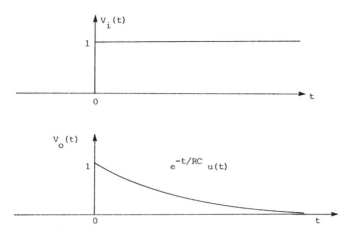

Fig. 2 Input and output voltage waveforms

The waveforms for input and output voltages are shown in Fig. 2.

• **PROBLEM 4-25**

Let $F(\omega)$ be the Fourier transform of a continuous function $f(t)$

Prove that $f(t) = f(t) * \frac{\omega}{\pi} \operatorname{sinc}(\omega t)$ for all $\omega > N$, provided that

$$F(\omega) = F(\omega) G(\omega) \quad \text{where}$$

$$G(\omega) = \begin{cases} 1, & |\omega| < N \\ 0, & \text{elsewhere} \end{cases}$$

(Hint: Fourier transform of $\frac{N}{\pi} \operatorname{sinc}(Nt) = u(\omega+N) - u(\omega-N)$

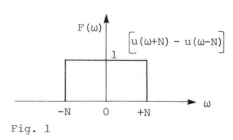

Fig. 1

<u>Solution</u>: With the information given, $F(\omega)$ can be written as

$$F(\omega) = F(\omega) [u(\omega+N) - u(\omega-N)] \tag{1}$$

Since $G(\omega)$ is bandlimited to N rad/sec, it can be represented by unit step functions, $u(\omega+N)$ and $u(\omega-N)$, which is shown in Fig. 1.

Now taking the inverse transform of Eq. (1),

$$f(t) = f(t) * \frac{N}{\pi} \sin C (Nt) \qquad (2)$$

where the sign * designates the operation of convolution.

Equation (2) follows from the fact that the inverse Fourier transform of a product of two time functions is the convolution of the respective functions in the time domain.

Also the inverse Fourier transform of $[u(\omega+N) - u(\omega-N)]$ is given by $\frac{N}{\pi} \sin C (Nt)$, from the symmetry theorem.

Hence, f(t) can be written as

$$f(t) = f(t) * \frac{N}{\pi} \text{sinc} (Nt)$$

or
$$f(t) = f(t) * \frac{\omega}{\pi} \text{sinc} (\omega t) \text{ for all } \omega > N,$$

Since $F(\omega)$ is zero for all $|\omega| > N$.

• **PROBLEM** 4-26

Given $F(\omega) = G(\omega-\omega_0) + G(\omega+\omega_0)$ and $G(\omega) = \begin{cases} \cos \omega, & |\omega| < \pi/2 \\ 0, & \text{elsewhere,} \end{cases}$

the Fourier transforms of f(t) and g(t) respectively,

(a) Determine g(t).

(b) Determine f(t).

(c) Find the frequency at which g(t) should be sampled for perfect reconstruction.

(d) Find A, ω_1, ω_2 so that y(t) = g(t) for the demodulation scheme given in Fig. 1.

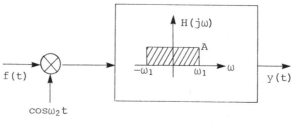

Fig. 1

Solution: (a) In this problem, note:

$$\cos \omega = \frac{e^{j\omega} + e^{-j\omega}}{2}$$

$$\sin \omega = \frac{e^{j\omega} - e^{-j\omega}}{2j}$$

and
$$\sin(t + 90°) = \cos(t)$$

$$\sin(t - 90°) = -\cos(t)$$

Now taking the inverse Fourier transform of $G(\omega)$,

$$g(t) = \frac{1}{2\pi} \int_{-\pi/2}^{\pi/2} \cos \omega \, e^{j\omega t} \, d\omega$$

$$= \frac{1}{4\pi} \int_{-\pi/2}^{\pi/2} (e^{j\omega} + e^{-j\omega}) e^{j\omega t} \, d\omega$$

$$= \frac{1}{4\pi} \int_{-\pi/2}^{\pi/2} e^{j\omega(t+1)} + e^{j\omega(t-1)} \, d\omega$$

$$= \frac{1}{4\pi} \left[\frac{1}{j(t+1)} e^{j\omega(t+1)} + \frac{1}{j(t-1)} e^{j\omega(t-1)} \right]_{-\frac{\pi}{2}}^{\frac{\pi}{2}}$$

$$= \frac{1}{4\pi} \left\{ \left[\frac{1}{j(t+1)} e^{j\frac{\pi}{2}(t+1)} + \frac{1}{j(t-1)} e^{j\frac{\pi}{2}(t-1)} \right] \right.$$

$$\left. - \left[\frac{1}{j(t+1)} e^{-j\frac{\pi}{2}(t+1)} + \frac{1}{j(t-1)} e^{-j\frac{\pi}{2}(t-1)} \right] \right\}$$

$$= \frac{1}{4\pi} \left\{ \left[\frac{1}{j(t+1)} e^{j\frac{\pi}{2}(t+1)} - \frac{1}{j(t+1)} e^{-j\frac{\pi}{2}(t+1)} \right] \right.$$

$$\left. + \left[\frac{1}{j(t-1)} e^{j\frac{\pi}{2}(t-1)} - \frac{1}{j(t-1)} e^{-j\frac{\pi}{2}(t-1)} \right] \right\}$$

$$= -\frac{1}{4\pi}\left\{\frac{2}{t+1}\left[\frac{e^{j\frac{\pi}{2}(t+1)} - e^{-j\frac{\pi}{2}(t+1)}}{2j}\right]\right.$$

$$\left.+ \frac{2}{t-1}\left[\frac{e^{j\frac{\pi}{2}(t-1)} - e^{-j\frac{\pi}{2}(t-1)}}{2j}\right]\right\}$$

$$= \frac{1}{2\pi}\left[\frac{(t-1)\sin\frac{\pi}{2}(t+1) + (t+1)\sin\frac{\pi}{2}(t-1)}{t^2 - 1}\right]$$

$$= \frac{1}{2\pi(t^2-1)}\left[(t-1)\underbrace{\sin(\tfrac{\pi}{2}t + 90°)}_{\cos(\pi/2\, t)} + (t+1)\underbrace{\sin(\tfrac{\pi}{2}t - 90°)}_{-\cos(\pi/2\, t)}\right]$$

$$= \frac{1}{2\pi(t^2-1)}\left[(t-1)\cos(\tfrac{\pi}{2}t) - (t+1)\cos(\tfrac{\pi}{2}t)\right]$$

$$= \frac{1}{2\pi}\cos(\tfrac{\pi}{2}t)\left[\frac{(t-1)-(t+1)}{(t^2-1)}\right]$$

$$= \frac{\cos(\tfrac{\pi}{2}t)}{\pi(1-t^2)}$$

(b) $F(\omega) = G(\omega-\omega_0) + G(\omega+\omega_0)$

By applying the frequency shifting property,

$$f(t) = \text{Fourier inverse of } F(\omega)$$

$$= g(t)\, e^{j\omega_0 t} + g(t)\, e^{-j\omega_0 t}$$

Hence, $f(t) = 2g(t)\cos\omega_0 t$ \hfill (1)

where $g(t)$ = Inverse Fourier transform of $G(\omega)$ obtained in part (a).

(c) The frequency f at which g(t) should be sampled for perfect reconstruction is given by $f \geq 2f_{max}$, where

$$2f_{max} = 2\left(\frac{\omega_{max}}{2\pi}\right) = \frac{2}{2\pi}\left(\frac{\pi}{2}\right) = \frac{1}{2}$$

Since $f = \frac{1}{T}$, therefore, for $f \geq \frac{1}{2}$, $T \leq 2$

(d) Knowing that $Z(t) = f(t) \cos \omega_2 t$ (2)

Substituting Eq. (1) into Eq. (2) and letting $\omega_2 = \omega_0$, $Z(t)$ becomes

$$Z(t) = 2 \cos^2 (\omega_0 t) \, g(t)$$

$$= [1 + \cos (2\omega_0 t)] \, g(t)$$

where $\cos^2 (\theta) = \dfrac{1 + \cos 2\theta}{2}$

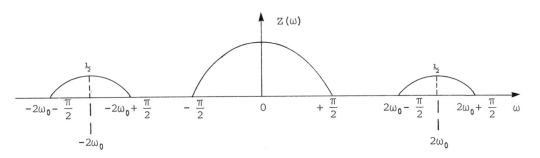

Fig. 2

$Z(\omega)$ = Frequency distribution of $Z(t)$ and its frequency spectrum is shown in Fig. 2.

Now in order to satisfy the condition $\dfrac{\pi}{2} < \omega_1 < 2\omega_0 - \dfrac{\pi}{2}$, $A = 1$ provided that $2\omega_0 > \pi$ or $\omega_0 > \dfrac{\pi}{2}$.

Thus one set of possible solution is:

$$\omega_1 = \dfrac{\pi}{2}$$

$$\omega_2 = \omega_0$$

and $A = 1$.

• **PROBLEM 4-27**

Given a modulation system as shown in Fig. 1. Plot the frequency distribution of $y(t)$, provided that

$$f(t) = 2 \cos 10t + 4 \cos 20t$$

and $m(t) = \cos 200t$

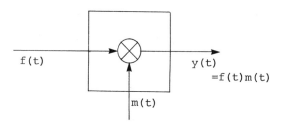

Fig. 1

Solution: From Fig. 1,

$$y(t) = f(t)\, m(t)$$
$$= (2 \cos 10t + 4 \cos 20t)(\cos 200t)$$
$$= 2 \cos 10t \cos 200t + 4 \cos 20t \cos 200t$$

Note: $\cos A \cos B = \tfrac{1}{2} \cos(A-B) + \tfrac{1}{2} \cos(A+B)$. Hence,

$$y(t) = [\cos 190t + \cos 210t] + 2[\cos 180t + \cos 220t]$$

since $Y(\omega)$ = Fourier transform of $y(t)$,

hence, $\bar{\bar{Y}}(\omega) = \pi[\delta(\omega-190) + \delta(\omega+190)] + \pi[\delta(\omega-210) + \delta(\omega+210)]$
$$+ 2\pi\big[\delta(\omega-180) + \delta(\omega+180)\big]$$
$$+ 2\pi\big[\delta(\omega-220) + \delta(\omega+220)\big]$$

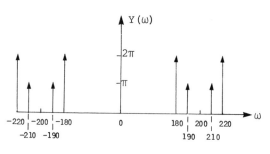

Fig. 2

The frequency spectrum of $y(t)$ is sketched in Fig. 2.

• **PROBLEM 4-28**

Given a system as shown in Fig. 1. Draw the frequency distribution of the output waveform and prove that it is linear.

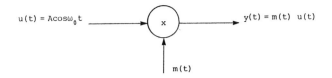

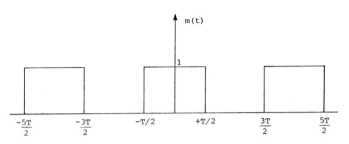

Fig. 1

Solution: From Fig. 1, the system output is $y(t) = m(t)u(t)$ and since the condition to be tested for linearity is chosen as $C_1 u_1(t) + C_2 u_2(t)$ such that $m(t)[C_1 u_1(t) + C_2 u_2(t)] = C_1 m(t) u_1(t) + C_2 m(t) u_2(t)$, this system is a linear time varying system.

The function $m(t)$ can be represented as an exponential Fourier series. In that case, $m(t)$ can be written as

$$m(t) = \sum_{n=-\infty}^{+\infty} C_n e^{jn\omega_m t}$$

Note that the time period of $m(t)$ is $2T$ rather than T.

Hence, $\quad \omega_m = \dfrac{2\pi}{(2T)} = \dfrac{\pi}{T}$

or $\quad \omega_m T = \pi$ \hfill (1)

The constant C_n is given by

$$C_n = \frac{1}{2T} \int_{-T}^{+T} m(t) e^{-jn\omega_m t}\, dt$$

$$= \frac{1}{2T} \int_{-T/2}^{+T/2} (1)\, e^{-jn\omega_m t}\, dt$$

$$= \frac{1}{2T} (2) \int_{0}^{T/2} \cos n\omega_m t\, dt$$

Since $e^{-j\phi} = \cos\phi - j\sin\phi$ and the odd function integrated with symmetric limits yields zero. Therefore,

$$C_n = \frac{1}{T} \left(\frac{\sin n\omega_m t}{n\omega_m} \right)_{t=0}^{t=T/2}$$

$$= \frac{1}{Tn\omega_m} \left(\frac{\sin \frac{n\omega_m T}{2}}{} \right)$$

$$= \frac{1}{Tn\omega_m} \left(\frac{Tn\omega_m}{2} \right) \left(\frac{\sin \frac{Tn\omega_m}{2}}{\frac{Tn\omega_m}{2}} \right)$$

$$= \frac{1}{2} \left(\frac{\sin \frac{n\omega_m T}{2}}{\frac{n\omega_m T}{2}} \right)$$

Since $\omega_m t = \pi$ (From Eq. 1) and $\dfrac{\operatorname{sinc}\left(\frac{n\pi}{2}\right)}{\frac{n\pi}{2}}$

$$C_n = \frac{1}{2} \operatorname{sinc}\left(\frac{n\pi}{2}\right)$$

Now,

$$y(t) = \Big[m(t)\Big]\Big[u(t)\Big]$$

$$= \left(\sum_{n=-\infty}^{+\infty} C_n e^{jn\omega_m t} \right) \left(A \cos\omega_0 t \right)$$

$$= \frac{A}{2} \left(\sum_{-\infty}^{+\infty} C_n e^{jn\omega_m t} \right) \left(e^{j\omega_0 t} + e^{-j\omega_0 t} \right)$$

$$= \frac{A}{2} \left\{ \sum_{-\infty}^{+\infty} C_n \left[e^{j(n\omega_m+\omega_0)t} + e^{j(n\omega_m-\omega_0)t} \right] \right\}$$

By applying the Frequency Shifting Theorem and noting that the Fourier transform of $[1] = 2\pi\delta(\omega)$,

$Y(\omega) = $ Fourier transform of $y(t)$

$$= \frac{A}{2} \left\{ \sum_{-\infty}^{+\infty} 2\pi C_n \left[\delta(\omega-\omega_0-n\omega_m) + \delta(\omega+\omega_0-n\omega_m) \right] \right\}$$

$$= \pi A \left\{ \sum_{-\infty}^{+\infty} C_n \left[\delta(\omega-\omega_0-n\omega_m) + \delta(\omega+\omega_0-n\omega_m) \right] \right\}$$

$$= \left\{ \sum_{-\infty}^{+\infty} \left[\frac{A\pi}{2} \operatorname{sinc} \frac{n\pi}{2}\right] \left[\delta(\omega-\omega_0-n\omega_m) + \delta(\omega+\omega_0-n\omega_m) \right] \right\}$$

where $C_n = \frac{1}{2} \operatorname{sinc}\left(\frac{n\pi}{2}\right)$

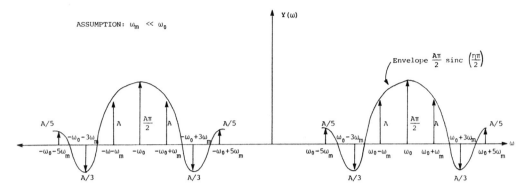

Fig. 2

Hence, the envelope of $Y(\omega) = \frac{A\pi}{2} \operatorname{sinc}\left(\frac{n\pi}{2}\right)$. The frequency distribution is plotted in Fig. 2.

● **PROBLEM** 4-29

Given the transfer function of a two port network
$H(\omega) = |H(\omega_0)| \exp(-j\phi)$.

Derive an expression for output as a function of time, for the single frequency input $v_i = \cos\omega_0 t$.

Solution: The output in the frequency domain can be written as

$$V_0(\omega) = H(\omega)\, V_i(\omega)$$

where $V_i(\omega)$ = Fourier transform of the input.

$$\begin{aligned}
V_i(\omega) &= \text{Fourier transform of } \cos\omega_0 t \\
&= \pi\left[\delta(\omega-\omega_0) + \delta(\omega+\omega_0)\right]
\end{aligned}$$

$v_0(t)$ = output in the time domain

$$= \frac{1}{2\pi} \int_{-\infty}^{+\infty} V_0(\omega)\, e^{j\omega t}\, d\omega$$

$$= \frac{\pi}{2\pi} \int_{-\infty}^{+\infty} H(\omega)\left[\delta(\omega-\omega_0) + \delta(\omega+\omega_0)\right] e^{j\omega t}\, d\omega$$

$$= \frac{1}{2}\left[H(\omega_0)e^{j\omega_0 t} + H(-\omega_0)e^{-j\omega_0 t}\right]$$

by applying the property of the delta function,

$$\text{i.e.} \quad \int_{-\infty}^{+\infty} f(t)\,\delta(t-a)\,dt = f(a)$$

Now, $\quad H(\omega_0) = |H(\omega_0)|\exp(-j\phi)$

$\quad\quad\quad H(-\omega_0) = |H(\omega_0)|\exp(j\phi)$

Because the magnitude spectrum of $H(\omega)$ is an even function of ω and the phase spectrum is an odd function of ω.

Hence,
$$v_0(t) = \frac{1}{2}|H(\omega_0)|\left(e^{-j\phi}e^{j\omega_0 t} + e^{j\phi}e^{-j\omega_0 t}\right)$$

$$= |H(\omega_0)|\left[\frac{e^{j(\omega_0 t - \phi)} + e^{-j(\omega_0 t - \phi)}}{2}\right]$$

$$= |H(\omega_0)|\cos(\omega_0 t - \phi)$$

CHAPTER 5

THE LAPLACE TRANSFORM

DEFINITION OF LAPLACE TRANSFORM

● **PROBLEM 5-1**

(a) Define the Laplace transform $L\{f(t)\}$ for a real valued function $f(t)$.

(b) State conditions under which $L\{f(t)\}$ exists.

(c) Show that $L\{f_1(t)\}$ exists where

$$f_1(t) = 2te^{t^2} \cos e^{t^2}$$

but that f_1 does not satisfy all conditions in (b).

Solution: (a) Let $f(t)$ be defined on some domain of the real axis containing $[0,\infty)$. Then the function $G(s)$ defined by

$$G(s) = \int_0^\infty e^{-st} f(t)dt \qquad (1)$$

is called the Laplace transform of $f(t)$ and we write

$$G(s) = L\{f(t)\}$$

to emphasize the point of view that $G(s)$ is the result of a certain operation (defined in (1)) performed on $f(t)$. In general, s can be complex but since most applications use real s, it will be assumed that s is real unless otherwise stated.

(b) The most widely applied existence theorem involves two definitions which will be stated now.

Definition 1: If f is defined on $a \leq t < \infty$, then f is piecewise continuous on $[a,\infty)$ if for every finite interval $a \leq t \leq b$ f has a finite number of discontinuities such that at each discontinuity $t = t_0$ the limits $f(t_0^+)$ and $f(t_0^-)$ exist.

Definition 2: A function $f(t)$ defined on $[a,\infty)$ is said to be exponential order $e^{\alpha t}$ on $[a,\infty)$ if

$$|e^{-\alpha t} f(t)| \leq M \quad (\alpha \text{ is real}) \quad (2)$$

for some real positive constant M and all $t \in [a,\infty)$.

Expressed more qualitatively, this means that $f(t)$ does not increase faster than $e^{\alpha t}$ as $t \to \infty$.

The existence theorem may now be stated.

Theorem: If $f(t)$ is piecewise continuous on $[0,\infty)$ and is of exponential order α, then the Laplace integral

$$G(s) = L\{f(t)\} = \int_0^\infty e^{-st} f(t) dt$$

converges for $s > \alpha$. Moreover, the integral is absolutely and uniformly convergent for $s \geq \alpha'$ where $\alpha' > \alpha$.

Note that if s is allowed to be complex, the theorem still holds but with the conditions

$$\text{Re } s > \alpha \quad \text{and} \quad \text{Re } s \geq \alpha'.$$

(c)
$$L\{f_1(t)\} = \int_0^\infty e^{-st} 2te^{t^2} \cos e^{t^2} dt. \quad (3)$$

Integrating by parts gives

$$L\{f_1(t)\} = e^{-st} \sin e^{t^2} \Big|_0^\infty + s \int_0^\infty e^{-st} \sin e^{t^2} dt,$$

$$= -\sin 1 + s \int_0^\infty e^{-st} \sin e^{t^2} dt. \quad (4)$$

Since

$$|\sin e^{t^2}| < 1 \quad , \quad |e^{-st} \sin e^{t^2}|$$

is bounded by e^{-st} for all $s > 0$, and the integral in (4) converges by the Weierstrass M-test for all

$s > 0$. Note however that $f_1(t)$ is not of any exponential order since

$$e^{t^2} > Me^{\alpha t}$$

for t suffiently large no matter how large M and α are. Thus, $f_1(t)$ illustrates the fact that the theorem of (b) provides sufficient conditions for the existence of the Laplace integral of (1), but not necessary conditions. However, almost all functions which appear in physical problems do satisfy the theorem.

● **PROBLEM 5-2**

(a) Prove that $f(t) = t^n$, $n > 0$, is of exponential order α on $[0,\infty)$ for all $\alpha > 0$.

(b) Prove that $f(t) = \sin kt$ is of exponential order α on $[0,\infty)$ for all $\alpha > 0$.

Solution: (a) The function $f(t)$ is said to be of exponential order α on $[0,\infty)$ if there exists a real constant α and a positive real constant M such that

$$e^{-\alpha t}|f(t)| < M \text{ for all } t > 0.$$

For positive t

$$|t^n| = t^n.$$

Thus,

$$e^{-\alpha t}|f(t)| = e^{-\alpha t} t^n.$$

Iterative use of L'Hospital's rule shows that, for $\alpha > 0$,

$$\lim_{t \to \infty} e^{-\alpha t} t^n = 0.$$

Arbitrarily setting $M = 1$, it follows from the definition of a limit that there exists a positive constant t_0 such that

$$e^{-\alpha t}|f(t)| < M, \text{ for all } t > t_0.$$

Now let

$$M_1 = \max \{e^{-\alpha t} t^n, t \in [0, t_0]\}$$

(such a max exists since the function $g(t) = e^{-\alpha t} t^n$ is bounded on the closed interval $[0, t_0]$). Then

$$e^{-\alpha t}|f(t)| < M'$$

for all $t > 0$, where

$$M' = \max\{M, M_1\}.$$

Hence, $f(t)$ is of exponential order α for any $\alpha > 0$. Therefore, by the existence theorem, the Laplace transform of f

$$L\{f(t)\} = \int_0^\infty e^{-st} f(t)\,dt$$

exists for all $s > 0$.

(b) The function $f(t)$ is said to be of exponential order α on $[0,\infty)$ if there exists a real constant α and a positive real constant M such that

$$e^{-\alpha t}|f(t)| < M \quad \text{for all } t > 0.$$

Notice that for $\alpha = 0$,

$$\lim_{t\to\infty} e^{-\alpha t} = 1,$$

and for $\alpha > 0$,

$$\lim_{t\to\infty} e^{-\alpha t} = 0.$$

Thus, for $\alpha \geq 0$,

$$\lim_{t\to\infty} e^{-\alpha t} < 2. \tag{1}$$

Also note that

$$|\sin kt| \leq 1, \tag{2}$$

for all t.

Multiplying inequalities (1) and (2) together, we obtain

$$\left(\lim_{t\to\infty} e^{-\alpha t}\right)|\sin kt| < 2,$$

for $\alpha \geq 0$.

It follows from the definition of a limit that there exists a positive constant t_0 such that

$$e^{-\alpha t}|\sin kt| < 2,$$

for all $t > t_0$. Since the function

$$g(t) = |e^{-\alpha t} \sin kt|$$

is bounded on the interval $[0, t_0]$ it has a maximum M' there, and we may take

$$M = \max \{2, M'\}$$

and conclude that

$$|g(t)| = e^{-\alpha t}|\sin kt| < M$$

for all $t > 0$.

Hence, $f(t)$ is of exponential order α for any $\alpha > 0$. Therefore, by the existence theorem, the Laplace transform of f

$$L\{f(t)\} = \int_0^\infty e^{-st} f(t) dt$$

exists for all $s > 0$.

● **PROBLEM 5-3**

(a) Prove that $f(t) = e^{at} \sin kt$ is of exponential order a on $[0,\infty)$.

(b) Prove that e^{t^2} is not of any exponential order on the interval $[c,\infty)$ for an arbitrary c.

Solution: (a) The function $f(t)$ is said to be of exponential order a on $[0,\infty)$ if there exists a real constant a and a positive real constant M such that

$$e^{-at}|f(t)| < M \text{ for all } t > 0.$$

Substituting our expression for $f(t)$,

$$e^{-at}|f(t)| = e^{-at}|e^{at} \sin kt|.$$

We can remove the absolute value sign from e^{at}, since e^{at} is positive for real a. Thus,

$$e^{-at}|f(t)| = e^{-at} e^{at}|\sin kt|$$

$$= |\sin kt|.$$

But

$$|\sin kt| \leq 1$$

for all t; hence,

$$e^{at}|f(t)| \leq 1 .$$

From this inequality it follows that

$$e^{-at}|f(t)| < 2$$

for all t > 0 . (1)

Hence, taking M = 2, (1) shows that f(t) is of exponential order a.

(b) It follows easily from the definition of exponential order already given that a function f(t) is of exponential order if there exists some number b such that

$$\lim_{t \to \infty} \frac{f(t)}{e^{bt}} = 0 .$$

In our case

$$f(t) = e^{t^2} .$$

Thus we examine

$$\lim_{t \to \infty} \frac{e^{t^2}}{e^{bt}} = \lim_{t \to \infty} e^{t(t-b)} .$$

For any value of b, t−b > 0 if t > b. Thus the exponent of e in $e^{t(t-b)}$ becomes and stays positive as t→∞ . Thus

$$\lim_{t \to \infty} e^{t(t-b)} = \infty \neq 0$$

which shows exp(t²) is not of any exponential order.

• **PROBLEM 5-4**

Determine those values of s for which the improper integral

$$G(s) = \int_0^\infty e^{-st} \, dt$$

converges, and find the Laplace transform of f(t) = 1.

Solution: By definition, the improper integral in the

problem is a limit of a proper integral

$$G(s) = \lim_{R \to \infty} \int_0^R e^{-st} \, dt \, . \tag{1}$$

For $s = 0$ this becomes

$$G(s) = \lim_{R \to \infty} \int_0^R e^{-(0)t} \, dt = \lim_{R \to \infty} \int_0^R dt$$

$$= \lim_{R \to \infty} t \Big|_{t=0}^R = \lim_{R \to \infty} R = \infty \, .$$

Hence, the integral $G(s)$ diverges for $s = 0$. For $s \neq 0$ equation (1) becomes

$$G(s) = \lim_{R \to \infty} -\left(\frac{1}{s} e^{-st}\right)\Big|_{t=0}^R$$

$$= \lim_{R \to \infty} \frac{1}{s}\left(1 - e^{-Rs}\right) = \frac{1}{s} - \lim_{R \to \infty} \frac{e^{-Rs}}{s} \, . \tag{2}$$

If $s < 0$, then $-Rs > 0$ for positive R; hence, e^{-Rs} approaches infinity as R approaches infinity, and the integral diverges. If $s > 0$, then $-Rs < 0$ for positive R; hence, e^{-Rs} approaches zero as R approaches infinity, and the integral converges to $1/s$.

Extending the domain of the Laplace transform to complex values of s, we evaluate the expression of e^{-Rs} by Euler's formula:

$$e^{-Rs} = e^{-R(\text{Re}\{s\}) - iR(\text{Im}\{s\})}$$

$$= e^{-R(\text{Re}\{s\})}\Big[\cos[-R(\text{Im}\{s\})] + i \sin[-R(\text{Im}\{s\})]\Big], \tag{3}$$

where $\text{Re}\{s\}$ is the real part of s, $\text{Im}\{s\}$ is the imaginary part of s, and $i \equiv \sqrt{-1}$ is the imaginary constant. The cosine and sine functions are bounded; hence expression (3) diverges, as $R \to \infty$, for $\text{Re}\{s\} < 0$ and converges to zero for $\text{Re}\{s\} > 0$. In the case where $s \neq 0$ and $\text{Re}\{s\} = 0$, we have

$$e^{-R(\text{Re}\{s\})} = 1;$$

hence,

$$e^{-Rs} = \cos[-R(\text{Im}\{s\})] + i \sin[-R(\text{Im}\{s\})],$$

which is a nonconstant periodic function, so e^{-Rs} does not converge to any value as $R \to \infty$.

Since e^{-Rs} is the only expression that varies with R in equality (2), its convergence properties for $R \to \infty$ determine the convergence properties of G(s). Thus, in general, G(s) converges to 1/s for Re{s} > 0 and diverges otherwise.

Using L as the Laplace transform operator, with

$$L\{f(t)\} = F(s) \equiv \int_0^\infty e^{-st} f(t)\, dt,$$

we take the Laplace transform of f(t) for $t \geq 0$:

$$L\{1\} = \int_0^\infty (e^{-st} \cdot 1)\, dt = \int_0^\infty e^{-st}\, dt,$$

which is the same integral as G(s) in the problem. As we already showed, this integral converges to 1/s when Re{s} > 0 and diverges otherwise. Thus, the required Laplace transform is

$$L\{1\} = \frac{1}{s}, \text{ for Re}\{s\} > 0.$$

Note that the function f(t) of which we take a Laplace transform needs only be defined for positive real values of its argument t since the integral,

$$\int_0^\infty e^{-st} f(t)\, dt,$$

is in a region in which $t \geq 0$.

● **PROBLEM 5-5**

Find the Laplace transform of

$$f(t) = t^n,$$

where n is a positive integer.

Solution: Using L as the Laplace transform operator with

$$L\{f(t)\} = \int_0^\infty e^{-st} f(t)\, dt,$$

and first considering real s, we find

$$L\{t^n\} = \int_0^\infty (e^{-st})(t^n)\,dt = \lim_{R\to\infty} \int_0^R t^n e^{-st}\,dt \ .$$

For $s = 0$,

$$L\{t^n\} = \lim_{R\to\infty} \int_0^R t^n e^{0t}\,dt = \lim_{R\to\infty} \int_0^R t^n\,dt$$

$$= \lim_{R\to\infty} \frac{R^{n+1}}{n+1} = \infty \ ;$$

hence,

$$L\{t^n\}$$

does not exist for $s = 0$. For $s \neq 0$, integrating by parts,

$$L\{t^n\} = \lim_{R\to\infty} \left\{ \left. \frac{-t^n e^{-st}}{s} \right|_{t=0}^R + \frac{n}{s} \int_0^R e^{-st} t^{n-1}\,dt \right\}$$

$$= \lim_{R\to\infty} \left\{ \frac{-R^n e^{-sR}}{s} + \frac{n}{s} \int_0^R e^{-st} t^{n-1}\,dt \right\}$$

$$= \lim_{R\to\infty} \left(\frac{-R^n e^{-sR}}{s} \right) + \frac{n}{s} \lim_{R\to\infty} \left(\int_0^R e^{-st} t^{n-1}\,dt \right)$$

$$= \lim_{R\to\infty} \left(\frac{-R^n e^{-sR}}{s} \right) + \frac{n}{s} L\{t^{n-1}\} \ . \tag{1}$$

For $s \leq 0$ the argument of the limit in expression (1) diverges as $R \to \infty$; hence,

$$L\{t^n\}$$

does not exist. For $s > 0$ rewrite (1) as

$$L\{t^n\} = \lim_{R\to\infty} \left(\frac{-R^n}{s e^{sR}} \right) + \frac{n}{s} L\{t^{n-1}\} \ . \tag{2}$$

Since both the numerator and denominator in the argument of the limit in equation (2) approach ∞ as $R \to \infty$, we can apply L'Hospital's rule:

$$\lim_{R \to \infty} \left(\frac{-R^n}{se^{sR}} \right) = \lim_{R \to \infty} \left[\frac{d/dR(-R^n)}{d/dR(se^{sR})} \right] = \lim_{R \to \infty} \left(\frac{-nR^{n-1}}{s^2 e^{sR}} \right) \quad (3)$$

As long as the numerator and denominator of the argument of our limit approach ∞ as $R \to \infty$, we can iteratively apply L'Hospital's rule to equality (3):

$$\lim_{R \to \infty} \left(\frac{-R^n}{se^{sR}} \right) = \lim_{R \to \infty} \left(\frac{-nR^{n-1}}{s^2 e^{sR}} \right) = \lim_{R \to \infty} \left[\frac{-n(n-1)R^{n-2}}{s^3 e^{sR}} \right]$$

$$= \ldots = \lim_{R \to \infty} \left[\frac{-n(n-1)(n-2)\ldots(2)(1)R^{n-n}}{s^{n+1} e^{sR}} \right]$$

$$= (-1) \lim_{R \to \infty} \left[\frac{n!}{s^{n+1} e^{sR}} \right] = 0.$$

Substituting this result in equation (2),

$$L\{t^n\} = \frac{n}{s} L\{t^{n-1}\} \quad \text{for } s > 0. \quad (4)$$

Substituting $(n-1)$ for n in equation (4), we find

$$L\{t^{n-1}\} = \frac{n-1}{s} L\{t^{n-2}\}.$$

Substituting this result back into equation (4),

$$L\{t^n\} = \frac{n}{s} \left(\frac{n-1}{s} L\{t^{n-2}\} \right) = \frac{n(n-1)}{s^2} L\{t^{n-2}\}.$$

By iterating this process, we obtain

$$L\{t^n\} = \frac{n(n-1)(n-2)\ldots(2)(1)}{s^n} L\{t^0\} = \frac{n!}{s^n} L\{1\}$$

$$= \frac{n!}{s^n} \lim_{R \to \infty} \int_0^R e^{-st}(1)dt,$$

which converges, since $s > 0$, to

$$L\{t^n\} = \frac{n!}{s^n} \lim_{R \to \infty} \left(\frac{1 - e^{-sR}}{s} \right) = \frac{n!}{s^n} \cdot \frac{1}{s}$$

$$= \frac{n!}{s^{n+1}}, \quad \text{for } s > 0. \quad (5)$$

202

If s is complex, a similar computation will give the same result, (5), except that the condition on s will now be Re s > 0.

● **PROBLEM 5-6**

Find the Laplace transforms of

(a) $f(t) = e^{kt}$,

where k is a complex constant of the form

$$k = \text{Re}\{k\} + i\,\text{Im}\{k\}$$

with Re{k} the real part of k, Im{k} the imaginary part of k, and

$$i \equiv \sqrt{-1}.$$

Use this Laplace transform to find the Laplace transforms of

$$f(t) = e^{-kt} \quad \text{and} \quad f(t) = 1.$$

(b) $f(t) = \sin kt$ where k is a real constant.

<u>Solution</u>: (a) Using L as the Laplace transform operator with

$$L\{f(t)\} = \int_0^\infty e^{-st} f(t)\,dt,$$

for complex s of the form $s = \text{Re}\{s\} + i\,\text{Im}\{s\}$, where Re{s} is the real part of s and Im{s} is the imaginary part of s,

$$L\{e^{kt}\} = \int_0^\infty e^{-st}(e^{kt})\,dt = \lim_{R\to\infty}\int_0^R e^{(k-s)t}\,dt.$$

We chose to solve the complex case in this problem because complex numbers are necessary when we determine

$$L\{\cos kt\} \quad \text{and} \quad L\{\sin kt\}$$

from $L\{e^{kt}\}$.

For $s = k$,

$$L\{e^{kt}\} = \lim_{R\to\infty}\int_0^R e^{(k-k)t}\,dt = \lim_{R\to\infty}\int_0^R dt$$

$$= \lim_{R\to\infty}\left(t\,\Big|_{t=0}^R\right) = \lim_{R\to\infty} R = \infty\,;$$

203

hence
$$L\{e^{kt}\}$$
does not exist for $s = k$. For $s \neq k$,

$$L\{e^{kt}\} = \lim_{R \to \infty} \int_0^R e^{(k-s)t} \, dt$$

$$= \lim_{R \to \infty} \left[\frac{e^{(k-s)t}}{k-s} \bigg|_{t=0}^R \right]$$

$$= \lim_{R \to \infty} \left[\frac{e^{(k-s)R} - 1}{k-s} \right]$$

$$= \frac{1}{s-k} + \frac{1}{k-s} \lim_{R \to \infty} e^{(k-s)R},$$

which diverges for
$$\text{Re}\{s\} \leq \text{Re}\{k\}$$
and converges to
$$\frac{1}{s-k}$$
for
$$\text{Re}\{s\} > \text{Re}\{k\}.$$

Thus, the Laplace transform of e^{kt} (for $t > 0$) is

$$L\{e^{kt}\} = \frac{1}{s-k} \quad \text{for } \text{Re}\{s\} > \text{Re}\{k\}. \tag{1}$$

Using the constant $(-k)$ in place of k in formula (1), we obtain a new formula:

$$L\{e^{-kt}\} = \frac{1}{s-(-k)} = \frac{1}{s+k}$$

for
$$\text{Re}\{s\} > -\text{Re}\{k\}.$$

If the special case $k = 0$ is used, equation (1) gives

$$L\{1\} = L\{e^{0t}\} = \frac{1}{s-0} = \frac{1}{s}, \quad \text{Re}\{s\} > 0.$$

(b) Using L as the Laplace transform operator with

$$L\{f(t)\} = \int_0^\infty e^{-st} f(t) \, dt,$$

and first considering real s,

$$L\{\sin kt\} = \int_0^\infty (e^{-st})(\sin kt)dt = \lim_{R\to\infty} \int_0^R e^{-st} \sin kt \, dt.$$

In the above equality, we substitute the exponential formula for the sine function:

$$\sin kt = \frac{e^{ikt} - e^{-ikt}}{2i}$$

where $i \equiv \sqrt{-1}$, so that

$$L\{\sin kt\} = \lim_{R\to\infty} \int_0^R e^{-st} \left(\frac{e^{ikt} - e^{-ikt}}{2i}\right) dt$$

$$= \frac{1}{2i} \lim_{R\to\infty} \left[\int_0^R e^{(ik-s)t} dt - \int_0^R e^{(-ik-s)t} dt\right]. \quad (2)$$

For $k = 0$ equation (2) gives us

$$L\{\sin kt\} = \frac{1}{2i} \lim_{R\to\infty} \left[\int_0^R e^{(0-s)t} dt - \int_0^R e^{(0-s)t} dt\right]$$

$$= \frac{1}{2i} \lim_{R\to\infty} [0] = 0.$$

When $k \neq 0$, s cannot equal $\pm ik$, since $\pm ik$ is nonzero imaginary; thus, equation (2) gives us

$$L\{\sin kt\} = \frac{1}{2i} \lim_{R\to\infty} \left[\frac{e^{(ik-s)R}}{ik-s} - \frac{1}{ik-s} + \frac{e^{(-ik-s)R}}{ik+s} - \frac{1}{ik+s}\right]$$

$$= \frac{1}{2i} \lim_{R\to\infty} \left[\frac{(ik+s)e^{(ik-s)R} - (ik+s) + (ik-s)e^{(-ik-s)R} - (ik-s)}{(ik-s)(ik+s)}\right]$$

$$= \frac{1}{2i} \lim_{R\to\infty} \left[\frac{(ik+s)e^{(ik-s)R} + (ik-s)e^{(-ik-s)R} - 2ik}{-s^2 - k^2}\right]$$

$$= \frac{k}{s^2+k^2} + \frac{1}{2i} \lim_{R\to\infty} \left[\frac{(ik+s)e^{(ik-s)R} + (ik-s)e^{(-ik-s)R}}{-s^2 - k^2}\right]. \quad (3)$$

When the real part of (ik-s) is negative (i.e., when

$-s < 0$) and the real part of $(-ik-s)$ is negative (i.e., when $-s < 0$), the argument of the limit in expression (3) approaches zero as $R \to \infty$; hence

$$L\{\sin kt\} = k/(s^2 + k^2) \text{ for } s > 0.$$

Otherwise, the argument of the limit diverges, and the Laplace transform of sin kt does not exist.

If s is complex, the result

$$L\{\sin kt\} = \frac{k}{s^2 + k^2}$$

still holds, but the condition on s becomes Re s > 0.

PROPERTIES OF LAPLACE TRANSFORM

• **PROBLEM 5-7**

Prove the following properties of the Laplace transform denoted by $L\{f(t)\}$

(a) $L\{c_1 f_1(t) + c_2 f_2(t) + \ldots + c_n f_n(t)\} = c_1 L\{f_1(t)\}$

$$+ c_2 L\{f_2(t)\} + \ldots + c_n L\{f_n(t)\},$$

where all c_j are constants.

(b) $L\{f^{(n)}(t)\} = s^n L\{f(t)\} - \sum_{k=1}^{n} s^{k-1} f^{(n-k)}(0)$

if $f^{(k)}(t)$ are of some finite exponential orders for $k = 1, 2, \ldots, n-1$ and if

$$L\{f^{(n)}(t)\}$$

exists.

(c) $L\{e^{-at} f(t)\} = G(s+a)$

where $G(s) = L\{f(t)\}$ and a is a real constant.

(d) $L\{t^n f(t)\} = (-1)^n \frac{d^n F}{ds^n}$

where $F(s) = L\{f(t)\}$.

(e) $L\{\frac{1}{t} f(t)\} = \int_{s}^{\infty} F(\sigma) d\sigma$

where

$$F(s) = L\{f(t)\}.$$

Solution: (a) This is called the linearity property and is the defining characteristic of the so-called linear operators of which differentiation, integration, and all integral operators are examples. To prove this property, simply use the definition of

$$L\{f(t)\} \text{ for } f(t) = c_1 f_1(t) + \ldots + c_n f_n(t)$$

to calculate

$$L\{f(t)\} = L\{c_1 f_1(t) + \ldots + c_n f_n(t)\}$$

$$= \int_0^\infty e^{-st}\left[c_1 f_1(t) + \ldots + c_n f_n(t)\right] dt$$

$$= c_1 \int_0^\infty e^{-st} f_1(t)\,dt + \ldots + c_n \int_0^\infty e^{-st} f_n(t)\,dt$$

$$= c_1 L\{f_1(t)\} + c_2 L\{f_2(t)\} + \ldots + c_n L\{f_n(t)\}.$$

Thus we conclude that L is a linear operator.

(b) This is the derivative property of Laplace transforms, and it is this property which makes Laplace transforms useful in solving differential equations and in many other applications. The property will be proven by induction. Thus, suppose $f(t)$ is continuous and integrate the Laplace integral of f by parts:

$$\int_0^\infty e^{-st} f(t)\,dt = \left(\frac{-1}{s}\right) e^{-st} f(t)\Big|_0^\infty + \frac{1}{s}\int_0^\infty e^{-st} f'(t)\,dt$$

or

(if $\lim_{R\to\infty} e^{-sR} f(R) = 0$)

$$s\int_0^\infty e^{-st} f(t)\,dt = f(0) + \int_0^\infty e^{-st} f'(t)\,dt.$$

Assuming that $L\{f'(t)\}$ exists, this yields

$$L\{f'(t)\} = s\,L\{f(t)\} - f(0). \qquad (1)$$

Now assume that the derivative property holds for an

integer k, i.e., assume that

$$L\{f^{(k)}(t)\} = s^k L\{f(t)\} - \sum_{i=1}^{k} s^{i-1} f^{(k-i)}(0).$$

Then (1) can be used to show that

$$L\{f^{(k+1)}(t)\} = L\{\left(f^{(k)}(t)\right)'\}$$

$$= s L\{f^{(k)}(t)\} - f^{(k)}(0)$$

$$= s\left[s^k L\{f(t)\} - \sum_{i=1}^{k} s^{i-1} f^{(k-i)}(0)\right] - f^{(k)}(0)$$

$$= s^{k+1} L\{f(t)\} - \sum_{i=1}^{k+1} s^{i-1} f^{(k+1-i)}(0). \quad (2)$$

Thus in (1) and (2) it has been shown that the derivative property holds for n=1 and that if it holds for n=k then it holds for n = k + 1. Therefore, by induction we conclude that the derivative property,

$$L\{f^{(n)}(t)\} = s^n L\{f(t)\} - \sum_{k=1}^{n} s^{k-1} f^{n-k}(0)$$

holds for all n provided that, for k = 0,1, ... , n-1,

$$\lim_{R \to \infty} f^{(k)}(R) e^{-sR} = 0$$

for s large enough (that is, $f^{(k)}(t)$ are of finite exponential orders), and that

$$L\{f^{(n)}(t)\}$$

exists for s large enough.

(c) This is known as the attenuation property or substitution property: If f(t) is "attentuated" by the exponential factor e^{-at}, then the transform is shifted (to the left) with respect to the variable s. To prove this, just recall the definition of the Laplace transform of a function f(t),

$$L\{f(t)\} = G(s) = \int_0^\infty e^{-st} f(t) dt,$$

and compute

$$L\{e^{-at} f(t)\} = \int_0^\infty e^{-st} e^{-at} f(t) dt$$

$$= \int_0^\infty e^{-(s+a)t} f(t) dt .$$

Setting $r = (s+a)$ yields

$$\int_0^\infty e^{-(s+a)t} f(t) dt = G(r) = G(s+a)$$

so that

$$L\{e^{-at} f(t)\} = G(s+a) .$$

Note that if the region of validity of $L\{f(t)\}$ is $\text{Re}\{s\} > \alpha$ for some α, then the region of validity of $L\{e^{-at} f(t)\}$ is $\text{Re}\{s\} > \alpha - a$. In this case (and almost all cases of "physical" functions) α is the smallest number such that f is of exponential order α.

(d) The expression

$$F(s) = \int_0^\infty e^{-st} f(t) dt \qquad (3)$$

is a uniformly convergent integral for a piecewise continuous function f and a suitable range of values of s (see problem 1 of this chapter). Therefore, interchange of differentiation and integration is allowed so that

$$\frac{dF(s)}{ds} = -\int_0^\infty t e^{-st} f(t) dt$$

or

$$\frac{dF}{ds} = - L\{tf(t)\} .$$

This formula can be generalized to find the nth derivative (provided $L\{t^n f(t)\}$ exists). Thus,

$$\frac{d^n F(s)}{ds^n} = (-1)^n \int_0^\infty t^n e^{-st} f(t) dt$$

or

$$L\{t^n f(t)\} = (-1)^n \frac{d^n F(s)}{ds^n},$$

and this will have the same region of validity, $s > \alpha$ for some α, as (3).

(e) Let $F(s) = L\{f(t)\}$ and consider

$$G(s) = \int_s^\infty F(\sigma) d\sigma \qquad (4)$$

By definition of $F(s)$,

$$G(s) = \int_s^\infty d\sigma \int_0^\infty e^{-\sigma t} f(t) dt .$$

Now, since F is uniformly convergent for all $\text{Re}\{s\} \geq \alpha'$ for any $\alpha' > \alpha$ (where $\text{Re}\{s\} > \alpha$ is the region of convergence of $L\{f(t)\}$), the order of integration may be interchanged to yield

$$G(s) = \int_0^\infty f(t) dt \int_s^\infty e^{-\sigma t} d\sigma = \int_0^\infty f(t) \frac{e^{-st}}{t} dt$$

$$= L\{\tfrac{1}{t} f(t)\} .$$

Therefore, recalling the definition of $G(s)$ in (4),

$$\int_s^\infty F(\sigma) d\sigma = L\{\tfrac{1}{t} f(t)\}$$

provided that the transform of $\tfrac{1}{t} f(t)$ exists.

• **PROBLEM 5-8**

Use the Laplace transform of

$$f(t) = e^{kt}, \qquad (1)$$

where k is a complex constant of the form

$$k = \text{Re}\{k\} + i\, \text{Im}\{k\}$$

with $\text{Re}\{k\}$ the real part of k, $\text{Im}\{k\}$ the imaginary part of k, and

$$i \equiv \sqrt{-1},$$

to find the Laplace transforms of

$f(t) = \cosh kt$, $\sinh kt$, $\cos kt$, and $\sin kt$.

<u>Solution</u>: It was found in problem 6 that

$$L\{e^{kt}\} = \frac{1}{s-k} \quad \text{for} \quad \text{Re}\{s\} > \text{Re}\{k\}. \tag{2}$$

This result could also be looked up in a table of Laplace transforms. In either case, we use the definitions

$$\cosh kt \equiv \frac{e^{kt} + e^{-kt}}{2},$$

and

$$\sinh kt \equiv \frac{e^{kt} - e^{-kt}}{2},$$

and the additional formula

$$L\{c_1 f_1(t) + c_2 f_2(t)\} =$$

$$c_1 L\{f_1(t)\} + c_2 L\{f_2(t)\}, \tag{3}$$

to find

$$L\{\cosh kt\} = L\left\{\frac{e^{kt} + e^{-kt}}{2}\right\}$$

$$= \frac{1}{2} (L\{e^{kt}\} + L\{e^{-kt}\})$$

$$= \frac{1}{2}\left(\frac{1}{s-k} + \frac{1}{s+k}\right) = \frac{1}{2}\left(\frac{2s}{s^2-k^2}\right)$$

$$= \frac{s}{s^2-k^2}, \quad \text{for} \quad \text{Re}\{s\} > |\text{Re}\{k\}|, \tag{4}$$

and

$$L\{\sinh kt\} = L\left\{\frac{e^{kt} - e^{-kt}}{2}\right\}$$

$$= \frac{1}{2} (L\{e^{kt}\} - L\{e^{-kt}\})$$

$$= \frac{1}{2}\left(\frac{1}{s-k} - \frac{1}{s+k}\right) = \frac{1}{2}\left(\frac{2k}{s^2-k^2}\right)$$

$$= \frac{k}{s^2-k^2}, \text{ for } \text{Re}\{s\} > |\text{Re}\{k\}|. \qquad (5)$$

The condition $\text{Re}\{s\} > |\text{Re}\{k\}|$ in formulas (4) and (5) comes from the fact that we derived those formulas for $L\{e^{kt}\}$ and $L\{e^{-kt}\}$, which require $\text{Re}\{s\} > \text{Re}\{k\}$ and $\text{Re}\{s\} > -\text{Re}\{k\}$, respectively. To insure that both $\text{Re}\{s\} > \text{Re}\{k\}$ and $\text{Re}\{s\} > -\text{Re}\{k\}$, it is necessary that $\text{Re}\{s\}$ be greater than the greater of $\text{Re}\{k\}$ and $-\text{Re}\{k\}$. Since one of these ($\text{Re}\{k\}$ or $-\text{Re}\{k\}$) must be positive and the other negative, the greater of the two is the positive one, which is equal to the absolute value of $\text{Re}\{k\}$.

Using the exponential formulas for the cosine and sine functions

$$\cos kt = \frac{e^{ikt} + e^{-ikt}}{2},$$

and

$$\sin kt = \frac{e^{ikt} - e^{-ikt}}{2i},$$

and again the addition formula (3), we find

$$L\{\cos kt\} = L\left\{\frac{e^{(ik)t} + e^{-(ik)t}}{2}\right\}$$

$$= \frac{1}{2}\left(L\{e^{(ik)t}\} + L\{e^{-(ik)t}\}\right),$$

and a similar expression holds for $L\{\sin kt\}$.

By substituting (ik) for k in formulas (1) and (2),

$$L\{\cos kt\} = \frac{1}{2}\left(\frac{1}{s-ik} + \frac{1}{s+ik}\right)$$

$$= \frac{1}{2}\left(\frac{2s}{s^2+k^2}\right)$$

$$= \frac{s}{s^2+k^2}, \qquad (6)$$

and

$$L\{\sin kt\} = L\left\{\frac{e^{(ik)t} - e^{-(ik)t}}{2i}\right\}$$

$$= \frac{1}{2i}(L\{e^{(ik)t}\} - L\{e^{-(ik)t}\})$$

$$= \frac{1}{2i}\left(\frac{1}{s-ik} - \frac{1}{s+ik}\right) = \frac{1}{2i}\left(\frac{2ik}{s^2+k^2}\right)$$

$$= \frac{k}{s^2+k^2} \tag{7}$$

Laplace transforms (6) and (7) are both subject to the same two existence conditions from the Laplace transforms of e^{kt} and e^{-kt} (which were the base of (6) and (7)). Since we used ik instead of k, the conditions are

$$Re\{s\} > Re\{ik\} \quad ,$$

and

$$Re\{s\} > -Re\{ik\} \quad .$$

Combining these two conditions as we did for the cosh and sinh Laplace transforms,

$$Re\{s\} > |Re\{ik\}| \quad .$$

But

$$|Re\{ik\}| = |Re\{i(Re\{k\} + iIm\{k\})\}|$$

$$= |Re\{-Im\{k\} + iRe\{k\}\}|$$

$$= |-Im\{k\}| = |Im\{k\}| \quad ;$$

hence, the condition for the existence of Laplace transforms (6) and (7) is

$$Re\{s\} > |Im\{k\}| \quad ,$$

which, for s and k real, is equivalent to $s > 0$.

• **PROBLEM 5-9**

For positive s, show that the Laplace transform

$$L\left[u(t-a)\right] = \frac{e^{-as}}{s}$$

where $u(t-a)$ is a unit step function.

Solution: The step function u(t-a) is defined as

$$u(t-a) = \begin{cases} 0 & \text{for } t < a \\ 1 & \text{for } t > a \end{cases}$$

Hence,

$$L[u(t-a)] = \int_0^\infty f(t)e^{-st}dt$$

$$= \int_0^a (0)e^{-st}dt + \int_a^\infty (1)e^{-st}dt$$

$$= \left(\frac{e^{-st}}{-s}\right)_a^\infty$$

Now, for positive s, $\lim_{t \to \infty} e^{-st} = 0$

Therefore,

$$L[u(t-a)] = \frac{-1}{s}\left(0 - e^{-as}\right)$$

$$= \frac{e^{-as}}{s} \quad \text{for } s > 0$$

SIMPLE FUNCTIONS

• **PROBLEM 5-10**

Find the Laplace transform of

$$f(t) = t, \text{ for } t > 0.$$

Solution: Using L as the Laplace transform operator, with

$$L\{f(t)\} = \int_0^\infty e^{-st} f(t)dt,$$

for real t and real s, we find

$$L\{t\} = \int_0^\infty (e^{-st})(t)dt = \lim_{R \to \infty} \int_0^R te^{-st}dt .$$

For s = 0,

$$L\{t\} = \lim_{R \to \infty} \int_0^R te^{-(0)t}dt = \lim_{R \to \infty} \int_0^R tdt$$

$$= \lim_{R \to \infty} \left(\frac{t^2}{2}\right)\Big|_{t=0}^R = \lim_{R \to \infty}\left(\frac{R^2}{2}\right) = \infty ;$$

hence $L\{t\}$ diverges for s = 0. For $s \neq 0$,

$$L\{t\} = \lim_{R \to \infty} \int_0^R te^{-st} dt.$$

Integrating by parts,

$$L\{t\} = \lim_{R \to \infty} \left[-\frac{e^{-st}}{s^2}(st+1)\right]_{t=0}^R$$

$$= \lim_{R \to \infty} \left[\frac{1}{s^2} - \frac{e^{-sR}}{s^2}(sR + 1) \right]$$

$$= \frac{1}{s^2} - \left(\frac{1}{s^2}\right) \lim_{R \to \infty} (e^{-sR})(sR + 1). \quad (a)$$

When $s < 0$, both e^{-sR} and $(sR + 1)$ diverge as $R \to \infty$; hence $L\{t\}$, in expression (a), diverges. When $s > 0$,

$$L\{t\} = \frac{1}{s^2} - \left(\frac{1}{s^2}\right) \lim_{R \to \infty} \frac{sR + 1}{e^{sR}}. \quad (b)$$

Since both the numerator and denominator in the argument of the limit approach ∞ as $R \to +\infty$, we can apply L'Hopital's rule:

$$\lim_{R \to \infty} \frac{sR + 1}{e^{sR}} = \lim_{R \to \infty} \frac{d/dR(sR+1)}{d/dR(e^{sR})}$$

$$= \lim_{R \to \infty} \frac{s}{se^{sR}} = \lim_{R \to \infty} e^{-sR} = 0.$$

Substituting this result into expression (b),

$$L\{t\} = \frac{1}{s^2} - \left(\frac{1}{s^2}\right)(0) = \frac{1}{s^2}.$$

Thus, the required Laplace transform is

$$L\{t\} = \frac{1}{s^2}, \text{ for } s > 0.$$

● **PROBLEM 5-11**

Find the Laplace transform of $f(t) = t^2$.

Solution: Using L as the Laplace transform operator, with

$$L\{f(t)\} = \int_0^\infty e^{-st} f(t)dt,$$

for real t and real s, we find

$$L\{t^2\} = \int_0^\infty (e^{-st})(t^2)dt = \lim_{R \to \infty} \int_0^R t^2 e^{-st} dt.$$

For $s = 0$,

$$L\{t^2\} = \lim_{R \to \infty} \int_0^R t^2 e^{-(0)t} dt = \lim_{R \to \infty} \int_0^R t^2 dt$$

$$= \lim_{R \to \infty} \left(\frac{t^3}{3}\right)\Big|_{t=0}^R = \lim_{R \to \infty} \left(\frac{R^3}{3}\right) = \infty;$$

hence $L\{t^2\}$ diverges for $s = 0$. For $s \neq 0$,

$$L\{t^2\} = \lim_{R \to \infty} \int_0^R t^2 e^{-st} dt.$$

Integrating by parts,

$$L\{t^2\} = \lim_{R \to \infty} \left[-\frac{t^2}{s}e^{-st} - \frac{2t}{s^2}e^{-st} - \frac{2}{s^3}e^{-st} \right]_{t=0}^R$$

215

$$= \lim_{R\to\infty}\left(-\frac{R^2}{s}e^{-sR} - \frac{2R}{s^2}e^{-sR} - \frac{2}{s^3}e^{-sR} + \frac{2}{s^3}\right)$$

$$= \frac{2}{s^3} - \lim_{R\to\infty}\left(\frac{R^2}{s} + \frac{2R}{s^2} + \frac{2}{s^3}\right)\left(e^{-sR}\right) \qquad (a)$$

When $s < 0$, both e^{-sR} and $\left(\frac{R^2}{s} + \frac{2R}{s^2} + \frac{2}{s^3}\right)$ diverge as $R \to \infty$; hence $L\{t\}$, in expression (a), diverges. When $s > 0$,

$$L\{t^2\} = \frac{2}{s^3} - \lim_{R\to\infty}\frac{(R^2/s + 2R/s^2 + 2/s^3)}{e^{sR}} \qquad (b)$$

Since both the numerator and denominator in the argument of the limit approach ∞ as $R \to +\infty$, we can apply L'Hopital's rule:

$$\lim_{R\to\infty}\frac{(R^2/s + 2R/s^2 + 2/s^3)}{e^{sR}} = \lim_{R\to\infty}\frac{d/dR(R^2/s + 2R/s^2 + 2/s^3)}{d/dR(e^{sR})}$$

$$= \lim_{R\to\infty}\frac{(2R/s + 2/s^2)}{se^{sR}} = \left(\frac{2}{s^2}\right)\lim_{R\to\infty}\frac{R+\frac{1}{s}}{e^{sR}} \qquad (c)$$

Here again, both the numerator and denominator in the argument of the limit approach either $+\infty$ or $-\infty$ as $R \to \infty$; hence, we can again apply L'Hopital's rule:

$$\lim_{R\to\infty}\frac{R+\frac{1}{s}}{e^{sR}} = \lim_{R\to\infty}\frac{d/dR\left(R+\frac{1}{s}\right)}{d/dR(e^{sR})} = \lim_{R\to\infty}\frac{1}{se^{sR}} = \left(\frac{1}{s}\right)\lim_{R\to\infty}(e^{-sR})$$

$$= \left(\frac{1}{s}\right)(0) = 0 .$$

Substituting this result into equality (c),

$$\lim_{R\to\infty}\frac{(R^2/s + 2R/s^2 + 2/s^3)}{e^{sR}} = \left(\frac{2}{s^2}\right)(0) = 0 .$$

Substituting this new result into expression (b),

$$L\{t^2\} = \frac{2}{s^3} - (0) = \frac{2}{s^3}$$

Thus, the required Laplace transform is

$$L\{t^2\} = \frac{2}{s^3}, \text{ for } s > 0 .$$

● **PROBLEM 5-12**

Find the Laplace transform of
$$f(t) = t^n ,$$
where n is a positive integer.

<u>Solution</u>: Using L as the Laplace transform operator, with

$$L\{f(t)\} = \int_0^\infty e^{-st} f(t)dt ,$$

for real t and real s, we find
$$L\{t^n\} = \int_0^\infty (e^{-st})(t^n)dt = \lim_{R\to\infty} \int_0^R t^n e^{-st} dt.$$
For $s = 0$,
$$L\{t^n\} = \lim_{R\to\infty} \int_0^R t^n e^{0t} dt = \lim_{R\to\infty} \int_0^R t^n dt$$
$$= \lim_{R\to\infty} \left(\frac{R^{n+1}}{n+1}\right) = \infty \;;$$
hence $L\{t^n\}$ diverges for $s = 0$. For $s \neq 0$,
$$L\{t^n\} = \lim_{R\to\infty} \int_0^R t^n e^{-st} dt .$$
Integrating by parts,
$$L\{t^n\} = \lim_{R\to\infty} \left\{ \left[\frac{-t^n e^{-st}}{s}\right]_{t=0}^R + \frac{n}{s} \int_0^R e^{-st} t^{n-1} dt \right\}$$
$$= \lim_{R\to\infty} \left\{ \frac{-R^n e^{-sR}}{s} + \frac{n}{s} \int_0^R e^{-st} t^{n-1} dt \right\}$$
$$= \lim_{R\to\infty} \left(\frac{-R^n e^{-sR}}{s}\right) + \frac{n}{s} \lim_{R\to\infty} \left(\int_0^R e^{-st} t^{n-1} dt \right)$$
$$= \lim_{R\to\infty} \left(\frac{-R^n e^{-sR}}{s}\right) + \frac{n}{s} L\{t^{n-1}\} . \tag{a}$$

For $s \leq 0$, the argument of the limit in expression (a) diverges as $R \to \infty$; hence, $L\{t^n\}$ diverges. For $s > 0$,
$$L\{t^n\} = \lim_{R\to\infty} \left(\frac{-R^n}{se^{sR}}\right) + \frac{n}{s} L\{t^{n-1}\} \tag{b}$$

Since both the numerator and denominator in the argument of the limit in equation (b) approach ∞ as $R \to \infty$, we can apply L'Hopital's rule:
$$\lim_{R\to\infty} \left(\frac{-R^n}{se^{sR}}\right) = \lim_{R\to\infty} \left[\frac{d/dR(-R^n)}{d/dR(se^{sR})}\right] = \lim_{R\to\infty} \left(\frac{-nR^{n-1}}{s^2 e^{sR}}\right) . \tag{c}$$

As long as the numerator and denominator of the argument of our limit approach ∞ as $R \to \infty$, we can iteratively apply L'Hopital's rule to equality (c):
$$\lim_{R\to\infty} \left(\frac{-R^n}{se^{sR}}\right) = \lim_{R\to\infty} \left(\frac{-nR^{n-1}}{s^2 e^{sR}}\right) = \lim_{R\to\infty} \left[\frac{-n(n-1)R^{n-2}}{s^3 e^{sR}}\right]$$
$$= \ldots = \lim_{R\to\infty} \left[\frac{-n(n-1)(n-2)\ldots(2)(1)R^{n-n}}{s^{n+1} e^{sR}}\right]$$
$$= (-1) \lim_{R\to\infty} \left[\frac{n!}{s^{n+1} e^{sR}}\right] = 0 .$$

Substituting this result in equation (b),
$$L\{t^n\} = \frac{n}{s} L\{t^{n-1}\}, \text{ for } s > 0 . \tag{d}$$

Substituting $(n-1)$ for n in equation (d), we find

$$L\{t^{n-1}\} = \frac{n-1}{s} L\{t^{n-2}\}.$$

Substituting this result back into equation (d),

$$L\{t^n\} = \frac{n}{s}\left(\frac{n-1}{s} L\{t^{n-2}\}\right) = \frac{n(n-1)}{s^2} L\{t^{n-2}\}.$$

By iterating this process, we obtain

$$L\{t^n\} = \frac{n(n-1)(n-2)\ldots(2)(1)}{s^n} L\{t^0\} = \frac{n!}{s^n} L\{1\} = \frac{n!}{s^n} \lim_{R\to\infty} \int_0^R e^{-st}(1)dt,$$

which converges, since $s > 0$, to

$$L\{t^n\} = \frac{n!}{s^n} \lim_{R\to\infty}\left(\frac{1-e^{-sR}}{s}\right) = \frac{n!}{s^n} \cdot \frac{1}{s}$$

$$= \frac{n!}{s^{n+1}}, \text{ for } s > 0.$$

● **PROBLEM** 5-13

Find the Laplace transform of
$$f(t) = e^{kt}, \text{ for } t > 0,$$
where k is a complex constant of the form $k = \text{Re}\{k\} + i\text{Im}\{k\}$, with $\text{Re}\{k\}$ the real part of k, $\text{Im}\{k\}$ the imaginary part of k, and $i = \sqrt{-1}$ the imaginary constant. Use this Laplace transform to find Laplace transforms of
$$f(t) = e^{-kt}, 1, \cosh kt, \sinh kt, \cos kt, \text{ and } \sin kt.$$

<u>Solution</u>: Using L as the Laplace transform operator, with

$$L\{f(t)\} = \int_0^\infty e^{-st} f(t) dt,$$

for real t and complex s of the form $s = \text{Re}\{s\} + i\text{Im}\{s\}$, where $\text{Re}\{s\}$ is the real part of s and $\text{Im}\{s\}$ is the imaginary part of s,

$$L\{e^{kt}\} = \int_0^\infty e^{-st}(e^{kt})dt = \lim_{R\to\infty}\int_0^R e^{(k-s)t} dt.$$

We chose to solve the complex case in this problem because complex numbers are necessary when we determine $L\{\cos kt\}$ and $L\{\sin kt\}$ from $L\{e^{kt}\}$.

For $s = k$,

$$L\{e^{kt}\} = \lim_{R\to\infty}\int_0^R e^{(k-k)t} dt = \lim_{R\to\infty}\int_0^R dt$$

$$= \lim_{R\to\infty} (t)\Big|_{t=0}^R = \lim_{R\to\infty} R = \infty;$$

hence $L\{e^{kt}\}$ diverges for $s = k$. For $s \neq k$,

$$L\{e^{kt}\} = \lim_{R\to\infty}\int_0^R e^{(k-s)t} dt$$

$$= \lim_{R \to \infty} \left[\frac{e^{(k-s)t}}{k-s} \right]_{t=0}^{R}$$

$$= \lim_{R \to \infty} \left[\frac{e^{(k-s)R} - 1}{k-s} \right]$$

$$= \frac{1}{s-k} + \frac{1}{k-s} \lim_{R \to \infty} e^{(k-s)R} ,$$

which diverges for $\text{Re}\{s\} \le \text{Re}\{k\}$ and converges to $\frac{1}{s-k}$ for $\text{Re}\{s\} > \text{Re}\{k\}$.

Thus, the Laplace transform of e^{kt} (for $t > 0$) is

$$L\{e^{kt}\} = \frac{1}{s-k}, \quad \text{for} \quad \text{Re}\{s\} > \text{Re}\{k\} \qquad (a)$$

Using the constant $(-k)$ in place of k in formula (a), we obtain a new formula:

$$L\{e^{-kt}\} = \frac{1}{s-(-k)} = \frac{1}{s+k}, \quad \text{for} \quad \text{Re}\{s\} > -\text{Re}\{k\}. \qquad (b)$$

If we take the special case of formula (a) when $k = 0$, we obtain

$$L\{1\} = L\{e^{0t}\} = \frac{1}{s-0} = \frac{1}{s}, \quad \text{for} \quad \text{Re}\{s\} > 0.$$

Using the definitions

$$\cosh kt \equiv \frac{e^{kt} + e^{-kt}}{2},$$

and

$$\sinh kt \equiv \frac{e^{kt} - e^{-kt}}{2},$$

and the addition formula

$$L\{c_1 f_1(t) + c_2 f_2(t)\} = \int_0^\infty e^{-st} [c_1 f_1(t) + c_2 f_2(t)] dt$$

$$= c_1 \int_0^\infty e^{-st} f_1(t) dt + c_2 \int_0^\infty e^{-st} f_2(t) dt$$

$$= c_1 L\{f_1(t)\} + c_2 L\{f_2(t)\}, \qquad (c)$$

we find

$$L\{\cosh kt\} = L\{\frac{e^{kt} + e^{-kt}}{2}\}$$

$$= \tfrac{1}{2}(L(e^{kt}) + L(e^{-kt}))$$

$$= \tfrac{1}{2}\left(\frac{1}{s-k} + \frac{1}{s+k}\right) = \tfrac{1}{2}\left(\frac{2s}{s^2-k^2}\right)$$

$$= \frac{s}{s^2-k^2}, \quad \text{for} \quad \text{Re}\{s\} > |\text{Re}\{k\}|, \qquad (d)$$

and

$$L\{\sinh kt\} = L\{\frac{e^{kt} - e^{-kt}}{2}\}$$

$$= \tfrac{1}{2}(L\{e^{kt}\} - L\{e^{-kt}\})$$

$$= \tfrac{1}{2}\left(\frac{1}{s-k} - \frac{1}{s+k}\right) = \tfrac{1}{2}\left(\frac{2k}{s^2-k^2}\right)$$

$$= \frac{k}{s^2-k^2}, \text{ for } \text{Re}\{s\} > |\text{Re}\{k\}|. \tag{e}$$

The condition $\text{Re}\{s\} > |\text{Re}\{k\}|$ in formulas (d) and (e) comes from the fact that we derived those formulas using the formulas for $L\{e^{kt}\}$ and $L\{e^{-kt}\}$, which require $\text{Re}\{s\} > \text{Re}\{k\}$ and $\text{Re}\{s\} > -\text{Re}\{k\}$, respectively. To insure that both $\text{Re}\{s\} > \text{Re}\{k\}$ and $\text{Re}\{s\} > -\text{Re}\{k\}$, it is necessary that $\text{Re}\{s\}$ be greater than the greater of $\text{Re}\{k\}$ and $-\text{Re}\{k\}$. Since one of these ($\text{Re}\{k\}$ or $-\text{Re}\{k\}$) must be positive and the other negative, the greater of the two is the positive one, which is equal to the absolute value of $\text{Re}\{k\}$.

Using the exponential formulas for the cosine and sine functions

$$\cos kt = \frac{e^{ikt} + e^{-ikt}}{2},$$

and

$$\sin kt = \frac{e^{ikt} - e^{-ikt}}{2i},$$

and again the addition formula (c), we find

$$L\{\cos kt\} = L\{\frac{e^{(ik)t} + e^{-(ik)t}}{2}\}$$

$$= \tfrac{1}{2}(L\{e^{(ik)t}\} + L\{e^{-(ik)t}\}),$$

and by substituting (ik) for k in formulas (a) and (b),

$$L\{\cos kt\} = \tfrac{1}{2}\left(\frac{1}{s-ik} + \frac{1}{s+ik}\right)$$

$$= \tfrac{1}{2}\left(\frac{2s}{s^2+k^2}\right)$$

$$= \frac{s}{s^2+k^2}, \tag{f}$$

and

$$L\{\sin kt\} = L\{\frac{e^{(ik)t} - e^{-(ik)t}}{2i}\}$$

$$= \frac{1}{2i}(L\{e^{(ik)t}\} - L\{e^{-(ik)t}\})$$

$$= \frac{1}{2i}\left(\frac{1}{s-ik} - \frac{1}{s+ik}\right) = \frac{1}{2i}\left(\frac{2ik}{s^2+k^2}\right)$$

$$= \frac{k}{s^2+k^2}. \tag{g}$$

Laplace transforms (f) and (g) are both subject to the same two existence conditions from the Laplace transforms (a) and (b) (which were the base of (f) and (g)). Since we used ik instead of k, the conditions are

$$\text{Re}\{s\} > \text{Re}\{ik\},$$

and

$$\text{Re}\{s\} > -\text{Re}\{ik\}.$$

Combining these two conditions as we did for the cosh and sinh Laplace transforms,

$$\text{Re}\{s\} > |\text{Re}\{ik\}|.$$

But

$$|\text{Re}\{ik\}| = |\text{Re}\{i(\text{Re}\{k\} + i\text{Im}\{k\})\}|$$

$$= |\text{Re}\{-\text{Im}\{k\} + i\text{Re}\{k\}\}|$$

$$= |-\text{Im}\{k\}| = |\text{Im}\{k\}|;$$

hence, the condition for the existence of Laplace transforms (f) and (g) is

$$\text{Re}\{s\} > |\text{Im}\{k\}|,$$

which, for s and k real, is equivalent to $s > 0$.

• **PROBLEM 5-14**

Find the Laplace transform of
$$f(t) = \sin kt, \text{ for } t > 0,$$
where k is a real constant.

<u>Solution</u>: Using L as the Laplace transform operator, with

$$L\{f(t)\} = \int_0^\infty e^{-st} f(t) \, dt,$$

for real t and real s,

$$L\{\sin kt\} = \int_0^\infty (e^{-st})(\sin kt) dt = \lim_{R \to \infty} \int_0^R e^{-st} \sin kt \, dt.$$

In the above equality, we substitute the exponential formula for the sine function:
$$\sin kt = \frac{e^{ikt} - e^{-ikt}}{2i}$$

where $i = \sqrt{-1}$, so that

$$L\{\sin kt\} = \lim_{R \to \infty} \int_0^R e^{-st} \left(\frac{e^{ikt} - e^{-ikt}}{2i} \right) dt$$

$$= \frac{1}{2i} \lim_{R \to \infty} \left[\int_0^R e^{(ik-s)t} dt - \int_0^R e^{(-ik-s)t} dt \right]. \quad (a)$$

For $k = 0$, equation (a) gives us

$$L\{\sin kt\} = \frac{1}{2i} \lim_{R \to \infty} \left[\int_0^R e^{(0-s)t} dt - \int_0^R e^{(0-s)t} dt \right]$$

$$= \frac{1}{2i} \lim_{R \to \infty} [0] = 0.$$

When $k \neq 0$, s cannot equal $\pm ik$, since $\pm ik$ is nonzero imaginary; thus, equation (a) gives us

$$L\{\sin kt\} = \frac{1}{2i} \lim_{R \to \infty} \left[\frac{e^{(ik-s)R}}{ik-s} - \frac{1}{ik-s} + \frac{e^{(-ik-s)R}}{ik+s} - \frac{1}{ik+s} \right]$$

$$= \frac{1}{2i} \lim_{R \to \infty} \left[\frac{(ik+s)e^{(ik-s)R} - (ik+s) + (ik-s)e^{(-ik-s)R} - (ik-s)}{(ik-s)(ik+s)} \right]$$

$$= \frac{1}{2i} \lim_{R \to \infty} \left[\frac{(ik+s)e^{(ik-s)R} + (ik-s)e^{(-ik-s)R} - 2ik}{-s^2 - k^2} \right]$$

$$= \frac{k}{s^2+k^2} + \frac{1}{2i} \lim_{R \to \infty} \left[\frac{(ik+s)e^{(ik-s)R} + (ik-s)e^{(-ik-s)R}}{-s^2 - k^2} \right]. \quad (b)$$

When the real part of (ik-s) is negative (i.e., when $-s < 0$) and the real part of (-ik-s) is negative (i.e., when $-s < 0$), the argument of the limit in expression (b) approaches zero as $R \to \infty$; hence $L\{\sin kt\} = k/(s^2+k^2)$, for $s > 0$. Otherwise, the argument of the limit diverges.

● **PROBLEM 5-15**

Find the Laplace transform of
$$f(t) = \cos kt, \text{ for } t > 0,$$
where k is a real constant.

<u>Solution</u>: Using L as the Laplace transform operator, with
$$L\{f(t)\} = \int_0^\infty e^{-st} f(t) \, dt,$$
for real t and real s, we find
$$L\{\cos kt\} = \int_0^\infty (e^{-st})(\cos kt) dt = \lim_{R \to \infty} \int_0^R e^{-st} \cos kt \, dt.$$

In the above equality, we substitute the exponential formula for the cosine function:
$$\cos kt = \frac{e^{ikt} + e^{-ikt}}{2},$$
so that
$$L\{\cos kt\} = \lim_{R \to \infty} \int_0^R e^{-st} \left(\frac{e^{ikt} + e^{-ikt}}{2}\right) et$$

$$= \tfrac{1}{2} \lim_{R \to \infty} \left[\int_0^R e^{(ik-s)t} dt + \int_0^R e^{(-ik-s)t} dt\right]. \quad (a)$$

For $s = k = 0$, equation (a) gives us
$$L\{\cos kt\} = \tfrac{1}{2} \lim_{R \to \infty} \left[\int_0^R e^{0t} dt + \int_0^R e^{0t} dt\right]$$

$$= \tfrac{1}{2} \lim_{R \to \infty} \left[2 \int_0^R dt\right] = \tfrac{1}{2} \lim_{R \to \infty}(2R) = \infty;$$

hence $L\{\cos kt\}$ diverges for $s = k = 0$. When it is not true that $s = k = 0$, we know that $s \neq ik$ and $s \neq -ik$, since s is real and $\pm ik$ are imaginary; hence equation (a) gives us

$$L\{\cos kt\} = \tfrac{1}{2} \lim_{R \to \infty} \left[\frac{e^{(ik-s)R}}{ik-s} - \frac{1}{ik-s} - \frac{e^{(-ik-s)R}}{ik+s} + \frac{1}{ik+s}\right]$$

$$= \tfrac{1}{2} \lim_{R \to \infty} \left[\frac{(ik+s)e^{(ik-s)R} - (ik+s) - (ik-s)e^{(-ik-s)R} + (ik-s)}{(ik-s)(ik+s)}\right]$$

$$= \tfrac{1}{2} \lim_{R \to \infty} \left[\frac{(ik+s)e^{(ik-s)R} - (ik-s)e^{(-ik-s)R} - 2s}{-s^2-k^2}\right]$$

$$= \frac{s}{s^2+k^2} + \tfrac{1}{2} \lim_{R \to \infty} \left[\frac{(ik+s)e^{(ik-s)R} - (ik-s)e^{(-ik-s)R}}{-s^2-k^2}\right] \quad (b)$$

When the real part of (ik-s) is negative (i.e., when $-s < 0$) and the real part of (-ik-s) is negative (i.e., when -s < 0), the argument of the limit in expression (b) approaches zero as R → ∞; hence L{cos kt} = $s/(s^2+k^2)$, for s > 0. Otherwise, the argument diverges.

• **PROBLEM 5-16**

Graph the function $f(t) = \alpha(t-2) - \alpha(t-3)$, where α is the unit step function:

$$\alpha(x) = \begin{bmatrix} 0, & x < 0 \\ 1, & x \geq 0 \end{bmatrix} \quad (a)$$

<u>Solution</u>: Using definition (a), we find that

$$\alpha(t-2) = \begin{bmatrix} 0, & t < 2 \\ 1, & t \geq 2 \end{bmatrix}$$

and

$$\alpha(t-3) = \begin{bmatrix} 0, & t < 3 \\ 1, & t \geq 3 \end{bmatrix}$$

To subtract $\alpha(t-3)$ from $\alpha(t-2)$, we must consider the values of their difference at various values of t. When $t < 2$, $\alpha(t-2) = 0$ and $\alpha(t-3) = 0$; hence,

$$f(t) = 0 - 0 = 0, \quad t < 2.$$

When $2 \leq t < 3$, $\alpha(t-2) = 1$ and $\alpha(t-3) = 0$; hence,

$$f(t) = 1 - 0 = 1, \quad 2 \leq t < 3.$$

When $t \geq 3$, $\alpha(t-2) = 1$ and $\alpha(t-3) = 1$; hence,

$$f(t) = 1 - 1 = 0, \quad t \geq 3.$$

Combining these results,

$$f(t) = \begin{bmatrix} 0, & x < 2 \\ 1, & 2 \leq x < 3 \\ 0, & x \geq 3 \end{bmatrix}$$

This function is graphed in the accompanying figure:

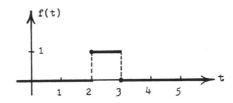

• **PROBLEM 5-17**

Give an analytic representation of the function f(t) graphed in the accompanying figure, using the unit step function α, where

$$\alpha(x) = \begin{bmatrix} 0, & x < 0 \\ 1, & x \geq 0 \end{bmatrix} \quad (a)$$

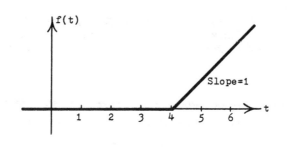

Solution: The part of the graph where $t \geq 4$ corresponds to a first-degree algebraic equation of the form

$$f(t) = mt + b, \tag{b}$$

where m is the slope and b is a constant. Since the slope is 1, equation (b) becomes

$$f(t) = t + b. \tag{c}$$

We observe from the graph that $f(4) = 0$ is on the part of the curve with slope 1. Substituting these values into equation (c),

$$0 = f(4) = 4 + b;$$

hence, $b = -4$ and

$$f(t) = t - 4, \quad t \geq 4.$$

To make $f(t)$ equal to zero for all values less than four, all we need to do is multiply $(t-4)$ by zero for $t < 4$ and by one for $t \geq 4$. Note that, based on definition (a), we have

$$\alpha(t-4) = \begin{bmatrix} 0, & t < 4 \\ 1, & t \geq 4 \end{bmatrix}$$

hence,

$$f(t) = (t-4)\,\alpha(t-4),$$

for all t

● **PROBLEM 5-18**

Find the Laplace transform $L\{g(t)\}$, where

$$g(t) = \begin{bmatrix} 0, & t < 4 \\ (t-4)^2, & t \geq 4 \end{bmatrix}$$

Solution: The function $g(t)$ can be expressed as $(t-4)^2 \alpha(t-4)$, where α is the unit step function, defined as follows:

$$\alpha(x) = \begin{bmatrix} 0, & x < 0 \\ 1, & x \geq 0 \end{bmatrix}$$

so that

$$\alpha(t-4) = \begin{bmatrix} 0, & t < 4 \\ 1, & t \geq 4 \end{bmatrix}$$

Using the property that $L\{f(t-c)\alpha(t-c)\} = e^{-cs} L\{f(t)\}$, where c is a nonnegative constant, provided that $L\{f(t)\}$ exists, we obtain (taking $c = 4$) $L\{g(t)\} = L\{(t-4)^2 \alpha(t-4)\}$

$$= e^{-4s} L\{t^2\}. \qquad (a)$$

In a table of Laplace transforms, we find that
$$L\{t^n\} = \frac{n!}{s^{n+1}}, \quad s > 0,$$
where n is a nonnegative integer constant; hence, taking n = 2, we find
$$L\{t^2\} = \frac{2}{s^3}$$
Substituting this result into equality (a),
$$L\{g(t)\} = e^{-4s}\left(\frac{2}{s^3}\right) = \frac{2e^{-4s}}{s^3}.$$

● **PROBLEM 5-19**

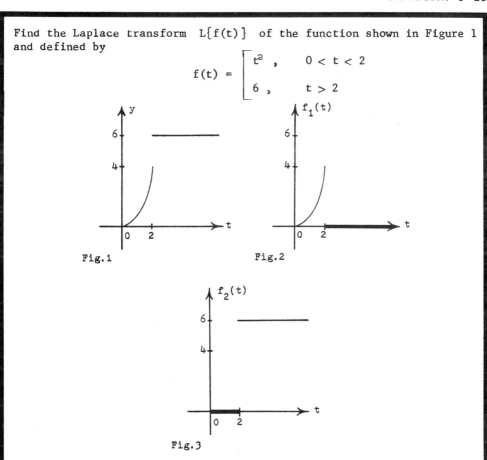

Find the Laplace transform $L\{f(t)\}$ of the function shown in Figure 1 and defined by
$$f(t) = \begin{cases} t^2, & 0 < t < 2 \\ 6, & t > 2 \end{cases}$$

Fig.1

Fig.2

Fig.3

Solution: We wish to express $f(t)$ in terms of the unit step function $\alpha(t)$, defined as
$$\alpha(t) = \begin{cases} 0, & t < 0 \\ 1, & t \geq 0 \end{cases}.$$
We can then use the property that $L\{g(t-c)\alpha(t-c)\} = e^{-cs}L\{g(t)\}$, where c is a nonnegative constant, provided that $L\{g(t)\}$ exists.

Let us build $f(t)$ out of continuous functions and unit step functions. First, we need a function $f_1(t)$ that is equal to t^2 when $0 < t < 2$ and zero when $t > 2$. The function $t^2 \alpha(t-2)$ is equal to t^2 when $t > 2$ and zero elsewhere, so that if we subtract it from t^2, we obtain the desired function (see Fig. 2):

$$f_1(t) = t^2 - t^2 \alpha(t-2).$$

Second, we need a function $f_2(t)$ that is equal to 6 when $t > 2$ and zero when $0 < t < 2$. We obtain this by multiplying the constant 6 by $\alpha(t-2)$, since $\alpha(t-2) = 0$ when $t < 2$ and $= 1$ when $t \geq 2$. Thus (see Fig. 3),

$$f_2(t) = 6\alpha(t-2).$$

The function $f(t)$ is obtained by adding together $f_1(t)$ and $f_2(t)$:

$$\begin{aligned} f(t) &= f_1(t) + f_2(t) \\ &= t^2 - t^2 \alpha(t-2) + 6\alpha(t-2) \\ &= t^2 + (6-t^2)\alpha(t-2). \end{aligned} \quad (a)$$

This is still not the form we require in order to use the property mentioned at the beginning of this solution. We need to express

$(6-t^2)$ as a function of $(t-2)$. We know that

$$(t-2)^2 = t^2 - 4t + 4;$$

hence,

$$\begin{aligned} 6 - t^2 &= -(t^2 - 4t + 4) - 4t + 10 \\ &= -(t-2)^2 - 4t + 10 \\ &= -(t-2)^2 - 4t + 8 + 2 \\ &= -(t-2)^2 - 4(t-2) + 2. \end{aligned}$$

Substituting this last expression into equality (a),

$$\begin{aligned} f(t) &= t^2 + [-(t-2)^2 - 4(t-2) + 2]\alpha(t-2) \\ &= t^2 - (t-2)^2 \alpha(t-2) - 4(t-2)\alpha(t-2) + 2\alpha(t-2). \end{aligned}$$

Taking the Laplace transform of this,

$$L\{f(t)\} = L\{t^2 - (t-2)^2 \alpha(t-2) - 4(t-2)\alpha(t-2) + 2\alpha(t-2)\}.$$

Since L is a linear operator,

$$L\{f(t)\} = L\{t^2\} - L\{(t-2)^2 \alpha(t-2)\} - 4L\{(t-2)\alpha(t-2)\} + 2L\{(1)\alpha(t-2)\}.$$

To the above equality we apply the property mentioned at the beginning of this solution:

$$L\{f(t)\} = L\{t^2\} - e^{-2s}L\{t^2\} - 4e^{-2s}L\{t\} + 2e^{-2s}L\{1\}. \quad (b)$$

From a table of Laplace transforms, we find that

$$L\{t^n\} = \frac{n!}{s^{n+1}}, \quad s > 0,$$

where n is a nonnegative integer constant. Taking $n = 0, 1,$ and 2, respectively, we obtain

$$L\{1\} = \frac{1}{s}, \quad L\{t\} = \frac{1}{s^2},$$

and

$$L\{t^2\} = \frac{2}{s^3}.$$

Substituting these results into equality (b), we obtain

$$L\{f(t)\} = \frac{2}{s^3} - \frac{2e^{-2s}}{s^3} - \frac{4e^{-2s}}{s^2} + \frac{2e^{-2s}}{s}.$$

• PROBLEM 5-20

Find the Laplace transform of the function $f(t)$ shown in the accompanying figure and defined by

$$f(t) = \begin{bmatrix} t, & 0 < t < 4 \\ 5, & t > 4 \end{bmatrix}.$$

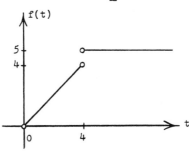

Solution: Note that $f(t)$ need not be defined at $t = 0$ or at $t = 4$, since such information is not essential in performing the integration to determine the Laplace transform.

Using L as the Laplace transform operator, with

$$L\{f(t)\} = \int_0^\infty e^{-st} f(t) \, dt,$$

for real t and real s, we find

$$L\{f(t)\} = \lim_{R \to \infty} \int_0^R f(t) e^{-st} \, dt$$

$$= \int_0^4 t e^{-st} \, dt + \lim_{R \to \infty} \int_4^R 5 e^{-st} \, dt.$$

For $s = 0$,

$$L\{f(t)\} = \int_0^4 t e^{0t} \, dt + \lim_{R \to \infty} \int_4^R 5 e^{0t} \, dt$$

$$= \int_0^4 t \, dt + \lim_{R \to \infty} \int_4^R 5 \, dt$$

$$= \frac{4^2}{2} + \lim_{R \to \infty} 5(R-4) = \infty \ ;$$

hence $L\{f(t)\}$ diverges for $s = 0$. For $s \neq 0$,

$$L\{f(t)\} = \int_0^4 t e^{-st} \, dt + \lim_{R \to \infty} \int_4^R 5 e^{-st} \, dt$$

$$= \int_0^4 t e^{-st} \, dt + \lim_{R \to \infty} \left(\frac{-5 e^{-sR} + 5 e^{-4s}}{s} \right)$$

$$= \int_0^4 t e^{-st} \, dt + \frac{5 e^{-4s}}{s} - \frac{5}{s} \lim_{R \to \infty} e^{-sR} \ .$$

Integrating by parts,

$$L\{f(t)\} = \left[-\frac{t}{s} e^{-st} - \frac{1}{s^2} e^{-st} \right]_{t=0}^4 + \frac{5 e^{-4s}}{s} - \frac{5}{s} \lim_{R \to \infty} e^{-sR}$$

227

$$= -\frac{4}{s}e^{-4s} - \frac{1}{s^2}e^{-4s} + \frac{1}{s^2} + \frac{5}{s}e^{-4s} - \frac{5}{s}\lim_{R\to\infty} e^{-sR}.$$

For $s > 0$, the above expression converges to

$$L\{f(t)\} = \frac{e^{-4s}}{s} - \frac{e^{-4s}}{s^2} + \frac{1}{s^2};$$

otherwise, $L\{f(t)\}$ diverges.

Thus, the required Laplace transform is

$$L\{f(t)\} = \frac{e^{-4s}}{s} - \frac{e^{-4s}}{s^2} + \frac{1}{s^2}, \text{ for } s > 0.$$

• **PROBLEM** 5-21

Determine those values of s for which the improper integral $G(s) = \int_0^\infty e^{-st} dt$ converges, and find the Laplace transform of $f(t) = 1$ for $t \geq 0$.

Solution: Converting the improper integral in the problem to a limit of a proper integral,

$$G(s) = \lim_{R\to\infty} \int_0^R e^{-st} dt \qquad (a)$$

For $s = 0$, this becomes

$$G(s) = \lim_{R\to\infty} \int_0^R e^{-(0)t} dt = \lim_{R\to\infty} \int_0^R dt$$

$$= \lim_{R\to\infty} t \Big|_{t=0}^R = \lim_{R\to\infty} R = \infty.$$

Hence, the integral $G(s)$ diverges for $s = 0$. For $s \neq 0$, equation (a) becomes

$$G(s) = \lim_{R\to\infty} \left(-\frac{1}{s}e^{-st}\right)\Big|_{t=0}^R$$

$$= \lim_{R\to\infty} \frac{1}{s}\left(1 - e^{-Rs}\right) = \frac{1}{s} - \lim_{R\to\infty} \frac{e^{-Rs}}{s}. \qquad (b)$$

If $s < 0$, then $-Rs > 0$ for positive R; hence, e^{-Rs} approaches infinity as R approaches infinity, and the integral diverges. If $s > 0$, then $-Rs < 0$ for positive R; hence, e^{-Rs} approaches zero as R approaches infinity, and the integral converges to $1/s$.

Extending the domain of the Laplace transform to complex values of s, we evaluate the expression e^{-Rs} by Euler's formula:

$$e^{-Rs} = e^{-R(\text{Re}\{s\}) - iR(\text{Im}\{s\})}$$

$$= e^{-R(\text{Re}\{s\})}\{\cos[-R(\text{Im}\{s\})] + i\sin[-R(\text{Im}\{s\})]\}, \qquad (c)$$

where $\text{Re}\{s\}$ is the real part of s, $\text{Im}\{s\}$ is the imaginary part of s, and $i \equiv \sqrt{-1}$ is the imaginary constant. The cosine and sine functions are bounded; hence expression (c) diverges, as $R \to \infty$, for $\text{Re}\{s\} < 0$ and converges to zero for $\text{Re}\{s\} > 0$. In the case where $s \neq 0$ and $\text{Re}\{s\} = 0$, we have

$$e^{-R(\text{Re}\{s\})} = 1; \text{ hence,}$$

$$e^{-Rs} = \cos[-R(\text{Im}\{s\})] + i \sin[-R(\text{Im}\{s\})],$$

which is a nonconstant periodic function, making e^{-Rs} diverge as $R \to \infty$.

Since e^{-Rs} is the only expression that varies with R in equality (b), its convergence properties for $R \to \infty$ determine the convergence properties of $G(s)$. Thus, in general, $G(s)$ converges to $1/s$ for $\text{Re}\{s\} > 0$ and diverges otherwise.

Using L as the Laplace transform operator, with

$$L\{f(t)\} = F(s) \equiv \int_0^\infty e^{-st} f(t) dt,$$

we take the Laplace transform of $f(t) = 1$ for $t \geq 0$:

$$L\{1\} = \int_0^\infty e^{-st} \cdot 1 \, dt = \int_0^\infty e^{-st} dt,$$

which is the same integral as $G(s)$ in the problem. As we already showed, this integral converges to $1/s$ when $\text{Re}\{s\} > 0$ and diverges otherwise. Thus, the required Laplace transform is

$$L\{1\} = \frac{1}{s}, \text{ for } \text{Re}\{s\} > 0.$$

Note that the function, $f(t)$, of which we take a Laplace transform need only be defined for positive real values of its argument, t, since the integral,

$$\int_0^\infty e^{-st} f(t) dt,$$

is in a region in which $t \geq 0$.

• **PROBLEM** 5-22

Find the Laplace transforms of

(a) $\quad g(t) = e^{-2t} \sin 5t$,

(b) $\quad h(t) = e^{-t} t \cos 2t$.

Solution: (a) We shall use the formula

$$L\{\sin kt\} = \frac{k}{s^2+k^2}, \tag{1}$$

where k is a real constant and s is a real variable. We also use the theorem that states that if $f(t)$ is defined for all nonnegative t, is piecewise continuous on every closed interval $[0,b]$ for $b > 0$, and is of exponential order $e^{\alpha t}$ then

$$L\{e^{at} f(t)\} = F(s-a) \tag{2}$$

for $s > \alpha + a$, where a is a real constant and

$$F(s) = L\{f(t)\} .$$

Using formula (1), with $f(t) = \sin 5t$, we find

$$f(s) = L\{\sin 5t\} = \frac{5}{s^2+5^2} = \frac{5}{s^2+25} .$$

In order to use the theorem associated with formula (2), we must first demonstrate that the function $f(t) = \sin 5t$ is of exponential order; i.e., there exists a real constant α and positive real constants M and t_0 such that

$$e^{-\alpha t}|f(t)| < M$$

for all $t > t_0$. For $\alpha > 0$,

$$\lim_{t \to \infty} e^{-\alpha t} = 0 ,$$

and for $\alpha = 0$,

$$\lim_{t \to \infty} e^{-\alpha t} = 1 .$$

Thus, for $\alpha \geq 0$,

$$\lim_{t \to \infty} e^{-\alpha t} < 2 . \qquad (3)$$

Also note that

$$|\sin 5t| \leq 1 , \qquad (4)$$

for all t. Multiplying inequalities (3) and (4) together, we obtain

$$\left(\lim_{t \to \infty} e^{-\alpha t}\right) |\sin 5t| < 2 ,$$

for $\alpha \geq 0$. It follows from the definition of a limit that there exists a positive constant t_0 such that, for $t > t_0$,

$$e^{-\alpha t}|\sin 5t| < 2 ,$$

which, taking M = 2, shows that $f(t) = \sin 5t$ is of exponential order $e^{\alpha t}$, for $\alpha \geq 0$.

Since $f(t) = \sin 5t$ is defined for all nonnegative t, is piecewise continuous on every closed interval [0,b] for $b > 0$, and is of exponential order $e^{\alpha t}$ for $\alpha \geq 0$, we can use formula (1) in formula (2);

$$L\{g(t)\} = L\{e^{-2t} f(t)\}$$

$$= F(s-a)$$

$$= \frac{5}{(s+2)^2 + 25}, \text{ for } s > -2.$$

(b) Using the property that

$$L\{e^{bt} f(t)\} = F(s-b),$$

where b is a constant and $F(s) = L\{f(t)\}$, provided that $L\{f(t)\}$ exists, we obtain

$$L\{e^{-t} t \cos 2t\} = G(s+1), \qquad (5)$$

where $G(s) = L\{t \cos 2t\}$.

We could find, in a table of Laplace transforms, that

$$L\{t \cos kt\} = \frac{s^2 - k^2}{(s^2 + k^2)^2}, \; s > 0.$$

However, this result may be established easily without consulting such a table. Using the fact that

$$L\{\cos kt\} = \frac{s}{s^2 + k^2}$$

which was established in a previous problem and the theorem which states that

$$L\{t^n f(t)\} = (-1)^n \frac{d^n}{ds^n} L\{f(t)\},$$

it is apparent that

$$L\{t \cos kt\} = (-1) \frac{d}{ds} L\{f(t)\}$$

$$= (-1) \frac{d}{ds} \left(\frac{s}{s^2 + k^2} \right)$$

$$= \frac{s^2 - k^2}{(s^2 + k^2)^2}$$

The region of validity of this formula is the same as that

for $L\{\cos kt\}$, i.e., $s > 0$ for real s.
Hence, taking $k = 2$, we find

$$L\{t \cos 2t\} = \frac{s^2 - 4}{(s^2 + 4)^2}.$$

Thus,

$$G(s) = \frac{s^2 - 4}{(s^2 + 4)^2},$$

and, from equality (5),

$$L\{e^{-t} t \cos 2t\} = \frac{(s+1)^2 - 4}{[(s+1)^2 + 4]^2}$$

● **PROBLEM 5-23**

Find the Laplace transform, $L\{f(t)\} = F(s)$, of

(a) $f(t) = 2 \sin t + 3 \cos 2t$

(b) $g(t) = \dfrac{1 - e^{-t}}{t}$

<u>Solution</u>: (a) We shall use the addition formula

$$L\{c_1 f_1(t) + c_2 f_2(t)\} = c_1 L\{f_1(t)\} + c_2 L\{f_2(t)\} \tag{1}$$

where c_1 and c_2 are constants, and the formulas which were derived in the previous problems:

$$L\{\sin kt\} = \frac{k}{s^2 + k^2} \quad \text{(for } s > 0\text{)} \tag{2}$$

$$L\{\cos kt\} = \frac{s}{s^2 + k^2} \quad \text{(for } s > 0\text{)} \tag{3}$$

where k is a real constant and s is a real variable.

By formula (1),

$$L\{2 \sin t + 3 \cos 2t\} = 2L\{\sin t\} + 3L\{\cos 2t\}.$$

Applying formulas (2) and (3) to the above equality,

$$L\{2 \sin t + 3 \cos 2t\} = 2\left(\frac{1}{s^2 + 1^2}\right) + 3\left(\frac{s}{s^2 + 2^2}\right)$$

$$= \frac{2}{s^2+1} + \frac{3s}{s^2+4}, \text{ for } s > 0.$$

(b) Expanding e^{-t} as an infinite series,

$$e^{-t} = \sum_{n=0}^{\infty} \frac{(-1)^n t^n}{n!};$$

hence,

$$\frac{1 - e^{-t}}{t} = \frac{1 - \left[\sum_{n=0}^{\infty} \frac{(-1)^n t^n}{n!}\right]}{t}$$

$$= \frac{1 - \left[1 + \sum_{n=1}^{\infty} \frac{(-1)^n t^n}{n!}\right]}{t}$$

$$= -\frac{1}{t} \sum_{n=1}^{\infty} \frac{(-1)^n t^n}{n!} = \sum_{n=1}^{\infty} \frac{(-1)^{n-1} t^{n-1}}{n!}.$$

Letting $k = n-1$ in the summation, this becomes

$$\frac{1 - e^{-t}}{t} = \sum_{k=0}^{\infty} \frac{(-1)^k t^k}{(k+1)!};$$

thus

$$L\left\{\frac{1 - e^{-t}}{t}\right\} = L\left\{\sum_{k=0}^{\infty} \frac{(-1)^k t^k}{(k+1)!}\right\} \tag{4}$$

Since L is a linear operator (i.e., since

$$L\{c_1 f_1(t) + c_2 f_2(t) + \ldots\} = c_1 L\{f_1(t)\}$$
$$+ c_2 L\{f_2(t)\} + \ldots, c_j = \text{constant}),$$

equality (4) becomes

$$L\left\{\frac{1 - e^{-t}}{t}\right\} = \sum_{k=0}^{\infty} \frac{(-1)^k}{(k+1)!} L\{t^k\}. \tag{5}$$

From the results of a previous problem, or from a table of Laplace transforms, we find that

$$L\{t^k\} = \frac{k!}{s^{k+1}}, \quad s > 0,$$

where k is a nonnegative integer. Substituting this result into equality (5), we obtain

$$L\left\{\frac{1-e^{-t}}{t}\right\} = \sum_{k=0}^{\infty} \frac{(-1)^k}{(k+1)!}\left(\frac{k!}{s^{k+1}}\right)$$

$$= \sum_{k=0}^{\infty} \frac{(-1)^k}{(k+1)}\left(\frac{1}{s}\right)^{k+1}$$

This last summation is the infinite series for the natural logarithm of $(1 + 1/s)$, where $|1/s| < 1$, i.e., where $|s| > 1$. Thus,

$$L\left\{\frac{1-e^{-t}}{t}\right\} = \log\left(1 + \frac{1}{s}\right)$$

An alternative method of solving this problem is to use the formula

$$L\left\{\frac{1}{t} f(t)\right\} = \int_s^{\infty} F(\sigma)\, d\sigma$$

where $F(s) = L\{f(t)\}$ (we obtained it in problem 7). Here

$$f(t) = 1 - e^{-t},$$

and

$$L\{1 - e^{-t}\} = L\{1\} - L\{e^{-t}\} = \frac{1}{s} - \frac{1}{s+1}$$

since

$$L\{e^{kt}\} = \frac{1}{s-k} \qquad \text{(for } k = 0, e^{kt} \equiv 1\text{)}.$$

Then

$$L\left\{\frac{1-e^{-t}}{t}\right\} = \int_s^{\infty} \left(\frac{1}{\sigma} - \frac{1}{\sigma+1}\right) d\sigma$$

$$= \left(\log|\sigma| - \log|\sigma + 1|\right)\Big|_s^{\infty}$$

$$= \log\left|\frac{\sigma}{\sigma+1}\right|\Big|_s^{\infty} = \log 1 - \log\frac{s}{s+1}$$

$$= \log\frac{s+1}{s} \qquad \text{(we have } s > 0\text{)}.$$

In fact, the condition $|s| > 1$, or $s > 1$ (since $s > 0$ for $L\{t^k\}$ to exist) is too restrictive, and the second solution shows that for any $s > 0$

$$L\left\{\frac{1 - e^{-t}}{t}\right\}$$

exists.

• **PROBLEM 5-24**

Find the Laplace transform of

(a) $$g(t) = te^{4t}$$

(b) $$f(t) = t^{7/2}$$

Solution: (a) We shall use the formula

$$L\{e^{kt}\} = \frac{1}{s-k}, \text{ for } s > k, \tag{1}$$

where k is a real constant, and s is a real variable. We also use the theorem that states that if $f(t)$ is defined for all nonnegative t, is piecewise continuous on every closed interval $[0,b]$ for $b > 0$, and is of exponential order $e^{\alpha t}$, then

$$L\{t^n f(t)\} = (-1)^n \frac{d^n}{ds^n} L\{f(t)\}, \tag{2}$$

for $s > \alpha$, where n is a positive integer.

Using formula (1), we find that

$$L\{e^{4t}\} = \frac{1}{s-4}, \text{ for } s > 4.$$

In order to use the theorem associated with formula (2), we must first demonstrate that the function

$$f(t) = e^{4t}$$

is of exponential order; i.e., there exists a real constant α and positive real constants M and t_0 such that

$$e^{-\alpha t}|f(t)| < M \text{ for all } t > t_0.$$

For $\alpha = 4$,

$$\lim_{t \to \infty} e^{-\alpha t} |e^{4t}| = \lim_{t \to \infty} e^{(4-\alpha)t} = 1.$$

For $\alpha > 4$,

$$\lim_{t \to \infty} e^{-\alpha t} |e^{4t}| = \lim_{t \to \infty} e^{(4-\alpha)t} = 0.$$

Thus, for $\alpha \geq 4$,

$$\lim_{t \to \infty} e^{-\alpha t} |e^{4t}| < 2.$$

It follows from the definition of a limit that there exists a positive constant t_0 such that, for $t > t_0$,

$$e^{-\alpha t} |e^{4t}| < 2,$$

which, taking $M = 2$, shows that $f(t) = e^{4t}$ is of exponential order $e^{\alpha t}$, for $\alpha \geq 4$.

Since $f(t) = e^{4t}$ is defined for all nonnegative t, is piecewise continuous on every closed interval $[0,b]$ for $b > 0$, and is of exponential order $e^{\alpha t}$ for $\alpha \geq 4$, we can substitute formula (1) for $L\{f(t)\}$ in formula (2):

$$L\{g(t)\} = L\{te^{4t}\}$$

$$= (-1) \frac{d}{ds} L\{e^{4t}\} = -\frac{d}{ds}\left(\frac{1}{s-4}\right)$$

$$= \frac{1}{(s-4)^2}, \text{ for } s > 4.$$

(b) Using the property that

$$L\{t^n f(t)\} = (-1)^n \frac{d^n}{ds^n} L\{f(t)\},$$

where n is a positive integer constant, provided that $L\{f(t)\}$ exists, we obtain

$$L\{t^{7/2}\} = L\{t^3 \sqrt{t}\}$$

$$= (-1)^3 \frac{d^3}{ds^3} L\{\sqrt{t}\}$$

$$= -\frac{d^3}{ds^3} L\{\sqrt{t}\} \tag{3}$$

Now $L\{\sqrt{t}\}$ may be found using the Γ function whose values are as well tabulated as, say, log x so that an answer expressible in terms of $\Gamma(x)$ is very useful. The definition of the Γ function is

$$\Gamma(k) = \int_0^\infty t^{(k-1)} e^{-t} dt$$

so that

$$\Gamma(k+1) = \int_0^\infty t^k e^{-t} dt \tag{4}$$

Now make the substitution $t = sx$ where s is some constant so that $dt = sdx$, $e^{-t} = e^{-sx}$, and $t^k = s^k x^k$. Then (d) becomes

$$\Gamma(k+1) = \int_0^\infty s^k x^k e^{-sx} s\, dx$$

$$= s^{k+1} \int_0^\infty x^k e^{-sx} dx \tag{5}$$

$$= s^{k+1} L\{x^k\}$$

so that

$$L\{x^k\} = \frac{\Gamma(k+1)}{s^{k+1}}, \quad k > -1, \; s > 0, \tag{6}$$

where the restrictions on s and k are imposed to insure the convergence of the integrals in (4) and (5) respectively. Therefore, using

$k = \frac{1}{2}$ in (6) gives

$$L(t^{\frac{1}{2}}) = L(\sqrt{t}) = \Gamma(3/2) s^{-3/2}$$

Now look up $\Gamma\{3/2\}$ in a table of $\Gamma(x)$ to find $\Gamma(3/2) = \sqrt{\pi}/2$ and

$$L(\sqrt{t}) = \frac{1}{2} \sqrt{\pi}\, s^{-3/2}$$

Substituting this result in equality (3),

$$L\{t^{7/2}\} = -\frac{d^3}{ds^3}\left(\frac{1}{2}\sqrt{\pi}s^{-3/2}\right)$$

$$= -\frac{d^2}{ds^2}\left(-\frac{3}{2}\cdot\frac{1}{2}\sqrt{\pi}\,s^{-5/2}\right)$$

$$= -\frac{d}{ds}\left[\left(-\frac{5}{2}\right)\left(-\frac{3}{2}\right)\cdot\frac{1}{2}\sqrt{\pi}\,s^{-7/2}\right]$$

$$= -\left(-\frac{7}{2}\right)\left(-\frac{5}{2}\right)\left(-\frac{3}{2}\right)\cdot\frac{1}{2}\sqrt{\pi}\,s^{-9/2}$$

$$= \frac{105}{16}\sqrt{\pi}\,s^{-9/2}.$$

INVERSE LAPLACE TRANSFORM

• **PROBLEM 5-25**

Determine the inverse Laplace transform of $F(s)$, where $F(s)$ is given by

$$F(s) = \frac{3s+1}{(s-1)(s^2+1)}$$

<u>Solution</u>: Using the partial fractions technique, $F(s)$ becomes

$$F(s) = \frac{3s+1}{(s-1)(s^2+1)} = \frac{A}{s-1} + \frac{(BS+C)}{s^2+1} \qquad (1)$$

where A, B, and C are constants.
To determine A, multiply $F(s)$ by $(s-1)$ and substitute $s=1$.

Hence,

$$A = \left.\frac{3s+1}{s^2+1}\right|_{s=1} = \frac{3(1)+1}{1+1} = \frac{4}{2} = 2$$

Now substitute the value of A into Eq. (1),

$$\frac{3s+1}{(S-1)(s^2+1)} = \frac{2}{s-1} + \frac{(Bs+C)}{(s^2+1)}$$

and multiplying throughout by $(s-1)(s^2+1)$ to obtain the constants B and C. Eq. (1) becomes

$$(3s+1) = 2(s^2+1) + (Bs+C)(s-1) \tag{2}$$

Comparing the coefficients of s^2 on both sides of Eq. (2), thus

$$0 = 2 + B$$

$$B = -2$$

Comparing the constants, thus

$$1 = 2 - C$$

$$C = 1$$

Therefore, $F(s) = \dfrac{2}{s-1} + \dfrac{(-2s+1)}{s^2+1}$

$$= \dfrac{2}{s-1} - 2\dfrac{s}{s^2+1} + \dfrac{1}{s^2+1}$$

Now, taking the Laplace inverse,

$$f(t) = 2e^t - 2\cos t + \sin t.$$

• **PROBLEM 5-26**

Evaluate the Laplace inverse of

$$F(s) = \dfrac{5s^2 - 15s + 7}{(s+1)(s-2)^3}$$

Solution: Applying partial fractions technique,

$$F(s) = \dfrac{5s^2-15s+7}{(s+1)(s-2)^3} = \dfrac{A}{(s+1)} + \dfrac{B}{(s-2)} + \dfrac{C}{(s-2)^2} + \dfrac{D}{(s-2)^3} \tag{1}$$

A can be determined by multiplying $F(s)$ by $(s+1)$ and substituting $s = -1$.

Hence,

$$A = \left.\dfrac{5s^2-15s+7}{(s-2)^3}\right|_{s=-1}$$

$$= \dfrac{5(1)-15(-1)+7}{(-1-2)^3} = \dfrac{5+15+7}{-27} = -1$$

Similarly, D can be determined by multiplying $F(s)$ by $(s-2)^3$ and substituting $s = 2$.

$$D = \left.\dfrac{5s^2-15s+7}{(s+1)}\right|_{s=2}$$

$$= \dfrac{5(4)-15(2)+7}{3}$$

$$= \dfrac{20-30+7}{3}$$

$$= \frac{-3}{3} = -1$$

Multiplying equation (1) by $(s+1)(s-2)^3$, thus

$$5s^2 - 15s + 7 = A(s-2)^3 + B(s+1)(s-2)^2 + C(s+1)(s-2) + D(s+1)$$

and substituting s=1

$$5 - 15 + 7 = -A + 2B - 2C + 2D$$

$$-3 = 1 + 2B - 2C - 2$$

$$-2 = 2B - 2C$$

$$-1 = B - C \tag{2}$$

Now for s = 0

$$7 = -8A + 4B - 2C + D$$

$$7 = 8 + 4B - 2C - 1$$

$$2B - C = 0 \tag{3}$$

Hence, from equations (2) and (3)

$$-1 = B - 2B$$

$$B = 1$$

$$C = 2$$

Therefore,

$$F(s) = \frac{-1}{(s+1)} + \frac{1}{(s-2)} + \frac{2}{(s-2)^2} - \frac{1}{(s-2)^3}$$

Now, taking the Laplace inverse and applying the shifting theorem,

$$f(t) = -e^{-t} + e^{2t} + 2te^{2t} - \frac{t^2}{2} e^{2t}$$

● **PROBLEM 5-27**

Find the inverse Laplace transform of

$$F(s) = \frac{2s^2-4}{(s-2)(s+1)(s-3)}$$

Solution: By using the partial fractions technique,

$$F(s) = \frac{2s^2-4}{(s-2)(s+1)(s-3)} = \frac{A}{(s-2)} + \frac{B}{(s+1)} + \frac{C}{(s-3)}$$

where A, B, and C are constants.
To determine A, multiply F(s) by (s-2) and substitute 2 for s,

$$A = \frac{2s^2-4}{(s+1)(s-3)}\Big|_{s=2}$$

$$= \frac{2(4)-4}{(2+1)(2-3)} = \frac{4}{(3)(-1)} = \frac{-4}{3}$$

Similarly, to get B, multiply F(s) by (s+1) and substitute s = -1,

$$B = \frac{2s^2-4}{(s-2)(s-3)}\Big|_{s=-1}$$

$$= \frac{2(1)-4}{(-3)(-4)} = \frac{-2}{(-3)(-4)} = \frac{-1}{6}$$

and also

$$C = \frac{2s^2-4}{(s-2)(s+1)}\Big|_{s=3}$$

$$= \frac{2(9)-4}{(1)(4)} = \frac{(14)}{(1)(4)} = \frac{7}{2}$$

Therefore,

$$F(s) = \frac{-4}{3}\frac{1}{(s-2)} + \left(\frac{-1}{6}\right)\frac{1}{s+1} + \frac{7}{2}\left(\frac{1}{s-3}\right)$$

Now, taking Laplace Inverse of F(s)

$$f(t) = \frac{-4}{3}e^{2t} - \frac{1}{6}e^{-t} + \frac{7}{2}e^{3t}.$$

• **PROBLEM 5-28**

Find the inverse Laplace transforms

(a) $L^{-1}\left\{\dfrac{1}{s^2 - 2s + 9}\right\}$,

(b) $L^{-1}\left\{\dfrac{s+1}{s^2 + 6s + 25}\right\}$.

Solution: (a) Completing the square in the denominator $(s^2 - 2s + 9)$,

$$\frac{1}{s^2 - 2s + 9} = \frac{1}{(s-1)^2 + (\sqrt{8})^2} = \frac{1}{\sqrt{8}}\left[\frac{\sqrt{8}}{(s-1)^2 + (\sqrt{8})^2}\right].$$

Thus,

$$L^{-1}\left\{\frac{1}{s^2 - 2s + 9}\right\} = L^{-1}\left\{\frac{1}{\sqrt{8}} \cdot \frac{\sqrt{8}}{(s-1)^2 + (\sqrt{8})^2}\right\}. \qquad (1)$$

Noting the property that

$$L^{-1}\{F(s-k)\} = e^{kt} L^{-1}\{F(s)\}$$

for any function F and constant k, provided that $L^{-1}\{F(s-k)\}$ and $L^{-1}\{F(s)\}$ exist, equality (1) becomes

$$L^{-1}\left\{\frac{1}{s^2 - 2s + 9}\right\} = e^t L^{-1}\left\{\frac{1}{\sqrt{8}} \cdot \frac{\sqrt{8}}{s^2 + (\sqrt{8})^2}\right\}$$

Since L^{-1} is a linear operator under the very weak restrictions on f(t) given in an earlier problem and which are assumed here,

$$L^{-1}\left\{\frac{1}{s^2 - 2s + 9}\right\} = \frac{1}{\sqrt{8}} e^t L^{-1}\left\{\frac{\sqrt{8}}{s^2 + (\sqrt{8})^2}\right\} \quad (2)$$

From a table of Laplace transforms or from the results of a previous problem we find that

$$L^{-1}\left\{\frac{b}{s^2 + b^2}\right\} = \sin bt$$

where b is a constant; hence, $b = \sqrt{8}$, we find

$$L^{-1}\left\{\frac{\sqrt{8}}{s^2 + (\sqrt{8})^2}\right\} = \sin(\sqrt{8}t)$$

Substituting this result into equality (2),

$$L^{-1}\left\{\frac{1}{s^2 - 2s + 9}\right\} = \frac{1}{\sqrt{8}} e^t \sin(\sqrt{8}t)$$

(b) Completing the square in the denominator $(s^2 + 6s + 25)$,

$$\frac{s+1}{s^2 + 6s + 25} = \frac{s+1}{(s+3)^2 + 4^2} = \frac{(s+3)-2}{(s+3)^2 + 4^2}$$

$$= \frac{(s+3)}{(s+3)^2 + 4^2} - \frac{1}{2}\left[\frac{4}{(s+3)^2 + 4^2}\right]$$

Thus, since L^{-1} is a linear operator under the restrictions mentioned earlier

$$L^{-1}\left\{\frac{s+1}{s^2 + 6s + 25}\right\} = L^{-1}\left\{\frac{(s+3)}{(s+3)^2 + 4^2}\right\} - \frac{1}{2} L^{-1}\left\{\frac{4}{(s+3)^2 + 4^2}\right\}$$

(3)

Noting the property that

$$L^{-1}\{F(s-k)\} = e^{kt} L^{-1}\{F(s)\}$$

for any function F and constant k, provided that $L^{-1}\{F(s-k)\}$ and $L^{-1}\{F(s)\}$ exist, equality (3) becomes

$$L^{-1}\left\{\frac{s+1}{s^2 + 6s + 25}\right\} = e^{-3t} L^{-1}\left\{\frac{s}{s^2 + 4^2}\right\} - \frac{1}{2} e^{-3t} L^{-1}\left\{\frac{4}{s^2 + 4^2}\right\} \quad (4)$$

From a table of Laplace transforms or from the results of a previous problem we find that

$$L^{-1}\left\{\frac{s}{s^2 + b^2}\right\} = \cos bt, \text{ and } L^{-1}\left\{\frac{b}{s^2 + b^2}\right\} = \sin bt,$$

where b is a constant; hence, taking b = 4, we find

$$L^{-1}\left\{\frac{s}{s^2 + 4^2}\right\} = \cos 4t, \text{ and } L^{-1}\left\{\frac{4}{s^2 + 4^2}\right\} = \sin 4t$$

Substituting these results into equality (4),

$$L^{-1}\left\{\frac{s+1}{s^2 + 6s + 25}\right\} = e^{-3t} \cos 4t - \frac{1}{2} e^{-3t} \sin 4t$$

● **PROBLEM 5-29**

Find and sketch the function g(t) which is the inverse Laplace transform

$$g(t) = L^{-1}\left\{\frac{3}{s} - \frac{4e^{-s}}{s^2} + \frac{4e^{-3s}}{s^2}\right\}$$

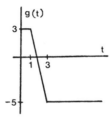

Solution: L^{-1} is a linear operator under the very weak restrictions on g(t) given in an earlier problem. Since these conditions are assumed to hold for all functions dealt with in this book,

$$g(t) = 3L^{-1}\left\{\frac{1}{s}\right\} - 4L^{-1}\left\{\frac{e^{-s}}{s^2}\right\} + 4L^{-1}\left\{\frac{e^{-3s}}{s^2}\right\}. \quad (1)$$

Recalling the property that

$$L^{-1}\{e^{-ks}F(s)\} = f(t-k)\alpha(t-k),$$

where $f(t) = L^{-1}\{f(s)\}$, k is nonnegative constant, and α is the unit step function

$$\alpha(x) = \begin{cases} 0, & x < 0 \\ 1, & x \geq 0 \end{cases}$$

provided that $L^{-1}\{e^{-ks}F(s)\}$ and $L^{-1}\{F(s)\}$ exist, equality (1) becomes

$$g(t) = \left[3L^{-1}\left\{\frac{1}{s}\right\}\bigg|_t - 4L^{-1}\left\{\frac{1}{s^2}\right\}\bigg|_{t-1} \cdot \alpha(t-1) + 4L^{-1}\left\{\frac{1}{s^2}\right\}\bigg|_{t-3} \times\right.$$
$$\left. \times \alpha(t-3)\right]. \quad (2)$$

From a table of Laplace transforms, or by recalling a previous problem, we find that

$$L^{-1}\left\{\frac{1}{s^n}\right\}\bigg|_x = \frac{x^{n-1}}{(n-1)!},$$

where n is a positive integer; hence,

$$L^{-1}\left\{\frac{1}{s}\right\}\bigg|_t = 1, \quad L^{-1}\left\{\frac{1}{s^2}\right\}\bigg|_{t-1} = t-1,$$

and

$$L^{-1}\left\{\frac{1}{s^2}\right\}\bigg|_{t-3} = t-3.$$

Substituting these results into equality (2),

$$g(t) = 3 - 4(t-1)\alpha(t-1) + 4(t-3)\alpha(t-3). \quad (3)$$

To remove the α function from expression (3), consider first the case $t < 1$. In that case,

$$\alpha(t-1) = \alpha(t-3) = 0,$$

so that

$$g(t) = 3-4(t-1)(0) + 4(t-3)(0) = 3 \ . \tag{4}$$

When $1 \leq t < 3$, we have $\alpha(t-1) = 1$ and $\alpha(t-3) = 0$, so that

$$g(t) = 3-4(t-1)(1) + 4(t-3)(0) = 7-4t \ . \tag{5}$$

When $t \geq 3$, we have $\alpha(t-1) = \alpha(t-3) = 1$, so that

$$g(t) = 3-4(t-1)(1) + 4(t-3)(1) = -5 \ . \tag{6}$$

Grouping together results (4), (5), and (6), we have

$$g(t) = \begin{cases} 3, & t < 1 \\ 7-4t, & 1 \leq t < 3 \\ -5, & t \geq 3 \end{cases} \ .$$

The graph of g(t) is shown in the figure.

• **PROBLEM 5-30**

Find the inverse Laplace transform

$$f(t) = L^{-1}\{F(s)\} = L^{-1}\left\{\frac{3s^2 + 17s + 47}{(s+2)(s^2 + 4s + 29)}\right\} \ .$$

Solution: Since the degree of the numerator polynomial is less than the degree of the denominator polynomial, factor the denominator and expand the rational fraction by partial fractions.

Thus,

$$F(s) = \frac{3s^2 + 17s + 47}{(s+2)(s+r_1)(s+r_2)}$$

where

$$r_1 = \frac{-4 + \sqrt{16-116}}{2} \ ; \quad r_2 = \frac{-4 - \sqrt{16-116}}{2}$$

$$r_1 = -2 + j5 \quad ; \quad r_2 = -2 - j5$$

and

$$j = \sqrt{-1} \ .$$

Expanding by partial fractions gives

$$F(s) = \frac{K_1}{s+2} + \frac{K_2}{s+2+j5} + \frac{K_2{}^*}{s+2-j5} \tag{1}$$

245

or, finding a common denominator for the right hand side of (1)

$$F(s) = \frac{K_1(s+2+j5)(s+2-j5) + K_2(s+2)(s+2-j5) + K_2^*(s+2)(s+2+j5)}{(s+2)(s+2+j5)(s+2-j5)}$$

(2)

But

$$F(s) = \frac{3s^2 + 17s + 47}{(s+2)(s+2+j5)(s+2-j5)}$$

and using this in (2) gives

$$3s^2 + 17s + 47 = K_1(s+2+j5)(s+2-j5) + K_2(s+2)(s+2-j5)$$
$$+ K_2^*(s+2)(s+2+j5) \ . \qquad (3)$$

Now (3) must hold for all s. In particular, when $s = -2$, (3) yields

$$(3s^2 + 17s + 47)\Big|_{s=-2} = K_1(s+2+j5)(s+2-j5)\Big|_{s=-2}$$

so that

$$K_1 = \frac{3s^2 + 17s + 47}{s^2 + 4s + 29}\Big|_{s=-2} = \frac{25}{25} = 1 \ .$$

Similarly at $s = -2 -j5$, (3) must hold so that

$$K_2 = \frac{3s^2 + 17s + 47}{(s+2)(s+2-j5)}\Big|_{s=-2-j5} = \frac{3(-21 + j20) - 34 - j85 + 47}{(-j5)(-j10)}$$

$$= \frac{-50 - j25}{-50} = 1 + j(0.5)$$

and

$$K_2^* = 1 - j(0.5) \ .$$

Substituting K_1, K_2, and K_2^* into the partial fraction expansion gives

$$F(s) = \frac{1}{s+2} + \frac{1+j0.5}{s+2+j5} + \frac{1-j0.5}{s+2-j5} \ .$$

Taking the inverse Laplace transform of F(s) gives

$$f(t) = e^{-2t} + (1+j0.5)e^{-2t-j5t} + (1-j0.5)e^{-2t+j5t}.$$

Multiplying out:

$$f(t) = e^{-2t} + e^{-2t}e^{-j5t} + j0.5e^{-2t}e^{-j5t} + e^{-2t}e^{j5t}$$

$$-j0.5e^{-2t}e^{j5t},$$

and factoring to obtain the form:

$$\cos \omega t = \frac{e^{j\omega t} + e^{-j\omega t}}{2}, \quad \sin \omega t = \frac{e^{j\omega t} - e^{-j\omega t}}{2j}$$

$$f(t) = e^{-2t} + 2e^{-2t}\left|\frac{e^{j5t} + e^{-j5t}}{2}\right| + (2j)(-j0.5e^{-2t})$$

$$\times \left|\frac{e^{j5t} - e^{-j5t}}{2j}\right|$$

$$f(t) = [e^{-2t} + 2e^{-2t}\cos 5t + e^{-2t}\sin 5t].$$

• **PROBLEM 5-31**

Find the inverse Laplace transforms

(a)
$$L^{-1}\left\{\frac{2s}{(s^2+1)^2}\right\},$$

(b)
$$L^{-1}\left\{\frac{1}{\sqrt{s}}\right\}$$

Solution: (a) In a table of Laplace transforms we find

$$L\{t \sin bt\} = \frac{2bs}{(s^2+b^2)^2},$$

where b is a constant; hence, taking b = 1, we find

$$L\{t \sin t\} = \frac{2s}{(s^2+1)^2}.$$

Therefore, by definition of the inverse Laplace transform,

$$L^{-1}\left\{\frac{2s}{(s^2+1)^2}\right\} = t \sin t .$$

(b) In a table of Laplace transforms, we find

$$L\left\{\frac{1}{\sqrt{t}}\right\} = \frac{\sqrt{\pi}}{\sqrt{s}} .$$

Thus, by definition of the inverse Laplace transform,

$$L^{-1}\left\{\frac{\sqrt{\pi}}{\sqrt{s}}\right\} = \frac{1}{\sqrt{t}}$$

Since L^{-1} is a linear operator,

$$L^{-1}\left\{\frac{1}{\sqrt{s}}\right\} = \frac{1}{\sqrt{\pi}} L^{-1}\left\{\frac{\sqrt{\pi}}{\sqrt{s}}\right\} = \frac{1}{\sqrt{\pi}} \cdot \frac{1}{\sqrt{t}}$$

● **PROBLEM 5-32**

Find the inverse Laplace transform

$$f(t) = L^{-1}\left\{\log \frac{s+1}{s-1}\right\}, \quad s > 1 .$$

Solution: From a table of infinite series, we can find

$$\log \frac{1+x}{1-x} = 2 \sum_{n=0}^{\infty} \frac{x^{2n+1}}{2n+1} , \tag{1}$$

for $|x| < 1$. We wish to put $\log[(s+1)/(s-1)]$ into some form for which the series (1) will be useful. Dividing the numerator and denominator by s does not change the value:

$$\log \frac{s+1}{s-1} = \log \frac{1+1/s}{1-1/s} .$$

Now, $|1/s| < 1$, since $s > 1$. Thus,

$$\log \frac{1+1/s}{1-1/2} = 2 \sum_{n=0}^{\infty} \frac{(1/s)^{2n+1}}{2n+1}$$

and

$$L^{-1}\left\{\log \frac{s+1}{s-1}\right\} = L^{-1}\left\{2 \sum_{n=0}^{\infty} \frac{(1/s)^{2n+1}}{2n+1}\right\} .$$

Since L^{-1} is a linear operator, under the very weak restrictions on f(t) given in an earlier problem and which are assumed here,

$$L^{-1}\left\{\log \frac{s+1}{s-1}\right\} = 2 \sum_{n=0}^{\infty} \left(\frac{1}{2n+1}\right) L^{-1}\left\{\left(\frac{1}{s}\right)^{2n+1}\right\}. \qquad (2)$$

From a table of Laplace transforms or from the result of a previous problem, it is found that

$$L^{-1}\left\{\left(\frac{1}{s}\right)^{k}\right\} = \frac{t^{k-1}}{(k-1)!},$$

where k is a positive integer. Substitution (2n+1) for k in the above formula,

$$L^{-1}\left\{\left(\frac{1}{s}\right)^{2n+1}\right\} = \frac{t^{2n}}{(2n)!},$$

where (2n+1) is a positive integer; i.e., where n is a nonnegative integer. Substituting this result into equality (2),

$$f(t) = L^{-1}\left\{\log \frac{s+1}{s-1}\right\} = 2 \sum_{n=0}^{\infty} \left(\frac{1}{2n+1}\right) \frac{t^{2n}}{(2n)!}$$

$$= 2 \sum_{n=0}^{\infty} \frac{t^{2n}}{(2n+1)!}$$

$$= \frac{2}{t} \sum_{n=0}^{\infty} \frac{t^{2n+1}}{(2n+1)!} = \frac{2}{t} \sinh t.$$

An alternative way of solving this problem is to use the equality

$$L\{tf(t)\} = -\frac{d}{ds} L\{f(t)\}$$

proved in problem 7. Differentiating our function,

$$\frac{d}{ds}\left(\log \frac{s+1}{s-1}\right) = -\frac{2}{s^2-1}.$$

If we find h(t) such that

$$L\{h(t)\} = -\frac{2}{s^2-1},$$

then

$$L\{h(t)\} = L\{t \cdot \frac{h(t)}{t}\} = -\frac{d}{ds} L\{\frac{h(t)}{t}\}$$

$$= -\frac{2}{s^2-1} = \frac{d}{ds}\left(\log\frac{s+1}{s-1}\right)$$

and

$$L\left\{-\frac{h(t)}{t}\right\} = \log\frac{s+1}{s-1} + \text{const.}$$

Constant we will show to be equal zero. We know

$$L\{\sinh t\} = \frac{1}{s^2-1}$$

so

$$L^{-1}\left\{-\frac{2}{s^2-1}\right\} = -2 \sinh t$$

and we have

$$L\left\{-\left(-\frac{2 \sinh t}{t}\right)\right\} = L\left\{\frac{2 \sinh t}{t}\right\} = \log\frac{s+1}{s-1} + \text{const.}$$

Now, as $s \to +\infty$ both

$$L\left\{\frac{2 \sinh t}{t}\right\} \quad \text{and} \quad \log\frac{s+1}{s-1} \to 0$$

so const = 0 and

$$L^{-1}\left\{\log\frac{s+1}{s-1}\right\} = \frac{2 \sinh t}{t}$$

• **PROBLEM 5-33**

Use the derivative property of Laplace transforms to solve the differential equation

$$y' - y = e^{-x} \qquad (1)$$

where $y(0) = 0$ is the initial value of y.

<u>Solution</u>: First multiply equation (1) by e^{-sx} and integrate from 0 to ∞ to get

$$\int_0^\infty e^{-sx} y'(x)\,dx - \int_0^\infty e^{-sx} y(x)\,dx = \int_0^\infty e^{-sx} \cdot e^{-x}\,dx \quad (2)$$

or

$$L\{y'(x)\} - L\{y(x)\} = L\{e^{-x}\} \quad (3)$$

Now the fact that $L\{e^{-ax}\} = \dfrac{1}{s+a}$ can be obtained from a previous problem in this chapter or can be looked up on a table of Laplace transforms and is valid for $s > a$. Also, the derivative property states that

$$L\{y'(x)\} = sL\{y(x)\} - y(0), \text{ so that}$$

denoting

$$Y(s) = L\{y(x)\}$$

and using (3), one obtains

$$[s\,Y(x) - y(0)] - Y(s) = \frac{1}{s+1}$$

or since

$$y(0) = 0, \quad (s-1)\,Y(s) = \frac{1}{s+1}$$

and

$$Y(s) = \frac{1}{(s+1)(s-1)} = \frac{1}{2}\frac{1}{s-1} - \frac{1}{2}\frac{1}{s+1}, \quad (4)$$

where we have used rational fraction decomposition of

$$\frac{1}{(s+1)(s-1)}$$

The equality in (4) is valid only for $s > 1$, but this will turn out to be insignificant since we wish to invert (4) anyway. Thus, recall that

$$L\{e^x\} = \frac{1}{s-1}, \quad L\{e^{-x}\} = \frac{1}{s+1}$$

so that (4) reads

$$L\{y(x)\} = \frac{1}{2}L\{e^x\} - \frac{1}{2}L\{e^{-x}\}$$

$$= L\left\{\frac{1}{2}e^x - \frac{1}{2}e^{-x}\right\} \quad (5)$$

since L is a linear operator. Thus $y(x)$ and

$$\frac{1}{2}e^x - \frac{1}{2}e^{-x}$$

have the same Laplace transforms. As will be discussed

later, under very general conditions, if two functions have the same Laplace transform, they are identical. This is not true for all functions, but is true for a class of almost all functions of any applicability, and we assume y(x) and

$$\frac{1}{2} e^x - \frac{1}{2} e^{-x}$$

to be such functions. Thus (5) implies that

$$y(x) = \frac{1}{2} e^x - \frac{1}{2} e^{-x} = \sinh x$$

is the solution to the problem. Questions concerning the validity of our "inversion" of the Laplace transform will be discussed more rigorously later in the chapter.

APPLICATIONS OF LAPLACE TRANSFORM

● **PROBLEM 5-34**

Solve the initial value problem

$$y''(t) + 2y'(t) + 5y(t) = H(t) \tag{1}$$

$$y(0) = y'(0) = 0,$$

where

$$H(t) = \begin{cases} 1, & 0 \le t < \pi \\ 0, & t \ge \pi, \end{cases}$$

as shown in the accompanying graph.

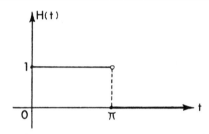

Solution: The function H(t) can be expressed as

$$H(t) = 1 - \alpha(t - \pi),$$

where α is the unit step function

$$\alpha(x) = \begin{cases} 0, & x < 0, \\ 1, & x \geq 0. \end{cases}$$

Thus, equation (1) is equivalent to

$$y''(t) + 2y'(t) + 5y(t) = 1 - \alpha(t - \pi).$$

Taking the Laplace transform of both sides,

$$L\{y''(t) + 2y'(t) + 5y(t)\} = L\{1 - \alpha(t - \pi)\}.$$

Since L is a linear operator,

$$L\{y''(t)\} + 2L\{y'(t)\} + 5L\{y(t)\} = L\{1\} - L\{\alpha(t - \pi)\}. \qquad (2)$$

From a table of Laplace transforms, we find that

$$L\{1\} = \frac{1}{s}, \qquad s > 0,$$

and

$$L\{\alpha(t - k)\} = \frac{e^{-ks}}{s}, \qquad s > 0,$$

where k is a nonnegative constant. Taking $k = \pi$, the latter result becomes

$$L\{\alpha(t - \pi)\} = \frac{e^{-\pi s}}{s}, \qquad s > 0.$$

Substituting these results into equation (2),

$$L\{y''(t)\} + 2L\{y'(t)\} + 5L\{y(t)\} = \frac{1}{s} - \frac{e^{-\pi s}}{s}. \qquad (3)$$

Using the properties that

$$L\{y'(t)\} = sL\{y(t)\} - y(0)$$

and

$$L\{y''(t)\} = s^2 L\{y(t)\} - sy(0) - y'(0),$$

provided that $y'(t)$, $y''(t)$, and $L\{y(t)\}$ exist, equation (3) becomes

$$[s^2 L\{y(t)\} - sy(0) - y'(0)] + 2[sL\{y(t)\} - y(0)]$$

$$+ 5L\{y(t)\} = \frac{1 - e^{-\pi s}}{s}.$$

Solving for $L\{y(t)\}$,

$$L\{y(t)\} = \frac{1 - e^{-\pi s} + s(s+2)y(0) + sy'(0)}{s(s^2 + 2s + 5)},$$

and since

$$y(0) = y'(0) = 0,$$

$$L\{y(t)\} = \frac{1 - e^{-\pi s}}{s(s^2 + 2s + 5)}.$$

Inverting the above equation,

$$y(t) = L^{-1}\left\{\frac{1 - e^{-\pi s}}{s(s^2 + 2s + 5)}\right\}.$$

Since L^{-1} is a linear operator under the conditions assumed in this chapter

$$y(t) = L^{-1}\left\{\frac{1}{s(s^2 + 2s + 5)}\right\} - L^{-1}\left\{\frac{e^{-\pi s}}{s(s^2 + 2s + 5)}\right\}.$$

Let

$$G(t) = L^{-1}\left\{\frac{1}{s(s^2 + 2s + 5)}\right\}.$$

Thus,

$$y(t) = G(t) - L^{-1}\left\{\frac{e^{-\pi s}}{s(s^2 + 2s + 5)}\right\}. \tag{4}$$

Using the property that

$$L^{-1}\{e^{-ks} F(s)\} = f(t - k)\alpha(t - k),$$

where k is a nonnegative constant and $f(t) = L^{-1}\{F(s)\}$, provided that $L^{-1}\{F(s)\}$ exists, we find (with $k = \pi$) that

$$L^{-1}\left\{\frac{e^{-\pi s}}{s(s^2 + 2s + 5)}\right\} = G(t - \pi)\alpha(t - \pi);$$

hence, equation (4) is equivalent to

$$y(t) = G(t) - G(t - \pi)\alpha(t - \pi). \tag{5}$$

We now evaluate the function $G(t)$. We use the method of partial fractions to decompose

$$\frac{1}{[s(s^2 + 2s + 5)]}.$$

Setting

$$\frac{1}{s(s^2 + 2s + 5)} = \frac{A}{s} + \frac{Bs + C}{s^2 + 2s + 5},$$

for all s, with A, B, and C constant, we find

$$A = \frac{1}{5}, \quad B = -\frac{1}{5}, \quad \text{and} \quad C = -\frac{2}{5}.$$

Thus,

$$G(t) = L^{-1}\left\{\frac{1}{5}\left(\frac{1}{s}\right) - \frac{1}{5}\left(\frac{s + 2}{(s^2 + 2s + 5)}\right)\right\}.$$

Completing the square in the denominator of the fraction

$$\frac{(s + 2)}{(s^2 + 2s + 5)}$$

above,

$$G(t) = L^{-1}\left\{\frac{1}{5}\left(\frac{1}{s}\right) - \frac{1}{5}\left[\frac{s + 2}{(s + 1)^2 + 2^2}\right]\right\}$$

$$= L^{-1}\left\{\frac{1}{5}\left(\frac{1}{s}\right) - \frac{1}{5}\left[\frac{s + 1}{(s + 1)^2 + 2^2}\right] - \frac{1}{10}\left[\frac{2}{(s + 1)^2 + 2^2}\right]\right\}.$$

Since L^{-1} is a linear operator,

$$G(t) = \frac{1}{5}L^{-1}\left\{\frac{1}{s}\right\} - \frac{1}{5}L^{-1}\left\{\frac{s + 1}{(s + 1)^2 + 2^2}\right\}$$

$$- \frac{1}{10}L^{-1}\left\{\frac{2}{(s + 1)^2 + 2^2}\right\}. \tag{6}$$

Using the property that

$$L^{-1}\{f(s - b)\} = e^{bt} L^{-1}\{f(s)\},$$

where b is a constant, provided that $L^{-1}\{f(s)\}$ exists, equation (6) becomes (with b = -1)

$$G(t) = \frac{1}{5} L^{-1}\left\{\frac{1}{s}\right\} - \frac{1}{5} e^{-t} L^{-1}\left\{\frac{s}{s^2 + 2^2}\right\}$$

$$- \frac{1}{10} e^{-t} L^{-1}\left\{\frac{2}{s^2 + 2^2}\right\} . \tag{7}$$

From a table of Laplace transforms, we find that

$$L^{-1}\left\{\frac{1}{s}\right\} = 1, \quad L^{-1}\left\{\frac{s}{s^2 + c^2}\right\} = \cos ct,$$

and

$$L^{-1}\left\{\frac{c}{s^2 + c^2}\right\} = \sin ct,$$

where c is a constant. Taking c = 2, the last two of the above results become

$$L^{-1}\left\{\frac{s}{s^2 + 2^2}\right\} = \cos 2t,$$

and

$$L^{-1}\left\{\frac{2}{s^2 + 2^2}\right\} = \sin 2t .$$

Substituting these results into equation (7),

$$G(t) = \frac{1}{5} - \frac{1}{5} e^{-t} \cos 2t - \frac{1}{10} e^{-t} \sin 2t, \tag{8}$$

and thus,

$$G(t - \pi) = \frac{1}{5} - \frac{1}{5} e^{-(t-\pi)} \cos 2(t - \pi)$$

$$- \frac{1}{10} e^{-(t-\pi)} \sin 2(t - \pi) . \tag{9}$$

Note that

$$\cos 2(t - \pi) = \cos (2t - 2\pi) = \cos 2t,$$

and

$$\sin 2(t - \pi) = \sin (2t - 2\pi) = \sin 2t;$$

therefore, equation (9) is equivalent to

$$G(t - \pi) = \frac{1}{5} - \frac{1}{5} e^{-(t-\pi)} \cos 2t - \frac{1}{10} e^{-(t-\pi)} \sin 2t. \quad (10)$$

Substituting equations (8) and (10) into equation (5),

$$y(t) = \frac{1}{5} - \frac{1}{5} e^{-t} \cos 2t - \frac{1}{10} e^{-t} \sin 2t$$

$$- \left[\frac{1}{5} - \frac{1}{5} e^{-(t-\pi)} \cos 2t - \frac{1}{10} e^{-(t-\pi)} \sin 2t \right] \alpha(t - \pi). \quad (11)$$

Since we made some assumptions about existence of $y(t), y'(t), y''(t)$ and $L(y(t))$, we must check our result (10) against the original problem. Differentiating twice, we obtain

$$y'(t) = \frac{1}{2} \left[1 - e^{\pi} \alpha(t - \pi) \right] e^{-t} \sin 2t$$

$$y''(t) = \frac{1}{2} \left[1 - e^{\pi} \alpha(t - \pi) \right] e^{-t} (-\sin 2t + 2 \cos 2t).$$

Checking in equation (1),

$$y''(t) + 2y'(t) + 5y(t)$$

$$= \frac{1}{2} [1 - e^{\pi} \alpha(t - \pi)] e^{-t} (-\sin 2t + 2 \cos 2t)$$

$$+ 2 \cdot \frac{1}{2} [1 - e^{\pi} \alpha(t - \pi)] e^{-t} \sin 2t$$

$$+ 5 \left[\frac{1}{5} - \frac{1}{5} e^{-t} \cos 2t - \frac{1}{10} e^{-t} \sin 2t \right.$$

$$\left. - \left(\frac{1}{5} - \frac{1}{5} e^{-(t-\pi)} \cos 2t - \frac{1}{10} e^{-(t-\pi)} \sin 2t \right) \alpha(t-\pi) \right]$$

$$= 1 - \alpha(t - \pi) = H(t),$$

$$y(0) = \frac{1}{5} - \frac{1}{5} e^0 \cos 0 - \frac{1}{10} e^0 \sin 0$$

$$- \left[\frac{1}{5} - \frac{1}{5} e \cos 0 - \frac{1}{10} e^{\pi} \sin 0 \right] \alpha(0 - \pi)$$

$$= 0$$

$$y'(0) = \frac{1}{2} [1 - e^{\pi} \alpha(0 - \pi)] e^0 \sin 0 = 0;$$

hence, equation (11) is the solution.

● **PROBLEM 5-35**

State a theorem which gives conditions under which two functions must be identical if they have the same Laplace transform, i.e., within what class of functions is the inverse Laplace transform unique? What does this imply about the functions

$$f(t) = L^{-1}\{aF_1(s) + bF_2(s)\}$$

and

$$\bar{f}(t) = aL^{-1}\{F_1(s)\} + bL^{-1}\{F_2(s)\} \tag{1}$$

where $F_1(s)$ and $F_2(s)$ are some Laplace transforms?

Solution: Consider all real valued functions which satisfy the following three conditions:

1) each function is piecewise smooth (i.e., has a piecewise continuous derivative) on $(0,\infty)$,

2) each function is of some exponential order (see problem 1 in this chapter),

3) each function is defined by

$$\frac{1}{2}[f(x_0+) + f(x_0-)]$$

at each jump discontinuity x_0.

Then two such functions possessing the same Laplace transform, $F(s)$, must be identical. In other words, the inverse Laplace transform is unique within the class of such functions.

Now the two functions in (1), $f(t)$ and $\bar{f}(t)$ have the same Laplace transform since L is a linear operator. That is

$$L\{f(t)\} = L\{L^{-1}\{aF_1(s) + bF_2(s)\}\} = aF_1(s) + bF_2(s)$$

and

$$L\{\bar{f}(t)\} = L\{aL^{-1}\{F_1(s)\} + bL^{-1}\{F_2(s)\}\}$$

$$= aL\{L^{-1}\{F_1(s)\}\} + bL\{L^{-1}\{F_2(s)\}\}$$

$$= aF_1(s) + bF_2(s)$$

so that

$$L\{f(t)\} = L\{\bar{f}(t)\} . \tag{2}$$

According to the theorem, if these functions satisfy the three conditions above, then (2) implies that they are equal, i.e.,

$$f(t) = \bar{f}(f)$$

or

$$L^{-1}\{aF_1(s) + bF_2(s)\} = aL^{-1}\{F_1(s)\} + bL^{-1}\{F_2(s)\} \quad (3)$$

Thus, under these conditions L^{-1} is a linear operator. The practical significance of this arises in the following way. Suppose that the Laplace transform has been applied to some functions in a particular problem, used for manipulative purposes, and is now to be inverted. In particular, suppose that we are faced with the problem of inverting a transform which can be split into two parts

$$F(s) = aF_1(s) + bF_2(s)$$

so that

$$f(t) = L^{-1}\{F(s)\} = L^{-1}\{aF_1(s) + bF_2(s)\}$$

Then in almost all cases, in order for the problem to be soluble, one would have to assume that L^{-1} is linear so that

$$L^{-1}\{aF_1(s) + bF_2(s)\} = aL^{-1}\{F_1(s)\} + bL^{-1}\{F_2(s)\}$$
$$= af_1(t) + bf_2(t) \quad (4)$$

and

$$f(t) = af_1(t) + bf_2(t).$$

Thus, assuming the linearity of L^{-1} (as is done almost without exception in Laplace transform problems) amounts to assuming that the original function, $f(t)$, satisfies the three conditions stated earlier. Since the great majority of physical problems deal with such functions, this assumption is usually justifiable.

• **PROBLEM 5-36**

The switch S is opened at t=0 in the circuit shown in Fig. 1. Determine an expression for the current i(t), assuming that the circuit had reached steady state before the switch S was opened.

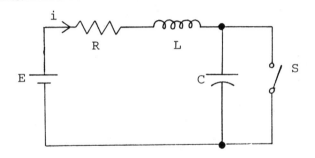

Solution: Since the current in an inductor and the voltage across a capacitor cannot change instantaneously, the values of the current in the inductor and the voltage across the capacitor will remain the same at the instant before and after the opening of the switch.

That means $i(0^-) = i(0^+) = E/R$

and $v_c(0^-) = v_c(0^+) = 0$

For $t > 0$, the differential equation for the circuit is

$$Ri(t) + L\frac{di}{dt}(t) + \frac{1}{C}\int_0^t i(t)dt + v_c(0) = E$$

Taking the Laplace transform and setting $v_c(0) = 0$

$$RI(s) + L\{sI(s) - i(0^-)\} + \frac{1}{C}\frac{I(s)}{s} = \frac{E}{s}$$

$$RI(s) + L\{sI(s) - \frac{E}{R}\} + \frac{1}{Cs}I(s) = \frac{E}{s}$$

$$sR^2C\,I(s) + s^2LRC\,I(s) - ELsC + RI(s) = ERC$$

$$(s^2LRC + sR^2C + R)\,I(s) = EC(R + Ls)$$

Solving for $I(s)$,

$$I(s) = \frac{ECL\left(s+\frac{R}{L}\right)}{LRC\left(s^2+\frac{sR^2C}{LRC}+\frac{R}{LRC}\right)}$$

$$= \frac{\left(\frac{E}{R}\right)\left(s+\frac{R}{L}\right)}{s^2 + \frac{R}{L}s + \frac{1}{LC}}$$

Now, $s^2 + \frac{R}{L}S + \frac{1}{LC} = \left(s+\frac{R}{2L}\right)^2 + \left(\frac{1}{LC} - \frac{R^2}{4L^2}\right)$

$$= (s+a)^2 + b^2$$

in which $a = \frac{R}{2L}$ and $b = \left(\frac{1}{LC} - \frac{R^2}{4L^2}\right)^{\frac{1}{2}}$ (assume $b > 0$)

Then $I(s)$ can be written as

$$I(s) = \left(\frac{E}{R}\right)\frac{(s+2a)}{(s+a)^2+b^2}$$

$$= \frac{E}{R}\left[\frac{(s+a)}{(s+a)^2+b^2} + \frac{a}{(s+a)^2+b^2}\right] \quad (1)$$

Taking the Laplace inverse of Eq. (1),

$$i(t) = \frac{E}{R}\left[e^{-at}\cos bt + \frac{a}{b}e^{-at}\sin bt\right]$$

where a and b are the same values as mentioned before.

● **PROBLEM 5-37**

A telephone line of length ℓ with capacitance C and resistance R is shown in Figure 1. At t=0, the switch S is turned on and assume the line is uncharged for t < 0. Determine the short circuited current at the receiving end.

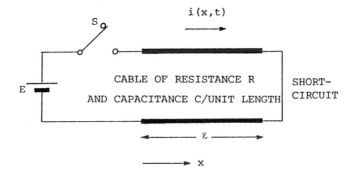

Solution: First, initial conditions must be considered. Assume v(x,t) to be the voltage at distance x from the sending end. Since the voltage across a capacitor cannot change instantaneously right after the switch is turned on,

$$v(x,t) = 0 \quad \text{for } t < 0$$

and $\quad v(x,t) = 0 \quad$ at $t = 0$

which imply v(x,0) = 0. (1)

Next, consider the boundary conditions:

At $x = 0$, $v(x,t) = E$

At $x = \ell$, $v(x,t) = 0$

therefore, $\bar{v}(0,s) = E/s$ and $\bar{v}(\ell,s) = 0$.

Now, also let i(x,t) be the current at distance x from the sending end. Then two related equations can be written for v(x,t) and i(x,t) as follows:

$$\left.\begin{array}{l}\dfrac{\partial v(x,t)}{\partial x} = -R\,i(x,t) \\[6pt] \dfrac{\partial i(x,t)}{\partial x} = \dfrac{-C\,\partial v(x,t)}{\partial t}\end{array}\right\} \quad (2)$$

Also, let the Laplace transform of $v(x,t) = \bar{v}(x,s)$

$$= \int_0^\infty v(x,t)e^{-st}\,dt$$

then because of the order of differentiation and integration is reversible,

$$\frac{\partial}{\partial x}\bar{v}(x,s) = \int_0^\infty \frac{\partial}{\partial x} v(x,t)e^{-st}\,dt. \qquad (3)$$

Using Eqs. (1) and (3), Eq. (2) becomes:

$$\frac{\partial}{\partial x}\bar{v}(x,s) = -R\bar{i}(x,s) \qquad (4)$$

where $\bar{i}(x,s)$ = Laplace transform of $i(x,t)$ and

$$\frac{\partial}{\partial x}\bar{i}(x,s) = -sc\,\bar{v}(x,s) \qquad (5)$$

Now, using Eqs. (4) and (5), $\bar{v}(x,s)$ and $\bar{i}(x,s)$ can be solved as follows:

Take partial derivative on both sides of Eq. (4) to obtain:

$$\frac{\partial^2}{\partial x^2}\bar{v}(x,s) = -R\frac{\partial}{\partial x}\bar{i}(x,s) \qquad (6)$$

Substitute Eq. (5) into Eq. (6)

$$\frac{\partial^2}{\partial x^2}\bar{v}(x,s) = -R\left[-Sc\bar{v}(x,s)\right]$$

Hence, $\quad -\dfrac{1}{R}\dfrac{\partial^2}{\partial x^2}\bar{v}(x,s) = -sC\,\bar{v}(x,s)$

$$\frac{\partial^2}{\partial x^2}\bar{v}(x,s) - sRC\,\bar{v}(x,s) = 0$$

The characteristic equation of this differential equation is $(D^2 - sRC) = 0$ having roots $\pm\sqrt{sRC}$.

Let the roots be $\pm a$ where $a = \sqrt{sRC}$, then the general solution for $\bar{v}(x,s)$ and $\bar{i}(x,s)$ can be written as:

$$\bar{v}(x,s) = A\exp(-ax) + B\exp(ax) \qquad (7)$$

and $\quad \bar{i}(x,s) = \left(\dfrac{a}{R}\right)\left[A\exp(-ax) - B\exp(ax)\right] \qquad (8)$

where A and B are constants.

Now, applying boundary conditions to Eqs. (7) and (8) A and B can be found as follows,

$$A + B = E/s \qquad (9)$$

$$A\exp(-a\ell) + B\exp(a\ell) = 0 \qquad (10)$$

Solving Eqs. (9) and (10),

$$A = \left(\frac{E}{s}\right) \frac{e^{a\ell}}{e^{a\ell}-e^{-a\ell}}$$

$$B = -\left(\frac{E}{s}\right) \frac{e^{-a\ell}}{e^{a\ell}-e^{-a\ell}}$$

Substitute the values of A and B into Eqs. (9) and (10) obtain:

$$\bar{i}(x,s) = -\frac{aE}{sR}\left[\frac{\exp(2a\ell-ax) + \exp(ax)}{1-\exp(2a\ell)}\right]$$

At $x = \ell$,

$$\bar{i}(\ell,s) = \frac{-aE}{sR}\left[\frac{2\exp(a\ell)}{1-\exp(2a\ell)}\right]$$

Applying the expansion theorem, the inverse Laplace transform of $\bar{i}(\ell,s)$ can be found.

Hence, since the denominator of $\bar{i}(\ell,s)$ has an infinite number of roots.

$$i(\ell,t) = \sum_{n=1}^{\infty} \frac{A(s)}{B'(s)} \exp(-st) + D$$

where $B'(s)$ = derivative of $B(s)$ with respect to s

$A(s) = \lambda E$

$B(s) = sR \sinh(a\ell)$

Setting $B(s) = 0$, the roots are $s = 0$ and $a = 0, \pm j\pi/\ell, \pm 2j\pi/\ell, \ldots, \pm jn\pi/\ell, \ldots$ for $\sinh(a\ell) = 0$.

The negative imaginary roots are irrelavant since 'a' is defined as $\sqrt{sRC}$

when $a = \frac{jn\pi}{\ell}$

$$s = \frac{-\pi^2 n^2}{\ell^2 CR}$$

Now, let $k = \frac{\pi^2}{\ell^2 CR}$

The roots of $B(s) = 0$ are $s = 0$ and $s = -n^2 k$, $n = 1, 2, \ldots$. Note: when $s = 0$, the numerator has a single zero and the denominator a double zero. This can be handled by splitting off the term D/s in the partial fraction expansion corresponding to this root. D is a constant to be determined.

It can be written from the expansion theorem,

$$i(\ell,t) = \sum_{n=1}^{\infty} \frac{A(-n^2k)}{B'(-n^2k)} \exp(-n^2kt) + D.$$

where $A(-n^2k) = jn\pi E/\ell : n = 1, 2, \ldots$.

$$B'(s) = R\sinh(a\ell) + sR\frac{d}{ds}\sinh(a\ell)$$

$$= R\sinh(a\ell) + sR\ell(\cosh a\ell)\left(\frac{da}{ds}\right)$$

$$= R\sinh(a\ell) + \frac{a^2\ell}{c}(\cosh a\ell)\left(\frac{CR}{2a}\right)$$

and $B'(-n^2k) = \left[R\sinh(a\ell) + \frac{a\ell R}{2}\cosh a\ell\right]_{a = jn\pi/2}$

Hence, $= \frac{jn\pi R}{2}(-1)^n \quad n = 1, 2, \ldots$

$$\frac{A(-n^2k)}{B'(-n^2k)} = \frac{2E}{R\ell}(-1)^n, \quad n = 1, 2, \ldots$$

Consider the expression given below to determine D

$$\frac{Ea}{sR\sinh(a\ell)} - \frac{D}{s} = \frac{Ea - RD\sinh(a\ell)}{sR\sinh(a\ell)}$$

Select D so that the numerator vanishes for $s = 0$.
    D is given by

$$Ea = RD\sinh(a\ell) \quad \text{if} \quad s = a = 0,$$

i.e. $D = \lim_{a \to 0} \frac{Ea}{R\sinh(a\ell)} = \frac{E}{R\ell}$

Hence,
$$i(\ell,t) = \frac{E}{R\ell}\left[1 + \sum_{n=1}^{\infty}(-1)^n \exp(-n^2kt)\right]$$

PERIODIC FUNCTIONS

● **PROBLEM** 5-38

Given a periodic waveform in Fig. 1., determine its Laplace transform.

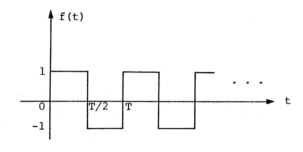

Solution: If f(t) is periodic with a period T,

$$F(s) = \text{Laplace transform of } f(t) = \frac{\int_0^T f(t)e^{-st}dt}{1-e^{-sT}}$$

Hence, with the equation given above, F(s) can be found as follows:

$$F(s) = \frac{1}{1-e^{-sT}}\left[\int_0^{T/2}(1)e^{-st}dt + \int_{T/2}^T(-1)e^{-st}dt\right]$$

$$= \frac{1}{1-e^{-sT}}\left[\left(\frac{e^{-st}}{-s}\right)_0^{T/2} + \left(\frac{e^{-st}}{s}\right)_{T/2}^T\right]$$

$$= \frac{1}{1-e^{-sT}}\left(\frac{1-e^{-\frac{sT}{2}}}{s} + \frac{e^{-sT}-e^{-\frac{sT}{2}}}{s}\right)$$

$$= \frac{1}{1-e^{-sT}}\left(\frac{1-2e^{-\frac{sT}{2}}+e^{-sT}}{s}\right)$$

$$= \frac{\left(1-e^{-\frac{sT}{2}}\right)^2}{(s)\left(1-e^{-\frac{sT}{2}}\right)\left(1+e^{-\frac{sT}{2}}\right)}$$

$$= \frac{1-e^{-\frac{sT}{2}}}{(s)\left(1+e^{-\frac{sT}{2}}\right)}$$

● **PROBLEM 5-39**

Find the Laplace transform L{f(t)}, where f(t) is the function shown in figures (a) and (b).

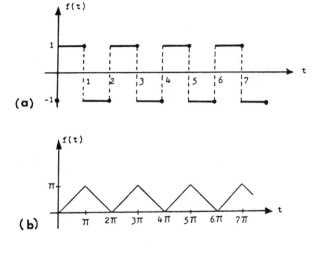

(a)

(b)

Solution: (a) Note that $f(t)$ is a periodic function with period $P = 2$; i.e., $f(t+2) = f(t)$ for all t. On the interval $0 < t \leq 2$, $f(t)$ can be expressed in analytic form:

$$f(t) = \begin{cases} 1, & 0 < t \leq 1 \\ -1, & 1 < t \leq 2 \end{cases}$$

Using the property that

$$L\{g(t)\} = \frac{\int_0^P e^{-st} g(t) dt}{1 - e^{-Ps}}$$

where $g(t)$ is periodic with period P, we obtain

$$L\{f(t)\} = \frac{\int_0^2 e^{-st} f(t) dt}{1 - e^{-2s}}$$

$$= \frac{\int_0^1 e^{-st}(1) dt + \int_1^2 e^{-st}(-1) dt}{1 - e^{-2s}}$$

$$= \frac{\left.\frac{-e^{-st}}{s}\right|_{t=0}^{1} + \left.\frac{e^{-st}}{s}\right|_{t=1}^{2}}{1 - e^{-2s}}$$

$$= \frac{\frac{1}{s}(1 - 2e^{-s} + e^{-2s})}{1 - e^{-2s}}$$

Factoring the numerator and denominator of this last fraction,

$$L\{f(t)\} = \frac{\frac{1}{s}(1 - e^{-s})^2}{(1+e^{-s})(1-e^{-s})}$$

$$= \frac{1}{s}\left(\frac{1 - e^{-s}}{1 + e^{-s}}\right)$$

Multiplying the numerator and denominator by $e^{s/2}$,

$$L\{f(t)\} = \frac{1}{s}\left(\frac{e^{s/2} - e^{-s/2}}{e^{s/2} + e^{-s/2}}\right)$$

$$= \frac{1}{s} \tanh \frac{s}{2}.$$

(b) Note that f(t) is a periodic function with period P = 2π; i.e., f(t+2π) = f(t) for all t. On the interval 0 ≤ t ≤ 2π, f(t) can be expressed in analytic form:

$$f(t) = \begin{cases} t, & 0 \le t \le \pi \\ 2\pi - t, & \pi < t \le 2\pi \end{cases}$$

Using the property that

$$L\{g(t)\} = \frac{\int_0^P e^{-st} g(t)\,dt}{1 - e^{-Ps}}$$

where g(t) is periodic with period P, we obtain

$$L\{f(t)\} = \frac{\int_0^{2\pi} e^{-st} f(t)\,dt}{1 - e^{-2\pi s}}$$

$$= \frac{\int_0^{\pi} e^{-st}(t)\,dt + \int_{\pi}^{2\pi} e^{-st}(2\pi - t)\,dt}{1 - e^{-2\pi s}}$$

$$= \frac{\int_0^{\pi} t e^{-st}\,dt - \int_{\pi}^{2\pi} t e^{-st}\,dt + 2\pi \int_{\pi}^{2\pi} e^{-st}\,dt}{1 - e^{-2\pi s}}.$$

Integrating by parts

$$L\{f(t)\} = \frac{\left(\dfrac{-t e^{-st}}{s} - \dfrac{e^{-st}}{s^2}\right)\bigg|_{t=0}^{\pi} - \left(\dfrac{-t e^{-st}}{s} - \dfrac{e^{-st}}{s^2}\right)\bigg|_{t=\pi}^{2\pi} - \left(\dfrac{2\pi e^{-st}}{s}\right)\bigg|_{t=\pi}^{2\pi}}{1 - e^{-2\pi s}}$$

$$= \frac{\left(\dfrac{-\pi e^{-\pi s}}{s} + \dfrac{1 - e^{-\pi s}}{s^2}\right) + \left(\dfrac{2\pi e^{-2\pi s} - \pi e^{-\pi s}}{s} + \dfrac{e^{-2\pi s} - e^{-\pi s}}{s^2}\right)}{1 - e^{-2\pi s}}$$

$$+ \frac{\left(\frac{2\pi e^{-\pi s} - 2\pi e^{-2\pi s}}{s}\right)}{1-e^{-2\pi s}}$$

$$= \frac{1 - 2e^{-\pi s} + e^{-2\pi s}}{s^2(1 - e^{-2\pi s})}$$

$$= \frac{\left(1 - e^{-\pi s}\right)^2}{s^2(1+e^{-\pi s})(1-e^{-\pi s})}$$

$$= \frac{1}{s^2}\left(\frac{1 - e^{-\pi s}}{1 + e^{-\pi s}}\right) .$$

Multiplying both the numerator and denominator in the last fraction by $e^{\pi s/2}$,

$$L\{f(t)\} = \frac{1}{s^2}\left(\frac{e^{\pi s/2} - e^{-\pi s/2}}{e^{\pi s/2} + e^{-\pi s/2}}\right)$$

$$= \frac{1}{s^2} \tanh \frac{\pi s}{2} .$$

● **PROBLEM 5-40**

Determine the Laplace transform of the half-wave rectified sine function shown in the figure.

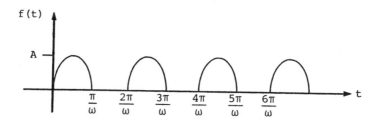

Solution: From the figure, the first cycle of the function can be recognized as a sine function, $A \sin \omega t$, between the interval $\left(0, \frac{\pi}{\omega}\right)$.

Since $f(t)$ is periodic with period $T = \frac{2\pi}{\omega}$, we can apply the property:

$$L\{f(t)\} = \frac{\int_0^T f(t)e^{-st}dt}{1-e^{-Ts}}$$

Hence,

$$L[f(t)] = \frac{A}{\left(1-e^{-\left(\frac{2\pi s}{\omega}\right)}\right)} \int_0^{\pi/\omega} \sin \omega t \, e^{-st} \, dt$$

$$= \left(\frac{A}{1-e^{-\left(\frac{2\pi s}{\omega}\right)}}\right) \frac{1}{2j} \int_0^{\pi/\omega} (e^{j\omega t} - e^{-j\omega t})e^{-st} \, dt$$

$$= \frac{A}{\left(1-e^{-\left(\frac{2\pi s}{\omega}\right)}\right)(2j)} \int_0^{\pi/\omega} \left(e^{-(s-j\omega)t} - e^{-(s+j\omega)t}\right)dt$$

$$= \frac{A}{\left(1-e^{-\left(\frac{2\pi s}{\omega}\right)}\right)(2j)} \left[\frac{e^{-(s-j\omega)t}}{-(s-j\omega)} - \frac{e^{-(s+j\omega)t}}{-(s+j\omega)}\right]_0^{\frac{\pi}{\omega}}$$

$$= \frac{-A}{\left(1-e^{-\left(\frac{2\pi s}{\omega}\right)}\right)(2j)} \left\{\left[\frac{e^{-(s-j\omega)\frac{\pi}{\omega}}}{(s-j\omega)} - \frac{e^{-(s+j\omega)\frac{\pi}{\omega}}}{(s+j\omega)}\right] - \left[\frac{1}{s-j\omega} - \frac{1}{(s+j\omega)}\right]\right\}$$

$$= \frac{-A}{\left(1-e^{-\left(\frac{2\pi s}{\omega}\right)}\right)(2j)} \left\{\frac{e^{-\frac{\pi s}{\omega}}}{s^2+\omega^2}\left[(s+j\omega)e^{j\pi} - (s-j\omega)e^{-j\pi}\right] - \frac{2j\omega}{s^2+\omega^2}\right\}$$

$$= \frac{-A}{(s^2+\omega^2)\left(1-e^{-\left(\frac{2\pi s}{\omega}\right)}\right)(2j)} \left\{e^{-\frac{\pi s}{\omega}}\left[(s+j\omega)(-1) - (s-j\omega)(-1)\right] - 2j\omega\right\}$$

$$= \frac{+A(2j\omega)}{(s^2+\omega^2)\left(1-e^{-\left(\frac{2\pi s}{\omega}\right)}\right)(2j)} \left(e^{-\frac{\pi s}{\omega}} + 1\right)$$

$$= \frac{A\omega}{s^2+\omega^2}\left[\frac{(1+e^{-\frac{\pi s}{\omega}})}{(1+e^{-\frac{\pi s}{\omega}})(1-e^{-\frac{\pi s}{\omega}})}\right]$$

$$= \frac{A\omega}{(s^2+\omega^2)} \left(\frac{1}{1-e^{-\pi s/\omega}} \right)$$

• **PROBLEM 5-41**

Find the Laplace transform $L\{h(t)\}$, where

$$h(t) = \begin{cases} 1, & 0 < t < c \\ -1, & c < t < 2c \end{cases}$$

and $h(t+2c) = h(t)$ for all t, with c a constant. (See Figure 1.)
Use the fact that

$$L\{g(t)\} = \frac{1}{s(1+e^{-cs})}, \tag{a}$$

where

$$g(t) = \begin{cases} 1, & 0 < t < c \\ 0, & c < t < 2c \end{cases}$$

and

$$g(t+2c) = g(t), \text{ for all } t.$$

(See Figure 2).

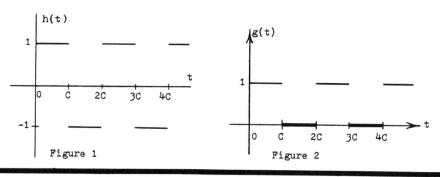

Figure 1

Figure 2

<u>Solution</u>: The crucial observation here is that $h(t) = 2g(t) - 1$, for all t. Since L is a linear operator,

$$L\{h(t)\} = L\{2g(t) - 1\}$$
$$= 2L\{g(t)\} - L\{1\}. \tag{b}$$

From a table of Laplace transforms, we find that

$$L\{t^n\} = \frac{n!}{s^{n+1}},$$

where n is a nonnegative integer constant; hence, taking $n = 0$,

$$L\{1\} = 1/s.$$

Substituting this result and equality (a) into equality (b), we find that

$$L\{h(t)\} = 2\left[\frac{1}{s(1+e^{-cs})}\right] - \frac{1}{s}$$

$$= \frac{1}{s}\left(\frac{2}{1+e^{-cs}} - 1\right)$$

$$= \frac{1}{s}\left[\frac{2-(1+e^{-cs})}{1+e^{-cs}}\right]$$

$$= \frac{1}{s}\left(\frac{1-e^{-cs}}{1+e^{-cs}}\right).$$

Multiplying the numerator and denominator by $e^{cs/2}$,

$$L\{h(t)\} = \frac{1}{s}\left(\frac{e^{cs/2}-e^{-cs/2}}{e^{cs/2}+e^{-cs/2}}\right)$$

$$= \frac{1}{s}\tanh\left(\frac{cs}{2}\right).$$

• **PROBLEM 5-42**

Prove that if
$$f(x+b) = -f(x) \tag{a}$$
for all x, where b is a constant, then

$$L\{f(t)\} = \frac{\int_0^b e^{-st}f(t)dt}{1 + e^{-bs}},$$

where L is the Laplace transform operator.

Solution: By an application of relation (a), setting $x = t + b$,

$$f(t+2b) = f([t+b]+b) = -f(t+b). \tag{b}$$

Next setting $x = t$ in relation (a),

$$-f(t+b) = -[-f(t)] = f(t). \tag{c}$$

Combining equalities (b) and (c),

$$f(t+2b) = f(t);$$

hence $f(t)$ is periodic with a period of $2b$. Using the property that

$$L\{g(t)\} = \frac{\int_0^P e^{-st}g(t)dt}{1 - e^{-Ps}},$$

where $g(t)$ is periodic with period P, we obtain

$$L\{f(t)\} = \frac{\int_0^{2b} e^{-st}f(t)dt}{1 - e^{-2bs}}$$

$$= \frac{\int_0^b e^{-st}f(t)dt + \int_b^{2b} e^{-st}f(t)dt}{1 - e^{-2bs}}. \tag{d}$$

Making the substitution $y + b = t$ in the second integral, we find that

$$\int_b^{2b} e^{-st}f(t)dt = \int_0^b e^{-s(y+b)}f(y+b)dy$$

$$= e^{-bs} \int_0^b e^{-sy}[-f(y)]dy$$

$$= -e^{-bs} \int_0^b e^{-sy}f(y)dy \ .$$

When we change the dummy variable from y to t in this last integral, the integral with its coefficient becomes

$$-e^{-bs} \int_0^b e^{-st}f(t)dt \ .$$

Substituting this result back into equality (d), we obtain

$$L\{f(t)\} = \frac{\int_0^b e^{-st}f(t)dt - e^{-bs}\int_0^b e^{-st}f(t)dt}{1 - e^{-2bs}}$$

$$= \frac{(1-e^{-bs}) \int_0^b e^{-st}f(t)dt}{1 - e^{-2bs}} \ .$$

Factoring the denominator,

$$L\{f(t)\} = \frac{(1-e^{-bs}) \int_0^b e^{-st}f(t)dt}{(1-e^{-bs})(1 + e^{-bs})}$$

$$= \frac{\int_0^b e^{-st}f(t)dt}{1 + e^{-bs}} \ .$$

• **PROBLEM 5-43**

Find the Laplace transform $L\{f(t)\}$, where $f(t)$ is the function graphed in the accompanying figure.

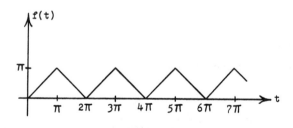

Solution: Note that $f(t)$ is a periodic function with period $P = 2\pi$; i.e., $f(t+2\pi) = f(t)$ for all t. On the interval $0 \leq t \leq 2\pi$, $f(t)$ can be expressed in analytic form:

$$f(t) = \begin{bmatrix} t, & 0 \leq t \leq \pi \\ 2\pi-t, & \pi < t \leq 2\pi \end{bmatrix} \ .$$

Using the property that

$$L\{g(t)\} = \frac{\int_0^P e^{-st}g(t)dt}{1 - e^{-Ps}},$$

where $g(t)$ is periodic with period P, we obtain

$$L\{f(t)\} = \frac{\int_0^{2\pi} e^{-st}f(t)dt}{1 - e^{-2\pi s}}$$

$$= \frac{\int_0^{\pi} e^{-st}(t)dt + \int_{\pi}^{2\pi} e^{-st}(2\pi - t)dt}{1 - e^{-2\pi s}}$$

$$= \frac{\int_0^{\pi} te^{-st}dt - \int_{\pi}^{2\pi} te^{-st}dt + 2\pi \int_{\pi}^{2\pi} e^{-st}dt}{1 - e^{-2\pi s}}.$$

Integrating by parts,

$$L\{f(t)\} = \frac{\left(\frac{-te^{-st}}{s} - \frac{e^{-st}}{s^2}\right)\Big|_{t=0}^{\pi} - \left(\frac{-te^{-st}}{s} - \frac{e^{-st}}{s^2}\right)\Big|_{t=\pi}^{2\pi} - \left(\frac{2\pi e^{-st}}{s}\right)\Big|_{t=\pi}^{2\pi}}{1 - e^{-2\pi s}}$$

$$= \frac{\left(\frac{-\pi e^{-\pi s}}{s} + \frac{1 - e^{-\pi s}}{s^2}\right) + \left(\frac{2\pi e^{-2\pi s} - \pi e^{-\pi s}}{s} + \frac{e^{-2\pi s} - e^{-\pi s}}{s^2}\right)}{1 - e^{-2\pi s}}$$

$$+ \frac{\left(\frac{2\pi e^{-\pi s} - 2\pi e^{-2\pi s}}{s}\right)}{1 - e^{-2\pi s}}$$

$$= \frac{1 - 2e^{-\pi s} + e^{-2\pi s}}{s^2(1 - e^{-2\pi s})}$$

$$= \frac{(1 - e^{-\pi s})^2}{s^2(1 + e^{-\pi s})(1 - e^{-\pi s})}$$

$$= \frac{1}{s^2}\left(\frac{1 - e^{-\pi s}}{1 + e^{-\pi s}}\right).$$

Multiplying both the numerator and denominator in the last fraction by $e^{\pi s/2}$,

$$L\{f(t)\} = \frac{1}{s^2}\left(\frac{e^{\pi s/2} - e^{-\pi s/2}}{e^{\pi s/2} + e^{-\pi s/2}}\right)$$

$$= \frac{1}{s^2}\tanh\frac{\pi s}{2}.$$

● **PROBLEM 5-44**

Given a saw-tooth waveform with period T in the figure. Determine its Laplace transform.

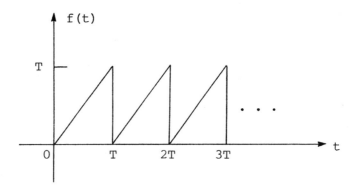

<u>Solution</u>: For a function f(t) which is periodic, i.e. $f(t+T) = f(t)$, the Laplace transform can be obtained as follows:

$$L[f(t)] = \frac{1}{1-e^{-sT}}\left[\int_0^T f(t)e^{-sT}\,dt\right]$$

$$= \frac{1}{1-e^{-sT}}\left(\int_0^T t\,e^{-st}\,dt\right)$$

$$= \frac{1}{1-e^{-sT}}\left(\frac{te^{-st}}{-s} - \int \frac{e^{-st}}{-s}\,dt\right)_0^T$$

$$= \frac{1}{1-e^{-sT}}\left[\frac{te^{-st}}{-s} + \frac{1}{s}\left(\frac{-1}{s}\right)e^{-st}\right]_0^T$$

$$= \frac{1}{1-e^{-sT}}\left[\frac{-t}{s}e^{-st} - \frac{1}{s^2}e^{-st}\right]_0^T$$

$$= \frac{1}{1-e^{-sT}}\left\{\left[-\frac{T}{s}e^{-sT} - \frac{1}{s^2}e^{-sT}\right] - \left[0 - \frac{1}{s^2}\right]\right\}$$

$$= \frac{1}{1-e^{-sT}}\left[-\frac{T}{s}e^{-sT} + \frac{1}{s^2}(1-e^{-sT})\right]$$

$$= \frac{1}{1-e^{-sT}}\left[\frac{-sTe^{-sT} + (1-e^{-sT})}{s^2}\right]$$

$$= \frac{1-(1+sT)e^{sT}}{s^2(1-e^{-sT})}$$

INITIAL AND FINAL VALUE THEOREMS

● **PROBLEM 5-45**

Given the Laplace transform of $i(t)$,

$$L[i(t)] = I(s) = \frac{E - V(0^+)}{R\left(s + \frac{1}{RC}\right)}$$

Determine $i(0^+)$.

Solution: $i(0^+)$ can be determined from $I(s)$ by using the Initial-Value Theorem.

i.e., $\quad \lim_{t \to 0^+} i(t) = i(0^+) = \lim_{s \to \infty} sI(s)$

Hence, $i(0^+) = \lim_{s \to \infty} \dfrac{s[E-v(0^+)]}{R\left(s + \frac{1}{RC}\right)}$

$\qquad = \lim_{s \to \infty} \dfrac{s[E-v(0^+)]}{Rs\left(1 + \frac{1}{sRC}\right)}$

$\qquad = \lim_{s \to \infty} \dfrac{[E-v(0^+)]}{R\left(1 + \frac{1}{sRC}\right)}$

$\therefore \quad i(0^+) = \dfrac{E-v(0^+)}{R}$

● **PROBLEM 5-46**

(a) Verify the equation $\quad I(s) = \dfrac{\frac{E}{s} + Li(0^+)}{Ls+R}$

(b) Determine $e_0(0^+)$ from the Laplace transform of $e_0(t)$, given by

$$E_0(s) = \frac{E - K}{s + \frac{1}{RC}}$$

Solution: (a) By applying the initial-value theorem

i.e. $i(0^+) = \lim_{s \to \infty} sI(s)$

$\lim_{s \to \infty} sI(s) = \lim_{s \to \infty} \dfrac{s\left[\frac{E}{s} + Li(0^+)\right]}{(Ls + R)}$

$$= \lim_{s \to \infty} \frac{s\left[\frac{E}{s} + Li(0^+)\right]}{s\left(L + \frac{R}{s}\right)}$$

$$= \lim_{s \to \infty} \frac{\left[\frac{E}{s} + Li(0^+)\right]}{L + \frac{R}{s}}$$

$$= \frac{Li(0^+)}{L} = i(0^+)$$

(b) $e_0(0^+)$ is obtained in the same way as above.

Hence, $$e_0(0^+) = \lim_{s \to \infty} sE_0(s)$$

$$= \lim_{s \to \infty} s \frac{(E-K)}{\left(s + \frac{1}{RC}\right)}$$

$$= \lim_{s \to \infty} \frac{s(E-K)}{s\left(1 + \frac{1}{sRC}\right)}$$

$$= E-K$$

CONVOLUTION

• **PROBLEM 5-47**

Let $F(k) = \Phi\{f(x)\}, G(k) = \Phi\{g(x)\}$ and suppose $F(k)G(k) = \Phi\{h(x)\}$. Prove the convolution theorem for Fourier transforms:

If $g(x)$ and $F(k)$ are absolutely integrable on $(-\infty, \infty)$ and if the Fourier inversion integral for $f(x)$ is valid for all x except possibly a countably infinite number of points, then

$$h(x) = (f * g),$$

where $(f * g)$ is the convolution of f and g defined by

$$(f * g) = \frac{1}{\sqrt{2\pi}} \int_{-\infty}^{\infty} f(\xi) \, g(x-\xi) \, d\xi \, . \tag{1}$$

Solution: Let Φ^{-1} denote the inverse Fourier transform. Then by definition

$$\Phi^{-1}\{F(k)G(k)\} = \frac{1}{\sqrt{2\pi}} \int_{-\infty}^{\infty} F(k)G(k) e^{-ikx} dk . \tag{2}$$

Using the definition of the Fourier transform of $g(\xi)$,

$$G(k) = \Phi\{g(\xi)\} = \frac{1}{\sqrt{2\pi}} \int_{-\infty}^{\infty} g(\xi) e^{ik\xi} d\xi$$

in (2) gives

$$\Phi^{-1}\{F(k)G(k)\} = \frac{1}{2\pi} \int_{-\infty}^{\infty} F(k) e^{-ikx} dk$$

$$\times \int_{-\infty}^{\infty} g(\xi) e^{ik\xi} d\xi . \tag{3}$$

The assumption that $g(x)$ and $F(k)$ are absolutely integrable on $(-\infty,\infty)$ means that the integrals

$$\int_{-\infty}^{\infty} |F(k)| dk , \quad \int_{-\infty}^{\infty} |g(x)| dx$$

are convergent so that

$$\int_{-\infty}^{\infty} F(k) e^{-ikx} dx , \quad \int_{-\infty}^{\infty} g(x) e^{-ikx} dx$$

are absolutely convergent (since

$$|F(k) e^{-ikx}| = |F(k)|$$

and

$$|g(x) e^{-ikx}| = |g(x)|).$$

Under these conditions the order of integration in (3) may be interchanged giving

$$\Phi^{-1}\{F(k)G(k)\} = \frac{1}{2\pi} \int_{-\infty}^{\infty} g(\xi) d\xi \int_{-\infty}^{\infty} F(k) e^{-ik(x-\xi)} dk . \tag{4}$$

Since the Fourier inversion integral is valid,

$$\frac{1}{\sqrt{2\pi}} \int_{-\infty}^{\infty} F(k) e^{-ik(x-\xi)} dk = f(x-\xi)$$

and using this result in (4) gives

$$\phi^{-1}\{F(k)G(k)\} = \frac{1}{\sqrt{2\pi}} \int_{-\infty}^{\infty} g(\xi) f(x-\xi) d\xi . \tag{5}$$

Noting that

$$\phi^{-1}\{F(k)G(k)\} = h(x)$$

and using the definition of convolution in (1), (5) becomes

$$h(x) = (f * g)$$

and the theorem is proved. The validity of (5) is assured even if the Fourier inversion integral for $f(x)$ has a countably infinite number of discrepancies with $f(x)$ since this will not affect the equality of the integrals

$$\frac{1}{\sqrt{2\pi}} \int_{-\infty}^{\infty} g(\xi) f(x-\xi) d\xi$$

and

$$\frac{1}{\sqrt{2\pi}} \int_{-\infty}^{\infty} g(\xi) \left(\frac{1}{\sqrt{2\pi}} \int_{-\infty}^{\infty} F(k) e^{-ik(x-\xi)} dk \right) d\xi$$

• PROBLEM 5-48

Prove that $f(t)*g(t) = g(t)*f(t)$, where the asterisk indicates convolution.

Solution: By the definition of convolution,

$$f(t)*g(t) = \int_0^t f(x)g(t-x)dx. \tag{a}$$

Define a new dummy variable $y = t-x$; hence $x = t-y$. Substituting $(t-y)$ for x in equality (a),

$$f(t)*g(t) = \int_{t-y=0}^{t-y=t} f(t-y)g(t-[t-y])d(t-y)$$

$$= \int_t^0 f(t-y)g(y)(-dy)$$

$$= -\int_t^0 g(y)f(t-y)dy$$

$$= \int_0^t g(y)f(t-y)dy$$

$$= g(t)*f(t).$$

• **PROBLEM 5-49**

Verify the theorem that $f(t)*g(t) = g(t)*f(t)$, where an asterisk indicates convolution, using
$$f(t) = e^{3t} \quad \text{and} \quad g(t) = e^{2t}$$

Solution: By the definition of convolution,

$$f(t)*g(t) = \int_0^t f(x)g(t-x)dx$$

$$= \int_0^t e^{3x} e^{2(t-x)} dx$$

$$= e^{2t} \int_0^t e^x dx = e^{2t} e^x \Big|_{x=0}^t$$

$$= e^{3t} - e^{2t}.$$

Again by the definition of convolution,

$$g(t)*f(t) = \int_0^t e^{2x} e^{3(t-x)} dx$$

$$= e^{3t} \int_0^t e^{-x} dx = -e^{3t} e^{-x} \Big|_{x=0}^t$$

$$= e^{3t} - e^{2t};$$

thus the theorem is verified.

• **PROBLEM 5-50**

Prove that
$$f(t)*[g(t) + h(t)] = [f(t)*g(t)] + [f(t)*h(t)],$$
where the symbol '*' indicates convolution.

Solution: By the definition of convolution,

$$f(t)*[g(t) + h(t)] = \int_0^t f(x)[g(t-x) + h(t-x)]dx$$

$$= \int_0^t [f(x)g(t-x) + f(x)h(t-x)]dx$$

$$= \int_0^t f(x)g(t-x)dx + \int_0^t f(x)h(t-x)dx$$

$$= [f(t)*g(t)] + [f(t)*h(t)].$$

● **PROBLEM** 5-51

Find $f(t)*g(t)$, where $f(t) = t$ and $g(t) = t^2$, and the asterisk indicates convolution.

<u>Solution</u>: By the definition of convolution,

$$f(t)*g(t) = \int_0^t f(x)g(t-x)dx$$

$$= \int_0^t x \cdot (t-x)^2 dx$$

$$= \int_0^t (t^2 x - 2tx^2 + x^3)dx$$

$$= t^2 \int_0^t x\,dx - 2t \int_0^t x^2\,dx + \int_0^t x^3\,dx$$

$$= \left(\frac{t^2 x^2}{2}\right)\Big|_{x=0}^t - \left(\frac{2tx^3}{3}\right)\Big|_{x=0}^t + \left(\frac{x^4}{4}\right)\Big|_{x=0}^t$$

$$= \frac{t^4}{2} - \frac{2t^4}{3} + \frac{t^4}{4} = \frac{t^4}{12}.$$

● **PROBLEM** 5-52

Find $f_1(t)*f_2(t)$ if $f_1(t) = 2e^{-4t}u(t)$ and $f_2(t) = 5\cos 3t\, u(t)$.

<u>Solution</u>: The convolution is represented by $f_1(t)*f_2(t)$ or "$f_1(t)$ asterisk $f_2(t)$" and it means

$$f_1(t)*f_2(t) = \int_{-\infty}^{+\infty} f_2(\tau)f_1(t-\tau)d\tau = \int_{-\infty}^{+\infty} f_1(\tau)f_2(t-\tau)d\tau \quad (1)$$

In this problem

$$f_1(t) = 2e^{-4t}u(t)$$

$$f_2(t) = 5\cos 3t\, u(t).$$

A function $g(t)$ when multiplied by the unit step function $u(t)$ means that $g(t)$ "turns on" at $t = 0$, or

$f_1(t) = 0$ for $t < 0$

$f_1(t) = 2e^{-4t}$ for $t \geq 0$

$f_2(t) = 0$ for $t < 0$

$f_2(t) = 5\cos 3t$ for $t \geq 0$

Now if the convolution of $f_1(t)$ and $f_2(t)$ is computed using (1),

$$f_1(t)*f_2(t) = \int_{-\infty}^{\infty} 5 \cos 3\tau \, u(\tau) \cdot 2e^{-4(t-\tau)} u(t-\tau) d\tau \quad (2)$$

and

$$u(t-\tau) = 1 \quad \text{for } (t-\tau) \geq 0 \text{ or } \tau \leq t \quad (3)$$
$$u(t-\tau) = 0 \quad \text{for } (t-\tau) < 0 \text{ or } \tau > t \quad (4)$$
$$u(\tau) = 1 \quad \text{for } \tau \geq 0 \quad (5)$$
$$u(\tau) = 0 \quad \text{for } \tau < 0 \quad (6)$$

Therefore, combining (2) through (6)

$$\begin{aligned} f_1(t)*f_2(t) &= \int_0^t 5 \cos 3\tau \cdot 2e^{-4(t-\tau)} d\tau \\ &= 10e^{-4t} \int_0^t e^{4\tau} \cos 3\tau \, d\tau \\ &= \frac{10e^{-4t}}{16+9} \left. e^{4\tau}(4 \cos 3\tau + 3 \sin 3\tau) \right|_0^t \\ &= 1.6 \cos 3t + 1.2 \sin 3t - 1.6e^{-4t} \end{aligned}$$

or

$$f_1(t)*f_2(t) = (1.6 \cos 3t + 1.2 \sin 3t - 1.6e^{-4t})u(t)$$

● **PROBLEM** 5-53

Find $f(t)*g(t)$ in Fig.1(a) and $f(t)*g(t)$ in Fig.1(b).

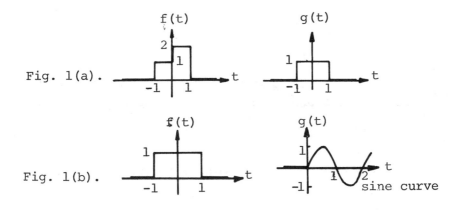

Fig. 1(a).

Fig. 1(b).

Solution: The definition of convolution of two functions is given as follows

$$f(t)*g(t) = \int_{-\infty}^{+\infty} f(u)g(t-u)du \qquad (1)$$

(a) By inspection of the given graph, it is known that

$f(u) = 0$	for	$-\infty < u < -1$
$f(u) = 1$	for	$-1 \leq u < 0$
$f(u) = 2$	for	$0 \leq u < 1$
$f(u) = 0$	for	$1 < u$
$g(t-u) = 0$	for	$-\infty < (t-u) < -1$
$g(t-u) = 1$	for	$-1 \leq (t-u) \leq +1$
$g(t-u) = 0$	for	$+1 < (t-u)$

Therefore, when Eq.(1) is calculated, by inspection there are following significant check points.

$t = -2$	$g(t-u)$ touches $f(u)$.
$t = -1$	a half of $g(t-u)$ overlaps with $f(u)$
$t = 0$	entire range of $g(t-u)$ coincides with $f(u)$
$t = +1$	a left half of $g(t-u)$ overlaps with $f(u)$
$t = +2$	the left edge of $g(t-u)$ touches with the right edge of $f(u)$

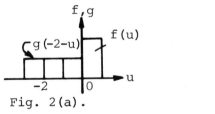
Fig. 2(a).

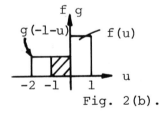
Fig. 2(b).

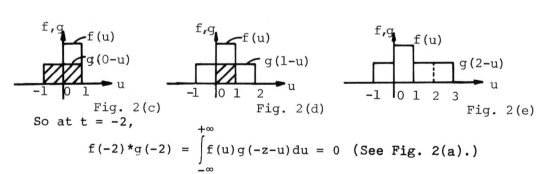
Fig. 2(c)   Fig. 2(d)   Fig. 2(e)

So at $t = -2$,

$$f(-2)*g(-2) = \int_{-\infty}^{+\infty} f(u)g(-z-u)du = 0 \quad \text{(See Fig. 2(a).)}$$

at $t = -1$,
$$f(-1)*g(-1) = \int_{-\infty}^{+\infty} f(u)g(-1-u)\,du = \int_{-1}^{0} 1 \times 1\,du = [u]_{-1}^{0} = 1$$
(See Fig. 2(b).)

at $t = 0$,
$$f(0)*g(0) = \int_{-\infty}^{\infty} f(u)g(0-u)\,du = \int_{-1}^{0} 1 \times 1\,du + \int_{0}^{1} 2 \times 2\,du$$
(See Fig. 2(c).)
$$= [u]_{-1}^{0} + [2u]_{0}^{1} = +1 + 2 = 3$$

at $t = 1$,
$$f(1)*g(1) = \int_{-\infty}^{\infty} f(u)g(1-u)\,du = \int_{0}^{1} 2 \times 1\,du = [2u]_{0}^{1} = 2$$
(See Fig. 2(d).)

at $t = 2$,
$$f(2)*g(2) = \int_{-\infty}^{+\infty} f(u)g(2-u)\,du = 0$$
(See Fig. 2(e).)

So the plot is given by

$f(-2)*g(-2)$ ... 0
$f(-1)*g(-1)$ ... 1
$f(0)*g(0)$ ... 3
$f(+1)*g(+1)$ ... 2
$f(+2)*g(+2)$ ... 0

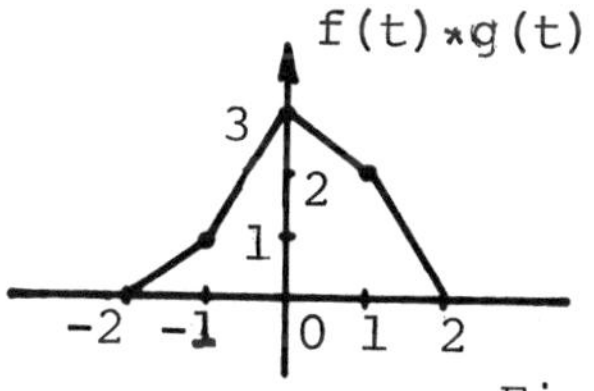

(See Fig. 3)

Fig. 3.

(b) By definition, the convolution of $f(t)$ and $g(t)$ is
$$f(t)*g(t) \equiv \int_{-\infty}^{+\infty} f(u)g(t-u)\,du \qquad (2)$$

By inspection of Fig. 1 (b)

$$f(u) = 0 \quad \text{for} \quad -\infty \le u < -1$$
$$\phantom{f(u) = 0 \quad \text{for} \quad } 1 < u \le +\infty \qquad (3)$$

$$f(u) = 1 \quad \text{for} \quad -1 \le u \le +1 \qquad (4)$$

$$g(t-u) = 0 \quad \text{for} \quad -\infty \le (t-u) \le 0 \qquad (5)$$

$$g(t-u) = \sin\pi(t-u) \quad \text{for} \quad 0 \le (t-u) \le +\infty \qquad (6)$$

The situation may be illustrated by Fig. 4.

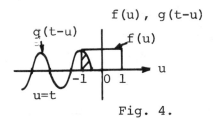
Fig. 4.

Therefore

$$f(t)*g(t) = \int_{-\infty}^{+\infty} f(u)g(t-u)\,du$$

$$= \int_{-\infty}^{-1} f(u)g(t-u)\,du + \int_{-1}^{+1} f(u)g(t-u)\,du + \int_{+1}^{+\infty} f(u)g(t-u)\,du$$

$$= \int_{-\infty}^{1} 0\cdot\sin\pi(t-u)\,du + \int_{-1}^{+t} 1\cdot\sin\pi(t-u)\,du + \int_{+1}^{+\infty} 0\cdot\sin\pi(t-u)\,du$$

$$= 0 + \left[\frac{-\cos\pi(t-u)}{-1}\right]_{-1}^{+t} + 0$$

$$= 0 + \frac{1}{\pi}[\cos\pi(t-t) - \cos\pi(t+1)] + 0 \qquad (7)$$

The convolution is zero for t<-1 and t>1 but in the range -1≤t<+1

$$f(t)*g(t) = \frac{1}{\pi}[1 - \cos\pi(t+1)]$$

This is an even function of t

t	Cosπ(t+1)	f(t)*g(t)
0	-1	$\frac{2}{\pi}$
1	+1	0

Fig. 5.

The sketch of f(t)*g(t) is shown in Fig. 5

● **PROBLEM 5-54**

Use the convolution integral to find the inverse Laplace transform of $F_1(s)F_2(s) =:$ (a) $(\frac{1}{s})(\frac{1}{s})$; (b) $(\frac{1}{s})(\frac{1}{s})^2$; (c) $(\frac{1}{s})(\frac{1}{s+1})$.

Solution: The convolution integral can be written as

$$\int_0^t f_1(\tau)f_2(t-\tau)d\tau = L^{-1}[F_1(s)F_2(s)]$$

where the operator L^{-1} indicates the inverse Laplace transform and

$$F_1(s) = L[f_1(t)]$$

$$F_2(s) = L[f_2(t)].$$

(a) $F_1(s)F_2(s) = \frac{1}{s}\frac{1}{s}$

Let

$$F_1(s) = \frac{1}{s}$$

$$F_2(s) = \frac{1}{s}.$$

The inverse transform of $\frac{1}{s}$ is 1 so

$$f_1(t) = 1 \quad \text{for} \quad t > 0$$

$$f_2(t) = 1 \quad \text{for} \quad t > 0$$

Then

$$L^{-1}[F_1(s)F_2(s)] = \int_0^t f_1(\tau)f_2(t-\tau)d\tau$$

$$L^{-1}[\frac{1}{s}\frac{1}{s}] = \int_0^t (1)(1)d\tau = \tau\Big|_0^t = t \quad \text{for } t > 0$$

or using step function notation, $u(t)$,

$$L^{-1}[\frac{1}{s}\frac{1}{s}] = t\, u(t) \quad \text{for all } t.$$

(b) $F_1(s)F_2(s) = \frac{1}{s}(\frac{1}{s})^2$

Let

$$F_1(s) = \frac{1}{s}$$

$$F_2(s) = \frac{1}{s^2}$$

Then

$$f_1(t) = 1$$

$$f_2(t) = t.$$

The inverse transform of $\frac{1}{s}(\frac{1}{s})^2$ is

$$L^{-1}[\frac{1}{s}(\frac{1}{s})^2] = \int_0^t f_1(\tau) f_2(t-\tau) d\tau$$

$$= \int_0^t (1)(t-\tau) d\tau$$

$$= (t\tau - \frac{\tau^2}{2})\Big|_0^t$$

$$= t^2 - \frac{1}{2} t^2$$

$$= \frac{1}{2} t^2 \quad \text{for } t > 0$$

$$= \frac{1}{2} t^2 u(t) \quad \text{for all } t.$$

(c) $F_1(s) F_2(s) = \frac{1}{s} \frac{1}{s+1}$.

Let

$$F_1(s) = \frac{1}{s}$$

$$F_2(s) = \frac{1}{s+1}.$$

Then

$$f_1(t) = 1$$

$$f_2(t) = e^{-t}.$$

The inverse transform of $\frac{1}{s} \frac{1}{s+1}$ is

$$L^{-1}[\frac{1}{s} \frac{1}{s+1}] = \int_0^t f_1(\tau) f_2(t-\tau) d\tau$$

$$= \int_0^t (1) e^{-(t-\tau)} d\tau$$

$$= e^{-t} \int_0^t e^\tau d\tau$$

$$= e^{-t} e^\tau \Big|_0^t$$

$$= e^{-t}(e^t - 1)$$

$$= 1 - e^{-t} \quad \text{for } t > 0$$

$$= (1 - e^{-t})u(t) \quad \text{for all } t.$$

• **PROBLEM** 5-55

Determine the Laplace inverse of

$$H(s) = \frac{10}{s(s+5)}$$

(a) by analytical convolution and

(b) by graphical convolution.

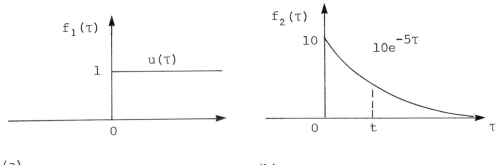

(a)   (b)

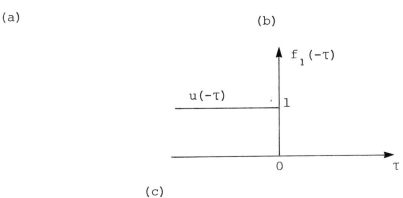

(c)

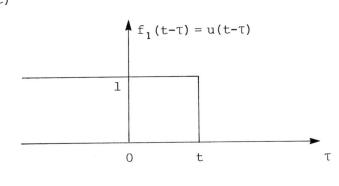

(d)

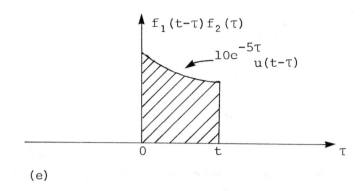

(e)

Solution: Let h(t) = Laplace inverse of H(s) and
H(s) = $F_1(s) F_2(s)$ where

$$F_1(s) = \frac{1}{s} \text{ and } F_2(s) = \frac{10}{s+5}$$

So that $f_1(t) = L^{-1}[F_1(s)] = L^{-1}\left[\frac{1}{s}\right] = u(t)$ \hfill (1)

where u(t) is a unit step function

$$f_2(t) = L^{-1}[F_2(s)] = L^{-1}\left[\frac{10}{s+5}\right] = 10e^{-5t} \quad (2)$$

(Note: Eqs.(1) and (2) are obtained by knowing Laplace transform pair.) Hence,

$$h(t) = L^{-1}[H(s)] = L^{-1}[F_1(s)F_2(s)] = f_1(t)*f_2(t)$$

where * designates convolution.

(a) Obtain h(t) analytically as follows:

$$h(t) = f_1(t) * f_2(t)$$

$$= \int_0^t f_1(\tau) f_2(t-\tau) d\tau$$

$$= \int_0^t f_1(t-\tau) f_2(\tau) d\tau$$

From eqs.(1) and (2),

$$f_1(\tau) = u(\tau)$$

and $$f_2(t-\tau) = 10e^{-5(t-\tau)}$$

Then,

$$h(t) = \int_0^t u(\tau) 10e^{-5(t-\tau)} d\tau$$

$$= 10 \int_0^t (1) e^{-5(t-\tau)} d\tau$$

$$= \left[(10) \frac{e^{-5(t-\tau)}}{5} \right]_{\tau=0}^{\tau=t}$$

$$= 2(1 - e^{-5t})$$

(b) Graphically, using the formula

$$h(t) = \int_0^t f_1(t-\tau) f_2(\tau) d\tau$$

The graphical construction of $f_1(t-\tau) f_2(t) = 10 e^{-5\tau} u(t-\tau)$ is shown in Fig. 1.

Since
$$h(t) = \int_0^t f_1(t-\tau) f_2(\tau) d\tau = \text{Area of the hatched region in Fig. 1(e)}.$$

Hence,
$$h(t) = \int_0^t 10 e^{-5\tau} dt$$

$$= 10 \left(\frac{e^{-5\tau}}{-5} \right)_0^t$$

$$= -2 \left(e^{-5t} - 1 \right)$$

$$= 2(1 - e^{-5t})$$

● **PROBLEM** 5-56

Find the inverse Laplace transform

$$L^{-1}\left\{\frac{1}{(s-1)^2}\right\},$$

using the convolution.

Solution: The function $1/(s-1)^2$ can be expressed as a product, $f(s)g(s)$, where $f(s) = g(s) = 1/(s-1)$.

From a table of Laplace transforms, we find that

$$L^{-1}\left\{\frac{1}{s-b}\right\} = e^{bt},$$

where b is a constant; hence, taking b = 1, we find

$$L^{-1}\left\{\frac{1}{s-1}\right\} = e^t.$$

Using the formula

$$L^{-1}\{f(s)g(s)\} = L^{-1}\{f(s)\} * L^{-1}\{g(s)\},$$

where the '*' symbol indicates convolution, provided that $L^{-1}\{f(s)\}$ and $L^{-1}\{g(s)\}$ exist, we obtain

$$\begin{aligned}
L^{-1}\left\{\frac{1}{(s-1)^2}\right\} &= L^{-1}\{f(s)g(s)\} \\
&= L^{-1}\{f(s)\} * L^{-1}\{g(s)\} \\
&= e^t * e^t \\
&= \int_0^t e^x e^{t-x}\, dx \\
&= e^t \int_0^t dx = xe^t\Big|_{x=0}^t = te^t
\end{aligned}$$

• **PROBLEM** 5-57

Find the inverse Laplace transform

$$L^{-1}\left\{\frac{1}{s(s^2+1)}\right\},$$

using the convolution.

Solution: The function $1/[s(s^2+1)]$ can be expressed as a product, $f(s)g(s)$, where $f(s) = 1/s$ and $g(s) = 1/(s^2+1)$. From a table of Laplace transforms we find that

$$L^{-1}\{1/s\} = 1 \quad \text{and} \quad L^{-1}\{1/(s^2+1)\} = \sin t.$$

Using the formula

$$L^{-1}\{f(s)g(s)\} = L^{-1}\{f(s)\} * L^{-1}\{g(s)\},$$

where the * symbol indicates convolution, provided that $L^{-1}\{f(s)\}$ and $L^{-1}\{g(s)\}$ exist, we obtain

$$\begin{aligned}
L^{-1}\left\{\frac{1}{s(s^2+1)}\right\} &= L^{-1}\{f(s)g(s)\} \\
&= L^{-1}\{f(s)\} * L^{-1}\{g(s)\} \\
&= (1) * (\sin t) \\
&= \int_0^t 1 \cdot \sin(t-x)\, dx.
\end{aligned} \qquad (a)$$

But the operation of convolution is commutative; hence,

$$\begin{aligned}
L^{-1}\left\{\frac{1}{s(s^2+1)}\right\} &= (\sin t)*(1) \\
&= \int_0^t (\sin x)(1)\, dx.
\end{aligned} \qquad (b)$$

We choose to evaluate integral (b), because it is slightly simpler than integral (a). Thus,

$$L^{-1}\left\{\frac{1}{s(s^2+1)}\right\} = \int_0^t (\sin x)(1)dx$$

$$= \int_0^t \sin x \, dx = -\cos x \Big|_{x=0}^t$$

$$= 1 - \cos t.$$

• **PROBLEM** 5-58

Find the inverse Laplace transform

$$L^{-1}\left\{\frac{1}{s(s^2+4)}\right\},$$

using the convolution.

Solution: The function $1/[s(s^2+4)]$ can be expressed as a product, $f(s)g(s)$, where $f(s) = 1/s$ and $g(s) = 1/(s^2+4)$.

From a table of Laplace transforms, we find that

$$L^{-1}\left\{\frac{1}{s^n}\right\} = \frac{t^{n-1}}{(n-1)!},$$

and

$$L^{-1}\left\{\frac{1}{s^2+b^2}\right\} = \frac{1}{b}\sin bt,$$

where b is a constant and n is a positive integer constant; hence, taking $b = 2$ and $n = 1$, we obtain

$$L^{-1}\left\{\frac{1}{s}\right\} = 1, \text{ and } L^{-1}\left\{\frac{1}{s^2+4}\right\} = \frac{1}{2}\sin 2t.$$

Using the formula

$$L^{-1}\{f(s)g(s)\} = L^{-1}\{f(s)\} * L^{-1}\{g(s)\},$$

where the '*' symbol indicates convolution, provided that $L^{-1}\{f(s)\}$ and $L^{-1}\{g(s)\}$ exist, we obtain

$$L^{-1}\left\{\frac{1}{s(s^2+4)}\right\} = L^{-1}\{f(s)g(s)\}$$

$$= L^{-1}\{f(s)\} * L^{-1}\{g(s)\}$$

$$= (1) * (\tfrac{1}{2}\sin 2t)$$

$$= \int_0^t 1 \cdot \tfrac{1}{2}\sin 2(t-x)dx. \quad (a)$$

But the operation of convolution is commutative; hence,

$$L^{-1}\left\{\frac{1}{s(s^2+4)}\right\} = (\tfrac{1}{2}\sin 2t) * (1)$$

$$= \int_0^t (\tfrac{1}{2}\sin 2x) \cdot (1)dx. \quad (b)$$

We choose to evaluate integral (b), because it is slightly simpler than integral (a). Thus,

$$L^{-1}\left\{\frac{1}{s(s^2+4)}\right\} = \int_0^t (\tfrac{1}{2} \sin 2x)dx$$

$$= -\tfrac{1}{4} \cos 2x \Big|_{x=0}^{t}$$

$$= \tfrac{1}{4}(1 - \cos 2t).$$

CHAPTER 6

SPECTRAL ANALYSIS

FOURIER SERIES REPRESENTATION

● **PROBLEM** 6-1

Determine the Fourier series and sketch the line frequency spectrum for each waveform shown in Fig. 1.

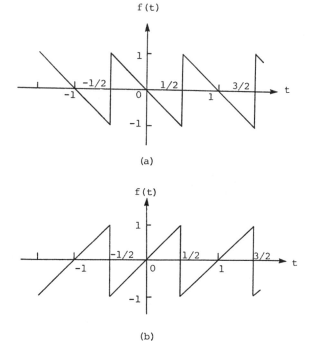

Fig. 1

Solution: Using the trigonometric form of the Fourier series, the function $f(t)$ can be written as $f(t) = a_o + \sum_{n=1}^{\infty}(a_n\cos n\omega_o t + b_n \sin \omega_o t)$, where the coefficients can be evaluated as follows:

$$a_o = \frac{1}{\tau}\int_0^{\tau} f(t)\, dt$$

$$a_n = \frac{2}{\tau}\int_0^{\tau} f(t)\cos n\omega_o t\, dt, \quad n = 1,2,3\ldots$$

$$b_n = \frac{2}{\tau}\int_0^{\tau} f(t)\sin n\omega_o t\, dt, \quad n = 1,2,3\ldots$$

Note that the function satisfies the Dirichlet conditions and can be expanded in a Fourier series. The function in Fig. 1a is an odd function since $f(-t) = -f(t)$ and is periodic with $\tau = 1$ and $\omega_1 = 2\pi$.

Therefore, for Fig.1(a), $f(t) = -2t, \left(-\frac{\tau}{2} = -\frac{1}{2}\right) \leq t \leq \left(\frac{\tau}{2} = \frac{1}{2}\right)$.

By symmetry, $a_o = a_n = 0$. Hence

$$b_n = 2\int_{-\frac{1}{2}}^{\frac{1}{2}}(-2t)\sin 2\pi nt\, dt.$$

Now compute the integral: from a table of integrals, the form is

$$\int x(\sin ax)dx = \frac{1}{a^2}\sin ax - \frac{x}{9}\cos ax.$$

Let $x = t$, $a = 2\pi n$.

Then, $-4\int_{-\frac{1}{2}}^{\frac{1}{2}} t\sin(2\pi nt)\, dt = -4\left[\frac{t}{(2\pi n)^2}\sin 2\pi nt \Big|_{-\frac{1}{2}}^{\frac{1}{2}} - \frac{t}{2\pi n}\cos 2\pi nt \Big|_{-\frac{1}{2}}^{\frac{1}{2}}\right]$

But, $t\sin 2\pi nt \Big|_{-\frac{1}{2}}^{\frac{1}{2}} = 0$ for all n

$t\cos 2\pi nt \Big|_{-\frac{1}{2}}^{\frac{1}{2}} = \begin{cases} -1, & n=1,3,5\ldots \\ 1, & n=2,4,6\ldots \end{cases}$

Therefore,

$$b_n = \begin{cases} -\dfrac{2}{\pi n}, & n=1,3,5\ldots \\ \dfrac{2}{\pi n}, & n=2,4,6\ldots \end{cases}$$

Hence,

$$f(t) = \frac{2}{\pi}\left(-\sin 2\pi t + \frac{1}{2}\sin 4\pi t - \frac{1}{3}\sin 6\pi t + \ldots\right). \tag{a}$$

Harmonic Amplitude	Frequency
$\dfrac{2}{\pi} = 0.64$	1
$\dfrac{1}{\pi} = 0.32$	2
$\dfrac{2}{3\pi} = 0.21$	3
$\dfrac{1}{2\pi} = 0.16$	4

Table 1.

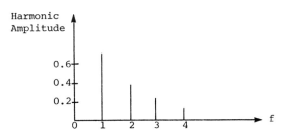

Fig. 2

For the line spectrum, $f_o = \dfrac{\omega_o}{2\pi} = 1$ and the harmonic amplitude versus the harmonic frequencies will be as shown in Table 1 or as sketched in Figure 2.

Harmonic Amplitude	Frequency
$\dfrac{2}{\pi} = 0.64$	1
$\dfrac{1}{\pi} = 0.32$	2
$\dfrac{2}{3\pi} = 0.21$	3
$\dfrac{1}{2\pi} = 0.16$	4

Table 1.
For Figure 1(b)

$f(t) = 2t$.

Therefore the procedure will be the same except for the sign, and

$$b_n = \begin{cases} \dfrac{2}{\pi n}, & n=1,3,5\ldots \\ -\dfrac{2}{\pi n}, & n=2,4,6\ldots \end{cases}$$

The line spectrum will be the same as Fig. 2.

● **PROBLEM 6-2**

Find the Fourier series over the interval $-\pi$ to π for the function x^2.

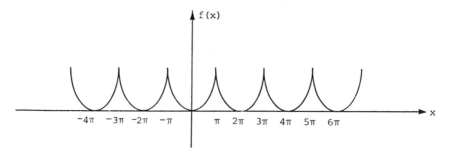

Solution: The Fourier series for x^2 in the interval $[-\pi, \pi]$ is

$$x^2 \simeq \frac{1}{2} a_0 + \sum_{n=1}^{\infty} (a_n \cos nx + b_n \sin nx) \quad (1)$$

where

$$a_n = \frac{1}{\pi} \int_{-\pi}^{\pi} x^2 \cos nx \, dx, \quad n = 0,1,2,\ldots \quad (2)$$

$$b_n = \frac{1}{\pi} \int_{-\pi}^{\pi} x^2 \sin nx \, dx \quad n = 1,2,3,\ldots \quad (3)$$

It is an interesting exercise to discuss the mathematical (not the physical) genesis of Fourier series. They arose out of the problem of expanding a given function $f(x)$ in an infinite series of orthonormal functions. Thus, assume that

$$f(x) = \sum_{n=1}^{\infty} c_n \phi_n(x) \quad (4)$$

where the functions $\phi_n(x)$ form an orthonormal system and the c_n are coefficients that must be determined. They are found to be given by the formula

$$c_n = \int_a^b f(x) \phi_n(x) r(x) dx, \quad (n = 1,2,3,\ldots) \quad (5)$$

where [a,b] is the interval over which the expansion exists and $r(x)$ is the weighting function. The series (4) with coefficients c_n as determined by (5) is called the Fourier series of $f(x)$ relative to the system $\{\phi_n\}$.

The trigonometric Fourier series form an important subset of the class of Fourier series. They are useful for expanding periodic functions and are in the solutions of various boundary-value problems in physics and engineering.

Since the given function, $f(x) = x^2$ is even, i.e., $f(x) = f(-x)$, the trigonometric Fourier series of $f(x)$ is considerably simplified. To show this, suppose that, in general, $f(x)$ is an even function. Consider

$$F_n(x) = f(x) \cos \frac{n\pi x}{L}, \quad -L \leq x \leq L.$$

$F_n(x)$ is the product of two even functions $f(x)$ and $\cos \frac{n\pi x}{L}$ (recall that $\cos(-\theta) = \cos(\theta)$). Thus $F_n(x)$ is even and hence

$$\int_{-L}^{L} F_n(x) dx = 2 \int_{0}^{L} F_n(x) dx, \quad (n = 0, 1, 2, \ldots)$$

The coefficients a_n are therefore

$$a_n = \frac{1}{L} \int_{-L}^{L} f(x) \cos \frac{n\pi x}{L} dx = \frac{2}{L} \int_{0}^{L} f(x) \cos \frac{n\pi x}{L} dx.$$

Let $G_n(x) = f(x) \sin \frac{n\pi x}{L}$. Since

$$G_n(-x) = f(-x) \sin \frac{n\pi(-x)}{L} = -f(x) \sin \frac{n\pi x}{L} = -G_n(x),$$

$G_n(x)$ is an odd function. Therefore,

$$\int_{-L}^{L} G_n(x) dx = 0, \quad (n = 1, 2, 3, \ldots)$$

and so

$$b_n = \frac{1}{L} \int_{-L}^{L} f(x) \sin \frac{n\pi x}{L} dx = 0, \quad (n = 1, 2, 3, \ldots).$$

Thus, the trigonometric Fourier series of an even function is given by

$$f(x) \simeq \frac{1}{2} a_0 + \sum_{n=1}^{\infty} a_n \cos \frac{n\pi x}{L} \tag{6}$$

where

$$a_n = \frac{2}{L} \int_{0}^{L} f(x) \cos \frac{n\pi x}{L} dx. \tag{7}$$

Letting $f(x) = x^2$, $L = \pi$ in (7)

$$a_n = \frac{2}{\pi}\int_0^\pi x^2 \cos nx \, dx, \quad (n = 0, 1, 2, \ldots) \tag{8}$$

For n = 0, the general method of solution of (8) would involve dividing by zero. Thus let n = 0 in (8) to obtain

$$a_0 = \frac{2}{\pi}\int_0^\pi x^2 \, dx = \frac{2\pi^2}{3} \tag{9}$$

For all other n, evaluate (8) using integration by parts. Thus,

$$a_n = \frac{2}{\pi}\int_0^\pi x^2 \cos nx \, dx$$

$$= \frac{2}{\pi}\left[\frac{x^2}{n}\sin nx\right]_0^\pi - \frac{4}{n\pi}\int_0^\pi x \sin nx \, dx$$

$$= \frac{4}{\pi n^2}\left[x \cos nx\right]_0^\pi - \frac{4}{\pi n^2}\int_0^\pi \cos nx \, dx$$

$$= \frac{4}{n^2}\left[\cos n\pi\right] - \frac{4}{\pi n^3}\left[\sin nx\right]_0^\pi$$

$$= \frac{4}{n^2}(-1)^n \tag{10}$$

Substituting (9) and (10) into (6), where $L = \pi$ and $f(x) = x^2$,

$$x^2 \sim \frac{\pi^2}{3} + 4\sum_{n=1}^\infty \frac{(-1)^n \cos nx}{n^2}, \quad -\pi \leq x \leq \pi. \tag{11}$$

The sum of the series (11) is sketched in the figure.

EXPONENTIAL FOURIER SERIES

• **PROBLEM 6-3**

Given a periodic square wave shown in Fig. 1, derive its exponential Fourier series.

Solution: In the interval $(0, 2\tau)$, the function $v(t)$ can be written as

$$v(t) = \begin{cases} 1 & \text{for } 0 < t < \tau \\ -1 & \text{for } \tau < t < 2\tau \end{cases}$$

The exponential Fourier series of $v(t)$ is given by:

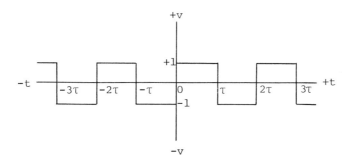

Fig. 1 Periodic square wave.

$$v(t) = \sum_{n=-\infty}^{+\infty} V_n e^{jn\omega_0 t}$$

where $\omega_0 = \frac{2\pi}{T}$

Since the period $T = 2\tau$, $\omega_0 = \frac{2\pi}{2\tau} = \frac{\pi}{\tau}$

Therefore,
$$v(t) = \sum_{n=-\infty}^{+\infty} V_n e^{jn\frac{\pi}{\tau}t}$$

The Fourier coefficients V_n is given by

$$V_n = \frac{1}{2\tau} \int_0^{2\tau} v(t) e^{-jn\omega_0 t} dt$$

$$= \frac{1}{2\tau} \left[\int_0^{\tau} (1) e^{-jn\frac{\pi}{\tau}t} dt + \int_{\tau}^{2\tau} (-1) e^{-jn\frac{\pi}{\tau}t} dt \right]$$

$$= \frac{1}{2\tau} \left[\left\{ \frac{e^{-jn\frac{\pi}{\tau}t}}{-jn\frac{\pi}{\tau}} \right\}_0^{\tau} - \left\{ \frac{e^{-jn\frac{\pi}{\tau}t}}{-jn\frac{\pi}{\tau}} \right\}_{\tau}^{2\tau} \right]$$

$$= \frac{1}{2\tau j \left(\frac{n\pi}{\tau}\right)} \left[-\left\{ e^{-jn\pi} - 1 \right\} + \left\{ e^{-j2n\pi} - e^{-jn\pi} \right\} \right]$$

$$= \frac{1}{j2n\pi} \left[1 - 2e^{-jn\pi} + e^{-j2n\pi} \right]$$

$$= \frac{1}{j2n\pi} \left[1 - e^{-jn\pi} \right]^2$$

Now knowing that $e^{-jn\pi} = \cos(n\pi) - j\sin(n\pi) = \cos n\pi = (-1)^n$

$$V_n = \frac{1}{2jn\pi} \left[1 - (-1)^n \right]^2$$

299

Three cases must be considered.

Case 1: For n = even,

$$V_n = \frac{1}{2jn\pi}\left[1 - 1\right]^2 = 0$$

Case 2: For n = odd,

$$V_n = \frac{1}{2jn\pi}\left[1 - (-1)\right]^2 = \frac{2}{jn\pi}.$$

Case 3: For n = 0, the expression for V_n becomes indeterminate. V_0 can be found directly as follows:

$$V_0 = \frac{1}{2\tau}\left[\int_0^{2\tau} v(t)dt\right] = \frac{1}{2\tau}\left[\int_0^{\tau} dt + \int_\tau^{2\tau} (-1)dt\right]$$

$$= 0.$$

Therefore,

$$V(t) = \begin{cases} \sum_{n=-\infty}^{+\infty} \frac{2}{jn\pi} e^{jn\frac{\pi}{\tau}t}, & n \text{ odd} \\ 0, & n \text{ even} \end{cases}$$

● **PROBLEM 6-4**

(a) Find the exponential Fourier series for the function g(t) shown in Fig. A.

(b) With the help of the solution obtained in part (a), find the Fourier coefficients F_n for the function f(t) shown in Fig. B.

(c) Find the smallest n (except n = 0) for which F_n is zero, given $\tau/T = 1/10$.

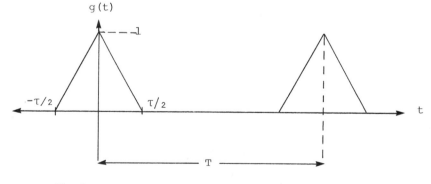

Fig. A

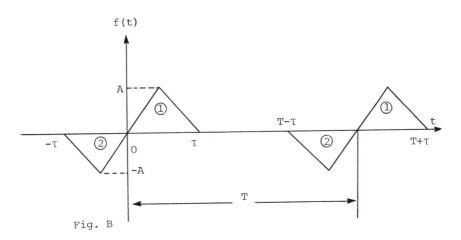

Fig. B

Solution: (a) The exponential Fourier series for $g(t)$ is

$$g(t) = \sum_{-\infty}^{+\infty} G_n e^{jn\omega_0 t}$$

where G_n is given by

$$G_n = \int_{-\tau/2}^{T-\frac{\tau}{2}} g(t) e^{-jn\omega_0 t} dt$$

From Fig. (A):

$$g(t) = \begin{cases} \frac{2}{\tau} t + 1 & , \quad -\frac{\tau}{2} < t < 0 \\ -\frac{2}{\tau} t + 1 & , \quad 0 < t < \frac{\tau}{2} \\ 0 & , \quad \frac{\tau}{2} < t < T - \frac{\tau}{2} \end{cases}$$

Therefore

$$G_n = \int_{-\tau/2}^{0} \left(\frac{2}{\tau} t + 1\right) e^{-jn\omega_0 t} dt + \int_{0}^{\tau/2} \left(-\frac{2}{\tau} t + 1\right) e^{-jn\omega_0 t} dt$$

$$= \frac{2}{\tau} \int_{-\tau/2}^{0} t e^{-jn\omega_0 t} dt - \frac{2}{\tau} \int_{0}^{\tau/2} t e^{-jn\omega_0 t} dt + \int_{-\tau/2}^{+\tau/2} e^{-jn\omega_0 t} dt$$

$$= \frac{2}{\tau} I_1 - \frac{2}{\tau} I_2 + I_3 \qquad (1)$$

where

$$I_1 = \int_{-\tau/2}^{0} t e^{-jn\omega_0 t} dt,$$

$$I_2 = \int_{0}^{\tau/2} t e^{-jn\omega_0 t} dt$$

and $\quad I_3 = \int_{-\tau/2}^{+\tau/2} e^{-jn\omega_0 t} \, dt$

Now notice that

$$\int_{-\frac{\tau}{2}}^{0} t e^{-jn\omega_0 t} \, dt = t\left(\frac{e^{-jn\omega_0 t}}{-jn\omega_0}\right) - (1)\left(\frac{e^{-jn\omega_0 t}}{-n^2\omega_0^2}\right)$$

$$= \frac{jn\omega_0 t}{n^2\omega_0^2} e^{-jn\omega_0 t} + \frac{1}{n^2\omega_0^2} e^{-jn\omega_0 t}$$

$$= \frac{1}{n^2\omega_0^2}\left(jn\omega_0 t \, e^{-jn\omega_0 t} + e^{-jn\omega_0 t}\right)$$

Hence,

$$I_1 = \frac{1}{n^2\omega_0^2}\left[jn\omega_0 t e^{-jn\omega_0 t} + e^{-jn\omega_0 t}\right]_{-\tau/2}^{0}$$

$$= \frac{1}{n^2\omega_0^2}\left[\{0 + 1\} - \left(-jn\omega_0 \frac{\tau}{2} e^{jn\omega_0 \frac{\tau}{2}} + e^{jn\omega_0 \frac{\tau}{2}}\right)\right]$$

$$= \frac{1}{n^2\omega_0^2}\left[1 + jn\omega_0 \frac{\tau}{2} e^{jn\omega_0 \frac{\tau}{2}} - e^{jn\omega_0 \frac{\tau}{2}}\right]$$

$$I_2 = \int_{0}^{\tau/2} t e^{-jn\omega_0 t} \, dt$$

$$= \frac{1}{n^2\omega_0^2}\left[jn\omega_0 t \, e^{-jn\omega_0 t} + e^{-jn\omega_0 t}\right]_{0}^{\tau/2}$$

$$= \frac{1}{n^2\omega_0^2}\left[\left\{jn\omega_0 \frac{\tau}{2} e^{-jn\omega_0 \frac{\tau}{2}} + e^{-jn\omega_0 \frac{\tau}{2}}\right\} - \{0 + 1\}\right]$$

$$= \frac{1}{n^2\omega_0^2}\left[-1 + jn\omega_0 \frac{\tau}{2} e^{-jn\omega_0 \frac{\tau}{2}} + e^{-jn\omega_0 \frac{\tau}{2}}\right]$$

$$I_3 = \int_{-\tau/2}^{+\tau/2} e^{-jn\omega_0 t} \, dt = \int_{-\tau/2}^{+\tau/2} \cos n\omega_0 t \, dt$$

$$= 2\int_{0}^{\tau/2} \cos n\omega_0 t \, dt$$

$$= 2\left[\frac{\sin n\omega_0 t}{n\omega_0}\right]_{0}^{\tau/2}$$

$$= 2\frac{\frac{\tau}{2} \sin n\omega_0 \frac{\tau}{2}}{n\omega_0 \frac{\tau}{2}}$$

$$= \tau \frac{\sin\left(n\omega_0 \frac{\tau}{2}\right)}{n\omega_0 \frac{\tau}{2}}$$

Substituting information obtained above into eq. (1)

$$G_n = \frac{2}{T} \frac{1}{n^2\omega_0^2}\left[2 + jn\omega_0\frac{\tau}{2}\left\{e^{jn\omega_0\frac{\tau}{2}} - e^{-jn\omega_0\frac{\tau}{2}}\right\} - \left\{e^{jn\omega_0\frac{\tau}{2}} + e^{-jn\omega_0\frac{\tau}{2}}\right\}\right]$$
$$+ \tau \frac{\sin n\omega_0 \frac{\tau}{2}}{n\omega_0 \frac{\tau}{2}}$$

$$= \frac{2}{T} \frac{1}{n^2\omega_0^2}\left[2 + jn\omega_0\frac{\tau}{2}\left\{2j\sin n\omega_0\frac{\tau}{2}\right\} - \left\{2\cos n\omega_0\frac{\tau}{2}\right\}\right] + \tau \frac{\sin n\omega_0\frac{\tau}{2}}{n\omega_0\frac{\tau}{2}}$$

$$= \frac{2}{T} \frac{\tau^2}{4} \frac{1}{(n\omega_0\frac{\tau}{2})^2} 2\left[(1-\cos n\omega_0\frac{\tau}{2}) - n\omega_0\frac{\tau}{2}\sin n\omega_0\frac{\tau}{2}\right] + \tau \frac{\sin n\omega_0\frac{\tau}{2}}{n\omega_0\frac{\tau}{2}}$$

$$= \tau \frac{1}{(n\omega_0\frac{\tau}{2})^2}\left[2\sin^2(n\omega_0\frac{\tau}{4}) - n\omega_0\frac{\tau}{2}\sin n\omega_0\frac{\tau}{2}\right] + \tau \frac{\sin n\omega_0\frac{\tau}{2}}{n\omega_0\frac{\tau}{2}}$$

$$= \tau \frac{1}{n^2\omega_0^2\frac{\tau^2}{4}}\left[2\sin^2(n\omega_0\frac{\tau}{4})\right]$$

$$= \frac{\tau}{2} \frac{4}{n^2\omega_0^2\frac{\tau^2}{4}} \sin^2(n\omega_0\frac{\tau}{4})$$

$$= \frac{\tau}{2} \frac{\sin^2(n\omega_0\frac{\tau}{4})}{(n\omega_0\frac{\tau}{4})^2}$$

where ω_0 = Fundamental angular frequency = $\frac{2\pi}{T}$

With $\quad n\omega_0\frac{\tau}{4} = n\left(\frac{2\pi}{T}\right)\left(\frac{\tau}{4}\right) = \frac{n\pi\tau}{2T}$

$$G_n = \frac{\tau}{2} \frac{\sin^2 \frac{n\pi\tau}{2T}}{\left(\frac{n\pi\tau}{2T}\right)^2} \qquad (1)$$

Part (B)

Comparing Figures A and B, f(t) = sum of the parts (1) and (2) indicated in Fig. B. Part (1) is similar to Fig. A except that it is translated to the right by $\tau/2$ and increased in magnitude by a factor 'A'. Hence, part (1) can be written as

$$Ag\left(t - \frac{\tau}{2}\right)$$

Part (2) is similar to Fig. B except for a translation towards the left by $\tau/2$ and an increase in magnitude by a factor $(-A)$. Hence, Part (2) can be written as

$$-Ag\left(t + \frac{\tau}{2}\right). \quad \text{So, } f(t) = Ag\left(t - \frac{\tau}{2}\right) - Ag\left(t + \frac{\tau}{2}\right)$$

Representing $f(t)$ and $g(t)$ by the exponential series and applying time shifting theorem,

$$\sum_{n=-\infty}^{+\infty} F_n e^{jn\omega_0 t} = A \sum_{n=-\infty}^{+\infty} e^{-jn\omega_0 \frac{\tau}{2}} G_n e^{jn\omega_0 t} - A \sum_{n=-\infty}^{+\infty} e^{jn\omega_0 \frac{\tau}{2}} G_n e^{jn\omega_0 t}$$

$$= \sum_{n=-\infty}^{+\infty} A \left(e^{-jn\omega_0 \frac{\tau}{2}} - e^{jn\omega_0 \frac{\tau}{2}}\right) \left[G_n e^{jn\omega_0 t}\right]$$

$$= \sum_{n=-\infty}^{+\infty} -2jA\sin\left(n\omega_0 \frac{\tau}{2}\right) \left[G_n e^{jn\omega_0 t}\right].$$

Note that the fundamental angular frequency is the same for both $f(t)$ and $g(t)$.

$$\sum_{n=-\infty}^{+\infty} F_n e^{jn\omega_0 t} = \sum_{n=-\infty}^{+\infty} -2Aj \left\{\sin\left(n\omega_0 \frac{\tau}{2}\right)\right\} G_n e^{jn\omega_0 t}$$

So, $F_n = -2Aj \left\{\sin\left(n\omega_0 \frac{\tau}{2}\right)\right\} G_n$ \hfill (2)

Substituting G_n from eq.(1) into eq.(2) gives

$$F_n = -2Aj \left\{\sin\left(n\omega_0 \frac{\tau}{2}\right)\right\} \frac{\tau}{2} \frac{\sin^2\left(\frac{n\pi\tau}{2T}\right)}{\left(\frac{n\pi\tau}{2T}\right)^2}$$

For $\omega_0 = \frac{2\pi}{T}$,

$$F_n = -j \left\{A\tau \sin\left(\frac{n\pi\tau}{T}\right)\right\} \left[\frac{\sin^2\left(\frac{n\pi\tau}{2T}\right)}{\left(\frac{n\pi\tau}{2T}\right)^2}\right]$$

Part C

If $\frac{\tau}{T} = \frac{1}{10}$, F_n first becomes zero for $n = 10$.

• **PROBLEM 6-5**

Given the function $f(t) = e^{-t}$, represent it as a complex Fourier series in the interval $(-1,1)$.

<u>Solution</u>: The function $f(t)$ is represented as an exponential Fourier series using the formula

$$f(t) = \sum_{n=-\infty}^{+\infty} C_n e^{jn\omega_0 t}$$

where

$$\omega_0 = \frac{2\pi}{T}.$$

Since the period $(T) = 2$,

$$\omega_0 = \frac{2\pi}{2} = \pi.$$

Therefore, $f(t) = \sum_{n=-\infty}^{+\infty} C_n e^{jn\pi t}$

where C_n is given by

$$C_n = \frac{1}{2}\int_{-1}^{+1}(e^{-t})e^{-jn\pi t}\, dt$$

$$= \frac{1}{2}\int_{-1}^{+1} e^{-(1+jn\pi)t}\, dt$$

$$= \frac{1}{2}\left[\frac{e^{-(1+jn\pi)t}}{-(1+jn\pi)}\right]_{-1}^{+1}$$

$$= -\frac{1}{2(1+jn\pi)}\left[e^{-(1+jn\pi)} - e^{(1+jn\pi)}\right]$$

$$= -\frac{1}{2(1+jn\pi)}\left[e^{(1+jn\pi)} - e^{-(1+jn\pi)}\right]$$

Now, $e^{jn\pi} = (\cos n\pi + j\sin n\pi) = \cos n\pi = (-1)^n$

$e^{-jn\pi} = (\cos n\pi - j\sin n\pi) = \cos n\pi = (-1)^n$

Hence, $C_n = \frac{(-1)^n}{2(1+jn\pi)}\left[e^1 - e^{-1}\right]$

$$= \frac{(-1)^n}{(1+jn\pi)}\left(\frac{e^1 - e^{-1}}{2}\right)$$

$$= \frac{(-1)^n}{(1+jn\pi)} \sinh(1)$$

$$= \frac{(-1)^n(1-jn\pi)}{(1+n^2\pi^2)} \sinh(1)$$

As a result,
$$f(t) = \sinh(1) \sum_{n=-\infty}^{+\infty} \frac{(-1)^n(1-jn\pi)}{(1+n^2\pi^2)} e^{jn\pi t}.$$

• **PROBLEM 6-6**

For the periodic gate function shown in Fig. 1, determine the spectral frequency distribution.

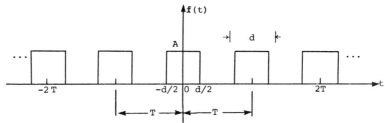

Fig. 1

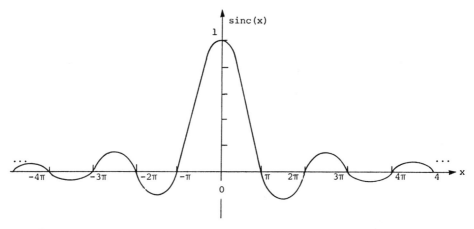

Fig. 2

Solution: The exponential Fourier series of $f(t)$ is given by

$$f(t) = \sum_{n=-\infty}^{+\infty} C_n e^{jn\omega_0 t}$$

where
$$\omega_0 = \frac{2\pi}{T}.$$

The Fourier coefficient C_n is given by

$$C_n = \frac{1}{T} \int_{-T/2}^{+T/2} f(t) e^{-jn\omega_0 t} dt$$

$$= \frac{1}{T}\int_{-d/2}^{+d/2} A\, e^{-jn\omega_0 t}\, dt$$

$$= \frac{2A}{T}\int_{0}^{d/2} \cos(n\omega_0 t)\, dt.$$

Note that f(t) is an even function. For an even function

$$C_n = \frac{2}{T}\int_{0}^{T/2} f(t)\cos(n\omega_0 t)\, dt.$$

Therefore,

$$C_n = \frac{2A}{T}\left[\frac{\sin(n\omega_0 t)}{n\omega_0}\right]_{0}^{d/2}$$

$$= \frac{2A}{T}\left[\frac{\sin(n\omega_0 \frac{d}{2})}{\frac{2}{d}n\omega_0 \frac{d}{2}}\right]$$

$$= \frac{Ad}{T}\left[\frac{\sin(n\omega_0 \frac{d}{2})}{(n\omega_0 \frac{d}{2})}\right]$$

Notice that $\frac{\sin x}{x} = \operatorname{sinc} x$ and $\omega_0 = \frac{2\pi}{T}$.

Hence,

$$C_n = \frac{Ad}{T}\left[\operatorname{sinc}(n\omega_0\, \frac{d}{2})\right]$$

$$= \frac{Ad}{T}\left[\operatorname{sinc}(n\, \frac{2\pi}{T}\, \frac{d}{2})\right]$$

$$= \frac{Ad}{T}\left[\operatorname{sinc}(n\pi\, \frac{d}{T})\right]$$

The frequency distribution can be expressed as

$$f(t) = \sum_{n=-\infty}^{+\infty}\left\{\frac{Ad}{T}\operatorname{sinC}(n\pi\, \frac{d}{T})\right\} e^{+jn\frac{2\pi}{T}t}$$

This is plotted in figure 2.

RESPONSE OF A LINEAR SYSTEM

• **PROBLEM 6-7**

Given a linear time invariant system

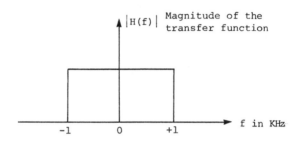

for which the transfer function and the Fourier transform of the input signal are shown in Fig. 1, determine the fraction of the energy that passes through the system.

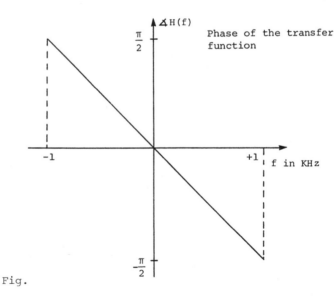

Fig.

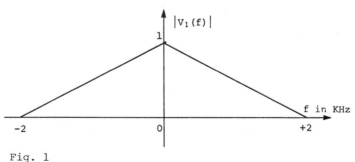

Fig. 1

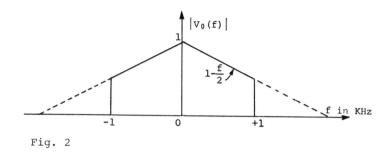

Fig. 2

Solution: For a linear time invariant system,

$$V_0(f) = H(f) V_i(f)$$

where $V_0(f)$ = Fourier transform of the output

$H(f)$ = Transfer function of the system

$V_i(f)$ = Fourier transform of the input.

The magnitude spectrum of $V_0(f)$ is shown in Fig. 2.

Applying Parseval's theorem, the energy in the signal $v_i(t)$ is given by

$$v_i(t) = \int_{-\infty}^{+\infty} |V_i(f)|^2 \, df.$$

Therefore, the fraction of the energy that passes through the system is

$$= \frac{\int_{-\infty}^{+\infty} |V_0(f)|^2 \, df}{\int_{-\infty}^{+\infty} |V_i(f)|^2 \, df}$$

$$= \frac{2\int_0^1 \left(1 - \frac{f}{2}\right)^2 df}{2\int_0^2 \left(1 - \frac{f}{2}\right)^2 df}$$

$$= \frac{\int_0^1 \left(1 - f + \frac{f^2}{4}\right) df}{\int_0^2 \left(1 - f + \frac{f^2}{4}\right) df}$$

$$= \frac{\left[f - \frac{f^2}{2} + \frac{f^3}{12}\right]_0^1}{\left[f - \frac{f^2}{2} + \frac{f^3}{12}\right]_0^2}$$

$$= \frac{\left[1 - \frac{1}{2} + \frac{1}{12}\right]}{\left[2 - 2 + \frac{8}{12}\right]} = \frac{\left[\frac{7}{12}\right]}{\left[\frac{8}{12}\right]}$$

$$= \frac{7}{8} = 0.875$$

Hence, 87.5% of the energy present in the input signal passes through the system.

• **PROBLEM** 6-8

A linear time invariant system such as a passive two port electrical network is indicated by the box in the figure. An exponential signal $v_i(t) = e^{-2t} u(t)$ is applied at the input port. Given the output in the frequency domain, how would you conclude that $V_0(\omega)$ is incorrect?

Given:

$$V_0(\omega) = \frac{3\omega}{2+j\omega}$$

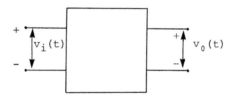

Solution: $V_0(\omega)$ = Fourier transform of the output

$$= \int_{-\infty}^{+\infty} v_0(t) e^{-j\omega t} \, dt$$

$$= \int_{-\infty}^{+\infty} v_0(t) \cos\omega t \, dt - j \int_{-\infty}^{+\infty} v_0(t) \sin\omega t \, dt$$

$$= R(\omega) - j X(\omega)$$

where

$$R(\omega) = \int_{-\infty}^{+\infty} v_0(t) \cos\omega t \, dt$$

and

$$X(\omega) = \int_{-\infty}^{+\infty} v_0(t) \sin\omega t \, dt$$

For $v_0(t)$ real, both $R(\omega)$ and $X(\omega)$ are real functions of ω. Since $R(-\omega) = R(\omega)$ and $X(-\omega) = -X(\omega)$, $R(\omega)$ is an even function of ω and $X(\omega)$ is an odd function of ω.

Now, splitting $V_0(\omega)$ into real and imaginary parts

$$V_0(\omega) = \frac{3\omega}{(2+j\omega)} \frac{(2-j\omega)}{(2-j\omega)} = \frac{6\omega - j3\omega^2}{4+\omega^2} = \frac{6\omega}{4+\omega^2} - j\frac{3\omega^2}{4+\omega^2}$$

In this case, the real part of $V_0(\omega)$ is an odd function of ω, and the imaginary part is an even function of ω. However, $V_0(t)$ is a real time function and hence it follows that $V_0(\omega) = \frac{3\omega}{2+j\omega}$ is invalid.

NORMALIZED POWER

• **PROBLEM 6-9**

An input signal $v(t)$ is applied to a low pass R-L circuit as shown in Fig. 1.

If $\quad v(t) = 2 + \cos 30t + \frac{1}{2}\sin 60t$,

(a) Determine the period of $v(t)$.

(b) Determine the mean square value of the first harmonic term of the input.

(c) Determine the mean square value of the input signal.

(d) Let the output be $y(t) = \sum_{n=-\infty}^{+\infty} Y_n e^{j30nt}$. Determine the magnitudes of Y_0, Y_1, and Y_2.

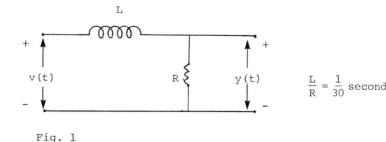

Fig. 1

Solution: (a) $v(t)$ can be written as

$$v(t) = 2 + \cos\omega_0 t + \frac{1}{2}\sin 2\omega_0 t$$

where $\omega_0 = 2\pi f_0$ and f_0 is the fundamental frequency. Since $\omega_0 = 30$, $2\pi f_0 = 30$ or

$$f_0 = \frac{30}{2\pi} = \frac{15}{\pi} \text{ Hz.}$$

Therefore, $\quad T = \text{period} = \dfrac{1}{f_0} = \dfrac{\pi}{15}$ seconds.

(b) The mean square value, or normalized power, of a sinusoidal waveform is given by one-half the square of the peak value.

Therefore, the mean square value of the first harmonic

$$= \dfrac{(1)^2}{2} = \dfrac{1}{2}.$$

(c) The total power of the input signal is the sum of the powers in each component. This is true because of the orthogonality of the terms.

The normalized power due to a d.c. term is equal to the square of that d.c. value.

Therefore, the normalized power of the input signal

$$= \underbrace{(2)^2}_{\substack{\text{d.c.}\\\text{term}}} + \underbrace{\dfrac{1}{2}(1)^2}_{\substack{\text{First}\\\text{Harmonic}\\\text{term}}} + \underbrace{\dfrac{1}{2}\left(\dfrac{1}{2}\right)^2}_{\substack{\text{Second}\\\text{Harmonic}\\\text{term}}}$$

$$= \left[4 + \dfrac{1}{2} + \dfrac{1}{8} = \dfrac{37}{8}\right] = 4.625.$$

(d) The transfer function changes the magnitude and phase of the input components.

The transfer function can be written as

$$H(jn\omega_0) = \dfrac{R}{R+jn\omega_0 L} = \dfrac{1}{1+jn\omega_0 \dfrac{L}{R}}$$

$$= \dfrac{1}{1+j\dfrac{n\omega_0}{30}} = \dfrac{1}{1+jn}$$

where ω_0 = Fundamental angular frequency = 30 $\dfrac{\text{radians}}{\text{seconds}}$

$H(j0) = 1; \quad H(j\omega_0) = \dfrac{1}{1+j}; \quad H(2j\omega_0) = \dfrac{1}{1+2j}$

$|H(J0)| = 1$

$|H(j\omega_0)| = \left|\dfrac{1}{1+j}\right| = \dfrac{1}{\sqrt{2}}$

$|H(2j\omega_0)| = \dfrac{1}{|1+2j|} = \dfrac{1}{\sqrt{5}}$

Now, $v(t)$ = Input = $2 + \cos 30t + \dfrac{1}{2}\sin 60t$

$$= 2 + \frac{1}{2}\left[e^{j30t} + e^{-j30t}\right] + \frac{1}{2(2j)}\left[e^{j60t} - e^{-j60t}\right]$$

$$= -\frac{1}{4j}e^{-j60t} + \frac{1}{2}e^{-j30t} + 2 + \frac{1}{2}e^{j30t} + \frac{1}{4j}e^{j60t}$$

$$= V_{-2}e^{-j60t} + V_{-1}e^{-j30t} + V_0 + V_1 e^{j30t} + V_2 e^{j60t}$$

Therefore, $V_0 = 2$; $|V_0| = \sqrt{2^2} = 2$

$V_{-1} = V_1 = \frac{1}{2}$; $|V_{-1}| = |V_1| = \sqrt{\frac{1}{2^2}} = \frac{1}{2}$

$V_{-2} = \frac{-1}{4j}$

$V_2 = \frac{1}{4j}$; $|V_{-2}| = |V_2| = \sqrt{\frac{1}{4^2}} = \frac{1}{4}$

Now, Output $= |Y_n| = |H(jn\omega_0)||V_n|$

Therefore, $|Y_0| = |H(j0)||V_0|$

$= (1)(2) = 2$

$|Y_1| = |H(j\omega_0)||V_1| = |H(j\omega_0)||V_{-1}|$

$= \frac{1}{\sqrt{2}} \cdot \frac{1}{2} = \frac{1}{2\sqrt{2}}$

$|Y_2| = |H(2j\omega_0)||V_2| = |H(2j\omega_0)||V_{-2}|$

$= \frac{1}{\sqrt{5}} \cdot \frac{1}{4} = \frac{1}{4\sqrt{5}}$.

• PROBLEM 6-10

A waveform m(t) is shown in Fig. 1. If it is given that $m(t) = m(t \pm T)$ and the Fourier expansion of $m(t) =$

$$\frac{2}{\pi} \sum_{n=1}^{\infty} \frac{(-1)^{n+1}}{n} \sin 2\pi n \frac{t}{T}$$

find the ratio of the normalized power contained in the first three harmonics to the total normalized power.

Solution: The Fourier expansion of the periodic waveform m(t) is given by

$$m(t) = \frac{2}{\pi} \sum_{n=1}^{\infty} \frac{(-1)^{n+1}}{n} \sin 2\pi n \frac{t}{T}$$

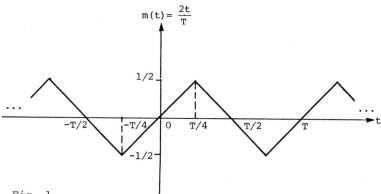

Fig. 1

$$= \frac{2}{\pi} \sum_{n=1}^{\infty} \frac{(-1)^{n+1}}{n} \frac{1}{2j} \left[e^{j2\pi n \frac{t}{T}} - e^{-j2\frac{\pi n t}{T}} \right]$$

$$= \frac{1}{\pi j} \sum_{\substack{n=-\infty \\ (n \neq 0)}}^{\infty} \frac{(-1)^{n+1}}{n} e^{j2\pi n t/T}$$

$$= \sum_{\substack{n=-\infty \\ (n \neq 0)}}^{+\infty} M_n e^{jn\omega_0 t}$$

where
$$M_n = \frac{1}{\pi j} \frac{(-1)^{n+1}}{n} \quad \text{and} \quad \omega_0 = \frac{2\pi}{T}$$

Now, S = Total normalized power = $\sum_{\substack{n=-\infty \\ (n \neq 0)}}^{+\infty} M_n M_n^*$

where M_n^* is the complex conjugate of the complex Fourier coefficient M_n

$$M_n M_n^* = \left[\frac{1}{\pi j} \frac{(-1)^{n+1}}{n} \right] \left[\frac{1}{-\pi j} \frac{(-1)^{n+1}}{n} \right]$$

$$= \frac{1}{\pi^2} \frac{\left[(-1)^{n+1}\right]^2}{n^2}$$

$$= \frac{1}{\pi^2 n^2}$$

Hence,
$$S = \sum_{\substack{n=-\infty \\ (n \neq 0)}}^{+\infty} M_n M_n^* = \sum_{\substack{n=-\infty \\ (n \neq 0)}}^{+\infty} \frac{1}{\pi^2 n^2}$$

$$= \frac{2}{\pi^2} \sum_{n=1}^{\infty} \frac{1}{n^2}$$

$$= \frac{2}{\pi^2}\left[\frac{1}{1} + \frac{1}{2^2} + \frac{1}{3^2} + \frac{1}{4^2} + \cdots\right]$$

Note: The series $\frac{1}{1^2} + \frac{1}{2^2} + \frac{1}{3^2} + \frac{1}{4^2} \cdots = \frac{\pi^2}{6}$

Therefore, $S = \frac{2}{\pi^2} \times \frac{\pi^2}{6} = \frac{1}{3} \cong 0.333$

Now for n = 1, The normalized power contained in the first harmonic is

$$S_1 = 2M_1 M_1^* = 2\frac{1}{\pi^2 1^2} = \frac{2}{\pi^2}$$

n = 2, S_2 = the normalized power contained in the second harmonic

$$= 2M_2 M_2^* = 2\frac{1}{\pi^2 2^2} = \frac{1}{2\pi^2}$$

and n = 3, S_3 = the normalized power contained in the third harmonic

$$= 2\frac{1}{\pi^2 3^2} = \frac{2}{9\pi^2}$$

Therefore, the fraction of the normalized power of this waveform which is contained in its first three harmonics

$$= \frac{S_1 + S_2 + S_3}{S} = \frac{\frac{2}{\pi^2} + \frac{1}{2\pi^2} + \frac{2}{9\pi^2}}{\frac{1}{3}}$$

$$= \frac{\frac{2}{\pi^2}\left(1 + \frac{1}{4} + \frac{1}{9}\right)}{1/3}$$

$$= 0.82 \text{ or } 82\%.$$

• **PROBLEM 6-11**

The total normalized power of a periodic signal is given by

$$S = \sum_{n=-\infty}^{n=+\infty} V_n V_n^*$$

where $V_n = \frac{1}{|n|} e^{-j\arctan(n/2)}$ for $n = \pm 1, \pm 2, \ldots$.

Calculate the ratio: $\frac{\text{normalized power(second harmonic)}}{\text{normalized power (first harmonic)}}$.

Solution: Since $V_n = \frac{1}{|n|} e^{-j\tan^{-1}(n/2)}$

$$V_n V_n^* = \left[\frac{1}{|n|} e^{-j\tan^{-1}(n/2)}\right]\left[\frac{1}{|n|} e^{j\tan^{-1}n/2}\right]$$

$$= \frac{1}{|n|^2}(1) = \frac{1}{|n|^2}.$$

Therefore,
$$S = \sum_{\substack{n=-\infty \\ (n \neq 0)}}^{n=+\infty} |V_n|^2 = \sum_{n=1}^{\infty} 2|V_n|^2 = \sum_{n=1}^{\infty} P_i$$

where P_i is the normalized power in the ith harmonic.

Thus,
$$\frac{P_2}{P_1} = \frac{2|V_2|^2}{2|V_1|^2} = \frac{2 \cdot \frac{1}{(2)^2}}{2 \cdot (1)} = \frac{1}{4} = 0.25.$$

● **PROBLEM 6-12**

For each waveform shown in Fig. 1, write an expression for the power spectral density $G(f)$.

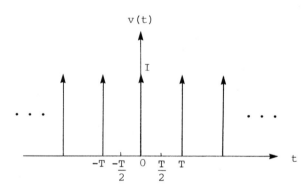

Fig. 1(A)

Solution: In Fig. 1a, an impulse train of strength I and period T is shown.

$v(t)$ can be represented as

$$v(t) = I \sum_{n=-\infty}^{+\infty} \delta(t-nT)$$

with power spectral density $G(f) = \sum_{n=-\infty}^{+\infty} |V_n|^2 \delta(f-nf_0)$

where V_n is the complex Fourier coefficient of $v(t)$.

Now, knowing that
$$V_n = \frac{1}{T}\int_{-T/2}^{T/2} v(t)e^{-jn\frac{2\pi}{T}t}\,dt$$

$$V_n = \frac{1}{T}\int_{-T/2}^{T/2} I\delta(t)e^{-jn\omega_0 t}\,dt = \frac{I}{T} \quad \text{where } \omega_0 = \frac{2\pi}{T}$$

= fundamental angular frequency

Therefore,
$$G(f) = \frac{I^2}{T^2}\sum_{n=-\infty}^{+\infty}\delta(f - nf_0).$$

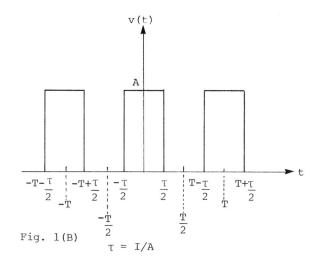

Fig. 1(B)

$\tau = I/A$

Similarly, G(f) is obtained for Fig. 1b as follows:

First, let $AP_\tau(t-a)$ represent a pulse of magnitude A, width τ and centered at $t = a$. Hence, v(t) can be represented as

$$v(t) = A\sum_{n=-\infty}^{+\infty} P_\tau(t-nT).$$

Now, V_n can be calculated as follows:

$$V_n = \frac{1}{T}\int_{-T/2}^{T/2} v(t)e^{-jn\omega_0 t}\,dt$$

$$= \frac{1}{T}\int_{-T/2}^{T/2} A\,P_\tau(t)e^{-jn\omega_0 t}\,dt$$

$$V_n = \frac{A}{T}\int_{-\tau/2}^{\tau/2} (1)e^{-jn\omega_0 t}\,dt$$

$$= \frac{A}{T}\int_{-\tau/2}^{\tau/2} (\cos n\omega_0 t - j\sin n\omega_0 t)\,dt$$

$$= \frac{A}{T}\left\{\int_{-\tau/2}^{\tau/2} \cos(n\omega_0 t)dt - j\int_{-\tau/2}^{\tau/2}\sin(n\omega_0 t)dt\right\}$$

Since the sine function is an odd function, integrating from $-\tau/2$ to $\tau/2$ gives zero; whereas, the cosine function is an even function, the integral can be thus simplified to:

$$V_n = 2\frac{A}{T}\int_0^{+\tau/2}\cos(n\omega_0 t)dt$$

$$= \frac{2A}{T}\left.\frac{\sin(n\omega_0 t)}{n\omega_0}\right|_0^{\tau/2}$$

$$= \frac{2A}{T}\left(\frac{\tau}{2}\frac{\sin(n\omega_0\tau/2)}{n\omega_0\tau/2}\right)$$

$$= \frac{A\tau}{T}\frac{\sin(n\omega_0\tau/2)}{n\omega_0\tau/2}$$

Substituting $I = A\tau$ (given),

$$V_n = \frac{I}{T}\frac{\sin(n\omega_0\tau/2)}{n\omega_0\tau/2}$$

Note that as $\tau \to 0$, V_n approaches the value obtained for Fig. 1a.

As a result,

$$G(F) = \sum_{n=-\infty}^{+\infty}|V_n|^2\delta(f-nf_0)$$

$$= \frac{I^2}{T^2}\sum_{n=-\infty}^{\infty}\left[\frac{\sin(n\omega_0\tau/2)}{n\omega_0\tau/2}\right]^2\delta(f-nf_0).$$

• **PROBLEM 6-13**

A periodic waveform $v_i(t)$ is applied to an ideal low-pass filter having a transfer function as shown in Fig. 1.

(a) Sketch the output power spectral density $G_0(f)$ of the filter.

(b) Determine the normalized power S_{in} of $v_i(t)$ and S_{out} of the filter output.

Note: f_0 = fundamental frequency = 1Hz.

Solution: (a) The power spectral density $G(f)$ is given by

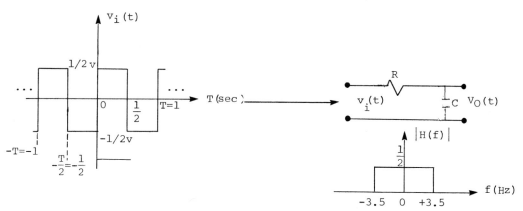

Fig. 1

$$G(f) = \sum_{n=-\infty}^{\infty} |V_n|^2 \delta(f-nf_0).$$

Hence,
$$G_0(f) = \sum_{n=-\infty}^{\infty} |V_{on}|^2 \delta(f-nf_0)$$

where the output coefficient V_{on} is related to the input coefficient by the transfer function H(f),

$$V_{on} = V_{in} H(f=nf_0).$$

Thus,
$$|V_{on}|^2 = |V_{in}|^2 |H(f=nf_0)|^2.$$

Now,
$$V_{in} = \frac{1}{T} \int_0^T v_i(t) e^{-jn\omega_0 t} \, dt$$

where $\omega_0 = 2\pi f_0 = 2\pi$ rad/sec = fundamental angular frequency.

Hence,
$$V_{in} = \frac{1}{T} \left[\int_0^{1/2} \frac{1}{2} e^{-jn\omega_0 t} dt + \int_{1/2}^1 -\frac{1}{2} e^{-jn\omega_0 t} dt \right]$$

$$= \frac{1}{2T} \left[\left(\frac{e^{-jn\omega_0 t}}{-jn\omega_0} \right)_0^{\frac{1}{2}} - \left(\frac{e^{-jn\omega_0 t}}{-jn\omega_0} \right)_{\frac{1}{2}}^1 \right]$$

$$= -\frac{1}{2jn\omega_0 T} \left[\left(e^{-j\frac{n\omega_0}{2}} - 1 \right) - \left(e^{-jn\omega_0} - e^{-j\frac{n\omega_0}{2}} \right) \right]$$

$$= \frac{1}{j(2n\omega_0 T)} \left[1 - 2e^{-j\frac{n\omega_0}{2}} + e^{-jn\omega_0} \right]$$

With $\omega_0 = \frac{2\pi}{T} = 2\pi$ radians/second.

Then,
$$e^{-jn\frac{\omega_0}{2}} = \cos \frac{n\omega_0}{T} - j \sin \frac{n\omega_0}{T}$$

$$= \cos n\pi - j \sin n\pi$$

and $\qquad e^{-jn\omega_0} = \cos n\omega_0 - j \sin n\omega_0$
$$= \cos n\pi = (-1)^n$$
$$= \cos 2\pi n - j \sin 2\pi n = 1 - 0 = 1$$

Therefore, $\quad V_{in} = \dfrac{1}{j(2n)(2\pi)} \left[1 - 2\cos n\pi + 1 \right]$

$$= \dfrac{1}{j(4n\pi)} \left[2(1 - \cos n\pi) \right] = \dfrac{(1 - \cos n\pi)}{j(2n\pi)}$$

$$= \dfrac{1 - (-1)^n}{j(2n\pi)}$$

Thus, $\qquad |V_{in}|^2 = \dfrac{[1-(-1)^n]^2}{4n^2\pi^2}$

Now, notice that:

$\qquad V_{in} = 0 \quad$ for $\quad n = $ even

for $\quad n = \pm 1, \quad |V_{in}|^2 = \dfrac{[1-(-1)]^2}{4\pi^2} = \dfrac{1}{\pi^2}$

and for $n = \pm 3, \quad |V_{in}|^2 = \dfrac{[1-(-1)]^2}{4(9)\pi^2} = \dfrac{1}{9\pi^2}$

and $\quad |V_{on}|^2 = 0 \quad$ for $\quad |n| \geq 4 \quad$ because $|H(f)| = 0$ for $|n| \geq 4$.

As a result, with all the information obtained above,

$$G_0(F) = \sum_{n=-\infty}^{\infty} |V_{in}|^2 |H(f=nf_0)|^2 \delta(f - nf_0)$$

$$= \sum_{\substack{n=-3 \\ (n=odd)}}^{+3} \left(\dfrac{1}{2}\right)^2 \dfrac{[1-(-1)^n]^2 \delta(f-nf_0)}{4n^2\pi^2}$$

$$= \dfrac{1}{16} \sum_{\substack{n=-3 \\ (n=odd)}}^{+3} \dfrac{\{1-(-1)^n\}^2}{n^2\pi^2} \delta(F-nf_0)$$

and the plot for $G_0(F)$ is shown in Fig. 2.

(b) The normalized power S_{in} of the input is given by

$$S_{in} = \sum_{n=-\infty}^{\infty} V_{in} V_{in}^* = \sum_{n=-\infty}^{\infty} |V_{in}|^2$$

$$= \sum_{n=-\infty}^{+\infty} \dfrac{(1-\cos n\pi)^2}{4n^2\pi^2}$$

$$= \sum_{n=-\infty}^{+\infty} \left\{ \dfrac{[1-(-1)^n]^2}{4n^2\pi^2} \right\}$$

$$= \sum_{\substack{n=-\infty \\ (n=\text{odd})}}^{+\infty} \frac{1}{n^2\pi^2}$$

$$= 2 \sum_{\substack{n=1 \\ (n=\text{odd})}}^{\infty} \frac{1}{n^2\pi^2}$$

$$= \frac{2}{\pi^2} \sum_{\substack{n=1 \\ (n=\text{odd})}}^{\infty} \frac{1}{n^2}$$

$$= \frac{2}{\pi^2} \left[\frac{1}{1^2} + \frac{1}{3^2} + \frac{1}{5^2} + \cdots \right]$$

Note: The series $\frac{1}{1^2} + \frac{1}{3^2} + \frac{1}{5^2} + \cdots = \frac{\pi^2}{8}$

Therefore, $S_{in} = \frac{2}{\pi^2} \frac{\pi^2}{8} = \frac{1}{4} = 0.25 \text{ volt}^2$.

Similarly, S_{out} can be calculated as follows:

Since, $V_{on} = V_{in} H(f = nf_0)$,

where $V_{in} = \frac{(1-\cos n\pi)}{j(2n\pi)}$

then $V_{on} = \frac{(1-\cos n\pi)}{j(2n\pi)} \times \frac{1}{2}$ for $|n| \leq 3$

Therefore, $S_{out} = \sum_{n=-3}^{+3} V_{on} V_{on}^*$

where $V_{on} V_{on}^* = \frac{1}{4} \frac{(1-\cos n\pi)}{4n^2\pi^2} = V_{in} V_{in}^*$.

Therefore, $S_{out} = \sum_{n=-3}^{+3} V_{on} V_{on}^* = \frac{1}{4} \sum_{\substack{n=-3 \\ (n=\text{odd})}}^{+3} \frac{[1-(-1)n]^2}{4n^2\pi^2}$

$$= \frac{1}{4} \times 2 \sum_{n=1}^{3} \frac{[1-(-1)^n]^2}{4n^2\pi^2}$$

$$= \frac{1}{2} \sum_{\substack{n=1 \\ (n=\text{odd})}}^{3} \frac{1}{n^2\pi^2}$$

$$= \frac{1}{2\pi^2} \sum_{\substack{n=1 \\ (n=\text{odd})}}^{3} \frac{1}{n^2}$$

$$= \frac{1}{2} \frac{1}{\pi^2} \left[\frac{1}{1^2} + \frac{1}{3^2} \right] = \frac{1}{2\pi^2} \left[1 + \frac{1}{9} \right] = 0.0563 \text{ volt}^2.$$

POWER CONTENT OF SIGNALS

• **PROBLEM** 6-14

Given a periodic square wave as shown in Fig. A, determine its exponential Fourier series and estimate the fraction of the total power contained within the first zero crossing of the spectrum envelope of f(t).

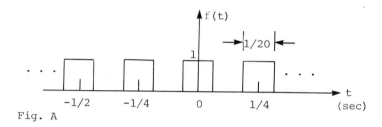

Fig. A

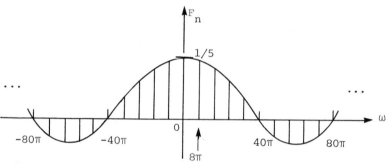

Fig. B

Solution: The exponential Fourier series of f(t) is,

$$f(t) = \sum_{n=-\infty}^{+\infty} C_n e^{jn\omega_0 t}$$

where T = Period = $\frac{1}{4}$ sec,

ω_0 = Fundamental angular frequency

$= \frac{2\pi}{T} = 8\pi$ radians/sec,

and d = pulsewidth = $\frac{1}{20}$ sec.

Therefore,

$$\frac{d}{T} = \frac{\frac{1}{20}}{\frac{1}{4}} = \frac{1}{5}$$

The Fourier coefficients C_n are given by

$$C_n = \frac{1}{T} \int_{-T/2}^{+T/2} f(t) e^{-jn\omega_0 t} \, dt$$

$$= \frac{1}{T} \int_{-d/2}^{+d/2} (1) \, e^{-jn\omega_0 t} \, dt$$

(Note: f(t) = 1 from figure A)

$$= \frac{1}{T} \left[\frac{e^{-jn\omega_0 t}}{-jn\omega_0} \right]_{-d/2}^{+d/2}$$

$$= \frac{1}{-jn\omega_0 T} \left[e^{-jn\omega_0 \frac{d}{2}} - e^{jn\omega_0 \frac{d}{2}} \right]$$

$$= \frac{1}{jn\omega_0 T} \left[e^{jn\omega_0 \left(\frac{d}{2}\right)} - e^{-jn\omega_0 \left(\frac{d}{2}\right)} \right]$$

$$= \frac{1}{jn\omega_0 T} \left[2j \sin \frac{n\omega_0 d}{2} \right]$$

$$= \frac{2}{n\omega_0 T} \sin \left(n\omega_0 \frac{d}{2} \right)$$

To obtain a desirable form, dividing and multiplying the above equation by d/2 yields,

$$C_n = \frac{d}{T} \frac{\sin \left(n\omega_0 \frac{d}{2}\right)}{\left(n\omega_0 \frac{d}{2}\right)} = \frac{d}{T} \text{sinc} \left(n\omega_0 \frac{d}{2}\right)$$

where $\frac{\sin(x)}{x} = \text{sinc}(x)$.

Now $\quad \frac{d}{T} = \frac{1}{5}$ and $\frac{n\omega_0 d}{2} = \frac{n\omega_0}{2} \frac{T}{5} = \frac{n\pi}{5}$

So, $\quad C_n = \frac{1}{5} \text{sinc} \frac{n\pi}{5}$

The frequency spectrum of f(t) is shown in Fig. B.

From Parseval's theorem, the total power of the signal is given by

$$P = \frac{1}{T} \int_{-T/2}^{+T/2} f^2(t) dt$$

$$= \frac{1}{T} \int_{-d/2}^{+d/2} (1) dt = \frac{2}{T} \frac{d}{2} = \frac{d}{T}$$

$$= \frac{1}{5} = 0.2$$

The first zero crossing of the spectrum of f(t) occurs at n=5. It corresponds to an angular frequency $\omega = n\omega_0 = 5 \times 8\pi = 40\pi$ radians/sec.

Now, Let P_1 = Power contained within the first zero crossing of the spectrum envelope of f(t).

Hence, $P_1 = C_0^2 + 2\{C_1 C_{-1} + C_2 C_{-2} + C_3 C_{-3} + C_4 C_{-4}\}$

with $C_n = \frac{1}{5} \text{sinc}\left(\frac{n\pi}{5}\right) = C_{-n}$,

$P_1 = C_0^2 + 2\{C_1^2 + C_2^2 + C_3^2 + C_4^2\}$

$= \left(\frac{1}{5}\right)^2 + \frac{2}{25}\left\{\text{sinc}^2\left(\frac{\pi}{5}\right) + \text{sinc}^2\left(\frac{2\pi}{5}\right) + \text{sinc}^2\left(\frac{3\pi}{5}\right) + \text{sinc}^2\left(\frac{4\pi}{5}\right)\right\}$

$= \frac{1}{25} + \frac{2}{25}\{0.875 + 0.573 + 0.255 + 0.055\}$

$\approx 0.04 + 0.141 = 0.181$.

Hence, the fraction of the total power contained within the first zero crossing of the spectrum envelope of f(t)

$= \frac{P_1}{P} = \frac{0.181}{0.20} = 0.905$ or 90.5%.

• **PROBLEM 6-15**

Find the ratio of the signal power at the fundamental frequency to the total signal power, for the periodic pulse x(t) shown in the figure.

x(t) is given by

$$x(t) = \sum_{n=-\infty}^{\infty} X_n e^{j\frac{n\pi t}{T_0}},$$

$$X_n = \frac{A\tau}{T_0} \frac{\sin \frac{n\pi\tau}{T_0}}{\frac{n\pi\tau}{T_0}} e^{-jn\pi t/T_0}, \quad \text{where } \tau/T_0 = \frac{1}{4}.$$

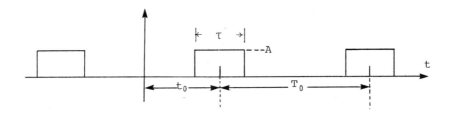

Solution: The component $f_0 = \frac{1}{T_0}$ corresponds to $n = \pm 1$ and is

$$X_1 e^{j \frac{\pi t}{T_0}} + X_{-1} e^{j \frac{\pi t}{T_0}} \tag{1}$$

Now, since $X_{-1} = X_1^*$ (complex conjugate pair), let

$$X_1 = A_1 + jB_1 \tag{2}$$

and

$$X_{-1} = A_1 - jB_1 \tag{3}$$

and substituting eq.(2) and (3) into eq.(1), eq.(1) becomes

$$(A_1 + jB_1) e^{j \frac{\pi t}{T_0}} + (A_1 - jB_1) e^{-j \frac{\pi t}{T_0}}$$

$$= A_1 (e^{j \frac{\pi t}{T_0}} + e^{-j \frac{\pi t}{T_0}}) + jB_1 (e^{j \frac{\pi t}{T_0}} - e^{-j \frac{\pi t}{T_0}})$$

$$= 2A_1 \cos \frac{\pi t}{T_0} - 2B_1 \sin \frac{\pi t}{T_0} \tag{4}$$

$\left(\text{Note: } \cos X = \frac{e^{jX} + e^{-jX}}{2} \text{ and } \sin X = \frac{e^{jX} - e^{-jX}}{2j} \right)$

Thus, the normalized power (S_0) = mean square value of eq. (4).

$$= \overline{\left[2A_1 \cos \frac{\pi t}{T_0} - 2B_1 \sin \frac{\pi t}{T_0} \right]^2}$$

$$= \frac{1}{2} (2A_1)^2 + \frac{1}{2} (2B_1)^2$$

$$= 2A_1^2 + 2B_1^2$$

$$= 2 |X_1| |X_{-1}|.$$

$$= 2 \left(\frac{A\tau}{T_0} \right)^2 \frac{\sin^2 \left(\frac{\pi \tau}{T_0} \right)}{\left(\frac{\pi \tau}{T_0} \right)^2} \left[e^{-j \frac{\pi t_0}{T_0}} e^{j \frac{\pi t_0}{T_0}} \right]$$

$$= 2 \left(\frac{A}{4} \right)^2 \frac{\sin^2 \left(\frac{\pi}{4} \right)}{\left(\frac{\pi}{4} \right)^2}$$

$$= 2 \frac{A^2}{16} \frac{(\frac{1}{2})}{\left(\frac{\pi^2}{16} \right)} = \frac{A^2}{\pi^2} = \left(\frac{A}{\pi} \right)^2$$

Total signal power = mean square value of x(t).

$$= S_T = \frac{1}{T_0} \int_{t_0 - \frac{\tau}{2}}^{t_0 + \frac{\tau}{2}} x^2(t) dt$$

$$= \frac{1}{T_0} \left[A^2 \tau \right] = A^2 \frac{\tau}{T_0} = \frac{A^2}{4} = \left(\frac{A}{2} \right)^2$$

Therefore, the ratio of the power at f_0 is:

$$\frac{S_\Omega}{S_T} = \frac{(A/\pi)^2}{(A/2)^2} = \left(\frac{2}{\pi}\right)^2 = \frac{4}{\pi^2}.$$

● **PROBLEM 6-16**

Given $f(t) = \int_{-\infty}^{+\infty} h(\sigma)\phi(t-\sigma)d\sigma$

determine the power spectral density of $f(t)$ in terms of the power spectral density of $\phi(t)$ and the Fourier transform of $h(t)$.

<u>Solution</u>: Realizing that the power spectral density and the autocorrelation function of a periodic waveform form a Fourier transform pair, let $R_f(\tau)$ and $G_f(\omega)$ be the autocorrelation function and the power spectral density of $f(t)$ respectively. Then $G_f(\omega)$ is the Fourier transform of $R_f(\tau)$. This can be represented by the equations

$$G_f(\omega) = \int_{\tau=-\infty}^{\tau=+\infty} R_f(\tau)e^{-j\omega\tau}\,d\tau$$

and

$$R_f(\tau) = \int_{t=-\infty}^{t=+\infty} f(t)\,f(t+\tau)dt.$$

So,

$$R_f(\tau) = \int_{t=-\infty}^{t=+\infty}\left[\int_{\sigma=-\infty}^{\sigma=+\infty} h(\sigma)\phi(t-\sigma)d\sigma\right]\left[\int_{\beta=-\infty}^{\beta=+\infty} h(\beta)\phi(t+\tau-\beta)d\beta\right]dt$$

$$= \int_{\sigma=-\infty}^{\sigma=+\infty}\int_{\beta=-\infty}^{\beta=+\infty} h(\sigma)h(\beta)\left[\int_{t=-\infty}^{t=+\infty}\phi(t-\sigma)\phi(t+\tau-\beta)dt\right](d\sigma)(d\beta)$$

$$= \int_{\sigma=-\infty}^{\sigma=+\infty}\int_{\beta=-\infty}^{\beta=+\infty} h(\sigma)h(\beta)\left[\int_{t=-\infty}^{t=+\infty}\phi(t-\sigma)\phi\{(t-\sigma)+(\tau-\beta+\sigma)\}d(t-\sigma)\right]d\sigma d\beta$$

$$= \int_{\sigma=-\infty}^{\sigma=+\infty}\int_{\beta=-\infty}^{\beta=+\infty} h(\sigma)h(\beta)\left[R_\phi(\tau-\beta+\sigma)\right]d\sigma d\beta$$

where $R_\phi(\tau-\beta+\sigma)$ is the autocorrelation function of $\phi(t)$.

Hence, $G_f(\omega)$ can be written as:

$$G_f(\omega) = \int_{\tau=-\infty}^{\tau=+\infty}\left[\int_{\sigma=-\infty}^{\sigma=+\infty}\int_{\beta=-\infty}^{\beta=+\infty} h(\sigma)h(\beta)\{R_\phi(\tau-\beta+\sigma)\}(d\sigma)(d\beta)\right]e^{-j\omega\tau}d\tau$$

Substituting $y = \tau - \beta + \sigma$ or $\tau = y + \beta - \sigma$ in this case $dy = d\tau$

So,

$$G_f(\omega) = \int_{y=-\infty}^{y=+\infty} \left[\int_{\sigma=-\infty}^{\sigma=+\infty} \int_{\beta=-\infty}^{\beta=+\infty} h(\sigma)h(\beta)\{R_\phi(y)\} d\sigma d\beta \right] e^{-j\omega(y+\beta-\sigma)} dy$$

$$= \left[\int_{\sigma=-\infty}^{\sigma=+\infty} h(\sigma) e^{-j(-\omega)\sigma} d\sigma \right] \left[\int_{\beta=-\infty}^{\beta=+\infty} h(\beta) e^{-j\omega\beta} d\beta \right] \left[\int_{y=-\infty}^{y=+\infty} R_\phi(y) e^{-j\omega y} dy \right]$$

$$= \left[H(-\omega) \right] \left[H(\omega) \right] \left[G_\phi(\omega) \right]$$

But $H(-\omega) = H^*(\omega) =$ complex conjugate of $H(\omega)$ and

$$H^*(\omega) H(\omega) = |H(\omega)|^2$$

Hence, $G_f(\omega) = |H(\omega)|^2 G_\phi(\omega)$.

FOURIER TRANSFORM REPRESENTATION

• **PROBLEM 6-17**

Let $A(\omega)$ be the Fourier transform of a signal $a(t)$. The magnitude of $A(\omega)$ is shown in Fig. 1.

(a) Plot the magnitude of $B(\omega)$, if

$$B(\omega) = \left[-j \, \text{Sgn}(\omega) \right] A(\omega)$$

(b) Plot the magnitude spectra of the signals
$f_1(t) = a(t) \cos \omega_0 t$ and $f_2(t) = b(t) \sin \omega_0 t$

(c) Let $g(t) = f_1(t) - f_2(t)$, where $f_1(t) = a(t)\cos\omega_0 t$ and $f_2(t) = b(t)\sin\omega_0 t$. Plot the magnitude of $G(\omega)$, where $G(\omega)$ is the Fourier transform of $g(t)$. Is $G(\omega)$ an upper or lower sideband?

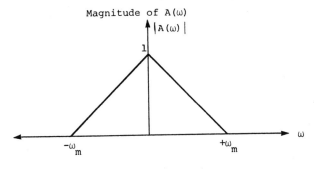

Fig. 1

Solution: (a) $B(\omega) = [-j \, \text{Sgn}(\omega)] A(\omega)$ is given and the function Sgn(ω) is defined as

$$\text{Sgn}(\omega) = \begin{cases} +1, & \omega > 0 \\ -1, & \omega < 0 \end{cases}.$$

Since $A(\omega)$ is purely real, it follows that $B(\omega)$ is purely imaginary. It can be shown that if the Fourier transform of a time function is complex, the real part and the imaginary part of the Fourier transform are, respectively, even and odd functions of ω.

Now, $|B(\omega)| = |-j\,\text{Sgn}(\omega)\,A(\omega)|$

$\qquad\qquad = \text{Sgn}(\omega)\,A(\omega) = $ Magnitude of $B(\omega)$.

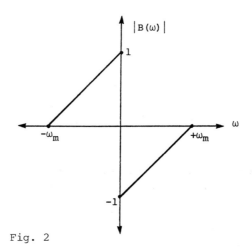

Fig. 2

The plot for $|B(\omega)|$ is shown in Fig. 2.

(b) $f_1(t) = a(t)\cos\omega_0 t$

$$= \frac{1}{2}(a(t))\left[e^{j\omega_0 t} + e^{-j\omega_0 t}\right] \qquad (1)$$

Applying the frequency shifting theorem to eq.(1) and with $F_1(\omega)$ = Fourier transform of $f_1(t)$,

$$F_1(\omega) = \frac{1}{2}\left[A(\omega-\omega_0)\right] + \frac{1}{2}\left[A(\omega+\omega_0)\right]$$

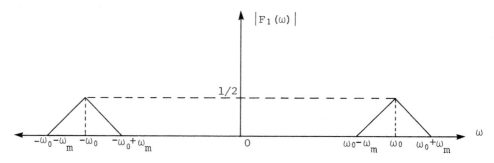

Fig. 3

Note: $A(\omega-\omega_0)$ is similar to $A(\omega)$ shifted by 'ω_0' units to the right. Similarly $A(\omega+\omega_0)$ corresponds to the translation of $A(\omega)$ by 'ω_0' units to the left. Also, $F_1(\omega)$ is purely real, hence $|F_1(\omega)|$ (magnitude of $F_1(\omega)$) can be plotted as shown in Fig. 3.

Similarly, $f_2(t) = b(t)\sin\omega_0 t$

$$= \frac{1}{2j} b(t) \left[e^{j\omega_0 t} - e^{-j\omega_0 t} \right].$$

From the Frequency-shifting theorem,

$$F_2(\omega) = \frac{1}{2j}\left[B(\omega-\omega_0)\right] - \frac{1}{2j}\left[B(\omega+\omega_0)\right],$$

since $B(\omega) = -j\,\text{Sgn}(\omega)\,A(\omega)$

$$F_2(\omega) = \frac{1}{2j}\left[-j\,\text{Sgn}(\omega-\omega_0)A(\omega-\omega_0)\right] - \frac{1}{2j}\left[-j\,\text{Sgn}(\omega+\omega_0)A(\omega+\omega_0)\right]$$

$$= \frac{1}{2}\text{Sgn}(\omega+\omega_0)\,A(\omega+\omega_0) - \frac{1}{2}\text{Sgn}(\omega-\omega_0)A(\omega-\omega_0).$$

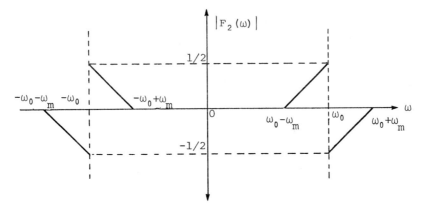

Fig. 4

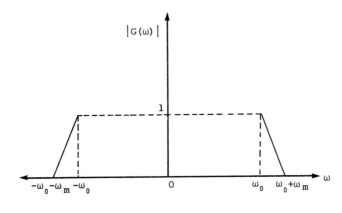

Fig. 5

Note that $F_2(\omega)$ is also purely real. $|F_2(\omega)|$ is plotted in Fig. 4.

(c) $g(t) = f_1(t) - f_2(t)$.

$G(\omega)$ = Fourier transform of $g(t)$

$= F_1(\omega) - F_2(\omega)$.

Using the solutions obtained in part (b), the plot for $|G(\omega)|$ is shown in Fig 5.

Note that this is an upper single-sideband signal.

• **PROBLEM 6-18**

Determine the Fourier transform of the function $f(t) = e^{-2t} \sin(6t - 18) u(t - 3)$ where $u(t)$ is the unit step function.

Note: $F\left[e^{-\alpha t}(\sin\omega_0 t)u(t)\right] = \dfrac{\omega_0}{(\alpha+j\omega)^2+\omega_0^2}$

Solution: $f(t) = e^{-2t}\sin(6t - 18) u(t - 3)$

$= e^{-2(t-3)}e^{-6}\{\sin 6(t-3)\} u(t-3)$

so that

$f(t) = e^{-6} f_1(t-3)$

where $f_1(t) = e^{-2t} \sin 6t\, u(t)$.

Now, $F[f(t)] = e^{-6} F[f_1(t-3)]$

and making use of the time shifting property

$F[f(t)] = e^{-6} e^{-3j\omega} F[f_1(t)]$.

The term $e^{-3j\omega}$ is due to a time delay of 3 units.

$F[f_1(t)] = F[e^{-2t} \sin 6t\, u(t)]$

Applying the frequency-shifting theorem,

$= \left[\dfrac{6}{(j\omega)^2+6^2}\right]_{j\omega \to (j\omega+2)}$

Because $F[\sin(at) u(t)] = \dfrac{a}{(j\omega)^2+a^2}$

$F[f_1(t)] = \dfrac{6}{(j\omega+2)^2+36}$.

Therefore $F[f(t)] = e^{-6} e^{-3j\omega} \dfrac{6}{(j\omega+2)^2+36}$

$$= 6 e^{-6} \frac{e^{-3j\omega}}{(j\omega+2)^2+36}$$

The term $e^{-3j\omega}$ is due to the 3 unit delay.

• **PROBLEM 6-19**

Find the Fourier transform of the function
$$f(t) = e^{-(2t+1)} \quad \text{for} \quad -\infty < t < +\infty$$

Solution: Let $F(\omega)$ be the Fourier transform of the function $f(t)$.

Note: This problem can be solved by applying the time-scaling and time-shifting properties as follows:

Time-scaling property:
$$F[y(at)] = \frac{1}{|a|} Y\left(\frac{\omega}{a}\right)$$

where $Y(\omega)$ = Fourier transform of $y(t)$.

Time-shifting property:
$$F[y(t-t_0)] = e^{-j\omega t_0} Y(\omega)$$

We can now approach the problem in two different ways.

Method I involves applying the time-scaling property first and then the time-shifting property. Alternatively, Method II involves applying the time-shifting property first and then the time-scaling property. Either method is correct.

Method I: Let $g(t) = e^{-|t|}$

and $h(t) = e^{-|2t|} = g(2t)$

Applying the time-scaling property,

$$H(\omega) = \text{Fourier transform of } h(t) = \frac{1}{2} G\left(\frac{\omega}{2}\right)$$

where $G(\omega)$ = Fourier transform of $g(t)$.

Since $f(t) = h\left(t + \frac{1}{2}\right)$, applying the time-shifting property,

$$F(\omega) = e^{j\frac{\omega}{2}} H(\omega) = \frac{e^{j\frac{\omega}{2}}}{2} G\left(\frac{\omega}{2}\right) \qquad (1)$$

Since, $G(\omega) = \int_{-\infty}^{+\infty} g(t) e^{-j\omega t} dt$

$$g(t) = \begin{cases} e^t, & t < 0 \\ e^{-t}, & t > 0. \end{cases}$$

Therefore,
$$G(\omega) = \int_{-\infty}^{0} e^t e^{-j\omega t} dt + \int_{0}^{\infty} e^{-t} e^{-j\omega t} dt$$

$$= \int_{-\infty}^{0} e^{-(j\omega-1)t} dt + \int_{0}^{\infty} e^{-(j\omega+1)t} dt$$

$$= \left[\frac{e^{-(j\omega-1)t}}{-(j\omega-1)}\right]_{-\infty}^{0} + \left[\frac{e^{-(j\omega+1)t}}{-(j\omega+1)}\right]_{0}^{\infty}$$

$$= \frac{-1}{(j\omega-1)}[1 - 0] - \frac{1}{j\omega+1}[0 - 1]$$

$$= \frac{1}{j\omega+1} - \frac{1}{(j\omega-1)} = \frac{j\omega-1-j\omega-1}{(-\omega^2-1)} = \frac{-2}{-(1+\omega^2)}.$$

Hence,
$$G\left(\frac{\omega}{2}\right) = \frac{2}{1+\left(\frac{\omega}{2}\right)^2}.$$

Now, substituting the value for $G\left(\frac{\omega}{2}\right)$ into eq.(1) gives

$$F(\omega) = \frac{e^{j\frac{\omega}{2}}}{2}\left[\frac{2}{1+\frac{\omega^2}{4}}\right] = \left[\frac{1}{1+\frac{\omega^2}{4}}\right] e^{j\frac{\omega}{2}}.$$

<u>Method II</u>: Let $x(t) = g(t+1) = e^{-|t+1|}$

and $f(t) = e^{-|2t+1|} = x(2t)$.

Applying the time-shifting property,

$$X(\omega) = e^{j\omega} G(\omega).$$

Applying the time-scaling property,

$$F(\omega) = \frac{1}{2}\left[X\left(\frac{\omega}{2}\right)\right] = \frac{1}{2} e^{j\frac{\omega}{2}} G\left(\frac{\omega}{2}\right) \qquad (2)$$

$G\left(\frac{\omega}{2}\right)$ can be found as in method I. Substituting the value of $G\left(\frac{\omega}{2}\right)$ into equation (2) gives

$$F(\omega) = \frac{1}{2} e^{j\frac{\omega}{2}} \left[\frac{2}{1+\frac{\omega^2}{4}}\right]$$

$$= \left[\frac{1}{1+\frac{\omega^2}{4}}\right] e^{j\frac{\omega}{2}}$$

Observe that either method leads to the same result.

PARSEVAL'S THEOREM

• **PROBLEM 6-20**

a) Use the concept of convergence in the mean to define another way in which the Fourier series of a function f can be said to satisfactorily represent f(x).

b) In this connection, state and prove Parseval's equality.

c) For what general class of functions does the Fourier series converge in the mean to its function?

Solution: (a) The Fourier series of a function f with period 2c may be considered to be an accurate representation of f on the interval (a, a+2c) if the total square deviation approaches zero

$$\lim_{n \to \infty} D_n = 0 \qquad (1)$$

where

$$D_n = \int_a^{a+2c} [f(x) - g_n(x)]^2 \, dx \qquad (2)$$

and $g_n(x)$ is the nth partial sum of the Fourier series

$$g_n(x) = \frac{a_0}{2} + \sum_{k=1}^{n} \left[a_k \cos \frac{k\pi x}{c} + b_k \sin \frac{k\pi x}{c} \right]. \qquad (3)$$

If (1) holds it is said that the Fourier series of f converges in the mean to f.

(b) It can be shown that the trigonometric polynomial g_n which minimized expression (2) (the mean square deviation) is the one with Fourier coefficients and that in this case (for f with period 2π)

$$[D_m]_{\min} = \int_{-\pi}^{\pi} [f(x)]^2 \, dx - \left[\frac{\pi}{2} a_0^2 + \pi \sum_{k=1}^{n} (a_k^2 + b_k^2) \right]. \qquad (4)$$

Now if the Fourier series for f converges in the mean to f, then by definition

$$\lim_{n \to \infty} [D_n]_{\min} = 0$$

so that from (4)

$$\frac{a_0^2}{2} + \sum_{k=1}^{\infty} (a_k^2 + b_k^2) = \frac{1}{\pi} \int_{-\pi}^{\pi} [f(x)]^2 \, dx. \qquad (5)$$

This is known as Parseval's equality and is useful in the summation of series.

(c) It can be proven that if f is defined on the interval (a,a+2c) and is piecewise continuous there, then the Fourier series of f converges in the mean to f. This requirement is less stringent than that imposed upon f in order that it be pointwise convergent and, of course, much less stringent than the uniform convergence conditions. This is so because the integral in (2) exists even if the integrand has a countably infinite number of discontinuities and its value is unaltered by any change in the functional values at these points. Thus, mean convergence is a statistical convergence and it must be kept in mind that it does not imply pointwise convergence. On the other hand, in most physical applications the shape of the function over the whole interval (or whole real axis) is the important consideration rather than the value at each point. Therefore, mean convergence is adequate in most physical situations.

• **PROBLEM 6-21**

a) Prove that if the functions g(x) and F(k) are absolutely integrable on $(-\infty, +\infty)$ and that the Fourier inversion integral for f(x) is valid for all x except possibly at a countably infinite number of points, then

$$\int_{-\infty}^{\infty} f(k) G(-k) dk = \int_{-\infty}^{\infty} f(x) g(x) dx \qquad (1)$$

where
$$F(k) = \Phi\{f(x)\}, \quad G(k) = \Phi\{g(x)\}.$$

This is known as the second Parseval theorem of Fourier transform theory.

b) From the above equation (1), prove the first Parseval theorem of Fourier transform theory,

$$\int_{-\infty}^{\infty} |F(k)|^2 \, dk = \int_{-\infty}^{\infty} |f(x)|^2 \, dx. \qquad (2)$$

Solution: a) The Fourier transform of a function f(x) is defined by

$$F(k) = \frac{1}{\sqrt{2\pi}} \int_{-\infty}^{\infty} f(x) e^{jkx} \, dx \qquad (3)$$

so that by definition

$$G(-k) = \frac{1}{\sqrt{2\pi}} \int_{-\infty}^{\infty} g(x) e^{-jkx} \, dx. \qquad (4)$$

Therefore,

$$\int_{-\infty}^{\infty} F(k)G(-k)dk = \int_{-\infty}^{\infty} F(k)dk \int_{-\infty}^{\infty} \frac{1}{\sqrt{2\pi}} g(x)e^{-jkx}dx. \quad (5)$$

Now $F(k)$ and $g(x)$ are absolutely convergent on $(-\infty, +\infty)$, that is, the integrals

$$\int_{-\infty}^{\infty} |F(k)|dk, \quad \int_{-\infty}^{\infty} |g(x)|dx$$

are convergent, so that

$$\int_{-\infty}^{\infty} F(k)e^{-jkx} dx, \quad \int_{-\infty}^{\infty} g(x)e^{-jkx}dx$$

are absolutely convergent since

$$|F(k)e^{-jkx}| = |F(k)| |e^{-jkx}| = |F(k)|$$

and

$$|g(x)e^{-jkx}| = |g(x)|.$$

Hence, the order of integration in (5) may be interchanged giving

$$\int_{-\infty}^{\infty} F(k)G(-k)dk = \int_{-\infty}^{\infty} g(x)dx \frac{1}{\sqrt{2\pi}} \int_{-\infty}^{\infty} F(k)e^{-jkx} dk. \quad (6)$$

Since the Fourier inversion integral is valid,

$$\frac{1}{\sqrt{2\pi}} \int_{-\infty}^{\infty} F(k)e^{-jkx} dk = f(x) \quad (7)$$

and using this result in (6) gives the second Parseval theorem:

$$\int_{-\infty}^{\infty} F(k)G(-k)dk = \int_{-\infty}^{\infty} g(x)f(x)dx. \quad (8)$$

The validity of (8) is insured even if the Fourier inversion integral for $f(x)$ has a countably infinite number of discrepancies with $f(x)$ since this will not affect the equality of the integrals

$$\int_{-\infty}^{\infty} g(x)(f(x))dx \quad \text{and} \quad \int_{-\infty}^{\infty} g(x) \frac{1}{\sqrt{2\pi}} \int_{-\infty}^{\infty} F(k)e^{-jkx}dk \, dx.$$

b) The first Parseval theorem is a corollary to the second Parseval theorem stated in equation (8) which follows by letting $f(x) = g(x)$ so that $F(k) = G(k)$ and recalling that (assuming $f = g$ real) $G(-k) = G^*(k)$ where $G^*(k)$ is the complex conjugate of $G(k)$. Noting that

$$G(k)G^*(k) = |G(k)|^2$$

and using these results in (8) gives

$$\int_{-\infty}^{\infty} G(k)G^*(k)\,dk = \int_{-\infty}^{\infty} [g(x)]^2\,dx$$

or

$$\int_{-\infty}^{\infty} |G(k)|^2\,dk = \int_{-\infty}^{\infty} |g(x)|^2\,dx.$$

Also,

$$\int_{-\varepsilon}^{+\varepsilon} \delta(x)\,dx = 1$$

for all values of ε. It appears then that letting $\varepsilon \to 0$, we have exactly

$$\int_{-\varepsilon}^{-\varepsilon} \delta(x)f(x)\,dx = f(0). \tag{9}$$

Note that the limits $-\varepsilon$ and $+\varepsilon$ may be replaced by any two numbers a and b provided that $a < 0 < b$. Now the integral in (9) is referred to as the sifting property of the delta function, that is $\delta(x)$ acts as a sieve, selecting from all possible values of $f(x)$ its value at the point $x = 0$.

• **PROBLEM 6-22**

Given the current pulse, $i(t) = te^{-bt}$:

(a) Find the total 1Ω energy associated with this waveform.

(b) What fraction of this energy is present in the frequency band from $-b$ to b rad/s?

Solution: The total 1-Ω energy associated with either a current or voltage waveform can be found by use of Parseval's theorem,

$$W_{1\Omega} = \frac{1}{2\pi} \int_{-\infty}^{\infty} |F(j\omega)|^2\,d\omega$$

where $F(j\omega)$ is the Fourier transform of the current or voltage waveform.

The Fourier transform of the current is

$$I(j\omega) = \int_0^{\infty} (t\,e^{-bt})e^{-j\omega t}\,dt$$

$$= \int_0^{\infty} t\,e^{-t(j\omega+b)}\,dt$$

$$= \left(\frac{-te^{-t(j\omega+b)}}{(j\omega+b)} - \int \frac{e^{-t(j\omega+b)}}{-(j\omega+b)} dt \right) \bigg|_0^\infty$$

$$= \left(-\frac{te^{-t(j\omega+b)}}{j\omega+b} - \frac{e^{-t(j\omega+b)}}{(j\omega+b)^2} \right) \bigg|_0^\infty$$

$$= \frac{1}{(j\omega+b)^2}$$

Then,

$$|I(j\omega)^2| = \frac{1}{(b^2-\omega^2)^2 + 4\omega^2 b^2} = \frac{1}{(b^2+\omega^2)^2}.$$

The total energy associated with the current is

$$W = \frac{1}{2\pi} \int_{-\infty}^{\infty} |I(j\omega)^2| d\omega.$$

Since

$$W = \frac{1}{\pi} \int_0^\infty |I(j\omega)|^2 d\omega$$

then

$$W = \frac{1}{\pi} \int_0^\infty \frac{1}{(b^2+\omega^2)^2} d\omega.$$

Using the trigonometric substitution,

$$\omega = b \tan \theta,$$

$$(b^2+\omega^2)^2 = (b^2 \tan^2\theta + b^2)^2 = (b^2 \sec^2\theta)^2.$$

Also,

$$d\omega = b\sec^2\theta d\theta.$$

Hence,

$$W = \frac{1}{\pi} \int_0^{\pi/2} \frac{b\sec^2\theta}{b^4\sec^4\theta} d\theta$$

$$W = \frac{1}{\pi} \int_0^{\pi/2} \frac{1}{b^3\sec^2\theta} d\theta$$

$$W = \frac{1}{\pi b^3} \int_0^{\pi/2} \cos^2\theta \, d\theta$$

$$W_\tau = \frac{1}{\pi b^3} \left(\frac{1}{2}\theta + \frac{1}{4} \sin 2\theta \right).$$

Since

$$\omega = b\tan\theta$$

and

$$\theta = \arctan \frac{\omega}{b}$$

then

$$W_T = \left\{\frac{1}{\pi b^3}\left(\frac{1}{2}\tan^{-1}\frac{\omega}{b}+\frac{1}{4}\sin 2(\tan^{-1}\frac{\omega}{b}))\right)\right\}\Big|_0^\infty$$

Since
$$\tan^{-1}\infty = \frac{\pi}{2}$$

then
$$W_T = \frac{1}{\pi b^3}\left(\frac{1}{2}\frac{\pi}{2}\right)$$

$$W_T = \frac{1}{4b^3}$$

(b) To find the energy present in the frequency band

$$-b < f < b$$

use Parseval's theorem and integrate from $-b$ to b:

$$W_b = \frac{1}{2\pi}\int_{-b}^{b}|I(j\omega)|^2\,d\omega$$

$$= \frac{1}{\pi}\int_0^b |I(j\omega)|^2\,d\omega$$

$$W_b = \left\{\frac{1}{\pi b^3}\left(\frac{1}{2}\tan^{-1}\frac{\omega}{b}+\frac{1}{4}\sin 2(\tan^{-1}\frac{\omega}{b}))\right)\right\}\Big|_0^b$$

$$= \left\{\frac{1}{\pi b^3}\left(\frac{1}{2}\tan^{-1}1+\frac{1}{4}\sin 2(\tan^{-1}1)\right)\right.$$
$$\left. -\left(\frac{1}{2}\tan^{-1}0+\frac{1}{4}\sin 2(\tan^{-1}0)\right)\right\}$$

$$= \frac{1}{\pi b^3}\left\{\frac{\pi}{8}+\frac{1}{4}-0-0\right\}$$

$$W_b = \left\{\frac{\pi+2}{8}\right\}\frac{1}{\pi b^3}$$

The fraction of energy present is

$$\frac{W_0}{W_T} = \frac{\left\{\frac{\pi+2}{8}\right\}\frac{1}{\pi b^3}}{\frac{1}{4b^3}} = \frac{\pi+2}{2\pi} = 0.818.$$

• PROBLEM 6-23

Derive Parseval's theorem $\int_{-\infty}^{\infty} f^2(t)dt = \frac{1}{2\pi}\int_{-\infty}^{\infty}|F(\omega)|^2 d\omega$
from the time convolution theorem.
Hint: If $f(t) \leftrightarrow F(\omega)$
       then $f(-t) \leftrightarrow F(-\omega)$
       and $f(t)*f(-t) \leftrightarrow F(\omega)F(-\omega)$

Solution: If $f(t) \leftrightarrow F(\omega)$, then using the time scaling property, we get $f(-t) \leftrightarrow F(-\omega)$ and $f(t)*f(-t) = F(\omega)F(-\omega) = |F(\omega)|^2$

Hence,
$$\int_{-\infty}^{\infty} f(\tau)\, f(\tau-t)d\tau \leftrightarrow |F(\omega)|^2$$

and
$$\int_{-\infty}^{\infty} f(\tau)f(\tau-t)d\tau = \frac{1}{2\pi}\int_{-\infty}^{\infty}|F(\omega)|^2 e^{j\omega t} d\omega$$

Now, letting t=0, the equality
$$\int_{-\infty}^{\infty} f^2(\tau)d\tau = \frac{1}{2\pi}\int_{-\infty}^{\infty}|F(\omega)|^2 d\omega$$
is proved.

• PROBLEM 6-24

Let v(t) be the inverse Fourier transform of V(f). Given that
$$V(f) = AT\left[\frac{\sin(2\pi fT)}{(2\pi fT)}\right]$$
determine the energy of v(t).

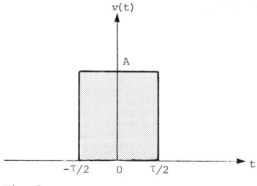

Fig. 1

Solution: First, a pulse v(t), as in Fig.(1), of strength A, and width τ, centered at t=0 can be represented by $P_\tau(t)$.

It is known that
$$A\,P_\tau(t) \leftrightarrow A\tau\,\frac{\sin\omega\,\tau/2}{\omega\,\tau/2},$$

where
$$\frac{\sin\omega\,\tau/2}{\omega\,\tau/2} \leftrightarrow \frac{1}{\tau}\,P_\tau(t)$$

form Fourier transform pairs. Therefore,
$$\frac{\sin\omega T}{\omega T} \leftrightarrow \frac{1}{2T}\,P_{2T}(t)$$

and
$$AT\,\frac{\sin\omega T}{\omega T} \leftrightarrow \frac{A}{2}\,P_{2T}(t).$$

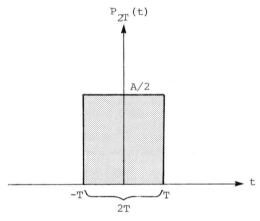

Fig. 2

Now, notice that $\frac{A}{2}\,P_{2T}(t)$ can be represented as in Fig. 2. Hence, using Parseval's theorem

$$E = \int_{-\infty}^{\infty} |V(f)|^2\,dF = \int_{-\infty}^{\infty} v^2(t)\,dt$$

$$= \int_{-T}^{T} \left(\frac{A}{2}\right)^2 dt$$

$$= \frac{A^2 T}{2}.$$

• **PROBLEM 6-25**

The Fourier transform of the waveform v(t) is shown in the figure below.

(a) By applying Parseval's theorem, determine the normalized energy E of the waveform.

(b) If only one-half of the normalized energy is contained in the interval from $-f_1$ to f_1, find the corresponding frequency f_1.

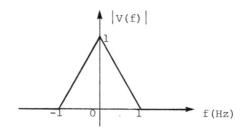

<u>Solution</u>: (a) By Parseval's theorem, the normalized energy E can be expressed as

$$E = \int_{-\infty}^{\infty} |V(f)|^2 \, df = \int_{-\infty}^{\infty} [v(t)^2] \, dt$$

Hence

$$E = \int_{-\infty}^{\infty} |V(F)|^2 \, df$$

$$= 2 \int_{0}^{\infty} |V(f)|^2 \, df$$

where

$$|V(f)| = \begin{cases} f + 1, & \text{for } -1 \leq f \leq 0 \\ -f + 1, & \text{for } 0 \leq f \leq 1 \end{cases}$$

hence,

$$|V(F)|^2 = \begin{cases} (f+1)^2, & \text{for } -1 \leq f \leq 0 \\ (-f+1)^2, & \text{for } 0 \leq f \leq 1 \end{cases}$$

Therefore,

$$E = 2 \int_{0}^{1} (1-f)^2 \, df$$

$$= 2 \left[\frac{-(1-f)^3}{3} \right]_{0}^{1} = \frac{2}{3} \text{ volt}^2\text{-sec.}$$

(b) To find frequency f_1 for $\frac{1}{2} E$,

341

set $\frac{1}{2} E = \frac{1}{2} \times \frac{2}{3} = \frac{1}{3} - \int_{-f_1}^{f_1} |V(f)|^2 df$

$$= 2 \left[\frac{-(1-f)^3}{3} \right]_0^{f_1}$$

$$= \frac{2}{3} [1-(1-f_1)^3]$$

Hence,

$$\frac{1}{3} = \frac{2}{3} [1-(1-f_1)^3]$$

$$\left(\frac{1}{3}\right)\left(\frac{3}{2}\right) = [1-(1-f_1)^3]$$

$$\frac{1}{2} = (1-f_1)^3$$

$$f_1 = 1 - \frac{1}{\sqrt[3]{2}} \cong 0.206 \text{ Hz}$$

• **PROBLEM 6-26**

For the low-pass circuit shown in the figure, a periodic signal y(t) of period T is applied at the input. The Fourier coefficients Y_n are given as follows:

$Y_0 = 0$, $Y_1 = (1+j1)$, $Y_2 = 2$, $Y_n = 0$ for $n > 2$

where $T = 2\pi Rc$.

(a) Determine the normalized power of y(t).

(b) Determine the normalized power of v(t) - y(t).

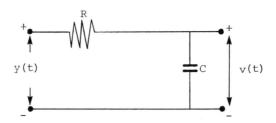

Solution: (a) By applying Parseval's theorem,

Normalized power of y(t) = Mean square value of y(t) = $\overline{y^2}$,

hence, $\overline{y^2} = \frac{1}{T} \int_0^T y^2(t) dt = \sum_{n=-\infty}^{\infty} Y_n Y_{-n}$.

Since the value $Y_n Y_{-n}$ is the same for either positive or negative values of n, and the only non-zero terms in the expansion for y(t) are Y_1, Y_2, Y_{-1}, and Y_{-2},

$$\overline{y^2} = Y_0^2 + 2\left[Y_1 Y_{-1} + Y_2 Y_{-2}\right]$$

$$= 0 + 2\left[(1+j)(1-j) + 2(2)\right]$$

$$= 2\left[1-(-1) + 4\right] = 2\left[2 + 4\right] = 12.$$

Therefore, the normalized power of y(t) = 12 V².

(b) Since $V(j\omega) = H(j\omega) Y(j\omega)$, the magnitude and phase of the sinusoidal components in the input are changed by the sinusoidal steady state transfer function.

The transfer function of the low-pass circuit is

$$H(j\omega_n) = \frac{(1/j\omega_n c)}{(1/j\omega_n c)+R} = \frac{1}{1+j\omega_n Rc}$$

where
$$\omega_n = n\omega_0 = n\left(\frac{2\pi}{T}\right)$$

hence,
$$\omega_n Rc = n\left(\frac{2\pi}{T}\right)(Rc) = n$$

and
$$H(j\omega_n) = \frac{1}{1+jn}$$

Now,
$$v(t) = \sum_{-\infty}^{+\infty} Y_n H(j\omega_n) e^{j\omega n t}$$

$$v(t) - y(t) = \sum_{-\infty}^{+\infty} Y_n H(j\omega_n) e^{j\omega n t} - \sum_{-\infty}^{+\infty} Y_n e^{j\omega n t}$$

$$= \sum_{-\infty}^{+\infty} Y_n \{H(j\omega_n)-1\} e^{j\omega n t}$$

$$= \sum_{-\infty}^{+\infty} Y_n' e^{j\omega n t}$$

where
$$Y_n' = Y_n \{H(j\omega_n)-1\}$$

$$= Y_n \left\{\frac{1}{1+jn} - 1\right\} = Y_n \left\{\frac{1-1-jn}{1+jn}\right\}$$

$$= \frac{-jn}{1+jn} Y_n$$

Thus
$$Y_0' = 0$$

343

$$Y_1' = \frac{-j}{1+j}(1+j) = -j$$

$$Y_2' = \frac{-2j}{1+j2}(2) = \frac{-4j(1-2j)}{5}$$

$$= -\frac{4}{5}(j+2) = -\frac{4}{5}(2+j)$$

Note: $Y_n' = 0$ for $n > 2$.

As in part (a) the normalized power of $y(t) - v(t)$ is

$$= Y_0' + 2Y_1' + Y_{-1}' + 2Y_2' + Y_{-2}'$$

$$= 0 + 2\left[(-j)(j) + \left(\frac{-4}{5}\right)(2+j)\left(\frac{-4}{5}\right)(2-j)\right]$$

$$= 2\left[1 + \frac{16}{25}(4+1)\right] = 2\left[1 + \frac{16(5)}{25}\right]$$

$$= 2\left[1 + \frac{16}{5}\right] = 8.4 \text{ V}^2$$

BANDLIMITING OF WAVEFORMS

• **PROBLEM 6-27**

A low-pass RC circuit is shown in Fig. 1. Determine the corresponding output $v_0(t)$ for inputs $v_1(t)$ and $v_2(t)$, which are shown in Fig. 2. Assume that $\tau \ll 1/f_2$ where f_2 is the 3-dB frequency of the filter.

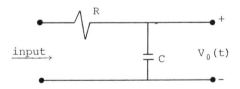

Fig. 1

Solution: The transfer function $H(\omega) = \dfrac{\frac{1}{j\omega C}}{\frac{1}{j\omega C}+R} = \dfrac{1}{1+j\omega RC}$

considering input $v_1(t)$ first, i.e. $v_1(t) = I\delta(t)$. The Fourier transform of the input is $V_1(\omega) = I(1) = I$.

The output is related to the input as follows:

$$V_0(\omega) = H(\omega) V_i(\omega) = \frac{1}{1+j\omega RC} \quad (I)$$

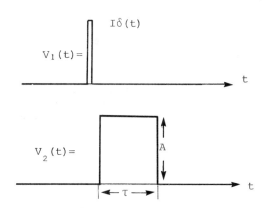

Fig. 2

$$= \frac{\frac{1}{RC}}{\frac{1}{RC} + j\omega} \quad (I)$$

Hence,

$$v_0(t) = \frac{1}{RC}(I)e^{-t/RC}$$

$$= \frac{I}{RC}e^{-t/RC} \text{ where } \frac{1}{a+j\omega} \leftrightarrow e^{-at}$$

Note: $\omega_2 = 2\pi f_2 = \frac{1}{RC}$ = cut-off frequency of the low-pass filter.

Next, consider input $v_2(t)$ similarly.

(Note: For a step input of amplitude A the output is

$$v_0(t) = A(1 - e^{-t/RC}) = A(1 - e^{-2\pi f_2 t}) \quad)$$

Now, realizing that the Fourier transform of the input $v_2(t)$ is $V_2(\omega) = A\tau \frac{\sin \omega\tau}{\omega\tau}$ and with $H(\omega) = \frac{1}{1+j\omega RC} = \frac{1}{1+jf/f_2} = \frac{1}{1+j\frac{\omega}{\omega_2}}$

$$V_0(\omega) = A\tau \frac{\sin \omega\tau}{\omega\tau} \frac{1}{1+j\frac{\omega}{\omega_2}}$$

$$= A\tau \frac{1}{1+j\frac{\omega}{\omega_2}}, \text{ for } \tau \ll \frac{1}{f_2}, \frac{\sin \omega\tau}{\omega\tau} = 1$$

$$= A\tau \frac{\omega_2}{\omega_2 + j\omega}$$

Hence,

$$v_0(t) = A\tau\omega_2 e^{-\omega_2 t} = \frac{A\tau}{RC} e^{-t/RC}.$$

If $\tau \ll 1/f_2$, the resulting output response is approximately the same as the one obtained by applying an impulse of strength $I' = A\tau$ as the input.

• **PROBLEM 6-28**

A high-pass RC circuit with a pulse input is shown in Fig. 1. Fig. 2 shows the response of the high-pass RC circuit. Verify that the area under the output waveform is zero.

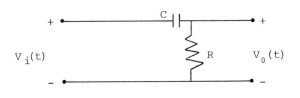

Fig. 1

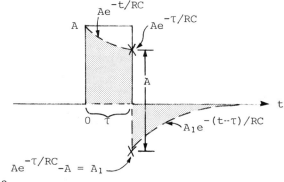

Fig. 2

Solution: The response of the high-pass circuit for $t < \tau$ decays exponentially to zero with time constant $\tau = \frac{1}{RC}$. This is the same as the response for a step input and is shown in Fig. 2.

At $t = \tau$, $v_0(t)$ drops abruptly by an amount A because, voltage across a capacitor cannot change instantaneously.

For $t > \tau$, the output again decays exponentially to zero with a time constant $\tau = \frac{1}{RC}$.

Therefore,

for $t < \tau$, $\quad v_0(t) = Ae^{\frac{-t}{RC}}$

$t = \tau$, $\quad v_0(t) = Ae^{-\tau/RC} - A = -A\left(1 - e^{-\tau/RC}\right) = A_1$

and $\quad t > \tau$, $\quad v_0(t) = A_1 e^{-(t-\tau)/RC}$.

Now, we can proceed to prove that the area under the curve $v_0(t)$ is zero, i.e. $\int_0^\infty v_0(t)dt = 0$

Hence,

$$\int_0^\infty v_0(t)dt = \int_0^\tau A e^{-t/RC} dt + \int_\tau^\infty A_1 e^{-(t-\tau)/RC} dt$$

$$= A \left[\frac{e^{-t/RC}}{-1/RC}\right]_0^\tau + A_1 \left[\frac{e^{-(t-\tau)/RC}}{-1/RC}\right]_\tau^\infty$$

$$= -A(RC)\left[e^{-\tau/RC} - 1\right] - A_1 RC\left[0-1\right]$$

$$= -ARC\, e^{-\tau/RC} + ARC - RC\left[A(1-e^{-\tau/RC})\right]$$

$$= -ARC\, e^{-\tau/RC} + ARC - ARC + ARC\, e^{-\tau/RC}$$

$$= 0$$

● **PROBLEM 6-29**

Given the power spectral density of a time function f(t) as follows:

$$G_X(\omega) = \begin{cases} \dfrac{P}{1+\left(\dfrac{\omega}{N}\right)^2} & \text{for } |\omega| < kN \\ 0 & \text{elsewhere} \end{cases}$$

where P, k, and N are positive real constants. Determine the root mean squared bandwidth of the power spectrum. What happens if $k \to \infty$?

Solution: The root mean squared bandwidth is a measure of the spread of a normalized power spectrum analogous to the standard deviation. The rms bandwidth is given by

$$\omega_{rms} = \frac{\int_{-\infty}^{+\infty} \omega^2\, G_X(\omega)\, d\omega}{\int_{-\infty}^{+\infty} G_X(\omega) d\omega}$$

$$= \frac{\displaystyle\int_{-kN}^{+kN} \omega^2 \frac{P}{1+\left(\frac{\omega}{N}\right)^2}\, d\omega}{\displaystyle\int_{-kN}^{+kN} \frac{P}{1+\left(\frac{\omega}{N}\right)^2}\, d\omega}$$

$$= \frac{2P \int_0^{kN} \frac{\omega^2}{1+\left(\frac{\omega}{N}\right)^2} d\omega}{2P \int_0^{kN} \frac{d\omega}{1+\left(\frac{\omega}{N}\right)^2}}$$

Now, substitute $y = \frac{\omega}{N}$

and notice: when $\omega = 0$, $y = 0$
when $\omega = kN$, $y = k$

Thus,

$$\omega_{rms} = \frac{\int_{y=0}^{y=k} \frac{N^2 y^2}{1+y^2} N dy}{\int_{y=0}^{y=k} \frac{1}{1+y^2} N dy}$$

$$= \frac{N^2 \int_0^k \frac{y^2}{1+y^2} dy}{\int_0^k \frac{dy}{1+y^2}}$$

Since, $\frac{y^2}{1+y^2}$ can be written as $\frac{y^2+1-1}{1+y^2} = 1 - \frac{1}{1+y^2}$

Hence,

$$\omega_{rms} = \frac{N^2 \int_0^k \left[1 - \frac{1}{1+y^2}\right] dy}{\int_0^k \frac{dy}{1+y^2}}$$

$$= \frac{N^2 \int_0^k dy}{\int_0^k \frac{dy}{1+y^2}} - \frac{N^2 \int_0^k \frac{dy}{1+y^2}}{\int_0^k \frac{dy}{1+y^2}}$$

where $\int_0^k \frac{dy}{1+y^2} = \tan^{-1} k$

$$\omega_{rms} = N^2 \left[\frac{k}{\tan^{-1} k} - 1\right] \quad (1)$$

Observing eq.(1), as $k \to \infty$, $\omega_{rms} \to \infty$.

AUTOCORRELATION

• **PROBLEM** 6-30

In general, the autocorrelation function $R(\tau)$ can be expressed as $R(\tau) = \lim_{T \to \infty} \frac{1}{T} \int_{-T/2}^{T/2} s(t)s(t+\tau)dt$. Show that $R(0) \geq R(\tau)$.

Solution: In order to prove that $R(0) \geq R(\tau)$, first consider the integral

$$I = \lim_{T \to \infty} \frac{1}{T} \int_{-T/2}^{T/2} [s(t) - s(t+\tau)]^2 dt. \qquad (1)$$

Multiplying out eq.(1)

$$I = \lim_{T \to \infty} \frac{1}{T} \left\{ \int_{-T/2}^{T/2} [s^2(t) + s^2(t+\tau) - 2s(t)s(t+\tau)]dt \right\}$$

$$= \lim_{T \to \infty} \frac{1}{T} \left\{ \int_{-T/2}^{T/2} s^2(t)dt + \int_{-T/2}^{T/2} s^2(t+\tau)dt - 2\int_{-T/2}^{T/2} s(t)s(t+\tau)dt \right\}$$

Now Let
$$I_1 = \lim_{T \to \infty} \frac{1}{T} \int_{-T/2}^{T/2} s^2(t)dt$$

$$I_2 = \lim_{T \to \infty} \frac{1}{T} \int_{-T/2}^{T/2} s^2(t+\tau)dt$$

and
$$I_3 = \lim_{T \to \infty} \frac{1}{T} \int_{-T/2}^{T/2} s(t)s(t+\tau)dt.$$

Hence, eq.(1) becomes: $I = I_1 + I_2 - 2I_3 \qquad (2)$

Now, evaluate I_1, I_2 and I_3 and express each result in desirable forms as follows:

$$I_1 = \lim_{T \to \infty} \frac{1}{T} \int_{-T/2}^{T/2} s^2(t)dt = \lim_{T \to \infty} \frac{1}{T} \int_{-T/2}^{T/2} s(t)s(t+0)dt$$

$$= R(0) \qquad (3)$$

$$I_2 = \lim_{T \to \infty} \frac{1}{T} \int_{-T/2}^{T/2} s^2(t+\tau)dt$$

letting $u = t+\tau$, $du = dt$

when $\quad t = -T/2, \quad u = -\dfrac{T}{2} + \tau$

when $\quad t = T/2, \quad u = T/2 + \tau$.

Hence,
$$I_2 = \lim_{T \to \infty} \frac{1}{T} \int_{u=-\frac{T}{2}+\tau}^{\frac{T}{2}+\tau} s^2(u) du$$

$$= \lim_{T \to \infty} \frac{1}{T} \int_{-\frac{T}{2}+\tau}^{\frac{T}{2}+\tau} s(u) \, s(u+0) du$$

$$= R(0) \tag{4}$$

Finally, $\quad I_3 = \lim_{T \to \infty} \dfrac{1}{T} \int_{-T/2}^{T/2} s(t)s(t+\tau)dt = R(\tau) \tag{5}$

Substituting eqs.(3), (4) and (5) into eq.(1), the following relationship is obtained.

$$I = 2\bigl[R(0) - R(\tau)\bigr]$$

Observe that

$$I = \lim_{T \to \infty} \frac{1}{T} \int_{-T/2}^{T/2} \bigl[s(t) - s(t+\tau)\bigr]^2 dt \quad \text{is greater}$$

than or equal to zero.

Therefore, $\quad I = 2\bigl[R(0) - R(\tau)\bigr] \geq 0$.

Hence, $\quad R(0) \geq R(\tau)$ is proved.

● **PROBLEM 6-31**

A periodic square wave m(t) is shown in Fig. 1. Determine the following:

(a) the autocorrelation function $R(\tau)$ for this waveform;

(b) the power spectral density $G(\omega)$.

Solution: Method I: Part (a) For the given periodic wave $R(\tau)$ = autocorrelation function

$$= \frac{1}{T} \int_{-T/2}^{T/2} m(t)m(t+\tau)d\tau. \tag{1}$$

Now, we must consider different cases for different time intervals.

Case 1: For $-T \leq \tau \leq -T/2$,

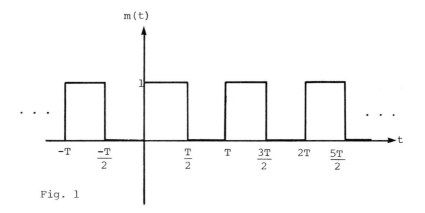

Fig. 1

$$R(\tau) = \frac{1}{T}\left[(1)(-\tau - \frac{T}{2})\right]$$

$$= -\frac{\tau}{T} - \frac{1}{2} \quad \text{(see Fig. 2.)}$$

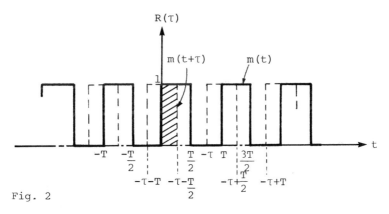

Fig. 2

Note: $R(\tau)$ for this time interval is obtained by applying eq.(1), namely, we multiply the waveforms $m(t)$ and $m(t+\tau)$ [which is $m(t)$ shifted to the left by τ]. The hatched area is exactly equal to the integral of $m(t) \times m(t+\tau)$ from $-\frac{T}{2}$ to $\frac{T}{2}$. Thus, $R(\tau)$ is found by calculating the hatched area and then dividing by the period T.

Similarly, case 2, 3 and 4 are calculated as follows:

Case 2: For $-T/2 < \tau \leq 0$,

$$R(\tau) = \frac{1}{T}\left[(1)\left(\frac{T}{2} - (-\tau)\right)\right]$$

$$= \frac{1}{T}\left[\frac{T}{2} + \tau\right]$$

$$= \frac{1}{2} + \frac{\tau}{T} \quad \text{(see Fig. 3.)}$$

Case 3: For $0 < \tau \leq T/2$,

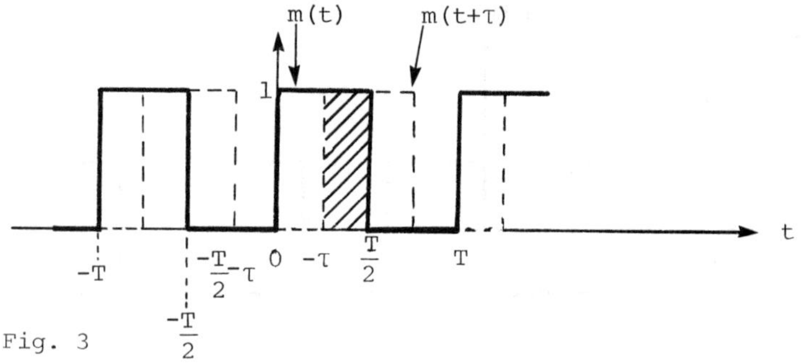

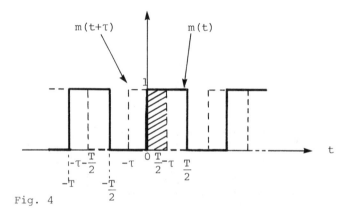

Fig. 3

$$R(\tau) = \frac{1}{T}\left[(1)\left(\left(\frac{T}{2}-\tau\right)-(0)\right)\right] = \frac{1}{2} - \frac{\tau}{T}$$

Fig. 4

Case 4: For $T/2 < \tau \leq T$,

$$R(\tau) = \frac{1}{T}\left\{(1)\left[\frac{T}{2} - (T-\tau)\right]\right\}$$

$$= \frac{1}{T}\left(-\frac{T}{2} + \tau\right)$$

$$= -\frac{1}{2} + \frac{\tau}{T}$$

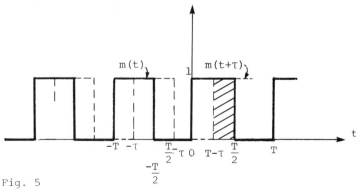

Fig. 5

Finally,

$$R(\tau) = \begin{cases} -\frac{1}{2} - \frac{\tau}{T} , & -T \leq \tau \leq -\frac{T}{2} \\ \frac{1}{2} + \frac{\tau}{T} , & -\frac{T}{2} < \tau \leq 0 \\ \frac{1}{2} - \frac{\tau}{T} , & 0 < \tau \leq \frac{T}{2} \\ -\frac{1}{2} + \frac{\tau}{T} , & \frac{T}{2} < \tau \leq T \end{cases}$$

and is plotted in Fig. (6).

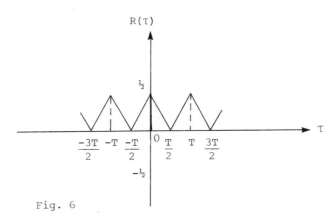

Fig. 6

Part (B) The autocorrelation function and the power spectral density of a periodic waveform are a Fourier transform pair. Hence,

$G(\omega)$ = Power Spectral Density of $m(t) = F[R(\tau)]$.

First determine $R(\tau)$.

$$R(\tau) = \sum_{n=-\infty}^{\infty} R_n e^{jn\omega_0 t} \quad \text{where} \quad \omega_0 = \frac{2\pi}{T}$$

and

$$R_n = \frac{1}{T} \int_{-T/2}^{T/2} R(\tau) e^{-jn\omega_0 \tau} d\tau$$

Thus,

$$R_n = \frac{1}{T} \left[\int_{-T/2}^{0} \left(\frac{1}{2} + \frac{\tau}{T}\right) e^{-jn\omega_0 \tau} d\tau + \int_{0}^{T/2} \left(\frac{1}{2} - \frac{\tau}{T}\right) e^{-jn\omega_0 \tau} d\tau \right]$$

$$= \frac{1}{T} \left[\int_{-T/2}^{T/2} \frac{1}{2} e^{-jn\omega_0 \tau} d\tau + \frac{1}{T} \int_{-T/2}^{0} \tau e^{-jn\omega_0 \tau} d\tau - \frac{1}{T} \int_{0}^{T/2} \tau e^{-jn\omega_0 \tau} d\tau \right]$$

$$= \frac{1}{2T} I_1 + \frac{1}{T^2} I_2 - \frac{1}{T^2} I_3 \tag{1}$$

in which

$$I_1 = \int_{-T/2}^{+T/2} e^{-jn\omega_0 \tau} d\tau \tag{2}$$

$$I_2 = \int_{-T/2}^{0} \tau e^{-jn\omega_0\tau} \, d\tau \tag{3}$$

and

$$I_3 = \int_{0}^{T/2} \tau e^{-jn\omega_0\tau} \, d\tau \tag{4}$$

Now, evaluating I_1, I_2 and I_3 respectively,

$$I_1 = \int_{-T/2}^{+T/2} e^{-jn\omega_0\tau} \, d\tau$$

$$= \int_{-T/2}^{+T/2} (\cos n\omega_0\tau) - j(\sin n\omega_0\tau) \, d\tau$$

for all n, $\sin \frac{2\pi n\tau}{T} = 0$ hence,

$$I_1 = \int_{-T/2}^{+T/2} \cos(n\omega_0\tau) \, d\tau = 2\int_{0}^{T/2} \cos(n\omega_0\tau) \, d\tau$$

$$= 2 \left[\frac{\sin(n\omega_0\tau)}{n\omega_0} \right]_{\tau=0}^{\tau=T/2}$$

$$= \frac{2}{n\omega_0} \left[\sin\left(\frac{n\omega_0 T}{2}\right) - 0 \right]$$

$$= \frac{2(T/2)}{n\omega_0 \frac{T}{2}} \sin\left(\frac{n\omega_0 T}{2}\right)$$

$$= T \frac{\sin\left(\frac{n\omega_0 T}{2}\right)}{\left(\frac{n\omega_0 T}{2}\right)} \tag{5}$$

Note: For an indefinite integral

$$\int \tau e^{-jn\omega_0\tau} d\tau,$$ we could do the following:

By applying the chain rule for integration,

$$\int u \, dv = uv_1 - u'v_2 + u''v_3 \ldots$$

where u' = First derivative of u
and v_1 = First integral of dv, etc.

Therefore we have,

$$\int \tau e^{-jn\omega_0\tau} d\tau = (\tau) \left[\frac{e^{-jn\omega_0\tau}}{(-jn\omega_0)} \right] - (1) \left[\frac{e^{-jn\omega_0\tau}}{(-jn\omega_0)^2} \right]$$

so,
$$I_2 = \int_{-T/2}^{0} \tau e^{-jn\omega_0\tau} d\tau = j\tau\left[\frac{e^{-jn\omega_0\tau}}{n\omega_0}\right] + \left[\frac{e^{-jn\omega_0\tau}}{n^2\omega_0^2}\right]$$

$$= \left[j\tau \frac{e^{-jn\omega_0\tau}}{n\omega_0} + \frac{e^{-jn\omega_0\tau}}{n^2\omega_0^2}\right]_{\tau=-T/2}^{\tau=0}$$

$$= \left[\frac{1}{n^2\omega_0^2} - \left(-j\frac{T}{2}\frac{e^{jn\omega_0\frac{T}{2}}}{n\omega_0} + \frac{e^{jn\omega_0 T/2}}{n^2\omega_0^2}\right)\right]$$

$$= \left[\frac{1}{n^2\omega_0^2} + j\frac{T}{2}\frac{e^{j\frac{n\omega_0 T}{2}}}{n\omega_0} + \frac{e^{j\frac{n\omega_0 T}{2}}}{n^2\omega_0^2}\right] \quad (6)$$

Finally,
$$I_3 = \int_{0}^{T/2} \tau e^{-jn\omega_0\tau} d\tau = \left[j\tau\frac{e^{-jn\omega_0\tau}}{n\omega_0} + \frac{e^{-jn\omega_0\tau}}{n^2\omega_0^2}\right]_{\tau=0}^{\tau=T/2}$$

$$= \left[\left(j\frac{T}{2}\frac{e^{-jn\omega_0 T/2}}{n\omega_0} + \frac{e^{-jn\omega_0 T/2}}{n^2\omega_0^2}\right) - \frac{1}{n^2\omega_0^2}\right]$$

$$= \left[\frac{-1}{n^2\omega_0^2} + j\frac{T}{2}\frac{e^{-jn\omega_0 T/2}}{n\omega_0} + \frac{e^{-jn\omega_0 T/2}}{n^2\omega_0^2}\right] \quad (7)$$

Now, substituting equations (5), (6) and (7) into equation (1) for R_n gives

$$R_n = \frac{1}{2T} T \frac{\sin\left(\frac{n\omega_0 T}{2}\right)}{\left(\frac{n\omega_0 T}{2}\right)}$$

$$+ \frac{1}{T^2}\left[\frac{1}{n^2\omega_0^2} + j\frac{T}{2}\frac{e^{jn\omega_0 T/2}}{n\omega_0} - \frac{e^{jn\omega_0 T/2}}{n^2\omega_0^2}\right]$$

$$- \frac{1}{T^2}\left[\frac{-1}{n^2\omega_0^2} + j\frac{T}{2}\frac{e^{-jn\omega_0 T/2}}{n\omega_0} + \frac{e^{-jn\omega_0 T/2}}{n^2\omega_0^2}\right]$$

$$= \frac{1}{2}\frac{\sin\left(\frac{n\omega_0 T}{2}\right)}{\left(\frac{n\omega_0 T}{2}\right)} + \frac{1}{T^2}\left(\frac{2}{n^2\omega_0^2}\right) + \frac{j}{2n\omega_0 T}\left(e^{jn\omega_0\frac{T}{2}} - e^{-jn\omega_0\frac{T}{2}}\right)$$

$$- \frac{1}{n^2\omega_0^2 T^2}\left(e^{jn\omega_0\frac{T}{2}} + e^{-jn\omega_0\frac{T}{2}}\right)$$

$$= \frac{1}{2}\frac{\sin\left(\frac{n\omega_0 T}{2}\right)}{\left(n\frac{\omega_0 T}{2}\right)} + \frac{2}{n^2\omega_0^2 T^2} + \frac{(-1)}{2n\omega_0 T}\left(2\sin\frac{n\omega_0 T}{2}\right) - \frac{2\cos n\omega_0 T/2}{n^2\omega_0^2 T^2}$$

$$R_n = \frac{2}{n^2\omega_0^2 T^2}\left[1 - \cos\frac{n\omega_0 T}{2}\right]$$

$$= \frac{2}{n^2\omega_0^2 T^2}\left[2\sin^2\frac{n\omega_0 T}{4}\right]$$

$$= \frac{1}{4}\left[\frac{\sin\left(\frac{n\omega_0 T}{4}\right)}{\left(\frac{n\omega_0 T}{4}\right)}\right]^2 = \frac{1}{4}\left[\frac{\sin x}{x}\right]^2 \tag{8}$$

where $\quad x = \dfrac{n\omega_0 T}{4} = \dfrac{n\left(\frac{2\pi}{T}\right)(T)}{4} = \dfrac{n\pi}{2}$

Therefore,
$$R(\tau) = \sum_{n=-\infty}^{\infty} R_n e^{jn\omega_0\tau} = \sum_{n=-\infty}^{\infty} \frac{1}{4}\left[\frac{\sin\left(\frac{n\pi}{2}\right)}{\left(\frac{n\pi}{2}\right)}\right]^2 e^{jn\omega_0\tau}$$

Since $G(\omega) = F[R(\tau)] =$ Fourier Transform of $R(\tau)$

$$= \int_{-\infty}^{\infty}\left\{\sum_{n=-\infty}^{\infty}\frac{1}{4}\left[\frac{\sin\left(\frac{n\pi}{2}\right)}{\left(\frac{n\pi}{2}\right)}\right]^2 e^{jn\omega_0\tau}\right\} e^{-j\omega\tau}\,d\tau$$

$$= \frac{\pi}{2}\sum_{n=-\infty}^{\infty}\left[\frac{\sin(n\pi/2)}{\frac{n\pi}{2}}\right]^2 \delta(\omega - n\omega_0)$$

where $c \leftrightarrow 2\pi c\delta(\omega)$ was used.

Thus,
$$G(\omega) = \sum_{n=-\infty}^{\infty}\frac{\pi}{2}\left[\frac{\sin\frac{n\pi}{2}}{\frac{n\pi}{2}}\right]^2 \delta(\omega - n\omega_0).$$

Method II:

a) For $m(t)$, a periodic waveform, $m(t)$ can be written as

$$m(t) = \sum_{n=-\infty}^{+\infty} M_n e^{j2\pi nt/T_0}$$

Now, using the autocorrelation function $R(\tau)$ in the form

$$R(\tau) = \frac{1}{T_0}\int_{-T_0/2}^{T_0/2} m(t)m(t+\tau)\,dt \quad \text{and}$$

substitute in $m(t)$ and $m(t+\tau)$, where $m(t+\tau)$ is simply equal to

$$\sum_{n=-\infty}^{+\infty} M_n e^{j2\pi n(t+\tau)/T_0}$$

Thus, $R(\tau)$ becomes

$$R(\tau) = \frac{1}{T_0}\int_{-T_0/2}^{T_0/2}\left[\sum_{m=-\infty}^{+\infty} M_m e^{j2\pi mt/T_0}\right]\left[\sum_{n=-\infty}^{+\infty} M_n e^{j2\pi n(t+\tau)/T_0}\right]dt \quad (1)$$

$$= \sum_{n,m=-\infty}^{+\infty}\frac{1}{T_0}\int_{-T_0/2}^{T_0/2}(M_m e^{j2\pi mt/T_0})\left[\sum_{n=-\infty}^{+\infty} M_n e^{j2\pi n(t+\tau)/T_0}\right]dt \quad (1)$$

Evaluating the integral separately, hence,

$$\frac{1}{T_0}\int_{-T_0/2}^{T_0/2}(M_m e^{j2\pi mt/T_0})(M_n e^{j2\pi n(t+\tau)/T_0})dt$$

$$= \frac{1}{T_0}e^{j2\pi n\tau/T_0}\int_{-T_0/2}^{T_0/2} M_m M_n e^{j2\pi(m+n)t/T_0}dt$$

$$= \frac{1}{T_0}e^{j2\pi n\tau/T_0}(M_m M_n)\int_{-T_0/2}^{T_0/2} e^{j2\pi(m+n)t/T_0}dt$$

$$= \frac{1}{T_0}e^{j2\pi n\tau/T_0}(M_m M_n)\frac{T_0}{j2\pi(m+n)}\left[e^{\frac{j2\pi(m+n)}{2}} - e^{\frac{-j2\pi(m+n)}{2}}\right]$$

$$= e^{\frac{j2\pi n\tau}{T_0}} M_m M_n \left[\frac{e^{j\pi(m+n)} - e^{-j\pi(m+n)}}{2j\pi(m+n)}\right]$$

$$= e^{\frac{j2\pi n\tau}{T_0}} M_m M_n \left[\frac{\sin\pi(m+n)}{\pi(m+n)}\right] \quad (2)$$

Notice that in Eq.(2), for all integers values m and n, eq. (2) will evaluate to zero except for m = -n since

$$\frac{\sin x}{x} \to 1 \quad \text{for } x \to 0.$$

Therefore, replacing m by -n and letting $\frac{\sin\pi(m+n)}{\pi(m+n)} = 1$ in Eq. (2), hence the result

$$M_{-n} M_n e^{\frac{j2\pi n\tau}{T_0}}$$

Now, going back to eq.(1), $R(\tau)$ becomes

$$R(\tau) = \sum_{n=-\infty}^{+\infty} M_n M_{-n} e^{j2\pi n\tau/T_0} = \sum_{n=-\infty}^{+\infty} |M_n|^2 e^{j2\pi n\tau/T_0}$$

$$= |M_0|^2 + 2\sum_{n=1}^{+\infty} |M_n|^2 \cos 2\pi n \frac{\tau}{T_0}$$

Finally,

$$R(\tau) = \frac{1}{T}\int_{-\frac{T}{2}}^{\frac{T}{2}} m(t)m(t+\tau)dt = \frac{1}{4} + \sum_{n=-\infty}^{+\infty}\left[\frac{1}{\pi(2n+1)}\right]^2 e^{\frac{j2\pi(2n+1)\tau}{T}}$$

$$= \frac{1}{4} + 2 \sum_{n=1}^{+\infty} \left[\frac{1}{\pi(2n+1)}\right]^2 \cos\left[2\pi(2n+1)\frac{\tau}{T}\right]$$

b) The relationship between the power spectral density $G(f)$ and the autocorrelation function $R(\tau)$ is as follows:

$$G(f) = F\left[R(\tau)\right] = \int_{-\infty}^{\infty} R(\tau) e^{-2\pi f \tau} d\tau$$

Hence, $G(f) = \int_{-\infty}^{+\infty} \left[\frac{1}{4} + \sum_{n=-\infty}^{\infty} \left[\frac{1}{\pi(2n+1)}\right]^2 e^{\frac{j2\pi(2n+1)\tau}{T}}\right] e^{-j2\pi f \tau} d\tau$

$$= \frac{1}{4} \int_{-\infty}^{+\infty} e^{-j2\pi f \tau} d\tau + \sum_{n=-\infty}^{+\infty} \left[\frac{1}{\pi(2n+1)}\right]^2 \times$$

$$\int_{-\infty}^{+\infty} e^{-j2\pi\left(f - \frac{2n+1}{T}\right)\tau} d\tau$$

$$= \frac{1}{4} \delta(f) + \sum_{n=-\infty}^{+\infty} \left[\frac{1}{\pi(2n+1)^2}\right]^2 \delta\left(f - \frac{2n+1}{T}\right)$$

● **PROBLEM 6-32**

(a) Determine the autocorrelation function for the rectangular pulse $y(t)$ shown in Fig. A.

(b) Determine the power spectral density of $y(t)$.

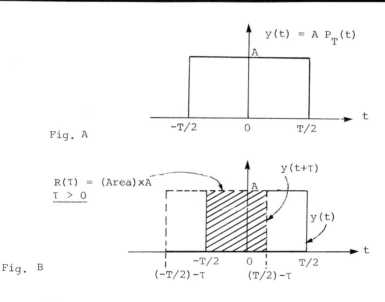

Fig. A

Fig. B

Solution: In Fig. A, the function $y(t)$ can be represented as $AP_T(t)$ where $P_T(t)$ is a pulse of unit magnitude, width T, and centered at $t = 0$.

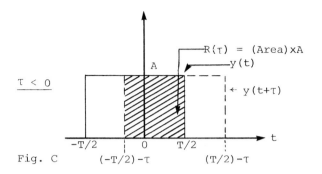

Fig. C

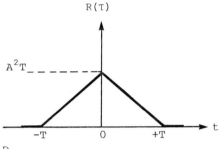

Fig. D

The autocorrelation function $R(\tau)$ for signals of finite energy is defined as

$$R(\tau) = \int_{-\infty}^{+\infty} y(t)\, y(t+\tau)\, dt$$

and is an even function of τ.

To determine $R(\tau)$, the product $y(t)y(t+\tau)$ is to be integrated with respect to t from $-\infty$ to $+\infty$. By observing the sketch in Fig. B, $R(\tau)$ is zero for $\tau > T$ because if τ is positive, $y(t+\tau)$ is equivalent to the translation of $y(t)$ to the left by τ units. By the same token, from Fig. C, $R(\tau)$ is zero for $\tau < -T$ because if τ is negative, $y(t)$ will be translated to the right by $|\tau|$ units. Hence, $R(\tau)$ is non-zero, only for values $|\tau| < T$. For $0 < \tau < T$, from Fig. B, $R(\tau)$ is equal to the area of the shaded portion multiplied with A, which is $A^2(T-\tau)$.

Similarly, from Fig. C, for $-T < \tau < 0$, $R(\tau) = A^2(T+\tau)$. Thus,

$$R(\tau) = \begin{cases} A^2(T-|\tau|), & \text{for } |\tau| < T \\ 0, & \text{elsewhere} \end{cases}$$

which is plotted in Fig. D.

(b) The power spectral density and the autocorrelation function of $y(t)$ are a Fourier transform pair. Hence,
$G(\omega)$ = Power spectral density of $y(t)$

$\qquad = F[R(\tau)]$

So,
$$G(\omega) = \int_{-\infty}^{+\infty} R(\tau) e^{-j\omega\tau} d\tau$$

$$= \int_{-T}^{0} A(T+\tau)e^{-j\omega\tau}d\tau + \int_{0}^{T} A(T-\tau)e^{-j\omega\tau}d\tau$$

$$= AT\int_{-T}^{T} e^{-j\omega\tau}d\tau + A\int_{-T}^{0} \tau e^{-j\omega\tau}d\tau - A\int_{0}^{T} \tau e^{-j\omega\tau}d\tau$$

$$G(\omega) = AT(I_1) + A(I_2) - A(I_3) \qquad (1)$$

where
$$I_1 = \int_{-T}^{+T} e^{-j\omega\tau} d\tau ,$$

$$I_2 = \int_{-T}^{0} \tau e^{-j\omega\tau} d\tau ,$$

and
$$I_3 = \int_{0}^{T} \tau e^{-j\omega\tau} d\tau$$

$$I_1 = \int_{-T}^{+T} e^{-j\omega\tau}d\tau = \int_{-T}^{T} \cos\omega\tau d\tau = 2\int_{0}^{T} \cos\omega\tau d\tau$$

$$= 2\left[\frac{\sin\omega\tau}{\omega}\right]_{0}^{T} = 2T\,\frac{\sin \omega T}{\omega T} .$$

$$I_2 = \int_{-T}^{0} \tau e^{-j\omega\tau}d\tau = \frac{1}{\omega^2}\left[(j\omega\tau e^{-j\omega\tau} + e^{-j\omega\tau})\right]_{-T}^{0}$$

Note:
$$\int \tau e^{-j\omega\tau}d\tau = (\tau)\left(\frac{e^{-j\omega\tau}}{-j\omega}\right) - (1)\left(\frac{e^{-j\omega\tau}}{-\omega^2}\right)$$

$$= \frac{j\omega\tau}{\omega^2} e^{-j\omega\tau} + \frac{1}{\omega^2} e^{-j\omega\tau}$$

$$= \frac{1}{\omega^2} (j\omega\tau\, e^{-j\omega\tau} + e^{-j\omega\tau})$$

$$I_2 = \frac{1}{\omega^2}\left[\{1\} - \{-j\omega T e^{j\omega T} + e^{j\omega T}\}\right]$$

$$= \frac{1}{\omega^2}\left[1 + j\omega T\, e^{j\omega T} - e^{j\omega T}\right].$$

$$I_3 = \int_{0}^{T} \tau e^{-j\omega\tau}d\tau = \frac{1}{\omega^2}\left[j\omega\tau e^{-j\omega\tau} + e^{-j\omega\tau}\right]_{0}^{T}$$

$$= \frac{1}{\omega^2}\left[\{j\omega T\, e^{-j\omega T} + e^{-j\omega T}\} - \{1\}\right]$$

$$= \frac{1}{\omega^2}\left[j\omega T\, e^{-j\omega T} + e^{-j\omega T} - 1\right]$$

Substituting the values for I_1, I_2, and I_3 into eq.(1)

$$G(\omega) = 2AT^2\,\frac{\sin \omega T}{\omega T} + A\,\frac{1}{\omega^2}\left[2 - (e^{j\omega T} + e^{-j\omega T}) + j\omega T(e^{j\omega T} - e^{-j\omega T})\right]$$

using $\quad e^{j\omega T} + e^{-j\omega T} = 2\cos \omega T$

$\qquad\qquad e^{j\omega T} - e^{-j\omega T} = 2j \sin \omega T$

and $\quad 1 - \cos X = 2\sin^2 \frac{X}{2}$

$$G(\omega) = 2AT^2\,\frac{\sin \omega T}{\omega T} + \frac{A}{\omega^2}\left[2 - 2\cos\omega T + j\omega T(2j\sin \omega T)\right]$$

$$= 2AT^2\,\frac{\sin \omega T}{\omega T} + \frac{2A}{\omega^2}\left[(1-\cos\omega T) - \omega T \sin \omega T\right]$$

$$= 2AT^2\,\frac{\sin \omega T}{\omega T} + \frac{2A}{\omega^2}(2\sin^2 \tfrac{\omega T}{2}) - 2AT^2\,\frac{\sin \omega T}{\omega T}$$

$$= \frac{AT^2}{\left(\frac{\omega^2 T^2}{4}\right)} \sin^2\left(\tfrac{\omega T}{2}\right) = AT^2\,\frac{\sin^2\left(\frac{\omega T}{2}\right)}{\left(\frac{\omega T}{2}\right)^2}$$

Since the energy density spectrum of the rectangular pulse in Fig. A is given by $(1/\pi)$ times its power spectral density, the energy density function $E(\omega) = \frac{1}{\pi} G(\omega)$

$$= \frac{1}{\pi} AT^2\,\frac{\sin^2\left(\frac{\omega T}{2}\right)}{\left(\frac{\omega T}{2}\right)^2}.$$

Note that $R(\tau)$ is an even function of τ and is maximum at $\tau = 0$.

• PROBLEM 6-33

Given a sinusoidal waveform $s(t) = \sin\omega_0 t$,

(a) write an expression for the autocorrelation function $R(\tau)$ of this waveform;

(b) determine its power spectral density $G(f)$.

Solution: (a) By definition, the autocorrelation function for any $s(t)$ is given as follows:

$$R(\tau) = \frac{1}{T} \int_{-T/2}^{T/2} s(t)s(t+\tau)dt$$

Hence, for the given waveform

$$R(\tau) = \frac{1}{T} \int_{-T/2}^{T/2} \sin(\omega_0 t)\sin\omega_0(t+\tau)dt$$

where T = Period of sinusoidal signal = $2\pi/\omega_0$.

Applying the trigonometric identity

$$\sin A \sin B = \frac{1}{2}[\cos(A-B) - \cos(A+B)],$$

$$\sin\omega_0(t+\tau)\sin\omega_0 t = \frac{1}{2}[\cos\omega_0(\tau) - \cos\omega_0(2t+\tau)]$$

Hence,

$$R(\tau) = \frac{1}{T} \int_{-T/2}^{T/2} \frac{1}{2}[\cos\omega_0(\tau) - \cos\omega_0(2t+\tau)]dt$$

$$= \frac{\cos\omega_0\tau}{2T} \int_{-T/2}^{T/2} dt - \frac{1}{2T} \int_{-T/2}^{T/2} \cos\omega_0(2t+\tau)dt$$

$$= \frac{\cos\omega_0\tau}{2T}(T) - \frac{1}{2T}\left[\frac{\sin\omega_0(2t+\tau)}{\omega_0(2)}\right]_{-T/2}^{T/2}$$

$$= \frac{1}{2}\cos\omega_0\tau - \frac{1}{4\omega_0 T}[\sin\omega_0(T+\tau) - \sin\omega_0(-T+\tau)]$$

$$= \frac{1}{2}\cos\omega_0\tau - \frac{1}{4\omega_0 T}[2\cos\omega_0\tau \sin\omega_0 T].$$

Now, notice that $\omega_0 T = 2\pi$ and $\sin 2\pi = 0$. Therefore, $R(\tau) = \frac{1}{2}\cos\omega_0\tau$.

(b) The power spectral density $G(\omega)$ can be easily obtained, since $G(\omega) = F[R(\tau)]$.

Hence,

$$G(\omega) = F\left[\frac{1}{2}\cos\omega_0\tau\right]$$

$$G(\omega) = F\left[\frac{1}{2}\left(\frac{e^{j\omega_0\tau} + e^{-j\omega_0\tau}}{2}\right)\right]$$

$$= F\left[\frac{1}{4}e^{j\omega_0\tau} + \frac{1}{4}e^{-j\omega_0\tau}\right]$$

$$= \int_{-\infty}^{\infty} (\tfrac{1}{4} e^{j\omega_0 \tau}) e^{-j\omega\tau} dt + \int_{-\infty}^{\infty} (\tfrac{1}{4} e^{-j\omega_0 \tau}) e^{-j\omega\tau} d\tau$$

$$= \int_{-\infty}^{\infty} \tfrac{1}{4} e^{-j(\omega-\omega_0)\tau} dt + \int_{-\infty}^{\infty} \tfrac{1}{4} e^{-j(\omega+\omega_0)\tau} d\tau$$

Using the Fourier transform pair $c \leftrightarrow 2\pi c \delta(\omega)$

$$G(\omega) = 2\pi \left(\tfrac{1}{4}\right) \delta(\omega-\omega_0) + 2\pi \left(\tfrac{1}{4}\right) \delta(\omega+\omega_0)$$

$$= \tfrac{\pi}{2} [\delta(\omega-\omega_0) + \delta(\omega+\omega_0)].$$

• **PROBLEM** 6-34

A nonperiodic waveform $v(t)$ with amplitude A ranging from $t = -t_0/2$ to $t = +t_0/2$ is shown in Fig. 1.

(a) Determine an expression for the correlation function of this waveform.

(b) The correlation function $R(\tau)$ and the energy spectral density are a Fourier transform pair. Hence, find the energy spectral density $G_E(\omega)$.

(c) Find $G_E(F)$ by applying Parseval's theorem.

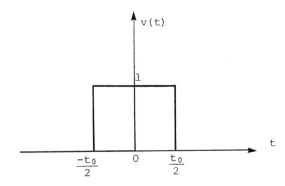

Fig. 1

<u>Solution</u>: (a) For a nonperiodic waveform of finite energy,

$$R(\tau) = \int_{-\infty}^{+\infty} v(t) v(t+\tau) dt.$$

Consider,

Case 1: For $t_0 \leq \tau \leq -t_0/2$. $R(\tau)$ is equal to the magnitude of the shaded region multiplyed with A. This multiplication by the factor A, accounts for the integration of the product $v(t) v(t+\tau)$.

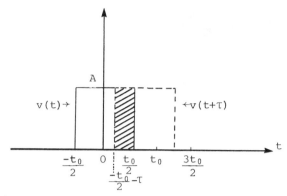

Fig. 2

From the figure,

$$R(\tau) = A^2\left[\left(\frac{t_0}{2}\right) - \left(\frac{-t_0}{2} - \tau\right)\right]$$

$$= A^2[t_0 + \tau]$$

Similarly, case 2, 3 and 4 are considered as follows:

Case 2: For $\frac{-t_0}{2} < \tau \leq 0$

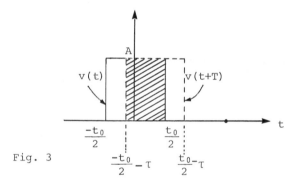

Fig. 3

$$R(\tau) = A^2\left[\frac{t_0}{2} - \left(-\frac{t_0}{2} - \tau\right)\right]$$

$$= A^2(t_0 + \tau)$$

Case 3: For $0 < \tau < t_0/2$

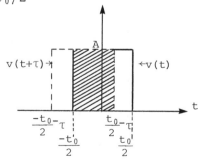

Fig. 4

$$R(\tau) = A^2\left[\left(\frac{t_0}{2}-\tau\right)-\left(\frac{-t_0}{2}\right)\right] = A^2\left[t_0-\tau\right]$$

Case 4: For $\frac{t_0}{2} < \tau < t_0$

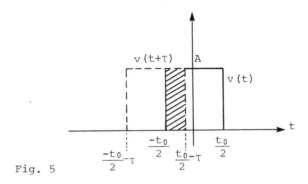

Fig. 5

$$R(\tau) = A^2\left[\left(\frac{t_0}{2}-\tau\right)-\left(\frac{-t_0}{2}\right)\right] = A^2\left[t_0-\tau\right]$$

Summarizing all four cases, the autocorrelation function $R(\tau)$ is

$$R(\tau) = \begin{cases} A^2\left[t_0+\tau\right], & -t_0 \leq \tau \leq 0 \\ A^2\left[t_0-\tau\right], & 0 < \tau \leq t \\ 0, & \text{elsewhere} \end{cases}$$

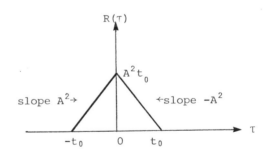

Fig. 6

and is plotted in Fig. 6.

(b) The correlation function and the energy spectral density form a Fourier transform pair.

Hence,

$$G_E(\omega) = F[R(\tau)] = \int_{-\infty}^{+\infty} R(\tau)e^{-j\omega\tau}d\tau$$

$$= \int_{-t_0}^{0} A^2(t_0+\tau)e^{-j\omega\tau}d\tau + \int_{0}^{t_0} A^2(t_0-\tau)e^{-j\omega\tau}d\tau$$

$$= A^2 t_0 \int_{-t_0}^{t_0} e^{-j\omega\tau} d\tau + A^2 \int_{-t_0}^{0} \tau e^{-j\omega\tau} d\tau - A^2 \int_{0}^{t_0} \tau e^{j\omega\tau} d\tau$$

$$= A^2 t_0 I_1 + A^2 I_2 - A^2 I_3 \tag{1}$$

where
$$I_1 = \int_{-t_0}^{t_0} e^{-j\omega\tau} d\tau = 2 \int_{0}^{t_0} \cos\omega\tau \, d\tau$$

$$= 2 \left[\frac{\sin \omega\tau}{\omega} \right]_0^{t_0}$$

$$= \frac{2t_0}{\omega t_0} [\sin\omega t_0] = 2t_0 \frac{\sin \omega t_0}{\omega t_0}$$

For I_2, first consider the integral $\int \tau e^{-j\omega\tau} d\tau$.

Applying the chain rule for integration:

$$\int \tau e^{-j\omega\tau} d\tau = \tau \left[\frac{e^{-j\omega\tau}}{-j\omega} \right] - (1) \left[\frac{e^{-j\omega\tau}}{(-j\omega)^2} \right]$$

$$= \frac{j\tau}{\omega} e^{-j\omega\tau} + \frac{e^{-j\omega\tau}}{\omega^2}$$

Thus,
$$I_2 = \int_{-t_0}^{0} \tau e^{-j\omega\tau} d\tau = \left[\frac{j\tau}{\omega} e^{-j\omega\tau} + \frac{e^{-j\omega\tau}}{\omega^2} \right]_{-t_0}^{0}$$

$$= \left[\left(\frac{1}{\omega^2}\right) - \left(\frac{-jt_0}{\omega} e^{j\omega t_0} + \frac{e^{j\omega t_0}}{\omega^2} \right) \right]$$

$$= \left[\frac{1}{\omega^2} + \frac{jt_0}{\omega} e^{j\omega t_0} - \frac{e^{j\omega t_0}}{\omega^2} \right]$$

and similarly, for I_3

$$I_3 = \int_{0}^{t_0} \tau e^{-j\omega\tau} d\tau = \left[\frac{j\tau}{\omega} e^{-j\omega\tau} + \frac{e^{-j\omega\tau}}{\omega^2} \right]_0^{t_0}$$

$$= \left[\left(\frac{jt_0}{\omega} e^{-j\omega t_0} + \frac{e^{-j\omega t_0}}{\omega^2} \right) - \left(\frac{1}{\omega^2} \right) \right]$$

$$= \left[-\frac{1}{\omega^2} + \frac{jt_0}{\omega} e^{-jt_0\omega} + \frac{e^{-j\omega t_0}}{\omega^2} \right]$$

Now, substituting I_1, I_2 and I_3 into eq.(1)

$$G_E(\omega) = A^2 t_0 (2t_0) \frac{\sin\omega t_0}{\omega t_0} + A^2 \left[\frac{2}{\omega^2} + \frac{jt_0}{\omega}(e^{j\omega t_0} - e^{-j\omega t_0}) \right.$$
$$\left. - \frac{1}{\omega^2}(e^{j\omega t_0} + e^{-j\omega t_0}) \right]$$

$$= \frac{2A^2 t_0}{\omega} \sin\omega t_0 + A^2 \left[\frac{2}{\omega^2} + \frac{jt_0}{\omega}(2j\sin\omega t_0) - \frac{2\cos\omega t_0}{\omega^2} \right]$$

$$= \frac{2A^2 t_0}{\omega} \sin\omega t_0 + \frac{2A^2}{\omega^2} - \frac{2A^2 t_0}{\omega} \sin\omega t_0 - \frac{2A^2 \cos\omega t_0}{\omega^2}$$

$$= \frac{2A^2}{\omega^2} (1 - \cos\omega t_0)$$

$$= \frac{2A^2}{\omega^2} 2\sin^2 \frac{\omega t_0}{2}$$

$$= \frac{4A^2 t_0^2}{\omega^2 t_0^2} \sin^2 \left(\frac{\omega t_0}{2}\right)$$

$$= A^2 t_0^2 \left[\frac{\sin\left(\frac{\omega t_0}{2}\right)}{\left(\frac{\omega t_0}{2}\right)} \right]^2$$

(c) By Parseval's theorem, for a nonperiodic pulse,

$$G_E(\omega) = |V(f)|^2 = |F[v(t)]|^2$$

$$= \left| \int_{-t_0/2}^{t_0/2} A e^{-j2\pi f t} dt \right|^2$$

$$= \left| A \frac{e^{-j\pi f t_0} - e^{j\pi f t_0}}{-j2\pi f} \right|^2$$

$$= A \left(\frac{\sin\pi f t_0}{\pi f} \right)^2$$

$$= A^2 t_0^2 \left[\frac{\sin\left(\frac{\omega t_0}{2}\right)}{\left(\frac{\omega t_0}{2}\right)} \right]^2 \qquad \text{where } \frac{\omega}{2} = \pi f.$$

CHAPTER 7

FREQUENCY RESPONSE IN LINEAR SYSTEMS

TRANSFER FUNCTION OF A SYSTEM

• **PROBLEM 7-1**

In the circuit of Fig. 1, derive the network function relating v_1 to v by voltage division.

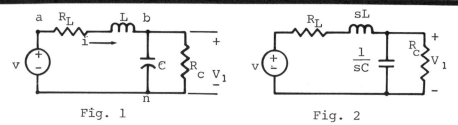

Fig. 1   Fig. 2

Solution: We use complex frequency s, as shown in Fig. 2, to express each impedance.

Combining $\frac{1}{sC}$ and R_C we obtain the circuit in Fig. 3.

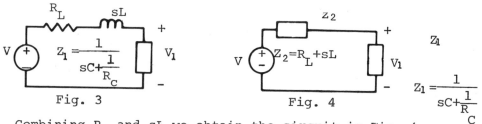

Fig. 3   Fig. 4

$$Z_1 = \frac{1}{sC + \frac{1}{R_C}}$$

Combining R_L and sL we obtain the circuit in Fig. 4.

Now by voltage division,

$$V_1 = \frac{Z_1 V}{Z_1 + Z_2}$$

thus, $$V_1 = \frac{\dfrac{V}{sC + \dfrac{1}{R_C}}}{R_L + sL + \dfrac{1}{sC + \dfrac{1}{R_C}}}$$

simplifying, $$V_1 = \frac{\dfrac{V R_C}{s R_C C + 1}}{R_L + sL + \dfrac{R_C}{s R_C C + 1}}$$

$$V_1 = \frac{\dfrac{V R_C}{s R_C C + 1}}{\dfrac{R_L (s R_C C + 1) + sL (s R_C C + 1) + R_C}{s R_C C + 1}}$$

yields $$\frac{V_1}{V} = H(s) = \frac{R_C}{s^2 (R_C L C) + s (R_L R_C C + L) + R_L + R_C}$$

which is the desired result.

• **PROBLEM 7-2**

Find the transfer function $\dfrac{V_2(s)}{V_1(s)}$ for the circuit in Fig. 1.

Solution: Write the loop equations for the loops shown in Fig. 1.

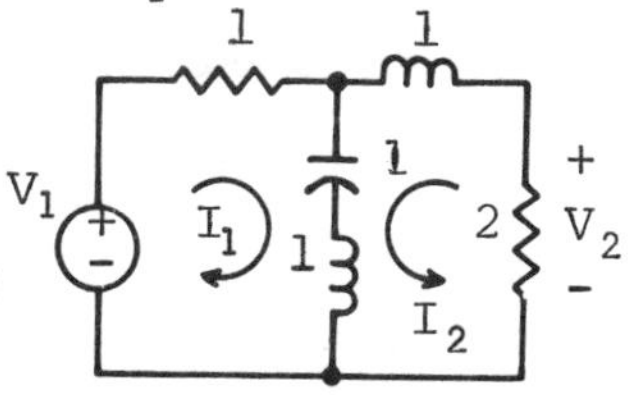

$$V_1 = I_1 (1 + \tfrac{1}{s} + s) + I_2 (\tfrac{1}{s} + s) \quad (1)$$

$$0 = I_1 (\tfrac{1}{s} + s) + I_2 (2 + 2s + \tfrac{1}{s}) \quad (2)$$

Fig. 1

Using **determinants** solve for I_2, since solving -or I_2 will give us V_1 times some impedance Z. Multiply $-I_2$ by the impedance across V_2 to give us V_2. $I_2 = YV_1$; $V_2 = -2YV_1$; $\dfrac{V_2}{V_1} = -2(Z)$.

Solving the equations (1) and (2) by Cramer's rule

$$I_2 = \frac{\begin{vmatrix} 1+\frac{1}{s}+s & V_1 \\ \frac{1}{s}+s & 0 \end{vmatrix}}{\begin{vmatrix} 1+\frac{1}{s}+s & \frac{1}{s}+s \\ \frac{1}{s}+s & 2+2s+\frac{1}{s} \end{vmatrix}} = \frac{-V_1(\frac{1}{s}+s)}{(1+\frac{1}{s}+s)(2+2s+\frac{1}{s}) - (\frac{1}{s}+s)^2}$$

$$I_2 = \frac{-V_1(\frac{1}{s}+s)}{\left|\frac{(s^2+s+1)}{s}\right|\left|\frac{2s^2+2s+1}{s}\right| - \left|\frac{(s^2+1)^2}{s^2}\right|} = \frac{-V_1(\frac{s^2+1}{s})}{\frac{s^3+4s^2+3s+3}{s}}$$

Simplifying gives

$$I_2 = \frac{-V_1(s^2+1)}{s^3 + 4s^2 + 3s + 3} \tag{1}$$

Since $V_2 = 2(-I_2)$ substituting into Eq.(1) gives

$$V_2 = \frac{2V_1(s^2+1)}{s^3 + 4s^2 + 3s + 3}$$

$$\frac{V_2(s)}{V_1(s)} = \frac{2(s^2+1)}{s^3 + 4s^2 + 3s + 3}$$

• **PROBLEM 7-3**

A rectangular voltage pulse, $v(t) = 10k[u(t)-u(t-\frac{1}{k})]V$, is applied in series with a 100-µF capacitor and a 20-kΩ resistor. Find the capacitor voltage at $t = 1$ as follows:
(a) if $k = 1$; (b) if $k = 10$; (c) if $k = 100$; (d) in the limit as $k \to \infty$.

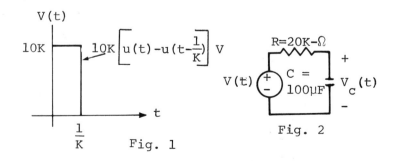

Fig. 1

Fig. 2

Solution: The pulse shown in Fig. 1 is applied to the circuit of Fig. 2.

For the capacitor voltage output shown the system function is

$$H(j\omega) = \frac{V_c(j\omega)}{V(j\omega)} = \frac{\frac{1}{j\omega C}}{\frac{1}{j\omega C} + R} = \frac{\frac{1}{Rc}}{\frac{1}{Rc} + j\omega} = \frac{(20\times 10^3)(100\times 10^{-6})}{j\omega + (20\times 10^3)(100\times 10^{-6})}$$

$$H(j\omega) = \frac{\frac{1}{2}}{j\omega + \frac{1}{2}}.$$

The Fourier transform of the input pulse is

$$\mathcal{F}\{10k[u(t) - u(t-\tfrac{1}{k})]\} = V(j\omega)$$

$$= 10k[\pi\delta(\omega) + \frac{1}{j\omega}] - 10k[\pi\delta(\omega) + \frac{1}{j\omega}]e^{-j\omega\frac{1}{k}}.$$

Hence, the output $V_c(t) = \mathcal{F}^{-1}\{V(j\omega) H(j\omega)\}$

$$= \mathcal{F}^{-1}\{\frac{\frac{1}{2}}{j\omega+\frac{1}{2}}[10k(\pi\delta(\omega) + \frac{1}{j\omega}) - 10k(\pi\delta(\omega) + \frac{1}{j\omega})e^{-j\omega\frac{1}{k}}]\}.$$

We can split the response into two parts:

$$v_{co}(t) = \mathcal{F}^{-1}\{\frac{\frac{1}{2}}{j\omega+\frac{1}{2}} 10k(\pi\delta(\omega) + \frac{1}{j\omega})\},$$

and $\quad v_{c1}(t) = \mathcal{F}^{-1}\{-\frac{\frac{1}{2}}{j\omega+\frac{1}{2}} 10k(\pi\delta(\omega) + \frac{1}{j\omega})e^{-j\omega\frac{1}{k}}\}.$

Evaluating $v_{co}(t)$ yields

$$v_{co}(t) = \mathcal{F}^{-1}\{\frac{\frac{1}{2}10k\pi\delta(\omega)}{j\omega+\frac{1}{2}} + \frac{\frac{1}{2}10k}{j\omega(j\omega+\frac{1}{2})}\}$$

$$v_{co}(t) = \mathcal{F}^{-1}\left(\frac{1-j\omega}{j\omega+\frac{1}{2}} 10K\Pi\delta(\omega) + \frac{\frac{1}{2}10K}{j(j\omega+\frac{1}{2})}\right)$$

$$= \mathcal{F}^{-1}\left(10K\Pi\delta(\omega) - \frac{j\omega}{(j\omega - \frac{1}{2})} 10K\Pi\delta(\omega) + \frac{10K}{j\omega} - \frac{10K}{j\omega+\frac{1}{2}}\right)$$

Making use of the property of delta function.

$$\phi(\omega)\delta(\omega) = \phi(0)\delta(\omega),$$

$$\frac{j\omega}{(j\omega+\frac{1}{2})} 10K\Pi \cdot \delta(\omega) = \left[10K\Pi \frac{j\omega}{j\omega+\frac{1}{2}}\right]_{\omega=0} \delta(\omega) = 0$$

Hence, $v_{co}(t) = F^{-1}\{10k\pi\delta(\omega) + \frac{10k}{j\omega} - \frac{10k}{j\omega+\frac{1}{2}}\}$

$$v_{co}(t) = 10k\, u(t) - 10k e^{-\frac{1}{2}t} u(t), \text{ hence}$$

Since $V_{c1}(j\omega) = -V_{co}(j\omega) e^{-j\omega\frac{1}{k}}$, from the time-shifting theorem,

$$V_{c1}(t) = -V_{co}(t) \quad t = (t-\frac{1}{k})$$

$$= -V_{co}(t-\frac{1}{k})$$

$$= -[10k\, u(t-\frac{1}{k}) - 10k\, e^{-\frac{1}{2}(t-\frac{1}{2})} u(t-\frac{1}{k})].$$

Hence, $V_c(t) = V_{co}(t) + V_{c1}(t)$

$$= 10k u(t) - 10k e^{-\frac{1}{2}t} u(t) - 10k u(t-\frac{1}{k}) + 10k e^{-\frac{1}{2}(t-\frac{1}{k})} u(t-\frac{1}{k})$$

(a) When $k = 1$,

$$v_c(t) = 10u(t) - 10e^{-\frac{1}{2}t} u(t) - 10u(t-1) + 10e^{-\frac{1}{2}(t-1)} u(t-1).$$

Hence, at $t = 1$

$$v_c(1) = 10 - 10e^{-\frac{1}{2}} - 10 + 10 = 3.93 \text{ V}.$$

(b) When $k = 10$,

$$v_c(t) = 100u(t) - 100e^{-\frac{1}{2}t} u(t) - 100u(t-\frac{1}{10})$$

$$+ 100e^{-\frac{1}{2}(t-\frac{1}{10})} u(t-\frac{1}{10})$$

and

$$v_c(1) = 100 - 100e^{-\frac{1}{2}} - 100 + 100e^{-\frac{9}{20}} = 3.11 \text{ V}.$$

(c) When $k = 100$,

$$v_c(t) = 1000u(t) - 1000e^{-\frac{1}{2}t} u(t) - 1000u(t-\frac{1}{100})$$

$$+ 1000e^{-\frac{1}{2}(t-\frac{1}{100})} u(t-\frac{1}{100})$$

and

$$v_c(1) = 1000 - 1000e^{-\frac{1}{2}} - 1000 + 1000e^{-\frac{99}{200}} = 3.04 \text{ V.}$$

For t>0 and as k→∞,

(d)
$$v_c(t) = 10k - 10ke^{-\frac{1}{2}t} + 10ke^{-\frac{1}{2}(t-\frac{1}{k})} - 10k$$

$$= 10k(-e^{-\frac{1}{2}t} + e^{-\frac{1}{2}(t-\frac{1}{k})})$$

$$= 10k\, e^{-\frac{1}{2}t} (e^{\frac{1}{2k}} - 1).$$

The Taylor series expansion for e^x is

$$e^x = 1 + x + \frac{x^2}{2} + \frac{x^3}{6} + \cdots$$

Substituting this expansion for $e^{\frac{1}{2k}}$ where $x = \frac{1}{2k}$

$$v_c(t) = 10k\, e^{-\frac{1}{2}t} (-1 + 1 + \frac{1}{2k} + \frac{1}{8k^2} + \frac{1}{24k^3} + \cdots)$$

$$= 10e^{-\frac{t}{2}} (\tfrac{1}{2} + 1/8k + 1/24k^2 \cdots)$$

In the limit as k→∞,

$$v_c(t) = 10e^{-t/2} (\tfrac{1}{2}) = 5e^{-t/2}$$

Therefore

$$v_c(1) = 5e^{-\frac{1}{2}} = 3.033 \text{ V.}$$

• **PROBLEM 7-4**

Obtain the Y-parameters of the network below.

Solution: Since the Y (admittance) parameters are required, first write the equations for I_1 and I_2 in terms of Y's, V_1, and V_2. Then, by individually setting V_1 and V_2 equal to

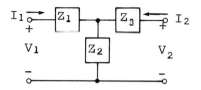

zero, solve for the Y's in terms of the Z's.

(a) The general equations for I_1 and I_2 can be written as functions of V_1 and V_2 as follows:

$$I_1 = Y_{11}V_1 + Y_{12}V_2 \tag{1}$$

$$I_2 = Y_{21}V_1 + Y_{22}V_2 . \tag{2}$$

The system of Y subscripts is chosen to identify each Y with the corresponding I and V. Thus, the first subscript is 1 where I_1 is involved and 2 where I_2 is involved. The second subscript is 1 where V_1 is involved and 2 where V_2 is involved. This notation is standardized.

(b) To solve for the various Y's, use the following scheme: note that

$$Y_{11} = I_1/V_1 \quad \text{for} \quad V_2 = 0 \tag{3}$$

in equation (1). Now, mathematically setting $V_2 = 0$ has the electrical counterpart of placing a short circuit across the two right-hand terminals. Placing this constraint on the network, redraw the circuit as shown in Fig. 1.

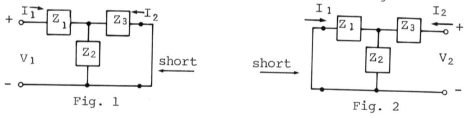

Fig. 1   Fig. 2

(c) The next task is solving for the ratio of I_1/V_1 which we recognize as the reciprocal of the impedance looking into the left set of terminals. Drawing upon our previous experience with circuits of this kind, we can write by inspection:

$$1/Y_{11} = Z_1 + \frac{Z_2 Z_3}{Z_2 + Z_3} \quad \text{ohms} \tag{4}$$

which becomes, with algebraic manipulation,

$$Y_{11} = \frac{Z_2 + Z_3}{Z_1 Z_2 + Z_2 Z_3 + Z_1 Z_3} \quad \text{mhos.} \tag{5}$$

(d) In a similar manner, choose from equation (2),

$$Y_{22} = I_2/V_2 \quad \text{for} \quad V_1 = 0 \tag{6}$$

and redraw the circuit with the <u>left</u> terminals shorted as seen in Figure 2.

Again, by similar analysis, we arrive at the solution

$$1/Y_{22} = Z_3 + \frac{Z_1 Z_2}{Z_1 + Z_2} \quad \text{ohms} \tag{7}$$

which becomes

$$Y_{22} = \frac{Z_1 + Z_2}{Z_1 Z_2 + Z_2 Z_3 + Z_1 Z_3} \quad \text{mhos.} \tag{8}$$

(e) Following the same general line of reasoning, set $V_1 = 0$ in equation (1), thereby solving for

$$Y_{12} = I_1/V_2 \quad \text{for} \quad V_1 = 0. \tag{9}$$

Since $V_1 = 0$, the circuit of Figure 2 still applies, but this time it is necessary to solve for I_1. Studying the network, we see that the current I_2 divides at the junction into two components, one of which is $(-I_1)$. This brings to mind the current division principle, which yields

$$-I_1 = I_2 \left(\frac{Z_2}{Z_1 + Z_2}\right). \tag{10}$$

Looking at our objective as seen in equation (9), recognize that Y_{12} can be obtained by dividing both sides of equation (10) by $(-V_2)$:

$$Y_{12} = I_1/V_2 = -\frac{I_2}{V_2} \left(\frac{Z_2}{Z_1 + Z_2}\right). \tag{11}$$

Looking back over the equations already obtained, recognize the I_2/V_2 factor in equation (11) has appeared before in equation (6), and, therefore, equation (11) can be rewritten as

$$Y_{12} = -(Y_{22}) \left(\frac{Z_2}{Z_1 + Z_2}\right). \tag{12}$$

Now, substituting equation (8) into (12),

$$Y_{12} = -\frac{Z_2}{Z_1 Z_2 + Z_2 Z_3 + Z_1 Z_3} \quad \text{mhos.} \tag{13}$$

(f) Using the same approach for Y_{21}, set $V_2 = 0$ in equation (2). This yields

$$Y_{21} = I_2/V_1 \quad \text{for} \quad V_2 = 0. \tag{14}$$

Since $V_2 = 0$, the circuit of Figure 1 still applies, and we must solve for I_2 in this circuit. Again, it is observed that the current division principle can be applied, resulting in

$$-I_2 = I_1 \left(\frac{Z_2}{Z_2+Z_3}\right). \qquad (15)$$

Dividing both sides of equation (15) by $(-V_1)$ to obtain Y_{21},

$$Y_{21} = I_2/V_1 = -\frac{I_1}{V_1}\left(\frac{Z_2}{Z_2+Z_3}\right). \qquad (16)$$

Now recognize the factor I_1/V_1 is Y_{11} and substitute equation (5) into equation (16):

$$Y_{21} = -\frac{Z_2+Z_3}{Z_1Z_2+Z_2Z_3+Z_1Z_3} \times \frac{Z_2}{Z_2+Z_3} \qquad (17)$$

for a final result of

$$Y_{21} = -\frac{Z_2}{Z_1Z_2+Z_2Z_3+Z_1Z_3} \quad \text{mhos.} \qquad (18)$$

(g) A quick glance over the four results indicates that $Y_{12} = Y_{21}$. This is worth noting for future reference.

● **PROBLEM 7-5**

Find the h-parameters for the network shown.

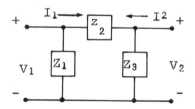

Solution: To solve this problem, it is important to know the definitions of h-parameters.

$$h_{11} \equiv \left.\frac{V_1}{I_1}\right|_{V_2=0} \quad \text{(self impedance at port-1)} \qquad (1)$$

By short circuiting Z_3, h_{11} is the parallel impedance of Z_1 and Z_2

$$h_{11} = \frac{Z_1 Z_2}{Z_1 + Z_2} \tag{2}$$

$$h_{22} \equiv \left.\frac{I_2}{V_2}\right|_{I_1=0} \quad \text{(self admittance at port-2)} \tag{3}$$

Now let $I_1 = 0$ (port 1 open). Thus h_{22} is the parallel admittance of Z_3 and series connection of Z_2 and Z_1

$$h_{22} = \frac{1}{Z_3} + \frac{1}{(Z_1+Z_2)} = \frac{Z_1 + Z_2 + Z_3}{(Z_1+Z_2)(Z_3)} \tag{4}$$

$$h_{12} \equiv \left.\frac{V_1}{V_2}\right|_{I_1=0} \quad \text{(voltage transfer ratio from Port 2 to Port 1)} \tag{5}$$

Using the current division principle, the current (I_2') (Z_1+Z_2) with $I_1=0$ is,

$$I_2' = \frac{Z_3}{Z_1+Z_2+Z_3} \times I_2 \tag{6}$$

Then, the voltage across Z_1 is,

$$V_1 = V_2 - Z_2 \left(I_2 \frac{Z_3}{Z_1+Z_2+Z_3}\right) = I_2' Z_1 \tag{7}$$

Solving equation (7) after substituting for I_2'

$$I_2 = \frac{V_2}{\frac{Z_3(Z_1+Z_2)}{Z_1+Z_2+Z_3}} = V_2 \frac{Z_1+Z_2+Z_3}{Z_3(Z_1+Z_2)} \tag{8}$$

Substituting Eq.(8) into Eq.(7)

$$V_1 = V_2 - Z_2 \left[V_2 \frac{Z_1+Z_2+Z_3}{Z_3(Z_1+Z_2)}\right]\left[\frac{Z_3}{Z_1+Z_2+Z_3}\right] = V_2 - \frac{Z_2}{Z_1+Z_2} V_2$$

$$= \left(1 - \frac{Z_2}{Z_1+Z_2}\right)V_2 = \frac{Z_1}{Z_1+Z_2} V_2 \tag{9}$$

So,
$$h_{12} \equiv \left.\frac{V_1}{V_2}\right|_{I_1=0} = \frac{Z_1}{Z_1+Z_2} \tag{10}$$

Now

$$h_{21} \equiv \left.\frac{I_2}{I_1}\right|_{V_2=0} \quad \text{(short circuit current transfer ratio from Port 1 to Port 2)} \tag{11}$$

Since $V_2=0$ or Port 2 is short circuited, I_2 flows in a direction opposite to that originally assigned. So, h_{21} is ordinarily negative. By the current division principle,

$$-I_2 = I_1 \frac{Z_1}{Z_1+Z_2} \tag{12}$$

Combining Eq. (12) with Eq. (11),

$$-h_{21} \equiv \left.\frac{I_2}{I_1}\right|_{V_2=0} = \frac{Z_1}{Z_1+Z_2} \tag{12}$$

h-parameters are also called hybrid parameters and originated from the following simultaneous equations.

$$\begin{aligned} V_1 &= h_{11}I_1 + h_{12}V_2 \\ I_2 &= h_{21}I_1 + h_{22}V_2 \end{aligned} \tag{13}$$

FREQUENCY DOMAIN ANALYSIS

● **PROBLEM 7-6**

Find $\frac{|I_2|}{|I_s|}$ in the circuit shown in fig. 1 if $\omega =$: (a) 200; (b) 2000; (c) 20×10^3 rad/s.

<u>Solution</u>: We can write an expression for $\frac{I_2}{I_s}$ by inspection if we recognize that the circuit is a current divider. Hence, the resistor and inductor can be combined to form an impedance z: Since $Z_L = sL$, $Z_R = R$, $Z_C = \frac{1}{sC}$,

where Z_L, Z_R, and Z_C represent the impedances offered by L, R and C respectively.

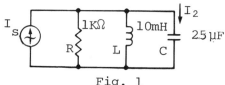

Fig. 1

$$Z = \frac{sRL}{sL + R}$$

then

$$I_2 = \frac{ZI_s}{Z + \frac{1}{sC}}$$

and

$$\frac{I_2}{I_s} = \frac{Z}{Z + \frac{1}{sC}}$$

$$\frac{I_2}{I_s} = \frac{\frac{sRL}{sL+R}}{\frac{sRL}{sL+R} + \frac{1}{sC}} = \frac{s^2}{s^2 + s\frac{1}{RC} + \frac{1}{LC}}.$$

Substituting $R = 1K\Omega$, $L = 10mH$ and $C = 25 \ \mu F$:

$$\frac{I_2}{I_s} = \frac{s^2}{s^2 + 40s + 4\times 10^6}.$$

Factoring the denominator we obtain the complex poles:

$$\frac{s^2}{(s + 20 - j2\times 10^3)(s + 20 + j2\times 10^3)}$$

poles: at $s = -20 + j2\times 10^3$

and $s = -20 - j2\times 10^3$

zeros: two zeros at $s = 0$.

The pole-zero diagram for $\frac{I_2}{I_s}$ is shown in fig. 2.

The magnitude of $\frac{I_2}{I_s}$, at a specific frequency ω, can be found quickly from the pole-zero diagram by finding the product of the distances for the zeros to the desired frequency ω on the $j\omega$-axis, and dividing by the product of the distances from the poles to the frequency ω on the $j\omega$-axis.

(a) To find $\frac{I_2}{I_s}$ at $\omega = 200 \ \frac{rad}{s}$ we find the distances shown in Fig. 3.

$$\left|\frac{I_2}{I_s}\right| = \frac{C^2}{AB} = \frac{200^2}{(1.8\times 10^3)(2.2\times 10^3)} = 0.0101$$

379

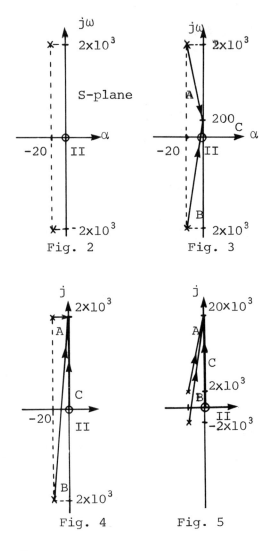

Fig. 2

Fig. 3

Fig. 4

Fig. 5

(b) To find $\left|\frac{I_2}{I_s}\right|$ at $\omega = 2 \times 10^3$ rad/s find the distances A, B, and C in fig. 4

$$\left|\frac{I_2}{I_s}\right| = \frac{C^2}{AB} = \frac{2000^2}{20(4 \times 10^3)} = 50$$

(c) To find $\left|\frac{I_2}{I_s}\right|$ at $\omega = 20 \times 10^3$ rad/s find the distances A, B, and C in fig. 5

$$\left|\frac{I_2}{I_s}\right| = \frac{C^2}{AB} = \frac{(20 \times 10^3)^2}{(18 \times 10^3)(22 \times 10^3)} = 1.01$$

• PROBLEM 7-7

Obtain the pole-zero plot for

$$H(s) = \frac{s(s^2 + \sqrt{2}s + 1)}{s^2 + 3s + 2}$$

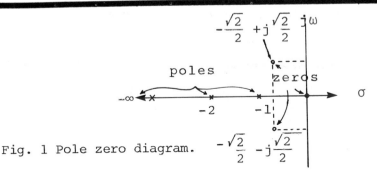

Fig. 1 Pole zero diagram.

Solution: The denominator and numerator polynomials must be factored into the form

$$H(s) = \frac{(s \pm a_1)(s \pm a_2)(s \pm a_3)\cdots}{(s \pm b_1)(s \pm b_2)(s \pm b_3)\cdots}$$

The poles are indicated in the denominator as the roots of that polynomial. The zeros are indicated in the numerator as the roots of that polynomial.

Factoring the denominator gives

$$(s^2 + 3s + 2) = (s + 1)(s + 2),$$

poles $= -1, -2$ and ∞. $S = \infty$ is a pole because as $S \to \infty$, $H(S) \to \infty$.

Factoring the numerator yields

$$s(s^2 + \sqrt{2}s + 1) = (s + 0)(s + (-z_1))(s + (-z_2))$$

$$z_1, z_2 = \frac{-b \pm \sqrt{b^2 - 4ac}}{2a} = \frac{-\sqrt{2} \pm \sqrt{2 - 4}}{2}$$

$$z_1, z_2 = \frac{-\sqrt{2}}{2} \pm j\frac{\sqrt{-2}}{2} = \frac{-\sqrt{2}}{2} \pm j\frac{\sqrt{2}}{2},$$

which gives $(s + 0)(s + \frac{\sqrt{2}}{2} - j\frac{\sqrt{2}}{2})(s + \frac{\sqrt{2}}{2} + j\frac{\sqrt{2}}{2}),$

zeroes $= 0, \frac{-\sqrt{2}}{2} + j\frac{\sqrt{2}}{2}, \frac{-\sqrt{2}}{2} - j\frac{\sqrt{2}}{2}.$

The pole-zero diagram is shown in fig. 1.

• **PROBLEM** 7-8

In a network, a current source $I(s)$ produces a response $V(s)$ such that

$$V(s) = \frac{s}{s^2 + s + 1} I(s).$$

Use the graphical method to determine the component of the response due to: (a) a sinusoidal source $i(t) = 10 \cos(t + 30°)A$; (b) $i(t) = 10e^{-t} \cos t$ A.

Solution: The procedure for the graphical solution is to plot the poles and zeros of the transfer function on the complex s plane and evaluate it for the desired frequency and then multiply this response by the input current function.

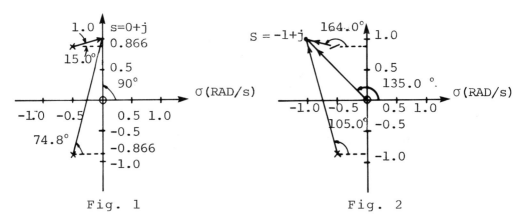

Fig. 1   Fig. 2

Complex s plane for part a   Complex s plane for part b

The transfer function $F(s)$ is

$$F(s) = \frac{s}{s^2 + s + 1} = \frac{s}{(s+0.5-j0.866)(s+0.5+j0.866)}.$$

A plot of the poles and zero is shown in Fig. 1.

(a) The current source is $i(t) = 10 \cos(t + 30°)A$ with an angular frequency of 1 rad/s. The response of the transfer function to $\omega = 1$ rad/s is found by graphically evaluating $F(s)$ at $s = j1$.

$$F(j1) = \frac{j1}{\{(0.5)+j(0.134)\}\{(0.5)+j(1.866)\}}$$

$$= \frac{1 \;/90°}{(0.5176 \;/15°)(1.9318 \;/75°)}$$

$$\cong 1 \;/0 = 1$$

382

The complex voltage is found by multiplying $F(j1)$ by the complex current, $10e^{j(t+30°)}$, giving

$$10e^{j(t+30°)} = 10\,\underline{/t+30°}$$

$v(t)$ is then determined by taking the real part:

$$v(t) = 10\cos(t+30°)\,V.$$

(b) The exponentially decaying current, $i(t) = 10e^{-t}\cos t$ A, will cause the voltage to decay also. In complex notation,

$$i(t) = 10e^{-t}e^{jt} = 10e^{(-1+j)t}\,A.$$

In this situation $F(s)$ is evaluated at $s = -1+j$ as shown in Fig. 2.

$$F(-1+j) = \frac{-1+j}{(-0.5+j0.134)(-0.5+j1.866)}$$

$$= \frac{(1.414)\,\underline{/105°}}{(0.517\,\underline{/165°})(1.93\,\underline{/105°})}$$

$$= 1.417\,\underline{/-135°}$$

Multiplying this by the original complex vector $10e^{(-1+j)t}$ gives

$$14.17e^{(-1+j)t}e^{-j135°} = 14.17e^{-t}e^{j(t-135°)}.$$

The final result for $v(t)$ is found by taking the real part:

$$v(t) = 14.17e^{-t}\cos(t-135°)\,V.$$

● **PROBLEM 7-9**

For a certain response-source relationship the transform network function is $H(s) = \dfrac{s+1}{(s+2)(s^2+2s+10)}$. Calculate graphically the component of the response due to each of the source functions: (a) $20e^{-3t}$; (b) $20\cos 3t$; (c) $20\cos(3t-30°)$; (d) $20e^{-3t}\cos t$.

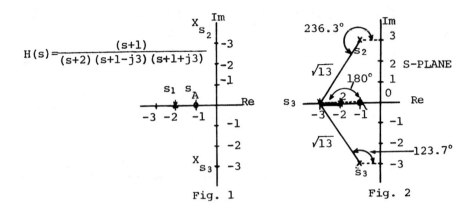

$$H(s) = \frac{(s+1)}{(s+2)(s+1-j3)(s+1+j3)}$$

Fig. 1

Fig. 2

Solution: If a certain source function $Ae^{s_g t}$ is impressed on a network, then if s_g is not a pole of the network function, the response component due to the source is

$$y_s(t) = \frac{N(s_g)}{D(s_g)} Ae^{s_g t} = H(s_g) Ae^{s_g t}$$

or

$$y_s(t) = \frac{(s_g - s_A)(s_g - s_B) \cdots (s_g - s_N)}{(s_g - s_1)(s_g - s_2) \cdots (s_g - s_M)} Ae^{s_g t}$$

Hence, the evaluation of the complete response depends on the evaluation of products of the form $\prod_k (s_g - s_k)$. Since each term $(s_g - s_k)$ may be a complex number, it is convenient to carry out the multiplication and division in polar form.

(a) The network function $H(s)$ has a zero at $s_A = -1$ and three poles, $s_1 = -2$, $s_2 = -1 + j3$, $s_3 = -1 - j3$, as shown in fig. 1.

Since the Laplace transform of $20e^{-3t} = \frac{1}{s+3}$,

The source function $20e^{-3t}$ can be represented by a pole at $s_g = -3$. Hence, the response due to the source is:

$$y_s(t) = \frac{|s_g - s_A| \,\underline{/\theta_{g-A}}}{|s_g - s_1||s_g - s_2||s_g - s_3| \,\underline{/\theta_{g-1}} \,\underline{/\theta_{g-2}} \,\underline{/\theta_{g-3}}} \, 20e^{-3t}.$$

The above calculation is illustrated graphically in fig. 2.

$$y_s(t) = \frac{2 \,\underline{/180°}}{1 \cdot \sqrt{13} \cdot \sqrt{13} \,\underline{/180°} \,\underline{/236.3°} \,\underline{/123.7°}} \, 20e^{-3t}$$

$$y_s(t) = \frac{2\,/180°}{13\,/540°} 20e^{-3t} = \frac{2}{13} 20e^{-3t}$$

$$y_s(t) = 3.08e^{-3t}$$

(b) Since the LaPlace transform of $e^{j3t} = \frac{1}{s-3j}$,
The source function $20\cos 3t$ can be represented by a pole at $S_g = j3$. The procedure for finding the response due to this source is illustrated below in Fig. 3.

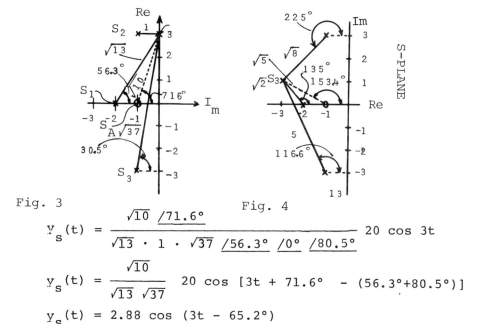

Fig. 3   Fig. 4

$$y_s(t) = \frac{\sqrt{10}\,/71.6°}{\sqrt{13}\cdot 1 \cdot \sqrt{37}\,/56.3°\,/0°\,/80.5°} 20\cos 3t$$

$$y_s(t) = \frac{\sqrt{10}}{\sqrt{13}\,\sqrt{37}} 20\cos[3t + 71.6° - (56.3° + 80.5°)]$$

$$y_s(t) = 2.88 \cos(3t - 65.2°)$$

(c) To find the response due to the source $20\cos(3t - 30°)$, we use the same procedure as in (b):

$$y_s(t) = \frac{\sqrt{10}}{\sqrt{13}\,\sqrt{37}} 20\cos[3t - 30° + 71.6° - (56.3° + 80.5°)]$$

$$y_s(t) = 2.88 \cos(3t - 95.2°).$$

(d) The source $20e^{-3t}\cos t$ can be represented by a pole at $s_g = -3 + j$. The procedure for finding the response due to the source is shown in fig. 4.

$$y_s(t) = \frac{\sqrt{5}\,/153.4°}{\sqrt{2}\cdot\sqrt{8}\cdot\sqrt{20}\,/135°\,/225°\,/116.6°} 20e^{-3t}\cos t$$

$$y_s(t) = \frac{\sqrt{5}}{\sqrt{2}\cdot\sqrt{8}\cdot\sqrt{20}} 20e^{-3t} \cos[t+153.4°-(135°+225°+116.6°)]$$

$$y_s(t) = 2.5e^{-3t} \cos(t + 36.8°)$$

• **PROBLEM 7-10**

Make an s-plane plot of the poles and zeros of the transfer admittance, $Y = \frac{I_1}{V_s}$, for the circuit shown in fig. 1.

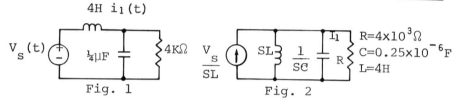

Fig. 1   Fig. 2

Solution: We can write the expression for the transfer admittance $Y = \frac{I_1}{V_s}$ by transforming the circuit in fig. 1 into the frequency domain and then transforming the voltage source into a current source and parallel impedance as shown in fig. 2.

Let Z_2 be the impedance of the parallel combination of the inductor and capacitor and Z_1 be the parallel combination of the inductor, capacitor and resister.

$$Y_2 = \frac{1}{Z_2} = \frac{1}{S_2} + Sc = \frac{S^2 LC+1}{SL} = \frac{C}{S}\left(S^2 + \frac{1}{LC}\right).$$

$$Y_1 = \frac{1}{Z_1} = \frac{1}{SL} + SC + \frac{1}{R} = \frac{R+S^2 LCR+SL}{SLR} = \frac{C}{S}\left(S + \frac{S^2}{RC} + \frac{1}{LC}\right)$$

The current I_1 is found by current division. Hence, from Fig. 2,

$$\frac{I_1}{\frac{V_s}{sL}} = \frac{Z_2}{Z_2+Z_1} = \frac{1}{1+\frac{Z_1}{Z_2}} = \frac{1}{1+\frac{Y_2}{Y_1}} = \frac{1}{1+\frac{S^2+\frac{1}{LC}}{S+\frac{S}{Rc}+\frac{1}{Lc}}}$$

$$= \frac{S^2 + \frac{S}{Rc} + \frac{1}{Lc}}{2S^2 + \frac{S}{Rc} + \frac{2}{Lc}}$$

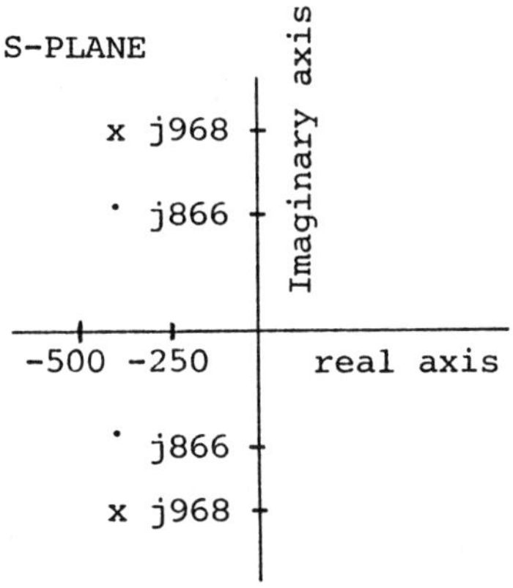

```
              S-PLANE        |
                    x j968   |
                      • j866 |
                             |
      ─────┼──────┼──────────┼──────────── real axis
          -500  -250         |
                      • j866 |
                    x j968   |
                             |
```

Fig. 3

Hence, $\dfrac{I_1}{V_s} = \dfrac{(S^2 + \dfrac{S}{Rc} + \dfrac{1}{Lc})}{2S1(S^2 + \dfrac{S}{2Rc} + \dfrac{1}{Lc})}$

$RC = 4 \times 10^3 \times 0.25 \times 10^{-6} = 10^{-3}; \quad \dfrac{1}{RC} = 10^3$

$LC = (4)(0.25 \times 10^{-6}) = 10^{-6}; \quad \dfrac{1}{LC} = 10^6$

$\dfrac{I_1}{V_s} = \dfrac{(S^2 + 10^3 S + 10^6)}{8S(S^2 + 500S + 10^6)}$

The zeros are at $S = S_1$ and $S = S_2$, where

$S_1, S_2 = \dfrac{-10^3 \pm \sqrt{10^6 - 4(1)(10^6)}}{2}$

$= \dfrac{-10^3}{2} \pm \dfrac{j\sqrt{3} \times 10^3}{2}$

$= -500 \pm j866$

The poles are at $S=0$, $S=S_3$, and $S=S_4$, where

$S_3, S_4 = \dfrac{-500 \pm \sqrt{(500)^2 - 4(1)(10^6)}}{2}$

$= \dfrac{-500 \pm 10^3 \sqrt{\tfrac{1}{4} - 4}}{2}$

$= -250 \pm j10^3 \dfrac{\sqrt{4 - \tfrac{1}{4}}}{2}$

$= -250 \pm j968$

• **PROBLEM 7-11**

In a certain low-loss network the response $v_2(t)$ is related to the source $v_1(t)$ by the equation

$$v_2(t) = \frac{1}{3} \frac{p(p^2 + 2p + 200)}{(p^2 + p + 50)(p^2 + 3p + 500)} v_1(t).$$

Use the approximate pole-zero-diagram method to sketch the amplitude response curve $|H(j\omega)| = h(\omega)$.

Solution: In this problem $|H(j\omega)| = \dfrac{v_2(t)}{v_1(t)}$ and

$$H(s) = \frac{s(s^2 + 2s + 200)}{3(s^2 + s + 50)(s^2 + 3s + 500)}.$$

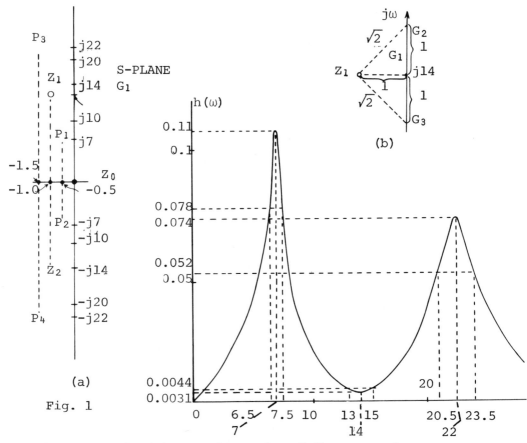

Fig. 1

The zeros of H(s) are the roots of the numerator:

$$s_1 = 0; \quad s_{2,3} = \frac{-2 \pm \sqrt{(2)^2 - 4(200)}}{2} \cong -1 \pm j14$$

The poles of H(s) are the roots of the denominator:

$$s_{12} = \frac{-1 \pm \sqrt{1 - 4(50)}}{2} \cong -0.5 \pm j7$$

$$s_{34} = \frac{-3 \pm \sqrt{9 - 4(500)}}{2} \cong -1.5 \pm j22$$

The pole-zero diagram is shown in fig. 1(a).

$$h(14) = \frac{(Z_0G_1)(Z_1G_1)(Z_2G_1)}{3(P_1G_1)(P_2G_1)(P_3G_1)(P_4G_1)}$$

where Z_0G_1 is the distance from Z_0 to the point $(0,j14)$, Z_1G_1 is the distance from Z_1 to $(0,j14)$, etc. $h(14)$ can be approximated as

$$h(14) \cong \frac{(14) \times (1) \times (28)}{3(7) \times (21) \times (8) \times (36)} = 0.0031.$$

A minimum also occurs at $\omega = 0$:

$$h(0) = \frac{(0) \times (14) \times (14)}{(7) \times (7) \times (22) \times (22)} = 0.$$

A maximum occurs at $\omega \cong 7$ and $\omega \cong 22$:

$$h(7) = \frac{(Z_0G_2)(Z_1G_2)(Z_2G_2)}{3(P_1G_2)(P_2G_2)(P_3G_2)(P_4G_2)}$$

$$\cong \frac{(7) \times (7) \times (21)}{3(0.5) \times (14) \times (1.5) \times (29)}$$

$$\cong 0.11$$

$$h(22) = \frac{(Z_0G_3)(Z_1G_3)(Z_2G_3)}{3(P_1G_3)(P_2G_3)(P_3G_3)(P_4G_3)}$$

$$\cong \frac{(22) \times (8) \times (36)}{3(15) \times (29) \times (1.5) \times (44)}$$

$$\cong 0.074$$

Now that all the maximum and minimum values are obtained, more values must be obtained to sketch $h(\omega)$. In the neighborhood of $\omega = 14$, the distance from Z_1 to $(0,j14)$ controls the magnitude of $h(\omega)$. The distances from all the other poles and zeros can be assumed constant.

From fig. 1b it is noted that

$$h(15) \cong h(13) \cong h(14) \times \sqrt{2} = 0.0044.$$

Similar conditions hold in the neighborhood of $\omega = 7$ and $\omega = 22$:

$$h(6.5) \cong h(7.5) = \frac{h(7)}{\sqrt{2}} = 0.078$$

$$h(20.5) \cong h(23.5) = \frac{h(22)}{\sqrt{2}} = 0.052.$$

For each ω we have the following values:

ω	0	6.5	7	7.5	13	14
$h(\omega)$	0	0.078	0.11	0.078	0.0044	0.0031

ω	15	20.5	22	23.5
$h(\omega)$	0.0044	0.052	0.074	0.052

The sketch of the response curve $h(\omega)$ is shown in fig. 1c.

• **PROBLEM 7-12**

In the circuit of fig. 1, $v(t) = V_m \cos\omega t\, V$ and, in the steady state, $v_{ab}(t) = (V_{ab})_m \cos(\omega t - \theta)\, V$. (a) Find the transform network function that relates v_{ab} to v. (b) Plot $(V_{ab})_m/V_m$ as a function of ω. (c) Plot θ as a function of ω.

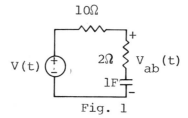

Fig. 1

Solution: (a) The transfer function $H(s) = \frac{V_{ab}(s)}{V(s)}$ is found by the concept of voltage division,

$$V_{ab}(s) = \frac{\frac{1}{s} + 2}{\frac{1}{s} + 12} V(s)$$

$$H(s) = \frac{V_{ab}(s)}{V(s)} = \frac{\frac{1}{s}+2}{\frac{1}{s}+12} = \frac{2s+1}{12s+1} = \frac{1}{6} \frac{(s+\frac{1}{2})}{(s+\frac{1}{12})}.$$

(b) By substituting $j\omega$ for s we can plot the transfer function $|H(j\omega)|$ as a function of ω.

$$H(j\omega) = \frac{(V_{ab})_m}{V_m} = \frac{1}{6} \frac{j\omega + \frac{1}{2}}{j\omega + \frac{1}{12}} = \frac{2j\omega + 1}{12j\omega + 1}$$

$$|H(j\omega)| = \frac{\sqrt{1 + 4\omega^2}}{\sqrt{1 + 144\omega^2}}$$

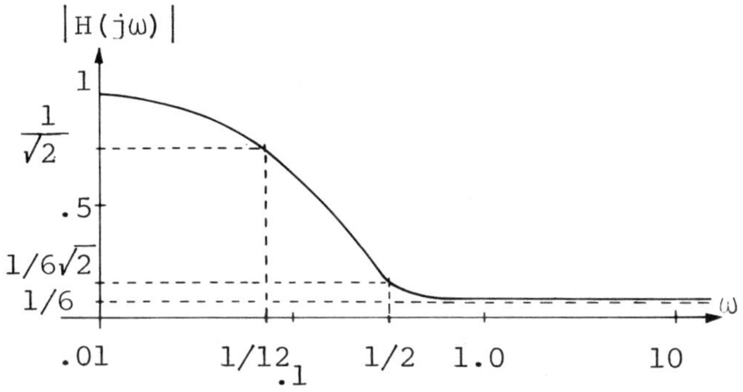

Fig. 2

(c) The plot of the phase angle, Θ, of the transfer function, $H(j\omega)$, is in the form

$$H(j\omega) = \frac{A(j\omega)}{B(j\omega)}$$

where

$$\Theta = \tan^{-1} \frac{\text{Im}[A(j\omega)]}{\text{Re}[A(j\omega)]} - \tan^{-1} \frac{\text{Im}[B(j\omega)]}{\text{Re}[B(j\omega)]} .$$

Hence, for

$$H(j\omega) = \frac{2j\omega + 1}{12j\omega + 1} \quad \text{we obtain}$$

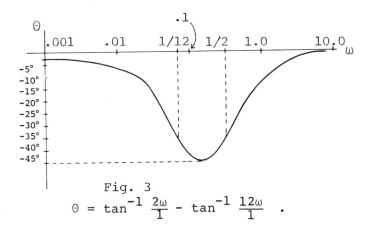

Fig. 3

$$\Theta = \tan^{-1} \frac{2\omega}{1} - \tan^{-1} \frac{12\omega}{1} .$$

Fig. 3 is the plot of Θ versus ω.

• **PROBLEM 7-13**

The drawing in fig. 1 shows a simple low-pass filter. (a) Calculate the steady-state response $v_R(t)$. (b) Determine the steady-state transfer function of the network, $H(j\omega)$, and sketch the amplitude and phase response.

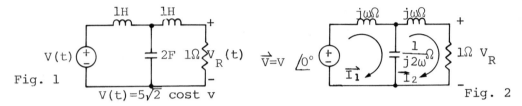

Fig. 1 $V(t) = 5\sqrt{2} \cos t$ Fig. 2

Solution: The circuit of fig. 1 is operating in the ac steady state. Once the transfer function is determined, it can be used to calculate the steady state response. Hence we will first derive $H(j\omega)$. The steady state equivalent circuit is shown in Figure 2. For derivation of $H(j\omega)$, note that one must assume an input of the complex phasor form shown. We must derive the phasor $\vec{V}_R$.

Now, since this circuit has two loops, we can use the loop currents indicated in Figure 2. The resulting loop equations are:

$$(j\omega + \frac{1}{j2\omega})\vec{I}_1 - \frac{1}{j2\omega}\vec{I}_2 = \vec{V} \tag{1}$$

$$-\frac{1}{j2\omega}\vec{I}_1 + (1 + j\omega + \frac{1}{j2\omega})\vec{I}_2 = 0. \tag{2}$$

Solving (2) for $\vec{I}_1$, we get

$$\vec{I}_1 = (2j\omega + 2(j\omega)^2 + 1)\vec{I}_2. \tag{3}$$

Now inserting this value of $\vec{I}_1$ in (1) we obtain:

$$[[j\omega + \frac{1}{j2\omega}][2j\omega + 2(j\omega)^2 + 1] - \frac{1}{j2\omega}]\vec{I}_2 = \vec{V} \tag{4}$$

or $$[2(j\omega)^3 + 2(j\omega)^2 + 2(j\omega) + 1]\vec{I}_2 = \vec{V}. \tag{5}$$

But $\vec{V}_R = \vec{I}_2$, so

$$H(j\omega) = \frac{\vec{V}_R}{\vec{V}} = \frac{\frac{1}{2}}{(j\omega)^3 + (j\omega)^2 + j\omega + \frac{1}{2}}, \tag{6}$$

which is the solution to part (b).

In order to compute the steady state response to $v(t) = 5\sqrt{2} \cos t$ V, simply let $\vec{V} = 5\sqrt{2} \underline{/0°}$ V in fig. 2

and note that $\omega = 1$ rad/s, hence

$$\vec{V}_R = \frac{\frac{1}{2}}{j^3 + j^2 + j + \frac{1}{2}} \times 5\sqrt{2} \underline{/0°} \qquad (7)$$

or $\quad \vec{V}_R = -5\sqrt{2} \underline{/0°}$ V. $\hfill (8)$

Hence $v_R(t) = \text{Re}\{V_R e^{jt}\} = -5\sqrt{2} \cos t$ V. $\hfill (9)$

The amplitude and phase response are obtained from $H(j\omega) = A(j\omega) e^{j\beta(j\omega)}$, where $A(j\omega)$ is the amplitude and $\beta(j\omega)$ is the phase. Therefore,

$$H(j\omega) = \frac{\frac{1}{2}}{(\frac{1}{2} - \omega^2) + j(\omega - \omega^3)} \qquad (10)$$

or

$$H(j\omega) = \frac{\frac{1}{2}}{\sqrt{(\frac{1}{2} - \omega^2)^2 + \omega^2(1 - \omega^2)^2}} \underline{/\tan^{-1} \frac{\omega(1-\omega^2)}{\frac{1}{2} - \omega^2}} \qquad (11)$$

so

$$A(j\omega) = \frac{1}{\sqrt{(\frac{1}{2} - \omega^2)^2 + \omega^2(1 - \omega^2)^2}} \qquad (12)$$

and

$$\beta(j\omega) = -\tan^{-1}\left[\frac{\omega \times (1 - \omega^2)}{\frac{1}{2} - \omega^2}\right] . \qquad (13)$$

The frequency response plots are shown in fig. 3.

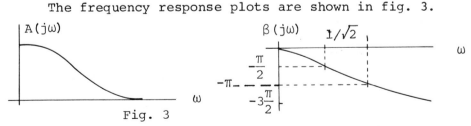

Fig. 3

Note that in sketching $\beta(j\omega)$ the inverse tangent function has several branches. Do not interpret (13) as the principal value.

• **PROBLEM 7-14**

Plot the sinusoidal-steady state amplitude response for the network shown in fig. 1. Let v(t) be the input and $v_R(t)$ be the response.

Solution: The steady state equivalent circuit is shown in fig. 2.

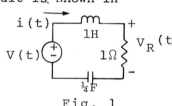

Fig. 1

Using KVL around the loop, we obtain

$$\vec{V} = \vec{I}\left[j\omega + 1 - \frac{j4}{\omega}\right].$$

Since $\vec{V}_R = \vec{I}$ then,

$$\vec{V} = \vec{V}_R\left[j\omega + 1 - \frac{j4}{\omega}\right]$$

$$H(j\omega) = \frac{\vec{V}_R}{\vec{V}} = \frac{1}{j\omega + 1 - \frac{j4}{\omega}}$$

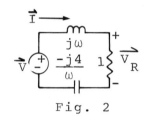

Fig. 2

$$H(j\omega) = \frac{j\omega}{(4-\omega^2) + j\omega}.$$

Hence, the magnitude of the response of the network is

$$|H(j\omega)| = \frac{\omega}{\sqrt{(4-\omega^2)^2 + \omega^2}}.$$

In order to plot the amplitude response as a function frequency, it is conventional to plot 20 times the logarithm of the magnitude, rather than the magnitude itself. This, of course, permits a wide variation of the function $|H(j\omega)|$ to be plotted on a compact graph.

Define the units of the function

$$20 \log_{10}|H(j\omega)| = 10 \log_{10}|H(j\omega)|^2$$

as decibels abbreviated as dB.

In general, the amplitude response of any function $H(j\omega)$ can be plotted by noting that there are only three types of terms that appear in the transfer function $H(j\omega)$. These terms are

1) s^n
2) $(s + a)^n$
3) $s^2 + 2\zeta\omega_n s + \omega_n^2$.

If s = jω then the three terms are:

1) $(j\omega)^n$ (1)

2) $(j\omega + a)^n$ (2)

3) $(\omega_n^2 - \omega^2) + j2\zeta_n\omega_n\omega$.   (3)

Since the magnitude function can be written

$$|H(j\omega)| = \left|\frac{B(j\omega)}{A(j\omega)}\right|,$$

then

$$20\log|H(j\omega)| = 20\log|B(j\omega)| - 20\log|A(j\omega)|.$$

It has been assumed that

$$B(j\omega) = B_1(j\omega)B_2(j\omega)\ldots B_m(j\omega)$$

and

$$A(j\omega) = A_1(j\omega)A_2(j\omega)\ldots A_n(j\omega)$$

where each $B_i(j\omega)$ and $A_i(j\omega)$ is given by one of the terms (1), - (3).

Thus, a plot of the function $20\log|H(j\omega)|$ consists of the addition and subtraction of plots corresponding to the magnitude of terms (1), - (3).

$$20\log|H(j\omega)| = 10\log\frac{\omega^2}{(4-\omega^2)+j\omega}$$

$$20\log|H(j\omega)| = 20\log|\omega| - 10\log|4-\omega^2+j\omega|$$

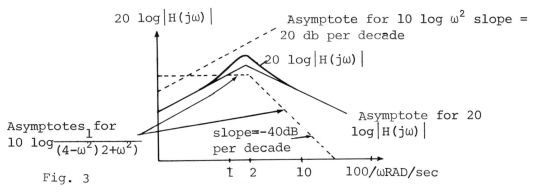

Fig. 3

The plot of this function is shown in fig. 3.

• PROBLEM 7-15

For the circuit shown in fig. 1, determine: the resonant frequency; the Q_o; the bandwidth; and sketch $|I|$ versus ω. What would the Q_o of the circuit be if the 20Ω resistor were short-circuited? What would it be if the 200 kΩ resistor were open-circuited?

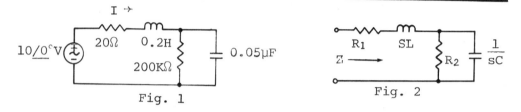

Fig. 1   Fig. 2

Solution: Since the circuit is an RLC circuit containing one inductor and one capacitor but two resistors we cannot use the formula for the resonant frequency in a series-parallel RLC circuit, $\frac{1}{\sqrt{LC}}$. Instead, we must write an expression for the impedance of the circuit in fig. 1 and compare it to the impedance of a series RLC circuit. Hence, from fig. 2

$$Z = R_1 + sL + \frac{\frac{R_2}{sC}}{R_2 + \frac{1}{sC}}$$

can be written

$$Z = \frac{s^2 + s\left(\frac{R_1}{L} + \frac{1}{R_2 C}\right) + \frac{R_1 + R_2}{R_2 LC}}{\frac{1}{L}\left(s + \frac{1}{R_2 C}\right)} \qquad (1)$$

Note that the impedance for a series RLC circuit in fig. 3 is of similar form to Eq. (1) above:

$$Z_s = \frac{s^2 + s\frac{R}{L} + \frac{1}{LC}}{\frac{1}{L}s} \qquad (2)$$

The resonant frequency of the series RLC circuit is $\frac{1}{\sqrt{LC}}$. Since $R_1 \ll R_2$ in the circuit of fig. 1 the resonant frequency is the square root of the last term in the numerator of Eq. (1). Hence,

$$\omega_o = \sqrt{\frac{R_1 + R_2}{R_2 LC}} \cong \frac{1}{\sqrt{LC}} = 10{,}000 \text{ rad/s}.$$

396

The Q_o of the series RLC circuit is $\frac{1}{\sqrt{LC}}\frac{L}{R} = \frac{\omega_o}{\frac{R}{L}}$. Note that the pole at $-\frac{1}{R_2C}$ in the impedance expression for the circuit in fig. 1 is -100 rad/s which is far enough away from its resonant frequency so as not to interfere with that portion of the response. Hence, the Q_o for the circuit is $\omega_o \left[\frac{1}{\frac{R_1}{L} + \frac{1}{R_2C}}\right] = 50$. The RLC circuit in fig. 1 will behave similarly to an equivalent series RLC circuit with $R_{eq} = 40\Omega$, $L_{eq} = .2H$ and $C_{eq} = 0.05\mu F$. The bandwidth of an RLC circuit is defined as $\frac{\omega_o}{Q_o} = \frac{10,000}{50} = 200$ rad/s.

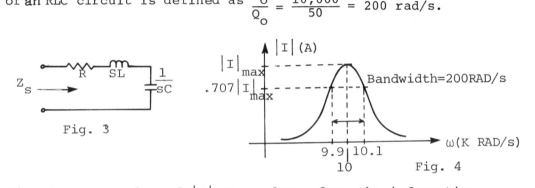

Fig. 3

Fig. 4

Fig. 4 shows a plot of $|I|$ vs. ω, drawn from the information found above.
The value for $|I|$ at the resonant frequency $|I|_{max}$ is found by the methods of sinusoidal steady-state analysis:

$$\frac{\vec{V}}{\vec{Z}} = \vec{I} = \frac{10 \underline{/0°}}{(20 + j\omega 0.2) + \dfrac{200\times 10^3}{200\times 10^3 + \dfrac{1}{j\omega 0.05\times 10^{-6}}}}$$

Letting $\omega = 10,000$, yields:

$$\vec{I}_{max} = \frac{10 \underline{/0°}}{(20 + j2000) + \dfrac{-j(200\times 10^3)(2\times 10^3)}{200\times 10^3 - j2\times 10^3}}$$

$$\vec{I}_{max} = \frac{10 \underline{/0°}}{\dfrac{(20 + j2\times 10^3)(200\times 10^3 - j2\times 10^3) - j4\times 10^8}{200\times 10^3 - j2\times 10^3}}$$

$$|\vec{I}|_{max} = \frac{1}{4} \text{ A}.$$

When the 20Ω resistor is short-circuited,

$$Z = \frac{s^2 + s \frac{1}{R_2 C} + \frac{1}{LC}}{\frac{1}{L}(s + \frac{1}{R_2 C})},$$

hence,

$$Q_o = \frac{\omega_o}{\frac{1}{R_2 C}} = \frac{1}{\sqrt{LC}} R_2 C = 100.$$

When the 200 kΩ resistor is open-circuited,

$$Z = \frac{s^2 + s \frac{R_1}{L} + \frac{1}{LC}}{\frac{1}{L} s},$$

the impedance of a series RLC circuit. Hence,

$$Q_o = \frac{\omega_o}{\frac{R_1}{L}} = 100.$$

● **PROBLEM 7-16**

If $Y(j\omega) = F(j\omega)H(j\omega)$, find $y(t)$ for $F(j\omega) = H(j\omega) = \frac{1}{1+\omega^2}$.

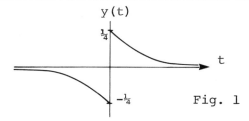

Fig. 1

Solution: The response $Y(j\omega)$ is the product of the system function, $H(j\omega)$, and the driving function, $F(j\omega)$. Since $F(j\omega) = H(j\omega) = \frac{1}{1+\omega^2}$, then $Y(j\omega) = \frac{1}{(1+\omega^2)^2}$.

Hence,

$$y(t) = F^{-1}\left\{\frac{1}{(1+\omega^2)^2}\right\}, \text{ the inverse Fourier transform.}$$

Since $1+\omega^2$ can be expanded into the form

$$-(j\omega+1)(j\omega-1)$$

then

$$y(t) = F^{-1}\left\{\frac{1}{(j\omega+1)^2(j\omega-1)^2}\right\}.$$

Partial fraction expansion yields

$$y(t) = F^{-1}\left\{\frac{\frac{1}{4}}{(j\omega+1)^2} + \frac{\frac{1}{4}}{(j\omega+1)} + \frac{\frac{1}{4}}{(j\omega-1)^2} - \frac{\frac{1}{4}}{(j\omega-1)}\right\}.$$

Noting that $F^{-1}\left\{\frac{\frac{1}{4}}{(j\omega+1)^2}\right\} = \frac{1}{4}te^{-t}u(t)$ and

$F^{-1}\left\{\frac{\frac{1}{4}}{j\omega+1}\right\} = \frac{1}{4}e^{-t}u(t)$, we obtain

$$y(t) = \frac{1}{4}te^{-t}u(t) + \frac{1}{4}e^{-t}u(t) + \frac{1}{4}te^{t}u(t) - \frac{1}{4}e^{t}u(t)$$

$$y(t) = \frac{1}{4}e^{-t}(t+1)u(t) + \frac{1}{4}e^{t}(t-1)u(t),$$

if we let $x(t) = \frac{1}{4}e^{-t}(t+1)u(t)$ and

$$z(t) = \frac{1}{4}e^{t}(t-1)u(t).$$

Hence,

$$y(t) = -z(-t) + z(t)$$

$$y(t) = x(t) - x(-t)$$

and $\quad y(t) = -z(-t) - x(-t) = x(t) + z(t).$

$y(t) = -y(-t)$ is an odd function (see Fig. 1).

• **PROBLEM 7-17**

Find the system function $H(j\omega) = \frac{V_i(j\omega)}{I_i(j\omega)}$, and impulse response of a one-port network if the input voltage, $v_i(t) = 100\cos\omega_o t$ V, produces the input current, $I_i(j\omega) = 100\pi[\delta(\omega+\omega_o) + \delta(\omega-\omega_o)](1-j/2\omega)$.

Solution: The Fourier transform of $v_i(t) = 100 \cos \omega_o t$ is

$$V_i(j\omega) = 100\pi [\delta(\omega+\omega_o) + \delta(\omega-\omega_o)].$$

Hence, the system function $H(j\omega)$ is

$$\frac{V_i(j\omega)}{I_i(j\omega)} = \frac{1}{1 - \frac{j}{2\omega}} = \frac{j2\omega}{1+j2\omega}.$$

The impulse response is found by taking the inverse Fourier transform of $H(j\omega)$. Hence,

$$h(t) = F^{-1}\left\{\frac{j2\omega}{1+j2\omega}\right\} = F^{-1}\left\{1 - \frac{1}{1+j2\omega}\right\}.$$

Using the transform pairs yields

$$\delta(t) \iff 1. \quad \text{Since } \delta(t-t_o) \iff e^{-j\omega t_o}$$

and $e^{-at} u(t) \iff \frac{1}{a+j\omega}$, we obtain

$$h(t) = \delta(t) - \frac{1}{2} e^{-\frac{1}{2}t} \mu(t).$$

STEADY STATE RESPONSE

• **PROBLEM** 7-18

Find the angle by which v_2 leads v_1 if

$v_1 = 4 \cos(1000t - 40°)V$ and $v_2 = $;

(a) $3 \sin(1000t - 40°)V$; (b) $-2 \cos(1000t - 120°)V$;
(c) $5 \sin(1000t - 180°)V$.

Solution: Although sines and cosines are actually two ways of describing the same wave of sinusoidal shape, the determination of phase angles between two sinusoids is less apt to result in mistakes if we first convert sines to cosines. Secondly, since we are asked specifically for the angle by which v_2 leads v_1 subtract the phase angle of v_1 from the phase angle of v_2. Thus, if v_2 actually lags v_1 the calculated angle will be negative.

(a) As a preliminary step, convert the sine function to a cosine function by subtracting 90 degrees from its phase angle. This yields

$$v_2 = 3 \sin(1000t - 40°) = 3 \cos(1000t - 40° - 90°)V \quad (1)$$

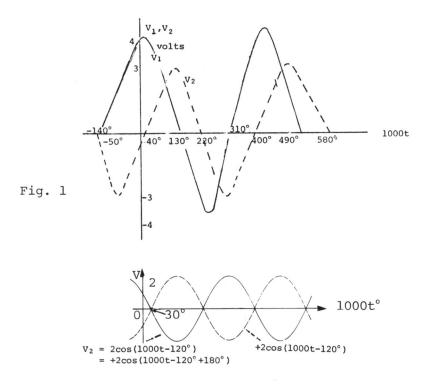

Fig. 1

$v_2 = 2\cos(1000t-120°)$
$= +2\cos(1000t-120°+180°)$

$+2\cos(1000t-120°)$

Fig. 2 Showing now adding 180° to phase angle eliminates minus sign in front of cosine

Since both voltages are cosines, subtract the angle of v_1 from the angle of v_2:

$$\theta_2 - \theta_1 = (-130°) - (-40°) = -90°. \tag{2}$$

The negative sign means that v_2 actually lags v_1. It is left in this form because the original statement of the problem asks for the amount by which v_2 leads v_1.

The relationship of these two sinusoids is shown in Fig. 1.

As a check, note that v_2 reaches its peak value when $1000t = 130°$ as sketched, and since a sine wave reaches its positive peak when its total angle is 90°, i.e.

$$v_2 = 3 \text{ when } \sin(1000t - 40°) = \sin 90° = 1, \tag{3}$$

then $1000t - 40° = 90°$ \hfill (4)

and $1000t = 130°$. \hfill (5)

(b) Conversion is unnecessary this time, but the minus sign in front of the cosine function should be interpreted as an additional 180° phase angle (either + or −, it doesn't matter). That is, the v_2 wave shown in Fig. 2 could be interpreted mathematically as the mirror image of

2 cos(1000t - 120°) (the minus sign does that), or it could also be interpreted as the 2 cos(1000t - 120°) shifted right or left by 180°.

Once v_2 has been modified, proceed as before to subtract phase angles:

$$v_2 = 2\cos(1000t - 120° + 180°) = 2\cos(1000t + 60°) V \qquad (6)$$

$$\theta_2 - \theta_1 = (+60°) - (-40°) = 100°. \qquad (7)$$

(c) Again it is necessary to convert the sine function to a cosine by subtracting 90 degrees, thus

$$v_2 = 5\sin(1000t - 180°) = 5\cos(1000t - 270°) V \qquad (8)$$

Subtracting phase angles gives

$$\theta_2 - \theta_1 = -270° - (-40°) = -230°. \qquad (9)$$

Although numerically correct, our answer will be modified by adding 360 because angles of lead or lag are customarily expressed as equal to or less than 180 in electrical engineering applications,

$$\theta_2 - \theta_1 = -230° + 360° = 130°. \qquad (10)$$

• **PROBLEM 7-19**

With $V_C(t)$ as the desired response, find the sinusoidal steady-state transfer function $H(j\omega)$ for the circuit of Fig. 1.

<u>Solution</u>: The transfer function $H(j\omega)$ is defined as

$$\frac{V_C(j\omega)}{V(j\omega)}.$$

Since $V_C(j\omega)$ can be found by voltage division as

$$V_C(j\omega) = \frac{V(j\omega) Z_C(j\omega)}{Z_R(j\omega) + Z_L(j\omega) + Z_C(j\omega)},$$

where Z's represent the impedances offered by resistor, inductor and capacitor.

the transfer function is

$$H(j\omega) = \frac{V_C(j\omega)}{V(j\omega)} = \frac{Z_C(j\omega)}{Z_R(j\omega) + Z_L(j\omega) + Z_C(j\omega)}.$$

Hence,

$$H(j\omega) = \frac{\frac{1}{j\omega 0.25}}{1 + j\omega + \frac{1}{j\omega 0.25}}$$

$$H(j\omega) = \frac{\frac{4}{j\omega}}{1 + j\omega + \frac{4}{j\omega}}$$

$$H(j\omega) = \frac{4}{(j\omega)^2 + j\omega + 4} = \frac{4}{4-\omega^2 + j\omega}.$$

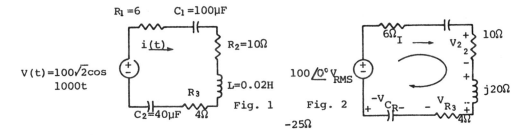

Fig. 1

• **PROBLEM 7-20**

In the series circuit shown in fig. 1, calculate i(t) in the steady-state and draw a voltage phasor diagram.

Solution: First, find the steady-state equivalent of the circuit by transforming, element-by-element;

$$100\sqrt{2} \cos 1000t \text{ V} \Rightarrow 100\underline{/0°} \text{ V rms}$$

The impedances offered by capacitors and the inductor are computed as follows:

$$C_1 = 100\mu F \Rightarrow \frac{1}{j\omega C} = \frac{1}{j(1000)(100\times 10^{-6})} = -j10\Omega$$

$$L = 0.02H = j\omega L = j1000(0.02) = j20$$

$$C_2 = 40\mu F \Rightarrow \frac{1}{j\omega C} = \frac{1}{j1000(40\times 10^{-6})} = -j25\Omega.$$

Hence, we have the circuit shown in fig. 2. Using KVL around the loop in fig. 2, we find

$$100\underline{/0°} = \vec{I}(6 - j10 + 10 + j20 + 4 - j25)$$

the phasor current is,

403

$$\vec{I} = \frac{100}{20 - j15}$$

$$\vec{I} = \frac{100(20 + j15)}{20^2 + 15^2} = \frac{20 + j15}{6.25}$$

$$\vec{I} = 3.2 + j2.4 \text{ A}$$

in polar form the current is,

$$\vec{I} = 4\underline{/36.87°} \text{ A}.$$

One can now calculate the voltage drop across each element,

$$\vec{V}_{R_1} = 6(4\underline{/36.87°}) = 24\underline{/36.87°} \text{ Vrms}$$

$$\vec{V}_{C_1} = -j10(4\underline{/36.87°})$$

$$= (10\underline{/-90°})(4\underline{/36.87°}) = 40\underline{/-53.13°} \text{ Vrms}$$

$$\vec{V}_{R_2} = 10(4\underline{/36.87°}) = 40\underline{/36.87°} \text{ Vrms}$$

$$\vec{V}_L = j20(4\underline{/36.87°})$$

$$= (20\underline{/90°})(4\underline{/36.87°}) = 80\underline{/126.87°} \text{ Vrms}$$

$$\vec{V}_{R_3} = 4(4\underline{/36.87°}) = 16\underline{/36.87°} \text{ Vrms}$$

$$\vec{V}_{C_2} = -j25(4\underline{/36.87°})$$

$$= (25\underline{/-90°})(4\underline{/36.87°}) = 100\underline{/-53.13°} \text{ Vrms}.$$

In rectangular form,

$$\vec{V}_{R_1} = 19.2 + j14.4 \text{ V}$$

$$\vec{V}_{C_1} = 24.0 - j32.0 \text{ V}$$

$$\vec{V}_{R_2} = 32.0 + j24.0 \text{ V}$$

$$\vec{V}_L = -48.0 + j64.0 \text{ V}$$

$$\vec{V}_{R_3} = 12.8 + j9.6 \text{ V}$$

$$\vec{V}_{C_2} = 60.0 - j80.0 \text{ V}$$

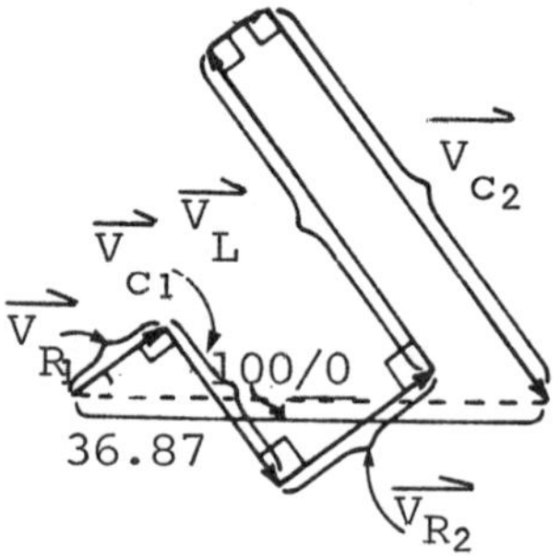

Fig.3

Adding the voltages in rectangular form gives 100 + j0 V,

the expected resulting source voltage.
   Using the voltage in polar form, add the phasor vectorially using the source voltage as a reference, Fig. 3 shows the vector addition of the voltage drops.

• **PROBLEM 7-21**

A sinusoidal driving function $V \sin \omega t$ is applied to a series RL network. Find the behavior of the current under steady state conditions. Sketch the magnitude and phase versus ω.

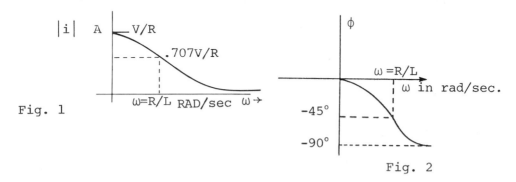

Fig. 1

Fig. 2

Solution: The differential equation which results from KVL applied to the R-L circuit is:

$$L \frac{di}{dt} + iR = V \sin \omega t$$

Since the response $i(t)$ must assume the time dependence of the forcing function, and since one term contains a derivative (with 90° shift), it would seem logical to assume:

$$i(t) = I_1 \sin \omega t + I_2 \cos \omega t \qquad (1)$$

where I_1 and I_2 are real constants which must depend on R, L, ω and V. A less cumbersome form results if we recognize that an equivalent form using a single amplitude constant and phase factor will give the same result.

$$i(t) = A \sin(\omega t + \phi) \qquad (2)$$

Putting (2) into the differential equation gives:

$$L[\omega A \cos(\omega t + \phi)] + R[A \sin(\omega t + \phi)] = V \sin \omega t.$$

Expanding the trigonometric forms gives:

$A\omega L[\cos \omega t \cos \phi - \sin \omega t \sin \phi]$

$\qquad + AR[\sin \omega t \cos \phi + \cos \omega t \sin \phi] = V \sin \omega t. \qquad (3)$

Now equate the coefficients of $\cos \omega t$:

$$A\omega L \cos \phi + AR \sin \phi = 0$$

from which

$$\tan \phi = \frac{\sin \phi}{\cos \phi} = \frac{-\omega L}{R}.$$

Also, from this result, it can be seen that

$$\sin \phi = \frac{-\omega L}{\sqrt{R^2 + \omega^2 L^2}} \quad \text{and} \quad \cos \phi = \frac{R}{\sqrt{R^2 + \omega^2 L^2}} \quad (4)$$

Now equate the coefficients of sin ωt in (3)

$$-A\omega L \sin \phi + AR \cos \phi = V.$$

Substituting the results of (4),

$$\frac{\omega^2 L^2 A}{\sqrt{R^2 + \omega^2 L^2}} + \frac{R^2 A}{\sqrt{R^2 + \omega^2 L^2}} = V$$

From which it is easy to obtain

$$A = \frac{\sqrt{R^2+\omega^2 L^2}}{R^2+\omega^2 L^2} V = \frac{V}{\sqrt{R^2+\omega^2 L^2}}$$

The solution is then written:

$$i(t) = \frac{V}{\sqrt{R^2 + \omega^2 L^2}} \sin\left(\omega t + \tan^{-1}\left(-\frac{\omega L}{R}\right)\right)$$

Figures 1 and 2 are sketches of magnitude phase of i versus ω.

● **PROBLEM 7-22**

The complete response of the circuit in fig. 1 was found to be $v_c(t) = \frac{1}{2}[e^{-t} - \cos t + \sin t]u(t)V$ for an input of $v(t) = \sin t \, u(t)V$. Use sinusoidal steady state methods to find the steady state voltage $V_{c_{ss}}(t)$ when the input is, $v(t) = \sin(t)V$. Compare the results with the expression for complete response, $v_c(t)$, above.

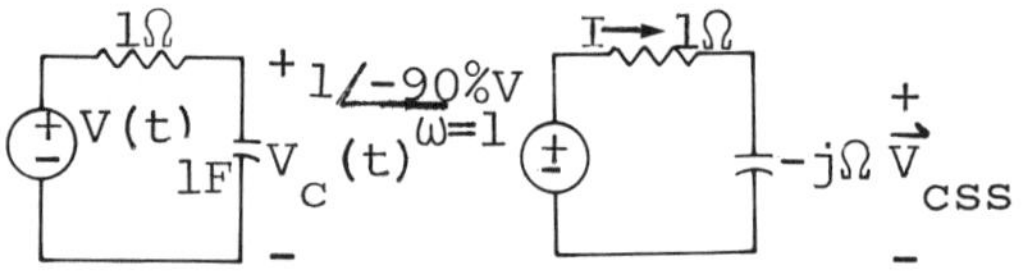

Fig. 1   Fig. 2

Solution: In order to perform steady state analysis, recall the convention for phasor representation of sinusoidals. It

was stated previously that
$$v(t) = A\cos(\omega t + \phi) = \text{Re}(\text{Im } e^{j(\omega t + \phi)}),$$
is represented by, $V = A e^{j\phi} = A\angle\phi$, the definition of the phasor.

Since, $\sin t = \cos(t - 90°)$, the phasor input becomes, $V = 1\angle -90°$ V. Fig. 2 shows the steady state phasor representation of the circuit in fig. 1.
Writing a KVL equation yields,
$$1\angle -90° = \vec{I}(1 - j)$$

$$\vec{V}_{C_{ss}} = \vec{I}(-j) = \frac{1\angle -90°}{1 - j}(-j)$$

in polar form:

$$\vec{V}_{C_{ss}} = \frac{(1\angle -90°)(1\angle -90°)}{\sqrt{2}\angle -45°} = \frac{1}{\sqrt{2}}\angle -135° \text{ V}$$

Using the convention stated above change the phasor $\vec{V}_{C_{ss}}$ to $v_{C_{ss}}(t)$ sinusoidal.

$$v_{C_{ss}}(t) = \frac{1}{\sqrt{2}}\cos(t - 135°) \text{V}. \qquad (1)$$

By using the angle difference relation,
$$\cos(\alpha - \beta) = \cos\alpha\cos\beta + \sin\alpha\sin\beta,$$
expand the expression for $V_{C_{ss}}$ into the sum of two sinusoids

$$v_{C_{ss}}(t) = \frac{1}{\sqrt{2}}\cos(t-135°) = \frac{1}{\sqrt{2}}[\cos t \cos(135°) + \sin t \sin(135°)]$$

$$V_{C_{ss}}(t) = \frac{1}{\sqrt{2}}\cos(t-135°) = -\frac{1}{2}\cos t + \frac{1}{2}\sin t \text{ V}. \qquad (2)$$

Explain the meaning of the results in equation (1) by comparing them to the complete response $V_c(t)$. Since the complete response contains both transient and steady state responses, the steady state response can be found by allowing, $t \to \infty$. At $t = \infty$ the term $\frac{1}{2}e^{-t}$ becomes zero and the steady state response becomes $-\frac{1}{2}\cos t + \frac{1}{2}\sin t$.

• **PROBLEM 7-23**

Evaluate the sinusoidal steady-state current for the circuit shown in Fig. 1 by replacing the circuit by its sinusoidal steady state equivalent.

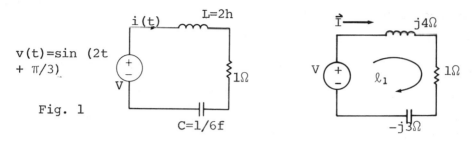

Fig. 1   Fig. 2

Solution: In order to replace the circuit in Fig. 1 by its sinusoidal steady state equivalent, note that from $v(t) = \sin(2t + \frac{\pi}{3})$ that $\omega = 2$ and that $v(t) = \sin(2t + 60°)$
$= \cos(2t+60 - 90°) = \cos(2t-30°) = \text{Re}\left[e^{j(2t-30°)}\right]$. This can be written in phasor form as $\text{Re}1\angle 2t-30°$.

$$\vec{Z}_L = j\omega 2 = j2(2) = j4\ \Omega$$

$$\vec{Z}_R = 1\Omega$$

$$\vec{Z}_C = \frac{1}{j\omega(\frac{1}{6})} = \frac{1}{j2(\frac{1}{6})} = \frac{-j}{\frac{1}{3}} = -j3\ \Omega.$$

Where Z's represent the impedances of L, R and C

Replace each of the elements in Fig. 1 by those found above. Fig. 2 shows the sinusoidal steady-state equivalent of the circuit in Fig. 1. Using KVL around the loop ℓ, solve for $\vec{I}$.

$$\left[1\angle 2t-30°\right] = \vec{I}(j4+1-j3)$$

$$\frac{1\angle 2t-30°}{1+j1} = \frac{1\angle 2t-30°}{\sqrt{2}\angle 45°}$$

$$I = \text{Re}\left[\frac{1}{\sqrt{2}}\angle 2t-75°\right] = \text{Re}\left[\frac{1}{\sqrt{2}}e^{j(2t-75°)}\right] = \frac{1}{\sqrt{2}}\cos(2t-75°)$$

$$I = \frac{1}{\sqrt{2}}\sin(2t - 75° + 90°) = \frac{1}{\sqrt{2}}\sin(2t + 15°)$$

$$I = \frac{1}{\sqrt{2}}\sin(2t + \frac{\pi}{12})\ \text{A}.$$

• **PROBLEM 7-24**

Find the steady-state current i(t) in the circuit shown in fig. 1. Assume

$$v(t) = 4 \sin(2t + \tfrac{\pi}{4}) \text{ V}.$$

Solution: Steady-state current i(t) is expressed in phasor form as

$$\vec{I} = I\, e^{j\phi_i} \tag{1}$$

where I is the rms value of i(t), ϕ_i is the initial phase angle of i(t), and j is equal to $\sqrt{-1}$. Steady-state voltage v(t) is expressed in a phasor form as

$$\vec{V} = V\, e^{j\phi_n} \tag{2}$$

where V is the rms value of v(t) and ϕ_v is the initial phase angle of v(t). Both $\vec{I}$ and $\vec{V}$ are related to each other by Ohm's law as follows

$$\vec{I} = \frac{\vec{V}}{\vec{Z}} \tag{3}$$

Here $\vec{Z}$ is the impedance of the circuit. If the circuit is formed of a series connection of a resistance R, inductance L and capacitance C then

$$\vec{Z} = R + j(\omega L - \tfrac{1}{\omega C})$$

$$= \sqrt{R^2 + (\omega L - \tfrac{1}{\omega C})^2}\, \exp\left[j\, \tan^{-1} \frac{\omega L - \tfrac{1}{\omega C}}{R}\right] \tag{4}$$

Now, $v(t) = 4 \sin(2t + \tfrac{\pi}{4})$

$$= 4 \cos\left[(2t + \tfrac{\pi}{4}) - \tfrac{\pi}{2}\right] = 4 \cos(2t - \tfrac{\pi}{4}) \tag{5}$$

$$\vec{V} = \frac{4}{\sqrt{2}}\, e^{j(-\tfrac{\pi}{4})} \text{ V} \tag{6}$$

Now $R = 1$, $L = 1$, $C = \tfrac{1}{4}$ and $\omega = 2$ so from Eq. (4)

$$\vec{Z} = \sqrt{1^2 + \left(2 \times 1 - \tfrac{1}{2 \times \tfrac{1}{4}}\right)^2}\, \exp\left[j\, \tan^{-1} \frac{2 \times 1 - \tfrac{1}{2 \times \tfrac{1}{4}}}{1}\right]$$

$$\vec{Z} = 1e^{j\,\tan^{-1}0} = 1e^{j0}\Omega. \tag{7}$$

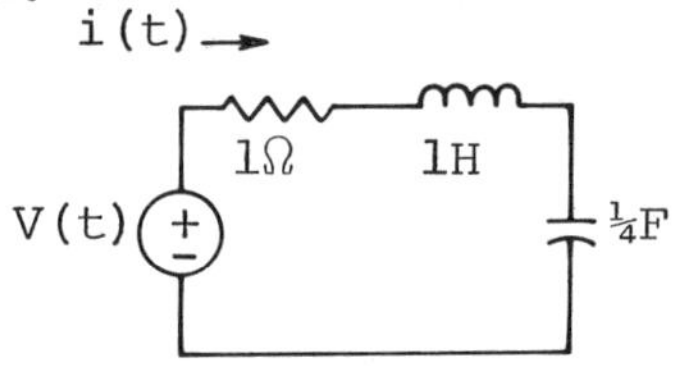

So, by Eq. (3),

$$\vec{I} = \frac{\frac{4}{\sqrt{2}}e^{-j\frac{\pi}{4}}}{1e^{j0}} = \frac{4}{\sqrt{2}}e^{-j\frac{\pi}{4}} A$$

Fig. 1

$$i_{ss}(t) = \sqrt{2}\,4I\cos(2t - \frac{\pi}{4})$$

$$= 4\sin(2t - \frac{\pi}{4} + \frac{\pi}{2}) = 4\sin(2t + \frac{\pi}{4})\,A \tag{8}$$

This problem can be solved by introducing a concept of power factor, or current lags voltage by an angle of

$$\theta = \tan^{-1}\frac{\omega L - \frac{1}{\omega C}}{R} \tag{9}$$

Then,

$$i_{ss}(t) = 4\sin(2t + \frac{\pi}{4} - 0)$$

$$= 4\sin(2t + \frac{\pi}{4} - 0)$$

$$= 4\sin(2t + \frac{\pi}{4})\,A \tag{10}$$

Note that all calculations using phasors here could have been done using only peak values given, then converting the answer if rms values were desired.

● **PROBLEM 7-25**

The R-L circuit shown has been connected to a sinusoidal source

$$v(t) = V_m \cos(\omega t + \phi)\,V$$

for a sufficiently long time so that the steady state is reached. At t = 0 the circuit is deenergized, solve for i(t) for all t ≥ 0⁺.

Solution: The first step of the solution is to find the steady-state current through the inductor which will then be the initial condition for the de-energized circuit.

The steady-state current with the switch up is found simply by

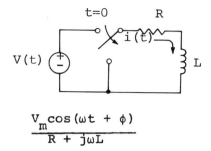

$$\frac{V_m \cos(\omega t + \phi)}{R + j\omega L}$$

when the switch is down the circuit equation is

$$Ri + L\frac{di}{dt} = 0$$

which has a solution of the form

$$i(t) = Ke^{-t/\tau}$$

where $\tau = L/R$.
The integration constant, K, is evaluated by setting the current equal to the original steady-state current for $t = 0$. Thus

$$K = \frac{V_m \cos \phi}{R + j\omega L}.$$

The solution is

$$i(t) = \frac{V_m \cos \phi}{R + j\omega L} e^{-t/\tau}.$$

Another form of the solution is obtained by setting $R + j\omega L = Z\underline{/\theta}$. In complex notation we have

$$\frac{V_m e^{j\phi}}{Z e^{j\theta}} e^{-t/\tau} = \frac{V_m}{Z} e^{j(\phi - \theta)} e^{-t/\tau} \text{ A}.$$

Taking the real part gives

$$i(t) = \left[\frac{V_m}{Z}\cos(\phi-\theta)\right] e^{-t/\tau} \quad \text{A}.$$

• **PROBLEM 7-26**

Calculate the steady-state voltage response of the network shown in Fig. 1.

Solution: Using phasor concepts, transform the circuit into its steady-state equivalent.

Doing so,

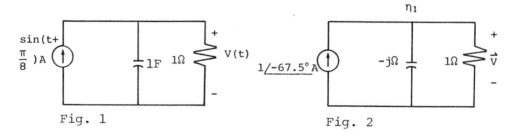

Fig. 1   Fig. 2

$$\sin(t + \tfrac{\pi}{8})A = 1 \underline{/- \tfrac{\pi}{2} + \tfrac{\pi}{8}} = 1\underline{/- 67.5°} \text{ A}$$

Since $\omega = 1$, the impedance of the capacitor will be $Z_c = \tfrac{1}{j\omega C}$ = $-j\Omega$, as shown in Fig. 2. At steady state, the current through the resistor and the capacitor are respectively $\tfrac{\vec{V}}{1}$ and $\tfrac{\vec{V}}{-j}$.

From KCL

$$1\underline{/- 67.5°} = \tfrac{\vec{V}}{1} - \tfrac{\vec{V}}{j} = \vec{V}(1 + j)$$

Solving for $\vec{V}$, the phasor branch voltage,

$$\vec{V} = \frac{1\underline{/- 67.5°}}{(1 + j)}$$

$$\vec{V} = \frac{1\underline{/- 67.5°}}{\sqrt{2}\,\underline{/45°}}$$

$$= \tfrac{1}{\sqrt{2}} \underline{/- 112.5°} \text{ V}$$

Converting from phasor form to cartesian form yields

$$V(t) = \tfrac{1}{\sqrt{2}} \cos(t - \tfrac{5}{8}\pi) \text{ V}.$$

● **PROBLEM 7-27**

Calculate the steady-state voltage, $v_c(t)$, for the network shown in Fig. 1.

Solution: Write a KCL equation for node n_1 in Fig. 1:

$$i(t) = \frac{V_1}{2} + \frac{V_1}{1 + \frac{1}{j\omega C}} \tag{1}$$

One can write

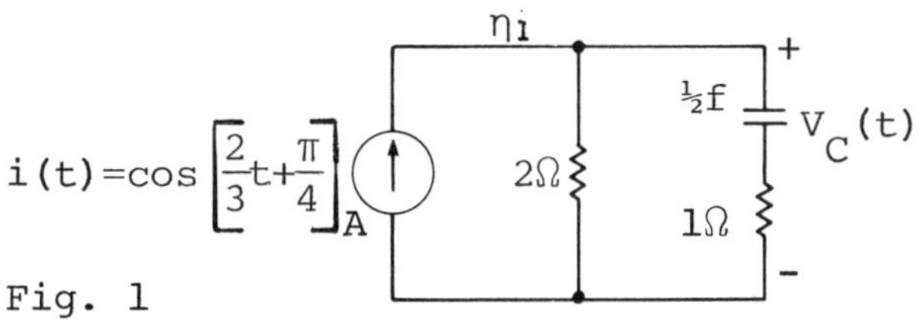

Fig. 1

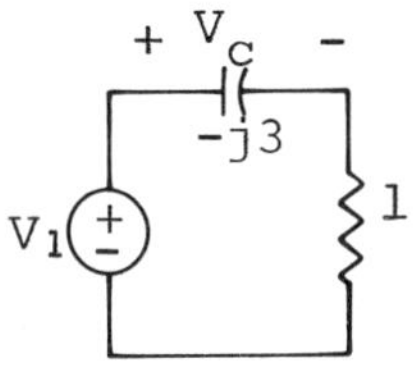

$i(t) = \cos\left[\frac{2}{3}t + \frac{\pi}{4}\right]$

Fig. 2

RC circuit in sinusoidal steady state.

$$i(t) = \cos\left(\frac{2}{3}t + \frac{\pi}{4}\right) = \text{Re}\left[e^{j\left(\frac{2}{3}t + \frac{\pi}{4}\right)}\right]$$

in polar form,

$$\vec{I} = 1\underline{/\frac{\pi}{4}} = 1\underline{/45°} \text{ A}$$

also, $\omega = \frac{2}{3}$ rad/s and $C = \frac{1}{2}$F. Substituting these values into Eq. (1) gives

$$1\underline{/45°} = \frac{\vec{V}_1}{2} + \frac{\vec{V}_1}{1 + j\left(\frac{2}{3}\right)\left(\frac{1}{2}\right)}$$

$$1\underline{/45°} = \vec{V}_1(.5) + \frac{\vec{V}_1}{1 - \frac{j}{\frac{1}{3}}}$$

$$1\underline{/45°} = \vec{V}_1\left(.5 + \frac{1}{1 - j3}\right)$$

Solving for V_1

$$\vec{V}_1 = \frac{1\underline{/45°}}{.5 + \frac{1}{\sqrt{10}\underline{/-71.57}}} = \frac{1\underline{/45°}}{.5 + \frac{\sqrt{10}}{10}\underline{/71.57}}$$

$$\vec{V}_1 = \frac{1\underline{/45°}}{0.5 + 0.1 + j(0.3)} = \frac{1\underline{/45°}}{(0.6) + j(0.3)}$$

$$\vec{V}_1 \frac{1\underline{/45°}}{\sqrt{0.45}\underline{/26.57°}} \text{ V.}$$

Using the concept of voltage division as illustrated in fig. 2,

$$V_C = \frac{V_1(-j3)}{1 - j3}$$

413

$$\vec{V}_C = \frac{\frac{1/45}{\sqrt{0.45}/26.57} \cdot 3/-90°}{\sqrt{10}/-71.56°}$$

$$V_C = \frac{3}{\sqrt{0.45}}\left(\frac{1}{\sqrt{10}}\right)/45 - 26.57 - 90 + 71.56$$

$$V_C = \sqrt{\frac{9}{0.45(10)}}/0°$$

$$V_C = \sqrt{\frac{9}{4.5}}/0° = \sqrt{2}/0° \text{ V}$$

$$V_C = \sqrt{2}\cos(\tfrac{2}{3}t)\text{V}$$

• **PROBLEM 7-28**

A sinusoidal current source, 10 cos 1000t A, is in parallel both with a 20-Ω resistor and the series combination of a 10-Ω resistor and a 10-mH inductor. Remembering the usefulness of Thévenin's theorem, find the voltage across the following: (a) inductor; (b) 10-Ω resistor; (c) current source.

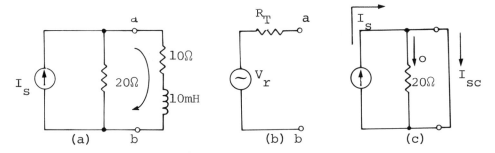

Solution: Following the suggestion, use Thevenin's theorem to convert the current source and 20-ohm resistor into a voltage source as shown in Figure 1(b). The resultant circuit shown in Figure 2 can now be solved for current and the current used to find the required voltages.

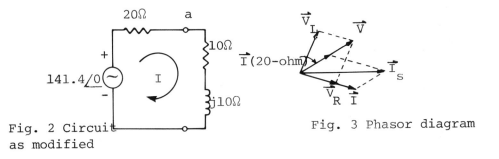

Fig. 2 Circuit as modified

Fig. 3 Phasor diagram

The original circuit as shown in Figure 1(a) will be modified by converting the current source and 20-ohm resistor to the left of a-b to an equivalent voltage source as shown in Figure 1(b). This is acomplished by first separating this part of the circuit from the remaining part and then determining the open-circuit voltage between a and b.

$$\text{Open-Circuit Voltage} = \vec{V}_t = 20\,\vec{I}_s \tag{1}$$

where I_s and V_t are phasor quantities (rms),

$$\vec{V}_t = 20 \times 10(.707)\underline{/0°} = 141.4\underline{/0°}\ \text{V}. \tag{2}$$

Next, determine the current through a hypothetical short-circuit between points a and b:

$$\text{Short-circuit Current } \vec{I}_{sc} = \vec{I}_s = 7.07\underline{/0°}\ \text{A}. \tag{3}$$

Finally, the Thevenin resistance R_t is found at the quotient

$$R_t = \frac{V_{ab}(\text{open circuit})}{I_{sc}(\text{short circuit})} = \frac{141.4\underline{/0°}}{7.07\underline{/0°}} = 20\Omega. \tag{4}$$

The resultant modified circuit is shown in Figure 2. Now we are ready to solve this simpler circuit.
First, solve for the inductive reactance

$$X_L = \omega L = 1000 \times .010 = 10\Omega. \tag{5}$$

and determine current with Kirchhoff's law:

$$\vec{I} = \frac{141.4\underline{/0°}}{(10+20)+j10} = \frac{141.4\underline{/0°}}{31.62\underline{/18.4°}} = 4.47\underline{/-18.4°}\ \text{A}. \tag{6}$$

Converted to an instantaneous value, the current is

$$i = 4.47\sqrt{2}\cos(1000t - 18.4°) = 6.32\cos(1000t - 18.4°)\text{A} \tag{7}$$

where $\sqrt{2}$ is included to convert the rms value to peak value.
The current can now be used to find out the voltage across the inductor:

$$\vec{V}_L = j\omega L\,\vec{I} = 10\underline{/90°} \times 4.47\underline{/-18.4°} = 44.7\underline{/71.6°}\ \text{V} \tag{8}$$

which converts to

$$v_L = 63.2\cos(1000t + 71.6°)\text{V}. \tag{9}$$

(d) The voltage across the 10-ohm resistor is

$$\vec{V}_R = \vec{I}R = 4.47\underline{/-18.4°} \times 10 = 44.7\underline{/-18.4°}\ \text{V} \tag{10}$$

which converts to

$$v_R = 63.2 \cos(1000t - 18.4°) \text{ V}. \tag{11}$$

(e) Finally, voltage across the current source is seen to be the voltage across a-b in Figure 1(a) which is

$$\vec{V} = \vec{I}\vec{Z} = (4.47)\underline{/-18.4°}(10+j10) = (4.47)(14.14)\underline{/-18.4°+45°} \tag{12}$$

$$\vec{V} = 63.2\underline{/26.6°} \text{ V}. \tag{13}$$

This converts to

$$v = 89.4 \cos(1000t + 26.6°) \text{V}. \tag{14}$$

The phasor diagram for this circuit is shown in Figure 3.

● **PROBLEM 7-29**

Solve for the steady-state current $i_L(t)$ in the circuit shown in fig. 1.

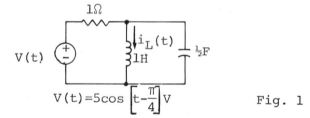

$$V(t) = 5\cos\left[t - \frac{\pi}{4}\right] \text{V}$$

Fig. 1

Solution: The phasor representation of the voltage v(t) is:

$$\vec{V} = 5\underline{/-45°} \text{ V}$$

with a frequency of 1 rad/sec. The impedance of each element is

$$\vec{Z}_R = 1 + j0 = 1\underline{/0°} \text{ } \Omega$$

$$\vec{Z}_L = 0 + j\omega L = 0 + j1 = 1\underline{/90°} \text{ } \Omega$$

$$\vec{Z}_C = \frac{1}{j\omega C} = \frac{-j}{(1)(\tfrac{1}{2})} = 0 - j2 = 2\underline{/90°} \text{ } \Omega.$$

Since the capacitor and inductor are in parallel, the equivalent impedance is:

$$Z_P = \frac{Z_C Z_L}{Z_L + Z_C} = \frac{1\underline{/90°}\ 2\underline{/-90°}}{+j1 - j2} = \frac{2\underline{/0°}}{1\underline{/-90°}} = 2\underline{/90°} = 0 + j2\Omega.$$

The total impedance is then:

$$\vec{Z}_T = \vec{Z}_R + \vec{Z}_P = 1 + j0 + 0 + j2 = 1 + j2 = \sqrt{5}/63.43° \ \Omega$$

It follows that the total current is

$$\vec{I}_T = \frac{\vec{V}}{\vec{Z}_T} = \frac{5/-45°}{\sqrt{5}/63.43°} = \sqrt{5}/-108.43° \ A.$$

The voltage across the parallel branches is

$$\vec{V}_P = \vec{I}_T \vec{Z}_P = \sqrt{5}/-108.43° \ \ 2/90°$$

$$= 2\sqrt{5}/-18.43° \ V.$$

The inductor current must be

$$\vec{I}_L = \frac{\vec{V}_P}{\vec{Z}_L} = \frac{2\sqrt{5}/-18.43°}{1/+90} = 2\sqrt{5}/-108.43° \ A.$$

In the time domain representation this is

$$i_L(t) = 2\sqrt{5} \cos(t - 108.43°) = 2\sqrt{5} \cos(t - 45 - 63.43°) A.$$

● **PROBLEM 7-30**

Write an expression for $v_c(t)$ in the circuit shown in fig. 1 by using a sinusoidal steady-state equivalent network.

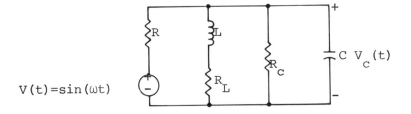

Solution: Replacing each element by its sinusoidal steady-state impedance produces the circuit in fig. 2.

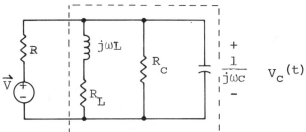

Fig. 2

By replacing the part of the circuit in fig. 2 which is enclosed by the dotted line, the voltage division concept can be applied to obtain V_C.

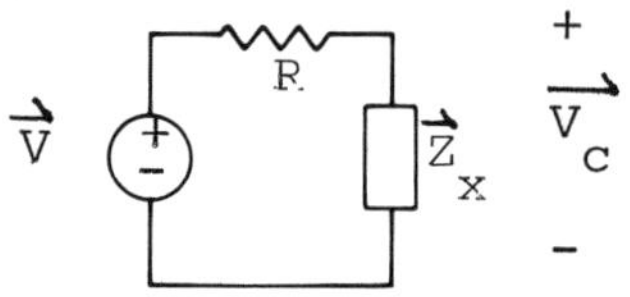

Fig. 3

$$\vec{Z}_X = (j\omega L + R_L) || R_C || \frac{1}{j\omega C}$$

$$\vec{Z}_X = \left[\frac{(j\omega L + R_L)R_C}{j\omega L + R_L + R_C}\right] || \frac{1}{j\omega C}$$

$$\vec{Z}_X = \frac{\dfrac{(j\omega L + R_L)R_C}{(j\omega L + R_L + R_C)j\omega C}}{\dfrac{(j\omega L + R_L)R_C}{j\omega L + R_L + R_C} + \dfrac{1}{j\omega C}}$$

Simplifying gives,

$$\vec{Z}_X = \frac{\dfrac{j\omega L R_C + R_L R_C}{j^2\omega^2 LC + j\omega C R_L + j\omega C R_C}}{\dfrac{(j\omega L + R_L)j\omega C R_C + j\omega L + R_L + R_C}{j^2\omega^2 LC + j\omega C R_L + j\omega C R_C}}$$

$$\vec{Z}_X = \frac{j\omega L R_C + R_L R_C}{(j\omega L + R_L)(j\omega C R_C + 1) + R_C}$$

$$\vec{Z}_X = \frac{R_C(j\omega L + R_L)}{(j\omega L + R_L)(j\omega C R_C + 1) + R_C}$$

since $\vec{V}_C = \dfrac{Z_X}{R + Z_X} V.$

$$\frac{\vec{Z}_X}{R + \vec{Z}_X} = \frac{\dfrac{R_C(j\omega L + R_L)}{(j\omega L + R_L)(j\omega C R_C + 1) + R_C}}{R + \dfrac{R_C(j\omega L + R_L)}{(j\omega L + R_L)(j\omega C R_C + 1) + R_C}}$$

Simplifying,

$$= \frac{\dfrac{R_C(j\omega L + R_L)}{(j\omega L + R_L)(j\omega C R_C + 1) + R_C}}{\dfrac{R(j\omega L + R_L)(j\omega C R_C + 1) + RR_C + R_C(j\omega L + R_L)}{(j\omega L + R_L)(j\omega C R_C + 1) + R_C}}$$

418

and dividing through by $R_C(j\omega L + R_L)$

$$= \cfrac{1}{\cfrac{R(j\omega CR_C + 1)}{R_C} + \cfrac{R}{j\omega L + R_L} + 1}$$

finally,

$$\vec{V}_c = \cfrac{V}{j\omega CR + \cfrac{R}{R_C} + \cfrac{R}{j\omega L + R_L} + 1}$$

• **PROBLEM 7-31**

Find the Thévenin equivalent circuit of the network shown in **Fig. 1** and use it to find I_{ab} through the 30Ω resistor.

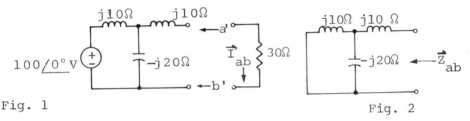

Fig. 1   Fig. 2

Solution: A Thévenin equivalent circuit is found by locating the open circuit voltage across the terminals of the desired network. This yields the Thévenin equivalent voltage source. The Thévenin equivalent impedance is found by determining the impedance across the terminals of the network when all independent voltage and/or current sources are "off." (i.e. voltages sources short circuited and current sources open circuited). The Thévenin equivalent circuit becomes the series combination of these two parameters.

In this example, the open circuit voltage $\vec{V}_{ab}$ is found by determining the voltage across the capacitor.
Hence,

$$\vec{V}_{ab} = \frac{(100\underline{/0°})(-j20)}{j10 - j20} = \frac{-j2000}{-j10}$$

$$\vec{V}_{ab} = 200\underline{/0°} \text{ V.}$$

The impedance across terminals ab is shown in Fig. 2.

$$\vec{Z}_{ab} = \frac{(j10)(-j20)}{j10 - j20} + j10$$

$$\vec{Z}_{ab} = \frac{200}{-j10} + j10$$

$$\vec{Z}_{ab} = j20 + j10 = j30\Omega.$$

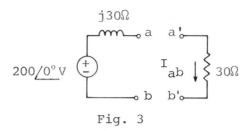

Fig. 3

Hence, the Thévenin equivalent circuit is shown in Fig. 3. The current $\vec{I}_{ab}$ is found by appling KVL:

$$200\underline{/0°} = \vec{I}_{ab}(j30 + 30)$$

$$\vec{I}_{ab} = \frac{200}{30 + j30} = \frac{200\underline{/0°}}{42.4\underline{/45°}} = 4.71\ \underline{/-45°}$$

$$\vec{I}_{ab} = 3.33 - j3.33\ A$$

• **PROBLEM 7-32**

Find the steady-state voltage $v_c(t)$ in the circuit of fig. 1.

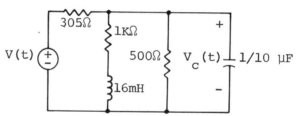

Fig. 1 $V(t) = \sin((2\pi \times 10^4)t+45°)V$

Solution: The equivalent steady-state network can be represented as three parallel admittances in series with an impedance and voltage source, as shown in Fig. 2.

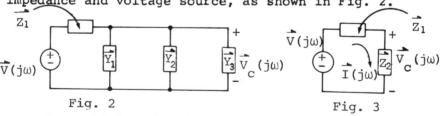

Fig. 2   Fig. 3

By combining the three parallel admittances, the total admittance is obtained

$$\vec{Y}_T = \vec{Y}_1 + \vec{Y}_2 + \vec{Y}_3$$

This admittance can be represented by a single impedance, $\frac{1}{Y_T} = Z_2$, shown in the equivalent circuit of Fig. 3.

The current around the single loop in the circuit of Fig. 3 is

$$\vec{I}(j\omega) = \frac{\vec{V}(j\omega)}{\vec{Z}_1 + \vec{Z}_2}$$

and $\vec{V}_C(j\omega) = \vec{I}(j\omega)Z_2 = \vec{V}(j\omega)\dfrac{\vec{Z}_2}{\vec{Z}_1 + \vec{Z}_2}$ \hfill (1)

Eq. (1) describes the concept of voltage division. In any series combination of impedances, the voltage across any particular element is the ratio of the impedance of the element to the total impedance in the series combination multiplied by the applied voltage.

Now, $Y_T = \dfrac{1}{1000+j1000} + 2 \times 10^{-3} + j\, 6.28 \times 10^{-3}$

Substituting the values of Z_1 and $Z_2 = \dfrac{1}{Y_T}$ in Eq. 1,

$$\vec{V}_C = 1\underline{/-45°}\left\{\dfrac{\dfrac{1}{\dfrac{1}{1000+j1000} + 2 \times 10^{-3} + j6.28 \times 10^{-3}}}{305 + \dfrac{1}{\dfrac{1}{1000+j1000} + 2 \times 10^{-3} + j6.28 \times 10^{-3}}}\right\}$$

$$\vec{V}_C = 1\underline{/-45°}\left\{\dfrac{\left[\dfrac{1}{0.5 - j0.5 + 2 + j6.28}\right]\dfrac{1}{10^{-3}}}{305 + \dfrac{1}{[0.5 - j0.5 + 2 + j6.28]}\dfrac{1}{10^{-3}}}\right\}$$

$$\vec{V}_C = 1\underline{/-45°}\left\{\dfrac{\left[\dfrac{1}{2.5 + j5.78}\right]\dfrac{1}{10^{-3}}}{305 + \left[\dfrac{1}{2.5 + j5.78}\right]\dfrac{1}{10^{-3}}}\right\}$$

$$\vec{V}_C = 1\underline{/-45°}\left\{\dfrac{\left(\dfrac{2.5 - j5.78}{6.3}\right)\dfrac{1}{10^{-3}}}{305 + \left[\dfrac{2.5 - j5.78}{6.3}\right]\dfrac{1}{10^{-3}}}\right\}$$

$$\vec{V}_C = 1\underline{/-45°}\left(\dfrac{397 - j917}{305 + 397 - j917}\right)$$

$$\vec{V}_C = 1\underline{/-45°}\ \dfrac{1000\underline{/\,66.6}}{1154.9\underline{/-52.6}}$$

$$\vec{V}_C = 0.866 \underline{/-59°} \text{ V}$$

$$v_c(t) = 0.866 \cos((2\pi \times 10^4) - 59°) \text{V}.$$

• **PROBLEM 7-33**

Use the superposition theorem to find the current through R_2 in Fig. 1. Both sources are of the same frequency.

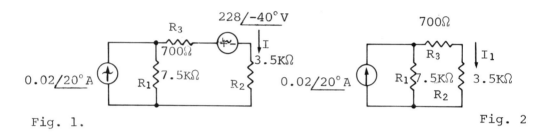

Fig. 1.   Fig. 2

Solution: First, consider the effect of the current source. With the voltage source "off," the circuit shown in Fig. 2 is obtained.
Using current division concepts produces

$$\vec{I}_1 = \frac{R_1}{R_1 + R_3 + R_2}(0.02\underline{/20°})$$

$$\vec{I}_1 = \frac{7.5K\Omega}{7.5K\Omega + 700\Omega + 3.5K\Omega}(0.02\underline{/20°})$$

$$\vec{I}_1 = \frac{7.5}{11.7}(0.02\underline{/20°})$$

$$\vec{I}_1 = 12.8\underline{/20°} \text{ mA}.$$

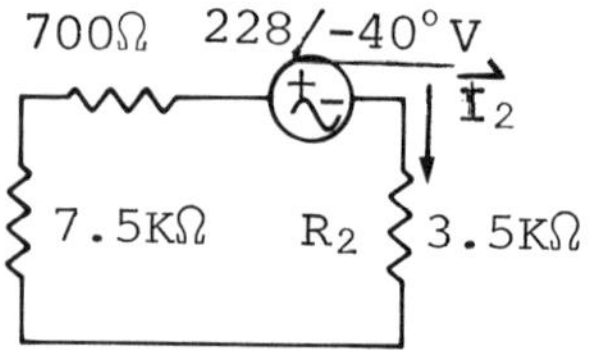

Fig. 3

Next, consider the effect of the voltage source. With the current source "off," the circuit shown in Fig. 3 is obtained.
Using KVL in the single loop circuit of Fig. 3, write the equation

$$-228\underline{/-40°} = \vec{I}_2(7.5K\Omega + 700\Omega + 3.5K\Omega)$$

where, $\vec{I}_2 = \dfrac{-228\underline{/-40°}}{11.7 \times 10^3}$

$$\vec{I}_2 = -19.5\underline{/-40°} \text{ mA}.$$

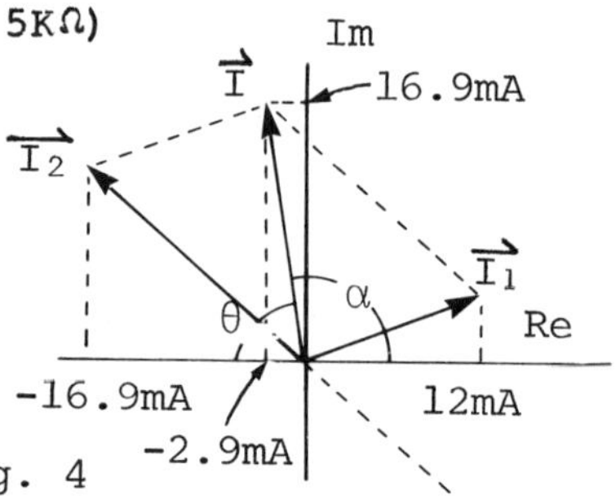

By superposition, we have

Fig. 4

$$\vec{I} = \vec{I}_1 + \vec{I}_2 = (12.8\underline{/20°} - 19.5\underline{/-40°})\text{mA}.$$

Change the currents to rectangular form and add to obtain the current I.

$$\vec{I} = 12.8 \cos 20° - 19.5 \cos(-40°)$$
$$+ j(12.8 \sin 20° - 19.5 \sin(-40°))$$
$$\vec{I} = (12 - 14.9) + j(4.4 + 12.5)$$
$$\vec{I} = -2.9 + j16.9 \text{ mA}$$
$$\vec{I} = \sqrt{2.9^2 + 16.9^2}\underline{/-\tan^{-1}\tfrac{16.9}{2.9} + 180°} = 17.1\underline{/99.7°} \text{ mA}.$$

Fig. 4 shows vectoriol representation of the addition of vectors $\vec{I}_1$ and $\vec{I}_2$.

CHAPTER 8

RANDOM VARIABLES AND PROCESSES

PROBABILITY

● **PROBLEM 8-1**

What is the probability of throwing a "six" with a single die?

Solution: The die may land in any of 6 ways:

1, 2, 3, 4, 5, 6

The probability of throwing a six,

$$P(6) = \frac{\text{number of ways to get a six}}{\text{number of ways the die may land}}$$

Thus $P(6) = \frac{1}{6}$.

● **PROBLEM 8-2**

A deck of playing cards is thoroughly shuffled and a card is drawn from the deck. What is the probability that the card drawn is the ace of diamonds?

Solution: The probability of an event occurring is

$$\frac{\text{the number of ways the event can occur}}{\text{the number of possible outcomes}}$$

In our case there is one way the event can occur, for there is only one ace of diamonds and there are fifty two possible outcomes (for there are 52 cards in the deck). Hence the probability that the card drawn is the ace of diamonds is 1/52.

• **PROBLEM 8-3**

In a single throw of a single die, find the probability of obtaining either a 2 or a 5.

<u>Solution:</u> In a single throw, the die may land in any of 6 ways:
$$1 \quad 2 \quad 3 \quad 4 \quad 5 \quad 6.$$

The probability of obtaining a 2,

$$P(2) = \frac{\text{number of ways of obtaining a 2}}{\text{numbers of ways the die may land}}, \quad P(2) = \frac{1}{6}.$$

Similarly, the probability of obtaining a 5,

$$P(5) = \frac{\text{number of ways of obtaining a 5}}{\text{number of ways the die may land}}, \quad P(5) = \frac{1}{6}.$$

The probability that either one of two mutually exclusive events will occur is the sum of the probabilities of the separate events. Thus the probability of obtaining either a 2 or a 5, P(2) or P(5), is

$$P(2) + P(5) = \frac{1}{6} + \frac{1}{6} = \frac{2}{6} = \frac{1}{3}$$

• **PROBLEM 8-4**

A box contains 7 red, 5 white, and 4 black balls. What is the probability of your drawing at random one red ball? One black ball?

<u>Solution:</u> There are 7 + 5 + 4 = 16 balls in the box. The probability of drawing one red ball,

$$P(R) = \frac{\text{number of possible ways of drawing a red ball}}{\text{number of ways of drawing any ball}}$$

$$P(R) = \frac{7}{16}.$$

Similarly, the probability of drawing one black ball

$$P(B) = \frac{\text{number of possible ways of drawing a black ball}}{\text{number of ways of drawing any ball}}$$

Thus,
$$P(B) = \frac{4}{16} = \frac{1}{4}$$

• **PROBLEM 8-5**

A penny is to be tossed 3 times. What is the probability there will be 2 heads and 1 tail?

Solution: We start this problem by constructing a set of all possible outcomes:

We can have heads on all 3 tosses: (HHH)
head on first 2 tosses, tail on the third: (HHT)   (1)
head on first toss, tail on next two: (HTT)
    •   (HTH)   (2)
    •   (THH)   (3)
        (THT)
    •   (TTH)
        (TTT)

Hence there are eight possible outcomes (2 possibilities on first toss x 2 on second x 2 on third = 2 x 2 x 2 = 8).

We assume that these outcomes are all equally likely and assign the probability 1/8 to each. Now we look for the set of outcomes that produce 2 heads and 1 tail. We see there are 3 such outcomes out of the 8 possibilities (numbered (1), (2), (3) in our listing). Hence the probability of 2 heads and 1 tail is 3/8.

• **PROBLEM 8-6**

A card is drawn at random from a deck of cards. Find the probability that at least one of the following three events will occur:

Event A : a heart is drawn.
Event B: a card which is not a face card is drawn.
Event C: the number of spots (if any) on the drawn card is divisible by 3.

Solution: Let $A \cup B \cup C$ = the event that at least one of the three events above will occur. We wish to find $P(A \cup B \cup C)$, the probability of the event $A \cup B \cup C$. Let us count the number of ways that at least A, B or C will occur. There are 13 hearts, 40 non-face cards, and 12 cards such that the number of spots is divisible by 3. (Cards numbered 3, 6, or 9 are all divisible by 3 and there are 4 suits each with 3 such cards, 3 × 4 = 12). If we add 40 + 13 + 12 we will have counted too many times. There are 10 cards which are hearts and non-face cards. 3 cards divisible by 3 and hearts, 12 cards which are non-face cards and divisible by 3. We must subtract each of these from our total of 40 + 13 + 12 giving 40 + 13 + 12 - 10 - 3 - 12. But we have subtracted too much; we have subtracted the 3 cards which are hearts and non-face cards and divisible by 3. We must add these cards to our total making

$$P(A \cup B \cup C) = \frac{40 + 13 + 12 - 10 - 3 - 12 + 3}{52} = \frac{43}{52}$$

Our counting technique used was called the principle of inclusion/exclusion and is useful for problems of this sort. Also look again at our answer,

$$P(A \cup B \cup C) = \frac{13 + 40 + 12 - 10 - 3 - 12 + 3}{52}$$

$$= \frac{13}{52} + \frac{40}{52} + \frac{12}{52} - \frac{10}{52} - \frac{3}{52} - \frac{12}{52} + \frac{3}{52}$$

Note that $P(A) = \dfrac{\text{number of hearts}}{\text{number of cards}} = \dfrac{13}{52}$

$P(B) = \dfrac{\text{number of non-face cards}}{\text{number of cards}} = \dfrac{40}{52}$

$P(C) = \dfrac{\text{number of cards divisible by 3}}{\text{number of cards}} = \dfrac{12}{52}$

$P(AB) = \dfrac{\text{number of hearts and non-face cards}}{\text{number of cards}} = \dfrac{10}{52}$

$P(AC) = \dfrac{\text{number of hearts and cards divisible by 3}}{\text{number of cards}}$

$= \dfrac{3}{52}$

● **PROBLEM 8-7**

In an experiment involving the toss of two dice, what is the probability that the sum is 6 or 7?

Solution: Let A be the event that a 6 is obtained and B be the event that a 7 is obtained in the toss of two dice. Since the events A and B are mutually exclusive (i.e. the occurence of one event precludes the occurence of another event). In other words,

$$P(A \text{ or } B) = P(A \cup B) = P(A) + P(B).$$

The union symbol "$\cup$" means that A and/or B can occur.

Now, number of ways to toss two dice = 6 × 6 = 36.

A = Set of the outcomes in which the sum is equal to 6

= { (1,5), (2,4), (3,3), (4,2), (5,1) }

B = Set of the outcomes in which the sum is equal to 7

= { (1,6), (2,5), (3,4), (4,3), (5,2), (6,1) }

So,
$P(A) = 5/36$

and $P(B) = 6/36 = 1/6$

and $P(A \cup B) = P(A) + P(B) = (5/36) + (6/36) = 11/36$.

• **PROBLEM 8-8**

Find the probability that a face card is drawn on the first draw and an ace on the second in two consecutive draws, without replacement, from a standard deck of cards.

Solution: This problem illustrates the notion of conditional probability. The conditional probability of an event, say event B, given the occurrence of a previous event, say event A, is written $P(B|A)$. This is the conditional probability of B given A.

$P(B|A)$ is defined to be $\frac{P(AB)}{P(A)}$, where $P(AB)$ = Probability of the joint occurrence of events A and B.

Let A = event that a face card is drawn on the first draw

B = event that an ace is drawn on the second draw.

We wish to find the probability of the joint occurrence of these events, $P(AB)$.

We know that $P(AB) = P(A) \cdot P(B|A)$.

$P(A)$ = probability that a face card is drawn on the first draw = $\frac{12}{52} = \frac{3}{13}$.

$P(B|A)$ = probability that an ace is drawn on the second draw given that a face card is drawn on the first

= number of ways an ace can be drawn on the second draw given a face card is drawn on the first divided by the total number of possible outcomes of the second draw.

= $\frac{4}{51}$; remember there will be only 51 cards left in the deck after the face card is drawn.

Thus $P(AB) = \frac{3}{13} \cdot \frac{4}{51} = \frac{4}{13 \times 17} = \frac{4}{221}$.

• **PROBLEM 8-9**

A coin is tossed 3 times. Find the probability that all 3 are heads,
(a) if it is known that the first is heads,
(b) if it is known that the first 2 are heads,
(c) if it is known that 2 of them are heads.

Solution: This problem is one of conditional probability.

If we have two events, A and B, the probability of event A given that event B has occurred is

$$P(A/B) = \frac{P(AB)}{P(B)}.$$

(a) We are asked to find the probability that all three tosses are heads given that the first toss is heads. The first event is A and the second is B.

P(AB) = probability that all three tosses are heads given that the first toss is heads

$= \dfrac{\text{the number of ways that all three tosses are heads given that the first toss is a head}}{\text{the number of possibilities resulting from 3 tosses}}$

$= \dfrac{\{H, HH\}}{\{\{H,H,H\}, \{H,H,T\}, \{H,T,H\}, \{H,T,T\}, \{T,T,T\}, \{T,T,H\}, \{T,H,T\}, \{T,H,H\}\}}$

$= \dfrac{1}{8}.$

P(B) = P(first toss is a head)

$= \dfrac{\text{the number of ways to obtain a head on the first toss}}{\text{the number of ways to obtain a head or a tail on the first of 3 tosses}}$

$= \dfrac{\{H,H,H\}, \{H,H,T\}, \{H,T,H\}, \{H,T,T\}}{8}$

$= \dfrac{4}{8} = \dfrac{1}{2}.$

$P(A/B) = \dfrac{P(AB)}{P(B)} = \dfrac{\frac{1}{8}}{\frac{1}{2}} = \dfrac{1}{8} \cdot \dfrac{2}{1} = \dfrac{1}{4}.$

To see what happens, in detail, we note that if the first toss is heads, the logical possibilities are HHH, HHT, HTH, HTT. There is only one of these for which the second and third are heads. Hence,

$P(A/B) = \dfrac{1}{4}.$

(b) The problem here is to find the probability that all 3 tosses are heads given that the first two tosses are heads.

P(AB) = the probability that all three tosees are heads given that the first two are heads

$= \dfrac{\text{the number of ways to obtain 3 heads given that the first two tosses are heads}}{\text{the number of possibilities resulting from 3 tosses}}$

$$= \frac{1}{8}.$$

P(B) = the probability that the first two are heads

$$= \frac{\text{number of ways to obtain heads on the first two tosses}}{\text{number of possibilities resulting from three tosses}}$$

$$= \frac{\{H,H,H\}, \{H,H,T\}}{8} = \frac{2}{8} = \frac{1}{4}.$$

$$P(A/B) = \frac{P(AB)}{P(B)} = \frac{\frac{1}{8}}{\frac{1}{4}} = \frac{4}{8} = \frac{1}{2}.$$

(c) In this last part, we are asked to find the probability that all 3 are heads on the condition that any 2 of them are heads.
Define:

A = the event that all three are heads

B = the event that two of them are heads

P(AB) = the probability that all three tosses are heads knowing that two of them are heads

$$= \frac{1}{8}.$$

P(B) = the probability that two tosses are heads

$$= \frac{\text{number of ways to obtain at least two heads out of three tosses}}{\text{number of possibilities resulting from 3 tosses}}$$

$$= \frac{\{H,H,T\},\{H,H,H\},\{H,T,H\},\{T,H,H\}}{8}$$

$$= \frac{4}{8}$$

$$= \frac{1}{2}$$

$$P(A/B) = \frac{P(AB)}{P(B)} = \frac{1/8}{1/2} = \frac{2}{8} = \frac{1}{4}$$

● **PROBLEM 8-10**

If 4 cards are drawn at random and without replacement from a deck of 52 playing cards, what is the chance of drawing the 4 aces as the first 4 cards?

Solution: We will do this problem in two ways. First we will use the classical model of probability which tells us

Probability = $\frac{\text{Number of favorable outcomes}}{\text{All possible outcomes}}$, assuming all outcomes are equally likely.

There are four aces we can draw first. Once that is gone any one of 3 can be taken second. We have 2 choices for third and only one for fourth. Using the Fundamental Principle of Counting we see that there are $4 \times 3 \times 2 \times 1$ possible favorable outcomes. Also we can choose any one of 52 cards first. There are 51 possibilitites for second, etc. The Fundamental Principle of Counting tells us that there are $52 \times 51 \times 50 \times 49$ possible outcomes in the drawing of four cards. Thus,

Probability = $\frac{4 \times 3 \times 2 \times 1}{52 \times 51 \times 50 \times 49} = \frac{1}{270,725} = .0000037.$

Our second method of solution involves the multiplication rule and shows some insights into its origin and its relation to conditional probability.

The formula for conditional probability $P(A|B) = \frac{P(A \cap B)}{P(B)}$ can be extended as follows:

$P(A|B \cap C \cap D) = \frac{P(A \cap B \cap C \cap D)}{P(B \cap C \cap D)}$; thus

$P(A \cap B \cap C \cap D) = P(A|B \cap C \cap D) \, P(B \cap C \cap D)$ but

$P(B \cap C \cap D) = P(B|C \cap D) \, P(C \cap D)$ therefore

$P(A \cap B \cap C \cap D) = P(A|B \cap C \cap D) \, P(B|C \cap D) \, P(C \cap D)$ but

$P(C \cap D) = P(C|D) \, P(D)$ hence

$P(A \cap B \cap C \cap D) = P(A|B \cap C \cap D) \, P(B|C \cap D) \, P(C|D) \, P(D).$

Let event D = drawing an ace on the first card

C = drawing an ace on second card

B = ace on third draw

A = ace on fourth card.

Our conditional probability extension becomes

P (4 aces) = P (on 4th|first 3) × P(3rd|first 2) × P(2nd|on first) × P(on first).

Assuming all outcomes are equally likely;

P(on 1st draw) = $\frac{4}{52}$. There are 4 ways of success in 52 possibilities. Once we pick an ace there are 51 remaining cards, 3 of which are aces. This leaves a probability of $\frac{3}{51}$ for picking a second ace once we have chosen the first. Once we have 2 aces there are 50 remaining cards, 2 of which are aces, thus P(on 3rd | first 2) = $\frac{2}{50}$. Similarly P(4th ace | first 3) = $\frac{1}{49}$. According to our formula above

$$P(4 \text{ aces}) = \frac{1}{49} \times \frac{2}{50} \times \frac{3}{51} \times \frac{4}{52} = .000037.$$

● **PROBLEM 8-11**

Twenty percent of the employees of a company are college graduates. Of these, 75% are in supervisory position. Of those who did not attend college, 20% are in supervisory positions. What is the probability that a randomly selected supervisor is a college graduate?

Solution: Let the events be as followed:

E : The person selected is a supervisor

Solution: Each question requires us to find the probability that a defective item was produced by a particular machine. Bayes' Rule allows us to calculate this using known (given) probabilities. First we define the necessary symbols: M_1 means the item was produced at A, M_2 means it was produced at B, and M_3 and M_4 refer to machines C and D, respectively. Let M mean that an item is defective. Using Bayes' Rule,

$$P(M_1|M) = \frac{P(M_1) P(M|M_1)}{P(M_1)P(M|M_1)+P(M_2)P(M|M_2)+P(M_3)P(M|M_3)+P(M_4)P(M|M_4)}$$

we substitute the given proportions as follows:

$$P(M_1|M) = \frac{(.1)(.001)}{(.1)(.001)+(.2)(.0005)+(.3)(.005)+(.4)(.002)}$$

$$= \frac{.0001}{.0001 + .0001 + .0015 + .0008} = \frac{.0001}{.0025} = \frac{1}{25}.$$

To compute $P(M_2|M)$ we need only change the numerator to $P(M_2)P(M|M_2)$. Substituting given proportions, we have $(.20)(.0005) = .0001$. We see that $P(M_2|M) = \frac{1}{25} = P(M_1|M)$. By the same procedure we find that $P(M_3|M) = \frac{3}{5}$ and $P(M_4|M) = \frac{8}{25}$.

To check our work, note that a defective item can be produced by any one of the 4 machines and that the four events "produced by machine i and defective" (i=1,2,3,4) are mutually exclusive. Thus

$$P(M) = \sum_{i=1}^{4} P(M \text{ and } M_i) \quad \text{or} \quad 1 = \sum_{i=1}^{4} \frac{P(M \text{ and } M_i)}{P(M)};$$

but

$$\frac{P(M \text{ and } M_i)}{P(M)} = P(M_i|M).$$

Thus $\sum_{i=1}^{4} P(M_i|M) = 1.$ Adding we see that

$$\frac{1}{25} + \frac{1}{25} + \frac{15}{25} + \frac{8}{25} = \frac{25}{25} = 1.$$

● **PROBLEM 8-12**

In a factory four machines produce the same product. Machine A produces 10% of the output, machine B, 20%, machine C, 30%, and machine D, 40%. The proportion of defective items produced by these follows: Machine A: .001; Machine B: .0005; Machine C: .005; Machine D: .002. An item selected at random is found to be defective. What is the probability that the item was produced by A? by B? by C? by D?

Solution: Each question requires us to find the probability that a defective item was produced by a particular machine. Bayes' Rule allows us to calculate this using known (given) probabilities. First we define the necessary symbols: M_1 means the item was produced at A, M_2 means it was produced at B, and M_3 and M_4 refer to machines C and D, respectively. Let M mean that an item is defective. Using Bayes' Rule,

$$P(M_1|M) = \frac{P(M_1) P(M|M_1)}{P(M_1)P(M|M_1)+P(M_2)P(M|M_2)+P(M_3)P(M|M_3)+P(M_4)P(M|M_4)}$$

we substitute the given proportions as follows:

$$P(M_1|M) = \frac{(.1)(.001)}{(.1)(.001)+(.2)(.0005)+(.3)(.005)+(.4)(.002)}$$

$$= \frac{.0001}{.0001 + .0001 + .0015 + .0008} = \frac{.0001}{.0025} = \frac{1}{25}.$$

To compute $P(M_2|M)$ we need only change the numerator to $P(M_2)P(M|M_2)$. Substituting given proportions, we have $(.20)(.0005) = .0001$. We see that $P(M_2|M) = \frac{1}{25} = P(M_1|M)$. By the same procedure we find that $P(M_3|M) = \frac{3}{5}$ and $P(M_4|M) = \frac{8}{25}$.

To check our work, note that a defective item can be produced by any one of the 4 machines and that the four events "produced by machine i and defective" (i=1,2,3,4) are mutually exclusive. Thus

$$P(M) = \sum_{i=1}^{4} P(M \text{ and } M_i) \quad \text{or} \quad 1 = \sum_{i=1}^{4} \frac{P(M \text{ and } M_i)}{P(M)} ;$$

but $\quad \dfrac{P(M \text{ and } M_i)}{P(M)} = P(M_i | M).$

Thus $\sum_{i=1}^{4} P(M_i | M) = 1.$ Adding we see that

$$\frac{1}{25} + \frac{1}{25} + \frac{15}{25} + \frac{8}{25} = \frac{25}{25} = 1.$$

PROBABILITY DENSITY FUNCTION OF RANDOM VARIABLES

• **PROBLEM 8-13**

Given that the continuous random variable X has distribution function $F(x) = 0$ when $x < 1$ and $F(x) = 1 - 1/x^2$ when $x \geq 1$, graph $F(x)$, find the density function $f(x)$ of X, and show how $F(x)$ can be obtained from $f(x)$.

<u>Solution</u>: To graph $F(x)$, observe that as x approaches 1 from the right side, $1 - 1/x^2$ approaches 0. As x approaches $+ \infty$, $1 - 1/x^2$ approaches 1 because $\lim_{x \to \infty} 1/x^2 = 0$.

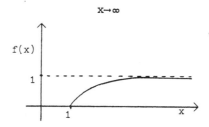

The curve is said to be asymptotic to the line $F(x) = 1$ because it comes closer and closer to it, as x increases without limit, but never touches it. $F(x)$ is a continuous function for all real numbers x because it satisfies 2 conditions: (1) $F(x)$ is defined for all values of x; (2) the function's value at any point c equals the left and right limits:

limit of $F(x) = F(c) =$ limit of $F(x)$
(as $x \to c$ from the left) (as $x \to c$ from the right).

Differentiating $F(x) = 1 - 1/x^2$ $(1 \leq x)$ yields
$$F'(x) = 0 - (-2x^{-3}) = 2/x^3$$

When $x < 1$, $F'(x) = d(0)/dx = 0.$ The derivative does not exist at $x = 1$ because

$$\text{limit of } F'(x) = 0 \neq 2 = \frac{2}{1^3} = \text{limit of } F'(x)$$
(as $x \to 1$ from left) \qquad\qquad $(x \to 1$ from right).

$$F'(x) = f(x) = \frac{2}{x^3} \text{ when } 1 \le x < \infty \quad = 0 \text{ when } x < 1$$

is the density function of X.

We can obtain $F(x)$ by integrating $f(x)$ from 1 to x when $x \ge 1$, and from $-\infty$ to x when $x < 1$. When $x \ge 1$,

$$F(x) = \int_1^x \frac{2}{t^3} dt = 2 \int_1^x t^{-3} dt = 2 \left. \frac{t^{-2}}{-2} \right]_1^x$$

$$= \left. \frac{-1}{t^2} \right]_1^x = -\frac{1}{x^2} - (-1) = -\frac{1}{x^2} + 1 = 1 - \frac{1}{x^2} = F(x).$$

When $x < 1$, $F(x) = \int_{-\infty}^x 0 \, dt = 0$

● PROBLEM 8-14

The cumulative distribution functions of the latitude angle $\theta(w)$ and the longitude angle $\phi(w)$ of the random orientation on the earth's surface are

$$F_\theta(\lambda) = \begin{cases} 0 & ; \lambda < 0 \\ 1-\cos\lambda & ; 0 \le \lambda \le \pi/2 \\ 1 & ; \lambda > \pi/2 \end{cases}$$

$$F_\phi(\lambda) = \begin{cases} 0 & ; \lambda < 0 \\ \lambda/2\pi & ; 0 \le \lambda \le 2\pi \\ 1 & ; \lambda > 2\pi \end{cases}$$

Find the corresponding density functions.

Solution: We know that $F(x) = \int_{-\infty}^x f(t) \, dt$. But by the Fundamental Theorem of Integral Calculus: $\frac{dF(x)}{dx} = f(x)$.

Hence $F_\theta(\lambda) = \frac{dF_\theta(\lambda)}{d\lambda} = \begin{cases} \frac{d}{d\lambda}(0) & ; \lambda < 0 \\ \frac{d}{d\lambda}(1-\cos\lambda) & ; 0 \le \lambda \le 2\pi \\ \frac{d}{d\lambda}(1) & ; \lambda > 2\pi \end{cases}$

$$= \begin{cases} 0 & ; \lambda < 0 \\ \sin\lambda & ; 0 \le \lambda \le 2\pi \\ 0 & ; \lambda > 2\pi \end{cases}$$

Also $F_\emptyset(\lambda) = \frac{dF(\lambda)}{d\lambda}$

$$= \begin{cases} \frac{d(0)}{d\lambda} & ; \quad \lambda < 0 \\ \frac{d}{d\lambda}\left(\frac{\lambda}{2\pi}\right) & ; \quad 0 \leq \lambda \leq 2\pi \\ \frac{d}{d\lambda}(1) & ; \quad \lambda > 2\pi \end{cases}$$

$$= \begin{cases} 0 & ; \quad \lambda < 0 \\ \frac{1}{2\pi} & ; \quad 0 \leq \lambda \leq 2\pi \\ 0 & ; \quad \lambda > 2\pi \end{cases}$$

• **PROBLEM 8-15**

Plot the probability density function of a random variable X, where X represents the number of heads that show in an experiment involving tossing of five fair coins.

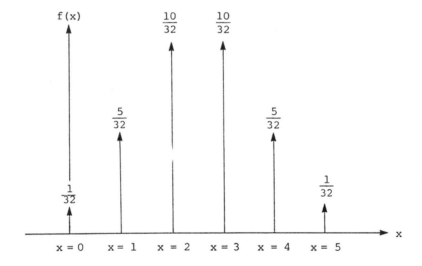

Solution: The event of getting x heads can occur in $\binom{5}{x}$ ways. So, the number of possible favorable outcomes $= \binom{5}{x}$. Since the sample space of the experiment contains 2^5 5-tuples, the total number of possible equally likely outcomes is $2^5 = 32$.

Probability density function $= \Pr(X=x) = f(x) = \frac{\binom{5}{x}}{2^5}$.

Therefore,
$\Pr(X=0) = \frac{\binom{5}{0}}{2^5} = \frac{1}{32} = f(0)$

$\Pr(X=1) = \frac{\binom{5}{1}}{2^5} = \frac{5}{32} = f(1)$

$$Pr(X=2) = \frac{\binom{5}{2}}{2^5} = \frac{10}{32} = f(2)$$

$$Pr(X=3) = \frac{\binom{5}{3}}{2^5} = \frac{\binom{5}{2}}{2^5} = \frac{10}{32} = f(3)$$

$$Pr(X=4) = \frac{\binom{5}{4}}{2^5} = \frac{\binom{5}{1}}{2^5} = \frac{5}{32} = f(4)$$

$$Pr(X=5) = \frac{\binom{5}{5}}{2^5} = \frac{\binom{5}{0}}{32} = \frac{1}{32} = f(5)$$

The probability density function is plotted in the Fig.

From the figure it is clear that the function is symmetrical, i.e. $Pr(X=0) = Pr(X=5) = 1/32$, where X=0 corresponds to "no heads", and X=5 corresponds to "all heads". Similarly, $Pr(X=1) = Pr(X=4)$, and so on.

• **PROBLEM 8-16**

Let X be the random variable equal to the sum of the values that appear in an experiment involving the tossing of two fair dice simultaneously. Determine the probability density function of X.

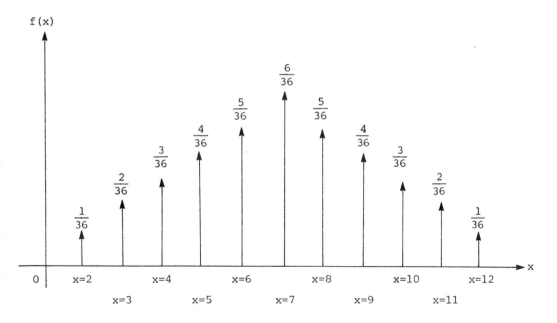

Solution: The random variable X, can take the values 2, 3, ..., 12.

Let $f(x) = \Pr(X=x)$ = Probability density function.

Now, $f(2) = f(12) = 1/36$,

$f(3) = f(11) = 2/36$,

$f(4) = f(10) = 3/36$,

$f(5) = f(9) = 4/36$,

$f(6) = f(8) = 5/36$, and

$f(7) = 7/36$.

The probability density function is shown in the figure in which the vertical lines indicate the probabilities for the value of x which the random variable assumes.

From the figure, it is observed that

$$f(x) = \Pr(X=x) = \begin{cases} \dfrac{6 - |x-7|}{36}, & x = 2,3,\ldots,12 \\ 0, & \text{otherwise} \end{cases}$$

It can be shown that $\sum_{x=2}^{12} f(x) = 1$

Consider the sum,

$$\sum_{x=2}^{12} f(x) = \sum_{x=2}^{12} \frac{6 - |7-x|}{36} = \sum_{x=2}^{12} \frac{6 - |x-7|}{36}$$

$= \dfrac{1}{36} \Big[(6-|7-2|) + (6-|7-3|) + (6-|7-4|) + (6-|7-5|)$

$\qquad + (6-|7-6|) + (6-|7-7|) + (6-|7-8|)$

$\qquad + (6-|7-9|) + (6-|7-10|) + (6-|7-11|) + (6-|7-12|) \Big]$

$= \dfrac{1}{36} \Big[(6-5) + (6-4) + (6-3) + (6-2) + (6-1) + (6-0)$

$\qquad + (6-1) + (6-2) + (6-3) + (6-4) + (6-5) \Big]$

$= \dfrac{1}{36} \Big(1 + 2 + 3 + 4 + 5 + 6 + 5 + 4 + 3 + 2 + 1 \Big)$

$= \dfrac{1}{36} (36)$

$= 1.$

● **PROBLEM 8-17**

(a) Show that the following properties are satisfied by a Rayleigh density given by

$$f(x) = \begin{cases} xe^{-x^2/2}, & x \geq 0 \\ 0, & x < 0 \end{cases}$$

(i) $f(x) \geq 0$ for all x

(ii) $\int_{-\infty}^{\infty} f(x)\,dx = 1.$

Solution:
(a) i) Since $e^{-x^2/2}$ is always positive and also $x \geq 0$, $f(x) \geq 0$ for all $x \geq 0$. For $x < 0$, $f(x) = 0$. Thus, the condition $f(x) \geq 0$ for all x is satisfied.

ii) $\int_{-\infty}^{\infty} f(x)\,dx = \int_{-\infty}^{0} f(x)\,dx + \int_{0}^{\infty} f(x)\,dx$

$$= \int_{-\infty}^{0} 0 \cdot dx + \int_{0}^{\infty} xe^{-x^2/2}\,dx \quad (1)$$

let $u = x^2/2$

Then, $du = \dfrac{2x\,dx}{2} = x\,dx.$

Equation (1) thus reduces to

$$\int_{-\infty}^{\infty} f(x)\,dx = \int_{0}^{\infty} e^{-u}\,du$$

$$= (-e^{-u})\Big|_{0}^{\infty}$$

$$= -[(0) - (1)] = 1.$$

Hence, the proof.

CUMULATIVE DISTRIBUTION FUNCTION

● **PROBLEM 8-18**

Let X be the random variable denoting the result of the single toss of a fair coin. If the toss is heads, $X = 1$. If the toss results in tails, $X = 0$.
What is the probability distribution of X?

Solution: The probability distribution of X is a function which assigns probabilities to the values X may assume.

This function will have the following properties if it defines a proper probability distribution. Let $f(x) = \Pr(X = x)$. Then $\Sigma \Pr(X = x) = 1$ and $\Pr(X = x) \geq 0$ for all x.

We have assumed that X is a discrete random variable. That is, X takes on discrete values.

The variable X in this problem is discrete as it only takes on the values 0 and 1.

To find the probability distribution of X, we must find $\Pr(X = 0)$ and $\Pr(X = 1)$.

Let $p_0 = \Pr(X = 0)$ and $\Pr(X = 1) = p_1$. If the coin is fair, the events X = 0 and X = 1 are equally likely. Thus $p_0 = p_1 = p$. We must have $p_0 > 0$ and $p_1 > 0$.

In addition,
$$\Pr(X = 0) + \Pr(X = 1) = 1$$
or
$$p_0 + p_1 = p + p = 1$$
or
$$2p = 1$$
and
$$p_0 = p_1 = p = \tfrac{1}{2},$$

thus the probability distribution of X is $f(x)$: where
$$f(0) = \Pr(X = 0) = \tfrac{1}{2} \text{ and}$$
$$f(1) = \Pr(X = 1) = \tfrac{1}{2}$$

f (anything else) = Pr(X = anything else) = 0. We see that this is a proper probability distribution for our variable X.

$$\Sigma f(x) = 1 \quad \text{and}$$
$$f(x) \geq 0.$$

● PROBLEM 8-19

Given that the random variable X has density function

$$f(x) = \begin{cases} 2x & 0 < x < 1 \\ 0 & \text{otherwise} \end{cases}$$

Find $\Pr(\tfrac{1}{2} < x < 3/4)$ and $\Pr(-\tfrac{1}{2} < x < \tfrac{1}{2})$.

Solution: Since $f(x) = 2x$ is the density function of a continuous random variable, $\Pr(\tfrac{1}{2} < x < 3/4)$ = area under $f(x)$ from $\tfrac{1}{2}$ to 3/4.

The area under $f(x)$ is the area of the triangle with vertices at (0,0), (1,0) and (1,2).

The area of this triangle is $A = \tfrac{1}{2} bh$ where b = base of the triangle and h is the altitude.

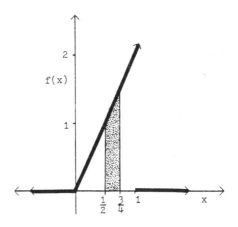

f(x) is indicated by the heavy line.

Thus, $A = \frac{1}{2}(1) \times 2 = \frac{2}{2} = 1$

proving that f(x) is a proper probability density function.

To find the probability that $\frac{1}{2} < x < 3/4$ we find the area of the shaded region in the diagram. This shaded region is the difference in areas of the right triangle with vertices (0,0), ($\frac{1}{2}$,0) and ($\frac{1}{2}$, f($\frac{1}{2}$)) and the area of the triangle with vertices (0,0), (3/4, 0) and (3/4, f(3/4)).

This difference is $\Pr(\frac{1}{2} < x < 3/4) = \frac{1}{2}(3/4)f(3/4) - \frac{1}{2}(\frac{1}{2})f(\frac{1}{2})$

$$= \frac{1}{2}[\frac{3}{4} \cdot \frac{6}{4} - \frac{1}{2} \cdot 1]$$

$$= \frac{1}{2}[\frac{9}{8} - \frac{1}{2}] = \frac{1}{2} \cdot \frac{5}{8} = \frac{5}{16} .$$

The probability that $-\frac{1}{2} < x < \frac{1}{2}$ is

$\Pr(-\frac{1}{2} < x < \frac{1}{2})$ = Area under f(x) from $-\frac{1}{2}$ to $\frac{1}{2}$.

Because f(x) = 0 from $-\frac{1}{2}$ to 0, the area under f(x) from $-\frac{1}{2}$ to 0 is 0. Thus

$\Pr(-\frac{1}{2} < x < \frac{1}{2}) = \Pr(0 < x < \frac{1}{2})$ = area under f(x) from 0 to $\frac{1}{2}$

$$= \frac{1}{2}(\frac{1}{2}) \ f(\frac{1}{2})$$

$$= \frac{1}{2}(\frac{1}{2}) \cdot 1 = \frac{1}{4} .$$

● **PROBLEM 8-20**

Let X be a continuous random variable. We wish to find probabilities concerning X. These probabilities are determined by a density function. Find a density function such that the probability that X falls in an interval (a,b) (0 < a < b < 1) is proportional to the length of the interval (a,b). Check that this is a proper probability density function.

Solution: The probabilities of a continuous random variable are computed from a continuous function called a density function in the following way. If f(x) is graphed and is

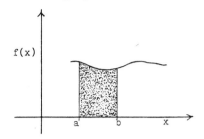

then, $Pr(a \leq X \leq b)$ = the area under the curve f(x) from a to b.

With this definition some conditions on f(x) must be imposed. f(x) must be positive and the total area between f(x) and the x-axis must be equal to 1.

We also see that if probability is defined in terms of area under a curve, the probability that a continuous random variable is equal to a particular value, $Pr(X = a)$ is the area under f(x) at the point a. The area of a line is 0, thus $Pr(X = a) = 0$. Therefore

$$Pr(a < X < b) = Pr(a \leq X \leq b)$$

To find a density function for $0 < X < 1$, such that $Pr(a < X < b)$ is proportional to the length of (a,b), we look for a function f(x) that is positive and the area under f(x) between 0 and 1 is equal to 1. It is reasonable to expect that the larger the interval the larger the probability that x is in the interval.

A density function that satisfies these criteria is

$$f(x) = \begin{cases} 1 & 0 < x < 1 \\ 0 & \text{otherwise} \end{cases}$$

A graph of this density function is

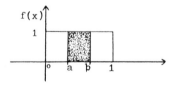

The probability that X is between a and b is the area of the shaded region. This is the area of a rectangle. The area of a rectangle is base × height.

Thus $Pr(a \leq X \leq b) = (b - a) \times 1 = b - a$. Similarly

$Pr(X \leq k) = (k - 0) \times 1 = k$ for $0 < k < 1$.

Often the density function is more complicated and integration must be used to calculate the area under the density function.

To check that this is a proper probability density function, we must check that the total area under f(x) is 1. The total area under this density function is $(1 - 0) \times 1 = 1$.

• **PROBLEM 8-21**

If $f(x) = 1/4$, $x = 0,1,2,3$ is a probability mass function, find $F(t)$, the cumulative distribution function and sketch its graph.

<u>Solution</u>: $F(t) = \sum_{x=0}^{t} f(x) = \Pr(X \le t)$. $F(t)$ changes for integer values of t. We have:

$$F(t) = 0 \qquad t < 0$$
$$F(t) = f(0) = 1/4, \quad 0 \le t < 1$$
$$F(t) = f(0) + f(1) = 1/4 + 1/4 = 1/2, \quad 1 \le t < 2$$
$$F(t) = f(0) + f(1) + f(2) \qquad 2 \le t < 3$$
$$= \tfrac{1}{4} + \tfrac{1}{4} + \tfrac{1}{4} = 3/4.$$
$$F(t) = \sum_{x=0}^{t} f(x) = 1 \qquad 3 \le t.$$

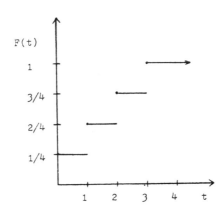

• **PROBLEM 8-22**

Consider the function, $g(x) = (1 + x^2)^{-1}$. Determine a constant k such that $f(x) = k(1 + x^2)^{-1}$ is a proper probability density for $-\infty < x < \infty$. Find $F(x) = \Pr(X \le x)$ if X is distributed with density function $f(x)$.

<u>Solution</u>: We see immediately that $f(x)$ is integrable, continuous and $f(x) > 0$ for all x.

To ensure that $f(x)$ is a proper probability function we must find a constant k such that $\int_{-\infty}^{\infty} f(x)\, dx = 1$.

Thus
$$\int_{-\infty}^{\infty} k(1 + x^2)^{-1}\, dx = k \int_{-\infty}^{\infty} \frac{1}{1 + x^2}\, dx$$

$$= k[\tan^{-1} x]_{-\infty}^{\infty}$$
$$= k[\tan^{-1} \infty - \tan^{-1} -\infty]$$
$$= k\pi.$$

Thus in order for $\int_{-\infty}^{\infty} f(x)\, dx$ to equal 1, k must be equal to $1/\pi$.

Then,
$$f(x) = \frac{1/\pi}{(1+x^2)} = \frac{1}{\pi(1+x^2)}$$

$$F(x) = \int_{-\infty}^{x} f(t)\, dt = \int_{-\infty}^{x} \frac{1}{\pi(1+t^2)}\, dt$$

$$= \frac{1}{\pi}[\tan^{-1} x - \tan^{-1}(-\infty)]$$

$$= \frac{1}{\pi}[\tan^{-1} x + \frac{\pi}{2}].$$

• **PROBLEM 8-23**

Let a distribution function F, be given by
$$F(x) = 0 \;;\quad x < 0$$
$$= (x+1)/2 \;;\quad 0 \leq x < 1$$
$$= 1 \;;\quad 1 \leq x.$$
Find $\Pr(-3 < x \leq \tfrac{1}{2})$ and sketch F.

Solution: First note that
$$\Pr(-3 < x \leq \tfrac{1}{2}) = \Pr(x \leq \tfrac{1}{2}) - \Pr(x \leq -3).$$

By definition, this is

$$F(\tfrac{1}{2}) - F(-3) = \frac{\tfrac{1}{2}+1}{2} - 0 = \frac{\tfrac{1}{2}+1}{2}$$

$$= \frac{3/2}{2} = \frac{3}{4}.$$

The graph of F follows:

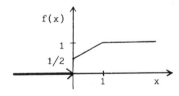

• **PROBLEM 8-24**

Let the probability density function for X be given as $f(x) = 1/k$ for $0 < x < k$ and 0 otherwise. Find $\Pr(X \leq t) = F(t)$ and sketch F(t).

Solution: $F(t) = \Pr(X \leq t)$ is defined as the cumulative distribution function, $F(t) = \int_{-\infty}^{t} f(x) \, dx$.

In this case $F(t) = \int_{-\infty}^{t} 1/k \, dx$ and

$$F(t) = \begin{cases} 0 & t \leq 0 \\ \int_{0}^{t} 1/k \, dx, & 0 < t < k \\ 1 & k \leq t \end{cases}$$

$$F(t) = \begin{cases} 0 & t \leq 0 \\ t/k & 0 < t < k \\ 1 & k \leq t \end{cases}$$

This function is graphed below:

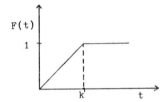

● **PROBLEM 8-25**

X is a continuous random variable with probability density function

$$f(x) = \begin{array}{ll} 1/k & 0 \leq x \leq k \\ 0 & \text{otherwise.} \end{array}$$

Show that this is a proper density function and find $\Pr(a \leq X \leq b)$ for $0 \leq a < b \leq k$.

Solution: To show that $f(x)$ is a proper probability density function we must show that $f(x) \geq 0$ for all x and that $\int_{-\infty}^{\infty} f(x) dx = 1$.

From the way $f(x)$ is defined, for any value of $k > 0$, $f(x) = 1/k \geq 0$.

Furthermore, $\int_{-\infty}^{\infty} f(x) \, dx = \int_{0}^{k} f(x) \, dx$

because $f(x) = 0$ for $x \geq k$ or $x \leq 0$.

$\int_{0}^{k} f(x) \, dx = \int_{0}^{k} 1/k \, dx = 1/k \cdot x\big]_{0}^{k} = 1/k[k - 0] = 1$.

$\Pr(a \leq X \leq b)$ for $0 \leq a < b \leq k$ is defined to be $\int_{a}^{b} f(x) \, dx$.

Thus $\Pr(a \leq X \leq b) = \int_{a}^{b} f(x) \, dx$

$$= \int_a^b 1/k \, dx = 1/k \, x \, \Big]_a^b$$

$$= \frac{b-a}{k}.$$

A graph of $f(x)$ is shown below:

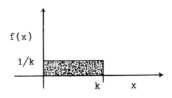

● **PROBLEM 8-26**

Consider the hardness of steel as a random variable, X, with values between 50 and 70 on the Rockwell B scale. We can assume that the hardness has density function

$$f(x) = 0 \quad \text{when} \quad x < 50,$$
$$f(x) = \frac{1}{20} \quad \text{when} \quad 50 \le x \le 70,$$
$$f(x) = 0 \quad \text{when} \quad x > 70.$$

Graph this density function. Compute the probability that the hardness of a randomly selected steel specimen is less than 65.

Solution: Graph of $f(x)$.

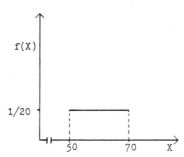

We use the cumulative distribution function $F(a) = P(X \le a)$ to compute $P(X < 65)$.

$$F(a) = \int_{-\infty}^a f(x)dx = \int_{50}^a \frac{1}{20} dx = \frac{1}{20} x \Big]_{50}^a = \frac{a-50}{20} = F(a)$$

when $50 \le a \le 70$. (When $b < 50$, $F(b) = \int_{-\infty}^b 0 \, dx = 0$. When $b > 70$,

$$F(b) = \int_{-\infty}^{50} f(x)dx + \int_{50}^{70} f(x)dx + \int_{70}^b f(x)dx = 0 + \frac{20}{20} + 0 = 1.)$$ We **find**
that $P(X < 65) = F(65) = \frac{65-50}{20} = \frac{15}{20} = \frac{3}{4}$.

• **PROBLEM 8-27**

Let x be the random variable representing the length of a telephone conversation. Let $f(x) = \lambda e^{-\lambda x}$, $0 \leq x < \infty$. Find the c.d.f., $F(x)$, and find $Pr(5 < x \leq 10)$.

Solution: The c.d.f. is the probability that an observation of a random variable will be less than or equal to x. The probability is equal to the following shaded area.

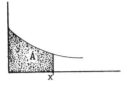

Area can be found by integration:

$$A = \int_0^x \lambda e^{-\lambda t} \, dt = - e^{-\lambda t}\Big|_{t=0}^{x} = 1 - e^{-\lambda x} = F(x).$$

$Pr(5 < x \leq 10)$ is represented by the following shaded area.

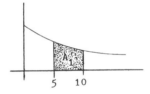

We find A_1 by integration,

$$A_1 = \int_{x=5}^{10} \lambda e^{-\lambda t} \, dt = -e^{-\lambda t}\Big|_{x=5}^{x=10} = e^{-5\lambda} - e^{-10\lambda}.$$

• **PROBLEM 8-28**

Let the random variable X represent the scores of an examination. It can be assumed that X has Gaussian probability density with mean, $m = 75$ and variance, $\sigma^2 = 64$.

What is the probability that

(a) X lies between 80 and 85,

(b) X will be greater than 85, and

(c) X will be less than 90.

A table, listing the values of cumulative normal distribution function for various values of x assumed by the random variable, is given.

Solution: Let $f(x)$ and $F(x)$ represent the probability density function (pdf) and the cumulative distribution function (cdf) respectively.

Since the random variable X has Gaussian density function,

$$f(x) = \frac{1}{\sigma\sqrt{2\pi}} \exp\left[\frac{-(x-m)^2}{2\sigma^2}\right]$$

where σ^2 is the variance and m is the mean of X.

Given $m = 75$ and $\sigma^2 = 64$

So,
$$f(x) = \frac{1}{8\sqrt{2\pi}} \exp\left[\frac{-(x-75)^2}{2(64)}\right]$$

$$= \frac{1}{8\sqrt{2\pi}} \exp\left[-\frac{1}{2}\left(\frac{x-75}{8}\right)^2\right]$$

(a) $\Pr(80 \leq X \leq 85) = \int_{x=80}^{x=85} f(x)dx$

$$= \frac{1}{8\sqrt{2\pi}} \int_{80}^{85} \exp\left[-\frac{1}{2}\left(\frac{x-75}{8}\right)^2\right] dx$$

Substituting $u = \frac{x-75}{8}$

$$du = \frac{dx}{8}$$

when $x = 85$, $u = \frac{85-75}{8} = \frac{10}{8} = 1.25$

when $x = 80$, $u = \frac{80-75}{8} = \frac{5}{8} = 0.625$

So, $\Pr(80 \leq X \leq 85)$

$$= \frac{1}{\sqrt{2\pi}} \int_{u=0.625}^{u=1.25} \exp\left(-\frac{1}{2}u^2\right) du$$

$$= \frac{1}{\sqrt{2\pi}} \int_{u=-\infty}^{u=1.25} \exp\left(\frac{-u^2}{2}\right) du - \frac{1}{\sqrt{2\pi}} \int_{u=-\infty}^{u=0.625} \exp\left(\frac{-u^2}{2}\right) du$$

$$= 0.8944 - 0.7340 \text{ (From the table)}$$

Hence, $\Pr(80 \leq X \leq 85) = 0.8944 - 0.7340 = 0.1604$

(b) $\Pr(X > 85) = \int_{x=85}^{x=\infty} f(x)dx$

$$= \frac{1}{8\sqrt{2\pi}} \int_{x=85}^{x=\infty} \exp\left[-\frac{1}{2}\left(\frac{x-75}{8}\right)^2\right] dx$$

With the same substitution as in part (a),

$$\Pr(X > 85) = \frac{1}{\sqrt{2\pi}} \int_{1.25}^{\infty} \exp\left(-\frac{1}{2} u^2\right) du$$

$$= \frac{1}{\sqrt{2\pi}} \int_{-\infty}^{+\infty} \exp\left(\frac{-u^2}{2}\right) du - \frac{1}{\sqrt{2\pi}} \int_{-\infty}^{1.25} \exp\left(\frac{-u^2}{2}\right) du$$

$$= 1 - 0.8944 = 0.1056 \quad \text{(From the table)}$$

(c) $\Pr(X \leq 90) = \frac{1}{8\sqrt{2\pi}} \int_{x=-\infty}^{x=90} \exp\left[-\frac{1}{2}\left(\frac{x-75}{8}\right)^2\right] dx$

With the same substitution as in parts (a) and (b),

$$\Pr(X \leq 90) = \frac{1}{\sqrt{2\pi}} \int_{u=-\infty}^{u=1.875} \exp\left(-\frac{1}{2} u^2\right) du$$

$$= 0.9696. \quad \text{(From the table)}$$

TABLE   CUMULATIVE NORMAL DISTRIBUTION

$$F(x) = \int_{-\infty}^{x} \frac{1}{\sqrt{2\pi}} e^{-t^2/2} dt$$

x	.00	.01	.02	.03	.04	.05	.06	.07	.08	.09
.0	.5000	.5040	.5080	.5120	.5160	.5199	.5239	.5279	.5319	.5359
.1	.5398	.5438	.5478	.5517	.5557	.5596	.5636	.5675	.5714	.5753
.2	.5793	.5832	.5871	.5910	.5948	.5987	.6026	.6064	.6103	.6141
.3	.6179	.6217	.6255	.6293	.6331	.6368	.6406	.6443	.6480	.6517
.4	.6554	.6591	.6628	.6664	.6700	.6736	.6772	.6808	.6844	.6879
.5	.6915	.6950	.6985	.7019	.7054	.7088	.7123	.7157	.7190	.7224
.6	.7257	.7291	.7324	.7357	.7389	.7422	.7454	.7486	.7517	.7549
.7	.7580	.7611	.7642	.7673	.7704	.7734	.7764	.7794	.7823	.7852
.8	.7881	.7910	.7939	.7967	.7995	.8023	.8051	.8078	.8106	.8133
.9	.8159	.8186	.8212	.8238	.8264	.8289	.8315	.8340	.8365	.8389
1.0	.8413	.8438	.8461	.8485	.8508	.8531	.8554	.8577	.8599	.8621
1.1	.8643	.8665	.8686	.8708	.8729	.8749	.8770	.8790	.8810	.8830
1.2	.8849	.8869	.8888	.8907	.8925	.8944	.8962	.8980	.8997	.9015
1.3	.9032	.9049	.9066	.9082	.9099	.9115	.9131	.9147	.9162	.9177
1.4	.9192	.9207	.9222	.9236	.9251	.9265	.9279	.9292	.9306	.9319
1.5	.9332	.9345	.9357	.9370	.9382	.9394	.9406	.9418	.9429	.9441
1.6	.9452	.9463	.9474	.9484	.9495	.9505	.9515	.9525	.9535	.9545
1.7	.9554	.9564	.9573	.9582	.9591	.9599	.9608	.9616	.9625	.9633
1.8	.9641	.9649	.9656	.9664	.9671	.9678	.9686	.9693	.9699	.9706
1.9	.9713	.9719	.9726	.9732	.9738	.9744	.9750	.9756	.9761	.9767
2.0	.9772	.9778	.9783	.9788	.9793	.9798	.9803	.9808	.9812	.9817
2.1	.9821	.9826	.9830	.9834	.9838	.9842	.9846	.9850	.9854	.9857
2.2	.9861	.9864	.9868	.9871	.9875	.9878	.9881	.9884	.9887	.9890
2.3	.9893	.9896	.9898	.9901	.9904	.9906	.9909	.9911	.9913	.9916
2.4	.9918	.9920	.9922	.9925	.9927	.9929	.9931	.9932	.9934	.9936

2.5	.9938	.9940	.9941	.9943	.9945	.9946	.9948	.9949	.9951	.9952
2.6	.9953	.9955	.9956	.9957	.9959	.9960	.9961	.9962	.9963	.9964
2.7	.9965	.9966	.9967	.9968	.9969	.9970	.9971	.9972	.9973	.9974
2.8	.9974	.9975	.9976	.9977	.9977	.9978	.9979	.9979	.9980	.9981
2.9	.9981	.9982	.9982	.9983	.9984	.9984	.9985	.9985	.9986	.9986
3.0	.9987	.9987	.9987	.9988	.9988	.9989	.9989	.9989	.9990	.9990
3.1	.9990	.9991	.9991	.9991	.9992	.9992	.9992	.9992	.9993	.9993
3.2	.9993	.9993	.9994	.9994	.9994	.9994	.9994	.9995	.9995	.9995
3.3	.9995	.9995	.9995	.9996	.9996	.9996	.9996	.9996	.9996	.9997
3.4	.9997	.9997	.9997	.9997	.9997	.9997	.9997	.9997	.9997	.9998

x	1.282	1.645	1.960	2.326	2.576	3.090	3.291	3.891	4.417
$F(x)$	.90	.95	.975	.99	.995	.999	.9995	.99995	.999995
$2[1 - F(x)]$	.20	.10	.05	.02	.01	.002	.001	.0001	.00001

• **PROBLEM 8-29**

The length of the time y for a long distance telephone call can be assumed to be a continuous random variable with an exponential probability density function f(y).

It is given that

$$f(y) = \begin{cases} \lambda e^{-\frac{1}{2}y}, & y > 0 \\ 0, & \text{elsewhere.} \end{cases}$$

Determine:

(a) the value of λ, if f(y) is a probability density function;

(b) the probability that the telephone call will last: (i) for a time between 10 and 15 minutes, (ii) for a time less than 10 minutes, (iii) for more than 15 minutes;

(c) the probability that the talk will last for a time less than or equal to 10 minutes, given that the call had lasted for a time between 8 and 13 minutes.

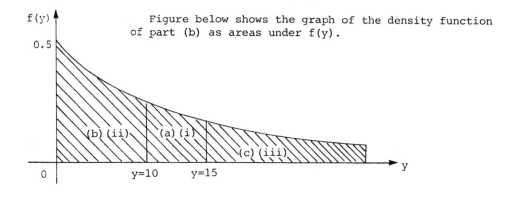

Figure below shows the graph of the density function of part (b) as areas under f(y).

Solution: (a) If $f(y)$ is a probability density function, the value of the definite integral $\int_{-\infty}^{+\infty} f(y)dy$ should be equal to unity.

Therefore,
$$\lambda \int_0^\infty e^{-\frac{1}{2}y} dy = 1$$

$$\lambda \left(\frac{e^{-\frac{1}{2}y}}{-\frac{1}{2}}\right)_0^\infty = 1$$

$$-2\lambda (0-1) = 1$$

$$2\lambda = 1$$

Hence, $\lambda = \dfrac{1}{2}$

(b) (i) $\Pr(10 \leq Y \leq 15) = \int_{10}^{15} f(y)dy$

$$= \frac{1}{2} \int_{10}^{15} e^{-\frac{1}{2}y} dy$$

$$= \frac{1}{2} \left(\frac{e^{-\frac{1}{2}y}}{-\frac{1}{2}}\right)_{10}^{15}$$

$$= -\left(e^{-\frac{15}{2}} - e^{-5}\right)$$

$$= e^{-5} - e^{-\frac{15}{2}}$$

$$= 0.00618$$

(ii) $\Pr(Y \leq 10) = \int_{-\infty}^{10} f(y)dy$

$$= \frac{1}{2} \int_0^{10} e^{-\frac{1}{2}y} dy$$

$$= \frac{1}{2} \left[\frac{e^{-(\frac{1}{2})y}}{-(\frac{1}{2})}\right]_0^{10}$$

$$= -\left(e^{-5} - 1\right)$$

$$= 1 - e^{-5}$$

$$= 0.993$$

(iii) $\Pr(Y \geq 15) = \int_{15}^{\infty} f(y)dy$

$$= \frac{1}{2} \int_{15}^{\infty} e^{-\frac{1}{2}y} dy$$

$$= \frac{1}{2}\left(\frac{e^{-\frac{1}{2}y}}{-\frac{1}{2}}\right)_{15}^{\infty}$$

$$= -\left(0 - e^{-\frac{15}{2}}\right) = e^{-\frac{15}{2}}$$

$$= e^{-\frac{15}{2}}$$

$$= 0.00055$$

(c) $\Pr\left[Y \leq 10 \;/\; 8 \leq Y \leq 13\right]$

$$= \frac{\Pr[8 \leq Y \leq 10]}{\Pr[8 \leq Y \leq 13]}$$

$$= \frac{\int_8^{10} f(y)\,dy}{\int_8^{13} f(y)\,dy}$$

$$= \frac{\frac{1}{2}\int_8^{10} e^{-\frac{1}{2}y}\,dy}{\frac{1}{2}\int_8^{13} e^{-\frac{1}{2}y}\,dy}$$

$$= \frac{\left(\dfrac{e^{-\frac{1}{2}y}}{-\frac{1}{2}}\right)_8^{10}}{\left(\dfrac{e^{-\frac{1}{2}y}}{-\frac{1}{2}}\right)_8^{13}}$$

$$= \frac{-\left(e^{-5} - e^{-4}\right)}{-\left(e^{-\frac{13}{2}} - e^{-4}\right)}$$

$$= \frac{e^{-4} - e^{-5}}{e^{-4} - e^{-\frac{13}{2}}}$$

$$= \frac{0.0115776}{0.0168121}$$

$$= 0.6886$$

• **PROBLEM 8-30**

Given the probability density function of a continuous random variable X. Determine its cumulative distribution function.

Probability density function = $f(x) = \begin{cases} \frac{3}{8}(x^2 - x + 1), & 0 \leq x \leq 2 \\ 0, & \text{otherwise.} \end{cases}$

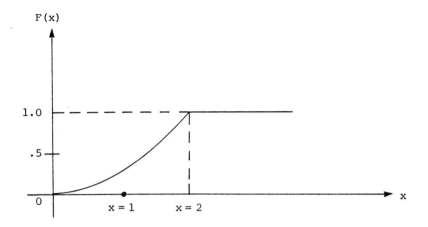

<u>Solution</u>: Let F(x) be the cumulative distribution function and it is defined by

$$F(x) = \Pr(X \le x) = \int_{-\infty}^{x} f(s)\,ds.$$

<u>(i) Case</u> For $x < 0$

$$F(x) = \int_{-\infty}^{x} f(x)\,dx = \int_{-\infty}^{x} 0 \cdot dx = 0$$

<u>(ii) Case</u> For $0 \le x \le 2$

$$F(x) = \int_{-\infty}^{x} f(x)\,dx$$

$$= \int_{-\infty}^{0} f(x)\,dx + \int_{0}^{x} f(x)\,dx$$

$$= F(0) + \int_{0}^{x} f(x)\,dx$$

$$= \int_{0}^{x} f(x)\,dx$$

$$= \int_{0}^{x} \frac{3}{8}(x^2 - x + 1)\,dx$$

$$= \frac{3}{8}\left[\frac{x^3}{3} - \frac{x^2}{2} + x\right]_0^x$$

$$= \frac{3}{8}\left(\frac{x^3}{3} - \frac{x^2}{2} + x\right)$$

$$= \frac{3}{8}\left(\frac{2x^3-3x^2+6x}{6}\right)$$

$$= \frac{1}{16} x(2x^2-3x+6)$$

(iii) Case   For x > 2,

$$F(x) = \int_0^2 f(x)dx + \int_2^\infty f(x)dx$$

$$= \int_0^2 \frac{3}{8}(x^2-x+1)dx$$

$$= \frac{3}{8}\left[\frac{x^3}{3} - \frac{x^2}{2} + x\right]_0^2$$

$$= \frac{3}{8}\left(\frac{8}{3} - \frac{4}{2} + 2\right)$$

$$= \frac{3}{8}\left(\frac{8}{3} - 2 + 2\right)$$

$$= 1$$

So,

$$F(x) = \begin{cases} 0, & x < 0, \\ \dfrac{x(2x^2-3x+6)}{16}, & 0 \le x \le 2, \\ 1, & x > 2. \end{cases}$$

F(x) is plotted in the figure.

JOINT DISTRIBUTION AND DENSITY FUNCTION

● **PROBLEM 8-31**

Use

$$f(x,y) = \begin{cases} e^{-x}e^{-y} & \begin{array}{l} x > 0 \\ y > 0 \end{array} \\ 0 & \text{otherwise} \end{cases}$$

to find the probability that $\{1 < X < 2 \text{ and } 0 < Y < 2\}$.

<u>Solution</u>: $\Pr(1 < X < 2 \text{ and } 0 < Y < 2)$ is the volume over the shaded rectangle:

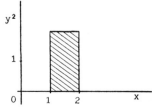

This volume over the rectangle and under f(x,y) is pictured below:

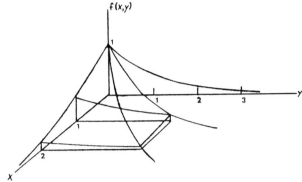

To find this volume we integrate X from 1 to 2 and Y from 0 to 2. Thus,

$$\Pr(1 < X < 2 \text{ and } 0 < Y < 2) = \int_0^2 \int_1^2 f(x,y)\,dx\,dy$$

$$= \int_0^2 e^{-y}\left(\int_1^2 e^{-x}\,dx\right)dy$$

$$= \int_0^2 e^{-y}dy\left[-e^{-x}\Big|_1^2\right] = \int_0^2 e^{-y}dy\,(e^{-1}-e^{-2})$$

$$= -e^{-y}\Big]_0^2 (e^{-1}-e^{-2})$$

$$= (e^0-e^{-2})(e^{-1}-e^{-2}) = (1-e^{-2})(e^{-1}-e^{-2})$$

$$= (.865)(.233) = .20 .$$

• **PROBLEM 8-32**

The joint probability density of two discrete random variables X and Y is given in Table 1. Determine the marginal densities of X and Y.

Table 1

X \ y	-2	0	1	4
-1	.3	.1	0	.2
3	0	.2	.1	0
5	.1	0	0	0

Solution: Let $f_X(x)$ and $f_Y(y)$ be the marginal probability densities of X and Y respectively.

$$f_X(x) = \sum_{y=-\infty}^{+\infty} f_{XY}(x,y)$$

where $f_{XY}(x,y)$ is the joint probability density.

$f_X(-1) = (0.3) + (0.1) + (0) + (0.2) = 0.6$

$f_X(3) = (0) + (0.2) + (0.1) + (0) = 0.3$

$f_X(5) = (0.1) + (0) + (0) + (0) = 0.1$

Similarly,
$$f_Y(y) = \sum_{x=-\infty}^{+\infty} f_{XY}(x,y)$$

$f_Y(-2) = (0.3) + (0) + (0.1) = 0.4$

$f_Y(0) = (.1) + (0.2) + (0) = 0.3$

$f_Y(1) = (0) + (0.1) + (0) = 0.1$

$f_Y(4) = (0.2) + (0) + (0) = 0.2$

Observe from the result obtained above, it is clear that $f_X(x)$ and $f_Y(y)$ satisfy the conditions of a probability density function.

i.e., $\int_{-\infty}^{\infty} p(q)dq = 1$ and $p(q) \geq 0$ for all q.

x	$f_X(x)$
-1	0.6
3	0.3
5	0.1

y	$f_Y(y)$
-2	0.4
0	0.3
1	0.1
4	0.2

Table 2

● **PROBLEM 8-33**

Determine the marginal densities of two dimensional, continuous random variables (X,Y). Given that f(x,y) = Joint probability density

$$= \begin{cases} \dfrac{x^3 y^3}{16}, & 0 \leq x \leq 2, \ 0 \leq y \leq 2. \\ 0, & \text{elsewhere.} \end{cases}$$

And, determine the cumulative distribution of X. Also sketch

$f_X(x)$, $f_Y(y)$, $F_X(x)$ and $F_Y(y)$.

Solution: Let $f_X(x)$ and $f_Y(y)$ represent the marginal densities of X and Y respectively.

$$f_X(x) = \int_{y=-\infty}^{y=+\infty} f(x,y)dy$$

$$= \int_{y=0}^{y=2} \frac{x^3 y^3}{16} dy$$

$$= \frac{x^3}{16} \int_{y=0}^{y=2} y^3 dy$$

$$= \frac{x^3}{16} \left(\frac{y^4}{4}\right)_0^2 = \frac{x^3}{(16)(4)}(2^4) = \frac{x^3}{4}$$

So, $\quad f_X(x) = \begin{cases} \frac{x^3}{4} & \text{for } 0 \leq x \leq 2 \\ 0, & \text{elsewhere} \end{cases}$

Similarly,
$$f_Y(y) = \int_{x=-\infty}^{x=+\infty} f(x,y)dx$$

$$= \int_{x=0}^{x=2} \frac{x^3 y^3}{16} dx$$

$$= \frac{y^3}{16} \left(\frac{x^4}{4}\right)_0^2 = \frac{y^3}{(16)(4)}(2^4) = \frac{y^3}{4}.$$

So, $\quad f_Y(y) = \begin{cases} \frac{y^3}{4} & \text{for } 0 \leq y \leq 2 \\ 0, & \text{elsewhere} \end{cases}$

Let $F_X(x)$ be the cumulative distribution function of X.
For $0 \leq x \leq 2$,
$$F_X(x) = \int_{\lambda=-\infty}^{\lambda=x} f_X(\lambda)d\lambda$$

$$= \int_{\lambda=0}^{\lambda=x} \frac{\lambda^3}{4} d\lambda$$

$$= \frac{1}{4}\left(\frac{\lambda^4}{4}\right)_0^x$$

$$= \frac{1}{16}\left(x^4\right)$$

457

$$= \frac{x^4}{16}$$

For $x > 2$

$$F_X(x) = \int_{\lambda=-\infty}^{\lambda=x} f_X(\lambda)d\lambda$$

$$= \int_{\lambda=0}^{\lambda=2} \frac{\lambda^3}{4} d\lambda$$

$$= \frac{1}{4}\left(\frac{\lambda^4}{4}\right)^2_{\lambda=0} = \frac{16}{16} = 1$$

So,
$$F_X(x) = \begin{cases} 0 , & x < 0 \\ \frac{x^4}{16} , & 0 \leq x \leq 2 \\ 1 , & x \geq 2 \end{cases}$$

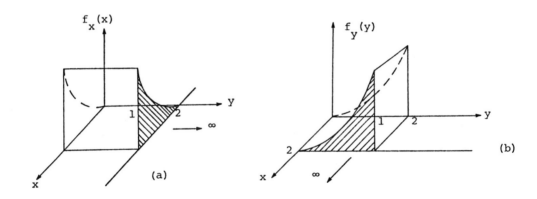

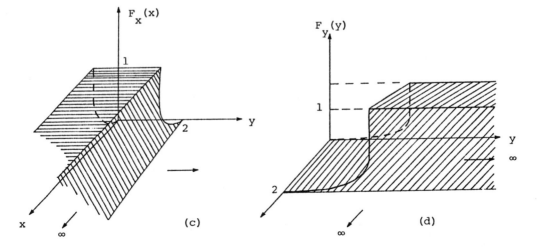

The graphical representation of $f_X(x)$, $f_Y(y)$, $F_X(x)$, and $F_Y(y)$ is shown in the Figures.

• **PROBLEM 8-34**

Given that the joint density function of X and Y is
$$f(x,y) = 6x^2 y \text{ where } \begin{cases} 0 < x < 1 \text{ and} \\ 0 < y < 1, \end{cases}$$
$$= 0 \text{ elsewhere,}$$
find $\Pr(0 < x < 3/4, 1/3 < y < 2)$.

Solution: The joint density function of two random variables represents a surface over the region on which it is defined. The volume under the surface is always 1. We need to find the volume over the region bounded by the given limits. We construct a double integral:

$$\int_{y=1/3}^{y=2} \int_{x=0}^{x=3/4} f(x,y) dx\, dy = \int_{y=1/3}^{y=1} \int_{x=0}^{x=3/4} 6x^2 y\, dx\, dy$$

$$+ \int_{y=1}^{2} \int_{x=0}^{x=3/4} 0\, dx\, dy\,;$$

because $f(x,y) = 0$ where $y \geq 1$), this becomes

$$\int_{y=1/3}^{y=2} \int_{x=0}^{x=3/4} f(x,y) dx\, dy = 6\int_{y=1/3}^{y=1} \frac{x^3}{3} y \Big]_0^{3/4} dy = 6\int_{y=1/3}^{y=1} \frac{9}{64} y\, dy$$

$$= \frac{54}{64} \frac{y^2}{2} \Big]_{1/3}^{1} = \frac{27}{32}\left(\frac{1}{2} - \frac{1}{18}\right)$$

$$= \frac{27}{32}\left(\frac{8}{18}\right) = \frac{27}{32}\left(\frac{4}{9}\right)$$

$$= \frac{3}{8} = \Pr(0 < x < 3/4, 1/3 < y < 2).$$

• **PROBLEM 8-35**

Given that the joint density function of the random variables X, Y and Z is $f(x,y,z) = e^{-(x+y+z)}$ when $0 < x < \infty$, $0 < y < \infty$, $0 < z < \infty$, find the cumulative distribution function $F(x,y,z)$ of X, Y, and Z.

Solution: $F(x,y,z)$ is defined to be $\Pr(X \leq x, Y \leq y, Z \leq z)$. Here we have extended the concept of cumulative probability from one random variable to three. The cumulative distribution function of one variable gives the area under the density curve to the left of the particular value of x. The distribution function of two variables gives the volume under the surface represented by the bivariate density function over the region bounded by the specified values of x and y.

This is obtained by integration with respect to each variable. The number given by a distribution function of three variables can be interpreted as a 4-dimensional volume. It is obtained by constructing an iterated triple integral and integrating with respect to each variable.

$$F(x,y,z) = \int_0^z \int_0^y \int_0^x e^{-u-v-w} \, du \, dv \, dw$$

$$= \int_0^z \int_0^y \int_0^x e^{-u}(e^{-v-w}) \, du \, dv \, dw$$

$$= \int_0^z \int_0^y (e^{-v-w})(1-e^{-x}) \, dv \, dw = (1-e^{-x}) \int_0^z e^{-w} \int_0^y e^{-v} \, dv \, dw$$

$$= (1-e^{-x}) \int_0^z e^{-w}(1-e^{-y}) \, dw$$

$$= (1-e^{-x})(1-e^{-y}) \int_0^z e^{-w} \, dw = (1-e^{-x})(1-e^{-y})(1-e^{-z})$$

for $x,y,z > 0$.

● **PROBLEM 8-36**

Suppose X has density function:

$$f(X) = [\pi(1+X^2)]^{-1} \qquad -\infty < x < \infty .$$

If $Y = X^2$, what is the density function of Y?

Solution: We will use the cumulative distribution technique to find the distribution of Y.

$$\Pr(Y \leq y) = \Pr(X^2 \leq y) .$$

But X^2 has two inverses. If $X^2 = y$, then $X = \sqrt{y}$ and $X = -\sqrt{y}$. Thus

$$\Pr(X^2 \leq y) = \begin{cases} \Pr(-\sqrt{y} \leq X \leq \sqrt{y}) & \text{if } y \geq 0 \\ 0 & \text{if } y < 0. \end{cases}$$

To see that this is true, if $-\sqrt{y} \leq X$,

$$(-\sqrt{y})(-\sqrt{y}) \geq (X)(X)$$

with the inequality reversed because there is multiplication by a negative number.

Thus,

$$(-\sqrt{y})(-\sqrt{y}) = (-1)^2 (\sqrt{y})^2 \geq X^2$$

or

$$y \geq X^2 .$$

Similarly,

$$X \leq \sqrt{y}$$

implies

$$(X)(X) \leq (\sqrt{y})(\sqrt{y})$$

or

$$X^2 \leq y .$$

But

$$\Pr(-\sqrt{y} \leq X \leq \sqrt{y}) = \Pr(X \leq \sqrt{y}) - \Pr(X \leq -\sqrt{y}) .$$

If $G(x) = \Pr(X \leq x)$, then

$$G(X) = \int_{-\infty}^{X} \frac{1}{\pi(1+t^2)} dt .$$

And
$$\Pr(Y \le y) = G(\sqrt{y}) - G(-\sqrt{y}) .$$

We know that
$$\frac{dG}{dX} = f(x)$$
the density function by the fundamental theorem of calculus.

We also know by the Chain Rule that
$$\frac{dG(f(y))}{dy} = G'(f(y)) \frac{df(y)}{dy} .$$

Thus, differentiating the $\Pr(Y \le y)$ with respect to y gives the density function of Y or

$$h(y) = \frac{d(\Pr(Y \le y))}{dy} = \frac{d}{dy}\left[G(\sqrt{y}) - G(-\sqrt{y})\right]$$

$$= f(\sqrt{y}) \frac{d(\sqrt{y})}{dy} - f(-\sqrt{y}) \frac{d(-\sqrt{y})}{dy}$$

$$= f(\sqrt{y}) \frac{1}{2\sqrt{y}} - f(-\sqrt{y}) \left(\frac{-1}{2\sqrt{y}}\right)$$

$$= \frac{1}{\pi(1+(\sqrt{y})^2)} \frac{1}{2\sqrt{y}} - \frac{1}{\pi(1+(-\sqrt{y})^2)} \left(\frac{-1}{2\sqrt{y}}\right)$$

$$= \frac{1}{\pi(1+y)} \left(\frac{1}{2\sqrt{y}}\right) + \frac{1}{\pi(1+y)} \left(\frac{1}{2\sqrt{y}}\right) .$$

Thus $h(y) = \begin{cases} \dfrac{1}{\pi\sqrt{y}(1+y)} & y > 0 \\ 0 & \text{otherwise.} \end{cases}$

● **PROBLEM 8-37**

Let (X,Y) have the distribution defined by the joint density function,
$$f(x,y) = \begin{cases} e^{-x-y} & x > 0 \\ & y > 0 \\ 0 & \text{otherwise .} \end{cases}$$
Find the marginal and conditional densities of Y and X. Are X and Y independent?

Solution: The marginal distributions of X and Y are:
$$g(x) = \int_0^\infty f(x,y) dy = \int_0^\infty e^{-x-y} dy$$
and
$$h(y) = \int_0^\infty f(x,y) dx = \int_0^\infty e^{-x-y} dx .$$

Thus
$$g(x) = e^{-x}(-e^{-y})\Big|_0^\infty = e^{-x} \qquad x > 0$$

and

$$h(y) = e^{-y}(-e^{-x})\Big|_0^\infty = e^{-y} \qquad y > 0.$$

We see that $f(x,y) = g(x)h(y)$ because $e^{-x-y} = e^{-x}e^{-y}$. Thus X and Y are independent.

The conditional densities of X and Y are defined by analogy to conditional probability.

$f(x|y) = $ conditional density of x given y is by definition

$$= \frac{f(x,y)}{h(y)}$$

for fixed y and

$f(y|x) = $ conditional density of y given x is

$$= \frac{f(x,y)}{g(x)}$$

for fixed y.

The conditional densities in this example are

$$f(x|y) = \frac{e^{-x-y}}{e^{-y}} \qquad x > 0$$

and

$$f(y|x) = \frac{e^{-x-y}}{e^{-x}} \qquad y > 0.$$

Hence

$$f(x|y) = g(x) = e^{-x} \quad \text{and} \quad f(y|x) = h(y) = e^{-y}.$$

Another alternate condition for independence is that the conditional density function is equal to the marginal density function.

● **PROBLEM 8-38**

In a bridge hand of 13 cards, what is the probability of

(a) having neither an ace nor a king,
(b) having no ace or one ace and exactly four kings.

Assume that the random variables X and Y represent the number of aces and kings respectively. Also assume that $\binom{52}{13}$ possible bridge hands are equally likely.

Solution: The joint probability density function (probability of having x aces and y kings in a bridge hand) is

$$\Pr(X=x, Y=y) = f(x,y) = \begin{cases} \binom{4}{x}\binom{4}{y}\binom{44}{13-x-y}, & x,y = 0,1,2,3,4 \\ 0, & \text{elsewhere} \end{cases}$$

Also,

$$\sum_{x=0}^{4} \sum_{y=0}^{4} f(x,y) = 1 \quad \text{and}$$

$$f(x,y) \geq 0 \quad \text{for all } x \text{ and } y.$$

(a) The probability of having neither an ace nor a king = $\Pr(X=0, Y=0)$

$$\Pr(X=0, Y=0) = f(0,0) = \frac{\binom{4}{0}\binom{4}{0}\binom{44}{13}}{\binom{52}{13}}$$

$$= \frac{51,915,526,432}{635,013,559,600} = 0.0818.$$

(b) The probability of having no ace or one ace and exactly four kings = $\Pr(X < 2, Y = 4)$

$$\Pr(X < 2, Y = 4) = \sum_{x=0}^{1} f(x, 4)$$

$$= \sum_{x=0}^{1} \frac{\binom{4}{x}\binom{4}{4}\binom{44}{9-x}}{\binom{52}{13}}$$

$$= \frac{\binom{4}{4}}{\binom{52}{13}} \left[\binom{4}{0}\binom{44}{9} + \binom{4}{1}\binom{44}{8} \right]$$

$$= \frac{1}{\binom{52}{13}} \left[\binom{44}{9} + 4\binom{44}{8} \right]$$

$$= 0.0022$$

• **PROBLEM 8-39**

Two random variables X and Y have a joint probability density function

$$f(x,y) = \begin{cases} ke^{-(x+y)}, & \text{for } 0 \le x \le \infty, \ 0 \le y \le \infty \\ 0, & \text{elsewhere} \end{cases}$$

(a) By applying the definition of a probability density function, determine the value of k.

(b) Determine f(x), the probability density of X, if X is an independent variable.

(c) Determine the probability that X lies in the range between 0 and 2, and Y lies in the range between 2 and 3.

(d) Prove that the random variables X and Y are independent.

Solution: (a) A necessary condition for f(x,y) to be a probability density function is that

$$\int_{-\infty}^{+\infty}\int_{-\infty}^{+\infty} f(x,y)\,dxdy = 1.$$

Thus, considering the function in this case

$$\int_{x=0}^{+\infty}\int_{y=0}^{+\infty} ke^{-(x+y)}\,dxdy = 1$$

Solving the above equation,

$$k\int_{x=0}^{\infty} e^{-x}\,dx \int_{y=0}^{\infty} e^{-y}\,dy = 1$$

$$k(-e^{-x})_0^{\infty}\,(-e^{-y})_0^{\infty} = 1$$

$$k(1)(1) = 1$$

Therefore, k=1.

(b) The probability density of the random variable X independent of the random variable Y is

$$f(x) = \int_{y=-\infty}^{\infty} f(x,y)\,dy.$$

Hence,

$$f(x) = \int_{y=-\infty}^{+\infty} e^{-(x+y)}\,dy$$

$$= e^{-x}\int_0^{\infty} e^{-y}\,dy$$

$$= e^{-x}(-e^{-y})_0^{\infty}$$

Therefore, $f(x) = \begin{cases} e^{-x}, & x \geq 0 \\ 0, & x < 0 \end{cases}$

(c) $P(0 \leq X \leq 2,\ 2 \leq Y \leq 3) = \int_{x=0}^{2}\int_{y=2}^{3} f(x,y)\,dxdy$

$$= \int_{x=0}^{2} \int_{y=2}^{3} e^{-(x+y)} \, dx \, dy$$

$$= \int_{0}^{2} e^{-x} \, dx \int_{2}^{3} e^{-y} \, dy$$

$$= (-e^{-x})_{0}^{2} (-e^{-y})_{2}^{3}$$

$$= \left[(-e^{-2}) - (-1)\right] \left[(-e^{-3}) - (-e^{-2})\right]$$

$$= (1 - e^{-2})(e^{-2} - e^{-3})$$

$$= e^{-2} - e^{-3} - e^{-4} + e^{-5}$$

$$= 0.135 - 0.049 - 0.0183 + 0.0067$$

$$= 0.0744$$

(d) From the solution obtained in part (b),

$$f(x) = e^{-x} \quad \text{for } x \geq 0.$$

Now, repeat part (b) for $f(y)$, i.e., probability density of the random variable Y independent of the random variable X, it follows that

$$f(y) = e^{-y}.$$

Therefore, $f(x) \cdot f(y) = e^{-(x+y)}$

$$= f(x,y)$$

Thus, the random variables X and Y are independent.

● **PROBLEM 8-40**

Given a Gaussian random variable X with average value (or expectation) of X equal to zero and variance $\sigma_x^2 = 1$, and Y, another random variable assuming values 1 or -1 with equal probability equal to 0.5.

(a) Determine $f(x,y)$, the joint density function.

(b) Prove that the probability density $f_Y(y)$, independent of the random variable X is

$$\int_{-\infty}^{\infty} f_{XY}(x,y) \, dx$$

Solution: (a) For a Gaussian random variable X, the probability density function is defined as

$$f(x) = \frac{1}{\sqrt{2\pi\sigma_x^2}} e^{-(x-m)^2/2\sigma_x^2}$$

where m is the average value or mean, and σ^2 is the variance associated with $f(x)$.

Since $m = E(x) = 0$ and $\sigma_x^2 = 1$

$$f(x) = \frac{1}{\sqrt{2\pi}} e^{-x^2/2}.$$

The variable Y is a discrete variable having the values 1 or -1, each with a probability of 0.5, i.e.,

$$P(Y=1) = 0.5 = f_Y(y)$$

$$P(Y=-1) = 0.5 = f_Y(y).$$

By definition, the probability density function $f_X(x)$ in terms of the cumulative distribution function $F_X(x)$ is

$$f_X(x) = \frac{d}{dx} F_X(x).$$

Hence, in this case, the joint density function $f_{XY}(x,y)$ is given by

$$f_{XY}(x,y) = \frac{\partial^2}{\partial x \partial y} F_{XY}(x,y)$$

where $F_{XY}(x,y) = P(X \leq x, Y \leq y) = P(X \leq x)P(Y \leq y).$

Hence, $F_{XY}(x,y) = \int_{-\infty}^{\infty} \frac{1}{\sqrt{2\pi}} e^{-\frac{x^2}{2}} dx \cdot \left[\tfrac{1}{2}u(y+1) + \tfrac{1}{2}u(y-1)\right]$

where $u(y)$ is a step function representing the density function of the random variable Y.

Therefore, $f_{XY}(x,y) = \frac{\partial^2}{\partial x \partial y} F_{XY}(x,y)$

$$= \frac{\partial^2}{\partial x \partial y} \int_{-\infty}^{\infty} \frac{1}{\sqrt{2\pi}} e^{-\frac{x^2}{2}} dx \cdot \left[\tfrac{1}{2}u(y+1) + \tfrac{1}{2}u(y-1)\right]$$

$$= \frac{1}{2} \frac{1}{\sqrt{2\pi}} e^{-\frac{x^2}{2}} \left[\delta(y+1) + \delta(y-1)\right].$$

(b) To prove that the probability density $f_Y(y)$ is independent of the random variable X, first, evaluate the probability density $f_Y(y)$ from the joint density function $f_{XY}(x,y)$ as

$$f_Y(y) = \int_{-\infty}^{+\infty} f_{XY}(x,y)dx$$

$$= \frac{1}{2\sqrt{2\pi}} \int_{-\infty}^{+\infty} e^{-\frac{x^2}{2}} dx.$$

Since $\int_{-\infty}^{+\infty} e^{-\alpha x^2} dx = \sqrt{\pi/\alpha}$

$$\int_{-\infty}^{+\infty} e^{-\frac{x^2}{2}} dx = \sqrt{2\pi}.$$

Therefore, $f_Y(y) = \frac{1}{2\sqrt{2\pi}} (\sqrt{2\pi})$

$$= \tfrac{1}{2}.$$

This density function is equal to the given density function (0.5) of Y. It thus follows that

$$f_Y(y) = \int_{-\infty}^{+\infty} f_{XY}(x,y)dx.$$

AVERAGE VALUE, MEAN AND VARIANCE

• **PROBLEM** 8-41

Given the probability distribution of the random variable X in the table below, compute E(X) and Var(X).

x_i	$\Pr(X = x_i)$
0	$\frac{8}{27}$
1	$\frac{12}{27}$
2	$\frac{6}{27}$
3	$\frac{1}{27}$

Solution:

$$E(X) = \sum_i x_i \Pr(X = x_i) \text{ and } \text{Var } X = E[(X - E(X))^2].$$

Thus, $E(X) = (0) \Pr(X = 0) + (1) \Pr(X = 1)$
$\qquad\qquad\qquad + (2) \Pr(X = 2) + (3) \Pr(X = 3)$

$$= (0)\frac{8}{27} + (1)\frac{12}{27} + (2)\frac{6}{27} + 3\left(\frac{1}{27}\right)$$

$$= 0 + \frac{12}{27} + \frac{12}{27} + \frac{3}{27} = \frac{27}{27} = 1.$$

$\text{Var } X = (0 - 1)^2 \Pr(X = 0) + (1 - 1)^2 \Pr(X = 1)$
$\qquad\qquad + (2 - 1)^2 \Pr(X = 2) + (3 - 1)^2 \Pr(X = 3)$

$$= (1^2)\frac{8}{27} + (0^2)\frac{12}{27} + (1^2)\frac{6}{27} + (2^2)\frac{1}{27}$$

$$= \frac{8}{27} + \frac{6}{27} + \frac{4}{27} = \frac{18}{27} = \frac{2}{3}.$$

● **PROBLEM 8-42**

Find the theoretical variance of the random variable with the following probability distribution.

x	$\Pr(X = x)$
0	$\frac{1}{4}$
1	$\frac{1}{2}$
2	$\frac{1}{8}$
3	$\frac{1}{8}$

Solution: The theoretical variance is the expected mean square error. The theoretical variance represents an idealized measure of the spread or dispersion of a probability distribution about its mean.

The variance of a random variable X is denoted σ_X^2 or Var X. It is defined as

$$\text{Var } X = E[(X - \mu)^2],$$

where $\mu = E(X)$.

In this example,

$E(X) = (0) \Pr(X = 0) + (1) \Pr(X = 1)$

$$+ (2)\Pr(X = 2) + 3\Pr(X = 3)$$

$$= 0 \cdot \tfrac{1}{4} + 1 \cdot \tfrac{1}{2} + 2 \cdot \tfrac{1}{8} + 3 \cdot \tfrac{1}{8}$$

$$= 0 + \tfrac{1}{2} + \tfrac{2}{8} + \tfrac{3}{8} = \tfrac{9}{8} \;.$$

We now compute the variance of X.

$$\text{Var } X = E[(X - \mu)^2]$$

$$= E\left[\left(X - \tfrac{9}{8}\right)^2\right]$$

$$= \left(0 - \tfrac{9}{8}\right)^2 \Pr(X = 0) + \left(1 - \tfrac{9}{8}\right)^2 \Pr(X = 1)$$

$$+ \left(2 - \tfrac{9}{8}\right)^2 \Pr(X = 2) + \left(3 - \tfrac{9}{8}\right)^2 \Pr(X = 3)$$

$$= \left(\tfrac{81}{64} \cdot \tfrac{1}{4}\right) + \left(\tfrac{1}{64} \cdot \tfrac{1}{2}\right) + \left(\tfrac{49}{64} \cdot \tfrac{1}{8}\right) + \left(\tfrac{225}{64} \cdot \tfrac{1}{8}\right)$$

$$= \frac{2(81) + 4 + 49 + 225}{64(8)}$$

$$= \frac{440}{(64)\,8} = \frac{55}{64} = .859\;.$$

A slightly less complicated formula for the theoretical variance is derived below,

$$E[(X - \mu)^2] = E[X^2 - 2X\mu + \mu^2]\;.$$

By the properties of expectation,

$$E[(X - \mu)^2] = E(X^2) - 2\mu E(X) + E(\mu^2)\;.$$

$E(\mu^2) = \mu^2$ because the expected value of a constant is a constant. Thus,

$$E[(X - \mu)^2] = E(X^2) - 2\mu E(X) + \mu^2$$

$E(X) = \mu$; thus

$$\text{Var } X = E(X^2) - 2\mu \cdot \mu + \mu^2$$

$$= E(X^2) - 2\mu^2 + \mu^2 = E(X^2) - \mu^2$$

or $\text{Var } X = E(X^2) - [E(X)]^2\;.$

Recomputing Var X we see that

$$\text{Var } X = 0^2 \cdot \Pr(X = 0) + 1^2 \Pr(X = 1) + 2^2 \Pr(X = 2)$$

$$+ 3^2 \Pr(X = 3) - \left[\tfrac{9}{8}\right]^2$$

$$= 0 + \frac{1}{2} + \frac{4}{8} + \frac{9}{8} - \left[\frac{9}{8}\right]^2$$

$$= \frac{(17)\ 8 - 81}{64} = \frac{136 - 81}{64} = \frac{55}{64} = .859$$

● **PROBLEM 8-43**

Find the variance of the random variable X + b where X has variance, Var X and b is a constant.

Solution: $\text{Var}(X + b) = E[(X + b)^2] - [E(X + b)]^2$

$$= E[X^2 + 2bX + b^2] - [E(X) + b]^2$$

$$= E(X^2) + 2bE(X) + b^2 - [E(X)]^2 - 2E(X)b - b^2,$$

thus $\text{Var}(X + b) = E(X^2) - [E(X)]^2 = \text{Var}\ X$.

● **PROBLEM 8-44**

Find the variance of the random variable, Z = X + Y if X and Y are not independent.

Solution: $\text{Var}\ Z = \text{Var}(X + Y) = E[((X + Y) - E(X + Y))^2]$

$$= E[(X - E(X) + Y - E(Y))^2]$$

(because $E(X + Y) = E(X) + E(Y)$),

$$= E[(X - E(X))^2 + 2(X - E(X))(Y - E(Y)) + (Y - E(Y))^2],$$

and by the properties of expectation,

$$= E[(X - E(X))^2] + 2E[(X - E(X))(Y - E(Y))] + E[Y - E(Y))^2]$$

Thus $\text{Var}\ Z = \text{Var}\ X + \text{Var}\ Y + 2E[(X - E(X))(Y - E(Y))]$.

If X and Y are independent, $E[(X - E(X))(Y - E(Y))] = 0$ but since X and Y are not independent, we may not assume that this cross product is zero.

$$E[(X - E(X))(Y - E(Y))]$$

is called the covariance of X and Y and is a measure of the linear relation between X and Y. It is a measure in the sense that , if X is greater than E(X) at the same time that Y is greater than E(Y) with high probability, then the covariance of X and Y will be positive. If X is below E(X) at the same time Y is above E(Y) with high probability, the covariance of X and Y will be negative.

Related to the covariance is the correlation coefficient defined as;

$$\rho = \frac{\text{Cov }(X, Y)}{\sqrt{\text{Var } X} \sqrt{\text{Var } Y}}$$

The correlation coefficient gives a clearer picture of the linear relation between X and Y because it takes account of the variation in the individual variables X and Y.

Other properties of covariance are: Cov (X, Y) = 0, if X and Y are independent. The converse is not true.

Cov (X, Y) = Cov (Y, X).

● PROBLEM 8-45

Find the variance of a random variable X that is uniformly distributed over the interval [0, 3].

Solution: The variance of X is by definition,

$$E([X - E(X)]^2) = E(X^2) - [E(X)]^2.$$

The density function of X is $f(x) = \begin{cases} \frac{1}{3} & 0 < x < 3 \\ 0 & \text{otherwise}. \end{cases}$

Thus, $E(X) = \int_0^3 xf(x)dx = \int_0^3 \frac{x}{3} dx = \left. \frac{x^2}{6} \right|_0^3 = \frac{9}{6} = \frac{3}{2}$

$E(X^2) = \int_0^3 x^2 f(x) dx = \int_0^3 \frac{x^2}{3} dx = \left. \frac{x^3}{9} \right|_0^3 = \frac{27}{9} = 3.$

And the variance of X is $\text{Var } X = 3 - \left(\frac{3}{2}\right)^2$

$$= 3 - \frac{9}{4} = \frac{12 - 9}{4} = \frac{3}{4}$$

● PROBLEM 8-46

Compute the conditional distribution of Y given X if X and Y are jointly distributed with density

$$f(x, y) = \begin{cases} x + y & \begin{matrix} 0 < x < 1 \\ 0 < y < 1 \end{matrix} \\ 0 & \text{otherwise}. \end{cases}$$

What is the conditional expectation of Y?

Solution: The conditional distribution of Y given X is defined by analogy with conditional probability to be:

$$f(y|x) = \frac{f(x, y)}{f(x)}$$

where $f(x, y)$ is the joint density of x and y and $f(x)$ is the marginal distribution of x.

In our example,
$$f(x) = \int_0^1 f(x, y)\, dy$$
$$= \int_0^1 (x + y)\, dy = \left[xy + \frac{y^2}{2}\right]_0^1$$
$$= x + \frac{1}{2} \qquad 0 < x < 1.$$

Thus $f(y|x) = \dfrac{f(x, y)}{f(x)} = \dfrac{x + y}{x + \frac{1}{2}} \qquad \begin{array}{l} 0 < y < 1 \\ 0 < x < 1 \end{array}.$

To see that $f(y|x)$ is a proper density function,

$$\int_0^1 f(y|x)\, dy = \int_0^1 \frac{x + y}{x + \frac{1}{2}}\, dy = \left(\frac{1}{x + \frac{1}{2}}\right) \int_0^1 (x + y)\, dy$$

$$= \left(\frac{1}{x + \frac{1}{2}}\right)\left(xy + \frac{y^2}{2}\right)_0^1 = \frac{x + \frac{1}{2}}{x + \frac{1}{2}} = 1.$$

The conditional expectation of Y given X is the expectation of y against the conditional density $f(y|x)$.

Thus, $\quad E(Y/X = x) = \displaystyle\int_{\text{all } y} y\, f(y/x)\, dy.$

For our example,

$$E(Y|X = x) = \int_0^1 y \left(\frac{x + y}{x + \frac{1}{2}}\right) dy = \left(\frac{1}{x + \frac{1}{2}}\right) \int_0^1 (xy + y^2)\, dy$$

$$= \frac{1}{x + \frac{1}{2}} \left[\frac{xy^2}{2} + \frac{y^3}{3}\right]_0^1 = \frac{\frac{x}{2} + \frac{1}{3}}{x + \frac{1}{2}}$$

$$= \frac{3x + 2}{3(2x + 1)} \qquad 0 < x < 1.$$

● **PROBLEM 8-47**

Conditional expectations may be used to find unconditional expectations. Show that

$$E[E(Y|X=x)] = E(Y).$$

Solution: Let $f(x)$, $f(y)$ and $f(x, y)$ be the marginal distribution of X, the marginal distribution of Y and the joint density of X and Y respectively.

$E(Y/X=x)$ is in general a function of x, say $E(Y|X=x) = h(x)$, then

$$E[E(Y|X=x)] = E[h(x)]$$

but the expected value of any function of a random variable is that function integrated against the density of X. Thus

$$E[E(Y|X=x)] = E(h(x)) = \int_{-\infty}^{\infty} h(x) f(x) dx$$

but $E(Y|X=x) = h(x) = \int_{-\infty}^{\infty} y f(y|x) dy$.

Thus

$$E[E(Y|X=x)] = \int_{-\infty}^{\infty} \left[\int_{-\infty}^{\infty} y f(y|x) dy \right] f(x) dx$$

$$= \int_{-\infty}^{\infty} \int_{-\infty}^{\infty} y f(y|x) f(x) dy dx.$$

Letting $f(y|x) = \dfrac{f(x, y)}{f(x)}$,

$$E\,E(Y/X=x) = \int_{-\infty}^{\infty} \int_{-\infty}^{\infty} \frac{y f(x, y)}{f(x)} f(x) dy dx$$

$$= \int_{-\infty}^{\infty} \int_{-\infty}^{\infty} y f(x, y) dy dx$$

$$= \int_{-\infty}^{\infty} y \left[\int_{-\infty}^{\infty} f(x, y) dx \right] dy$$

and

$$\int_{-\infty}^{\infty} f(x, y) dx = f(y).$$

Thus

$$E[(Y|X=x)] = \int_{-\infty}^{\infty} y\, f(y)\, dy = E(Y)$$

• **PROBLEM 8-48**

Given that the random variable X has density function

$$f(x) = \frac{1}{2}(x+1) \text{ when } -1 < x < 1$$

and $f(x) = 0$ elsewhere

calculate the mean value or expected value of X and the variance of X.

Solution: Recall that when X is a discrete random variable, its mean value

$$\mu = E(X) = \Sigma x\, f(x), \text{ where } f(x) = \Pr(X = x).$$

This sum of products is a weighted average of the values of X.

In this problem X is a continuous random variable. Therefore we must integrate $xf(x)$ from -1 to $+1$, since $f(x) = 0$ when $x \leq -1$ or $x \geq 1$.

$$\mu = \int_{-\infty}^{\infty} xf(x)\, dx = \int_{-1}^{1} x\, \frac{x+1}{2}\, dx$$

(we have substituted $\frac{x+1}{2}$ for $f(x)$)

$$= \int_{-1}^{1}\left(\frac{x^2}{2} + \frac{x}{2}\right) dx = \frac{1}{2}\int_{-1}^{1} x^2\, dx + \frac{1}{2}\int_{-1}^{1} x\, dx$$

$$= \frac{1}{2}\frac{x^3}{3}\bigg]_{-1}^{1} + \frac{1}{2}\frac{x^2}{2}\bigg]_{-1}^{1}$$

$$= \frac{1}{2}\left(\frac{1}{3} + \frac{1}{3}\right) + \frac{1}{2}\left(\frac{1}{2} - \frac{1}{2}\right) = \frac{1}{2}\left(\frac{2}{3}\right) = \frac{1}{3}$$

The variance of X is defined to be

$$\sigma^2 = E[(X-\mu)^2]. \quad \text{Since}$$

$$(X-\mu)^2 = (X-\mu)(X-\mu) = X^2 - 2\mu X + \mu^2,$$

we can write,

$$\sigma^2 = E[(X-\mu)^2] = E(X^2 - 2\mu X + \mu^2)$$

$$= \int_{-1}^{1} (X^2 - 2\mu X + \mu^2) f(x) \, dx$$

$$= \int_{-1}^{1} X^2 f(x) \, dx - 2\mu \int_{-1}^{1} x f(x) \, dx + \mu^2 \int_{-1}^{1} 1 \cdot f(x) \, dx$$

$$= E(X^2) - 2\mu E(X) + \mu^2 E(1)$$

$$= E(X^2) - 2\mu^2 + \mu^2 \quad [E(1) = 1 \text{ because } f(x) \text{ is a density}$$

function and must satisfy the condition that $\int_{-\infty}^{\infty} f(x) \, dx = 1$].

$$E(X^2) - 2\mu^2 + \mu^2 = E(X^2) - \mu^2$$

$$= \int_{-\infty}^{\infty} X^2 f(x) \, dx - \left(\frac{1}{3}\right)^2$$

$$= \int_{-1}^{1} X^2 \left(\frac{x+1}{2}\right) dx - \frac{1}{9}$$

$$= \int_{-1}^{1} \frac{x^3}{2} \, dx + \int_{-1}^{1} \frac{x^2}{2} \, dx - \frac{1}{9}$$

$$= \left. \frac{x^4}{8} \right|_{-1}^{1} + \left. \frac{x^3}{6} \right|_{-1}^{1} - \frac{1}{9}$$

$$= \left(\frac{1}{8} - \frac{1}{8}\right) + \frac{1}{6} - \left(\frac{-1}{6}\right) - \frac{1}{9} = \frac{1}{3} - \frac{1}{9} = \frac{2}{9}$$

● **PROBLEM 8-49**

Briefly discuss the Central Limit Theorem.

<u>Solution</u>: The theorem has to do with the means of large (greater than 30) samples. As the sample size increases, the distribution of the sample mean, $\bar{X}$, has a distribution which is approximately normal. This distribution has a mean equal to the population mean and a standard deviation equal to the population standard deviation divided by the square root of the sample size.

Since $\bar{X}$ is approximately normal, $\dfrac{\bar{X} - E(\bar{X})}{\sigma_{\bar{X}}} = \dfrac{\bar{X} - \mu}{\sigma/\sqrt{n}} = \dfrac{\sqrt{n}(\bar{X} - \mu)}{\sigma}$

will have a standard normal distribution.

● **PROBLEM 8-50**

Find ρ for X and Y if

$$f(x, y) = x + y \quad \text{for} \quad 0 < x < 1$$
$$0 < y < 1$$

is the joint density of X and Y.

Solution: The correlation coefficient, ρ, is defined to be

$$\rho = \frac{\text{Cov }(X, Y)}{\sqrt{\text{Var } X} \sqrt{\text{Var } Y}}$$

where $\text{Cov }(X, Y) = E[(X - E(X))(Y - E(Y))]$

and $\text{Var } X = E(X^2) - [E(X)]^2$

$\text{Var } Y = E(Y^2) - [E(Y)]^2$

$$\begin{aligned}\text{Cov }(X, Y) &= E[(X - E(X))(Y - E(Y))]\\ &= E[XY - E(X)Y - E(Y)X + E(X)E(Y)]\\ &= E(XY) - E(X)E(Y) - E(Y)E(X) + E(X)E(Y)\end{aligned}$$

by the properties of expectation. Thus,

$$\text{Cov. }(X, Y) = E(XY) - E(X)E(Y).$$

In our problem $E(X) = \iint_R x\, f(x, y)\, dx\, dy$,

$$E(X) = \int_0^1 \int_0^1 x(x + y)\, dx\, dy = \int_0^1 \left[\frac{x^3}{3} + \frac{x^2 y}{2}\right]_0^1 dy$$

$$= \int_0^1 \left[\frac{1}{3} + \frac{y}{2}\right] dy = \frac{y}{3} + \frac{y^2}{4}\bigg|_0^1 = \frac{1}{3} + \frac{1}{4} = \frac{7}{12}.$$

$$E(X^2) = \int_0^1 \int_0^1 x^2(x + y)\, dx\, dy = \int_0^1 \int_0^1 (x^3 + yx^2)\, dx\, dy$$

$$= \int_0^1 \left|\frac{x^4}{4} + \frac{yx^3}{3}\right|_0^1 dy = \int_0^1 \left(\frac{1}{4} + \frac{y}{3}\right) dy = \left(\frac{y}{4} + \frac{y^2}{6}\right)\bigg|_0^1$$

$$= \frac{1}{4} + \frac{1}{6} = \frac{10}{24} = \frac{5}{12}.$$

Similarly, $E(Y^2) = \frac{5}{12}$ and $E(Y) = \frac{7}{12}$.

Finally, $E(XY) = \int_0^1 \int_0^1 xy(x+y)\,dx\,dy$

$= \int_0^1 \int_0^1 (x^2 y + y^2 x)\,dx\,dy = \int_0^1 \left[\frac{x^3 y}{3} \frac{y^2 x^2}{2} \right]_0^1 dy$

$= \int_0^1 \left(\frac{y}{3} + \frac{y^2}{2} \right) dy = \left[\frac{y^2}{6} + \frac{y^3}{6} \right]_0^1 = \frac{1}{6} + \frac{1}{6} = \frac{1}{3}$.

Thus, $\rho = \dfrac{\text{Cov}(X,Y)}{\sqrt{\text{Var } X}\sqrt{\text{Var } Y}} = \dfrac{E(XY) - E(X)E(Y)}{\sqrt{E(X^2) - [E(X)]^2}\sqrt{E(Y^2) - [E()]^2}}$

$= \dfrac{\frac{1}{3} - \left(\frac{7}{12}\right)\left(\frac{7}{12}\right)}{\sqrt{\frac{5}{12} - \left(\frac{7}{12}\right)^2}\sqrt{\frac{5}{12} - \left(\frac{7}{12}\right)^2}} = \dfrac{\frac{1}{3} - \frac{49}{144}}{\frac{5}{12} - \left(\frac{7}{12}\right)^2}$

$= \dfrac{\frac{48}{144} - \frac{49}{144}}{\frac{60}{144} - \frac{49}{144}} = \dfrac{-\frac{1}{144}}{\frac{11}{144}} = \dfrac{-1}{11}$.

● **PROBLEM** 8-51

Given a Gaussian random variable X with density function $f_X(x) = \dfrac{1}{\sqrt{2\pi}} e^{-x^2/2}$, prove the following for $n = 1, 2, 3, \ldots$

(a) $E(X^{2n}) = 1 \cdot 3 \cdot 5 \ldots (2n-1)$ and,

(b) $E(X^{2n-1}) = 0$.

Solution:

(a) The general expression for the expected value of a random variable X is

$$E(X) = \int_{-\infty}^{+\infty} x\, f_X(x)\,dx.$$

Therefore, $E(X^{2n}) = \displaystyle\int_{-\infty}^{+\infty} x^{2n} f_X(x)\,dx$

$$= \frac{1}{\sqrt{2\pi}} \int_{-\infty}^{+\infty} x^{2n} e^{-x^2/2} dx$$

$$= \frac{1}{\sqrt{2\pi}} \int_{-\infty}^{+\infty} (x^2)^n e^{-x^2/2} dx$$

Since the integral is an even function of x

$$E(X^{2n}) = \frac{2}{\sqrt{2\pi}} \int_{0}^{\infty} (x^2)^n e^{-x^2/2} dx \qquad (1)$$

Let $u = x^2/2$

Hence, $x^2 = 2u$, $x^{2n} = 2^n u^n$

$x = \sqrt{2u}$, $dx = \sqrt{2} \frac{1}{2\sqrt{u}} du$

$$= \frac{1}{\sqrt{2u}} du.$$

With the above substitutions, equation (1) reduces to

$$E(X^{2n}) = \frac{2}{\sqrt{2\pi}} \int_{0}^{\infty} (2^n u^n) e^{-u} \left(\frac{1}{\sqrt{2u}}\right) du$$

$$= \frac{2}{\sqrt{2\pi}} \cdot \frac{2}{\sqrt{2}} \int_{0}^{\infty} u^{n-\frac{1}{2}} e^{-u} du$$

$$= \frac{2^n}{\sqrt{\pi}} \int_{0}^{\infty} u^{(n+\frac{1}{2})-1} e^{-u} du$$

$$= \frac{2^n}{\sqrt{\pi}} \Gamma(n+\tfrac{1}{2}) \qquad (2)$$

where $\Gamma(n+\tfrac{1}{2}) = \int_{0}^{\infty} u^{(n+\frac{1}{2})-1} e^{-u} du$, which follows from

the equality

$$\Gamma(n) = \int_{0}^{\infty} e^{-t} t^{n-1} dt.$$

The value of $\Gamma(n+\tfrac{1}{2})$ can be evaluated using the relations

$$\Gamma(n+1) = n\Gamma(n) \qquad \text{for } n > 0,$$

and $\Gamma(\tfrac{1}{2}) = \sqrt{\pi}$.

Thus, $\Gamma(n+\tfrac{1}{2}) = \Gamma(n-\tfrac{1}{2}+1) = (n-\tfrac{1}{2})\,\Gamma(n-\tfrac{1}{2})$

$$= (n-\tfrac{1}{2})\Gamma(n-\tfrac{3}{2}+1)$$

$$= (n-\tfrac{1}{2})(n-\tfrac{3}{2})\Gamma(n-\tfrac{3}{2})$$

$$= (n-\tfrac{1}{2})(n-\tfrac{3}{2})\cdots\cdots(\tfrac{3}{2})(\tfrac{1}{2})\Gamma(\tfrac{1}{2})$$

$$= \frac{(2n-1)(2n-3)\cdots\cdots[2n-(2n-3)][2n-(2n-1)](\sqrt{\pi})}{2^n}$$

$$= \frac{1\cdot 3\cdot 5\cdots\cdots(2n-1)(\sqrt{\pi})}{2^n}.$$

Substituting for $\Gamma(n+\tfrac{1}{2})$ into equation (2)

$$E(x^{2n}) = \frac{2^n}{\sqrt{\pi}}\left[\frac{1\cdot 3\cdot 5\cdots\cdots(2n-1)}{2^n}\sqrt{\pi}\right]$$

$$= 1\cdot 3\cdot 5\cdots\cdots(2n-1)$$

Hence, the proof.

(b) $\qquad E(X^{2n-1}) = \int_{-\infty}^{\infty} x^{2n-1} f_X(x)\,dx$

$$= \frac{1}{\sqrt{2\pi}}\int_{-\infty}^{\infty} x^{2n-1} e^{-x^2/2}\,dx \qquad (3)$$

Since x^{2n-1} is an odd function of x and $e^{-x^2/2}$ is an even function of x, their product $x^{2n-1} e^{-x^2/2}$ is an odd function of x. The integral of an odd function with limits symmetric to the origin is zero. Thus, the integral in equation (3) yields a zero value, and therefore, $E(X^{2n-1}) = 0$.

● **PROBLEM 8-52**

Given the following density functions of a random variable X:

(a) $f_{X_1}(x) = \dfrac{1}{\sqrt{2\pi}} e^{-(x-m)^2/2}$, for all x

(b) $f_{X_2}(x) = \begin{cases} xe^{-x^2/2}, & \text{for } x \geq 0 \\ 0, & \text{elsewhere} \end{cases}$

Determine the maximum $f_X(x)$ and the expectation of the random variable X.

Solution:

(a) Consider the density function $f_{X_1}(x) = \dfrac{1}{\sqrt{2\pi}} e^{-(x-m)^2/2}$ (1)

It is obvious from equation (1) that $f_{X_1}(x)$ will be maximum when x = m. Therefore, the most probable value of X is m.
Average value of X = E(X) = Expected value of X

where $E(X) = \displaystyle\int_{-\infty}^{+\infty} x\, f_X(x)\, dx$

$= \dfrac{1}{\sqrt{2\pi}} \displaystyle\int_{-\infty}^{+\infty} x\, e^{-(x-m)^2/2}\, dx.$

Let u = x−m, therefore du = dx, hence,

$E(X) = \dfrac{1}{\sqrt{2\pi}} \displaystyle\int_{-\infty}^{+\infty} (m+u)\, e^{-u^2/2}\, du$

$= \dfrac{1}{\sqrt{2\pi}} \displaystyle\int_{-\infty}^{+\infty} m e^{-u^2/2}\, du + \dfrac{1}{\sqrt{2\pi}} \displaystyle\int_{-\infty}^{+\infty} u e^{-u^2/2}\, du$

$= I_1 + I_2$

where $I_1 = \dfrac{m}{\sqrt{2\pi}} \displaystyle\int_{-\infty}^{+\infty} e^{-u^2/2}\, du$.

Since $\int_{-\infty}^{+\infty} e^{-\alpha x^2} dx = \sqrt{\pi/\alpha}$

$\int_{-\infty}^{+\infty} e^{-u^2/2} du = \sqrt{2\pi}$

Therefore, $I_1 = \dfrac{m}{\sqrt{2\pi}} \times \sqrt{2\pi} = m$

$I_2 = 0$ because the integrand is an odd function and the limits are symmetric with respect to the origin.

Thus, $E(X) = I_1 + I_2$

$\qquad = m$

(b) Now, consider the density function

$$f_{X_2}(x) = \begin{cases} xe^{-x^2/2}, & \text{for } x \geq 0 \\ 0, & \text{elsewhere} \end{cases}$$

The most probable value of X can be evaluated by differentiating $xe^{-x^2/2}$ with respect to x, and equating the result to zero

i.e., $\dfrac{d}{dx}(xe^{-x^2/2}) = 0$

$x\left[e^{-x^2/2}(-\tfrac{1}{2} \times 2x)\right] + e^{-x^2/2} = 0$

$-x^2 e^{-x^2/2} + e^{-x^2/2} = 0$

$e^{-x^2/2}(-x^2 + 1) = 0$

since $e^{-x^2/2} \neq 0$, $-x^2 + 1 = 0$

Hence, $x^2 = 1$

$\qquad x = \pm 1$

since $x > 0$, $x = 1$

Therefore, $f_{X_2}(x)$ is a maximum when $x = 1$.

Now, the average value of $X = E(X)$

$$E(X) = \int_{-\infty}^{+\infty} x \, f_{X_2}(x) \, dx$$

$$= \int_{0}^{\infty} x \cdot x e^{-x^2/2} \, dx$$

$$= \int_{0}^{\infty} x^2 e^{-x^2/2} \, dx$$

let $u = x^2/2$, $x^2 = 2u$

$x = \sqrt{2u}$, $dx = \sqrt{2} \, \dfrac{1}{2\sqrt{u}} \, du = \dfrac{1}{\sqrt{2}\sqrt{u}} \, du$

Therefore, $E(X) = \int_{0}^{\infty} (2u)(e^{-u})\left(\dfrac{1}{\sqrt{2}\sqrt{u}}\right) du$

$$= \sqrt{2} \int_{0}^{\infty} u^{\frac{1}{2}} e^{-u} \, du$$

$$= \sqrt{2} \int_{0}^{\infty} u^{\frac{3}{2} - 1} e^{-u} \, du$$

Using the identity $\Gamma(n) = \int_{0}^{\infty} t^{n-1} e^{-t} \, dt$, we have

$$E(X) = \sqrt{2} \, \Gamma(3/2) \qquad (2)$$

The value of $\Gamma(3/2)$ can be obtained from the relations $\Gamma(n+1) = n\Gamma(n)$ and $\Gamma(\frac{1}{2}) = \sqrt{\pi}$.

Therefore, $\Gamma(3/2) = \Gamma(\frac{1}{2}+1)$

$$= \tfrac{1}{2}\Gamma(\tfrac{1}{2}) = \tfrac{1}{2}\sqrt{\pi}$$

Substituting for $\Gamma(3/2)$ into equation (2)

$$E(X) = \sqrt{2} \times \tfrac{1}{2}\sqrt{\pi} = \sqrt{\pi/2}$$

Thus, average value of $X = \sqrt{\pi/2}$

• **PROBLEM 8-53**

Determine the variance for each of the density functions given below:

(a) The Gaussian density $f_{X_1}(x) = \frac{1}{\sqrt{2\pi}} e^{-(x-m)^2/2}$, $-\infty \le x \le 0$.

(b) The Rayleigh density $f_{X_2}(x) = xe^{-x^2/2}$, $x \ge 0$ and

(c) The uniform density $f_{X_3}(x) = 1/a$, $-a/2 \le x \le a/2$.

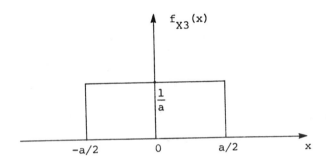

Solution:

(a) Consider the Gaussian density

$$f_{X_1}(x) = \frac{1}{\sqrt{2\pi}} e^{-(x-m)^2/2}, \quad -\infty \le x \le \infty.$$

The variance or second moment of a random variable X_1 is defined as

$$\sigma_{X_1}^2 \equiv E\left[(X - m_x)^2\right]$$

where E denotes the expected value and m is the mean or average value of the random variable.

Since $m_x = m$,

$$\sigma_{X_1}^2 = E\left[(X-m)^2\right] = \int_{-\infty}^{\infty} (x-m)^2 f_{X_1}(x) dx$$

$$= \frac{1}{\sqrt{2\pi}} \int_{-\infty}^{\infty} (x-m)^2 e^{-(x-m)^2/2} dx$$

let $u = x-m$, $du = dx$

Therefore, $\sigma_{X_1}^2 = \frac{1}{\sqrt{2\pi}} \int_{-\infty}^{\infty} u^2 e^{-u^2/2} du$

483

Since the integrand is an even function

$$\sigma_{X_1}^2 = \frac{2}{\sqrt{2\pi}} \int_0^\infty u^2 e^{-u^2/2} \, du.$$

Let $y = u^2/2$,

then $u^2 = 2y$ and $u = \sqrt{2y}$ (since $u > 0$).

Thus, $du = \sqrt{2} \, \frac{1}{2\sqrt{y}} \, dy = \frac{1}{\sqrt{2}\sqrt{y}} \, dy$

Hence $\sigma_{X_1}^2 = \frac{2}{\sqrt{2\pi}} \int_0^\infty (2y)(e^{-y})\left(\frac{1}{\sqrt{2}\sqrt{y}}\right) dy$

$$= \frac{2}{\sqrt{\pi}} \int_0^\infty y^{\frac{1}{2}} e^{-y} \, dy$$

$$= \frac{2}{\sqrt{\pi}} \int_0^\infty y^{\frac{3}{2}-1} e^{-y} \, dy$$

Using the identity $\Gamma(n) = \int_0^\infty t^{n-1} e^{-t} \, dt,$ \hfill (a)

$$\sigma_{X_1}^2 = \frac{2}{\sqrt{\pi}} \Gamma(3/2) \hfill (1)$$

The value of $\Gamma(3/2)$ can be evaluated from the relations $\Gamma(n+1) = n\Gamma(n)$ and $\Gamma(\frac{1}{2}) = \sqrt{\pi}$.

Therefore, $\Gamma(3/2) = \Gamma(\frac{1}{2}+1)$

$$= \tfrac{1}{2}\Gamma(\tfrac{1}{2}) = \tfrac{1}{2} \cdot \sqrt{\pi}$$

Substituting for $\Gamma(3/2)$ into equation (1)

$$\sigma_{X_1}^2 = \frac{2}{\sqrt{\pi}} \cdot \frac{\sqrt{\pi}}{2} = 1.$$

Thus, the variance of the random variable X_1 is '1'. Note that the variance $\sigma_{X_1}^2$ can be directly evaluated from the expression for the density function. In the present case, comparing the given density function with a standard Gaussian density function, it follows,

$$\frac{1}{\sqrt{2\pi}} e^{-(x-m)^2/2} = \frac{1}{\sqrt{2\pi\sigma^2}} e^{-(x-m)^2/2\sigma^2}$$

It is thus evident that the variance $\sigma_{X_1}^2 = 1$.

(b) Given the Rayleigh density
$$f_{X_2}(x) = xe^{-x^2/2}$$

The variance can be expressed as

$$\sigma_{X_2}^2 = E(X_2^2) - \left[E(X_2)\right]^2. \qquad (2)$$

So,
$$E(X_2^2) = \int_{-\infty}^{\infty} x^2 \, f_{X_2}(x) \, dx$$

$$= \int_{0}^{\infty} (x^2)(xe^{-x^2/2}) \, dx$$

$$= \int_{0}^{\infty} x^2 \, e^{-x^2/2} \, (xdx).$$

Let $y = x^2/2$

$$dy = \frac{2xdx}{2} = xdx.$$

Hence,
$$E(X_2^2) = \int_{0}^{\infty} (2y)(e^{-y}) \, dy$$

$$= 2 \int_{0}^{\infty} y \, e^{-y} \, dy$$

$$= 2 \int_{0}^{\infty} y^{2-1} \, e^{-y} \, dy$$

As before, using the identity (a)

$$E(X_2^2) = 2\,\Gamma(2) = 2\Gamma(1+1)$$
$$= 2 \times 1 \times \Gamma(1) = 2.$$

Now,
$$E(X_2) = \int_{-\infty}^{\infty} x \, f_{X_2}(x) \, dx$$

$$= \int_{0}^{\infty} x \cdot xe^{-x^2/2} \, dx$$

$$= \int_{0}^{\infty} x \, e^{-x^2/2} \, (xdx)$$

Let $u = x^2/2$, $du = xdx$

and since $x > 0$, $x = \sqrt{2} \cdot \sqrt{u}$

Therefore, $E(X_2) = \int_0^\infty \sqrt{2} \, u^{\frac{1}{2}} e^{-u} \, du$

$$= \sqrt{2} \int_0^\infty u^{\frac{3}{2} - 1} e^{-u} \, du$$

Using identity (a) again

$$E(X_2) = \sqrt{2} \, \Gamma(3/2)$$

$$= \sqrt{2} \times \tfrac{1}{2}\Gamma(\tfrac{1}{2})$$

$$= \sqrt{2} \cdot \tfrac{1}{2} \cdot \sqrt{\pi}$$

$$= \sqrt{\tfrac{\pi}{2}}.$$

Substituting for $E(X_2^2)$ and $E(X_2)$ into equation (2)

$$\sigma_{X_2}^2 = (2) - \left(\sqrt{\tfrac{\pi}{2}}\right)^2$$

$$= 2 - \tfrac{\pi}{2}.$$

Thus, the variance of the random variable $X_2 = 2 - \tfrac{\pi}{2} \cong 0.43$

(c) Consider the uniform density function

$$f_{X_3}(x) = 1/a, \quad -a/2 \leq x \leq a/2$$

The variance can be evaluated from the identity

$$\sigma_{X_3}^2 = E(X_3^2) - \left[E(X_3)\right]^2$$

where $E(X_3^2) = \int_{-\infty}^\infty x^2 \, f_{X_3}(x) \, dx$

$$= \int_{-a/2}^{a/2} x^2 \left(\tfrac{1}{a}\right) dx$$

$$= \tfrac{2}{a} \int_0^{a/2} x^2 \, dx$$

$$= \tfrac{2}{a} \left(\tfrac{x^3}{3}\right)_0^{a/2}$$

$$= \frac{2}{a} \times \frac{1}{3} \times \frac{a^3}{8}$$

$$= \frac{a^2}{12}$$

and $E(X_3) = \int_{-\infty}^{\infty} x\, f_{X_3}(x)\, dx$

$$= \int_{-a/2}^{a/2} x \left(\frac{1}{a}\right) dx = \frac{1}{a} \int_{-a/2}^{a/2} x\, dx = 0$$

Therefore, $\sigma_{X_3}^2 = \frac{a^2}{12}$

• **PROBLEM** 8-54

The Cauchy density function is given by

$$f(x) = \frac{K}{1+x^2}, \quad -\infty \leq x \leq \infty .$$

Determine (a) the value of K such that f(x) is a probability density function.
(b) the expected value of X.
(c) the variance of X.

<u>Solution</u>: (a) Consider the Cauchy density function

$$f(x) = \frac{K}{1+x^2}$$

The necessary condition for f(x) to be a density function is that

$$\int_{-\infty}^{\infty} f(x)\, dx = 1 .$$

Hence, $\int_{-\infty}^{\infty} \frac{K}{1+x^2} dx = 1 .$

Solving the above integral,

$$2K \int_{0}^{\infty} \frac{1}{1+x^2}\, dx = 1$$

$$2K(\tan^{-1}x)_0^\infty = 1$$

$$2K[(\pi/2) - (0)] = 1$$

$$K\pi = 1 \quad \text{or} \quad K = \frac{1}{\pi}.$$

Hence, $f(x) = \dfrac{1/\pi}{1+x^2}$, $-\infty \leq x \leq \infty$.

(b) $E(X)$ = Expected value of X

$$= \int_{-\infty}^{\infty} x\, f_X(x)\, dx = \int_{-\infty}^{\infty} (x) \cdot \frac{1/\pi}{1+x^2}\, dx$$

Since the integrand is an odd function and the limits are symmetric with respect to the origin,

$$\int_{-\infty}^{\infty} (x)\, \frac{1/\pi}{1+x^2}\, dx = 0.$$

Therefore, $E(X) = 0$.

(c) $\text{Var}(X) = E(X^2) - [E(X)]^2$

since $E(X) = 0$,

$$\text{Var}(X) = E(X^2) = \int_{-\infty}^{\infty} x^2\, f_X(x)\, dx$$

$$= \int_{-\infty}^{\infty} x^2\, \frac{1/\pi}{(1+x^2)}\, dx$$

$$= \frac{1}{\pi} \int_{\infty}^{\infty} \frac{x^2}{(1+x^2)}\, dx$$

Now, $\dfrac{x^2}{1+x^2} = \dfrac{x^2+1-1}{1+x^2} = 1 - \dfrac{1}{(1+x^2)}$

$$\text{Var}(X) = \frac{2}{\pi} \int_0^\infty \left(1 - \frac{1}{1+x^2}\right) dx$$

$$= \frac{2}{\pi} \int_0^\infty dx - \frac{2}{\pi} \int_0^\infty \frac{dx}{1+x^2}$$

$$= I_1 - I_2$$

where $I_2 = \frac{2}{\pi} \int_0^\infty \frac{dx}{1+x^2} = \frac{2}{\pi}(\tan^{-1} x)\Big|_0^\infty = \frac{2}{\pi}(\pi/2) = 1$.

As far as integral I_1 is concerned

$$I_1 = \lim_{x \to \infty} \frac{2}{\pi} x$$

Thus, it follows that, $I_1 \to \infty$, and so does the variance of X.

• **PROBLEM** 8-55

A random variable X has a Gaussian density function

$$f_X(x) = \frac{1}{\sqrt{2\pi\sigma^2}} e^{-(x-m)^2/2\sigma^2}$$

For $n = 1, 2, \ldots$, prove that

(a) $E\left[(X-m)^{2n-1}\right] = 0$ and,

(b) $E\left[(X-m)^{2n}\right] = 1 \cdot 3 \cdot 5 \cdots (2n-1)\sigma^{2n}$.

Solution: For a random variable X, the general expression for the expected value is

$$E(X) = \int_{-\infty}^\infty x \, f(x) \, dx.$$

Therefore, $E\left[(X-m)^{2n-1}\right] = \int_{-\infty}^\infty (x-m)^{2n-1} f(x) \, dx$

$$= \int_{-\infty}^\infty (x-m)^{2n-1} \frac{1}{\sqrt{2\pi\sigma^2}} e^{-(x-m)^2/2\sigma^2} dx$$

$$= \frac{1}{\sqrt{2\pi\sigma^2}} \cdot \int_{-\infty}^\infty (x-m)^{2n-1} \cdot e^{-(x-m)^2/2\sigma^2} dx.$$

In the above expression, the terms $(x-m)^{2n-1}$ and $e^{-(x-m)^2/2\sigma^2}$ are odd and even respectively. Thus, the integrand is an odd function, and since the limits are symmetric with respect to the origin, the above expression evaluates to zero.

i.e., $E\left[(X-m)^{2n-1}\right] = 0$.

(b)
$$E\left[(X-m)^{2n}\right] = \int_{-\infty}^{\infty} (x-m)^{2n} f(x) \, dx$$

$$= \frac{1}{\sqrt{2\pi\sigma^2}} \int_{-\infty}^{\infty} (x-m)^{2n} e^{-(x-m)^2/2\sigma^2} \, dx$$

let $u = x-m$, $du = dx$

$$E\left[(X-m)^{2n}\right] = \frac{1}{\sqrt{2\pi\sigma^2}} \int_{-\infty}^{\infty} u^{2n} e^{-u^2/2\sigma^2} \, du.$$

Since both u^{2n} and $e^{-u^2/2\sigma^2}$ are even, the integrand is an even function.

Therefore,
$$E\left[(x-m)^{2n}\right] = \frac{2}{\sqrt{2\pi\sigma^2}} \int_{0}^{\infty} u^{2n} e^{-u^2/2\sigma^2} \, du \qquad (1)$$

let $y = u^2/2\sigma^2$

$u^2 = 2\sigma^2 y$

Since $u > 0$, $u = \sigma\sqrt{2y}$

$$du = \sigma\sqrt{2} \cdot \frac{1}{2\sqrt{y}} \, dy = \frac{\sigma}{\sqrt{2}} y^{-\frac{1}{2}} \, dy.$$

Substituting the above values in equation (1)

$$E\left[(X-m)^{2n}\right] = \frac{2}{\sqrt{2\pi\sigma^2}} \int_{0}^{\infty} (2^n \sigma^{2n} y^n) \, e^{-y} \left(\frac{\sigma}{\sqrt{2}} y^{-\frac{1}{2}}\right) dy$$

$$= \frac{2}{\sigma\sqrt{2\pi}} \cdot (2^n \sigma^{2n}) \left(\frac{\sigma}{\sqrt{2}}\right) \int_{0}^{\infty} y^{n-\frac{1}{2}} e^{-y} \, dy$$

$$= \frac{2^n \sigma^{2n}}{\sqrt{\pi}} \int_{0}^{\infty} y^{n+\frac{1}{2}-1} e^{-y} \, dy$$

From the general rule $\int_{0}^{\infty} t^{n-1} e^{-t} \, dt = \Gamma(n)$ \qquad (a)

(where Γ represents the Gamma function) it follows

$$E\left[(X-m)^{2n}\right] = \frac{2^n \sigma^{2n}}{\sqrt{\pi}} \Gamma(n+\tfrac{1}{2}), \qquad (2)$$

Since for $n > 0$ $\Gamma(n+1) = n\Gamma(n)$ and $\Gamma(\tfrac{1}{2}) = \sqrt{\pi}$

$$\Gamma(n+\tfrac{1}{2}) = \Gamma(n-\tfrac{1}{2}+1) = (n-\tfrac{1}{2})\ \Gamma(n-\tfrac{1}{2})$$

$$= (n-\tfrac{1}{2})(n-3/2)\ \Gamma(n-3/2)$$

$$= (n-\tfrac{1}{2})(n-3/2) \cdots \left(\tfrac{3}{2}\right)\cdot\left(\tfrac{1}{2}\right)\Gamma(\tfrac{1}{2})$$

$$= \left\{n-\tfrac{[2(1)-1]}{2}\right\}\left\{n-\tfrac{[2(2)-1]}{2}\right\}\cdots\left\{n-\tfrac{[2(n-1)-1]}{2}\right\}$$

$$\left\{n-\tfrac{[2(1)-1]}{2}\right\}\sqrt{\pi}$$

$$= \left[\frac{(2n-1)(2n-3)\quad\cdots\quad 5\cdot 3\cdot 1}{2^n}\right]\sqrt{\pi}$$

Substituting for $\Gamma(n+\tfrac{1}{2})$ into equation (2)

$$E\left[(X-m)^{2n}\right] = \frac{2^n \sigma^{2n}}{\sqrt{\pi}}\left[\frac{1\cdot 3\cdot 5 \cdots (2n-3)(2n-1)}{2^n}\right]\sqrt{\pi}$$

$$= 1\cdot 3\cdot 5 \cdots (2n-1)\ \sigma^{2n}$$

● **PROBLEM** 8-56

Two random variables X and Y have a joint density function

$$f(x,y) = \frac{1}{\pi\sqrt{3}}\ e^{-2(x^2 - xy + y^2)/3}$$

Prove that the density function of each random variable is Gaussian and independent of the other.

Solution: Consider the joint density function

$$f(x,y) = \frac{1}{\pi\sqrt{3}}\ e^{-2(x^2 - xy + y^2)/3}$$

The density function f(x) of the random variable X, independent of Y is

$$f(x) = \int_{-\infty}^{+\infty} f(x,y)\ dy$$

$$= \frac{1}{\pi\sqrt{3}} \int_{-\infty}^{+\infty} e^{-2(x^2 - xy + y^2)/3}\ dy$$

$$= \frac{1}{\pi\sqrt{3}} \int_{-\infty}^{\infty} e^{-2x^2/3}\ e^{-2(y^2 - xy)/3}\ dy.$$

Since x is a constant in the above integral

$$f(x) = \frac{1}{\pi\sqrt{3}} e^{-2x^2/3} \int_{-\infty}^{\infty} e^{-2(y^2 - xy)/3} \, dy. \qquad (1)$$

Now, $y^2 - xy = (y - \frac{x}{2})^2 - \frac{x^2}{4}$

$-\frac{2}{3}(y^2 - xy) = -\frac{2}{3}(y - \frac{x}{2})^2 + \frac{1}{6}x^2$

Thus, $e^{-\frac{2}{3}(y^2 - xy)} = e^{-\frac{2}{3}(y - \frac{x}{2})^2} \cdot e^{\frac{1}{6}x^2}$

Substituting the above value into equation (1)

$$f(x) = \frac{1}{\pi\sqrt{3}} e^{-2x^2/3} \int_{-\infty}^{+\infty} e^{\frac{1}{6}x^2} e^{-\frac{2}{3}(y - \frac{x}{2})^2} \, dy$$

$$= \frac{1}{\pi\sqrt{3}} e^{x^2(-\frac{2}{3} + \frac{1}{6})} \int_{-\infty}^{+\infty} e^{-\frac{2}{3}(y - \frac{x}{2})^2} \, dy$$

let $t = y - \frac{x}{2}$

$$f(x) = \frac{1}{\pi\sqrt{3}} e^{-\frac{x^2}{2}} \int_{-\infty}^{\infty} e^{-\frac{2}{3}t^2} \, dt$$

Since $\int_{-\infty}^{\infty} e^{-\alpha y^2} = \sqrt{\pi/\alpha}$

$$f(x) = \frac{1}{\pi\sqrt{3}} e^{-\frac{x^2}{2}} \sqrt{\frac{3\pi}{2}}$$

$$= \frac{e^{-\frac{x^2}{2}}}{\sqrt{2\pi}} \qquad (2)$$

The general expression for a Gaussian density function is

$$f(x) = \frac{1}{\sqrt{2\pi\sigma^2}} e^{-(x-m)^2/2\sigma^2} \qquad (3)$$

in which σ^2 is the variance and m is the mean of X.

Comparing equations (2) and (3) it is obvious that equation (2) represents a Gaussian density function with zero mean (m=0) and variance equal to one ($\sigma^2=1$)

Similarly, it can be proved that if the random variable Y is considered independent of X, its density function too will be Gaussian with zero mean (m=0) and variance = 1.

• **PROBLEM** 8-57

A random variable R has a Rayleigh density

$$f(r) = \begin{cases} \dfrac{r}{\alpha^2} e^{-r^2/2\alpha^2} &, 0 \le r \le \infty \\ 0 &, r < 0 \end{cases}$$

Determine the following:
(a) the expected value or R, $E[R]$
(b) the expected value of R^2, $E[R^2]$
and (c) the variance σ^2 of R

Solution: (a)

$$f(r) = \begin{cases} \dfrac{r}{\alpha^2} e^{-r^2/2\alpha^2} &, 0 \le r \le \infty \\ 0 &, r < 0 \end{cases}$$

The mean value, or the expected value of R is

$$E[R] = \int_{-\infty}^{+\infty} r\, f(r)\, dr$$

$$= \int_0^{\infty} r \left(\dfrac{r}{\alpha^2}\right) e^{-r^2/2\alpha^2}\, dr$$

$$= \dfrac{1}{\alpha^2} \int_0^{\infty} r^2 e^{-r^2/2\alpha^2}\, dr \qquad (1)$$

let $u = r^2/2\alpha^2$

Hence, $r^2 = 2\alpha^2 u$

Since $r > 0$, $r = \sqrt{2} \cdot \alpha \cdot \sqrt{u}$

and $dr = \sqrt{2}\, \alpha \dfrac{1}{2\sqrt{u}}\, du = \dfrac{\alpha}{\sqrt{2}\sqrt{u}}\, du$.

Substituting these values into equation (1)

$$E[R] = \dfrac{1}{\alpha^2} \int_0^{\infty} (2\alpha^2 u)\, e^{-u} \dfrac{\alpha}{\sqrt{2}\sqrt{u}}\, du$$

$$= (\sqrt{2} \cdot \alpha) \int_0^{\infty} u^{(3/2)-1} e^{-u}\, du$$

493

Using the identity, $\Gamma(n) = \int_0^\infty t^{n-1} e^{-t} dt$. (2)

$$\int_0^\infty u^{(3/2)-1} e^{-u} du = \Gamma(3/2).$$

Since $\Gamma(n+1) = n\Gamma(n)$ and $\Gamma(\frac{1}{2}) = \sqrt{\pi}$,

$$\Gamma(3/2) = \Gamma(\tfrac{1}{2} + 1)$$
$$= \tfrac{1}{2}\Gamma(\tfrac{1}{2})$$
$$= \frac{\sqrt{\pi}}{2}$$

Therefore, $E[R] = (\sqrt{2}\alpha) \cdot \frac{\sqrt{\pi}}{2} = \left(\sqrt{\frac{\pi}{2}}\right) \cdot \alpha$

(b) Mean square value of $R = E[R^2] = \int_{-\infty}^\infty r^2 f(r) dr$

$$E[R^2] = \int_0^\infty (r^2) \left(\frac{r}{\alpha^2}\right) e^{-r^2/2\alpha^2} dr$$

$$= \frac{1}{\alpha^2} \int_0^\infty r^3 e^{-r^2/2\alpha^2} dr$$

$$= \frac{1}{\alpha^2} \int_0^\infty r^2 e^{-r^2/2\alpha^2} (r\, dr). \tag{3}$$

If $u = r^2/2\alpha^2$ then, $r^2 = 2\alpha^2 u$

$$du = \frac{1}{2\alpha^2} 2r\, dr = \frac{1}{\alpha^2} r\, dr$$

$$r\, dr = \alpha^2 du$$

Substituting these values into equation (3)

$$E[R^2] = \frac{1}{\alpha^2} \int_0^\infty (2\alpha^2 u) e^{-u} \alpha^2 du$$

$$= 2\alpha^2 \int_0^\infty u e^{-u} du$$

$$= 2\alpha^2 \int_0^\infty u^{2-1} e^{-u} du.$$

As before, using identity (2)

$$\int_0^\infty u^{2-1} e^{-u} du = \Gamma(2) = \Gamma(1+1) = 1\Gamma(1)$$

i.e., $E[R^2] = 2\alpha^2 \cdot 1\Gamma(1) = 2\alpha^2$.

Thus, the mean square value of $R = 2\alpha^2$.

(c) The variance of R is defined as

$$\sigma_r^2 = \text{(Mean square value)} - \text{(Mean value)}^2$$

$$= 2\alpha^2 - (\sqrt{\tfrac{\pi}{2}} \cdot \alpha)^2$$

$$= 2\alpha^2 - \tfrac{\pi}{2}\alpha^2$$

$$= \alpha^2 (2 - \tfrac{\pi}{2})$$

$$\cong \alpha^2 (0.43)$$

RANDOM PROCESSES

● **PROBLEM** 8-58

Let Θ be a uniformly distributed random variable on the interval $(0, 2\pi)$. If A and ω_0 are constants, prove that the random process $X(t) = A\cos(\omega_0 t + \Theta)$ is wide-sense stationary.

Solution: A random process is said to be wide-sense stationary, provided the following two conditions are satisfied.

(i) The expected value or mean of X(t) is a constant.

(ii) The autocorrelation function is independent of time.

Expected value of $X(t) = E[X(t)]$

$$= \int_0^{2\pi} \tfrac{1}{2\pi} A\cos(\omega_0 t + \theta) d\theta$$

$$= \tfrac{A}{2\pi} \Big[\sin(\omega_0 t + \theta) \Big]_0^{2\pi}$$

$$= \frac{A}{2\pi}\left[\sin(\omega_0 t + 2\pi) - \sin(\omega_0 t)\right]$$

$$= \frac{A}{2\pi}\left[\sin\omega_0 t - \sin\omega_0 t\right] = 0$$

The autocorrelation function is defined by

$$R_{XX}(t_1,t_2) = \left[E\ X(t_1)\ X(t_2)\right]$$

where $t_1 = t$ and $t_2 = t + \tau$

Now,
$$R_{XX}(t,t+\tau) = E\left[A\cos(\omega_0 t+\theta)A\cos(\omega_0 t+\omega_0\tau+\theta)\right]$$

$$= \frac{A^2}{2}\ E\left[\cos(\omega_0\tau)+\cos(2\omega_0 t+\omega_0\tau+2\theta)\right]$$

$$= \frac{A^2}{2}\cos(\omega_0\tau) + \frac{A^2}{2}\ E\left[\cos(2\omega_0 t+\omega_0\tau+2\theta)\right]$$

$$= \frac{A^2}{2}\cos(\omega_0\tau) + \frac{A^2}{2}\int_0^{2\pi}\frac{1}{2\pi}\cos(2\omega_0 t+\omega_0\tau+2\theta)d\theta$$

$$= \frac{A^2}{2}\cos(\omega_0\tau) + \frac{A^2}{2\times 2\pi}(0)$$

$$= \frac{A^2}{2}\cos(\omega_0\tau)$$

Thus, $R_{XX}(t,t+\tau)$ is independent of t and the expected value is a constant, so X(t) is wide-sense stationary.

• **PROBLEM** 8-59

Let X and Y be two random variables of zero mean and zero correlation coefficient. $N_1(t)$ and $N_2(t)$ are two random processes given by

$N_1(t) = X\cos(\omega_0 t) + Y\sin(\omega_0 t)$

$N_2(t) = Y\cos(\omega_0 t) - X\sin(\omega_0 t)$

where ω_0 is a constant. Prove that $N_1(t)$ and $N_2(t)$ are jointly wide-sense stationary.

Solution: The cross-correlation $R_{12}(t,t+\tau)$ between two processes $N_1(t)$ and $N_2(t)$ is defined as

$$R_{12}(t,t+\tau) = E\left[N_1(t)\ N_2(t+\tau)\right]$$

$$= E\left[\{X\cos(\omega_0 t)+Y\sin(\omega_0 t)\}\{Y\cos(\omega_0 t+\omega_0\tau)-X\sin(\omega_0 t+\omega_0\tau)\}\right]$$

$$= E\left[XY\cos(\omega_0 t)\cos(\omega_0 t+\omega_0\tau)-X^2\cos(\omega_0 t)\sin(\omega_0 t+\omega_0\tau)\right.$$
$$\left.+Y^2\sin(\omega_0 t)\cos(\omega_0 t+\omega_0\tau)-XY\sin(\omega_0 t)\sin(\omega_0 t+\omega_0\tau)\right]$$

$$= E\left[XY\{\cos(\omega_0 t)\cos(\omega_0 t+\omega_0\tau)-\sin(\omega_0 t)\sin(\omega_0 t+\omega_0\tau)\}\right.$$
$$\left.-X^2\cos(\omega_0 t)\sin(\omega_0 t+\omega_0\tau)+Y^2\sin(\omega_0 t)\cos(\omega_0 t+\omega_0\tau)\right]$$

$$= E[XY][\cos(\omega_0 t+\omega_0 t+\omega_0\tau)]$$
$$-E[X^2]\cos(\omega_0 t)\sin(\omega_0 t+\omega_0\tau)$$
$$+E[Y^2]\sin(\omega_0 t)\cos(\omega_0 t+\omega_0\tau)$$

Since X and Y are zero-mean, uncorrelated random variables $E[XY] = 0$. Also, since X and Y have equal variances, $E[X^2] = E[Y^2] = \sigma^2$.

So, $R_{12}(t,t+\tau) = -\sigma^2\left[\sin(\omega_0 t+\omega_0\tau)\cos(\omega_0 t)-\cos(\omega_0 t+\omega_0\tau)\sin(\omega_0 t)\right]$
$$= -\sigma^2\left[\sin(\omega_0 t+\omega_0\tau-\omega_0 t)\right]$$
$$= -\sigma^2 \sin(\omega_0\tau)$$

Since $R_{12}(t,t+\tau)$ does not depend on t, $N_1(t)$ and $N_2(t)$ are jointly wide-sense stationary.

• **PROBLEM 8-60**

Determine the first and second moments of the random process X(t), both as time averages and statistical averages. It is given that the random process with sample functions is:

$X(t) = A \cos(\omega_0 t + \Theta)$, where ω_0 is a constant

and Θ a random variable with uniform density on the interval $(0,2\pi)$.

Solution: Computed as time averages,

$\langle X(t)\rangle$ = First moment

$$= \lim_{T\to\infty} \frac{1}{2T} \int_{-T}^{+T} A \cos(\omega_0 t+\Theta)dt$$

$$= \lim_{T\to\infty} \frac{1}{2T} A \left[\frac{\sin(\omega_0 t+\Theta)}{\omega_0}\right]_{-T}^{+T}$$

$$= \lim_{T\to\infty} \frac{A}{2\omega_0 T} \left[\sin(\omega_0 T+\Theta) - \sin(-\omega_0 T+\Theta)\right]$$

$$= \lim_{T\to\infty} \frac{A}{2\omega_0 T} \left[\sin(2\pi+\Theta) - \sin(-2\pi+\Theta)\right]$$

$$= \frac{A}{2(2\pi)} \left[\sin(2\pi+\Theta) + \sin(2\pi-\Theta)\right]$$

$$= \frac{A}{2(2\pi)} \left[\sin\theta - \sin\theta \right] = 0$$

and
$$\langle X^2(t) \rangle = \text{Second moment}$$

$$= \lim_{T \to \infty} \frac{1}{2T} \int_{-T}^{+T} A^2 \cos^2(\omega_0 t + \theta) dt$$

$$= \lim_{T \to \infty} \frac{A^2}{2T(2)} \int_{-T}^{+T} [1 + \cos 2(\omega_0 t + \theta)] dt$$

$$= \lim_{T \to \infty} \frac{A^2}{4T} \left[\int_{-T}^{T} dt + \int_{-T}^{+T} \cos 2(\omega_0 t + \theta) dt \right]$$

$$= \lim_{T \to \infty} \frac{A^2}{4T} \left[(2T + 0) \right]$$

$$= \frac{A^2}{2}$$

Computed as statistical averages,

$$E[X(t)] = \text{First moment} = \int_{-\infty}^{+\infty} A\cos(\omega_0 t + \theta) f_\Theta(\theta) d\theta$$

$f_\Theta(\theta)$ = Probability density function of the random variable Θ

$$= \begin{cases} \frac{1}{2\pi}, & 0 < \theta < 2\pi \\ 0, & \text{elsewhere} \end{cases}$$

So, $E[X(t)] = \frac{A}{2\pi} \int_0^{2\pi} \cos(\omega_0 t + \theta) d\theta$

$$= 0$$

and,
$$E\left[X^2(t)\right] = \text{Second moment}$$

$$= \frac{1}{2\pi} \int_0^{2\pi} A^2 \cos^2(\omega_0 t + \theta) d\theta = \frac{A^2}{2\pi} (\pi) = \frac{A^2}{2}$$

Note that from the above problem, the random process is stationary and ergodic because $\langle X(t) \rangle$ and $\langle X^2(t) \rangle$ are constants.

CHAPTER 9

AMPLITUDE MODULATION SYSTEMS (AM)

AMPLITUDE MODULATION

● **PROBLEM 9-1**

$v(t) = 12 (1 + 0.1 \cos 2\pi 1000t) \cos (2\pi 10,000t)$ is an amplitude-modulated signal.

(a) Find the sum and difference frequency components, their corresponding amplitude, and modulation index (m).

(b) Determine the power dissipated across a resistor ($R = 15\Omega$) due to the modulated signal.

Solution:

(a) First, in order to examine the sum and difference frequency components, we have to represent $v(t)$ in a more desirable form.

Hence, multiplying $v(t)$ out,

$v(t) = (12 + 1.2 \cos 2\pi 1000t) \cos (2\pi 10,000t)$

$= 12 \cos(2\pi 10^4 t) + 1.2 \cos(2\pi 10^4 t) \cos(2\pi 10^3 t)$ (1)

Applying the trigonometric identity,

$\cos A \cos B = \tfrac{1}{2} \cos (A-B) + \tfrac{1}{2} \cos (A+B)$, to Eq. (1).

$v(t) = \underset{v_C}{\underline{12}} \cos(2\pi 10^4 t) + \underset{v_L}{\underline{0.6}} \cos 2\pi(10^4 - 10^3)t$

$\quad + \underset{v_U}{\underline{0.6}} \cos 2\pi(10^4 + 10^3)t$ (2)

499

Hence, the sum frequency = (10^4+10^3)Hz = 11kHz

the difference frequency = (10^4-10^3)Hz = 9kHz

their corresponding amplitude =

$$v(f_c + f_m) = v(f_c - f_m) = 0.6v$$

(Note: $f_c=10^4$Hz=carrier frequency, $f_m=10^3$Hz=modulating frequency.)

The modulation index (m) is obtained by observing the original equation for v(t). Hence m = 0.1.

(b) From Eq. (2), we have,

$$v_c \text{ (carrier)}=12v, \quad v_{Lower} \text{ and } v_{Upper} \text{ (sideband)}=0.6v$$

With the above values, power across the 15Ω resistor can be calculated as follows:

$$\text{Carrier power} = \frac{(v_c)^2}{2} \times \frac{1}{R} = \frac{(12)^2}{2(15)} = 4.8W$$

$$\text{Sideband power} = \frac{(v_L)^2}{2R} = \frac{(v_U)^2}{2R} = \frac{(0.6)^2}{2(15)} = 0.012W$$

Thus, total sideband power = 0.012W × 2 = 0.024W.

• **PROBLEM** 9-2

For m(t), a single tone, such that
$$m(t) = a\cos \omega_o t$$
show that the sideband power cannot exceed one-third of the total transmitted power in Amplitude Modulation.

Solution: Since m(t) is a single tone,
$$m(t) = a\cos \omega_o t,$$
the average power of $m_{ac}(t)$ is $\frac{a^2}{2}$, where $m_{ac}(t)$ is m(t) without the dc term m_o. Thus,

$$P_{m_{ac}} = \frac{a^2}{2}.$$

When c(t), the continuous wave carrier tone, is double sideband modulated by m(t), the harmonics are

$$S_{DSB}(t) = \frac{Aa}{2}\cos(\omega_c+\omega_o)t + \frac{Aa}{2}\cos(\omega_c-\omega_o)t$$

The average power of the above is, by inspection, equal to

$\frac{(Aa)^2}{4}$, which is seen to be $A^2/2$ times the average power of $m_{ac}(t)$ (which is $a^2/2$). For Amplitude Modulation (modulation of $\cos \omega_c t$ by $a\cos \omega_m t$), the amplitude modulated signal is

$$S_{AM}(t) = A\left[1+m_a \cos \omega_o t\right]\cos \omega_c t \quad \text{where } m_a = \frac{a}{A}.$$

The average power is given by

$$P = \frac{1}{T_c}\int_{-\frac{T_c}{2}}^{\frac{T_c}{2}} A^2\left[1+m_a m_o\right]^2 \cos^2\omega_c t\, dt + \frac{m_a^2}{\Delta T}\int_{T_1}^{T_2} m_{ac}^2(t)\, A^2\cos^2\omega_c t\, dt$$

$$+ \frac{A^2}{2}\frac{m_a^2}{\Delta T}\int_{T_1}^{T_2} m_{ac}^2(t)\, dt + P_{SB}\ .$$

The first term is the average power in the residual pilot carrier, and the second term is the average power in the information sidebands. m_o is the dc component of $m(t)$, m_a is the amplitude modulation index in volts per volt and P_{SB} is the average power of the modulation signal sidebands. Thus the average power of $S_{AM}(t)$ follows from the last three listed equations as

$$P = \left[\frac{A^2}{2} + \frac{m_a^2 A^2}{4}\right] = P_{carrier} + P_{SB} \qquad (1)$$

It is seen from this that the sideband power is limited to

$$P_{SB} = \frac{m_a^2}{2} P_c$$

The maximum sideband power occurs with 100 percent modulation which implies that,

$$0 \leq P_{SB} \leq P_c/2$$

Thus, the total average power can be estimated from equation (1) and the above constraint as

$$P_c \leq P \leq 3/2\, P_c$$

It can therefore be inferred that, in AM, even with 100 percent modulation, the sideband power is limited to one-third of the total transmitted power.

Note that both in single sideband, and double sideband, the entire average power is represented by the sidebands.

• **PROBLEM** 9-3

A sinusoidal carrier is amplitude modulated by a baseband signal m(t) as shown in the figure. With the data given, calculate the percentage modulation in each case.

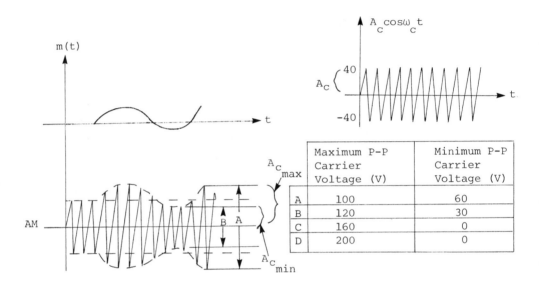

	Maximum P-P Carrier Voltage (V)	Minimum P-P Carrier Voltage (V)
A	100	60
B	120	30
C	160	0
D	200	0

Solution: The extent to which a carrier is amplitude modulated is expressed in terms of a percentage modulation (P) given by the formula

$$P = \frac{A_{c(max)} - A_{c(min)}}{2A_c} \times 100 \% \quad (1)$$

where $A_c(max) = \frac{\text{maximum P-P carrier voltage}}{2}$

and $A_c(min) = \frac{\text{minimum P-P carrier voltage}}{2}$

Hence, using Eq. (1), the percentage modulation for cases A to D are calculated as follows:

A. $P = \frac{A_{c(max)} - A_{c(min)}}{2A_c} \times 100 \%$

$= \frac{50 - 30}{2(40)} \times 100 \%$

= 25 %

B. $P = \dfrac{60 - 15}{2(40)} \times 100\% = 56.25\%$

C. $P = \dfrac{80 - 0}{80} \times 100\% = 100\%$

D. In this case, since the peak value of the modulated carrier is too high, overmodulation occurs.

● **PROBLEM** 9-4

An AM signal is transmitted through an antenna. The magnitude of the antenna current is 15A when an unmodulated signal is trnasmitted and, it increases to 17A for a modulated signal. Determine the percentage modulation.

Solution: For AM transmission, in order to maintain a high efficiency, since the sidebands of the transmitted signal contain the information and have maximum power at 100% modulation, the percentage modulation (5m) must be chosen as high as possible but not overmodulated.

A useful relationship between power and modulation index (m) is:
$$P_{SC} = P_C\left(1 + \dfrac{m^2}{2}\right)$$

where P_{SC} = total transmitted power of sidebands and carrier, (1)

P_C = carrier power.

Since $P = I^2R$, Eq. (1) can be solved in terms of the current.
Therefore,
$$P_{SC} = P_C\left(1 + \dfrac{m^2}{2}\right)$$

$$I_{SC}^2 R = I_C^2 R\left(1 + \dfrac{m^2}{2}\right)$$

$$I_{SC} = \sqrt{I_C^2\left(1 + \dfrac{m^2}{2}\right)}$$

$$= I_C\sqrt{1 + \dfrac{m^2}{2}} \quad (2)$$

where I_{SC} = total transmitted current,

I_C = carrier current.

Substituting information given into Eq. (2), % modulation can be calculated as follows:

$$17 = 15\sqrt{1 + \frac{m^2}{2}}$$

$$\left(\frac{17}{15}\right)^2 = 1 + \frac{m^2}{2}$$

$$m = \sqrt{2\left[\left(\frac{17}{15}\right)^2 - 1\right]}$$

Hence, %m = 0.75 × 100% = 75%.

• **PROBLEM 9-5**

An auxiliary signal with carrier frequency (f_c=1.0 MHz) is modulated by a signal bandlimited to the frequency range 25 Hz (f_1) to 20 kHz (f_2). Sketch the resulting frequency spectrum and find the upper and lower sideband frequencies.

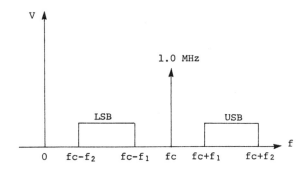

Solution: A frequency spectrum of the AM modulator's output is sketched below to provide a better understanding on the calculation of the upper and lower sideband frequencies.

Therefore, with the aid of the figure, the upper sideband (USB) will include the frequencies from

$fc + f_1$ = 1,000,000 Hz + 25 Hz = 1,000,025 Hz

to $fc + f_2$ = 1,000,000 Hz + 20,000 Hz = 1,020,000 Hz.

The lower sideband (LSB) will include the frequencies from

$fc - f_2$ = 1,000,000 Hz - 20,000 Hz = 980,000 Hz

to $fc - f_1$ = 1,000,000 Hz - 25 Hz = 999,975 Hz.

• **PROBLEM 9-6**

A chopper-type modulator used for AM signal generation is shown in Fig. (1). v(t) and $A\cos \omega_c t$ are represented in Fig. (2) and (3) respectively; sketch the waveforms at aa', bb', and cc'. In order to have an AM signal output at cc', what conditions must amplitude A satisfy?

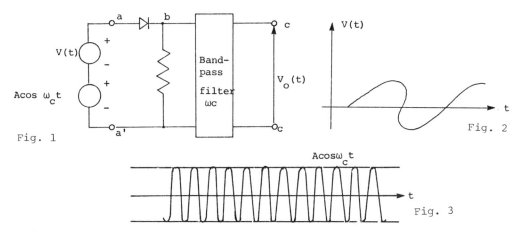

Fig. 1

Fig. 2

Fig. 3

Solution: Since v(t) and $A\cos\omega_c t$ are given, the waveforms at aa', bb' and cc' are sketched below.

At aa': $v_{aa'}(t) = v(t) + A\cos\omega_c t$

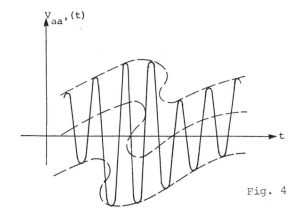

Fig. 4

At bb':

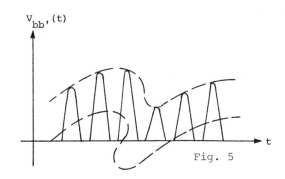

Fig. 5

At cc':

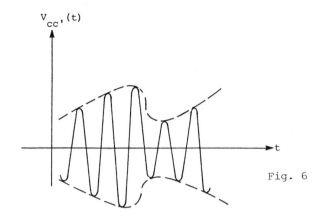

Fig. 6

The amplitude A of the sinusoidal signal must be chosen such that v(t) + A >>0. If A becomes too small or negative, the zero crossings of the signal v(t) + Acos ω_ct will not be periodic. Hence, in order to have periodic zero crossings, the condition v(t) + ωA >>0 must be satisfied. In this case, rectification is equivalent to multiplication by a square wave.

● **PROBLEM 9-7**

(a) By using amplitude modulation, demonstrate and explain the modulation of a sinusoidal carrier by a sinusoidal signal. Write an expression for the amplitude-modulated signal in the time-domain.
(b) Write the time-domain expression in another form so that it will show the steady-state sinusoidal components, and draw the spectrum of the AM signal.

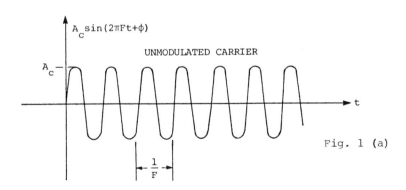

Fig. 1 (a)

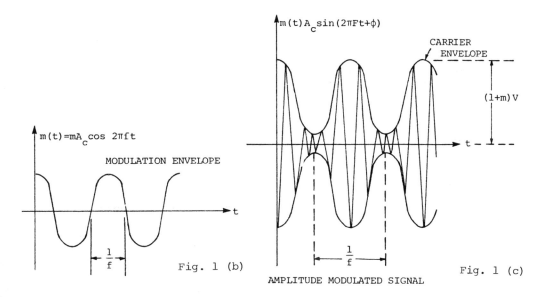

Fig. 1 (b)

Fig. 1 (c)

AMPLITUDE MODULATED SIGNAL

Solution:

(a) The process of amplitude-modulation is illustrated in Fig. (1). The figure provides a better understanding on how a sinusoidal carrier-V_c is modulated by another sinusoidal waveform m(t).

Note: the modulation index (m) must be chosen in the range $0 < m \leq 1$ in order to avoid distortion .

In Fig. (1A) and (1B), the carrier and the modulating waveform are expressed in the time domain as

$V_c = A_c \sin(2\pi Ft+\phi)$ and $m(t) = mA_c \cos(2\pi ft)$ respectively

where A_c = maximum carrier voltage
ϕ = carrier phase
m = modulation index

The resultant waveform in Fig. (1c) is one in which the carrier is modulated in amplitude. It is expressed in the time domain as

$$V(t) = A_c(1 + m \cos 2\pi ft) \sin(2\pi ft + \phi) \qquad (1)$$

(b) Now in order to show the steady-state sinusoidal components, some manipulation has to be done.

From equation (1) it follows that,

$$V(t) = A_c \sin(2\pi Ft + \phi) + mA_c \cos(2\pi ft)\sin(2\pi ft + \phi) \qquad (2)$$

Applying the trigonometric identity

$\cos B \sin A = \tfrac{1}{2} \sin(A+B) + \tfrac{1}{2} \sin(A-B)$

Eq. (2) becomes

$$V(t) = A_c \sin(2\pi Ft + \phi) + \frac{mA_c}{2}\sin\left[2\pi(F+f)t + \phi\right] + \frac{mA_c}{2}\sin\left[2\pi(F-f)t + \phi\right] \quad (3)$$

In Eq. (3), the three components forming the equation are the steady-state sinusoidal components which can be identified as: the carrier component $A_c \sin(2\pi Ft + \phi)$, the upper sideband $\frac{mA_c}{2}\sin\left[2\pi(F-f)t + \phi\right]$, and the lower sideband

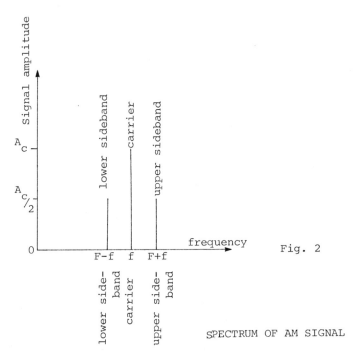

Fig. 2

SPECTRUM OF AM SIGNAL

$\frac{mA_c}{2}\sin\left[2\pi(F-f)t + \phi\right]$. The spectrum of the amplitude modulation signal is shown in Fig.(2).

• **PROBLEM 9-8**

Use an adder circuit and a multiplier to demonstrate amplitude modulation. With the information given below, determine the peak voltage of the carrier term and the product term of the amplitude modulated signal.
Given:
$V_1(t) = V_1 \sin \omega_1 t$ = modulating signal
$V_2(t) = V_2 \sin \omega_2 t$ = carrier signal
where $V_1 = 10v$, $V_2 = 5v$, $\omega_1 = 2\pi$, $\omega_2 = 2\pi \times 10^4$

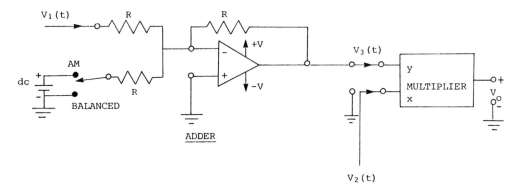

Solution: An adder-multiplier circuit is shown in the figure. The values of resistance R and the dc voltage supply are assumed to be 10KΩ and 5v, respectively.

Since the input voltages of the adder are $V_1(t) = -10\sin(2\pi t)$ and 5v (d-c), the resulting output is $V_3(t) = 5 = 10\sin(2\pi t)$. The output of the adder $V_3(t)$ is then multiplied by the carrier signal $V_2(t) = 5\sin(2\pi \times 10^4 t)$. The resulting output is

$$V_0 = \frac{V_2^2 \sin(\omega_2 t)}{10} + \frac{V_1 V_2}{10}(\sin \omega_2 t)(\sin \omega_1 t). \qquad (1)$$

Hence, the peak voltage for the carrier term is

$$\frac{V_2^2}{10} = 2.5v \ .$$

and the peak voltage for the product term is

$$\frac{V_1 V_2}{10} = 5v.$$

Note: the upper and lower sideband frequency components can be obtained by multiplying out Eq. (1).

• **PROBLEM 9-9**

Two signals are transmitted on the same carrier. With the aid of Fig. (1), show that the signals can be recovered by quadrature multiplexing.

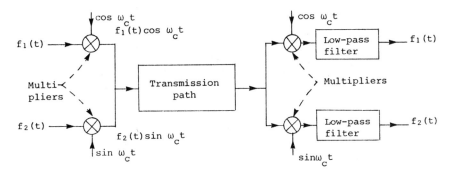

Solution: The received signal r(t) is

$$r(t) = f_1(t)\cos\omega_c t + f_2(t)\sin\omega_c t.$$

Multiply r(t) by $\cos\omega_c t$, to obtain

$$r(t)\cos\omega_c t = [f_1(t)\cos\omega_c t + f_2(t)\sin\omega_c t]\cos\omega_c t$$
$$= f_1(t)\cos^2\omega_c t + f_2(t)\cos\omega_c t \sin\omega_c t$$
$$= \tfrac{1}{2}f_1(t) + \tfrac{1}{2}[f_1(t)\cos\omega_c t + f_2(t)\sin 2\omega_c t]$$

The terms in the bracket can be filtered out by a low pass filter and $f_1(t)$ is then obtained. Similarly, $f_2(t)$ can be obtained from $r(t)\sin\omega_c t$. Therefore, this shows that signals can be recovered by a method known as quadrature multiplexing.

AM DEMODULATORS

• **PROBLEM** 9-10

A sinusoidal carrier c(t) is tone modulated by a co-sinusoidally varying signal m(t). Plot the frequency spectrum and determine the channel bandwidth.

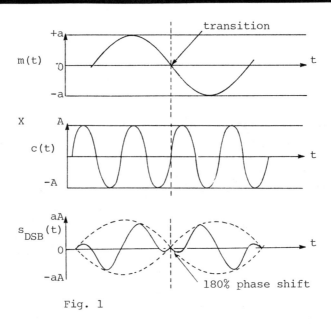

Fig. 1

Waveform structure-tone modulation of a sinusoidal carrier.

Solution: Let the modulating signal be

$$m(t) = a \cos \omega_m t, \quad \omega_m \ll \omega_c$$

Figure (A) shows the waveform m(t), the sinusoidal carrier c(t) and the resulting tone modulated carrier.
It can be seen that, at zero crossing, as shown by the vertical dotted line in Fig. (B), the tone modulated waveform experiences a 180° phase shift, resulting in the transformation of the amplitude excursions of m(t) into both amplitude variation and phase shift of c(t). In other words the resultant signal is modulated both in amplitude and phase and has two side band components, an upper side band and a lower side band. These side bands are the single tone (frequency) analogs of the continuous spectrum as shown in figure (2).

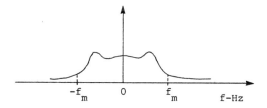

Fig. 2   Amplitude frequency spectrum-baseband information signal.

Considering the expanded form of the upper and lower side bands, the Fourier transform of the modulated signal can be evaluated as

$$\frac{1}{2\pi} \int_{-\infty}^{\infty} \delta(f_c \pm f_m) e^{j\omega t} d\omega = e^{\pm j(\omega_c \pm \omega_m)t} \quad (1)$$

(The above step follows from the sifting property of delta functions).

From equation (1), it can be inferred that the spectrum consists of four delta functions at $\pm(f_c \pm f_m)$ as shown in figure (c).

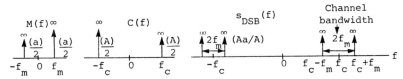

Fig. 3   Frequency spectrum tone modulation of a sinusoidal carrier.

The channel bandwidth can be directly determined from the above spectrum which reveals that for proper transmission, the channel bandwidth is twice the bandwidth of the modulating signal.
Thus, channel bandwidth = 2 × f_m.

• PROBLEM 9-11

A simple broadcasting system is represented in Fig. (1). If the Fourier transform of $f'(t) \leftrightarrow F'(\omega)$ and $f''(t) \leftrightarrow F''(\omega)$, where $F'(\omega)$ and $F''(\omega)$ are shown in Fig. (2), sketch the transform of $f(t)$ and $g(t)$ indicated in figure (1).

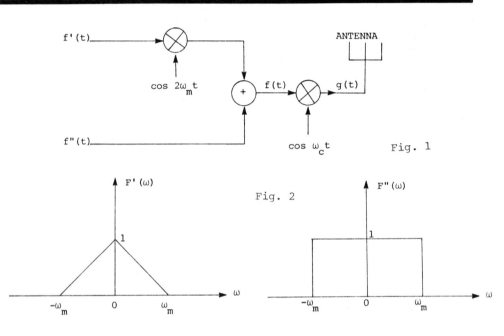

Fig. 1

Fig. 2

Solution:

With the aid of Fig. (1), an expression for $f(t)$ can be obtained.

Hence, $f(t) = f'(t)\cos 2\omega_m t + f''(t)$.

Now, since $F'(\omega)$ and $F''(\omega)$ are given, and knowing that
$$F\left[m(t)\cos \omega_x t\right] = \tfrac{1}{2}\left[M(\omega + \omega_x) + M(\omega - \omega_x)\right], \quad (1)$$
the transform of $f(t)$ is sketched in Fig. (3).

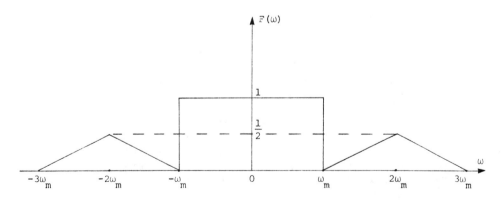

Fig. 3

Note: Equation (1) indicates that when a signal m(t) is multiplied by a sinusoidal waveform, the signal is shifted to the left and right by the angular frequency (ω_x), and with a decrease in amplitude by a factor ½.

Now to sketch the transform of g(t), apply Eq. (1) again and repeat the same procedure as above, but this time with g(t) = f(t)cos ω_xt.

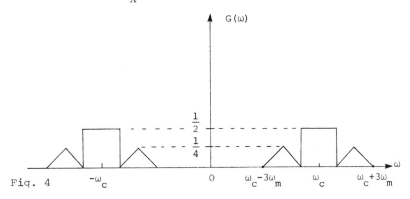

Fig. 4

Thus, shifting the waveform obtained in Fig. (3) to the left and to the right by ω_c, and with a decrease in amplitude by a factor ½, the sketch of G(ω) is obtained in Fig. (4)

• **PROBLEM 9-12**

A transistor square-law demodulator circuit and its input/output characteristics are shown in Fig.(A) and (B).

Show that the modulating signal is completely recovered at the output if a single-sideband signal, together with a carrier are applied to the input.

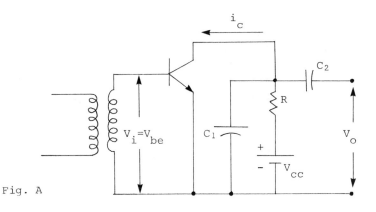

Fig. A

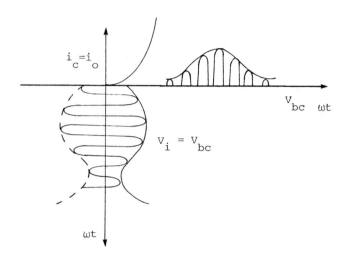

Solution: Let the input signal be
$$v_i(t) = \underbrace{v_m \cos(\omega_c + \omega_m)t}_{\text{SSB signal}} + \underbrace{v_c \cos \omega_c t}_{\text{carrier}} \tag{1}$$

For a nonlinear system, the relation between the output current (i_o) and the input voltage (v_i) is expressed as

$$i_o = a_o + a_1 v_i + a_2 v_i^2. \tag{2}$$

Substituting Eq. (2) into Eq. (1)

$$i_o(t) = a_o + a_1 \left[v_m \cos(\omega_c + \omega_m)t \ v_c \cos \omega_c t \right] +$$
$$a_2 \left[v_m \cos(\omega_c + \omega_m)t + v_c \cos \omega_c t \right]^2$$
$$= a_o + a_1 v_m \cos(\omega_c + \omega_m)t + a_1 v_c \cos \omega_c t + a_2 v_m^2 \cos^2(\omega_c + \omega_m)t$$
$$+ a_2 v_c^2 \cos^2 \omega_c t + 2 a_2 v_m v_c \cos(\omega_c + \omega_m)t \cos \omega_c t$$
$$= a_o + a_1 v_m \cos(\omega_c + \omega_m)t + a_1 v_c \cos \omega_c t + (a_2 v_m^2/2)\Big[1$$
$$+ \cos 2(\omega_c + \omega_m)t\Big] + (a_2 k_2^2/2)(1 + \cos 2\omega_c t)$$
$$+ a_2 v_c v_m \Big[\cos(2\omega_c + \omega_m)t + \cos \omega_m t\Big].$$

Notice that, after all the high frequency components and dc terms are filtered out or rejected, only the modulating signal remains.

Hence, $i_o(t) = a_2 v_m v_c \cos \omega_m t$.

Therefore, no distortion occurs, and the input signal is completely recovered.

BALANCED MODULATORS

• **PROBLEM** 9-13

A block diagram of the gated demodulator is shown in the figure below, where $G(t)$ is a periodic waveform given by the Fourier series expansion.

$$G(t) = \sum_{i=0}^{\infty} a_i \cos i\omega_c t$$

and $f_m(t) = f(t) \cos \omega_c t$.

Sketch the waveforms for $i = 0$, 1 and 2 at position (1) and determine the output signal at position (2).

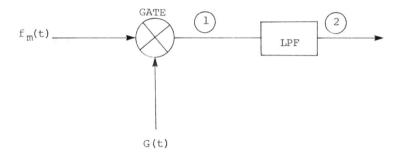

Solution: The signal at (1) is a product of $f_m(t)$ and $G(t)$.

Thus, $f_m(t) G(t) = f(t) \cos \omega_c t \sum_{i=0}^{\infty} a_i \cos i\omega_c t$

$$= f(t) \sum_{i=0}^{\infty} a_i \cos \omega_c t \cos i\omega_c t$$

$$= f(t) \sum_{i=0}^{\infty} a_i \tfrac{1}{2} \Big[\cos(i-1)\omega_c t + \cos(i+1)\omega_c t\Big] \quad (1)$$

Now, the waveforms for $i = 0$, 1 and 2 can be sketched by using Eq. (1) and substituting in appropriate values.

Hence, for

Notice that for $i > 1$, waveforms at position (1) are shifted to some multiple of ω_c, except for the term $\tfrac{1}{2} a_1 f(t)$, which after passing through the low-pass filter will be filtered out. Hence, the output signal at (2) is $\tfrac{1}{2} a_1 f(t)$ and therefore demodulation is accomplished.

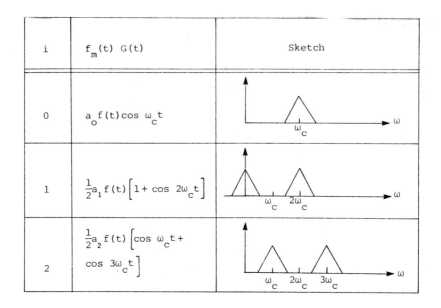

● **PROBLEM 9-14**

The figure below shows a transmission system with a two-tone modulator. If the system is operating with a 10 MHz carrier, evaluate the different transmitter output frequencies and, determine the bandwidth of operation for this system.

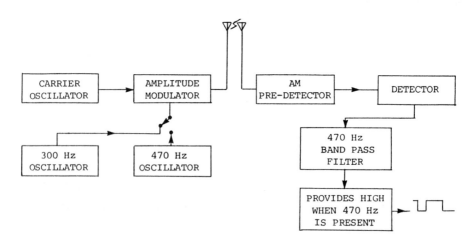

Solution: Since this is an amplitude modulation system, the output frequencies of the transmitter will be, 10 MHz ± 300 Hz corresponding to the 300 Hz modulating signal, and 10 MHz and 10 MHz ± 470 Hz corresponding to the 470 Hz modulating signal.

Since the maximum frequency deviation for the given transmitter is ± 470 Hz, the required bandwidth of operation will be 2 x 470 Hz = 940 Hz. Thus, a 1 kHz channel would be adequate.

This problem illustrates the advantage of a two-tone modulation system with regard to bandwidth utilization. It indicates the possibility of packing one hundred 1-kHz channels in the frequency spectrum from 10 Mhz to 10.1 MHz. Other advantages include the elimination of gain control problems at the receiver, and the reduction in ionospheric fading of the transmitted signal due to the presence of three different frequencies (carrier, USB and LSB).

• **PROBLEM 9-15**

A multiplier shown in the figure is used as a balanced modulator. The inputs to terminals X and Y are sinusoidal signals: $V_m(t)$ = modulating signal and $V_c(t)$ = carrier signal, as indicated in the figure.

(a) Write an expression for the output voltage, $V_o(t)$.

(b) Calculate the peak voltage and the frequency of the sum and the difference frequency components.

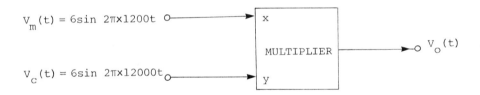

Solution:

(a) The output of the multiplier is the product of the input signals, $V_m(t)$ and $V_c(t)$. Thus, the output can be represented as follows:

$$V_o(t) = CXY = CV_{mp}V_{cp}(\sin 2\pi f_c t)(\sin 2\pi f_m t) \qquad (1)$$

where: $C \approx \frac{1}{10}$ is a constant scale factor,

V_{mp} and V_{cp} = the peak voltage of the modulating and the carrier signals.

f_m and f_c = the modulating frequency and the carrier frequency.

Substituting information given into Eq. (1), Vo(t) becomes

$$Vo(t) = \frac{6 \times 6}{10} \left[\sin 2\pi \times 12{,}000 t\right]\left[\sin 2\pi \times 1{,}200 t\right]$$

$$= 3.6 \left[\sin 2\pi \times 12{,}000 t\right]\left[\sin 2\pi \times 1{,}200 t\right] \qquad (2)$$

(b) In order to represent Eq. (2) in a commonly used form in communication, apply the trigonometric identity

$(\sin A)(\sin B) = \frac{1}{2}\left[\cos(A-B) - \cos(A+B)\right]$ to Eq. (2)

$$V_o(t) = \frac{3.6}{2}\cos 2\pi(f_c-f_m)t - \frac{3.6}{2}\cos 2\pi(f_c+f_m)t$$

$$= 1.8\cos 2\pi(12{,}000-1{,}200)t - 1.8\cos 2\pi(12{,}000+1{,}200)t$$

$$= 1.8\cos 2\pi(10{,}800)t - 1.8\cos 2\pi(13{,}200)t \qquad (3)$$

Therefore, from Eq. (3),

The peak values for both the sum and the difference frequency components are 1.8 volts,
where the sum frequency $(f_c + f_m) = 1.32$ kHz
and the difference frequency $(f_c - f_m) = 1.08$ kHz.

AMPLITUDE MODULATION RECEIVER

• **PROBLEM 9-16**

A balanced modulator circuit is shown in Fig.(1) and the output-versus-input characteristic of the diode is shown in Fig. (2). Write an expression for the output $V_o(t)$ indicated in the figure and show that $g(t)$ can be recovered from $V_o(t)$.

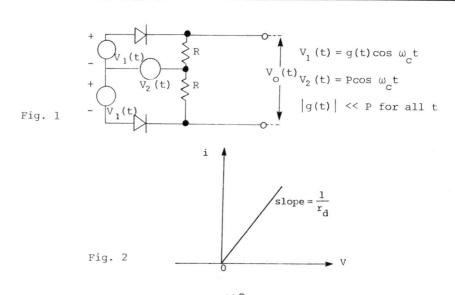

Fig. 1

$V_1(t) = g(t)\cos \omega_c t$
$V_2(t) = P\cos \omega_c t$
$|g(t)| \ll P$ for all t

Fig. 2

slope $= \frac{1}{r_d}$

Solution: Notice that the diodes of the circuit in Fig.(1) are functioning as a square chopper and are on and off every other half cycle. The relationship, for $|g(t)| \ll P$, between the diodes and the output voltage $V_o(t)$ is as follows:

	diodes operation	$V_o(t)$ output voltage
When $\cos \omega_c t$ is positive →	(SHORT) ON $\frac{2R}{R+r_d}$	$V_1(t)$
When $\cos \omega_c t$ is negative →	(OPEN) OFF	0

Thus, the output $V_o(t)$ is $\frac{2R}{R+r_d} V_1(t) = \frac{2R}{R+r_d} g(t) \cos \omega_c t$.

Now, to prove that $g(t)$ can be recovered from $V_o(t)$, we set $V_o(t) = (\frac{2R}{R+r_d}) g(t) \cos \omega_c t \, S(t)$ where $S(t)$ is a square wave or frequency ω_c.

Thus, using relationships:

$$f(t) \, p(t) \longleftrightarrow \tfrac{1}{2} F(\omega) + \frac{1}{\pi} \sum_{\substack{n=-\infty \\ (n=\text{odd})}}^{\infty} \frac{(-1)^{\frac{n-1}{2}}}{n} F(\omega - n\omega_c) \qquad (1)$$

$$V_o(t) \longleftrightarrow (\frac{2R}{R+r_d}) \left[\tfrac{1}{2} \bar{V}_1(\omega) + \frac{1}{\pi} \sum_{n=\text{odd}} \frac{(-1)^{(\frac{n-1}{2})}}{n} \bar{V}_1(\omega - n\omega_c) \right]$$

where $\bar{V}_1(\omega) = \tfrac{1}{2} \left[G(\omega + \omega_c) + G(\omega - \omega_c) \right]$

It can be seen that $V_o(t)$ contains terms $G(\omega)$ (when $n = \pm 1$) The remaining terms represent $G(\omega)$ shifted to $\pm 2\omega_c, \pm 3\omega_c \ldots$ etc.

Hence, $G(t)$ can be recovered from $V_o(t)$ by passing $V_o(t)$ through a low pass filter with a cut off frequency $= \pm \omega_c$. In conclusion, the balanced modulator circuit can be used as a synchronous detector.

• **PROBLEM 9-17**

A multiplier is shown in Fig. (1). The input to the x and y terminals of the multiplier is a sinusoidal waveform, $V_i(t) = 8 \sin 2\pi \times 12,000 t$.

(a) Write a time domain expression for the output

$V_o(t)$ indicated in Fig. (1).

(b) Calculate the dc and ac output voltage components if
V_{ip}(input peak voltage)=20V

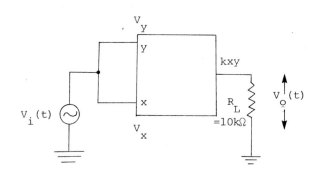

Fig. 1

Solution:

(a) The input-output voltage relationship of the multiplier can be written as
$$V_o = Cxy \quad (1)$$
where $C \simeq \frac{1}{10}$ is a constant scale factor.

Such a constant scale factor (C) is chosen such that for the best output performance, the input voltages x or y should not exceed ±10V with respect to the ground.

Now, expressing Eq. (1) in the time-domain and substituting the corresponding values,

$$V_o(t) = \frac{1}{10} V_x(t) V_y(t)$$

$$= \frac{1}{10} (8\sin 2\pi \times 12,000t)^2 \quad \text{since } V_x(t) = V_y(t)$$

$$= 6.4 (\sin^2 2\pi \times 12,000t).$$

Using trigonometric identity:
$$\sin^2 2\pi f t = \frac{1}{2} - \frac{\cos 2\pi(2f)t}{2}$$

Vo(t) becomes $V_o(t) = 6.4 \left[\frac{1}{2} - \frac{\cos 2\pi \times 24,000t}{2} \right]$

$$= 3.2 - 3.2\cos 2\pi \times 24,000t \quad (2)$$

In Eq. (2), the first term (3.2V) is the constant d.c. term which can be easily removed by a coupling capacitor. The second term is the ac component with 3.2V peak voltage.

(b) Inspect Eq. (2) carefully, the relation between the output dc and ac peak values, and the input peak voltage is as follows:

dc output voltage $= \frac{(V_{ip})^2}{20} = A_c$ peak output voltage with output frequency doubled the input frequency.

Hence, for $V_{ip} = 20V$ at 1.5 kHz, the output dc voltage $= \frac{(20)^2}{20} = 20V = A_c$ peak output voltage at a frequency $= 1.5$ kHz $\times 2$ kHz $= 3$ kHz.

• **PROBLEM 9-18**

Find the image frequency for the receiver as shown in Fig. (1). Draw a diagram to show how double conversion provides image frequency response rejection.

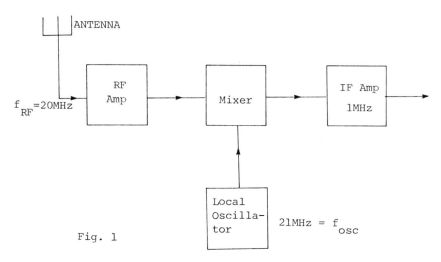

Fig. 1

Solution: By using the relationship $f_{IF} = (f_{osc} - f_{RF})$,

(21 - 20) MHz = 1 MHz.

Hence, the image frequency is the frequency which, when mixed with the local oscillator frequency, will produce the same 1 MHz output. Therefore, 22 MHz is chosen as the image frequency.

The response of the radio frequency and mixer tuned circuit is shown in Fig. (2).

Note that, due to the closeness of the image frequency (22 MHz) to the desired signal (20 MHz), the tuned RF and mixer circuits will partially attenuate the image frequency. However, if double conversion is used, the image frequency will be completely attenuated.

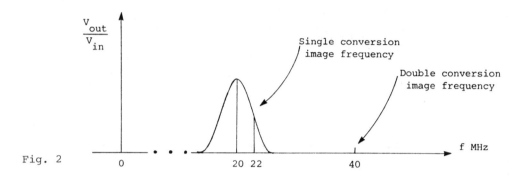

Fig. 2

• **PROBLEM 9-19**

A double conversion block diagram of a receiver is shown below. Find the image frequency for mixer (1).

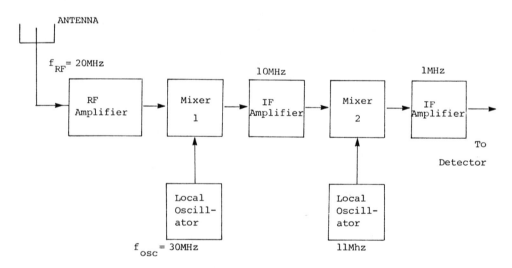

Solution: From the diagram, an input signal is furnished by the receiving antenna. Then it is amplified with a radio-frequency (RF) amplifier and passed on to a mixer. In the mixer, the modulated RF carrier is multiplied with a sinusoidal waveform generated by a local oscillator which operates at a frequency (f_{osc}), producing the sum and difference frequencies.

The sum-frequency will be rejected by the built-in filter inside the mixer. The difference-frequency carrier is called the intermediate frequency (IF) which can be calculated by using the relationship $f_{IF} = f_{osc} - f_{RF}$

Hence, $f_{IF} = f_{oc} - f_{RF}$
$= (30 - 20)$ MHz
$= 10$ MHz

Now, the image frequency is defined as the frequency which, when mixed with the 30 MHz local oscillator frequency, will produce the same 10 MHz output.

Therefore, the 40 MHz "image frequency" will satisfy the requirement, since $(40-30)$ MHz $= 10$ MHz.

SINGLE SIDEBAND MODULATION

• **PROBLEM** 9-20

An AM receiver is shown below.
(a) Write an expression for X(t) and find the output signal $\hat{m}(t)$ if the low-pass filter rejects double frequency components.
(b) Determine the maximum phase θ if 87% of the maximum possible value of the signal is recovered.
(c) Find minimum value of f_o so that the baseband signal m(t) with bandwidth (B) 4kHz will be recovered without distortion.

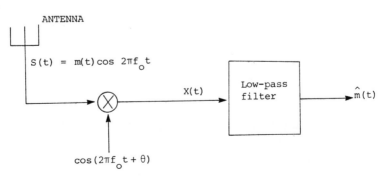

Solution:
(a) X(t) is the product of S(t) and $\cos_o(2\pi f\, t + \theta)$.

Hence, $X(t) = S(t)\cos(2\pi f_o t + \Theta)$
$= m(t)\cos(2\pi f_o t)\cos(2\pi f_o t + \Theta)$.

Applying trigonometric identity $\cos A \cos B = \tfrac{1}{2}\cos(A-B) + \tfrac{1}{2}\cos(A+B)$

$X(t) = m(t)\ \tfrac{1}{2}\cos\Theta + \tfrac{1}{2}\cos(4\pi f_o t + \Theta)$
$= \tfrac{1}{2}\cos\Theta\, m(t) + \tfrac{1}{2}m(t)\cos(4\pi f_o t + \Theta)$.

Since $X(t)$ is transmitted through a low-pass filter, the output signal is $\hat{m}(t) = \tfrac{1}{2}m(t)\cos\Theta$. Notice that the double-frequency component is rejected by the filter.

(b) From the information given, we can set up equations as follows:

$$\tfrac{1}{2}m(t)\cos\Theta = 87\%\ \left[(\tfrac{1}{2})m(t)\cos\Theta\right]_{\max\Theta} \qquad (1)$$

Rearranging Eq. (1): $\dfrac{(\tfrac{1}{2})m(t)\cos\Theta}{\left[\tfrac{1}{2}m(t)\cos\Theta\right]_{\max\Theta}} = 0.87$

Since for max. Θ, $\cos\Theta = 1$.

Therefore, $\dfrac{\tfrac{1}{2}m(t)\cos\Theta}{\tfrac{1}{2}m(t)(1)} = \cos\Theta = 0.87$.

As a result: $\Theta = \cos^{-1} 0.87 = 29.54°$.

(c) In order to recover $m(t)$ by filtering the product $v(t)\cos(2\pi f_o t + \Theta)$, the lowest frequency contained in the second term of $X(t)$ should be greater than the highest frequency contained in the first term.

Therefore, $2f_o - BW > BW$
$2f_o > 2BW$
$f_o > BW$ where $BW = 4$ kHz

Hence, min $f_o = 4$ kHz for distortionless signal.

● **PROBLEM 9-21**

Figure (1) below shows a schematic circuit of a frequency shifter. For each signal component at the output of the multiplier at point P, determine the frequency and the corresponding peak value. Given that

At input A:

Frequency	upper side-band $(f_c + f_m) = 1025$ kHz	carrier $f_c = 1020$ kHz	lower side-band $(f_c - f_m) = 1015$ kHz
Peak Amplitude	2V	5V	2V

where modulating frequency $f_m = 5$ kHz and $f_c =$ carrier frequency

And at input B: Is a sinusoidal waveform produced by a local oscillator of 6V peak voltage and with oscillating frequency f_o = 1445 kHz due to the intermediate frequency.

Solution: The output of the multiplier is a product of the A and B input components and can be expressed as

$$E_o = \frac{E_c^2}{10} \sin 2\pi f_c t + \frac{E_c E_m}{10} (\sin 2\pi f_c t)(\sin 2\pi f_m t)$$

          (Carrier term)   (Product term)

where 1/10 is a scale factor.

    E_c = peak carrier voltage

    E_m = peak voltage of the modulating signal

Since $(\sin A)(\sin B) = \frac{1}{2}[\cos(A-B) - \cos(A+B)]$,

$$E_o = \frac{E_c^2}{10} \sin 2\pi f_c t + \frac{E_c E_m}{20} \cos 2\pi(f_c - f_m)t$$

    (carrier term)   (lower side frequency)

$$- \frac{E_c E_m}{20} \cos 2\pi(f_c + f_m)t$$

    (upper side frequency)

The above expression denotes the sum and difference frequencies together with their peak amplitudes as a result of the product of the peak amplitude of the A input and the peak amplitude of the local oscillator frequency.

The table below shows the frequencies and corresponding peak values at the output of the multiplier.

INPUT FREQUENCIES AT A (kHz)	MULTIPLIER PEAK AMPLITUDE (Volts)	OUTPUT FREQUENCY (kHz)
1025	$\frac{2 \times 6}{20} = 0.6v$	1445+1025=2470 1445-1025=420
1020	$\frac{5 \times 6}{20} = 1.5v$	1445+1020=2465 1445-1020=425
1015	$\frac{2 \times 6}{20} = 0.6v$	1445+1015=2460 1445-1015=430

The multiplier is followed by a low-pass filter which passes only the three lower intermediate frequencies of 420 kHz, 425 kHz and 430 kHz. The other frequency components present at the input of the filter corresponding to the upper band intermediate frequencies of 2460, 2465 and 2470 are usually removed by the filter. Figure (2) shows the frequencies in the output signal prior to filtering.

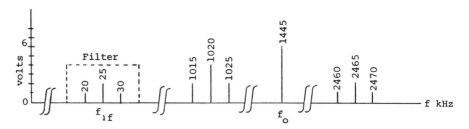

Fig. (2)   Frequency spectrum at the output of the multiplier.

From the frequency spectrum, it is seen that the 'A' input frequencies are transferred to different upper and lower intermediate frequencies, of which the desired intermediate frequency can be obtained by the use of a band-pass filter.

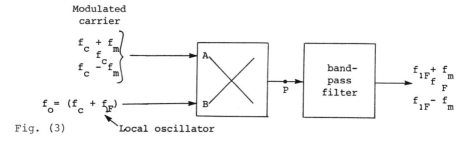

Fig. (3)

● **PROBLEM** 9-22

A LC tuned circuit with output/input-versus-frequency characteristic is shown in the figure below.

(a) If the resonant frequency (f_R) of the tuned circuit is changed from (f_{R_1}) 400 kHz to (f_{R_2}) 1000 kHz, determine the maximum and minimum values of the capacitor. Note: L = constant = 8.5 µH.

(b) Calculate the quality factor Q, provided that a practical 10-kHz bandwidth occurs at f_R = 900 kHz.

(c) A tuned radio frequency receiver is to be designed with the above tuned circuit, determine the bandwidth, BW, of this receiver at f_{R_1} and f_{R_2}.

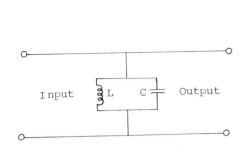
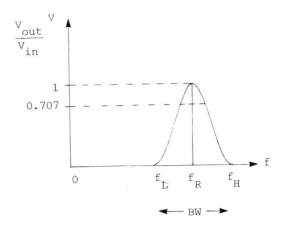

Solution:

(a) From the figure, since the output is maximum at the resonant frequency, a useful and accurate relationship between the resonant frequency (f_R) and capacitance is given by

$$f_R = \frac{1}{2\pi\sqrt{LC}} \quad \text{where} \quad \frac{1}{\sqrt{LC}} = \omega_R \tag{1}$$

Now, given L, f_{R_1} and f_{R_2}, the maximum and minimum values of the capacitor can be found by using Eq. (1). Since, at f_{R_2} = 1000 kHz, C = C_{min} and at f_{R_1} = 400 kHz, C = C_{max}.

Substituting given values into Eq. (1)

$$f_{R_2} = \frac{1}{2\pi\sqrt{LC_{min}}} \quad \Big| \quad f_{R_1} = \frac{1}{2\pi\sqrt{LC_{max}}}$$

$$1000 \times 10^3 = \frac{1}{2\pi\sqrt{10\times 10^{-6}C_{min}}} \quad \Big| \quad 400 \times 10^3 = \frac{1}{2\pi\sqrt{10\times 10^{-6}C_{max}}}$$

Therefore, C_{min} = 2.54 × 10^{-9} F = 2.54 nF

and C_{max} = 1.58 × 10^{-8} F = 15.8 nF

Hence, the values ranging from 2.54 nF to 15.8 nF are values required for the capacitor.

(b) The quality factor Q provides a measure of the tuned circuit's frequency selectivity. As Q increases, the bandwidth will narrow down. A useful relationship between Q, BW and f_R is

$$Q = \frac{f_R}{BW} \tag{2}$$

Therefore, with BW = 10 kHz and f_R = 900 kHz

$$Q = \frac{f_R}{BW} = \frac{900 \times 10^3}{10 \times 10^3} = 90 \ .$$

Note: a circuit with high Q discriminates sharply but a low-Q circuit is relatively unselective.

(c) Using Eq. (2) with Q = 90, the bandwidth of the receiver at 400 kHz (f_{R_1}) and 1000 kHz (f_{R_2}) can be calculated.

Hence, $BW_1 = \dfrac{f_{R_1}}{Q} = \dfrac{400 \text{ kHz}}{90} = 4.44 \text{ kHz}$

and $BW_2 = \dfrac{f_{R_2}}{Q} = \dfrac{1000 \text{ kHz}}{90} = 11.11 \text{ kHz}.$

● **PROBLEM 9-23**

For an AM system with a modulation index m = 0.6, compare the AM signal power to the SSB signal power.

Solution: Equations for the power of a modulated wave dissipated in a resistance R are:

Carrier power = $V_c^2/(2R)$

Sideband power = $V_c^2 m^2/(8R)$

Total sideband power = $V_c^2 m^2/(4R)$ where V_c is the carrier peak voltage.

Now, let P_1 = sideband power = $V_c^2 m^2/(8R)$

P_2 = power in both sidebands and carrier

$= \dfrac{V_c^2}{2R} + \dfrac{V_c^2 m^2}{4R}$

$= \dfrac{V_c^2 \, m^2 (2 + 4/m^2)}{8R}$

Therefore, $\dfrac{P_1}{P_2} = \dfrac{\left(\dfrac{V_c^2 m^2}{8R}\right)}{\dfrac{V_c^2 m^2 (2+4/m^2)}{8R}} = \dfrac{1}{(2 + 4/m^2)}$

For m = 0.6, $\dfrac{P_1}{P_2} = \dfrac{9}{118}$

● **PROBLEM 9-24**

A SSB-SC signal is given as follows:

$$v(t) = \sum_{n=1}^{N} \left[\cos \omega_c t \cos(\omega_n t + \theta_n) - \sin \omega_c t \sin(\omega_n t + \theta_n)\right]$$

where $\omega_c \gg \omega_n$.

(a) Determine whether this is the upper or the lower sideband.
(b) Write expressions for the missing sideband and the total DSB-SC signal.

Solution: Apply the trigonometric identities:
$$\cos A \cos B = \tfrac{1}{2}\cos(A-B) + \tfrac{1}{2}\cos(A+B)$$
and $$\sin A \sin B = \tfrac{1}{2}\cos(A-B) - \tfrac{1}{2}\cos(A+B)$$
to $v(t)$, and let $v(t) = v_1(t)$

$$v_1(t) = \sum_{n=1}^{N} \left[\cos \omega_c t \cos(\omega_n t + \theta_n) - \sin \omega_c t \sin(\omega_n t + \theta_n)\right]$$

$$= \sum_{n=1}^{N} \left\{\tfrac{1}{2}\cos\left[(\omega_c - \omega_n)t - \theta_n\right] + \tfrac{1}{2}\cos\left[(\omega_c + \omega_n)t + \theta_n\right] - \tfrac{1}{2}\cos\left[(\omega_c - \omega_n)t - \theta_n\right] + \tfrac{1}{2}\cos\left[(\omega_c + \omega_n)t + \theta_n\right]\right\}$$

$$= \sum_{n=1}^{N} \cos\left[(\omega_c + \omega_n)t + \theta_n\right].$$

Hence, the result is the sum frequency component which constitutes the upper sideband of an SSB-SC signal.

(b) An expression for the lower sideband can be written as

$$v_2(t) = \sum_{n=1}^{N} \left[\cos \omega_c t \cos(\omega_c t + \theta_n) + \sin \omega_c t \sin(\omega_n t + \theta_n)\right]$$

Simplifying as in part (a)

$$v_2(t) = \sum_{n=1}^{N} \cos\left[(\omega_c - \omega_n)t - \theta_n\right].$$

As a result, for the total DSB-SC signal:

$$v_{12}(t) = v_1(t) + v_2(t)$$

$$= \sum_{n=1}^{N} \cos\left[(\omega_c + \omega_n)t + \theta_n\right] + \sum_{n=1}^{N} \cos\left[(\omega_c - \omega_n)t - \theta_n\right]$$

$$= \sum_{n=1}^{N} \cos \omega_c t \cos(\omega_n t + \theta_n)$$

$$= \left[2 \sum_{n=1}^{N} \cos(\omega_n t + \theta_n)\right] \cos \omega_c t.$$

SQUARE LAW MODULATION

• **PROBLEM 9-25**

$$v(t) = [A + m(t)]\cos \omega_c t - \hat{m}(t)\sin \omega_c t \qquad (1)$$

is a single-sideband modulated signal with carrier term added.

(a) Write $v(t)$ in the form:
$$v(t) = A(t)\cos[\omega_c t + \theta(t)],$$
showing that the signal has a time-varying phase and a time-varying envelope. Write expressions for $A(t)$ and $\theta(t)$ in terms of A, $m(t)$ and $\hat{m}(t)$.

(b) If $v(t)$ is applied to an envelope detector and a low pass filter, give the conditions for a good approximation to $m(t)$.

<u>Solution</u>: Applying a trigonometric identity to $\cos[\omega_c t + \theta]$,
$\cos[\omega_c t + \theta] = \cos \theta \cos \omega_c t - \sin \theta \sin \omega_c t$.
Therefore, $v(t)$ contains both $\sin \omega_c t$ and $\cos \omega_c t$ terms.
Now, if we can normalize so that the coefficient of $\cos \omega_c t$ is $\cos \theta$ and the coefficient of $\sin \omega_c t$ is $\sin \theta$, we will obtain $v(t)$ in the desired form.

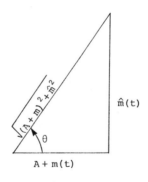

Using Pythagoras' theorem,

$$[A + m]^2 + [\hat{m}]^2 = H^2$$
$$H = \sqrt{[A + m]^2 + \hat{m}^2}$$

Multiply and divide Eq. (1) by H:

$$v(t) = \left[\underbrace{\frac{A + m(t)}{\sqrt{(A+m)^2 + \hat{m}^2}}}_{\cos\theta} \cos\omega_c t - \underbrace{\frac{\hat{m}(t)}{\sqrt{(A+m^2) + \hat{m}^2}}}_{\sin\theta} \sin\omega_c t\right] \sqrt{(A+m)^2 + \hat{m}^2}$$

v(t) is now in the form

$$[\cos\theta\cos\omega_c t - \sin\theta\sin\omega_c t] \sqrt{(A+m)^2 + \hat{m}^2} .$$

thus, $v(t) = \underbrace{\sqrt{(A+m)^2 + \hat{m}^2}}_{A(t)} \cos(\omega_c t + \theta)$

where $\theta \longleftrightarrow \theta(t) = \tan^{-1}\left[\frac{\hat{m}(t)}{A+m(t)}\right]$.

b) The envelope of v(t) is A(t).
Factoring out A,

$$A(t) = A\sqrt{(1 + \frac{m}{A})^2 + (\frac{\hat{m}}{A})^2} .$$

For $x \ll 1$, use the approximation $(1+x)^{\frac{1}{2}} \cong 1 + \frac{1}{2}x$

$$A(t) \cong A\left[1 + \frac{1}{2}(\frac{2m}{A} + \frac{m^2}{A^2} + \frac{\hat{m}^2}{A^2})\right]$$

The squared terms can be neglected under the condition $\frac{m}{A} \ll 1$ and $\frac{\hat{m}}{A} \ll 1$, hence, $A(t) \simeq A + m(t)$.

An envelope detector with low-pass filter will have A(t) as output. Under the condition $m(t) \ll A$ and $\hat{m}(t) \ll A$, this output will be approximately proportional to m(t) after the dc bias term A is removed.

• **PROBLEM 9-26**

$$v(t) = \sum_{n=1}^{N} \left[\cos\omega_c t\cos(\omega_n t+\theta_n) - \sin\omega_n t\sin(\omega_n t+\theta_n)\right]$$

is a SSB signal.

(a) If the signal is multiplied by $\cos\omega_c t$ and then passed through a low-pass filter with cutoff frequency f_o, where $f_m < f_o < 2f_c$, determine the output waveform of the filter. (Note: f_m = modulating frequency, f_c = carrier frequency.)

(b) If the multiplying signal has a phase offset θ, what is the output of the low-pass filter?

(c) If the multiplying signal has an angular frequency offset $\Delta\omega$, determine the recovered signal.
(Note: $\Delta\omega << \omega$)

Solution:

(a) $v(t) \times \cos\omega_c t$

$$= \sum_{n=1}^{N} \left[\cos\omega_c t \cos(\omega_n t + \theta_n) - \sin\omega_c t \sin(\omega_n t + \theta_n)\right] \cos\omega_c t$$

$$= \left\{\sum_{n=1}^{N} \left\{\tfrac{1}{2}\cos\left[(\omega_c - \omega_n)t - \theta_n\right] + \tfrac{1}{2}\cos\left[(\omega_c + \omega_n)t + \theta_n\right] - \tfrac{1}{2}\cos\left[(\omega_c - \omega_n)t - \theta_n\right]\right.\right.$$
$$\left.\left. + \tfrac{1}{2}\cos\left[(\omega_c + \omega_n)t + \theta_n\right]\right\}\right\} \cos\omega_c t$$

$$= \left\{\sum_{n=1}^{N} \cos\left[(\omega_c + \omega_n)t + \theta_n\right]\right\} \cos\omega_c t$$

$$= \sum_{n=1}^{N} \left\{\tfrac{1}{2}\cos\left[(2\omega_c + \omega_n)t + \theta_n\right] + \tfrac{1}{2}\cos(\omega_n t + \theta_n)\right\}$$

$$= \sum_{n=1}^{N} \tfrac{1}{2}\cos\left[(2\omega_c + \omega_n)t + \theta_n\right] + \sum_{n=1}^{N} \tfrac{1}{2}\cos(\omega_n t + \theta_n).$$

Since the double frequency component is rejected, the output waveform of the filter is $\sum_{n=1}^{N} \tfrac{1}{2}\cos(\omega_n t + \theta_n)$. The modulation is completely recovered.

(b) In SSB, the phase offset $\cos(\omega_c t + \theta)$ of the carriers produce a phase change but not an amplitude change.

Thus, $v(t) \cos(\omega_c t + \theta)$

$$= \sum_{n=1}^{N} \cos\left[(\omega_c + \omega_n)t + \theta_n\right] \cos(\omega_c t + \theta)$$

$$= \tfrac{1}{2} \sum_{n=1}^{N} \cos\left[(2\omega_c+\omega_n)t+(\theta_n+\theta)\right] + \tfrac{1}{2}\sum_{n=1}^{N} \cos\left[\omega_n t+(\theta_n-\theta)\right].$$

Hence, after passing through a low-pass filter, the output is $\tfrac{1}{2} \sum_{n=1}^{N} \cos\left[\omega_n t+(\theta_n-\theta)\right]$.

(c) If the local oscillator carrier has an angular frequency offset $\Delta\omega$, and is of the form $\cos(\omega_c+\Delta\omega)t$, then in SSB, the amplitude remains fixed, but the frequency of the recovered signal is in error by an amount $\Delta\omega$.

Thus, $v(t)\cos(\omega_c+\Delta\omega)t$

$$= \sum_{n=1}^{N} \cos\left[(\omega_c+\omega_n)t+\theta_n\right]\cos(\omega_c+\Delta\omega)t$$

$$= \tfrac{1}{2}\sum_{n=1}^{N} \cos\left[(2\omega_c+\omega_n+\Delta\omega)t+\theta_n\right] + \tfrac{1}{2}\sum_{n=1}^{N}\cos\left[(\omega_n-\Delta\omega)t+\theta_n\right]$$

The recovered signal is $\tfrac{1}{2}\sum_{n=1}^{N}\cos\left[(\omega_n-\Delta\omega)t+\theta_n\right]$.

• **PROBLEM 9-27**

Given the envelope modulated signal
$$y(t) = A\left[1+2\alpha\sin\omega_A t + \alpha\cos 2\omega_A t\right]\cos\omega_c t$$
where ω_c is the carrier frequency, ω_A and $2\omega_A$ are two modulating signal frequency components.

(a) What is the largest value of α for which envelope detection could be performed on $y(t)$ to recover the modulating signal without distortion?

(b) If $\alpha = 0.2$, what would be the percentage saving in power transmitted if instead, a suppressed carrier version,

$A\left[2\alpha\sin\omega_A t + \alpha\cos 2\omega_A t\right]\cos\omega_c t$, was sent?

Solution:

(a) The most negative value of $2\alpha\sin\omega_A t + \alpha\cos 2\omega_A t$ is

-3α (see sketch).

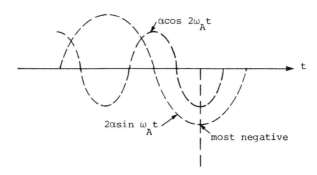

The condition for no distortion is $3\alpha < 1$,

hence, $\alpha_{max} = \frac{1}{3}$

(b) Using the trigonometric formula for sin acos b and cosa cosb in terms of sum and difference angles,

$$y(t) = A\cos \omega_c t + \alpha A \sin(\omega_c + \omega_A)t$$
$$= \alpha A \sin(\omega_c - \omega_A)t + \frac{\alpha A}{2}\cos(\omega_c + \omega_A)t$$
$$+ \frac{\alpha A}{2}\cos(\omega_c - \omega_A)t.$$

Powers in the sinusoids at different frequencies add: where $\alpha = 0.2$,

$$P = \frac{A^2}{2} + \frac{(\alpha A)^2}{2} + \frac{(\alpha A)^2}{2} + \frac{1}{2}\left(\frac{\alpha A}{2}\right)^2 + \frac{1}{2}\left(\frac{\alpha A}{2}\right)^2$$

$$P = \underbrace{\left(\frac{A^2}{2}\right.}_{\text{(carrier)}} + \underbrace{\left.0.05A^2\right)}_{\text{(sidebands)}} = 0.55A^2 .$$

The percentage saving by suppressing the carrier term is thus,

$$\frac{A^2/2}{0.55A^2} \times 100 = 90.9 \text{ percent}$$

● **PROBLEM 9-28**

A square-law diode modulator is shown in Fig. (1). The input voltage is $v_i(t) = m(t) + \cos 2\pi f_c t$ (1) and is assumed to be small. Write an expression for the output voltage $v_o(t)$ and sketch the spectral density of $v_o(t)$.

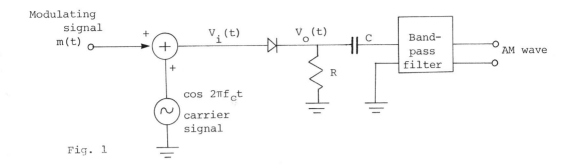

Fig. 1

Solution: The output voltage across the resistor (R) can be obtained using power series of the form $v_o = \sum_{n=1}^{\infty} a_1 v_i^n$. (2)

Since the input voltage is small, we can approximate the power series using only the first two terms.

$$v_o = a_1 v_i + a_2 v_i^2 .$$ (3)

Now, substituting Eq. (1) into Eq. (3) in the time domain, the output $v_o(t) = a_1 \left[m(t) + \cos 2\pi f_c t \right] + a_2 \left[m(t) + \cos 2\pi f_c t \right]^2$

$= a_1 m(t) + a_1 \cos 2\pi f_c t + a_2 m(t)^2 + a_2 m(t) \cos 2\pi f_c t + a_2 \cos^2 2\pi f_c t$

$= a_1 m(t) + a_2 m^2(t) + a_2 \cos^2 2\pi f_c t + a_1 \left[1 + \frac{2a_2}{a_1} m(t) \right] \cos 2\pi f_c t$ (4)

With some substitution, we can express Eq. (4) as a recognizable amplitude modulated waveform.

Hence, let $A_c \equiv a_1$, $n \equiv 2a_2/a_1$ and $\cos^2 \theta = 1 + \frac{\cos 2\theta}{2}$,

Eq. (4) becomes

$$v_o(t) = A_c m(t) + a_2 m^2(t) + \frac{a_2}{2}(1 + \cos 4\pi f_c t) + A_c \left[1 + n m(t) \right] \cos 2\pi f_c t .$$

Now, in order to sketch the spectral density of $v_o(t)$, for square-law modulation, the carrier frequency (f_c) must be chosen such that $f_c > 3\omega$ where ω is the bandwidth of the modulating signal.

The spectral diagram of $v_o(t)$ is shown in Fig. (2)

Note: In order to filter out the amplitude modulated signal without distortion, the bandwidth for the band-pass filter is chosen to be 2ω centered at f_c.

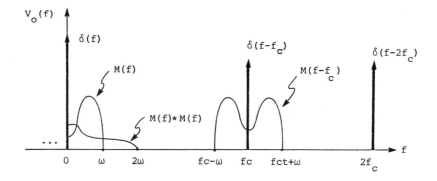

NOTE: $V_o(f)$ is symmetric about the vertical axis.   Fig. 2

• **PROBLEM 9-29**

Show that, in amplitude modulation, use of a square-law demodulator to recover the baseband signal will result in distortion of the original signal.

Solution: First, assume that for a square-law device, the relationship between the input (x) and output (y) is $y = Kx^2$ where K is a constant.

Let the applied signal $x = A_o + A_c \left[1 + m(t) \right] \cos \omega_c t$

where A_o = constant dc term, A_c = amplitude of unmodulated signal, and $m(t)$ = baseband signal.

Then, the output is:

$$y = Kx^2$$
$$= K \left(A_o + A_c \left[1 + m(t) \right] \cos \omega_c t \right)^2$$
$$= K \left(A_o^2 + 2A_o A_c \left[1+m(t) \right] \cos \omega_c t + A_c^2 \left[1+m(t) \right]^2 \cos^2 \omega_c t \right).$$

Dropping dc terms and considering that a low-pass filter, at the output, will reject spectral components near ω_c and $2\omega_c$.

Hence, $y = K \left(A_o^2 + 2A_o A_c \left[1+m(t) \right]^2 \cos \omega_c t + A_c^2 \left[1+m(t) \right]^2 \cos^2 \omega_c (t) \right)$

$$= K A_c^2 \left[1 + m(t) \right]^2 \left[\frac{1 + \cos 2\omega_c(t)}{2} \right]$$

536

$$= \frac{K A_c^2}{2} \underbrace{\left[1 + 2m(t) + m^2(t)\right]}_{\text{(dc term)}} \underbrace{\left[1 + \cos 2\omega_c(t)\right]}_{\text{(dc term) } (2\omega_c \text{ term})}$$

$$= K A_c^2 \left[m(t) + \tfrac{1}{2}m^2(t)\right].$$

Looking at the result, not only $m(t)$ is recovered but $m^2(t)$. As a result distortion will occur. However, if $\tfrac{1}{2}m^2(t) \ll |m(t)|$, the small distortion is acceptable.

• **PROBLEM** 9-30

A function $e_i(t) = \sin 2\omega_1 t + \cos 5\omega_1 t + \cos \omega_c t$ is applied to a non-linear device having characteristic

$$e_o(t) = a_1 e_i(t) + a_2 \left[e_i(t)\right]^2.$$

Assume $\omega_c \gg 5\omega_1$.

(a) List the frequency components present in the output $e_o(t)$.

(b) Indicate which frequencies should be eliminated by a filter to obtain a double-sideband AM signal, where ω_c is the carrier frequency and $\sin 2\omega_1 t + \cos 5\omega_1 t$ is the modulating signal.

Solution: (a) Substituting $e_i(t)$ in the given equation, the following terms appear in $e_o(t)$ after squaring:

(1) $\sin 2\omega_1 t$
(2) $\cos 5\omega_1 t$ } from the linear term
(3) $\cos \omega_c t$
(4) $\sin^2 2\omega_1 t$
(5) $\cos^2 5\omega_1 t$ } square
(6) $\cos^2 \omega_c t$
(7) $\sin 2\omega_1 t \cos 5\omega_1 t$ } cross product terms
(8) $\sin 2\omega_1 t \cos \omega_c t$ } resulting from squaring $e_i(t)$.
(9) $\cos 5\omega_1 t \cos \omega_c t$

Now we can make use of the trigonometric formula:
$\sin a \cos b = \tfrac{1}{2}\sin(a+b) + \tfrac{1}{2}\sin(a-b);$ $\sin^2 x = \tfrac{1}{2} - \tfrac{1}{2}\cos 2x$
$\cos a \cos b = \tfrac{1}{2}\cos(a+b) + \tfrac{1}{2}\cos(a-b);$ $\cos^2 x = \tfrac{1}{2} + \tfrac{1}{2}\cos 2x$

The frequencies in the various terms are then as follows:

term   angular frequencies

(1) $2\omega_1$
(2) $5\omega_1$
(3) ω_c
(4)   d.c., $4\omega_1$
(5)   d.c., $10\omega_1$
(6)   d.c., $2\omega_c$
(7) $3\omega_1, 7\omega_1$ ⎫
(8) $\omega_c-2\omega_1, \omega_c+2\omega_1$ ⎬ sum and difference frequencies
(9) $\omega_c+5\omega_1, \omega_c-5\omega_1$ ⎭

An enumeration of the frequencies present is: $\overset{x}{2\omega_1}$, $\overset{x}{5\omega_1}$, ω_c, $\overset{x}{4\omega_1}$, $\overset{x}{10\omega_1}$, $\overset{x}{2\omega_c}$, $\overset{x}{3\omega_1}$, $\overset{x}{7\omega_1}$, $\omega_c-2\omega_1$, $\omega_c+2\omega_1$, $\omega_c-5\omega_1$, $\omega_c+5\omega_1$.

(b)   A double-sideband signal should have the form:

$$(A + B\sin 2\omega_1 t + B\cos 5\omega_1 t)\cos \omega_c t =$$
$$\underbrace{A\cos \omega_c t}_{\text{term (3)}} + \underbrace{B\sin 2\omega_1 t\cos \omega_c t}_{\text{term (8)}} + \underbrace{B\cos 5\omega_1 t\cos \omega_c t}_{\text{term (9)}}$$

A check will show that terms (8) and (9) do have the same amplitude coefficient. If the frequencies in the enumeration with an x above them are eliminated by a filter, the result will be the desired double sideband signal.

CHAPTER 10

FREQUENCY MODULATION SYSTEMS (FM)

ANGLE MODULATION

• **PROBLEM 10-1**

The signal y(t) is a two-tone angle-modulated wave given by

$$y(t) = A\cos\left[2\pi(10^6)t + \beta_1\sin2\pi(200)t + \beta_2\sin2\pi(125)t\right]$$

with carrier frequency $f_c = 10^6$ Hz. Find the sideband frequencies nearest to the carrier.

Solution: Euler's identity can be used to rewrite y(t). Euler's identity states that

$$(\cos\theta + j\sin\theta) = e^{j\theta} \quad \text{or} \quad \cos\theta = \text{Re}\left[e^{j\theta}\right].$$

Therefore,

$$y(t) = A\text{Re}\left[e^{j[2\pi(10^6)t+\beta_1\sin2\pi(200)t+\beta_2\sin2\pi(125)t]}\right]$$

$$= A\text{Re}\left[e^{j\omega_c t}\, e^{j\beta_1\sin2\pi(200)t}\, e^{j\beta_2\sin2\pi(125)t}\right] \quad (1)$$

where $\omega_c = 2\pi(10^6)$ = carrier angular frequency in radians/second. Each of the last two terms in eq.(1) is periodic, and so, can be written as a Fourier series. The first term in eq.(1) has a fundamental frequency of 200 Hz and its nth harmonic is 200 n. The second term has a fundamental frequency of 125 Hz and the kth harmonic is 125k. Thus,

$$y(t) = A\text{Re}\left\{e^{j\omega_c t}\left[\sum_{n=-\infty}^{\infty} C_n e^{j2\pi(200n)t}\right]\left[\sum_{k=-\infty}^{\infty} D_k e^{j2\pi(125k)t}\right]\right\}$$

539

The (k,n) term is a sinusoid at the frequency

$$f_c + 200n + 125k$$ where the ranges of values for k and n are:

$$-\infty \leq k \leq \infty$$

$$-\infty \leq n \leq \infty .$$

The 200n terms are: $0, \pm200, \pm400, \pm600, \pm800, \ldots$. The 125k terms are: $0, \pm125, \pm250, \pm375, \pm500, \pm625, \pm750, \ldots$. Notice that the smallest magnitude difference between a 200n term and a 125k term (other than zero) is 25 Hz for $n = \pm3$ and $k = \pm5$ (e.g. ±600 and ±625). Thus, the sidebands nearest to the carrier are at $(10^6 \pm 25)$Hz.

• **PROBLEM 10-2**

An angle-modulated signal $v(t)$ is given as

$$v(t) = A \cos\left[\omega_c t + 2\cos60\pi t + 5\cos40\pi t\right]$$

with carrier frequency $f_c = \dfrac{\omega_c}{2\pi}$ Hertz.

(A) Find the maximum phase deviation in radians.

(B) At $t = \dfrac{1}{30}$ second, find the instantaneous frequency deviation in Hertz.

Solution: (A) An angle-modulated signal $x(t)$, can be written in the form $x(t) = A \cos[\omega_c t + \theta(t)]$ where $\theta(t)$ explicitly represents the phase deviation in radians. For this problem, $\theta(t) = 2\cos60\pi t + 5\cos40\pi t$. Thus, the maximum value of $\theta(t) = 2(1) + 5(1) = 7$, occuring at $t = 0$. Hence, the maximum phase deviation is 7 radians.

(B) For the instantaneous frequency deviation (f_d) in Hertz, differentiate the radian phase deviation with respect to time and divide by 2π radians,

$$f_d = \frac{1}{2\pi} \frac{d\theta(t)}{dt}$$

Therefore,

$$f_d = \frac{1}{2\pi}\left[\frac{d}{dt}(2\cos60\pi t + 5\cos40\pi t)\right]$$

$$= \frac{1}{2\pi}(-120\pi\sin60\pi t - 200\pi\sin40\pi t)$$

$$= -60\sin60\pi t - 100\sin40\pi t .$$

For $t = \dfrac{1}{30}$ second,

$$f_d = -60\sin 2\pi - 100\sin \frac{4}{3}\pi$$

$$= 0 - 100\left(\frac{-\sqrt{3}}{2}\right)$$

$$= 50\sqrt{3}.$$

Thus, $f_d = 86.6$ Hertz.

● **PROBLEM 10-3**

Find an expression for the instantaneous amplitude and phase deviation of the angle-modulated signal

$$v(t) = \cos\omega_c t - \beta\sin\omega_c t \sin\omega_m t$$

where ω_c is the carrier's angular frequency.

Solution: An angle-modulated signal $x(t)$ can be written in the form $x(t) = a(t)\cos[\omega_c t + \theta(t)]$ where $\theta(t)$ explicitly represents the instantaneous phase deviation and $a(t)$ the instantaneous amplitude.

In order to get $v(t)$ in the form $a(t)\cos[\omega_c t + \theta(t)]$, first use Euler's identity to write

$$a(t)\cos[\omega_c t + \theta(t)] = \text{Re}\left[a(t)e^{j\theta(t)} e^{j\omega_c t}\right].$$

Similarly,

$$\cos\omega_c t - \beta\sin\omega_c t \sin\omega_m t = \text{Re}\left[e^{j\omega_c t}\right] - \beta\text{Re}\left[(-j)\sin\omega_m t e^{j\omega_c t}\right]$$

$$= \text{Re}\left[e^{j\omega_c t} + j\beta\sin\omega_m t\, e^{j\omega_c t}\right]$$

$$= \text{Re}\left[(1 + j\beta\sin\omega_m t)e^{j\omega_c t}\right].$$

This can be verified by multiplying out the complex numbers and taking the real part of the result.

Hence,

$$(1+j\beta\sin\omega_m t)e^{j\omega_c t} = (1+j\beta\sin\omega_m t)(\cos\omega_c t + j\sin\omega_c t)$$

$$= \cos\omega_c t - \beta\sin\omega_m t \sin\omega_c t + j\beta\sin\omega_m t\cos\omega_c t + j\sin\omega_c t$$

where the real part is $\cos\omega_c t - \beta\sin\omega_m t\sin\omega_c t$. Therefore, the complex phasor $1 + j\beta\sin\omega_m t$ corresponds to $a(t)e^{j\theta(t)}$, where the instantaneous amplitude $a(t) = \sqrt{1^2+\beta^2\sin^2\omega_m t}$ and the instantaneous phase deviation $\theta(t) = \tan^{-1}(\beta\sin\omega_m t)$,

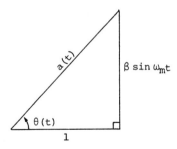

as shown in the phasor diagram.

FREQUENCY MODULATION

• **PROBLEM** 10-4

A square wave, $f(t)$, and a triangular wave, $g(t)$, are shown in Fig. 1 and 2, respectively. Sketch the frequency modulated and amplitude-modulated suppressed carrier waveforms for both signals.

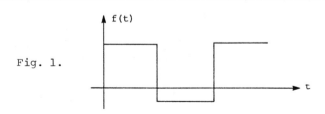

Fig. 1.

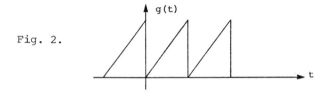

Fig. 2.

<u>Solution</u>: An amplitude-modulated suppressed carrier waveform is one in which the carrier is modulated in amplitude.

The AMSC waveforms for $f(t)$ and $g(t)$ are shown in Fig. 3A and B.

On the other hand, the frequency-modulated signal changes frequency whenever the modulation changes level with amplitude at constant level.

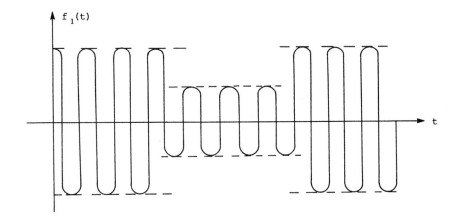

Fig. 3(a). AMSC

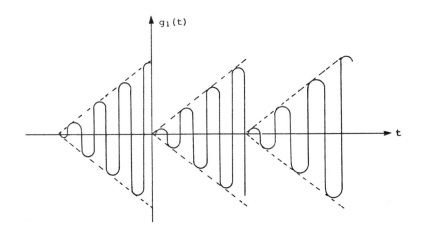

Fig. 3(b). AMSC

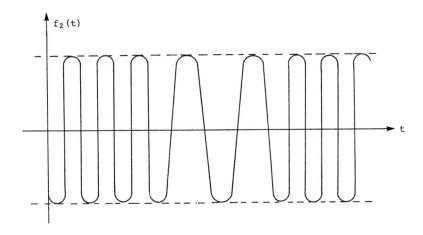

Fig. 4(a). FM

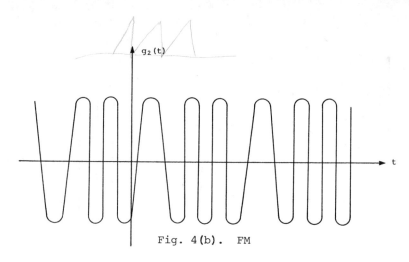

Fig. 4(b). FM

The FM waveforms for f(t) and g(t) are shown in Fig. 4A and B.

● **PROBLEM** 10-5

Let g(t) and h(t) represent two waveforms given by

$$g(t) = \begin{cases} \cos t, & t < 1 \\ \cos 2t, & 1 \leq t \leq 2 \\ \cos 3t, & 2 < t \end{cases}$$

and $h(t) = 10\cos(1000t + \sin 5t)$.

Determine the instantaneous frequency for each waveform.

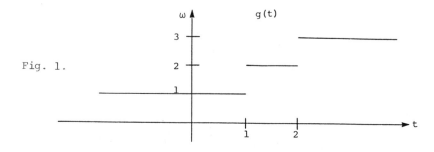

Fig. 1.

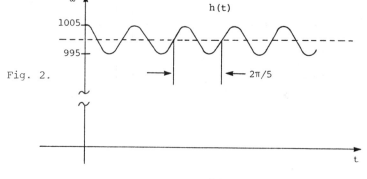

Fig. 2.

Solution: For the waveform g(t),

$$\theta(t) = \begin{cases} t, & t < 1 \\ 2t, & 1 \leq t \leq 2 \\ 3t, & 2 < t \end{cases}$$

Instantaneous frequency is given by

$$\omega = \frac{d}{dt}\theta(t)$$

Therefore,

$$\omega = \begin{cases} 1, & t < 1 \\ 2, & 1 \leq t \leq 2 \\ 3, & 2 < t \end{cases}$$

Similarly, for the waveform h(t),

$$\theta(t) = 1000t + \sin 5t$$

$$\omega = \text{Instantaneous angular frequency} = \frac{d\theta}{dt}$$

$$= 1000 + 5\cos 5t.$$

The instantaneous frequencies for g(t) and h(t) are shown in figures 1 and 2 respectively.

● **PROBLEM 10-6**

A frequency-modulated signal with a carrier frequency of 100MHz, is fluctuating at a rate of 100 times per second between the frequencies 100.001 MHz and 99.999 MHz.

(a) Find f_m, the modulating frequency and calculate Δf, frequency deviation.

(b) If the frequency swing is now between the frequencies 100.002 MHz and 99.998 MHz, state if there is any effect.

(c) Determine the modulation index β in each case.

Solution: (a) Since the swing rate is given as 100 times per second, the modulating frequency (f_m) is equal to 100Hz.

(b) The frequency deviation Δf is equal to the difference between the frequencies 100.001 MHz and 99.999 MHz.

Hence, Δf = 100.001 MHz - 99.999 MHz

= 2kHz

(b) Now, similarly for swing frequencies between 100.002 MHz and 99.998 MHz,

$$\Delta f_2 = 100.002 \text{ MHz} - 99.998 \text{ MHz} = 4\text{kHz}.$$

Since the frequency deviation is doubled, the modulation amplitude will also be doubled.

If the signal is f(t), then f(t) can be represented as

$$f(t) = A \cos(\omega_c t + \frac{\Delta f}{f_m} \sin \omega_m t)$$

where $\frac{\Delta f}{f_m}$ gives the amplitude of the modulating sinusoidal signal and is called the modulation index β.

(c) Hence, the modulation index is defined as

$$\beta = \frac{\Delta f}{f_m}$$

Therefore,

$$\beta_1 = \frac{2\text{kHz}}{100\text{Hz}} = 20$$

and

$$\beta_2 = \frac{4\text{kHz}}{100\text{Hz}} = 40.$$

● **PROBLEM 10-7**

An FM signal ranging from $f_{max} = 100.02$ MHz to $f_{min} = 99.98$ MHz is modulated by a 3kHz sine wave. Find

(A) The carrier frequency f_c of the signal.

(B) The carrier swing for the signal.

(C) The frequency deviation Δf of the signal.

(D) The index of modulation β.

Solution: (A) The carrier frequency is the arithmetic mean of the maximum frequency and minimum frequency reached by the modulated wave. Thus

$$f_c = \frac{f_{max} + f_{min}}{2}$$

$$= \frac{(100.02 \times 10^6 \text{Hz}) + (99.98 \times 10^6 \text{Hz})}{2}$$

$$= 100 \text{ MHz}.$$

(B) The carrier swing is defined as the difference between the maximum and minimum signal frequencies reached by the modulated wave. Thus,

$$\text{carrier swing} = (100.02 \times 10^6 \text{Hz}) - (99.98 \times 10^6 \text{Hz})$$
$$= 40 \times 10^3 \text{ Hz}$$
$$= 40 \text{ kHz.}$$

(C) Frequency deviation (in Hertz) expresses the amount above or below the carrier frequency by which the modulated signal varies. Thus, the frequency deviation must be ½ of the carrier swing or 20 kHz.

(D) The modulation index is the frequency deviation divided by the modulating signal frequency.

Hence,
$$\beta = \frac{\Delta f}{f_m}$$
$$= \frac{20 \text{kHz}}{3 \text{kHz}} = 6.667$$

● **PROBLEM 10-8**

The frequency range of a commercially broadcast FM signal is 88 to 108 MHz, with carrier swing of 125 kHz. Find the percentage modulation of the signal.

Solution: The frequency deviation Δf of a frequency modulated signal is equal to one-half of the carrier swing, i.e. $\Delta f = \frac{(\text{carrier swing})}{2}$. Therefore, the frequency deviation $= \frac{125 \text{kHz}}{2} = 62.5$ kHz. According to Federal Communications Commission regulations, the maximum frequency deviation of the FM broadcast signal is 75 kHz, i.e. $\Delta f_{max} = 75$ kHz

Therefore, percentage modulation

$$= \frac{\Delta f_{actual}}{\Delta f_{maximum}} = \Delta f \times 100 = \frac{62.5}{75} \times 100 = 83.3\%$$

● **PROBLEM 10-9**

Find the carrier swing and the frequency deviation Δf of a TV signal, given that the percentage modulation of the audio portion is 80%.

Solution: According to Federal Communications Commission regulations, the maximum allowable frequency deviation of the audio portion of the TV signal is 25 kHz. The percentage modulation is equal to the ratio of the actual frequency deviation Δf to the maximum frequency deviation Δf_{max} multiplied by 100 %

Therefore,
$$\frac{\Delta f_{actual}}{\Delta f_{max}} \times 100 \% = 80\%$$

so,
$$\Delta f = \frac{80 \times 25 kHz}{100} = 20 \ kHz$$

The carrier swing is equal to twice the frequency deviation.

Therefore, carrier swing = $2\Delta f$

$$= 2(20 kHz) = 40 \ kHz.$$

● **PROBLEM** 10-10

Determine the frequency deviation (Δf) and the carrier frequency (f_c) at points (a), (b), and (c) of the block diagram of an FM transmitter.

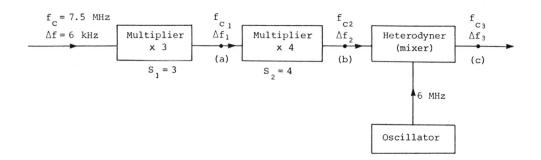

Solution: The first multiplier multiplies both frequency deviation and carrier frequency by a factor 3. Hence, at point a:

$$f_{c1} = S_1 f_c$$

$$= 3(7.5 MHz) = 22.5 MHz$$

and
$$\Delta f = S_1 \Delta f$$

$$= 3(6 kHz) = 18 kHz.$$

Similarly, at point b:

$$f_{c2} = S_2 f_{c1}$$

$$= 4(22.5 \text{MHz}) = 90 \text{MHz}$$

and $\quad \Delta f_2 = S_2 \Delta f_1$

$$= 4(18 \text{kHz}) = 72 \text{kHz}.$$

At the mixer, the carrier frequency will either be incremented or decremented by an amount equal to the frequency of the oscillator. Let's assume that there is an increment in carrier frequency at point c:

Hence $\quad f_{c_3} = f_{c_2} + f_{osc}$

$$= 90 \text{MHz} + 6 \text{MHz} = 96 \text{MHz}$$

since heterodyning due to the mixer and the oscillator has no effect on frequency deviation.

Therefore,

$$\Delta f_3 = \Delta f_2 = 72 \text{kHz}.$$

• **PROBLEM 10-11**

A simplified block diagram for a frequency modulation transmitter is given below with values as shown:

If for tolerable distortion, the phase deviation at point (1) must be less than or equal to 0.5 radians:

(A) Find the value of x_1, if x_1 is made up of stages each of which may have a maximum multiplication of 4. Specify the minimum number of multiplier stages which can be used for x_1 as well as the multiplication value of each stage.

(B) What local oscillator freuqency (f_0) must be used for x_1 as found in part (A)?

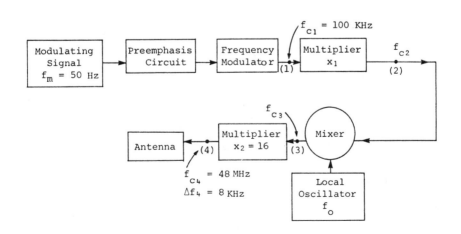

Solution: (A) By definition,

$$\text{phase deviation}(\Delta\theta) = \frac{\text{frequency deviation}(\Delta f)}{\text{modulating frequency}(f_m)}.$$

For $\Delta\theta_1$ (the phase deviation at point 1) to be a maximum, Δf_1 must be a maximum and the modulating frequency used should be a minimum. From the distortion constraint, x_1 is derived as follows:

Since $\Delta f_4 = x_1 x_2 \Delta f_1$,

$$\Delta\theta_{1max} = \frac{\Delta f_{1max}}{f_{mMIN}} = \frac{\Delta f_{4max}}{x_1 x_2 f_{mMIN}} \leq 0.5 \text{ rad}$$

Hence,

$$x_1 \geq \frac{2\Delta f_{4max}}{x_2 f_{mMIN}} = \frac{(2)8kHz}{(16)50Hz} = 20.$$

If it is not possible to exactly match the value of x_1 to the types of multiplier stages used, always choose the closest possible value larger than the calculated x_1 value. The best choice in this case for the multiplier x_1 would consist of 3 stages with multiplications 2, 3 and 4 making $x_1 = 24$. This is the closest high-side value to 20. (Note: with $x_1 = 24$, $\Delta\theta_{1max} = 0.417$ rad.).

(B) With x_1 known:

$$f_{C2} = x_1 f_{C4}$$

$$= 24(100kHz)$$

$$= 2.4 \text{ MHz}$$

$$f_{C3} = f_{C4}/x_2$$

$$= 48 \text{ MHz}/16$$

$$= 3 \text{ MHz}$$

and $\Delta f_{C2} = \Delta f_{C3} = \Delta f_{C4}/x_2$

$$= 8kHz/16$$

$$= 500 \text{ Hz}$$

Knowing the characteristic of the mixer, the local oscillator frequency can be found as:

$$f_{C3} = |f_{C2} \pm f_{osc}|$$

or

$$f_{osc} = 3 \text{ MHz} - 2.4 \text{ MHz}$$

$$= 600 \text{ kHz}.$$

• **PROBLEM 10-12**

The frequency deviation of a 5MHz frequency modulated signal is 4kHz. With this input signal, how can a frequency modulated output with a signal frequency of 50MHz and a frequency deviation of 24kHz be obtained?

Solution: To obtain the desired output frequency deviation, the input signal should first be passed through a frequency multiplier. The frequency multiplication required can be found from the relation

$$\Delta f_{out} = S \, \Delta f_{in}$$

where
Δf_{out} = frequency deviation of the output

Δf_{in} = frequency deviation of the input

S = amount of frequency multiplication

Hence,
$$S = \frac{24 \text{kHz}}{4 \text{kHz}} = 6$$

(Note: Since it is not common to have a single 6x multiplier, a cascade arrangement of a 2x and a 3x multiplier is required).

The frequency multiplication used does not necessarily cause the output carrier frequency to have the required value. After passing through the multipliers, the frequency of the F.M. carrier signal f_c will be six times its original value or $6 \times 5\text{MHz} = 30\text{MHz}$. In order to get the desired output carrier frequency without changing the frequency deviation, the signal should be passed through a heterodyne section made up of a mixer and an oscillator. A block diagram of the system used to get the desired output is shown below:

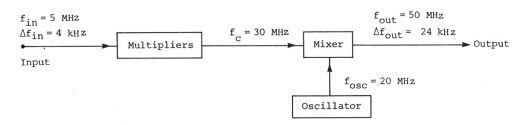

The frequency of the oscillator for the desired output is found from
$$f_c + f_{osc} = f_{out}$$

$$30\text{MHz} + f_{osc} = 50\text{MHz}$$

Thus,
$$f_{osc} = 20\text{MHz}.$$

• **PROBLEM 10-13**

A frequency-multiplier circuit is shown in Fig. 1. The collector current of the transistor has the waveform shown in Fig. 2.

(a) Write an expression of the resonant frequency f_0 in terms of L.

Then calculate (b) the inductance L,

                (c) the resonant impedance R, and

                (d) the amplitude of the third harmonic voltage across the tank circuit

under the following signal conditions:

The input signal frequency is 1 MHz, the multiplication factor is 3, the capacitance is 200 pF, and the inductor Q is 30.

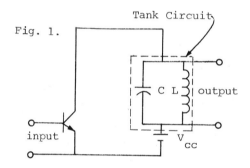

Fig. 1.

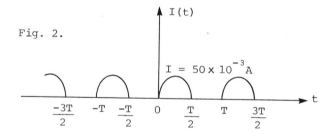

Fig. 2.

$I = 50 \times 10^{-3}$ A

Solution: (a) $\quad f_0 = 3 \times 1 \times 10^6 = \dfrac{1}{2\pi\sqrt{LC}} = \dfrac{1}{2\pi\sqrt{L \times 2 \times 10^{-10}}}$

(b) $\quad 3 \times 10^6 = \dfrac{1}{2\pi\sqrt{L \times 2 \times 10^{-10}}}$

$\quad\quad 2\pi(3 \times 10^6) = \dfrac{1}{\sqrt{L \times 2 \times 10^{-10}}}$

$$[2\pi(3\times10^6)]^2 = \frac{1}{2\times10^{-10} L}$$

Therefore, $L = \dfrac{1}{2\times10^{-10}[2\pi(3\times10^6)]^2} = 14.07 \; \mu H$

(c) $Q = 30 = \dfrac{R}{\omega_0 L} = \dfrac{R}{2\pi\times3\times10^6\times1.4\times10^{-5}}$

Therefore, $R = 7.9 \times 10^3 \; \Omega$

(d) The Fourier series expansion of the collector current waveform is given by

$$I(t) = \frac{I}{\pi} + \frac{I}{2}\sin2\pi ft - \frac{2I}{\pi}\sum_{n=1}^{\infty}\frac{1}{4n^2-1}\cos4\pi nft$$

Hence, by observation, the amplitude of the third harmonic voltage across the tank circuit is zero volts.

• **PROBLEM 10-14**

For broadcast FM, the maximum frequency deviation is equal to 75 kHz. If the modulating frequencies range from 30 Hz to 15 kHz,

(a) calculate the maximum allowable range in modulation index β;

(b) repeat part (a) for a narrow-band frequency modulation (NBFM) system which permits a maximum frequency deviation Δf of 10 kKz and modulating frequencies from 100 Hz to 3 kHz.

Solution: (a) The equation for calculating β is given by

$$\beta = \frac{\Delta f \text{(frequency deviation)}}{f_m \text{(modulating frequency)}}$$

Therefore, the maximum allowable range of β is from $\dfrac{75\text{kHz}}{30\text{Hz}} = 2500$ to $\dfrac{75\text{kHz}}{15\text{kHz}} = 5$.

(b) Using the same formula in part (a) but with different values for Δf and f_m, hence, in this case

β ranges from $\dfrac{10\text{kHz}}{100\text{Hz}} = 100$ to $\dfrac{10\text{kHz}}{3\text{kHz}} = 3.33$

• **PROBLEM 10-15**

A frequency modulated wave has an angular carrier frequency ω_c = 5000 rad/sec and the modulation index β = 10. Determine the bandwidth (BW) and the upper and lower sideband frequencies if the modulating signal m(t) is m(t) = 20 cos5t.

Solution: The bandwidth using Carson's rule is given by

$$BW = 2(\beta+1)f_m.$$

Therefore, in this case

$$BW = 2(10+1)f_m \quad \text{where} \quad f_m = \text{modulating frequency}$$

$$= \frac{\omega_m}{2\pi}$$

$$= \frac{5}{2\pi}$$

Thus, $\quad BW = 2(10+1)\frac{5}{2\pi} = 17.51$ Hz

Now, since the carrier frequency is given, the band of frequencies occupied by this FM wave can be calculated as follows:

$$\text{Lower-sideband} = \frac{5000}{2\pi} - \frac{110}{2\pi} = \frac{4890}{2\pi} \cong 778.3 \text{Hz}$$

$$\text{Upper-sideband} = \frac{5000}{2\pi} + \frac{110}{2\pi} \cong 813.3 \text{Hz}$$

• **PROBLEM 10-16**

The modulation index (m) of a 10-kw FM transmitter is 0.25. Find the normalized power in the sidebands and the carrier, given that the Bessel coefficients for m=0.25 are J_0 = 0.98 and J_1 = 0.12.

Solution: For a modulation index = 0.25, the amplitude of the modulated carrier is equal to 0.98 times its unmodulated amplitude. There is only one significant sideband whose relative amplitude is equal to J_1 = 0.12.

Therefore, the normalized power of the carrier =

$(0.98)^2 \times 10\text{kw} = 9.604$ kw.

The power of each sideband = $(0.12)^2 \times 10$ kw

$$= 0.144 \text{ kw}$$

Therefore, the normalized power in the carrier and sidebands

$$= 9.604 + 2(0.144)$$

$$= 9.892 \text{ kw}$$

$$\cong 10 \text{ kw}$$

From the result of this problem, it can be seen that a frequency-modulated waveform varies only in its frequency but never in its amplitude, and therefore regardless of the level of modulation, the total power remains a constant.

• **PROBLEM** 10-17

Given the bandwidth $BW = (2\beta+1)f_m$ used for a space communication system, where β is the modulation index. If $\beta = 0.2$ and 5, determine the fraction of the signal power passed by the filter.

Solution: In order to have a total power transmission, usually 98% power transmission for a sinusoidally modulated FM signal, the bandwidth BW is given by $BW = 2(\beta+1)f_m$.

Thus, for $\beta = 0.2$, the fraction of the signal power passed by the filter can be calculated as follows:

$$\frac{(2\beta+1)f_m}{2(\beta+1)f_m} \times 0.98 = \frac{[2(0.2)+1]}{2(0.2+1)} \times 0.98$$

$$= 0.57 \quad \text{or} \quad 57\%$$

Similarly, for $\beta = 5$, $\quad \frac{[2(5)+1]}{2(5+1)} \times 0.98$

$$\cong 0.90 \quad \text{or} \quad 90\%$$

• **PROBLEM** 10-18

(a) Determine the bandwidth occupied by a frequency modulated waveform which results from modulation of a carrier signal by a sinusoidal modulating signal of frequency 3 kHz. The frequency deviation of the modulated signal was found to be equal to 5 kHz.

(b) What would be the new bandwidth if the frequency of the modulating signal is increased to 4 kHz and the amplitude is decreased by a factor 2?

Solution: (a) The bandwidth using Carson's rule is given by

$$BW = 2(\Delta f + f_m)$$

since $\Delta f = 5kHz$ and $f_m = 3kHz$

$$BW = 2(5k+3k)Hz = 16 \text{ kHz}$$

(b) Δf is related to β by the equation

$$\Delta f = \beta f_m$$

Hence, if β is decreased by a factor 2, ΔF will also decrease by the same amount. Therefore, the new values for Δf and f_m are $\Delta f = 2.5kHz$ and $f_m = 4kHz$.

Thus, the bandwidth in this case is

$$BW = 2(\Delta f + f_m)$$

$$= 2(2.5+4) = 13kHz.$$

(Note: For FM, the amplitude of the modulating signal is proportional to the frequency deviation Δf, i.e. an increase of amplitude by a factor will cause the increase in Δf by the same factor.)

● **PROBLEM 10-19**

A carrier of 100 MHz is frequency-modulated by f(t) shown in Fig. 1. The modulating system constant k_p is 10^6. Find and sketch the spectrum of the modulated carrier. (Hint: The FM carrier can be expressed as a sinusoid of one frequency over a part of the cycle and a sinusoid of another frequency over the remaining cycles. This is equivalent to the sum of two sinusoids, each multiplied by a periodic gate function.)

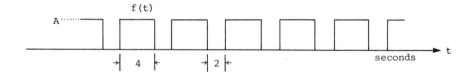

Solution: The modulated carrier is:

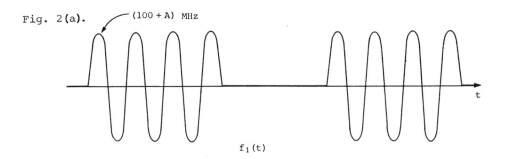

Fig. 2(a). (100 + A) MHz

$f_1(t)$

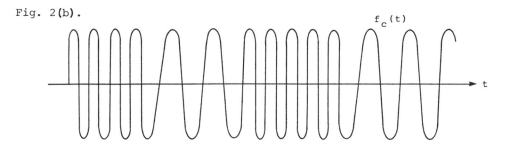

Fig. 2(b). $f_c(t)$

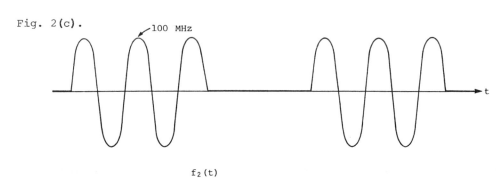

Fig. 2(c). 100 MHz

$f_2(t)$

$f_c(t) = f_1(t) + f_2(t)$ where $f_1(t)$ and $f_2(t)$ are both modulated periodic gate functions. The carrier in $f_1(t)$ has a frequency 100 MHz + ak_f = (100+A)MHz and that in $f_2(t)$ is 100 MHz.

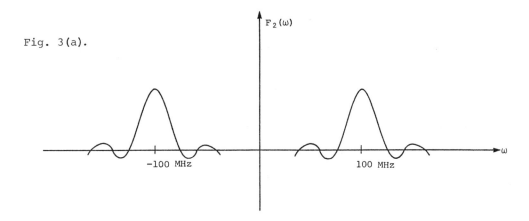

Fig. 3(a).

Fig. 3(b).

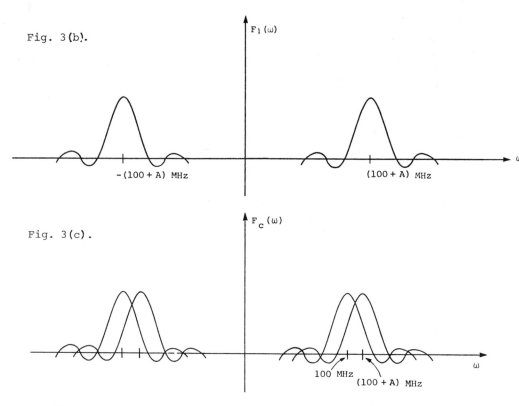

Fig. 3(c).

The spectrum of $f_1(t)$ is shown in figure 3. $f_2(t)$ has a similar spectrum. In each case, the spectrum is a sampling function shifted to a carrier frequency.

● **PROBLEM 10-20**

Consider $\theta(t) = 2\pi f_d \int_{t_0}^{t} m(\lambda)d\lambda + \theta_0$, the phase deviation of a frequency modulated carrier, with $m(t) = \cos\omega_m t$ in an FM system. Determine the output of the modulator and sketch the spectrum of the output signal.

(Note: f_d = frequency deviation constant of the modulator and θ_0 = phase deviation at $t = t_0$.)

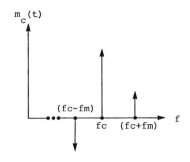

Solution: The instantaneous phase deviation at $t = t_0 = 0$ is

$$\theta(t) = 2\pi f_d \int_0^t \cos\omega_m \lambda\, d\lambda$$

$$= \frac{2\pi f_d}{\omega_m} (\sin\omega_m \lambda)\Big|_0^t$$

$$= \frac{2\pi f_d}{\omega_m} (\sin\omega_m t)$$

$$= \frac{2\pi f_d}{2\pi f_m} \sin\omega_m t \qquad \text{(Note: } \omega_m = 2\pi f_m\text{)}$$

$$= \frac{f_d}{f_m} \sin\omega_m t$$

In general the modulator output $m_c(t)$ can be expressed as

$$m_c(t) \cong A_c \left[\cos\omega_c t - \theta(t)\sin\omega_m t\right]. \qquad (1)$$

Hence, substitute $\theta(t)$ into eq. (1),

$$m_c(t) \cong A_c \left[\cos\omega_c t - \frac{f_d}{f_m} \sin\omega_m t\right]$$

$$\cong A_c \cos\omega_c t + \frac{A_c}{2} \frac{f_d}{f_m} \left[\cos(\omega_c+\omega_m)t - \cos(\omega_c-\omega_m)t\right] \qquad (2)$$

Now, in order to plot the spectrum of $M_c(t)$, express eq.(1) in a more convenient form as

$$m_c(t) = \text{Re}\left\{A_c e^{j\omega_c t}\left[1 + \frac{f_d}{2f_m}(e^{j\omega_m t} - e^{-j\omega_m t})\right]\right\}.$$

The spectrum of $m_c(t)$ is shown in Fig. 1. Notice the phase relationship between upper and lower sidebands.

● **PROBLEM 10-21**

A carrier with amplitude 4 volts and frequency $f_c = 2\text{MHz}$ is frequency modulated by a modulating signal $m(t) = \sin 1000\pi t$. The amplitude of the modulating signal is 2 volts and the frequency deviation was found to be equal to 1 kHz. If the amplitude and frequency of the modulating signal are increased to 8 volts and 2 kHz respectively, write an expression for the new modulated signal.

Solution: The frequency modulated signal can be represented as:

$$v(t) = B\cos(2\pi f_c t + \frac{kA}{f_m} \sin 2\pi f_m t)$$

where f_c = carrier frequency = 2×10^6 Hz

B = carrier amplitude = 4 volts

Δf = kA = Frequency deviation = 1 kHz

A = Amplitude of the modulating signal
  = 2 volts

Hence, $k = \frac{\Delta f}{A} = \frac{1 \text{ kHz}}{2 \text{ volts}} = 500$ Hz/volt.

When A = 8 volts and f_m = 2 kHz,

$$\frac{kA}{f_m} = \frac{(500 \text{Hz/volt})(8 \text{ volts})}{2000 \text{ Hz}} = 2$$

Therefore the new modulated signal can be represented as:

$v(t) = 4\cos(4\pi \times 10^6 t + 2 \sin 2\pi \times 2000 t)$.

• **PROBLEM 10-22**

The output of an F.M. modulator is given as

$$v(t) = 40\cos\left[\omega_c t + (2\pi f_D) \int_o^t m(\lambda) d\lambda \right]$$

where f_D = 10 Hertz/volt and m(t) is shown in fig. 1.

(A) Find the phase deviation $\phi(t)$ as a function of time and sketch.

(B) Find the frequency deviation Δf as a function of time and sketch.

(C) Find the peak frequency deviation in Hertz.

(D) Find the peak phase deviation in radians.

(E) What is the normalized modulator output power?

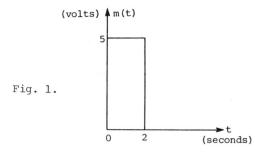

Fig. 1.

Solution: (A) The given output can be expressed in the general form: $x(t) = a(t)\cos[\omega_c t + \phi(t)]$ where $\phi(t)$ explicitly represents the phase deviation.

Thus,
$$\phi(t) = 2\pi(10)\int_0^t m(\lambda)d\lambda = 20\pi \int_0^t m(\lambda)d\lambda.$$

The integral can be evaluated as the area from 0 to t under the curve m(t) in figure 1. Hence,

$$\phi(t) = \begin{cases} 100\pi t & \text{for } 0 \leq t \leq 2 \\ 200\pi & \text{for } t > 2 \\ 0 & \text{elsewhere} \end{cases}$$

Fig. 2.

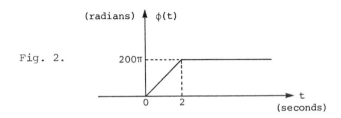

(B) To determine the frequency deviation in Hertz, differentiate the phase deviation with respect to time and divide by 2π. Therefore,

$$\Delta f = \frac{1}{2\pi} \frac{d\phi(t)}{dt} = f_D m(t)$$

Thus,
$$\Delta f = \begin{cases} \frac{100\pi}{2\pi} = 50 & \text{for } 0 \leq t \leq 2 \\ 0 & \text{elsewhere} \end{cases}$$

Fig. 3.

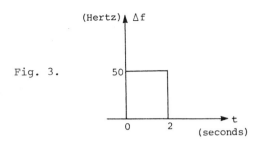

(C) From the result obtained in part (B), it can be seen that 50 Hz is the peak frequency deviation.

(D) From the result obtained in part (A) it can be seen that 200π radians is the peak phase deviation.

(E) Normalized output power is not affected by the phase modulation; if v(t) represents the voltage across 1Ω resistor, the normalized output power is one-half the square of the peak signal amplitude a(t), or

$$\frac{(40)^2}{2} = 800 \text{ watts.}$$

● **PROBLEM** 10-23

A frequency deviation of 3 kHz is produced when a cosinusoidal carrier of frequency 50 MHz is frequency-modulated using a sinusoid of frequency 10 kHz. Prove that the frequency spectrum of the modulated carrier consists of a carrier and two pairs of sidebands. Also, determine the ratios of the amplitudes of those components with respect to the amplitude of the unmodulated carrier.

Solution: Let v(t) be the modulated carrier. Hence, v(t) can be written as

$$v(t) = A\cos(\omega_c t + \beta \sin\omega_m t)$$

where ω_c = Frequency of the unmodulated carrier = 50 MHz

ω_m = Frequency of the modulating signal = 10 kHz

β = Modulation index = $\dfrac{\text{Frequency deviation}}{\text{Modulating signal frequency}}$

$$= \frac{\Delta f}{f_m} = \frac{3 \text{ kHz}}{10 \text{ kHz}} = 0.3$$

Expanding v(t) gives

$$v(t) = A\left[\cos(\omega_c t)\cos(\beta\sin\omega_m t) - \sin(\omega_c t)\sin(\beta\sin\omega_m t)\right] \quad (1)$$

Expanding $\cos(\beta\sin\omega_m t)$ and $\sin(\beta\sin\omega_m t)$ using the infinite series expansion results in,

$$\cos(\beta\sin\omega_m t) = 1 - \frac{(\beta\sin\omega_m t)^2}{2!} + \cdots$$

and since $\beta < 1$, this can be approximated as

$$\cos(\beta\sin\omega_m t) \cong 1 - \frac{\beta^2 \sin^2\omega_m}{2} \cdot \quad (2)$$

Similarly,

$$\sin(\beta\sin\omega_m t) = \beta\sin\omega_m t - \frac{(\beta\sin\omega_m t)^3}{3!} + \cdots$$

$$\cong \beta\sin\omega_m t. \quad (3)$$

Substituting equations (2) and (3) into (1) gives

$$v(t) = A\left[\cos(\omega_c t)\left(1 - \frac{\beta^2 \sin^2 \omega_m}{2}\right) - \sin(\omega_c t)(\beta \sin \omega_m t)\right]$$

$$= A\left\{\cos(\omega_c t)\left[1 - \frac{\beta^2}{2}\frac{(1-\cos 2\omega_m t)}{2}\right] - \beta \sin(\omega_c t)\sin(\omega_m t)\right\}$$

$$= A\left\{\left(\frac{1-\beta^2}{4}\right)\cos\omega_c t + \frac{\beta^2}{4}\cos(\omega_c t)\cos(2\omega_m t) - \beta\sin(\omega_c t)\sin(\omega_m t)\right\}$$

$$= A\left\{\left(1 - \frac{\beta^2}{4}\right)\cos\omega_c t + \frac{\beta^2}{4}\cdot\frac{1}{2}\left[\cos(\omega_c - 2\omega_m)t + \cos(\omega_c + 2\omega_m)t\right]\right.$$
$$\left. - \beta\frac{1}{2}\left[\cos(\omega_c - \omega_m)t - \cos(\omega_c + \omega_m)t\right]\right\}$$

Thus, the modulated carrier consists of a carrier of amplitude $\left(1 - \frac{\beta^2}{4}\right)A$, a sideband pair of frequency $(\omega_c \pm \omega_m)$ and amplitude $\frac{\beta^2 A}{8}$, and a second sideband pair of frequency $(W_c \pm 2W_m)$ and amplitude $\frac{1}{2}\beta A$. Note, as β increases, the number of significant sidebands increases. The word significant is usually taken to mean those sidebands which have a magnitude of at least 1% of the unmodulated carrier magnitude.

With respect to the amplitude of the unmodulated carrier, the amplitudes of the components in the modulated carrier are as follows:

Relative amplitude of the carrier $= 1 - \frac{\beta^2}{4} = 1 - \frac{(0.3)^2}{4}$

$= 0.9775$.

Relative amplitude of the 1st sideband pair $= \frac{\beta}{2} = \frac{0.3}{2}$

$= 0.15$

Relative amplitude of the 2nd sideband pair $= \frac{\beta^2}{8}$

$= \frac{(0.3)^2}{8}$

$\cong 0.0113$.

• **PROBLEM 10-24**

Consider a FM signal represented by

$$v(t) = \cos\left[2\pi f_c t + k \int_{-\infty}^{t} m(\lambda)d\lambda\right]$$

(a) determine the power spectral density $G_v(f)$ of $v(t)$ if the probability density of the amplitude of $m(t)$ is Rayleigh:

$$f(m) = \begin{cases} me^{-m^2/2} & m \geq 0 \\ 0 & \text{elsewhere} \end{cases}$$

(b) repeat part (a) if the probability density of $m(t)$ is $f(m) = \frac{1}{2} e^{-|m|}$.

Solution: (a) Since $G(v)$ (the power spectral density) (with $v \equiv f - f_c$) is proportional to $f(m)$, $G(v) = \alpha f(m)$ where α is a constant of proportionality.

Since $v(t) = km(t)/2\pi$,

$$G(v) = \begin{cases} \dfrac{2\pi \alpha v}{k} e^{-(2\pi v)^2/2k^2}, & v \geq 0 \\ 0, & v < 0 \end{cases}$$

Replacing v by $f \pm f_c$, and expressing the power spectral density for both positive and negative frequencies,

$$G_v(f) = \begin{cases} \dfrac{\pi \alpha (|f|-f_c)}{k} e^{-2\pi^2(|f|-f_c)^2/k^2} & |f| \geq f_c \\ 0 & |f| < f_c \end{cases}$$

To evaluate α, note that the power of the FM waveform is $A^2/2$. Hence,

$$\int_{-\infty}^{\infty} G_v(f) df = \alpha \int_{f_c}^{\infty} \frac{2\pi(f-f_c)}{k} e^{-2\pi^2(f-f_c)^2/k^2} df$$

$$= \alpha/2\pi \int_0^{\infty} \frac{\omega}{k} e^{-\omega^2/2k^2} d\omega$$

$$= \alpha k/2\pi = A^2/2$$

therefore, $\alpha = \pi A^2/k$

Hence,

$$G_v(f) = \begin{cases} \dfrac{\pi^2 A^2 ||f|-f_c|}{k^2} e^{-2\pi^2(|f|-f_c)^2/k^2} & |f| \geq f_c \\ 0 & |f| < f_c \end{cases}$$

(b) Repeat the same procedure as (a)

$$G(v) = \alpha f(m) = \frac{\alpha}{2} e^{-|m|} = \frac{\alpha}{2} e^{-2\pi |v|/k} \text{ where } v = f \pm f_c,$$

$v(t) = km(t)/2\pi$ and α is a constant of proportionality.

Replacing ν by $f \pm f_c$,

$$G_\nu(f) = \frac{\alpha}{4} e^{-2\pi|f-f_c|/k} + \frac{\alpha}{4} e^{-2\pi|f+f_c|/k}$$

Evaluate α:

$$\int_{-\infty}^{\infty} G_\nu(f)df = \int_{-\infty}^{\infty} G(\nu)d\nu = \alpha \int_0^{\infty} e^{-2\pi\nu/k} d\nu$$

$$= \frac{\alpha}{2\pi} \int_0^{\infty} e^{-\omega/k} d\omega = \frac{\alpha k}{2\pi} = \frac{A^2}{2}$$

therefore, $\quad \alpha = \pi A^2/k$

Hence,
$$G_\nu(f) = \frac{\pi A^2}{4k} e^{-2\pi|f-f_c|/k} + \frac{\pi A^2}{4k} e^{-2\pi|f+f_c|/k}$$

• **PROBLEM** 10-25

Determine the bandwidth of a frequency modulated signal $v(t)$, given by

$$v(t) = 10\cos(2 \times 10^7 \pi t + 20\cos 1000\pi t)$$

Solution: Generally, a frequency modulated waveform is represented in the form

$$A\cos(2\pi f_c t + \beta \cos\omega_m t).$$

For this problem the carrier frequency

$$f_c = 10^7 \text{Hz} = 10\text{MHz} \quad \text{and} \quad \omega_m = 2\pi f_m = 1000\pi$$

Therefore,
$$fm = \frac{1000\pi}{2\pi} = 500 \text{ Hz}.$$

The maximum frequency deviation (Δf) is given by

$$\Delta f = \beta f_m$$

where β, the modulation index, is given above as 20.

Hence, $\quad \Delta f = 20(500\text{Hz}) = 10 \text{ kHz}$

and according to Carson's rule the bandwidth (BW) of the frequency modulated waveform is given as

$$BW = 2(\Delta f + f_m)$$

$$= 2(10 \text{ kHz} + 500 \text{ Hz})$$

$$= 2(10.5 \text{ kHz}) = 21 \text{ kHz}$$

Thus, BW = 21 kHz.

● **PROBLEM** 10-26

(a) An FM signal has a maximum frequency deviation $\Delta f = 20$ kHz. Find the bandwidth required to transmit the FM signal if the modulating signal frequency is 10 kHz.

(b) Repeat part (a) with $f_m = 5$ kHz.

TABLE
FM Side Frequencies from Bessel Functions

x (β)	J_0	J_1	J_2	J_3	J_4	J_5	J_6	J_7	J_8	J_9	J_{10}	J_{11}	J_{12}	J_{13}	J_{14}	J_{15}	J_{16}
0.00	1.00	—	—	—	—	—	—	—	—	—	—	—	—	—	—	—	—
0.25	0.98	0.12	—	—	—	—	—	—	—	—	—	—	—	—	—	—	—
0.5	0.94	0.24	0.03	—	—	—	—	—	—	—	—	—	—	—	—	—	—
1.0	0.77	0.44	0.11	0.02	—	—	—	—	—	—	—	—	—	—	—	—	—
1.5	0.51	0.56	0.23	0.06	0.01	—	—	—	—	—	—	—	—	—	—	—	—
→2.0	0.22	0.58	0.35	0.13	0.03	—	—	—	—	—	—	—	—	—	—	—	—
2.5	−0.05	0.50	0.45	0.22	0.07	0.02	—	—	—	—	—	—	—	—	—	—	—
3.0	−0.26	0.34	0.49	0.31	0.13	0.04	0.01	—	—	—	—	—	—	—	—	—	—
→4.0	−0.40	−0.07	0.36	0.43	0.28	0.13	0.05	0.02	—	—	—	—	—	—	—	—	—
5.0	−0.18	−0.33	0.05	0.36	0.39	0.26	0.13	0.05	0.02	—	—	—	—	—	—	—	—
6.0	0.15	−0.28	−0.24	0.11	0.36	0.36	0.25	0.13	0.06	0.02	—	—	—	—	—	—	—
7.0	0.30	0.00	−0.30	−0.17	0.16	0.35	0.34	0.23	0.13	0.06	0.02	—	—	—	—	—	—
8.0	0.17	0.23	−0.11	−0.29	−0.10	0.19	0.34	0.32	0.22	0.13	0.06	0.03	—	—	—	—	—
9.0	−0.09	0.24	0.14	−0.18	−0.27	−0.06	0.20	0.33	0.30	0.21	0.12	0.06	0.03	0.01	—	—	—
10.0	−0.25	0.04	0.25	0.06	−0.22	−0.23	−0.01	0.22	0.31	0.29	0.20	0.12	0.06	0.03	0.01	—	—
12.0	0.05	−0.22	−0.08	0.20	0.18	−0.07	−0.24	−0.17	0.05	0.23	0.30	0.27	0.20	0.12	0.07	0.03	0.01
15.0	−0.01	0.21	0.04	−0.19	−0.12	0.13	0.21	0.03	−0.17	−0.22	−0.09	0.10	0.24	0.28	0.25	0.18	0.12

Solution: (a) A frequency modulated wave can be represented as

$$v(t) = A\sin(\omega_c t + \beta\sin\omega_m t)$$

in which ω_c = Angular frequency of the carrier

ω_m = Modulating signal frequency

β = Modulation index of the FM signal

A = Amplitude of the unmodulated carrier

note: $\beta = \dfrac{\Delta f}{f_m} = \dfrac{20 \text{ kHz}}{10 \text{ kHz}} = 2$.

From the table, with β = 2, the following significant components are obtained:

J_0, J_1, J_2, J_3, J_4.

This means that besides the carrier, J_1 will exist at ±10kHz around the carrier, J_2 at ±20kHz, J_3 at ±30kHz, and J_4 at ±40 kHz. Therefore, the total required bandwidth is

$$2 \times 40 \text{ kHz} = 80 \text{ kHz}.$$

(b) In this case

$$\beta = \frac{\Delta f}{f_m} = \frac{20 \text{ kHz}}{5 \text{ kHz}} = 4.$$

From the table with $\beta = 4$, the highest significant side-frequency component is J_7. Since J_7 is at 7×5 kHz around the carrier, the required BW is

$$2 \times 35 \text{ kHz} = 70 \text{ kHz}.$$

• **PROBLEM** 10-27

The input resistance h_{ie} of a transistor reactance modulator used in an FM transmitter is equal to 600Ω. The beta (β) of the transistor is 65. The modulator is shown in the figure.

(a) Find the equivalent capacitance due to this circuit in parallel with the tank circuit.

(b) What is the value of the capacitor range if the beta of the transistor is changed from 50 to 75?

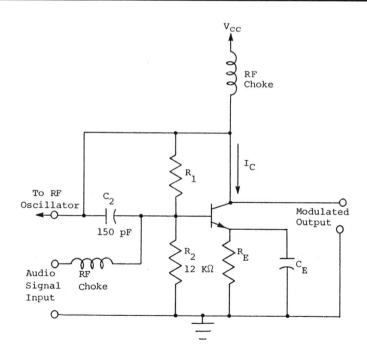

Solution: (a) The equivalent capacitance of a transistor reactance modulator is given by

$$C_{eq} = \frac{\beta R_2 C_2}{h_{ie} + R_2}$$

$$= \frac{(65)(12000)(150 \times 10^{-12})}{(600+12000)}$$

$$\cong 9.29 \; 10^{-9} \; \text{Farad}$$

$$= 9.29 \; \text{nF}.$$

(b) For $\beta = \beta_1 = 50$,

$$C_{eq1} = \frac{\beta_1 R_2 C_2}{h_{ie} + R_2}$$

$$= \frac{(50)(12000)(150 \times 10^{-12})}{(600+12000)}$$

$$= 7.14 \times 10^{-9} \; \text{Farad}$$

$$= 7.14 \; \text{nF}.$$

For $\beta = \beta_2 = 75$,

$$C_{eq2} = \frac{\beta_2 R_2 C_2}{h_{ie} + R_2}$$

$$= \frac{(75)(12000)(150 \times 10^{-12})}{600+12000}$$

$$= 1.07 \times 10^{-8} \; \text{Farad}$$

$$= 10.7 \; \text{nF}.$$

Therefore, the equivalent capacitance swings from 7.14 nanofarads to 10.7 nanofarads, while the beta changed from 50 to 75.

● **PROBLEM 10-28**

Find the range of capacitance offered by a reactance tube modulator shown in the figure. The range of the transconductance g_m of the remote cutoff tube is from 2500 μmhos to 3500 μmhos.

Solution: The equivalent capacitance of a reactance tube modulator is given by

$$C_{eq} = g_m RC$$

in which g_m varies due to the variation of the audio signal.

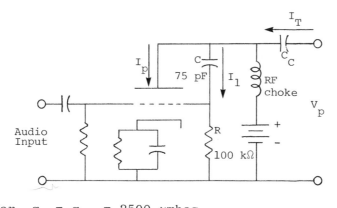

For $g_m = g_{m1} = 2500$ μmhos

$$C_{eq1} = (2500 \times 10^{-6})(100 \times 10^3)(75 \times 10^{-12})$$

$$= 1.875 \times 10^{-8} \text{ Farad}$$

$$= 18.75 \times 10^{-9} \text{ Farad}$$

$$= 18.75 \text{ nanoFarads.}$$

For $g_m = g_{m2} = 3500$ μmhos

$$C_{eq2} = (3500 \times 10^{-6})(100 \times 10^3)(75 \times 10^{-12})$$

$$= 2.625 \times 10^{-8} \text{ F}$$

$$= 26.25 \times 10^{-9} \text{ F}$$

$$= 26.25 \text{ nanoFarads.}$$

Hence, the range of capacitance presented by the reactance tube modulator is 18.75 nF to 26.25 nF.

PHASE MODULATION

• **PROBLEM 10-29**

Illustrate the relationship between phase and frequency modulation by using block diagrams.

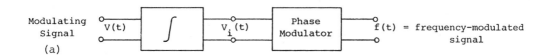

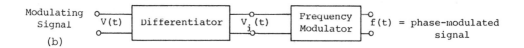

(b)

Solution: Simplified block diagrams for phase and frequency modulation are shown in Fig. 1A and 1B.

In fig. 1(A), the time domain expression for the output signal f(t) can be written as

$$f(t) = F\cos\left[\omega_c t + M_1 V_i(t)\right]$$

which is a carrier, phase-modulated by the input signal $V_i(t)$ with M_1 being a constant.

Since $V_i(t)$ results from integrating $V(t)$, $V_i(t)$ can be written in terms of $V(t)$ as follows:

$$V_i(t) = M_2 \int_{-\infty}^{t} V(t)dt \quad \text{in which } M_2 \text{ is also a constant.}$$

Then with $M = M_1 M_2$,

$$f(t) = F\cos\left[\omega_c t + M \int_{-\infty}^{t} V(t)dt\right].$$

Hence, the instantaneous angular frequency (ω_i) is

$$\omega_i = \frac{d}{dt}\left[\omega_c t + M \int_{-\infty}^{t} V(t)dt\right] = \omega_c + MV(t)$$

and the frequency deviation (Δf) is given by $(f - f_c)$ where $f_c = \frac{\omega_c}{2\pi}$ is the carrier frequency.

Thus, $\quad \Delta f \equiv f - f_c = \frac{M}{2\pi} V(t)$

Note: Since the instantaneous frequency is directly proportional to the modulating signal, the combination shown in Fig. (1A) is a device for producing a frequency-modulated output. Similarly the combination of fig. 1(b) of the differentiator and frequency modulator generates a phase-modulated output, i.e., a signal whose phase departure from the carrier is proportional to the modulating signal.

• **PROBLEM** 10-30

The received signal at the end of a communication network is given by

$$h(t) = 10 \cos(10^6 t + 200 \sin 500t).$$

Determine   a)   the bandwidth (BW) of $h(t)$;

b) the modulating signal m(t) if h(t) is an FM signal

c) m(t) if h(t) is a phase-modulated wave.

Solution: (a) Using Carson's rule the bandwidth is given by

$$BW = 2(\Delta f + f_m) = 2(\Delta \omega + \omega_m)$$

where Δf and $\Delta \omega$ = frequency and angular frequency deviation, respectively.

f_m and ω_m = modulating signal, frequency and angular frequency, respectively.

Thus, $f_m = \dfrac{\omega_m}{2\pi} = \dfrac{500}{2\pi} = 79.58$ Hz

and $\Delta f \cdot \beta f_m = (200)(79.58) = 15923$ Hz

Therefore, $BW = 2(\Delta f + f_m) = 2(15923 + 79.58) \cong 32$ kHz

(Note: The above waveform would be considered a wideband waveform since the bandwidth of a narrowband case is approximately 159 hz.)

(b) If h(t) were an FM wave, then h(t) could be written as

$$h(t) = A\cos\left[\omega_c t + \beta \sin \omega_m t\right] \qquad (1)$$

where $\omega_c = 10^6$ rad/sec = carrier angular frequency

and β = modulation index = 200

Now, letting $\theta(t)$ = the instantaneous phase

$$= \omega_c t + \beta \sin \omega_m t,$$

the instantaneous frequency f is given by

$$f = \dfrac{1}{2\pi} \dfrac{d}{dt} \theta(t) = \dfrac{1}{2\pi}\left[\omega_c t + \beta \sin \omega_m t\right]$$

$$= \dfrac{\omega_c}{2\pi} + \dfrac{\beta \omega_m}{2\pi} \cos \omega_m t \qquad (2)$$

or, since $\omega = 2\pi f$, eq.(2) can be expressed as

$$\omega = \omega_c + \beta \omega_m \cos \omega_m t$$

$$= 10^6 + 200 \times 500 \cos 500t$$

$$= 10^6 + 10^5 \cos 500t$$

Therefore, the modulating signal m(t) is easily recognized as m(t) = cos 500t.

(Note: In general, for FM, $h(t) = A\cos\left[\omega_c t + K\int_{-\infty}^{t} m(t)dt\right]$.)

(c) If h(t) is a phase modulated (P.M.) wave, the general form is given by eq. (1).
Similarly, from part (b),

$$\theta(t) = \omega_c t + \beta \sin\omega_m t$$

$$= 10^6 t + 200 \sin\omega_m t$$

where the information signal m(t) = sin 500t is easily obtained.

Note: The essential guideline in solving problems involving PM is to remember that the instantaneous phase is directly proportional to the modulating signal. Unlike PM, FM is the case where the time derivative of the instantaneous phase $\theta(t)$ is proportional to the waveform of the message.

The difference between cos 500t and sin 500t is not really a severe problem, the only difference being that each is 90° out of phase with respect to the other and essentially the shape of the modulating signal is the same. However, if the modulating signal was any continuous time function other than a sinusoidal waveform, then differentiation or integration of it would generate a severe distortion of the information.

FREQUENCY DEMODULATOR

● **PROBLEM** 10-31

What is the difference in phase between the signals E_X and E_Y shown in the figure, given

$$V_{o\ dc} = 1.25 \text{ volt.}$$

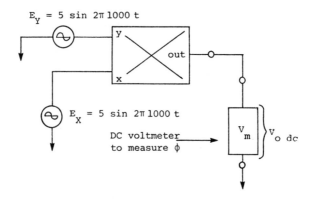

Solution: Let ϕ be the phase difference between E_X and E_Y. Then ϕ is related to the peak voltages, E_{Xpeak} and E_{Ypeak} by

$$V_{o\,dc} = \frac{E_{Xpeak}E_{Ypeak}}{20}\cos(\phi)$$

so,
$$\cos\phi = \frac{20(1.25)}{(5)(5)} = 1$$

hence,
$$\phi = \cos^{-1}(1) = 0°.$$

Therefore, there is no phase shift, i.e. $\phi = 0°$.

● **PROBLEM 10-32**

A narrowband frequency modulator is illustrated in the figure below. The output FM signal is applied to an FM demodulator.

(a) Determine the demodulator output $v_0(t)$ assuming that the frequency deviation is small.

(b) Calculate the maximum allowable frequency deviation Δf if the normalized power of the third harmonic is to be within 0.95 percent of the power associated with the fundamental frequency. Assume the modulating frequency to be equal to 50 Hz.

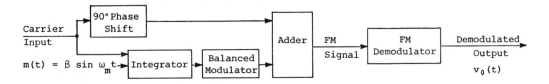

Solution: (a) For narrowband modulation, i.e. $m(t) \ll 1$,

$$\cos\left[\omega_c t + m(t)\right] \cong \cos\omega_c t - m(t)\sin\omega_c t$$
$$= \cos\omega_c t - \sin\omega_c t \left[\int_0^t \Delta\omega \cos\omega_m \lambda d\lambda\right] \quad (1)$$

where $\Delta\omega$ = angular frequency deviation and

$$\int_0^t \Delta\omega \cos\omega_m \lambda d\lambda = \frac{\Delta\omega}{\omega_m} \sin\omega_m t$$

$$= \frac{\beta\omega_m}{\omega_m} \sin\omega_m t$$

$$= \beta \sin\omega_m t$$

(Notice: $\Delta\omega = \beta\omega_m$ by definition.)

Therefore, eq.(1) becomes

$$\cos\omega_c t - \sin\omega_c t(\beta\sin\omega_m t)$$

$$= \sqrt{1 + \beta^2\sin^2\omega_m t} \cos\left[\omega_c t + \tan^{-1}(\beta\sin\omega_m t)\right]$$

(Also note: $\cos(\omega_c t + \beta\sin\omega_m t) = \cos\omega_c t \cos(\beta\sin\omega_m t)$

$$- \sin\omega_c t \sin(\beta\sin\omega_m t)$$

$$\cong \cos\omega_c t - \beta\sin\omega_m t \sin\omega_c t$$

as obtained above.)

Now, the demodulator output is

$$v_0(t) = \frac{d\theta(t)}{dt} = \frac{d}{dt}\left[\tan^{-1}(\beta \sin\omega_m t)\right]$$

$$= \frac{d}{dt}\left[\beta\sin\omega_m t - \frac{1}{3}\beta^3\sin^3\omega_m t + \frac{1}{5}\beta^5\sin^5\omega_m t \ldots\right]$$

$$= \beta\omega_m \cos\omega_m t - \beta^3\omega_m \cos\omega_m t \sin^2\omega_m t$$

$$+ \beta^5\omega_m \cos\omega_m t \sin^4\omega_m t \ldots$$

$$\cong \Delta\omega\left(\cos\omega_m t - \frac{\beta^2}{4}\cos\omega_m t + \frac{\beta^2}{2}\cos3\omega_m t - \frac{5\beta^4}{16}\cos3\omega_m t + \ldots\right) \quad (2)$$

Note: If the frequency deviation is not kept small, the odd harmonics will also appear at the demodulator output.

(b) Observing, in eq.(2),

the normalized power of the third harmonic $= \left(\frac{\beta^2}{2}\right)^2$

and the fundamental power $= 1^2 = 1$.

Hence, $\left(\frac{\beta^2}{2}\right)^2 = 0.0095(1^2)$

$$\frac{\beta^4}{4} = 0.0095$$

$$\beta = 0.44$$

Therefore, $\Delta f_{max} = \beta f_m = 0.44 \times 50\text{Hz} = 22\text{Hz}$

● PROBLEM 10-33

(a) An FM demodulator is shown in Fig. 1. The 3dB-frequency of the RC integrating network is

$$f' = \frac{1}{2\pi RC}.$$

Determine the maximum change of output over change in input frequency. (i.e. sensitivity.) Assume the center frequency of the FM waveform to be equal to f_c.

(b) With the result obtained in part (a) and information given below, calculate the change in demodulator output.

Given:

f_c = carrier frequency of FM wave = 1MHz and the change in input frequency = 1Hz.

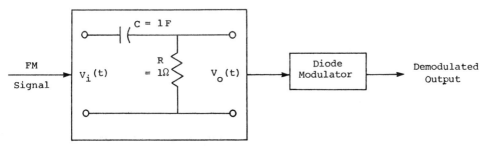

Frequency Selective Network

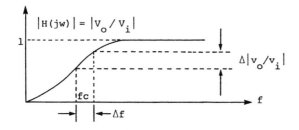

Solution: (a) The transfer function of the RC integrating network is given by $\frac{1}{1+jf/f'}$.

Hence,
$$|H(f)| = \left| \frac{1}{1+jf/f'} \right|$$

$$= \frac{1}{\sqrt{1+(f/f')^2}}$$

575

$$= \frac{1}{\sqrt{1+x^2}}\bigg|_{x=f/f'} = \left|\frac{V_o}{V_i}\right|$$

Thus, the maximum change in output over input frequency is calculated as follows:

$$\frac{d^2|H(f)|}{dx^2} = \frac{d^2}{dx^2}\frac{1}{\sqrt{1+x^2}}$$

$$= \frac{d}{dx}\frac{0-\frac{1}{2}(2x)(1+x^2)^{-\frac{1}{2}}}{(1+x^2)}$$

$$= -\frac{d}{dx}\frac{x}{(1+x^2)^{\frac{3}{2}}}$$

$$= -\frac{(1+x^2)^{\frac{3}{2}}(1) - x\left(\frac{3}{2}\right)(2x)(1+x^2)^{\frac{1}{2}}}{(1+x^2)^3}$$

$$= \frac{-(1+x^2)^{\frac{3}{2}}+3x^2(1+x^2)^{\frac{1}{2}}}{(1+x^2)^3} \qquad (1)$$

Now, setting eq.(1) equal to zero to obtain the maximum value

$$\frac{-(1+x^2)^{\frac{3}{2}} + 3x^2(1+x^2)^{\frac{1}{2}}}{(1+x^2)^3} = 0$$

$$-(1+x^2)^{\frac{3}{2}} = -3x^2(1+x^2)^{\frac{1}{2}}$$

$$\frac{(1+x^2)^{\frac{3}{2}}}{(1+x^2)^{\frac{1}{2}}} = 3x^2$$

$$1 + x^2 = 3x^2$$

$$1 = 2x^2$$

$$\frac{1}{2} = x^2$$

Therefore,

$$x = \frac{f_c}{f'}\bigg|_{f=f_c} = \frac{1}{\sqrt{2}} \quad \text{or} \quad f' = 1.414\ f_c$$

(b) The change in demodulator output is calculated by

$$\frac{d|H(f)|}{dx}\bigg|_{f=f_c} = \frac{-x}{(1+x^2)^{\frac{3}{2}}}\bigg|_{x=\frac{f_c}{f'}}$$

$$= \frac{-(f_c/f')}{[1+(f_c/f')^2]^{\frac{3}{2}}} \qquad (2)$$

Since $f_c = 1 \times 10^6 \, \text{Hz}$ and $f' = 1.414 f_c = 1.414 \times 10^6 \, \text{Hz}$, eq.(2) is equal to

$$\frac{-(1 \times 10^6 / 1.414 \times 10^6)}{[1 + (1 \times 10^6 / 1.414 \times 10^6)^2]^{\frac{3}{2}}} = \frac{-0.707}{1.84} = 0.38 \, \text{volt/Hz}.$$

BESSEL'S EQUATION

• **PROBLEM** 10-34

Find the general solution to Bessel's equation of order zero.

Solution: Bessel's equation of order p is

$$x^2 y'' + y' + (x^2 - p^2) y = 0. \tag{1}$$

The solutions to (1) are power series and are known as Bessel's function of order p. When $p = 0$, (1) reduces to

$$x^2 y'' + xy' + x^2 y = 0$$

or,

$$xy'' + y' + xy = 0 \tag{2}$$

if $x \neq 0$. Equation (2) is Bessel's equation of order zero. Since $x = 0$ is a regular singular point of (2), we seek solutions for $0 < x < R$ of the form

$$y = \sum_{n=0}^{\infty} c_n x^{n+r}, \quad c_0 \neq 0 \tag{3}$$

Then,

$$y' = \sum_{n=0}^{\infty} (n+r) c_n x^{n+r-1} \quad \text{and}$$

$$y'' = \sum_{n=0}^{\infty} (n+r)(n+r-1) c_n x^{n+r-2}.$$

Substituting these results into (2),

$$\sum_{n=0}^{\infty} (n+r)(n+r-1) c_n x^{n+r-1} + \sum_{n=0}^{\infty} (n+r) c_n x^{n+r-1}$$

$$+ \sum_{n=0}^{\infty} c_n x^{n+r+1} = 0. \tag{4}$$

Equation (4) may be simplified to give

$$\sum_{n=0}^{\infty} (n+r)^2 c_n x^{n+r-1} + \sum_{n=0}^{\infty} c_{n-2} x^{n+r-1} = 0$$

or, writing down the first two terms for the first series on the left,

$$r^2 c_0 x^{r-1} + (1+r)^2 c_1 x^r + \sum_{n=2}^{\infty}\left[(n+r)^2 c_n + c_{n-2}\right] x^{n+r-1} = 0. \tag{5}$$

If (5) is to hold, the coefficient of the lower power in x must be zero. Thus, $r^2 = 0$ and the roots of the indicial equation are $r_1 = r_2 = 0$. Similarly we obtain

$$(1 + r)^2 c_1 = 0 \tag{6}$$

and the recurrence formula

$$(n + r)^2 c_n + c_{n-2} = 0,$$

from which

$$c_n = -\frac{c_{n-2}}{(n+r)^2}, \quad n \geq 2. \tag{7}$$

Letting $r = 0$ in (6) and (7) we find that

$$c_1 = 0, \quad \text{and for n odd,}$$

$$c_1 = c_3 = c_5 = \ldots = c_{2n+1} = \ldots = 0.$$

For n even $c_2 = -\frac{c_0}{2^2}$,

$$c_4 = -\frac{c_2}{4^2} = \frac{c_0}{4^2 2^2}$$

and in general,

$$c_{2n} = \frac{(-1)^n c_0}{2^2 \cdot 4^2 \cdot 6^2 \ldots (2n)^2} = \frac{(-1)^n c_0}{(n!)^2 2^{2n}}, \quad n \geq 1.$$

Thus one solution of (2) is

$$y = \sum_{n=0}^{\infty} c_n x^{n+r} = c_0 \sum_{n=0}^{\infty} \frac{(-1)^n}{(n!)^2} \left(\frac{x}{2}\right)^{2n}. \tag{8}$$

Letting $c_0 = 1$ in (8) we obtain the Bessel function of the first kind of order zero. Thus,

$$J_0(x) = \sum_{n=0}^{\infty} \frac{(-1)^n}{(n!)^2} \left(\frac{x}{2}\right)^{2n} \tag{9}$$

$$J_0(x) = 1 - \left(\frac{x}{2}\right)^2 + \frac{1}{(2!)^2}\left(\frac{x}{2}\right)^4 - \frac{1}{(3!)^2}\left(\frac{x}{6}\right)^6 + \ldots$$

$$= 1 - \frac{x^2}{4} + \frac{x^4}{64} - \frac{x^6}{2304} + \ldots$$

We have found one solution, but the general solution of a second order differential equation contains two arbitrary constants. Hence we seek another solution. It has been shown using advanced complex analysis that when the roots of the indicial equation are both zero, then a second solution of (1) which is linearly independent of $J_0(x)$ is given by

$$y(x) = x \sum_{n=0}^{\infty} a_n x^n + J_0(x) \ln x. \tag{10}$$

How are we to find the values of the coefficients a_n? The method of reduction of order assures us that given one solution, another linearly independent solution is of the form

$$y_2(x) = J_0(x) \int \frac{e^{-\int dx/x}}{[J_0(x)]^2} dx .$$

Thus,
$$y_2(x) = J_0(x) \int \frac{dx}{[J_0(x)]^2 x} .$$

From (9), $[J_0(x)]^2 = 1 + \frac{x^2}{2} + \frac{3x^4}{32} - \frac{5x^6}{576} + \ldots$

and hence $\frac{1}{[J_0(x)]^2} = 1 + \frac{x^2}{2} + \frac{5x^4}{32} + \frac{23x^6}{576} + \ldots$

Thus $y_2(x) = J_0(x) \int \left[\frac{1}{x} + \frac{x}{2} + \frac{5x^3}{32} + \frac{23x^5}{576} + \ldots \right] dx$

$$= J_0(x) \left[\ln x + \frac{x^2}{4} + \frac{5x^4}{128} + \frac{23x^6}{3456} + \ldots \right]$$

$$= J_0(x) \ln x + \left[1 - \frac{x^2}{4} + \frac{x^4}{64} - \frac{x^6}{2304} + \ldots \right]$$

$$\cdot \left[\frac{x^2}{4} + \frac{5x^4}{128} + \frac{23x^6}{3456} + \ldots \right]$$

$$= J_0(x) \ln x + \frac{x^2}{4} - \frac{3x^4}{128} + \frac{11x^6}{13284} + \ldots \tag{11}$$

Examining (11) we see that it would be similar to (10) if we could find some general formula for the series

$$\frac{x^2}{4} - \frac{3x^4}{128} + \frac{11x^6}{13284} + \ldots$$

Observe that

$$(-1)^2 \frac{1}{2^2(1!)^2} (1) = \frac{1}{2^2} = \frac{1}{4}$$

$$(-1)^3 \frac{1}{2^4(2!)^2} \left(1+\frac{1}{2}\right) = \frac{-3}{2^4 \cdot 2^2 \cdot 2} = \frac{-3}{128}$$

$$(-1)^4 \frac{1}{2^6(3!)^2} \left(1+\frac{1}{2}+\frac{1}{3}\right) = \frac{11}{2^6 \cdot 6^2 \cdot 6} = \frac{11}{13824} .$$

Hence, $y_2(x)$ may be written as

$$y_2(x) = J_0(x) \ln x + \frac{x^2}{2^2} - \frac{x^4}{2^4(2!)^2}\left(1 + \frac{1}{2}\right)$$

$$+ \frac{x^6}{2^6(3!)^2}\left(1 + \frac{1}{2} + \frac{1}{3}\right) + \ldots$$

In general, the coefficient a_{2n} is given by

$$a_{2n} = \frac{(-1)^{n+1}}{2^{2n}(n!)^2}\left(1 + \frac{1}{2} + \frac{1}{3} + \ldots + \frac{1}{n}\right), \quad n \geq 1.$$

We therefore write the second solution as

$$y_2(x) = J_0(x) \ln x + \sum_{n=1}^{\infty} \frac{(-1)^{n+1} x^{2n}}{2^{2n}(n!)^2}$$

$$\cdot \left(1 + \frac{1}{2} + \frac{1}{3} + \ldots + \frac{1}{n}\right)$$

and the general solution is

$$y = J_0(x) + y_2(x).$$

Another form of the general solution uses a linear combination of J_0 and $y_2(x)$ to obtain the second solution of (2). This combination is

$$\frac{2}{\pi}[y_2(x) + (\gamma - \ln 2)J_0(x)] = y_0 \quad (12)$$

where $\gamma = \lim_{n\to\infty}\left(1 + \frac{1}{2} + \frac{1}{3} + \ldots + \frac{1}{n} - \ln n\right) \approx 0.5772.$

If we choose y_0 as the second solution of (2), the general solution for $0 < x < R$ is given by

$$y = c_1 J_0(x) + c_2 y_0(x).$$

● **PROBLEM 10-35**

Find the general solution of Bessel's equation of order one.

Solution: Bessel's equation of order one is

$$x^2 y'' + xy' + (x^2 - 1)y = 0 \quad (1)$$

If we write this in the form of

$$y'' + P(x)y' + Q(x)y = 0 \quad (2)$$

we find
$$P(x) = x^{-1}$$
$$Q(x) = 1 - x^{-2}$$

and $x = 0$ is a regular singular point. By the method of Frobenius let
$$y = x^\lambda \sum_{n=0}^{\infty} a_n x^n.$$

Then,
$$y' = \sum_{n=0}^{\infty} (n + \lambda) a_n x^{n+\lambda-1}$$

$$y'' = \sum_{n=0}^{\infty} (n + \lambda)(n + \lambda - 1) a_n x^{n+\lambda-2}$$

Substituting in (2)
$$y'' + \frac{y'}{x} + y - \frac{y}{x^2} = 0$$

$$\sum_{n=0}^{\infty} a_n x^\lambda \left| (n+\lambda)(n+\lambda-1)x^{n-2} + (n+\lambda)x^{n-2} + x^n + x^{n-2} \right| = 0,$$

$$\sum_{n=0}^{\infty} x^\lambda \left[[(\lambda+n)^2 - 1] a_n + a_{n-2} \right] x^n = 0.$$

Thus,
$$(\lambda^2 - 1) a_0 = 0 \tag{3}$$
$$[(x+1)^2 - 1] a_1 = 0 \tag{4}$$

and for $n \geq 2$
$$[(x+n)^2 - 1] a_n + a_{n-2} = 0$$

or
$$a_n = \frac{-1}{(\lambda+n)^2-1} a_{n-2} \quad (n \geq 2). \tag{5}$$

From (3) the indicial equation is
$$\lambda^2 - 1 = 0$$

which has solutions $\lambda = \lambda_1 = 1$ and $\lambda = \lambda_2 = -1$. Now $\lambda_1 - \lambda_2 = 2$ which is a positive integer. By the method of Frobenius we obtain the solution
$$y_2(x) = d_{-1} y_1(x) \ln x + x^{\lambda_2} \sum_{n=0}^{\infty} d_n(\lambda_2) x^n \tag{6}$$

where d_{-1}, d_n are constants and
$$y_1(x) = x^{\lambda_1} \sum_{n=0}^{\infty} a_n(\lambda_1) x^n. \tag{7}$$

Substituting $\lambda = \lambda_1 = 1$ into (4) and (5) we obtain

$$a_1 = 0$$

and

$$a_n = \frac{-1}{n(n+2)} a_{n-2}. \qquad (n \geq 2)$$

Since $a_1 = 0$ it must follow that $a_3 = a_5 = a_7 = \ldots = 0$. Furthermore,

$$a_2 = \frac{-1}{2(4)} a_0 = \frac{-1}{2^2 1! 2!} a_0,$$

$$a_4 = \frac{-1}{4(6)} a_2 = \frac{1}{2^4 2! 3!} a_0,$$

$$a_6 = \frac{-1}{6(8)} a_4 = \frac{-1}{2^6 3! 4!} a_0$$

and in general

$$a_{2k} = \frac{(-1)^k}{2^{2k} k! (k+1)!} a_0$$

for $k = 1, 2, 3, \ldots$

Substituting this in (7)

$$y_1(x) = a_0 x \sum_{n=0}^{\infty} \frac{(-1)^n}{2^{2n} n! (n+1)!} x^{2n}. \qquad (8)$$

Now let us use $\lambda = \lambda_2 = -1$. Substituting this in (5)

$$a_n = \frac{-1}{n(n-2)} a_{n-2}.$$

We cannot let $n = 2$ for we will get zero in the denominator. We must use the modified Frobenius method. From (5)

$$a_2 = \frac{-1}{(\lambda+3)(\lambda+1)} a_0,$$

$$a_4 = \frac{1}{(\lambda+5)(\lambda+3)^2(\lambda+1)} a_0, \quad \text{etc.}$$

Hence we define

$$y(\lambda, x) = a_0 \left[x^\lambda - \frac{x^{\lambda+2}}{(\lambda+3)(\lambda+1)} + \frac{x^{\lambda+4}}{(\lambda+5)(\lambda+3)^2(\lambda+1)} + \ldots \right].$$

Since $\lambda - \lambda_2 = \lambda + 1$

$$(\lambda - \lambda_2) y(\lambda, x) =$$

$$a_0 \left[(\lambda+1) x^\lambda - \frac{x^{\lambda+2}}{(\lambda+3)} + \frac{x^{\lambda+4}}{(\lambda+5)(\lambda+3)^2} + \ldots \right]$$

and
$$\frac{\partial}{\partial x}\left[(\lambda-\lambda_2)y(\lambda,x)\right] = a_0\left[x^\lambda + (\lambda+1)x^\lambda \ln x\right.$$
$$+ \frac{1}{(\lambda+3)^2}x^{\lambda+2} - \frac{1}{(\lambda+3)}x^{\lambda+2}\ln x$$
$$- \frac{x^{\lambda+4}}{(\lambda+5)^2(\lambda+3)^2} - \frac{2}{(\lambda+5)(\lambda+3)^3}\cdot x^{\lambda+4}$$
$$\left. + \frac{1}{(\lambda+5)(\lambda+3)^2}x^{\lambda+4}\ln x + \ldots\right].$$

The modified Frobenius method says that

$$y_2(x) = \frac{\partial}{\partial x}\left[(\lambda-\lambda_2)y(\lambda,x)\right]\Big|_{\lambda=\lambda_2=-1}$$

which yields immediately

$$y_2(x) = a_0\left[x^{-1} + 0 + \frac{x}{4} - \frac{x \ln x}{2} - \frac{1}{64}x^3 - \frac{2}{32}x^3 + \frac{x^3 \ln x}{16} + \ldots\right]$$

$$= -\frac{1}{2}(\ln x)a_0 x\left[1 - \frac{x^2}{8} + \ldots\right] + a_0\left[\frac{1}{x} + \frac{x}{4} - \frac{5x^3}{64} + \ldots\right]$$

or $\quad y_2(x) = -\frac{1}{2}(\ln x)y_1(x) + \frac{a_0}{x}\left[1 + \frac{x^2}{4} - \frac{5}{64}x^4 + \ldots\right].$ (9)

Equations (8) and (9) represent the general solution

$$y(x) = c_1 y_1(x) + c_2 y_2(x)$$

of the first order Bessel equation.

● **PROBLEM 10-36**

Find solutions of Bessel's equation
$$x^2 y'' + xy' + (x^2 - p^2)x = 0 \qquad (1)$$
when p is not an integer.

Solution: Note that $x = 0$ is a regular singular point of (1). Hence we are interested in constructing solutions of (1) valid in a neighborhood around $x = 0$. Using the method of Frobenius, assume

$$y = x^\lambda(a_0 + a_1 x + \ldots a_r x^r + \ldots)$$
$$= \sum_{n=0}^{\infty} a_n x^{n+\lambda}$$
(2)

is a solution of (1).

The problem is to determine λ and the coefficients a_0, a_1, ... so that (2) satisfies the differential equation (1).

Since (2) is a solution, substitution of it and its derivatives into (1) should result in zero. Differentiating $y(x)$ twice,

$$y' = \sum_{n=0}^{\infty} (n+\lambda)a_n x^{n+\lambda-1};$$

$$y'' = \sum_{n=0}^{\infty} (n+\lambda)(n+\lambda-1)a_n x^{n+\lambda-2}.$$

Substituting into (1)

$$\sum_{n=0}^{\infty} (n+\lambda)(n+\lambda-1)a_n x^{n+\lambda} + \sum_{n=0}^{\infty} (n+\lambda)a_n x^{n+\lambda}$$

$$+ \sum_{n=0}^{\infty} a_n x^{n+\lambda+2} - \sum_{n=0}^{\infty} p^2 a_n x^{n+\lambda} = 0.$$

Combining terms with like powers of x,

$$\sum_{n=0}^{\infty} \{(n+\lambda)(n+\lambda-1) + (n+\lambda) - p^2\} a_n x^{n+\lambda} + \sum_{n=0}^{\infty} a_n x^{n+\lambda+2} = 0$$

Replacing n by $(n-2)$ in the second summation and simplifying the first,

$$\sum_{n=0}^{\infty} \left[(n+\lambda)^2 - p^2\right] a_n x^{n+\lambda} + \sum_{n=2}^{\infty} a_{n-2} x^{n+\lambda} = 0 \qquad (3)$$

From (3),

$$(\lambda^2 - p^2)a_0 x^\lambda + \left[(\lambda+1)^2 - p^2\right]a_1 x^{\lambda+1}$$

$$+ \sum_{n=2}^{\infty} \left\{\left[(n+\lambda)^2 - p^2\right]a_n + a_{n-2}\right\} x^{n+\lambda} = 0 \qquad (4)$$

If (4) is to hold, the coefficient of each power of x must be zero. Thus,

$$(\lambda^2 - p^2)a_0 = 0 \qquad (5)$$

$$[(\lambda+1)^2 - p^2]a_1 = 0 \qquad (6)$$

$$[(n+\lambda)^2 - p^2]a_n = -a_{n-2} \qquad (7)$$

Equation (5), the contribution of the $n = 0$ term, is called the indicial equation. From (5), since we assume $a_0 \neq 0$,

$$\lambda = \pm p. \tag{8}$$

Then, from (6), $a_1 = 0$. Examining (7), see that this forces all the odd coefficients to be zero. Thus, $a_1 = a_3 = \ldots = a_{2n+1} = \ldots = 0$. When n is even, from (7) again,

$$a_n = \frac{-1}{(\lambda+n)^2 - p^2} a_{n-2}. \tag{9}$$

We have found the values of λ and discovered a recursive relation for the a_n. Hence, we are ready to construct the general solution. From (8), when $\lambda = -p$, substitution into (2) gives the solution.

$$y_1(x) = a_0 x^p \{1 - \frac{1}{2(2p+2)} x^2 + \frac{1}{4(2p+4)2(2p+2)} x^4 - \ldots\}, \tag{10}$$

while letting $\lambda = -p$ yields

$$y_2(x) = a_0' x^{-p} \{1 + \frac{1}{2(2p-2)} x^2 + \frac{1}{4(2p-4)2(2p-2)} x^4 + \ldots\} \tag{11}$$

where the terms in the denominator were obtained by substituting $\lambda = p$ and $\lambda = -p$ respectively into (9).

The general solution is a linear combination of (10) and (11).

We offer a brief justification for why we restricted p to non-integral values. Assume, contrary to hypothesis that p is a positive integer. Then, from (9), letting $\lambda = -p$,

$$a_n = \frac{1}{n(2p-n)} a_{n-2}, \quad n \geq 2. \tag{12}$$

Since p is a positive integer, for some n we must have $(2p - n) = 0$ which means a_n is undefined. When p is a negative integer, letting $\lambda = p$ yields a similar result from (9).

● **PROBLEM 10-37**

Find one solution of Bessel's equation of order p using the method of Frobenius.

Solution: Bessel's equation is a second order differential equation with variable coefficients. Thus, it is of the form

$$y'' + p(x)y' + a(x)y = 0. \tag{1}$$

In particular, Bessel's equation of order p is:

$$x^2 y'' + xy' + (x^2 - p^2)y = 0 \tag{2}$$

or, writing (2) to conform with (1),

585

$$y'' + \frac{1}{x} y' + \frac{x^2-p^2}{x^2} y = 0, \qquad (3)$$

where p is a parameter which denotes the order of the equation. Examining (3) we see that the point $x = 0$ is a singular point, i.e. the coefficients $p(x)$, $a(x)$ are not analytic there, but since $xp(x)$ and $x^2 a(x)$ are analytic, the singularity is weak and hence $x = 0$ is a regular singular point. A solution of (3) around $x = 0$ will be a valid solution in the interval $|x| < R$.

We now use the method of Frobenius to obtain a solution of (3). Assume that

$$y(x) = x^m(a_0 + a_1 x + \ldots + a_n x^n + \ldots) \qquad (4)$$

$$= x^m \sum_{n=0}^{\infty} a_n x^n$$

$$= \sum_{n=0}^{\infty} a_n x^{m+n} .$$

Then, differentiating (4),

$$y' = ma_0 x^{m-1} + (m+1)a_1 x^m + \ldots + (n+m)a_n x^{m+n-1} + \ldots$$

$$= \sum_{n=0}^{\infty} (m+n)a_n x^{m+n-1}, \qquad (5)$$

and $y'' = m(m-1)a_0 x^{m-2} + (m+1)(m)a_1 x^{m-1} + \ldots$

$$= \sum_{n=0}^{\infty} (m+n)(m+n-1)a_n x^{m+n-2} . \qquad (6)$$

Substitute these expressions for y, y', y'' into (2). We thus obtain

$$x^2 y'' = \sum_{n=0}^{\infty} (n+m-1)(n+m)a_n x^{n+m}$$

$$xy' = \sum_{n=0}^{\infty} (n+m)a_n x^{n+m}$$

$$x^2 y = \sum_{n=0}^{\infty} a_n x^{m+n+2} = \sum_{n=2}^{\infty} a_{n-2} x^{n+m}$$

$$-p^2 y = \sum_{n=0}^{\infty} - p^2 a_n x^{n+m}, \quad \text{or}$$

$$\sum_{n=0}^{\infty} [(n+m)(n+m-1) + (n+m) - p^2] a_n x^{n+m}$$

$$+ \sum_{n=2}^{\infty} a_{n-2} x^{n+m} = 0 \tag{7}$$

where we have combined the coefficients of like powers of x. Equation (7) may be rewritten as

$$[m(m-1) + m - p^2] a_0 x^m + [(m+1)m + (m+1) - p^2] a_1 x^{m+1}$$

$$+ \sum_{n=2}^{\infty} [((n+m)(n+m-1) + (n+m) - p^2) a_n + a_{n-2}] x^{n+m} = 0 \tag{8}$$

Since, by assumption $a_0 \neq 0$ (if it were, then (4) would take a different form), we find the indicial equation is

$$m^2 - p^2 = 0. \tag{9}$$

The roots of (9) are $m_1 = p$ and $m_2 = -p$. Further, we have, from (8),

$$[(m+1)^2 - p^2] a_1 = 0, \tag{10}$$

and in general

$$[(m+n)^2 - p^2] a_n + a_{n-2} = 0, \quad n \geq 2$$

or, $\quad a_n = -\dfrac{1}{(m+n)^2 - p^2} a_{n-2} \quad n \geq 2. \tag{11}$

Equation (11) is the general recurrence formula. Since $m = \pm p$, (10) implies that,

$$a_1 = 0, \tag{12}$$

$$a_n = \frac{-1}{n(2p+n)} a_{n-2} \quad (n \geq 2).$$

Hence, $a_1 = a_3 = a_5 = \ldots = 0$ from (11) and (12) and

$$a_2 = \frac{-1}{2(2p+2)} a_0 = \frac{-1}{2^2 1!(p+1)} a_0,$$

$$a_4 = \frac{-1}{2^2 \cdot 2(p+2)} a_2 = \frac{1}{2^4 2!(p+2)(p+1)} a_0,$$

$$a_6 = \frac{-1}{2^2 3(p+3)} a_4 = \frac{-1}{2^6 3!(p+3)(p+2)(p+1)} a_0$$

and generally,

$$a_{2k} = \frac{(-1)^k}{2^{2k} k!(p+k)(p+k-1) \ldots (p+1)} a_0, \quad k \geq 1.$$

Therefore, the solution to (2) is

$$y_1(x) = x^m \sum_{n=0}^{\infty} a_n x^n = x^p \left[a_0 + \sum_{k=1}^{\infty} a_{2k} x^{2k} \right]$$

$$= a_0 x^p \left[1 + \sum_{k=1}^{\infty} \frac{(-1)^k x^{2k}}{2^{2k} k! (p+k)(p+k-1) \cdots (p+2)(p+1)} \right].$$

If we let $a_0 = \dfrac{1}{2^p \Gamma(p+1)}$, we may express the solution as

$$y_1(x) = \frac{1}{2^p \Gamma(p+1)} x^p + \sum_{k=1}^{\infty} \frac{(-1)^k x^{2k+p}}{2^{2k+p} k! \Gamma(p+k+1)}$$

where we have used the property of the gamma function that

$$\Gamma(p+k+1) = (p+k)(p+k-1) \cdots (p+2)(p+1)\Gamma(p+1).$$

Thus, the final solution is

$$y_1(x) = \sum_{k=0}^{\infty} \frac{(-1)^k x^{2k+p}}{2^{2k+p} k! \, \Gamma(p+k+1)} \, J_p(x).$$

CHAPTER 11

PULSE-MODULATION SYSTEMS (PM)

THE SAMPLING THEOREM

● **PROBLEM 11-1**

> The signal $u(t) = \cos(3\pi t) + 0.125 \cdot \cos(8\pi t)$ is periodically sampled every T_s seconds,
>
> (a) What is the maximum value of the sampling time T_s.
>
> (b) Given a sampling signal $S(t) = 4\sum_{n=-\infty}^{\infty} \delta(t-0.125n)$ and the sampled signal $u_s(t) = u(t)S(t)$ where $u_s(t) = \sum_{n=-\infty}^{\infty} I_n \delta(t-0.125n)$, determine I_0, I_1, I_2 and prove that $I_{n+4} = I_n$.
>
> (c) Determine the minimum bandwidth for a low-pass filter so that the recovered signal will be distortionless.

Solution:

(a) From the sampling theorem, in order to sample a signal without overlapping, the condition $T_s \leq 1/(2f_m)$ must be satisfied.

Hence, for maximum T_s,
$$T_{s(max)} = \frac{1}{2f_{m(max)}} \qquad (1)$$

(where $f_{m(max)}$ = maximum modulating frequency.)

Thus, since $\omega_{m(max)} = 2\pi f_{m(max)}$

589

$$8\pi = 2\pi f_{m(max)}$$

$$f_{m(max)} = 4 \text{Hz}$$

Now, substituting $f_{m(max)}$ into Eq. (1),

$$T_{S(max)} = \frac{1}{2f_{m(max)}} = \frac{1}{2(4)} = 0.125 \text{ sec}$$

(b) Since the sampled signal is given by $u_s(t) = u(t)S(t)$, substituting for $u(t)$ and $S(t)$,

$$u_s(t) = (\cos 3\pi t + 0.125 \cos 8\pi t) \times 4 \sum_{n=-\infty}^{\infty} I_n \delta(t-0.123n),$$

$$= \sum_{n=-\infty}^{\infty} 4(\cos 3\pi t + 0.125 \cos 8\pi t) \cdot \delta(t-0.125n)$$

$$= \sum_{n=-\infty}^{\infty} 4(\cos 3\pi n + 0.125 \cos 8\pi n) \cdot \delta(t-0.125n)$$

$$= \sum_{n=-\infty}^{\infty} I_n \cdot \delta(t-0.125n),$$

(where I_n = strength of each impulse.)

Since $I_n = 4(\cos 3\pi n + 0.125 \cos 8\pi n)$,

$$I_0 = 4(1+0.125) = 4.5$$

$$I_1 = 4(\cos 3\pi + 0.125 \cos 8\pi) = -3.5$$

and $I_2 = 4(\cos 6\pi + 0.125 \cos 16\pi) = 4.5$

Proof: $I_{n+4} = 4\left[\cos 3\pi(n+4) + 0.125 \cos 8\pi(n+4)\right]$

$$= 4\left[\cos(3\pi n + 12\pi) + 0.125 \cos(8\pi n + 24\pi)\right]$$

$$= 4\left[\cos 3\pi n + 0.125 \cos 8\pi n\right] = I_n$$

(c) Since $f_{m(max)}$ the highest frequency component of $u(t)$ is equal to 4Hz, the minimum bandwidth necessary for the filter is 4Hz in order to reconstruct the signal without distortion.

• **PROBLEM 11-2**

A periodic train of pulses given by $S(t) = I \sum_{n=-\infty}^{\infty} \delta(t-nT_s)$ is used to sample a bandpass signal $u(t) = \cos(11\omega_0 t) + \cos(12\omega_0 t) + \cos(13\omega_0 t)$. The sampling operation is performed in a multiplier as shown in the figure.

(a) Determine the minimum sampling frequency f_s.

(b) Write an expression for $u_s(t)$.

(c) Write expressions for $u_o(t)$ if the low-pass filter's bandwidth, BW = $2f_o$ and also if BW = $4f_o$.

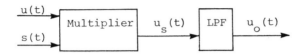

Solution:

(a) To ensure no overlapping between spectrum patterns, the minimum sampling frequency given by $f_s = 2(f_H - f_L)$ is allowable for either f_H or f_L, where f_H and f_L are harmonics of f_s.

Hence, $f_s = 2(f_H - f_L) + \varepsilon_1$

$\qquad = \varepsilon_1 + 2(13f_o - 11f_o)$

$\qquad = 4f_o + \varepsilon_1$

where ε_1 is an arbitrary small positive number.

(b) Due to the multiplier, $u_s(t)$ is equal to the product of $u(t)$ and $S(t)$.

Now, notice that $S(t)$ can be written as an exponential Fourier series as follows:

Let $f(t) = S(t) = \sum_{n=-\infty}^{\infty} \delta(t - nT_S)$

Then $f(t) = \sum_{n=-\infty}^{\infty} F_n e^{(jn\omega_o t)}$

where $F_n = \frac{1}{T_S} \int_{-T/2}^{T_S/2} f(t) e^{-(jn\omega_o t)} dt = \frac{1}{T_S} \int_{-T/2}^{T_S/2} S(t) e^{(-jn\omega_o t)} dt$

$\qquad = \frac{1}{T_S} \left(e^{-jn\omega_o t} \right)_{t=0} = \frac{1}{T_S}$

(Note: $S(t)$ is a train of impulses with strength I and perios T_S.)

Therefore, $f(t) = \sum_{n=-\infty}^{\infty} \frac{1}{T_S} e^{(jn\omega_o t)} = \frac{1}{T_S} \sum_{n=-\infty}^{\infty} (\cos n\omega_o t + j\sin n\omega_o t)$

$$= \frac{1}{T_s} \cdot 2 \cdot \sum_{n=0}^{\infty} \cos n\omega_o t$$

$$= \frac{1}{T_s} + \left[\frac{2}{T_s} \sum_{n=1}^{\infty} \cos(n\omega_o t)\right]$$

$$= S(t)$$

Now, $u_s(t)$ can be written as

$$u_s(t) = u(t)S(t) = \left[(\cos 11\omega_o t + \cos 12\omega_o t + \cos 13\omega_o t)\right] \times$$

$$\frac{1}{T_s}\left[1 + 2\sum_{n=1}^{\infty} \cos n\omega_o t\right]$$

$$= (4f_o + \varepsilon_1)\left\{\cos 10\omega_o t + \cos 11\omega_o t + \cos 12\omega_o t + \right.$$

$$\sum_{n=1}^{\infty}\left[\cos(10 + 4n + \varepsilon_2 n)\omega_o t + \cos(10 - 4n - \varepsilon_2 n)\omega_o t\right] +$$

$$\sum_{n=1}^{\infty}\left[\cos(11 + 4n + \varepsilon_2 n)\omega_o t + \cos(11 - 4n - \varepsilon_2 n)\omega_o t\right] +$$

$$\left.\sum_{n=1}^{\infty}\left[\cos(12 + 4n + \varepsilon_2 n)\omega_o t + \cos(12 - 4n - \varepsilon_2 n)\omega_o t\right]\right\}$$

where $\varepsilon_2 = \varepsilon_1/f_o$ and $T_s = \frac{1}{f_s} = \frac{1}{4f_o + \varepsilon_1}$

(c) For a filter with BW = $2f_o$

Observe: The components in $u_o(t)$ are those of $u_s(t)$ which have a bandwidth $\leq 2f_o$, since higher frequency components are rejected by the filter. Considering individual components of $u_s(t)$ for different values of n, it is obvious that the components of interest are,

(i) $\cos(2 - 2\varepsilon_2)\omega_o t$ corresponding to $\cos(10 - 4n - \varepsilon_2 n)\omega_o t$ for n = 2,

(ii) $\cos(1 + 3\varepsilon_2)\omega_o t = \cos(-1 - 3\varepsilon_2)\omega_o t$ corresponding to $\cos(11 - 4n - \varepsilon_2 n)\omega_o t$ for n = 3,

and (iii) $\cos 3\varepsilon_2 \omega_o t = \cos(-3\varepsilon_2)\omega_o t$ corresponding to $\cos(12 - 4n - \varepsilon_2 n)\omega_o t$ for n = 3.

Hence, $u_o(t) = (4f_o + \varepsilon_1)\left[\cos(2 - 2\varepsilon_2)\omega_o t + \cos(1 + 3\varepsilon_2)\omega_o t + \cos 3\varepsilon_2 \omega_o t\right]$

$$\simeq 4f_o(\cos 2\omega_o t + \cos\omega_o t + 1)$$

Similarly, for a filter with BW = $4f_o$, as in the previous case, evaluating the components of interest, we have,

$$u_o(t) = \left(4f_o + \varepsilon_1\right)\cdot\cos(2 + 3\varepsilon_2)\omega_o t + \cos(3 - 2\varepsilon_2)\omega_o t$$
$$+ \cos(4 - 2\varepsilon_2)\omega_o t$$

$$\simeq 4f_o(\cos 2\omega_o t + \cos 3\omega_o t + \cos 4\omega_o t)$$

• **PROBLEM 11-3**

Prove the sampling theorem following the procedures given below:

(a) Write a Fourier series expression for the sampling waveform S(t) shown in Fig. (1).

(b) The frequency spectrum of the signal m(t) which is to be sampled is shown in Fig. (2). The two signals S(t) and m(t) are then applied to a multiplier where m(t) is sampled at intervals T_s. Write an expression for the product S(t)m(t) and sketch its spectrum representation. Assume that m(t) is bandlimited to f_m.

(c) Show that in order to have a distortionless sampled signal, the sampling time must be chosen as

$$T_s = \frac{1}{f_s} < \frac{1}{2f_m}.$$

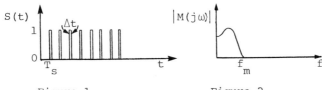

Figure 1.   Figure 2.

Solution:

(a) The trigonometric Fourier Series for a periodic pulse train is given by

$$S(t) = C_o + \sum_{n=1}^{\infty} C_n \cos\left(\frac{2\pi n t}{T_s} + \phi_n\right)$$

where

$$C_o = \frac{1}{T_s}\int_0^{T_s} S(t)dt = \frac{\Delta\pi}{T}$$

$$a_n = \frac{2}{T_s}\int_0^{T_s} S(t)\cos\left(\frac{2\pi nt}{T}\right)dt = \frac{1}{n}\sin\left(\frac{2\pi\Delta T}{T_s}n\right)$$

$$b_n = \frac{2}{T_s}\int_0^{T_s} S(t)\sin\left(\frac{2\pi nt}{T}\right)dt = \frac{1}{\pi n}\left[1-\cos\left(\frac{2\pi\Delta T}{T_s}n\right)\right]$$

Thus,

$$C_n = \sqrt{a_n^2 + b_n^2} = \sqrt{\frac{2}{(\pi n)^2}\left[1 - \cos\left(\frac{2\pi\Delta T}{T_s}n\right)\right]}$$

and

$$\phi_n = -\tan^{-1}\left(\frac{b_n}{a_n}\right) = -\tan^{-1}\left[\frac{1 - \cos\left(\frac{2\pi\Delta T}{T_s}n\right)}{\sin\left(\frac{2\pi\Delta T}{T_s}n\right)}\right]$$

(b) The product $S(t)m(t)$ sampled at intervals of T_s is given by the Fourier series for $S(t)$ multiplied by $m(t)$ as follows:

$$S(t)m(t) = C_o m(t) + \sum_{n=1}^{\infty} C_n \cdot m(t) \cdot \cos\left(\frac{2\pi n}{T_s}t + \phi_n\right).$$

Let the Fourier transform of $S(t)m(t)$ be $H(f)$, the spectral range of which can be seen from the diagram below:

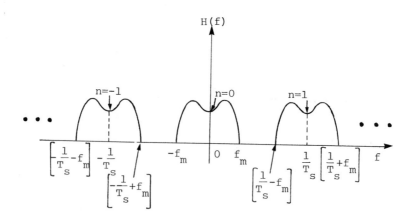

(c) Observe from the spectrum diagram above, for no overlap between the $n = 0$ and $n=1$ terms, it must be true that

$$\left(\frac{1}{T_s} - f_m\right) > f_m \qquad \text{thus} \qquad \frac{1}{T_s} > 2f_m$$

Hence, $f_s > 2f_m$ (where $\frac{1}{T_s} = f_s$) and so on. No matter which

term is chosen, the condition for no overlapping is $f_s > 2f_m$,

hence, $m(t)$ is recoverable.

• **PROBLEM 11-4**

For the following signals, calculate the Nyquist rate and the Nyquist interval.

(a) $S_a(200t)$

(b) $[S_a(200t)]^2$

Solution: The sampling theorem states that a signal bandlimited to a frequency f_m cps is uniquely determined by its values at uniform intervals less than $\frac{1}{2f_m}$ seconds apart. Thus, if the Fourier transform of a signal $m(t)$ is zero for frequency larger than f_m, then the complete information about $m(t)$ is contained in its samples spaced uniformly at a distance less than $\frac{1}{2f_m}$ seconds.

In other words, $M(\omega)$ will repeat periodically without distortion as long as

$$\omega_0 \geq 2\omega_m \quad \text{or} \quad \frac{2\pi}{T} \geq 2(2\pi f_m)$$

$$\text{where } T \leq \frac{1}{2f_m} \quad \text{(Nyquist interval)} \tag{1}$$

(a) Given the sampling function $S_a(200t)$, $\omega_m = 200 \frac{\text{rad}}{\text{sec}}$

Hence, the minimum sampling rate or Nyquist rate =

$$f_m = \frac{\omega_m}{2\pi} = \frac{200}{2\pi} = \frac{100}{\pi} \text{ cps.}$$

Now, the Nyquist interval is determined by the relationship in Eq. (1).

$$T_{Nyq.} = \frac{1}{2f_m} = \frac{1}{2\left(\frac{100}{\pi}\right)} = \frac{\pi}{200} \text{ sec.}$$

(b) Similarly, for $S_a^2(200t)$,

Note: $S_a(x) = \frac{\sin x}{x}$, therefore $S_a^2(x) = \frac{\sin^2 x}{x^2} = \frac{(1-\cos 2x)}{2x^2}$

Hence, $\omega_m = 2(200) = 400 \frac{\text{rad}}{\text{sec}}$ and $f_m = \frac{400}{2\pi} = \frac{200}{\pi}$ cps.

Thus, $T_{Nyq.} = \frac{1}{2f_m} = \frac{1}{2\left(\frac{200}{\pi}\right)} = \frac{\pi}{400}$ sec.

Note: An increase in sampling rate above the Nyquist rate increases the width of the guard band, thereby easing the problem of filtering.

• **PROBLEM 11-5**

A) Given a sampling function $S_a(200\pi t)$, sketch its spectrum representation.

B) A train of pulses represented by the function $\delta_T(t)$ is used to sample the above signal. Sketch the spectrum of $\delta_T(t)$ for

$$T = \frac{1}{400}, \frac{1}{200} \text{ and } \frac{1}{100} \text{ second, respectively.}$$

C) In each of the three cases in part (B), sketch the spectrum of the sampled function by graphical convolution.

Solution:

(A) The spectrum of $S_a(200\pi t)$:

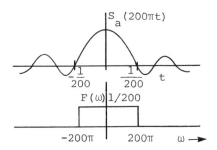

(B) The spectrum of $\delta_T(t)$ for $T = \frac{1}{400}, \frac{1}{200}, \frac{1}{100}$ sec.:

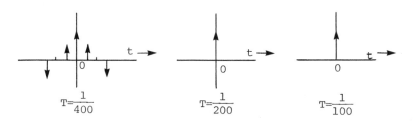

(C) The spectrum of the sampled function by graphical convolution and the sampled function in each case as in (B) are shown below:

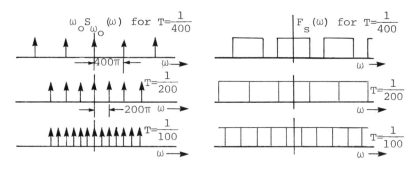

● **PROBLEM** 11-6

A singer's performance is to be recorded by sampling and storing the sample values. Assuming the highest frequency tone to be recorded is 15,800 Hertz, what is the minimum sampling frequency that can be used? How many samples would be required to store a three minutes performance? If each sample is quantized into 128 levels, how many binary digits (bits) would be required to store the three minutes performance?

Solution: An important and useful result in signal analysis is the property that a band-limited signal can be uniquely represented by a set of samples taken at time intervals spaced $\frac{1}{2BW}$ seconds apart, where BW is the signal bandwidth in Hertz. Thus, for a signal whose highest frequency tone is 15,800 Hertz, the minimum sampling frequency must be 2 x 15,800 Hz = 3.16×10^4 Hertz.

(Note: Minimum sampling rate = 3.16×10^4 samples/second)

For a three minutes performance, the number of samples is:

$3.16 \times 10^4 \frac{\text{samples}}{\text{second}} \times 60 \frac{\text{second}}{\text{minute}} \times 3 \text{ minutes} = 5.688 \times 10^6$ samples.

For 128 levels per sample,

$(2)^7 = 128$. Thus, seven bits per sample are used to quantify the signal.

Therefore, 7 bits/sample x 5.688×10^6 samples = 3.9816×10^7 bits are used to quantize the three minutes performance of a singer whose highest frequency tone is 15,800 Hertz.

● **PROBLEM** 11-7

Design a rotating commutator system for the signals $F_1(t) = \cos \omega_0 t$, $F_2(t) = 0.5 \cos \omega_0 t$, $F_3(t) = 2 \cos 2\omega_0 t$, and $F_4(t) = \cos 4\omega_0 t$ by referring to the time-division multiplexing system shown in the figure.

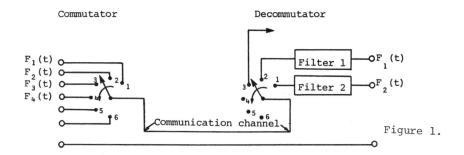

Figure 1.

Solution: To design a rotating commutator, we must first calculate the minimum allowable sampling rate for each signal.

From the sampling theorem, in order that a given signal is sampled, the sampling rate (f_s) must be at least twice the signal frequency.

Thus, for the signals $F_1(t)$, $F_2(t)$, $F_3(t)$ and $F_4(t)$, when considered individually, the corresponding sampling rates are

$$
\begin{aligned}
\text{for } F_1(t) &\quad f_s > 2 f_o \\
F_2(t) &\quad f_s > 2 f_o \\
F_3(t) &\quad f_s > 4 f_o \\
F_4(t) &\quad f_s > 8 f_o
\end{aligned}
$$

where f_s is the sampling rate and f_o is the base frequency of ω_0.

The rotating commutator system is shown in Fig. (2)

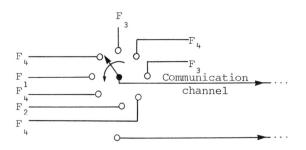

Figure 2.

To facilitate proper sampling of all 4 signals, the suitable sampling rate should be twice the highest frequency among the given signal frequencies. Thus, the minimum sampling rate is $8 f_o$. Therefore, the commutator speed is chosen to be $8 f_o$ revolutions per second

• **PROBLEM 11-8**

Write an expression for f(t) using the "reconstitution formula" and approximate the maximum value of f(t), given that the time samples for f(t) taken at Nyquist interval are {...0, 0, 10, 30, 50, 50, -40, -10,...}.

$\left(\text{Hint: } f(t) = \sum_n f(nT) \text{ sinc}\left[\omega_n (t - nT)\right] \text{ (reconstitution formula)} \right)$

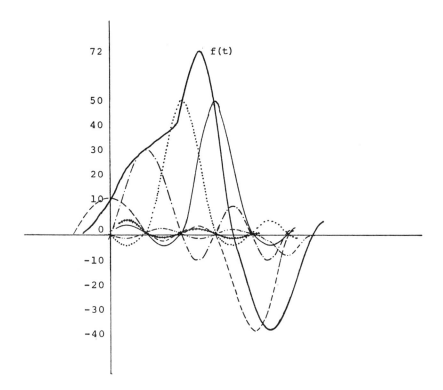

Solution: Given the time samples {...0, 0, 10, 30, 50, 50, -40, -10,...}, f(t) can be reconstructed via the sampling theorem using the "reconstitution formula",

$$f(t) = \sum_n f(nt) \, \text{sinc}\left[\omega_n (t - nT)\right]$$

It can be assumed (without loss of generality) that 10 is the sample at t = 0; f(t) can be written as

$$f(t) = 10 \, \text{sinc}\left(\omega_m t\right) + 30 \, \text{sinc}\left[\omega_m (t - T)\right] + 50 \, \text{sinc}\left[\omega_m (t - 2T)\right]$$

$$+ 50 \, \text{sinc}\left[\omega_m (t - 3T)\right] - 40 \, \text{sinc}\left[\omega_m (t - 4T)\right]$$

$$- 10 \, \text{sinc}\left[\omega_m (t - 5t)\right] \quad (1)$$

where $\omega_m = \pi/T$.

To estimate the maximum value if f(t), sketch of an approximate waveform corresponding to equation (1) is shown below.

Now, to find f(t), add the shifted sinc functions together. The maximum value of f(t) occurs at $t = \frac{5T}{2}$ with a value of approximately 72. One could compute the maximum value from the formula given before for f(t).

• **PROBLEM 11-9**

A train of pulses having duration δ, separated by the sampling time T_s are transmitted through a communication channel with bandwidth of 200 kHz. Determine T_s. How should δ be chosen for maximum transmission?

Solution: The time waveform may be expressed as

$$f(t) = \sum_{n=-\infty}^{\infty} a_n \, p(t - nT_s)$$

where a_n are the pulse heights and $p(t)$ is a typical pulse. Let $a(t)$ be a waveform which is band-limited to 200 kHz and such that $a(nT_s) = a_n$. Then $f(t)$ may be represented as

$$f(t) = \left[a(t) \cdot \sum_{n=-\infty}^{\infty} \delta(t - nT_s) \right] * p(t)$$

with transform

$$F(j\omega) = \frac{P(j\omega)}{T_s} \sum_{n=-\infty}^{\infty} A\left(j\omega - \frac{n 2\pi}{T_s} \right)$$

as sketched:

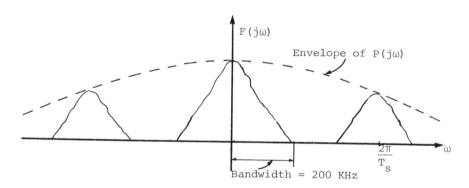

In order to avoid spectral overlap, which would distort the transmitted information, it is necessary that the sampling frequency

$$\frac{1}{T_s} \leq 2 f_m = 2 \times 200 \times 10^3 \text{ Hz}.$$

Hence, the sampling time must be $T_s \leq 2.5 \, \mu$ seconds.

And to minimize spectral distortion from $P(j\omega)$, the pulse width δ should be chosen as small as practically possible.

• **PROBLEM 11-10**

Given the equalities

$$\sum_{k=0}^{2n} e^{j\frac{2\pi nk}{2N+1}} = \sum_{k=0}^{2n} e^{j\frac{2\pi n}{T}\frac{kT}{(2n+1)}} \begin{cases} = 2N+1 \text{ if n is an integral multiple of } (2N+1), \text{ including zero.} \\ = 0 \text{ for other values of n.} \end{cases}$$

and

$$\sum_{n=-N}^{+N} e^{j\frac{2\pi n}{T}x} = (1) + 2 \cdot \sum_{n=1}^{N} \cos\frac{2\pi n}{T}x$$

$$= \frac{\sin\left[\frac{(2N+1)\pi}{T}x\right]}{\sin\left(\frac{\pi}{T}x\right)} \quad \text{(from C.R. Cahn's sampling theorem).}$$

If $f(t)$ is periodic with period T and considering Fourier coefficients up to the Nth harmonic only, prove that

$$f(t) = \sum_{k=0}^{2N} f\left(\frac{kT}{2N+1}\right) \frac{\sin\left[(2N+1)\frac{\pi}{T}\left(t - \frac{kT}{2N+1}\right)\right]}{(2N+1)\sin\left[\frac{\pi}{T}\left(t - \frac{kT}{2N+1}\right)\right]}$$

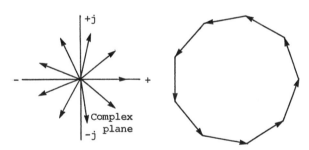

(a) The individual vectors   (b) Vectorial addition of vectors in (a)

Solution: Since only the Fourier coefficients up to the Nth harmonic are under consideration, the periodic function can be expressed as

$$f(t) = \sum_{n=-N}^{N} C_n e^{j\frac{2\pi n}{T}t} \qquad (1)$$

Now, consider the first equality given

$$\sum_{k=0}^{2N} e^{j\frac{2\pi nk}{2N+1}} = \sum_{k=0}^{2N} e^{j\frac{2\pi n}{T}\left(\frac{kT}{2N+1}\right)} = \begin{cases} 2N+1 \text{ if n=0 or a multiple of } (2N+1). \\ 0, \text{ for other values of n.} \end{cases}$$

601

The above equality relates the sum of all the individual vectors of the form $e^{j\frac{2\pi nk}{2N+1}}$ and their phases. It introduces the constraint that unless the (2N+1) vectors are in phase, their resulting sum would be zero, as illustrated in the figure.

For $t = \frac{kT}{2N+1}$, equation (1) evaluates to

$$f\left(\frac{kT}{2N+1}\right) = \sum_{n=-N}^{+N} C_n \, e^{j\frac{2\pi nk}{2N+1}} \tag{2}$$

Using the first equality, the sum of all the vector components of the periodic function is

$$\sum_{k=0}^{2N} f\left(\frac{kT}{2N+1}\right) e^{-j\frac{2\pi mk}{2N+1}} = \sum_{n=-N}^{N} C_n \sum_{k=0}^{2N} e^{j\frac{2\pi}{2N+1} k(n-m)} \tag{3}$$

The above equation can be evaluated by the method used to determine the Fourier coefficients C_n of an infinite Fourier series.

For an infinite Fourier series it follows that

$$f(t) = \sum_{n=-\infty}^{+\infty} C_n \, e^{j(2\pi nt/T)}$$

Multiplying both sides by $e^{-j(2\pi nt/T)}$ and integrating from o to T yields

$$\int_0^T f(t) \, e^{-j(2\pi nt/T)} \, dt = C_n T$$

where $-N \leq n \leq N$

Comparing the first equality and equation (3) we have

$$\sum_{k=0}^{2N} f\left(\frac{kT}{2N+1}\right) e^{-j\frac{2\pi nk}{2N+1}} = (2N+1)C_n \tag{4}$$

Substituting for C_n in equation (1) yields

$$f(t) = \sum_{n=-N}^{N} \left[\frac{1}{2N+1} \sum_{k=0}^{2N} f\left(\frac{kT}{2N+1}\right) e^{-j\frac{2\pi nk}{2N+1}} \right] e^{j\frac{2\pi nt}{T}}$$

$$= \sum_{k=0}^{2N} f\left(\frac{kT}{2N+1}\right) \frac{1}{2N+1} \cdot \sum_{n=-N}^{N} e^{j\frac{2\pi n}{T}\left(t - \frac{kT}{2N+1}\right)} \tag{5}$$

Using the standard identity,

$$\sum_{n=-N}^{N} e^{j\frac{2\pi n}{T} x} = 1 + 2 \cdot \sum_{n=1}^{N} \cos\frac{2\pi n}{T} x = \frac{\sin (2N+1) \frac{\pi}{T} x}{\sin \frac{\pi}{T} x},$$

equation (5) can be expanded to yield the sampling formula as shown below.

$$f(t) = \sum_{k=0}^{2N} f\left(\frac{kT}{2N+1}\right) \frac{\sin\left[(2N+1) \frac{\pi}{T} \left(t - \frac{kT}{2N+1}\right)\right]}{(2N+1) \sin\left[\frac{\pi}{T}\left(t - \frac{kT}{2N+1}\right)\right]}$$

PULSE-AMPLITUDE MODULATION

• **PROBLEM 11-11**

Detection of an amplitude-modulated suppressed carrier signal, $g(t) \cdot \cos(\omega_c t)$, may be done by multiplying it by a pulse train $q(t)$ with frequency ω_c, where $q(t)$ has Fourier transform $q(t) \leftrightarrow \pi\delta(\omega) + 2\sum_{\substack{n=-\infty \\ (n=\text{odd})}}^{\infty} \frac{(-1)^{\frac{n-1}{2}}}{n} \delta(\omega - n\omega_c)$.
Derive the spectrum density function for the resulting signal $g(t) \cdot \cos(\omega_c t) \cdot q(t)$ analytically.

Solution: First, let the function $\phi(t) = g(t) \cos \omega_c t$, then the Fourier transform of the function $\phi(t)$ is easily obtained as $\Phi(\omega) = \frac{1}{2}\left[G(\omega + \omega_c) + G(\omega - \omega_c)\right]$. Now, since the Fourier transform of $q(t)$ is given,

$$g(t) \cos(\omega_c t) q(t) = \phi(t) q(t) \leftrightarrow \frac{1}{2\pi}\left[\pi\Phi(\omega) + 2\sum_{n \text{ odd}} \frac{(-1)^{\frac{n-1}{2}}}{n} \Phi(\omega - n\omega_c)\right]$$

$$= \frac{1}{2}\Phi(\omega) + \frac{1}{\pi} \sum_{n \text{ odd}} \frac{(-1)^{\frac{n-1}{2}}}{n} \Phi(\omega - n\omega_c). \quad (1)$$

Note: It can be easily verified that when $n = \pm 1$, the right-hand side of Eq.(1) yields $G(\omega)$, from which the signal $G(t)$ can be obtained, because

$$\Phi(\omega \pm n\omega_c) = \frac{1}{2}\left[G(\omega + \omega_c \pm n\omega_c) + G(\omega - \omega_c \pm n\omega_c)\right]$$

$$= \frac{1}{2}\left[G(\omega + 2\omega_c) + G(\omega)\right] \quad \text{for } n = 1$$

$$= \frac{1}{2}\left[G(\omega) + G(\omega - 2\omega_c)\right] \quad \text{for } n = -1$$

Notice that the original signal can be recovered by using a low-pass filter which rejects double-frequency components, i.e., $\pm 2\omega_c$.

• **PROBLEM 11-12**

A carrier waveform shown in Fig.(1) is pulse amplitude modulated by an information signal $Sa(\pi t)$. Sketch the process and the output $M_S(\omega)H(\omega)$.

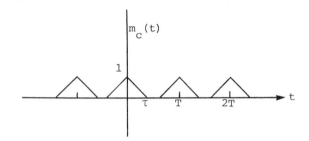

Figure 1.

Solution: In Fig. (2), the state by state process of the pulse amplitude modulation is shown

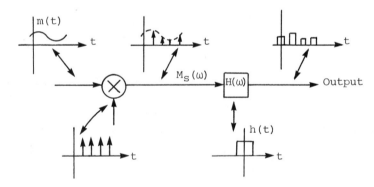

Figure 2

Since the output $M_S(\omega)$ of an ideal impulse sampler has a transform $M_S(\omega) = \frac{1}{T} \sum_{K=-\infty}^{\infty} M(\omega - K\omega_0)$ and $Sa(\pi t)$ can be written as $\frac{\sin \pi t}{\pi t} = m(t)$, the transform of $m(t)$ is a pulse shown in Fig. 3(a).

In this case, the filter would have to change each impulse into a triangular pulse. Its impulse response is therefore a single triangular pulse which has as its transform

$$H(\omega) = \frac{4 \sin^2(\omega\tau/2)}{\tau\omega^2}$$

Finally, the transform of the PAM waveform is given by the product of $M_s(\omega)$ with $H(\omega)$ as shown in Fig. (3b).

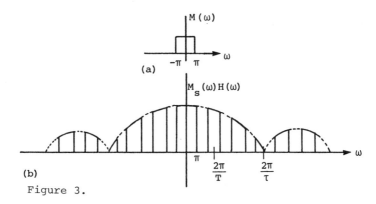

Figure 3.

• **PROBLEM 11-13**

A signal g(t) and a pulse train $P_T(t)$ are input to a multiplier as shown in Fig. (1). If there are two possible outputs which the multiplier can produce, namely, (1) $O_a(t)$ with modulated pulses of constant amplitudes and (2) $O_b(t)$ where the pulses follow the signal g(t),

(a) Write expressions for $O_a(t)$ and $O_b(t)$ in terms of g(t) and P(t), where the impulse train is given by

$$\delta_T(t) = \sum_{k=-\infty}^{\infty} \delta(t-KT).$$

(b) For $G(j\omega)$ given in Fig. (2), determine and sketch the spectrum of $O_a(j\omega)$ and $O_b(j\omega)$.

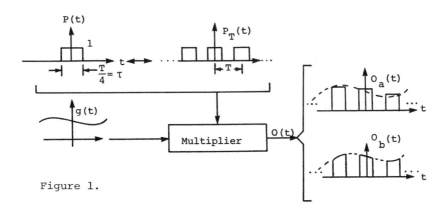

Figure 1.

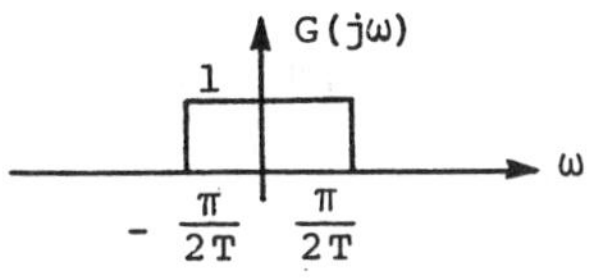

Figure 2.

Solution:

(a) The time waveforms $O_a(t)$ and $O_b(t)$ in terms of $g(t)$ and $P(t)$ can be obtained by using the operations of multiplication and convolution.

Hence,

$$O_a(t) = \sum_{-\infty}^{\infty} g(nt) P(t-nT) = \left[g(t) \cdot \Sigma \delta(t-nT)\right] * P(t-nT)$$

$$= \left[g(t) \, \delta_T(t)\right] * P(t-nT) \quad \text{and}$$

$$O_b(t) = \sum_{-\infty}^{\infty} g(t) P(t-nT) = g(t) \sum_{-\infty}^{\infty} P(t-nT)$$

$$= g(t) \left[\sum_{-\infty}^{\infty} \delta(t-nT) * P(t)\right]$$

$$= g(t) \left[\delta_T(t) * P(t)\right]$$

with

$\delta_T(t) \longleftrightarrow \omega_o \cdot \delta_{\omega_o}(\omega)$, where $\omega_o = \frac{2\pi}{T}$, form a Fourier transform pair.

(b) Thus,

$$O_a(j\omega) = \left[\frac{G(j\omega)}{2\pi} * \omega_o \delta_{\omega_o}(\omega)\right] P(j\omega)$$

$$= \frac{1}{T} P(j\omega) \cdot \sum_{-\infty}^{\infty} G(j\omega - n\omega_o)$$

and

$$O_b(j\omega) = \frac{G(j\omega)}{2\pi} * \left[\omega_o \delta_{\omega_o}(\omega) \cdot P(j\omega)\right]$$

$$= \frac{1}{T} G(j\omega) * \sum_{-\infty}^{\infty} P(n\omega_o) \delta(\omega - n\omega_o)$$

$$= \frac{1}{T} \sum_{-\infty}^{\infty} P(n\omega_o) G(j\omega - n\omega_o)$$

Both frequency spectra are sketched below:

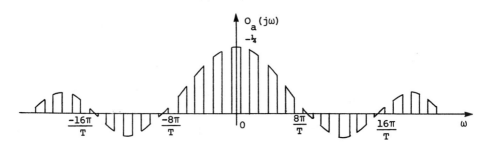

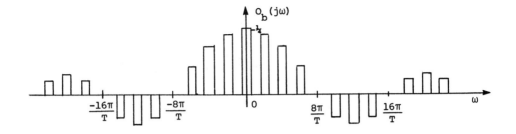

• **PROBLEM 11-14**

The generation of a SSB signal is shown in Fig. (1). Show by analytical and graphical methods, how to recover the amplitude modulated single-sideband (AMSSB) signal by synchronous demodulation.

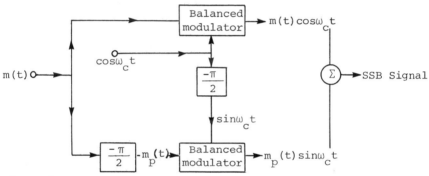

Figure 1.

Solution: From Fig. (1), the output SSB-signal can be expressed as $m_{SSB}(t) = m(t)\cos\omega_c t + m_p(t)\sin\omega_c t$ where $m_p(t)$ is the response of a phase shifter ($\pi/2$ phase shifted) reference to $m(t)$.
Now, for synchronous demodulation, the AM-SSB signal is multiplied by a sinusoidal signal of carrier frequency ω_c.

Hence,

$$m_{SSB}(t)\cos\omega_c t = s(t) = \left[m(t)\cos\omega_c t + m_p(t)\sin\omega_c t\right]\cos\omega_c t \quad (1)$$

Applying trigonometric identities: $\cos^2 x = \tfrac{1}{2}(1+\cos 2x)$
and $\sin(x)\cos(y) = \tfrac{1}{2}\sin(x+y) + \tfrac{1}{2}\sin(x-y)$ to Eq. (1).

$$\begin{aligned}
s(t) &= m(t)\cos^2\omega_c t + m_p(t)\sin\omega_c t \cdot \cos\omega_c t \\
&= m(t)\left[\tfrac{1}{2}(1+\cos 2\omega_c t)\right] + m_p(t)\left[\tfrac{1}{2}\sin 2\omega_c t + \tfrac{1}{2}\sin(0)\right] \\
&= \tfrac{1}{2}m(t) + \tfrac{1}{2}m(t)\cos 2\omega_c t + \tfrac{1}{2}m_p(t)\sin 2\omega_c t + 0 \\
&= \tfrac{1}{2}m(t) + \tfrac{1}{2}\left[m(t)\cos 2\omega_c t + m_p(t)\sin 2\omega_c t\right] \quad (2)
\end{aligned}$$

Now, observe that in Eq. (2), the signal $\frac{1}{2} m(t)$ is the desired signal and can be obtained easily by filtering out the double frequency components.

Next, s(t) can also be obtained by the graphical convolution method as shown by the spectral diagrams in Fig. (2).

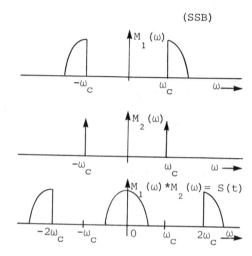

Figure 2.

NATURAL AND FLAT-TOPPED SAMPLING

• **PROBLEM 11-15**

In Fig. (1), the block diagram represents an information signal m(t) sampled by a pulse train $P_T(t)$ of rectangular pulses having duration τ, repeating every T seconds. If m(t) and M(ω) are given in Fig. (2),

(a) Sketch $P_T(t)$ and $P_T(\omega)$.

(b) Sketch the sampled signal $m_s(t)$ i.e., natural sampling and write an expression for $M_s(\omega)$ and sketch its spectrum.

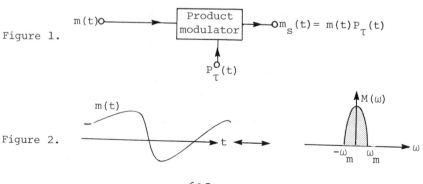

Figure 1.

Figure 2.

Solution:

(a) $P_T(t)$ and $P_T(\omega)$ are sketched below:

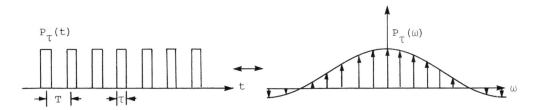

(b) The naturally sampled signal $m_s(t)$ is sketched in Fig. (3a).

Observe that the sampled signal follows the waveform of the signal $m(t)$. Note that since the generation of true impulses is impossible, usually the sampling is performed by very narrow pulses of finite width. As a result, sampling occurs over a finite time interval.

Now, since $m_s(t) = m(t) \cdot P_T(t)$, using the frequency convolution property namely,

$$m_1(t) \leftrightarrow M_1(\omega) \text{ and } m_2(t) \leftrightarrow M_2(\omega)$$

then $m_1(t) \, m_2(t) \leftrightarrow \dfrac{1}{2\pi} \displaystyle\int_{-\infty}^{\infty} M_1(u) \, M_2(\omega - u) \, du$

That is, $m_1(t) \, m_2(t) \leftrightarrow \dfrac{1}{2\pi} \left[M_1(\omega) * M_2(\omega) \right]$

In this case, $M_s(\omega) = \dfrac{1}{2\pi} M(\omega) * P_T(\omega)$ \hfill (1)

In order to determine $P_T(\omega)$, assume $P_T(t)$ to be a periodic gate function and then expand it by the exponential Fourier series as follows:

$$P_T(t) = \begin{cases} A & -\tau/2 < t < \tau/2 \\ 0 & \text{elsewhere} \end{cases}$$

where A is the constant amplitude of the pulses.

Hence,

$$P_n = \frac{1}{T} \int_{-\tau/2}^{\tau/2} P_T(t) \cdot e^{-jn\omega_0 t} \, dt$$

$$= \frac{1}{T} \int_{-\tau/2}^{\tau/2} A \, e^{-jn\omega_0 t} \, dt$$

$$= -\frac{A}{T(jn\omega_0)} e^{-jn\omega_0 t} \bigg|_{-\tau/2}^{\tau/2}$$

$$= -\frac{A}{Tn\omega_0} \frac{(2)}{(2)} \frac{e^{jn\omega_0(\tau/2)} - e^{-jn\omega_0(\tau/2)}}{j}$$

$$= \frac{2A}{n\omega_0 T} \sin\left[(n\omega_0 \tau/2)\right]$$

$$= \frac{2A}{n\omega_0 T} (n\omega_0 \cdot \tau/2) \frac{\sin(n\omega_0 \cdot \tau/2)}{(n\omega_0 \cdot \tau/2)}$$

$$= \frac{A\tau}{T} \frac{\sin(n\omega_0 \cdot \tau/2)}{n\omega_0 \cdot \tau/2}$$

$$= \frac{A\tau}{T} S_a(n\omega_0 \cdot \tau/2)$$

Therefore, $P_\tau(t) = \sum_{n=-\infty}^{\infty} P_n e^{jn\omega_0 t} = \frac{A\tau}{T} \sum_{n=-\infty}^{\infty} S_a\left(\frac{n\pi}{T}\tau\right) e^{jn\omega_0 t}$

where $\omega_0 = \frac{2\pi}{T} = 2\omega_m$ and $\frac{n\omega_0 \tau}{2} = \frac{n\pi\tau}{T}$

Thus, $F\left[P_\tau(t)\right] = P_\tau(\omega) = 2A\tau\omega_m \sum_{n=-\infty}^{\infty} S_a(n\tau\omega_m) \delta(\omega - 2n\omega_m)$ (2)

Substituting Eq. (2) into Eq. (1).

$$M_s(\omega) = \frac{A\tau\omega_m}{\pi} M(\omega) * \sum_{n=-\infty}^{\infty} S_a(n\tau\omega_m) \delta(\omega - 2n\omega_m)$$

$$= \frac{A\tau}{T} \sum_{n=-\infty}^{\infty} S_a(n\tau\omega_m) M(\omega) * \delta(\omega - 2n\omega_m)$$

$$= \frac{A\tau}{T} \sum_{n=-\infty}^{\infty} S_a(n\tau\omega_m) M(\omega - 2n\omega_m)$$

The spectrum of $M_s(t)$ is sketched in Fig. (3b).

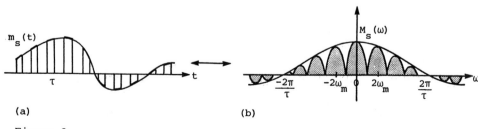

(a)   (b)

Figure 3.

• **PROBLEM 11-16**

A signal of the form $m(t) = \cos\omega_0 t + \cos 10\omega_0 t$ is naturally sampled by the sampling signal

$$S(t) \begin{cases} 1 & kT - \tau/2 \leq t \leq kT + \tau/2, \text{ k is an integer} \\ 0 & \text{elsewhere} \end{cases}$$

with pulses of duration τ and period T_S.

(a) Calculate f_S, the minimum sampling rate.

(b) Sketch the sampled signal $m_S(t) = S(t)m(t)$ for $\tau = 1/64f_0$ and $\tau = 1/640f_0$.

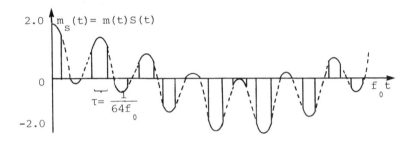

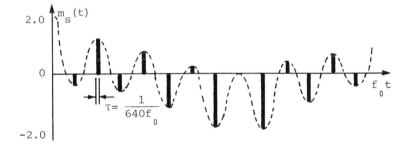

Solution:

(a) By the definition of the sampling theorem, the sampled signal can be recovered exactly when

$$T_S \leq \frac{1}{2f_m}.$$

Hence, the minimum sampling rate $f_S = \frac{1}{T_S} = 2f_m = 2 \times 10f_0$

$$= 20f_0$$

(b) The approximate plots for $\tau = 1/64f_0$ and $\tau = 1/640f_0$ of $m_S(t)$ are shown in the figures above.

● **PROBLEM** 11-17

An input signal $v(t) = I\delta(t)$ (impulse function) is passed through a filter having function $(1 - e^{-j\omega\tau})/j\omega$ where τ is the stretched umpulse width. Determine the output of the filter.

Solution: Since the input is an impulse function, the output of the filter can be obtained easily by taking the inverse Fourier transform of $I(1 - e^{-j\omega\tau})/j\omega$.

Hence, $v_o(t) = F^{-1}\left[I(1 - e^{-j\omega\tau})/j\omega\right]$

$$= \frac{1}{2\pi}\int_{-\infty}^{\infty} \frac{I(1 - e^{-j\omega\tau})}{j\omega} \cdot e^{j\omega\tau} d\omega$$

$$= \frac{1}{2\pi}\int_{-\infty}^{\infty} I(1 - e^{-j\omega\tau}) \cdot \int_{-\infty}^{t} e^{j\omega\sigma} d\sigma\, d\omega$$

$$= \int_{-\infty}^{t} d\sigma \frac{1}{2\pi} \cdot \int_{-\infty}^{\infty} I\left[e^{j\omega\sigma} - e^{j\omega(\sigma-\tau)}\right] d\omega$$

$$= \int_{-\infty}^{t} I\left[\delta(\sigma) - \delta(\sigma-\tau)\right] d\sigma$$

$$= I\left[u(t) - u(t-\tau)\right]$$

Note: The result obtained above is equal to integrating the impulse for τ sec.

i.e., $v_o(t) = I\left[u(t) - u(t-\tau)\right] = \begin{cases} I & 0 \le t \le \tau \\ 0 & \text{elsewhere} \end{cases}$

$$= \begin{cases} \int_0^t I\delta(t)\, dt & 0 \le t \le \tau \\ 0 & \text{elsewhere} \end{cases}$$

● **PROBLEM** 11-18

Flat-topped sampling is used for the signal $m(t)$ in Fig. (1).

(a) Write an expression for the sampled signal $m_s(t)$ and prove that it is a periodic function. (Note: signal is sampled at Nyquist rate.)

(b) Assuming τ = 0.05 sec., obtain the Fourier series representation of $m_s(t)$.

(c) Write an expression for the output $m_o(t)$ in Fig. (1).

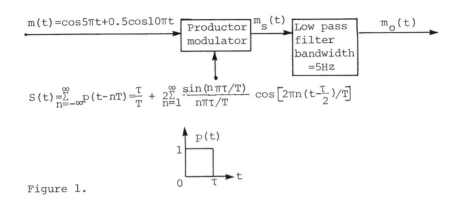

Figure 1.

Solution:

(a) The sampled signal $m_s(t)$ can be written as

$$m_s(t) = m(t)S(t)$$
$$= (\cos 5\pi t + 0.5 \cos 10\pi t) \cdot \sum_{n=-\infty}^{\infty} p(t-nT)$$

(Note: T = period of pulses = $\frac{1}{f_s} = \frac{1}{2f_m} = \frac{1}{2 \times 5} = 0.1$ sec.

where $f_m = \frac{\omega_m}{2\pi} = \frac{10\pi}{2\pi} = 5$ and f_s = sampling rate or Nyquist rate from the sampling theorem.)

Hence, $m_s(t) = (\cos 5\pi t + 0.5 \cos 10\pi t) \sum_{n=-\infty}^{\infty} \delta(t-0.1n)$ (1)

Now, observe in Eq. (1), the period $T_1 = \frac{2\pi}{\omega} = \frac{2\pi}{5\pi} = \frac{2}{5}$ sec. (for the 1st cos term) and the period $T_2 = \frac{2\pi}{10\pi} = \frac{1}{5}$ sec (for the 2nd cos term). Hence, in order to satisfy Eq. (1), we let $\frac{2}{5}$ sec. to be the period of Eq. (1).

Next, in order to prove that $m_s(t)$ is periodic, proceed as follows:

It is to be shown that,

$m_s(t+kT) = m_s(t)$ where k is any integer.

Then, $m_s(t+0.4k) = m(t+0.4k)S(t+0.4k)$

$$= \left[\cos 5\pi(t+0.4k) + 0.5 \cos 10\pi(t+0.4k)\right] \times$$

613

$$\sum_{n=-\infty}^{\infty} p(t+0.4k-0.1n)$$

$$= \left[\cos(5\pi t+2\pi k) + 0.5 \cos(10\pi t+4\pi k)\right] \times \sum_{n=-\infty}^{\infty} \delta(t-0.1n + 0.4k)$$

Notice that since $\cos(5\pi t+2\pi k) = \cos 5\pi t$

and $\cos(10\pi t+4\pi k) = \cos 10\pi t$

Also for delta function, $\sum_{n=-\infty}^{+\infty} \delta(t-0.1n+0.4k)$

$$= \sum_{n=-\infty}^{+\infty} \delta\left[t-0.1(n-4k)\right]$$

$$= \sum_{-\infty}^{+\infty} \left[t-0.1n\right]$$

Therefore, $m_S(t)$ is periodic.

(b) Given $S(t) = \frac{\tau}{T} + 2 \sum_{n=1}^{\infty} \frac{\sin(n\pi\tau/T)}{n\pi\tau/T} \cos\left[2\pi n(t-\tau/2)T\;(2)\right]$

(2)

and $\tau = 0.05$ sec., and also from part (a), T is calculated to be 0.1 sec.

Thus, Eq. (2) becomes

$$S(t) = \frac{1}{2} + 2 \sum_{n=1}^{\infty} \frac{\sin(n\pi/2)}{n\pi/2} \cos\left(20\pi nt - \frac{n\pi}{2}\right)$$

Now, the Fourier series representation of $m_S(t)$ can be obtained as follows:

$m_S(t) = m(t) S(t)$

$$= (\cos 5\pi t+0.5 \cos 10\pi t) \times \left[\frac{1}{2} +2 \sum_{n=1}^{\infty} \frac{\sin(n\pi/2)}{n\pi/2} \cos(20\pi nt - \frac{n\pi}{2})\right]$$

$$= \frac{1}{2}(\cos 5\pi t+0.5 \cos 10\pi t) + 2 \sum_{n=1}^{\infty} \frac{\sin(n\pi/2)}{n\pi/2} \left[\cos(20\pi nt-\frac{n\pi}{2}) \times\right.$$

$(\cos 5\pi t + 0.5 \cos 10\pi t)$

$$= \frac{1}{2}(\cos 5\pi t+0.5 \cos 10\pi t) + 2 \sum_{n=1}^{\infty} \frac{\sin(n\pi/2)}{n\pi/2} \left[\cos(20\pi nt-\frac{n\pi}{2}) \times\right.$$

$\left. \cos 5\pi t + \cos(20\pi nt - \frac{n\pi}{2}) \times 0.5 \cos 10\pi t\right]$

(Applying trigonometric identity: $\cos A \cos B = \frac{1}{2} \cos(A-B) + \frac{1}{2} \cos(A+B)$)

$$= \tfrac{1}{2}(\cos 5\pi t + 0.5 \cos 10\pi t) +$$

$$2 \sum_{n=1}^{\infty} \frac{\sin(n\pi/2)}{n\pi/2} \left[\tfrac{1}{2} \cos(20\pi nt - \tfrac{n\pi}{2} - 5\pi t) + \right.$$

$$\tfrac{1}{2} \cos(20\pi nt - \tfrac{n\pi}{2} + 5\pi t) +$$

$$(0.5) \tfrac{1}{2} \cos(20\pi nt - \tfrac{n\pi}{2} - 10\pi t) +$$

$$\left. (0.5) \tfrac{1}{2} \cos(20\pi nt - \tfrac{n\pi}{2} + 10\pi t) \right]$$

$$= \tfrac{1}{2}(\cos 5\pi t + 0.5 \cos 10\pi t) +$$

$$2 \times \tfrac{1}{2} \sum_{n=1}^{\infty} \frac{\sin(n\pi/2)}{n\pi/2} \left\{ \cos\left[(20n+5)\pi t - \tfrac{n\pi}{2}\right] + \right.$$

$$\cos\left[(20n-5)\pi t - \tfrac{n\pi}{2}\right] +$$

$$0.5 \cos\left[(20n+10)\pi t - \tfrac{n\pi}{2}\right] +$$

$$\left. 0.5 \cos\left[(20n-10)\pi t - \tfrac{n\pi}{2}\right] \right\} \quad (3)$$

(c) The low-pass filter has a bandwidth BW = 5Hz, which is equivalent to 10π rad/s.

Therefore, in equation (3), all components larger than 10π rad/s will be filtered out due to the low-pass filter.

Hence, the output $m_o(t) = \tfrac{1}{2}(\cos 5\pi t + 0.5 \cos 10\pi t) + \tfrac{1}{\pi} \sin 10\pi t$

$$= 0.5 \cos 5\pi t +$$
$$\sqrt{(0.25)^2 (1/\pi)^2} \sin(10\pi t + \tan^{-1}\tfrac{\pi}{4})$$

$$= 0.5 \cos 5\pi t + 0.4 \sin(10\pi t + 38°)$$

SIGNAL RECOVERY - HOLDING AND CROSSTALK

• **PROBLEM** 11-19

A holding circuit shown in Fig. (1) is commonly used for recovery of a bandlimited signal m(t) from a sampled signal $m_s(t)$, sampled at every $\frac{1}{2f_m}$ seconds interval. A) If m(t) is given in Fig. (2), sketch the input and output waveforms of the circuit. B) Write an expression for the transfer function of this circuit and sketch its frequency response. Show that it closely resembles an ideal low-pass filter.

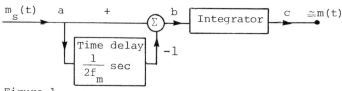

Figure 1.

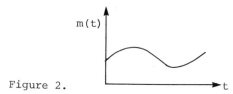

Figure 2.

Solution:

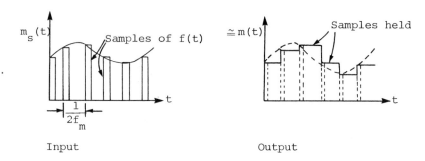

Input   Output

B) In order to determine the transfer function of the circuit, first find h(t), the unit impulse response of the system.

Hence, if $\delta(t)$ is applied at point (a) the signal at point (b) is given by $\delta(t) - \delta(t - \frac{1}{2f_m})$ and the signal at c is given by $\int \left[\delta(t) - \delta(t - \frac{1}{2f_m}) \right] dt = u(t) - u(t - \frac{1}{2f_m})$. This is of course the impulse response h(t).

Therefore, $h(t) = u(t) - u(t - \frac{1}{2f_m})$ and

$$H(s) = \frac{1}{s} - \frac{1}{s} e^{-s/2f_m}$$

$$= \frac{1}{s}(1 - e^{-s/2f_m})$$

$$H(j\omega) = \frac{1}{j\omega}(1 - e^{-j\omega/2f_m})$$

The frequency response is given by $|H(j\omega)|$

$$|H(j\omega)| = \frac{1}{\omega} \left| 1 - e^{-j\omega/2f_m} \right|$$

$$= \frac{1}{\omega} \left| 1 - \cos\frac{\omega}{2f_m} + j\sin\frac{\omega}{2f_m} \right|$$

$$= \frac{1}{2f_m} \frac{\sin(\omega/4f_m)}{(\omega/4f_m)}$$

$$= \frac{1}{2f_m} Sa(\frac{\omega}{4f_m})$$

This is obviously a crude form of a low pass filter. The frequency response is shown in Fig. (3).

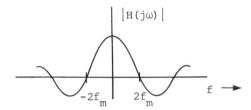

Figure 3.

● **PROBLEM 11-20**

Two thousand signals are sampled at Nyquist rate by the sampling function $S(t) = I \sum_{n=-\infty}^{\infty} (t - \frac{1}{f_s} n)$ where f_s is the sampling rate. The signals are then time-division multiplexed. If each signal has a bandwidth of 5 kHz, calculate the sampling time T_s and determine τ_{max} for pulses of duration τ. (i.e., stretched impulses).

Solution: From the sampling theorem, the sampling time is given by $T_s \leq \frac{1}{2f_m}$.

Hence, in this case, T_s can be calculated as follows:

$$T_s = \frac{1}{f_s} = \frac{1}{2f_m} = \frac{1}{2 \times 5000} = 1 \times 10^{-4} \text{ sec}.$$

Therefore, the maximum duration for each pulse is,

$$\tau_{max} = \frac{T_s}{2000} = \frac{1 \times 10^{-4}}{2000} = 0.5 \times 10^{-7} \text{ sec}.$$

• **PROBLEM 11-21**

Typically, for voice message transmission on a telephone line, the voice signal is bandlimited to $f_m = 3.3$ kHz and the sampling frequency is chosen to be 8.0 kHz.

(a) Calculate the sampling time T_s.

(b) Determine the guard band according to the sampling theorem and give a brief explanation.

(c) If 10 speech channels are to be multiplexed in a PAM system, allocate each sample interval together with its associated guard space.

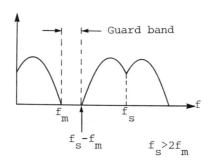

Figure 1.

Solution:

(a) Since the sampling frequency $f_s = 8.0$ kHz,

$$T_s = \frac{1}{f_s} = \frac{1}{8000} = 125 \text{ μsec is easily obtained.}$$

(b) In Fig. (1), a plot of the spectrum of a sampled signal will help to understand this problem.

According to the sampling theorem, the sampling rate f_s must be larger than or equal to $2f_m$ in order to reconstruct the signal without ditortion. In practical cases, a lowpass filter used during demodulation does not have an infinitely sharp cut off. Hence, for this reason, a guard band indicated in Fig. (1) is always required where f_s must be larger than $2f_m$.

Now, observing the diagram in Fig. (1), the guard band is $f_s - 2f_m$, hence, in this case,

guard band = 8.0 - 2 × 3.3 = 1.4 kHz.

(c) For 10 channel multiplexing, each sample interval together with its associated guard space may be allocated

$$\frac{125 \ \mu sec}{10} = 12.5 \ \mu sec.$$

● **PROBLEM** 11-22

The block diagram in Fig. (1) represents the transmission of two signals $V_1(t)$ and $V_2(t)$ over a transmission channel.

The transmission channel behaves like a high-pass filter shown in Fig. (2). If there is no guard band between each sampled pulse and the Nyquist rate for each signal is 20 kHz, (a) calculate the maximum time slot (τ_{max}). (b) First, approximate the accumulate-tilt (Δ) due to the initially flat top of the pulse, then approximate the crosstalk signal and the crosstalk ratio K.

Given: $V_1(t) = 20$ V and $V_2(t) = 0$ during a specific time interval lasting for 100 samples.

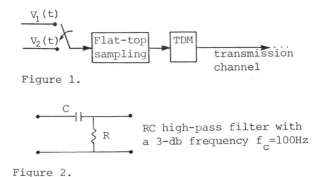

Figure 1.

RC high-pass filter with a 3-db frequency $f_c = 100$Hz

Figure 2.

Solution:

(a) If K signals are to be multiplexed, then the maximum time slot τ_{max} is given by $\tau_{max} = \frac{T_s}{K}$ where T_s is the sampling time.

Hence, since 2 signals are multiplexed and $f_s = 20$ kHz,

$$\tau_{max} = \frac{1}{2f_s K} = \frac{1}{2 \times 2 \times 20 \times 10^3} = 1.25 \times 10^{-5} \ sec$$

(b) The pulse distortion and overlapping between two time slots due to the bandwidth restriction of the channel is shown in Fig. (3). The accumulate tilt (Δ) is given by

$$\Delta = V(1 - e^{-\tau/\tau_c}) \quad (1)$$

where $\tau_c = RC$ is the channel time constant. Notice that the 3dB cut off frequency of the channel is

$$f_c = \frac{1}{2\pi\tau_c}$$

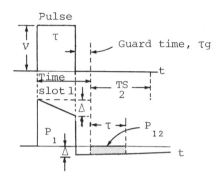

Figure 3.

Now, in order to reduce crosstalk, it is required that $\tau \ll \tau_c$. Hence, the exponential tilt in Eq. (1) is approximately linear.

Therefore, let $\tau_{max} = \tau$, and $\Delta \simeq V\dfrac{\tau}{\tau_c} \simeq \dfrac{20(1.25 \times 10^{-5})}{\dfrac{1}{2\pi f_c}}$

$$\simeq 20(1.25 \times 10^{-5}) \times 2\pi \times 100$$

$$\simeq 0.157 \text{ volts.}$$

In this case, since $\tau_g \ll \tau_c$, the area P_{12} which measures the crosstalk signal can be approximated as follows:

$$P_{12} \simeq \Delta\tau = 0.157 \times 1.25 \times 10^{-5} = 1.96 \times 10^{-6} \text{ volt sec.}$$

The crosstalk ratio K is given by $K = \dfrac{P_{12}}{P_1}$

where $P_1 \simeq V\tau$

Therefore, $K \simeq \dfrac{P_{12}}{V\tau} \simeq \dfrac{1.96 \times 10^{-6}}{20(1.25 \times 10^{-5})} \simeq 0.00784 \simeq -42 \text{ dB}$

METHODS OF GENERATING PULSE-TIME AND PULSE-DURATION MODULATION SIGNALS

● **PROBLEM** 11-23

Show how a pulse-time-modulated (PTM) signal can be generated by first generating a PAM waveform.

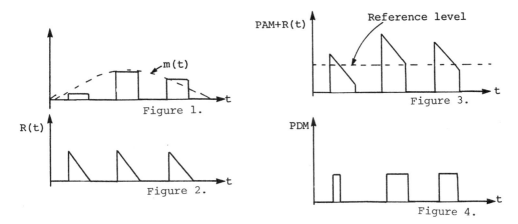

Solution: First, a baseband signal m(t) is flat-top sampled, thereby generating a pulse-amplitude modulated wavetrain as shown in Fig. (1).

In Fig. (2), a linear ramp waveform R(t) is generated synchronously with the PAM samples.

In Fig. (3), the PAM samples and the synchronized ramp are added and then passed through a comparator with the reference level as shown.

Note: The comparator's output has two voltage level characteristics under the following conditions:

1. When the input signal is less than the reference level.

2. When the input signal is larger than the reference level.

Finally, in Fig. (4), a pulse-duration-modulated signal is generated. Notice that the leading edge of a pulse output of the comparator is the first crossing of the reference level by the waveform PAM + R(t) and the trailing edge is the second crossing of the reference level.

• **PROBLEM 11-24**

Show how to generate a PTM waveform without first generating a PAM waveform.

Solution: Instead of generating a pulse amplitude modulated waveform first, generate a pulse time modulated (PTM) waveform by simply adding the baseband signal to a ramp waveform and pass the sum through a comparator. The procedures are shown as follows:

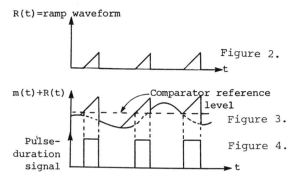

Figure 2.

Figure 3.

Figure 4.

Note: In this problem the PTM signal can be obtained by modulating the leading edge of a pulse shown in Fig. (3) & (4).

● **PROBLEM 11-25**

Use simplified block diagrams to illustrate the circuits involved in the generation of a pulse-duration-modulated signal (PDM).

Solution:

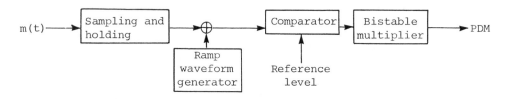

● **PROBLEM 11-26**

Show how baseband signal recovery can be done by converting a PDM waveform to PAM.

Solution: In order to illustrate the process of recovering the baseband signal by converting a PDM to PAM, it is of advantage to understand before hand, the process of conversion of a baseband signal to a PDM. This is illustrated in figures (A) to (D).

To recover the baseband signal, the pulses in the PDM waveform are used to generate a linear ramp waveform (Fig. (E)), wherein the height attained by the ramp is proportional to the pulse duration of the PDM. The level attained by the ramp is sustained for a time (i.e., the capacitor of the ramp forming circuit is not permitted to discharge) after which the voltage is returned to its initial level at some arbitrary time.

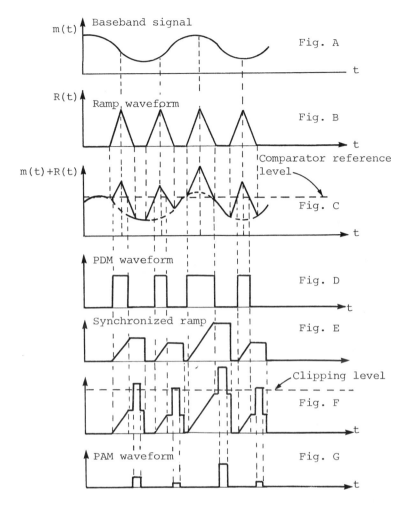

Fig. A — Baseband signal m(t)
Fig. B — Ramp waveform R(t)
Fig. C — m(t)+R(t), Comparator reference level
Fig. D — PDM waveform
Fig. E — Synchronized ramp
Fig. F — Clipping level
Fig. G — PAM waveform

A sequence of pulses of fixed amplitude and duration generated at the demodulator are now added to the ramp waveform resulting in the waveform in figure (F). This waveform when applied to a clipping circuit will transmit only the portion of the waveform above the reference level as in Fig. (G) and thus, yields a PAM waveform. The baseband signal can now be recovered by passing this PAM waveform through an appropriate low-pass filter.

• **PROBLEM 11-27**

Choosing suitable parameters, design a ramp generator analog to digital converter (A.D.C.) to transform $f(t) = \sin 2\pi t$ into a digital signal of 4 bits.

Solution: The sampling theorem states that, if for a time function, the Fourier transform is zero for $|\omega| > \omega_m$ and the time function has definite values for $t = nT_s$, where $T_s < \pi/\omega_m$, then the time function has definite values for all values of t.

In general, for the problem at hand,

$$T_s < \frac{\pi}{\omega_m} < \frac{\pi}{2\pi} < \frac{1}{2},$$

or, $\quad f_s > \dfrac{1}{T_s} > 2.$

That is, to comply with the sampling theorem, the sampling rate must be greater than 2 samples/sec. Therefore, choose a sampling rate of 3 samples/sec.

Since the ADC required is a ramp generator ADC, it is necessary that the slope of the ramp must be considerably greater than the maximum slope of the signal. For the given signal of slope 2π, choose a ramp slope of 100 V/sec., such that to attain the maximum value of the signal (which is unity in the present case), the ramp function will require 0.01 second. Since a 4-bit digital signal is required, the counter should be able to count from 0000 to 1111 (a maximum of 16 numbers) in 0.01 second or one count in 0.01/16 = 0.000625 second.

Thus, the counting rate of the clock should be

= 1 count/0.000625 second

= 1,600 counts/second.

CHAPTER 12

PULSE-CODE MODULATION (PCM)

QUANTIZATION OF SIGNALS

● **PROBLEM** 12-1

(a) If a voltage of 1 volt rms is applied at the sending end of a telephone cable of length 1000 miles, determine the voltage at the receiving end. Assume that the attenuation in the cable is 1 dB/mile.

(b) It is desired to obtain a 1 volt rms signal at the receiving end. This can be accomplished by using repeaters. Assume that repeaters yield a maximum rms output of 1 volt and have a voltage gain of 100. Determine the number of repeaters needed and the spacing between them.

Solution: (a) The voltage at the receiving end will be attenuated by 1000 dB because the attenuation in the cable is 1 dB/mile and the cable is 1000 miles long.

Now, let the received (or output) voltage be v_0 volts. Then by definition,

$$-1000 \text{ dB} = 20 \log_{10}\left(\frac{v_0}{v_{in}}\right) = 20 \log\left(\frac{v_0}{1}\right)$$

Thus, $v_0 = 10^{-50}$ volts rms

(b) The voltage gain of the repeater is given as 100, therefore, $20\log(100) = 40$ dB.

Since the attenuation in the cable is 1 db/mile, as a result, each repeater is set for every 40 miles and the number of repeaters required is equal to

$$\frac{1000}{40} = 25.$$

● **PROBLEM 12-2**

The probability density of a signal is given by

$$f(x) = \begin{cases} K\,e^{-|x|} & \text{for } |x| < 4 \\ 0 & \text{otherwise} \end{cases}$$

(a) What is the step size S if there are four quantization levels.

(b) Find the value of the constant K.

(c) Assume that there are four quantization levels and that f(x) is not constant over each level. Determine the variance of the quantization error.

<u>Solution</u>: (a) Since x ranges from -4 to +4 volts, there are 4 - (-4) = 8 levels.

Hence, the step size S is calculated as

$$S = \frac{8}{4(\text{quantization levels})} = 2$$

Thus, the quantization levels are:

$$x_1 = -3 \text{ volts}$$
$$x_2 = -1 \text{ volt}$$
$$x_3 = 1 \text{ volt}$$
$$x_4 = 3 \text{ volts}$$

(b) By definition, the integral of f(x) with respect to x from x = -∞ to x = +∞ should be equal to 1 (i.e., the probability density over the entire range of the variable x is unity). Hence,

$$\int_{-\infty}^{+\infty} f(x)\,dx = 1$$

$$K \int_{-4}^{+4} e^{-|x|}\,dx = 1$$

626

Since f(x) is symmetric about the x-axis:

$$2K \int_0^4 e^{-x} \, dx = 1$$

$$2K \left[\frac{e^{-x}}{-1} \right]_0^4 = 1$$

$$-2K \left[e^{-4} - 1 \right] = 1$$

Therefore,

$$K = \frac{1}{2(1-e^{-4})} \cong 0.5093$$

(c) The variance of the quantization error is

$$= \text{var}\left[e(x)\right]$$

$$= \int_{-4}^{-2} (-3-x)^2 Ke^{-|x|} \, dx + \int_{-2}^{0} (-1-x)^2 Ke^{-|x|} \, dx$$

$$+ \int_0^2 (1-x)^2 Ke^{-|x|} \, dx + \int_2^4 (3-x)^2 Ke^{-|x|} \, dx$$

$$= 2K \int_0^2 (1-x)^2 e^{-x} \, dx + 2K \int_2^4 (3-x)^2 e^{-x} \, dx$$

$$= 2K \left[\{(1-x)^2\}\{-e^{-x}\} - \{-2(1-x)\}\{e^{-x}\} + \{2\}\{-e^{-x}\} \right]_0^2$$

$$+ 2K \left[\{(3-x)^2\}\{-e^{-x}\} - \{-2(3-x)\}\{e^{-x}\} + \{2\}\{-e^{-x}\} \right]_2^4$$

$$= 2K \left[e^{-x}\{-(1-x)^2 + 2(1-x) - 2\} \right]_0^2$$

$$+ 2K \left[e^{-x}\{-(3-x)^2 + 2(3-x) - 2\} \right]_2^4$$

$$= 2K \left[e^{-x}\{-1 - x^2 + 2x + 2 - 2x - 2\} \right]_0^2$$

$$+ 2K \left[e^{-x}\{-9 - x^2 + 6x + 6 - 2x - 2\} \right]_2^4$$

$$= 2K \left[e^{-x}\{-1 - x^2\} \right]_0^2 + 2K \left[e^{-x}\{-5 - x^2 + 4x\} \right]_2^4$$

627

$$= 2K\left[e^{-2}(-5) - (-1)\right] + 2K\left[e^{-4}(-5-16+16) - e^{-2}(-5-4+8)\right]$$

$$= 2K(1-5e^{-2}) + 2K(e^{-2}-5e^{-4})$$

$$= 2K(1 - 5e^{-2} + e^{-2} - 5e^{-4})$$

$$= \frac{2(1-4e^{-2} - 5e^{-4})}{2(1-e^{-4})}.$$

Thus, the variance of the quantization error is

$$= \frac{(1-4e^{-2} - 5e^{-4})}{(1-e^{-4})} \cong 0.3739.$$

● **PROBLEM 12-3**

In practice, logarithmic companding is commonly used for speech telephony. It is specified as µ-law companding and has the form

$$v_0 = \begin{cases} \dfrac{v_{max}\ln(1+\mu\, v/v_{max})}{\ln(1+\mu)} & 0 \leq v < v_{max} \\ -v_{max}\dfrac{\ln(1-\mu v/v_{max})}{\ln(1+\mu)} & -v_{max} \leq v \leq 0 \end{cases}$$

(a) By selecting $\mu = 0, 3, 100$, plot the resulting compressor and expander characteristics.

(b) If the quantization level is chosen to be 32, write an expression for the step size S in terms of input voltage v.

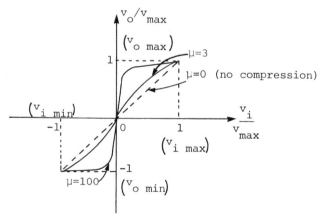

COMPRESSOR CHARACTERISTICS

(i)

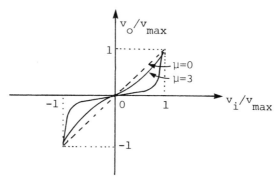

EXPANDER CHARACTERISTICS

(ii)

Solution: (a) The resulting compressor and expander characteristics are shown in the figures

(b) Let v_m be the mth quantization level of the input signal v.

Then

$$(v_0)_m = m\left(\frac{2v_{max}}{32}\right) = \begin{cases} v_{max} \dfrac{\ln(1+\mu v_m/v_{max})}{\ln(1+\mu)} & 0 \le v_m \le v_{max} \\ -v_{max} \dfrac{\ln(1-\mu v_m/v_{max})}{\ln(1+\mu)} & -v_{max} \le v_k \le 0 \end{cases}$$

Solving the above equation for v_m gives

$$\frac{mv_m}{16} = v_{max} \frac{\ln(1+\mu v_m/v_{max})}{\ln(1+\mu)}$$

$$\frac{m}{16} = \frac{\ln(1+\mu v_m/v_{max})}{\ln(1+\mu)}$$

$$\frac{m}{16} \times \ln(1+\mu) = \ln(1+\mu\, v_m/v_{max})$$

$$\ln(1+\mu)^{\frac{m}{16}} = \ln(1+\mu v_m/v_{max}).$$

Therefore,

$$(1+\mu)^{\frac{m}{16}} = 1+\mu v_m/v_{max}$$

$$\frac{\mu v_m}{v_{max}} = (1+\mu)^{\frac{m}{16}} - 1.$$

Hence,

$$v_m = \pm \frac{v_{max}}{\mu}\left[(1+\mu)^{\frac{m}{16}} - 1\right]. \quad (1)$$

Since the step size, S_m is given by

$$S_m = (v_{m+1} - v_m), \quad (2)$$

Substituting values v_{m+1} and v_m obtained from eq.(1) into eq.(2) gives

$$S_m = \frac{v_{max}}{\mu}\left[(1+\mu)^{\frac{m+1}{16}} - 1\right] - \frac{v_{max}}{\mu}\left[(1+\mu)^{\frac{m}{16}} - 1\right]$$

$$= \frac{v_{max}}{\mu}\left[(1+\mu)^{\frac{m+1}{16}} - (1+\mu)^{\frac{m}{16}} - 1 + 1\right]$$

$$= \frac{v_{max}}{\mu}\left[(1+\mu)^{\frac{m+1}{16}} - (1+\mu)^{\frac{m}{16}}\right]$$

● **PROBLEM** 12-4

Consider a binary PCM system, in which $r(t)$ is a rectangular pulse of height A and width T. If the impulse response of the matched filter is given by $h(t) = r(T-t)$, determine the probability of error P_e for $K = 10$ (rms value of noise). It is given that the energy E of $r(t)$ is $E = A^2T$ and $A = K\sigma_n$ where σ_n is the rms value of the noise signal. (i.e., $\sigma_n^2 = \overline{n^2(t)} = N_i$ (input noise power).

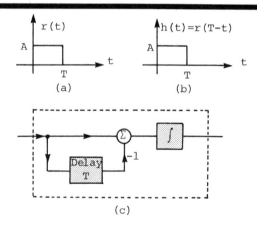

Figure 1.

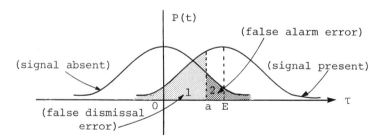

Figure 2.

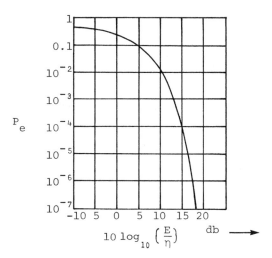

Figure 3.

Solution: Observe, in Fig. (1), the matched filter is represented by the block diagram and the function h(t) is equal to r(t) because h(t) can be obtained by shifting the mirror image of r(t) to the right by T units.

Hence, $h(t) = r(T-t) = r(t)$.

Now, calculate P_e. Since the maximum rate of transmission for a binary pulse is 1 bit per second, then for a system with bandwidth = BHz, 2B pulses/second can be transmitted.

In this case, since the rate of transmisson is $\frac{1}{T}$ pulses/sec., where T is the pulse width,

$$\frac{1}{T} = 2B \ \frac{\text{pulses}}{\text{sec}}.$$

Therefore, $B = \frac{1}{2T}$ Hz is the required transmission bandwidth.

Now, assume white channel noise with power spectral density $\eta/2$ (η=constant) and the input noise power N_i for a single channel given by $N_i = \eta$ BW (for white noise).

Thus, $N_i = \eta \ BW = \frac{\eta}{2T} = \sigma_n^2$

$$\eta = 2T\sigma_n^2$$

Since the probability of error is given by

$$P_e = \frac{1}{\sqrt{\pi \eta E}} \int_a^\infty e^{-r^2/\eta E} \ dr = \int_a^\infty p(r)dr \qquad (1)$$

to ease the problem of integration, the error function erf(μ) is

$$\text{erf}(\mu) = \frac{1}{\sqrt{2\pi}} \int_{-\infty}^{\mu} e^{-y^2/2} \, dy \quad \text{and the complementary}$$

error function erfc(μ) is

$$\text{erfc}(\mu) = \frac{1}{\sqrt{2\pi}} \int_{\mu}^{\infty} e^{-y^2/2} \, dy \qquad (2)$$

Now, comparing eqs.(1) and (2):

Let $\quad \dfrac{r^2}{\eta E} = \dfrac{y^2}{2}$

Hence, $\quad r^2 = \dfrac{y^2 \eta E}{2} \quad$ and $\quad dr = \sqrt{\dfrac{\eta E}{2}} \, dy$

Thus,

$$P_e = \frac{1}{\sqrt{\pi \eta E}} \int_{\frac{a}{\sqrt{\eta E/2}}}^{\infty} e^{-\frac{y^2 \eta E}{2\eta E}} \cdot \sqrt{\frac{\eta E}{2}} \, dy$$

$$= \frac{1}{\sqrt{\pi \eta E}} \sqrt{\frac{\eta E}{2}} \int_{\frac{a}{\sqrt{\eta E/2}}}^{\infty} e^{-y^2/2} \, dy$$

$$= \frac{1}{\sqrt{2\pi}} \int_{\frac{a}{\sqrt{\eta E/2}}}^{\infty} e^{-y^2/2} \, dy$$

$$= \text{erfc}\left(\frac{a}{\sqrt{\eta E/2}}\right).$$

Note: $a = \dfrac{E}{2}$ is defined as the optimum threshold. (see Fig. 2).

In the figure, the error probability is given by the mean of the 2 shaded areas and the sum of the areas is minimum if $a = E/2$.

Hence, $\quad P_e = \text{erfc}\left(\sqrt{\dfrac{E}{2\eta}}\right)$

where $\quad \dfrac{E}{\eta} = \dfrac{A^2 T}{2T\sigma_n^2} = \dfrac{A^2}{2\sigma_n^2} = \dfrac{K^2 \sigma_n^2}{2\sigma_n^2} = \dfrac{K^2}{2}$

Therefore, $P_e = \text{erfc}\left(\sqrt{\frac{K^2}{2} \cdot \frac{1}{\sqrt{2}}}\right) = \text{erfc}\left(\frac{K}{2}\right).$

For $K = 10$, $P_e = \text{erfc}(5)$

Using the approximation given by

$$\text{erfc}(\mu) \cong \frac{1}{\mu\sqrt{2\pi}}\left(1 - \frac{1}{\mu^2}\right)e^{-\mu^2/2} \quad \text{for } \mu > 2$$

gives

$$P_e \cong 2.85 \times 10^{-7} \quad \text{(an acceptable error)}$$

or 16.9 dB. (from Fig. 3)

● **PROBLEM** 12-5

For a quantized signal, the mean-square quantization error is defined as

$$\overline{e^2} = \sum_{i=1}^{M} \int_{m_i - \frac{S}{2}}^{m_i + \frac{S}{2}} f(m)(m-m_i)^2 \, dm \quad \text{where } S = \text{the}$$

step size, m_i = quantization level location for $i = 1, 2, 3\ldots$ and M = the number of quantization levels. Assuming M to be small, determine $\overline{e^2}$ if the signal is distributed uniformly.

<u>Solution</u>: Let the peak to peak voltage of the signal be equal to MS.

Thus, $f(m)dm$, the probability that $m(t)$ lies in the voltage range $m_i - \frac{S}{2} \leq m \leq m_i + \frac{S}{2}$ for uniform distribution is

$f(m) = \frac{1}{MS}$.

Hence,
$$\overline{e^2} = \sum_{i=1}^{M} \int_{m_i - \frac{S}{2}}^{m_i + \frac{S}{2}} f(m)(m-m_i)^2 \, dm$$

$$= \frac{1}{MS} \sum_{i=1}^{M} \int_{m_i - \frac{S}{2}}^{m_i + \frac{S}{2}} (m-m_i)^2 \, dm$$

$$= \frac{1}{MS} \sum_{i=1}^{M} \left[\frac{(m-m_i)^3}{3}\right]_{m_i - \frac{S}{2}}^{m_i + \frac{S}{2}}$$

$$= \frac{1}{MS} \sum_{i=1}^{M} \frac{(m_i + \frac{S}{2} - m_i)^3 - (m_i - \frac{S}{2} - m_i)^3}{3}$$

$$= \frac{1}{MS} \sum_{i=1}^{M} \frac{1}{3}\left[\left(\frac{S}{2}\right)^3 - \left(-\frac{S}{2}\right)^3\right]$$

$$= \frac{1}{MS} \sum_{i=1}^{M} \frac{1}{3}\left(2 \cdot \frac{S^3}{8}\right)$$

$$= \frac{1}{MS} \sum_{i=1}^{M} \frac{S^3}{12}$$

Therefore, for small M, $\overline{e^2} \simeq \frac{S^2}{12}$

● **PROBLEM 12-6**

Determine the mean-square quantization error during the quantization of a signal. Assume that the error (equally likely) lies in the range $-S/2$ to $+S/2$, where S is the step size.

Solution: Let e represent the error. Then, the error is defined as the difference between the original signal and the quantized signal. i.e., $e = m(t) - m_q(t)$.

Since the error is uniformly distributed and lies in the range $-\frac{S}{2}$ to $\frac{S}{2}$, P_e = error probability = $\dfrac{1}{\left[\frac{S}{2} - \left(-\frac{S}{2}\right)\right]}$

$$= \frac{1}{S}$$

Now, the mean-square quantization error is equal to the expected value of the error squared.

So, $\overline{e^2}$ = Mean-square quantization error

$$= E[e^2] = \text{Expected value of the error squared}$$

$$= \int_{-S/2}^{+S/2} e^2 \, P_e \, de$$

$$= \frac{1}{S} \int_{-S/2}^{+S/2} e^2 \, de$$

$$= \frac{2}{S} \int_{0}^{S/2} e^2 \, de$$

$$= \frac{2}{S} \left[\frac{e^3}{3}\right]_{0}^{S/2}$$

$$= \frac{2}{3S}\left(\frac{S}{2}\right)^3$$

$$= \frac{2}{3S} \cdot \frac{S^3}{2^3}$$

$$= \frac{S^2}{12}$$

THE PCM SYSTEM

● **PROBLEM** 12-7

An information signal is pulse-code modulated using eight levels quantization. If the signal is given as $\cos 4\pi t$ with sampling rate $f_s = 8Hz$, and the sampling signals are a train of pulses of unit height and period $T_p = \frac{n}{8}$ sec, $-\infty < n < \infty$,

sketch the sampled message signal and determine the binary representation of each sample. What is the number of bits per sample required?

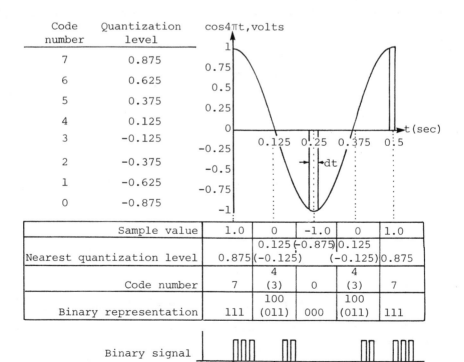

Solution: Since 8 quantization levels are employed and the peak to peak voltage of the signal is 2 volts, the quantization levels are located at -0.875, -0.625, 0.375, -0.125, 0,... +0.875 volts as indicated in the figure.

The binary representation can be easily obtained as shown in the table.

Note that the sampling time $T_s = \frac{1}{f_s} = \frac{1}{8\text{Hz}} = 0.125$ sec and the pulse width is assumed to be dt.

The, number of bits per sample is $\log_2 8 = 3$.

• **PROBLEM 12-8**

An analog signal is quantized and encoded using pulse code modulation (PCM).

(a) If the number of quantization levels M = 16 and the word length is 400 μsec determine the minimum channel bandwith (f_c) required to transmit the binary PCM signal.

(b) If the channel has a bandwidth half of what is found in (a), determine the new number of quantization levels so that the PCM signal can pass through this channel.

Solution: (a) Let each word have N time slots each of width τ with $M = 2^N$, then, the word length is equal to Nτ, which is equal to 400 μsec in this case.

Thus, using the equation $M = 2^N$, and substituting M = 16, $16 = 2^N$ gives N = 4.

Therefore, the duration τ of each time slot can be calculated as follows:

$$N\tau = 400 \times 10^{-6}$$

$$\tau = \frac{400 \times 10^{-6}}{4} = 100 \text{ μsec}$$

Now, since $\omega_c > 1.59$ must be satisfied so as to eliminate intersymbol interference, the minimum channel bandwidth can be evaluated as

$$\omega_c = \frac{1.59}{\tau} = \frac{1.59}{100 \times 10^{-6}} = 15900 \text{ rad/sec.}$$

Therefore, $f_c \equiv \dfrac{\omega_c}{2\pi} \geq \dfrac{15900}{2\pi} = 2.53 \times 10^{+3}\,\text{Hz}$

Hence, $2.53 \times 10^{+3}\,\text{Hz}$ is the minimum channel bandwidth required.

(b) Assume for part (a) the channel bandwidth $= f_{c_1}$ and for part (b) the channel bandwidth $= f_{c_2}$, then, $f_{c_2} = \tfrac{1}{2} f_{c_1}$.

Now, since the word length is a constant (i.e., $N_1 \tau_1 = N_2 \tau_2$) and $M = 2^N$, the following equalities are obtained.

$$\log_2 M_1 = N_1 \log_2 2 \tag{1}$$
$$\log_2 M_2 = N_2 \log_2 2$$

$$\dfrac{N_2}{N_1} = \dfrac{\tau_2}{\tau_1} = \dfrac{\log_2 M_2}{\log_2 M_1} \tag{2}$$

and since $\tau_1 f_{c_1} = \tau_2 f_{c_2}$,

$$\dfrac{\tau_2}{\tau_1} = \dfrac{f_{c_1}}{f_{c_2}} \tag{3}$$

From eqs.(2) and (3) it follows,

$$\dfrac{f_{c_2}}{f_{c_1}} = \dfrac{N_2}{N_1} = \dfrac{\log_2 M_2}{\log_2 M_1}$$

Thus, $\dfrac{f_{c_2}}{f_{c_1}} = \dfrac{\log_2 M_2}{\log_2 M_1}$

$$\dfrac{f_{c_2}}{f_{c_1}} \log_2 M_1 = \log_2 M_2$$

$$M_1^{(f_{c_2}/f_{c_1})} = M_2$$

$$M_1^{(\frac{1}{2} f_{c_1}/f_{c_1})} = M_2$$

$$M_1^{\frac{1}{2}} = M_2$$

$$16^{\frac{1}{2}} = M_2$$

$$4 = M_2 = \text{new number of quantization levels}$$

• **PROBLEM 12-9**

Consider a PCM system shown below:

(a) Determine the maximum sampling period allowed to reconstruct the analog signal from the digital samples without distortion.

(b) Calculate the quantization noise due to the 4 bit A/D converter.

(c) What happens to the quantization noise if a 5 bit A/D converter were used? Compare the advantages and disadvantages of using a 5 bit A/D or a 4 bit A/D converter in terms of the channel bandwidth required.

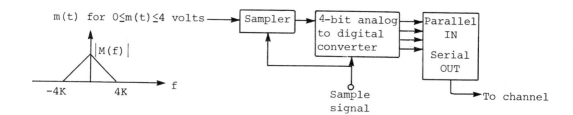

Solution: (a) From the sampling theorem, the sampling period T_s must be less than or equal to $1/2f_m$, where f_m = modulating signal frequency, for reconstruction of the signal without distortion. With f_m = 4kHz, the maximum sampling period $T_s = \frac{1}{2(4k)} = 1.25 \times 10^{-4}$ sec.

(b) Assume M, the number of quantization levels to be large such that the step size (S) will be small. With the above assumption, the quantization noise (or the mean-square quantization error) $\overline{e^2} \cong \frac{S^2}{12}$.

Now, since $M = 2^N$ and $N = 4$ (i.e., 4 bit A/D), the number of quantization levels $M = 2^4 = 16$. Thus, the step size

$S = \frac{\text{peak to peak voltage of m(t)}}{M}$

$= \frac{4}{16}$ volts

$= 0.25$ volts.

Therefore, $\overline{e^2} \cong \frac{S^2}{12} \cong \frac{(0.25)^2}{12} = 0.0052$, which is very small.

(c) Similar to part (b), for a 5 bit A/D converter,

$M = 2^5 = 32$,

$$S = \frac{4}{32} \text{ volts} = 0.125 \text{ volts}.$$

Thus, $\overline{e^2} = \frac{(0.125)^2}{12} = 0.0013$.

Therefore, by using a 5 bit A/D converter, the quantization noice is reduced by $\frac{0.0013}{0.0052} \times 100\% = 25\%$

Expressing in terms of the bandwidth (BW),

$$\frac{BW_{(5 \text{ BIT})}}{BW_{(4 \text{ BIT})}} = \frac{N_5}{N_4} = \frac{\log_2 M_5}{\log_2 M_4} = \frac{\log_2 32}{\log_2 16} = \frac{5}{4}$$

or

$$BW_{(5 \text{ BIT})} = \frac{5}{4} BW_{(4 \text{ BIT})}$$

Therefore, the required channel bandwidth for a 5 bit A/D converter is larger than the 4 bit A/D converter.

• **PROBLEM** 12-10

In a PCM system, the following binary code numbers are transmitted:

001, 101, 111, 110.

(a) Sketch the pulse and voltage representation of such coded samples. Assume guard band = τ_g.

(b) If the bandwidth of the communication channel is restricted, distortion will occur in the transmitting signal. Sketch and explain such a problem. (Hint: $d(t) = V(1 - 2e^{-\omega_c t})$ $0 \leq t \leq \tau$ is the distorted waveform, where ω_c is the angular cutoff frequency of the low-pass filter.) Then, calculate the minimum cutoff frequency f_c for $\tau = 150$ nsec to obtain a correct decision. (i.e., channel bandwidth).

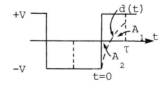

Figure 1.

Solution: (a)

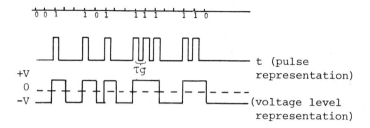
t (pulse representation)

(voltage level representation)

(b) Due to restriction of the bandwidth in the channel distortion (or intersymbol interference) occurs which is produced by the high frequency cutoff of the channel. Such interference can be represented by the waveform shown in Fig. 1 (i.e., having the characteristics of a low-pass filter).

As a result, areas A_1 and A_2 are produced.

Given $d(t) = V(1 - 2e^{-\omega_c t})$ for $0 \leq t \leq \tau$, the total area $A_1 + A_2$ can be calculated as,

$$\text{total area} = \int_0^\tau V(1 - 2e^{-\omega_c t}) dt$$

$$= V \int_0^\tau (1 - 2e^{-\omega_c t}) dt$$

$$= V \left[t + \frac{2}{\omega_c} e^{-\omega_c t} \right]_0^\tau$$

$$= \frac{V}{\omega_c} (\omega_c \tau + 2e^{-\omega_c \tau} - 2).$$

Since a positive total area is desired,

$$\frac{V}{\omega_c} (\omega_c \tau + 2e^{-\omega_c \tau} - 2) > 0$$

or

$$\omega_c \tau > 2(1 - e^{-\omega_c \tau})$$

Hence, $\omega_c \tau > 1.59$ must be satisfied.

Therefore, for $\tau = 150$ nsec

$$\omega_c (150 \times 10^{-9}) > 1.59$$

$$2\pi f_c (150 \times 10^{-9}) > 1.59$$

$$f_c > \frac{1.59}{(150 \times 10^{-9})(2\pi)}$$

$$f_c > 1.69 \text{ MHz}$$

Thus, $\quad f_{c,\min} = 1.69 \text{ MHz}$

● **PROBLEM 12-11**

> Consider a pulse code modulation (PCM) system. If the bandwidth (BW) of the transmission channel is f Hz and the information signal is bandlimited to f_m Hz where $f_m < f$, determine the maximum number of pulses that can be placed between the time samples.

Solution: The problem is to sample a time signal representing information, quantize the samples and then transmit a binary code for each of the quantized samples. Since the time samples are spaced $\frac{1}{2f_m}$ seconds apart, it is required to determine the number of binary pulses that can be placed between the time samples.

The more the number of binary samples sent between time samples, the more number of levels that can be coded, and thus, the finer the quantization. The bandwidth of the system determines how short the pulses are to be.

The signals below form a Fourier transform pair:

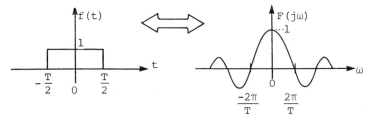

WHERE THE PULSE BANDWIDTH EQUALS $\frac{1}{T}$.

If the bandwidth of the pulse were to be defined as the first zero crossing in its spectrum, the pulse bandwidth is $\frac{1}{T}$ Hertz. Thus, for a bandwidth of f Hz, pulses can be transmitted with a width of $\frac{1}{f}$ seconds which is the width for the maximum number of pulses and is sketched below.

● **PROBLEM 12-12**

A telephone transmission line with allowable signal range 300 Hz to 3000 Hz is used to transmit a binary sequence of pulses at 9600 bits/second. If 12.5 percent sinusoidal roll-off shaping is used, prove that with 16-state QAM (quadrature amplitude modulation), the desired bit rate is achieved. If the carrier frequency is 16.5 kHz, determine the 6-dB bandwidth about the carrier.

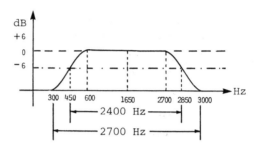

Solution: In QAM, a binary sequence of pulses is transmitted as a sequence of samples each sample consisting of n bits. The number of bits (n) per sample is given by

$$16 = 2^n$$

Hence, $n = 4$

Therefore, 9600 bits/second = 2400 samples/second,

$$= \frac{1}{T} \text{ (the sampling rate)}.$$

The transmission bandwidth BW_T with QAM is given by

$$BW_T = 2 \times \frac{(1+r)}{2T} = \frac{(1+r)}{T}$$

where $r = \frac{12.5}{100} = 0.125$ is the sinusoidal roll-off factor.

Thus, $BW_T = (1 + 0.125) \cdot (2400)$

$$= 2700 \text{ Hz}$$

This bandwidth of 2700 Hz is in the desired range of 300 to 3000 Hz. Hence, with 16 state QAM, the desired bit rate is achieved for the given transmission line.

From the figure above, observe that the 6dB bandwidth about the carrier is

$$(2850 - 450) = 2400 \text{ Hz}$$

• **PROBLEM** 12-13

In Fig. 1, a PCM system is time division multiplexed with a data source. Then the output is fed into a PSK modulator, after which the signal is transmitted over a channel of bandwidth $BW_T = 1$ MHz with center frequency at 100 MHz.

(a) Determine the maximum number of quantization levels required in this system.

(b) Now, if the data source is doubled and the number of quantization levels is fixed at 256, with all other requirements as above, what modification is needed for the modulator in this situation?

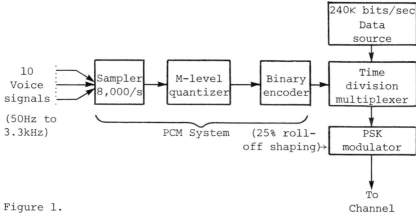

Figure 1.

Solution: (a) Since BW_T = transmission bandwidth = $2BW$, where BW is the baseband bandwidth BW = 500 kHz. For a modulator with a roll-off = r, the baseband bandwidth is given by

$$BW_T = \frac{1}{2T}(1+r)$$

where the bit rate $\frac{1}{T}$ can be evaluated as

$$\frac{1}{T} = \frac{2BW_T}{(1+r)} \quad \text{for r = 25\% roll-off = 0.25}$$

$$= \frac{2(500k)}{(1+0.25)}$$

$$= 800k \text{ bits/sec.}$$

Thus, at the output of the modulator, the signal is a sequence of pulses at 800k bits/second. Since the number of pulses from the data source is 240k bits/sec, the number of pulses from the binary encoder will be

$$= (800k - 240k)$$

$$= 560k \text{ bits/sec.}$$

The sequence of pulses at the output of the PCM system is a direct consequence of the sampler output. With 8000 samples/sec, and 10 voice signals at the input, the total sampler output is

$$= 8000 \cdot (10) = 80 \text{ k samples/sec.}$$

The number of bits per sample (n) at the PCM system output is therefore,

$$n = \frac{560k}{80k} = 7.$$

For binary signals, the number of quantization levels M is given by

$$M = 2^n = 2^7 = 128 \text{ levels.}$$

(b) For $M = 256$, the number of bits per sample is $n = 8$ (where $256 = 2^8$). With 80k samples per second and 8 bits per sample, the output of the binary encoder will be

$$= (80k \times 8) = 640k \text{ bits/second.}$$

Thus, the output bit rate of the multiplier is

$$= 640k + 2(240k)$$

$$= 1120k \text{ bits/second}$$

$$= 1.12 \text{ Megabits/second}$$

In the present case, let us assume that the PSK modulator in Figure 1 is a $4-\phi$ PSK such that its bit rate is a constant and equal to $\frac{1}{T} = 560k$ bits/sec.

From the equality

$$2BW_T = \frac{1}{T}(1+r)$$

$$2 \cdot (500k) = 560k(1+r)$$

$$r = 0.78$$

Thus, if 1120k bits/second is the bit rate of the multiplier output, a $4-\phi$ PSK modulator with a 78% roll-off is required.

• **PROBLEM** 12-14

A time-division multiplexing system using PCM is used to multiplex telephone conversations over a single communications channel. It is found that a minimum interval of 1 μs must be allowed for reliable identification of bits at the receiving end. If the allowable voice bandwidth for telephone line transmission is 3kHz and the quatization level is given as 16, approximate the number of voice signals that can be multiplexed.

Solution: The use of 16 levels means $(2)^4$ levels; thus, 4 bits are needed for each signal.

For a bandwidth of 3,000 Hertz the period is

$$\frac{1}{3000 \text{ Hertz}} \text{ or } 333 \text{ } \mu \text{ seconds}$$

Thus, the number of signals that can be multiplexed using 4 bits/signal is:

$$\frac{333 \text{ } \mu s}{4 \text{ } \mu s/\text{signal}} \cong 83 \text{ signals}$$

Note: in practical systems, the telephone line's transmission capabilities are not pushed to their theoretical limits. Thus, about 75 signals may be used in a case as described above.

• **PROBLEM 12-15**

Design a PCM multiplexing system using a 256-level signal quantizer for the transmission of three signals m_1, m_2 and m_3, bandlimited to 5kHz, 10kHz and 5kHz, respectively. Assuming that each signal is sampled at its nyquist rate and 8 bits are transmitted simultaneously, Compute:

(a) the maximum bit duration,

(b) the channel bandwidth required to pass the PCM signal,

(c) the commutator speed in rpm and then sketch the system,

(d) the increase in the channel bandwidth, if 512 quantization levels are used,

(e) Modify the scheme if the commutator sends only one bit of each sample.

Solution: (a) Maximum bit duration

$$= \frac{1}{\text{Nyquist rate}} \times \frac{1}{4} \times \frac{1}{8 \text{ bits}}$$

$$= \frac{1}{10^4} \times \frac{1}{4} \times \frac{1}{8}$$

$$= 3.125 \text{ } \mu\text{sec.}$$

(b) Channel bandwidth

$$= \frac{1}{2 \times \text{max bit duration}}$$

$$= 160 \text{ kHz}$$

(c) Commutator speed

$$= 10^4 \text{ rps}$$

$$= 6 \times 10^5 \text{ rpm}$$

The system is shown in Fig. 1 by using block diagram.

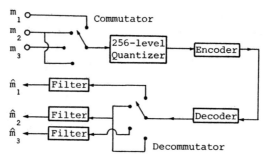

Figure 1.

(d) When the number of quantized levels of the message signal are increased, it affects the channel bandwidth also. If M_1 quantized levers are used, then each word will have N_1 time slots of duration τ_1 such that $M_2 = 2^{N_1}$ if the word length is kept constant and the number of levers are increased to M_2, then each word will have N_2 slots of duration τ_2 such that $M_2 = 2^{N_2}$.

Since the word length is a constant,

$$N_1 \tau_1 = N_2 \tau_2$$

If the bandwidth in the first case is f_1 and in the second case is f_2, then for the same degree of error immunity

so
$$\tau_1 f_1 = \tau_2 f_2$$

$$\frac{f_2}{f_1} = \frac{N_2}{N_1} = \frac{\log_2 M_2}{\log_2 M_1}$$

$$= \frac{\log_2 512}{\log_2 256} = \frac{9}{8}$$

Hence, $f_2 = 1.125 \, f_1$

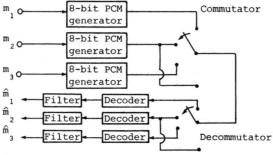

MODIFIED SCHEME FOR PART (e).

Figure 2.

(e) The modified scheme is shown in figure 2. In this case, all the quantities except the commutator speed will remain as is.

new commutator speed

$$= \frac{1}{4} \times \frac{1}{\text{max. bit duration}} = 80{,}000 \text{ rps.}$$

DELTA MODULATION

• **PROBLEM 12-16**

Let m(t), the modulating signal be a sinusoid of amplitude A and frequency f_0 in a delta modulator.

(a) If the step size S is larger than 2A, show that $\hat{m}(t)$, the approximation to the input signal m(t) is no longer valid. Verify the above (i.e., show step-size limiting).

(b) In order to avoid slope-overload in delta modulation, prove that the maximum amplitude is $A = Sf_s/2\pi f_0$, where f_s is the sampling frequency.

(c) Show that in order to avoid both step-size limiting and slope-overload, the condition $f_s > 3f_0$ must be satisfied.

Solution: (a) Observe that in the sketch below when S > 2A, the waveform $\hat{m}(t)$ is no longer responsive to any changes in m(t).

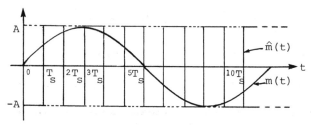

(b) In delta modulation, overload occurs when m(t) changes between samplings, by an amount greater than the size of a step. Therefore, such overload is determined by the slope but not by the amplitude of m(t) as is common in other modulating schemes.

For the given signal $m(t) = A \sin 2\pi f_0 t$, to avoid slope-overload,

$$\left. \frac{d}{dt} m(t) \right|_{t=0} < \frac{S}{T_s} \quad \text{where } T_s \text{ is the sampling time.}$$

That means,
$$\frac{d}{dt}(A\sin 2\pi f_0 t)\bigg|_{t=0} < \frac{S}{T_s}$$

$$A\, 2\pi f_0 \cos 2\pi f_0 t\bigg|_{t=0} < \frac{S}{T_s}$$

since $\frac{1}{T_s} = f_s$ $\qquad A 2\pi f_0 < S f_s$

$$A < \frac{S f_s}{2\pi f_0}$$

Therefore, the maximum signal amplitude is $Sf_s/2\pi f_0$ and also note that the point of slope-overload occurs when $2\pi f_0 A = Sf_s$ (the rates of rising or the slopes are equal).

(c) In order to avoid both step-size limiting and slope-overload, the conditions:

$$S < 2A \quad \text{and} \quad A = Sf_s/2\pi f_0 \text{ must be satisfied.}$$

Thus, from the above constraints,

$$S < 2\,\frac{Sf_s}{2\pi f_0}$$

therefore, $\qquad f_s > \pi f_0 > 3 f_0.$

• **PROBLEM 12-17**

Consider the delta PCM system shown in the following figure. Compute the following graphically if, $\tau = \frac{1}{8}$ sec. and the input signal $s(t) = 0.05 \sin(2\pi t)$

(a) $\tilde{s}(t)$, the signal at the output of the ingegrator, and

(b) input to the delta modulator i.e., $\Delta(t)$.

Also sketch the compatible receiver for the above system.

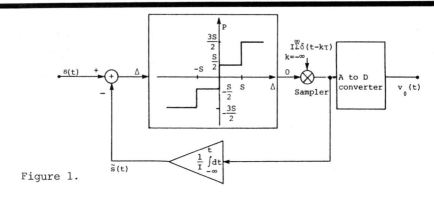

Figure 1.

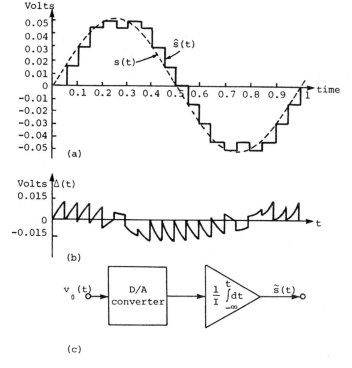

Figure 2.

Solution: This is a four-level delta modulator. The incoming signal is s(t), and Δ(t) will depend on $\tilde{s}$ which is the approximate signal.

Δ(t) is positive for $s(t) > \tilde{s}(t)$

Δ(t) is negative for $s(t) < \tilde{s}(t)$

The output, O, of the modulator could have only four possible levels depending on the incoming signal Δ(t) as shown in figure 1. The signal O is sampled, and integrated to obtain the approximate signal $\tilde{s}(t)$. The sampled signal of O is encoded and transmitted. The input to the delta modulator Δ(t) and $\tilde{s}(t)$ is sketched in Figs. (2a) and (b). The corresponding receiver is shown in Fig. (2) by using simple block diagram.

• **PROBLEM 12-18**

The block diagram of a delta-modulator is shown below. The input to the system is a signal m(t) which is sampled by the sampling frequency f_s. Show that:

(a) If the input is $m(t) = k \cdot t$, then slope-overload occurs when k exceeds $S \cdot f_s$, where S is the step-size;

(b) If the input $m(t) = M \cdot \sin\omega t$ and $2M < S$, then the signal $m(t)$ cannot be recovered; and

(c) If the input is $M \cdot \sin\omega_0 t$, then $f_s > 3 \cdot f_0$ to avoid slope-overload and step-size limiting.

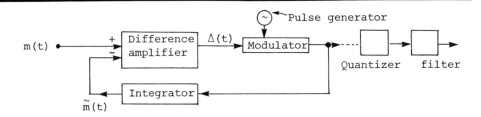

A DELTA-MODULATION SYSTEM

Figure 1.

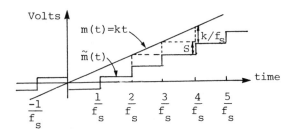

SLOPE OVERLOAD SITUATION

Figure 2.

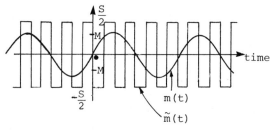

Figure 3.

Solution: (a) Fig. (2) shows the slope-overload situation. In this figure, the input signal $m(t) = k \cdot t$ and the output of the integrator $\bar{m}(t)$ are both plotted against time 't'. The output of the integrator $\bar{m}(t)$ is an approximation to the input signal $m(t)$. The overload occurs when the modulating signal changes between samplings by an amount greater than the step-size. So from Fig. (2), slope-overload occurs when

$$\frac{k}{f_s} > S \quad \text{or} \quad k > S \cdot f_s.$$

(b) From Fig.(3), it follows that

when $\bar{m}(k) = \frac{S}{2}$ then

$$\bar{m}(k+1) = \bar{m}(k) + S \cdot \text{Sgn}\left[m(k) - \bar{m}(k)\right]$$

$$= \frac{S}{2} + S \cdot \text{Sgn}\left[M \cdot \sin\omega kT - \frac{S}{2}\right]$$

$$= \frac{S}{2} - S$$

$$= \frac{-S}{2}$$

Similarly, when $\bar{m}(k) = \frac{-S}{2}$ then

$$\bar{m}(k+1) = \frac{-S}{2} + S \cdot \text{Sgn}\left[M \cdot \sin\omega kT + \frac{S}{2}\right]$$

$$= \frac{-S}{2} + S$$

$$= \frac{S}{2}$$

Thus, it can be seen that $\bar{m}(t)$ is a squarewave with period $1/f_{S_1}$ and hence, the original signal cannot be recovered.

(c) The necessary condition for avoiding slope-overload and step-size limiting is as follows:

$$2\pi \cdot f_0 \cdot M < S \cdot f_S$$

and

$$S < 2M$$

Hence,

$$2\pi \cdot f_0 \cdot M < S \cdot f_S < 2M \cdot f_S$$

Therefore:

$$3f_0 < \pi \cdot f_0 < f_S$$

• **PROBLEM** 12-19

A sinusoidal signal $m(t) = 0.1 \sin(2\pi \times 10^3)t$ is modulated by a delta-modulator as shown below. Assuming that at $t = 0$, $S = m(t) - \bar{m}(t)$, where S = the step-size = 20mv and $\bar{M}(t)$ is a stepped approximation of $m(t)$.

(a) Plot $m(t)$, $\bar{m}(t)$ and $E(t)$ (the energy) as a function of time.

(b) Prove the best step-size (optimum step-size) $S_{optimum} \approx 2\pi A/(f_S/f_m)$, where A is the peak amplitude of the sine wave, f_S is the clock rate and f_m

is the frequency of the sine wave. Then calculate S_{opt}.

(c) For S = 4mv and S = 60mv does any slope-overloading occur?

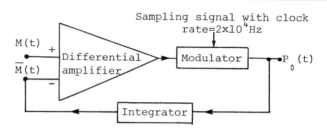

DELTA MODULATOR

Figure 1.

Solution: (a) The waveforms for m(t), $\tilde{m}(t)$ and E(t) are plotted in Fig. 2.

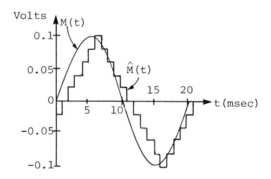

Figure 2.

(b) If $m(t) = A \sin\omega_m t$, then slope-overload will occur when

$$\left.\frac{d}{dt} m(t)\right|_{t=0} \geq \frac{S}{T_s}$$

$$\left.\frac{d}{dt} A \sin\omega t\right|_{t=0} \geq \frac{S}{T_s}$$

$$\omega_m A \geq \frac{S}{T_s}$$

Therefore $\quad S_{opt} \cong T_s 2\pi f_m A$ (1)

$$\cong \frac{2\pi A}{(f_s/f_m)}$$

Substituting $f_s = 20\text{kHz}$ and $f_m = 1 \times 10^3 \text{Hz}$ into eq.(1)

$$S_{opt} \cong \frac{2\pi A}{(f_s/f_m)} \cong \frac{2\pi \times 0.1}{20 \times 10^3/1 \times 10^3} \cong 30\text{mv}$$

(c) Note that slope-overload occurs when the step-size $S \geq S_{opt}$.

Thus for a step-size of 4mv, there will not be any slope-overload. However for a step-size of $S = 60\text{mv}$, $S \geq S_{opt}$ which will result in a greater slope and hence in slope-overloading.

• **PROBLEM** 12-20

The equation,

$$S\left[(k+1)T_s\right] = |S(kT_s)| E(kT_s) + S_0 E\left[(k-1)T_s\right]$$

characterizes an adaptive delta-modulating system.

Assuming that $M(t)$ remains constant within the step size S, (which together with $\bar{M}(t)$ are shown in the figure below), prove that the frequency at which $\bar{M}(t)$ fluctuates about $M(t)$ is one-half the frequency that would be encountered in linear delta-modulation.

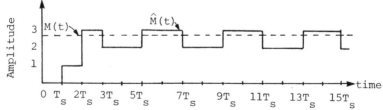

Solution:

In a linear delta-modulator, the frequency at which $\bar{M}(t)$ fluctuates about $M(t)$ is given by

$$f_{OSC} = \frac{1}{2} T_s \quad \text{where } T_s \text{ is the sampling period}$$

Observe from the figure for the adaptive delta-modulator, the oscillation period, $T_{OSC} = 4T_s$

Hence, $\quad f_{OSC} = \dfrac{1}{4T_s}$

Therefore, the frequency of oscillation of an adaptive delta-modulator system is one-half of the frequency encountered in a linear delta-modulator.

BINARY COMMUNICATIONS - ON-OFF KEYING, PSK, DPSK AND FSK

● **PROBLEM** 12-21

The difference equation

$$S\left[(k+1)T_s\right] = |S(kT_s)| \, E(kT_s) + S_0 E\left[(k-1)T_s\right]$$

characterizes an adaptive delta-modulator. A waveform $M(t)$ shown in Fig. 1 below is applied to the modulator. Assuming $\hat{M}(t) = 0$ at $t = 0$, with a minimum step-size $S_0 = 0.5$ volts, plot $M(t)$ and $\hat{M}(t)$ from $t = 0$ to $36T_s$.

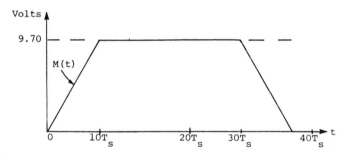

Figure 1.

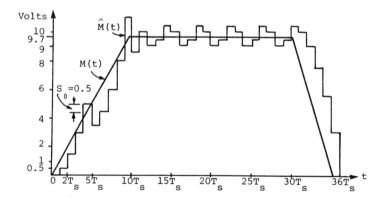

Figure 2.

Solution: The plot is shown in Fig. 2.

• **PROBLEM** 12-22

The following binary sequence is given.

1 1 0 1 0 0 0 1 1 0 0

(a) If phase shift keying (PSK) is used to transmit this signal, determine the transmitted sequence and the corresponding carrier phase.

(b) What is the sequence that is transmitted and the corresponding carrier phase if Differential Phase Shift Keying (DPSK) is used to transmit the sequence?

(c) State one advantage and one disadvantage of DPSK over PSK.

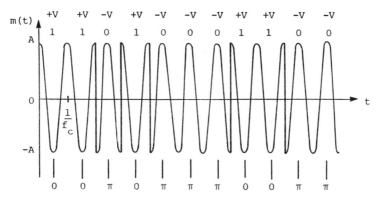

Figure 1.

Solution: (a) In a phase-shift keying system, let's assume $m(t)$ to be a modulating binary signal, which has values $m(t) = +V$ or $m(t) = -V$.

Then, the PSK waveform is given as

$$m_{PSK}(t) = A\cos\left[\omega_0 t + \phi(t)\right] = \frac{m(t)}{V} A\cos\omega_0 t \quad (1)$$

where A = constant amplitude,

and $\phi = 0$ for $m(t) = +V$
$\phi = \pi$ for $m(t) = -V$ (assumption)

(Notice that the waveform in eq.(1) is generated by applying the signal $m(t)$ and the carrier $\cos\omega_0 t$ to a balanced modulator.)

Hence, the received signal is,

$$m_{PSK}(t) = \frac{m(t)}{V} A\cos(\omega_0 t + \theta)$$

where θ is a phase angle corresponding to the length of the transmission channel.

Now, let +V correspond to the binary bit 1 and -V correspond to 0, then the transmitting sequence and its corresponding phase angle is shown in Fig. 1. (Assume m(t) has duration $T = 1/f_c$.)

(b) Since differential phase-shift keying is used, let m(t) be the given sequence, then an auxiliary binary sequence m'(t) is generated as shown in Fig. 2.

$$m(t) \rightarrow 1\ 1\ 0\ 1\ 0\ 0\ 0\ 1\ 1\ 0\ 0$$
$$m'(t) \rightarrow 1\ 1\ 1\ 0\ 0\ 1\ 0\ 1\ 1\ 1\ 0\ 1$$
$$\text{phase} \rightarrow 0\ 0\ 0\ \pi\ \pi\ 0\ \pi\ 0\ 0\ 0\ \pi\ 0$$

Figure 2.

Note: From the basis rule, the first 1 of m'(t) is arbitrarily chosen. Then, if m(t) is 1, m'(t) remains the same value as before. On the other hand, if m(t) is 0, m'(t) is changed.

The block diagram used to generate m'(t) is shown in Fig. 3. Next, the signal together with the carrier is applied to a balanced modulator. As a result, a signal $m_{DPSK}(t)$ is transmitted similar to part (a).

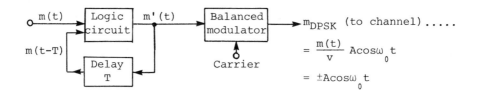

Figure 3.

Note: The logic circuit operates as follows:

If m(t) and m(t-T) are the same, then m'(t) = 1. Otherwise, m'(t) = 0.

(c) One advantage of a DPSK coherent system over PSK is that, the circuit involved in the generation of a local carrier at the DPSK receiver is less complicated. On the other hand, the error rate during transmission in DPSK is greater than PSK because bit determination is based on the signal received in two successive bit intervals. Hence, noise in one bit interval will produce errors for two bit determinations.

• **PROBLEM 12-23**

A binary encoded message which is represented by the binary sequence b'(t) = 11001100111100100, is transmitted using a DPSK transmitter. Find the auxiliary bit stream b(t) generated by the transmitter and show how the original stream b'(t) can be recovered by a DPSK receiver.

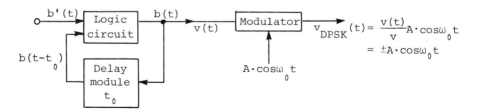

DPSK TRANSMITTER

Figure 1.

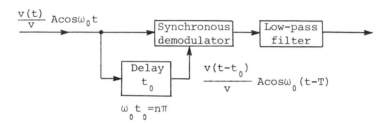

DPSK RECEIVER

Figure 2.

Solution: The input binary sequence is 11001100111100100. When this binary sequence is fed to the logic circuit as shown in Fig. 1, an auxiliary sequence b(t) is generated. This sequence has one extra digit than the input stream. The extra bit added by the circuitry is 1 (arbitrarily chosen) and subsequent digits in b(t) are either 0 or 1 depending on whether b'(t) is 1 or 0 during some bit interval. If b'(t) is a 1, then b(t) in that interval does not change from its value in the preceding interval, but if b'(t) is 0, b(t) does change.

Applying this rule, the following auxiliary bit stream b(t) results.

```
b'(t)    =   1 1 0 0 1 1 0 0 1 1 1 1 0 0 1 0 0
b(t)     = 1 1 1 0 1 1 1 0 1 1 1 1 0 1 1 0 1
b(t-t₀)  = 0 1 1 1 0 1 1 1 0 1 1 1 1 0 1 1 0
```

When this stream b(t) is used to modulate a carrier signal $A\cos\omega_0 t$, an output signal of maximum amplitude $\frac{v(t)}{V} A = A$ results. This signal is processed in the receiver shown in Fig. 2 using a delay module. Both signals, the received signal and the delayed signal are fed into a synchronous demodulator which acts as an exclusive-OR gate. The output of this modulator is thus,

$$b(t) \oplus b(t-t_0) = 1\ 1\ 1\ 0\ 1\ 1\ 1\ 0\ 1\ 1\ 1\ 1\ 0\ 1\ 1\ 0\ 1$$

$$\oplus\ 0\ 1\ 1\ 1\ 0\ 1\ 1\ 1\ 0\ 1\ 1\ 1\ 1\ 0\ 1\ 1\ 0$$

$$=\ 1\ 1\ 0\ 0\ 1\ 1\ 0\ 0\ 1\ 1\ 1\ 1\ 0\ 0\ 1\ 0\ 0$$

$$= b'(t)$$

Note how $b(t-t_0)$ is obtained above. When $b'(t) = b(t)$, then $b(t-t_0) = 1$, else equal to zero.

• **PROBLEM** 12-24

A communication channel is used for transmitting audio signals in the range 600 to 3000 Hz, using a 1800 Hz carrier wave. If phase-shift keying technique is used, show that,

(a) Using a four-phase PSK with raised cosine shaping, it is possible to transmit data at a rate of 2400 bits per second.

(b) It is possible to transmit data at a rate of 4800 bits per second using eight-phase PSK with 50% sinusoidal roll-off.

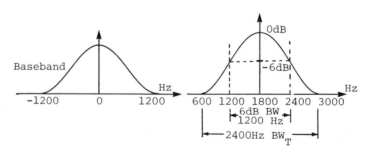

Figure 1.

Solution: (a) For raised cosine shaping, r = 1 (roll-off factor) and with four-phase PSK (i.e., $4 = 2^n$, n = 2), the required bandwidth is given as

$$BW = \frac{1}{T} = \frac{2400 \text{ bits/sec}}{2 \text{ bits/symbols}} = 1200 \text{ Hz} \quad (\text{see Fig. 1})$$

Since, the transmission bandwidth is given as follows,

$$BW_T = 2BW$$

$$BW_T = 2400 \text{ Hz} \quad \text{as shown in Fig. 1.}$$

Note that in Fig. 1, the 6dB bandwidth about the carrier is 1200Hz which is $1/2$ BW_T.

(b) Similarly, for 8ϕ PSK, with data rate = 4800 bits per second

$$\frac{1}{T} = \frac{4800}{3} = 1600 \text{ symbols per second.}$$

(See fig. 2).

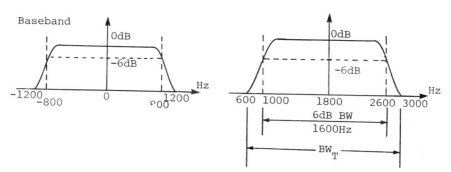

Figure 2.

Since 8ϕ corresponds to n = 3, $\frac{1}{T} = \frac{\text{data rate}}{n}$

and r = 0.5 (50% sinusoidal roll-off).

Hence, the required bandwidth for sinusoidal roll-off spectrum is given as

$$BW = \frac{1}{2T}(1+r)$$

$$= 800\,(1.5)$$

$$= 1200 \text{Hz}$$

Therefore, $\quad BW_T = 2BW = 2400 \text{Hz}$

Note that the 6-dB bandwidth is now 1,600Hz as indicated in Fig. 2.

• **PROBLEM** 12-25

Consider a Pulse Code Modulation (PCM) system with binary encoder output of 2×10^6 bits/second. Determine the transmission bandwidth (BW_T) required and sketch the spectrum for each of the following cases:

(a) On-off keying (OOK) transmission with amplitude modulation of a sine wave carrier.

(b) Frequency-shift keying (FSK) with (i) switching between two sine waves of frequencies 100 and 104 MHz. ii) Same as (i) but with frequencies 100 and 120 MHz.

Also assume raised cosine shaping is used for the baseband pulse.

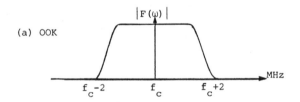
(a) OOK

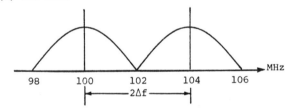
(b) FSK
(i) $2\Delta f = 4$MHz

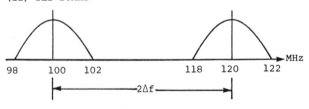
(ii) $2\Delta f = 20$MHz

<u>Solution</u>: (a) The transmission bandwidth BW_T for OOK is given by $BW_T = 2BW$ where $BW = \frac{1}{2T}(1+r)$ = baseband bandwidth. Since $\frac{1}{T} = 2 \times 10^6$ bits/sec and

$r = 1$ = roll-off factor for raised cosine,

$BW = \frac{1}{2T}(2) = \frac{1}{T} = 2$ MHz

and $\quad BW_T = 2BW = 4$ MHz

The OOK spectrum is shown in Fig. (1a).

(b) (i) Since the approximate transmission bandwidth for FSK is given by

$$BW_T = (2\Delta f + 2BW)$$

$$BW_T = [(104M - 100M) + 2(2M)] \text{ Hz} = 8\text{MHz}$$

(ii) Note: $2\Delta f = 4$MHz. See Fig. (1b).

For $2\Delta f = (120M - 100M)$Hz $= 20$MHz (See Fig. 1b)

$$BW_T = 2\Delta f + 2BW$$
$$= 20\text{MHz} + 4\text{MHz}$$
$$= 24\text{MHz}.$$

The frequency spectrum for each of the above cases is shown in the figure

CHAPTER 13

MATHEMATICAL REPRESENTATION OF NOISE

FREQUENCY DOMAIN REPRESENTATION AND SPECTRAL CHARACTERISTICS OF NOISE

● **PROBLEM** 13-1

Given a square waveform g(t) of peak to peak voltage equal to 2V, where V is a uniformly distributed random variable, as shown in Fig. (1).

(a) Plot the power spectrum of the waveform.

(b) Determine the normalized power of the waveform contained in the frequency range -3×10^3 to $+3 \times 10^3$ Hz in percentage.

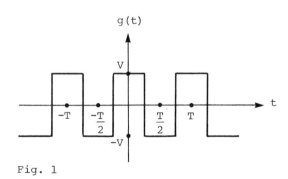

Fig. 1

Solution: Since the waveform g(t) is periodic, it can be expanded in a Fourier series.

Hence, $g(t) = \sum_{k=1}^{\infty} g_k \cos 2\pi k \frac{t}{T}$ a periodic waveform,

where g_k is known as the constant coefficient of the spectral term and is obtained as follows,

$$g_k = \frac{2}{T}\int_{-T/2}^{-T/4} -V\cos 2\pi k\frac{t}{T}\,dt + \frac{2}{T}\int_{-T/4}^{T/4} V\cos 2\pi k\frac{t}{T}\,dt$$

$$+ \frac{2}{T}\int_{T/4}^{T/2} -V\cos 2\pi k\frac{t}{T}\,dt$$

$$g_k = \frac{-2V}{T}\left[\frac{\sin\left(\frac{2\pi\cdot k\cdot t}{T}\right)\cdot T}{2\pi\cdot k}\right]_{-T/2}^{-T/4} + \frac{2V}{T}\left[\frac{\sin\left(\frac{2\pi\cdot k\cdot t}{T}\right)\cdot T}{2\pi\cdot k}\right]_{-T/4}^{T/4}$$

$$+ \frac{-2V}{T}\left[\sin\left(\frac{2\pi\cdot k\cdot t}{T}\right)\right]_{T/4}^{T/2}$$

$$= \frac{-V}{\pi\cdot k}\left[\sin\left(\frac{-2\pi\cdot k}{T}\times\frac{T}{4}\right) - \sin\left(-\frac{2\pi\cdot k}{T}\times\frac{T}{2}\right)\right]$$

$$+ \frac{2V}{2\pi\cdot k}\left[\sin\frac{\pi k}{2} + \sin\frac{\pi k}{2}\right] - \frac{2V}{2\pi\cdot k}\left[\sin(\pi\cdot k) - \sin\left(\frac{\pi k}{2}\right)\right]$$

$$= \frac{V}{\pi k}\left[\sin\left(\frac{\pi k}{2}\right) - \sin(-\pi k) + \sin\left(\frac{\pi k}{2}\right) + \sin\left(\frac{\pi k}{2}\right) - \sin(\pi k)\right.$$

$$\left. + \sin\left(\frac{\pi k}{2}\right)\right]$$

Since $\sin(-\theta) = -\sin(\theta)$, above expression simplifies to

$$g_k = \frac{4V}{\pi k}\cdot\sin\frac{\pi k}{2}$$

Thus, the expression for the given square waveform is

$$g(t) = \sum_{k=0}^{\infty}\frac{(-1)^k 4V}{\pi(2k+1)}\cos 2\pi(2k+1)\frac{t}{T}$$

$$= \sum_{k=0}^{\infty}\frac{(-1)^k 4V}{\pi(2k+1)}\cos 2\pi(2k+1)\Delta f\cdot t.$$

Since V is a uniform random variable between 1V and 2V, its variance can be expressed as

$$\overline{V^2} = \int_{-\infty}^{\infty} V^2 f(V)\,dV$$

$$= \int_1^2 V^2 dV$$

$$= \frac{7}{3}$$

Now, the power spectral density at the frequency $(2k+1)\Delta f$ will be

$$G_n\left[(2k+1)\Delta f\right] \equiv \frac{\overline{g_{2k+1}^2}}{4\Delta f}$$

g_{2k+1} can be evaluated from equation (1) as

$$g_{2k+1} = \frac{4V}{\pi(2k+1)} \sin \frac{\pi(2k+1)}{2}$$

$$= \frac{4V}{\pi(2k+1)} \sin \pi(k+\tfrac{1}{2})$$

$$= \frac{4V}{\pi(2k+1)} \left[(\sin \pi k \cdot \cos \pi/2) + (\cos \pi k \cdot \sin \pi/2)\right]$$

$$= \frac{4V}{\pi(2k+1)} \left[0 + (-1)^k\right]$$

For k is even,

$$g_{2k+1} = \frac{4V}{\pi(2k+1)}$$

Thus,

$$G_n\left[(2k+1)\Delta f\right] = \frac{4\overline{V}^2}{\pi^2(2k+1)^2 \Delta f}$$

$$= \frac{28}{3\pi^2(2k+1)^2 \Delta f}$$

and

$$G_n\left[2k\Delta f\right] = 0$$

A plot of the power spectrum for the given waveform is shown in figure (2) below.

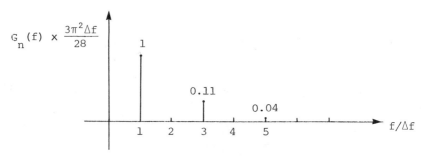

Fig. 2

(b) From the power spectrum shown in figure (2), it is observed that there is no component of power in the negative half of the spectrum. Thus, in effect, the power contained in the frequency range -3×10^3 to $+3 \times 10^3$ Hz is the same as the power contained in the range 0 to $+3 \times 10^3$ Hz, which is equal to 1 corresponding to $f/\Delta f = 1$.

Expressing this component of power in terms of the total power in the spectrum, we have

$$\frac{G_n(\Delta f) + G_n(-\Delta f)}{\sum_{k=-\infty}^{\infty} G_n[k\Delta f]} = \frac{1 + 0}{\sum_{k=0}^{\infty} \frac{1}{(2k+1)^2}} = \frac{1}{(\pi^2/8)} = 0.81$$

or $\simeq 81\%$

• **PROBLEM 13-2**

For a noise $n(t)$ with a power spectral density $G_n(f)$ as shown in figure (1),

(a) Determine the normalized power P in terms of η and f_m. Evaluate P for $\eta = 2\mu$ V²/Hz and $f_m = 8$ kHz.

(b) Determine and plot the autocorrelation function of the noise.

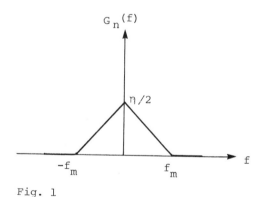

Fig. 1

Solution:
(a) The normalized power P of the noise can be expressed as

$$P = \int_{-\infty}^{\infty} G_n(f) df$$

From Fig. (1), $G_n(f)$ can be expressed as

$$G_n(f) = \frac{\eta}{2f_m} (f_m - f)$$

Hence, the normalized power is calculated as follows:

$$P = \int_{-f_m}^{f_m} \frac{\eta}{2f_m}(f_m - f)df$$

$$= 2\int_{0}^{f_m} \frac{\eta}{2f_m}(f_m - f)df$$

$$= \frac{\eta}{f_m}\left[-\frac{(f_m-f)^2}{2}\right]_0^{f_m}$$

$$= \frac{\eta}{f_m} \cdot \frac{f_m^2}{2}$$

$$= \frac{\eta f_m}{2}$$

For $\eta = 2\ \mu V^2/Hz$ and $f_m = 8$ kHz,

$$P = \frac{(2 \times 10^{-6})(8 \times 10^3)}{2}$$

$$= 8 \times 10^{-3} \text{ watt}$$

$$= 8 \text{ m watt}.$$

(b) The autocorrelation function of the noise is defined as

$$R_n(\tau) = \int_{-\infty}^{\infty} G_n(f) e^{j\omega\tau}\, df$$

$$= \int_{-\infty}^{\infty} \frac{\eta}{2f_m}(f_m - f)\cos(2\pi f\tau)\, df$$

$$= \frac{\eta}{f_m}\int_0^{f_m}(f_m - f)\cos 2\pi f\tau\, df$$

$$= \frac{\eta}{f_m}\left(\int_0^{f_m} f_m \cdot \cos(2\pi f\tau)\, df - \int_0^{f_m} f \cdot \cos(2\pi f\tau)\, df\right)$$

$$= \frac{\eta}{f_m}\left\{f_m\left[\frac{\sin 2\pi f\tau}{2\pi\tau}\right]_0^{f_m} - f\cdot\left[\frac{\sin 2\pi f\tau}{2\pi\tau}\right]_0^{f_m}\right.$$

$$\left. - \int_0^{f_m} \frac{\sin 2\pi f\tau}{2\pi\tau}\cdot df\right\}$$

$$= \frac{\eta}{f_m} \left(f_m \cdot \frac{\sin 2\pi f_m \tau}{2\pi\tau} - f_m \cdot \frac{\sin 2\pi f_m \tau}{2\pi\tau} \right.$$

$$\left. + \int_0^{f_m} \frac{\sin 2\pi f \tau}{2\pi\tau} df \right)$$

$$= \frac{\eta}{2\pi f_m \tau} \int_0^{f_m} \sin 2\pi f \tau \cdot df$$

$$= \frac{\eta}{2\pi f_m \tau} \left[\frac{-\cos 2\pi f \tau}{2\pi\tau} \right]_0^{f_m}$$

$$= \frac{\eta}{(2\pi\tau)^2 f_m} (1 - \cos 2\pi f_m \tau)$$

$$= \frac{\eta}{(2\pi\tau)^2 f_m} \left[2(\sin \pi f_m \tau)^2 \right]$$

$$= \frac{\eta f_m}{2} \left(\frac{\sin \pi f_m \tau}{\pi \tau f_m} \right)^2$$

The plot of the autocorrelation function is shown in figure (2) below.

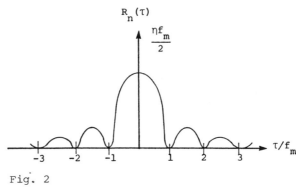

Fig. 2

● **PROBLEM 13-3**

If the autocorrelation function $R_n(\tau)$ of the noise $n(t)$ is

$$R_n(\tau) = A \cos \omega_0 \tau,$$

(a) Find the power spectral density $G_n(f)$ of the noise,

(b) Sketch both $R_n(\tau)$ and $G_n(f)$.

(c) Repeat part (a) and (b) if $R_n(\tau) = N e^{-\omega_0 |\tau|}$

Solution:
(a) By definition, the power spectral density $G_n(f)$ of $n(t)$ is the Fourier transform of the autocorrelation $R_n(\tau)$.

Hence,
$$G_n(f) = F[R_n(\tau)] = \int_{-\infty}^{\infty} R_n(\tau)e^{-j\omega\tau}d\tau$$

$$= \int_{-\infty}^{\infty} A(\cos\omega_0\tau)\cdot e^{-j\omega\tau}d\tau$$

$$= \frac{A}{2}\int_{-\infty}^{\infty}(e^{j\omega_0\tau} + e^{-j\omega_0\tau})e^{-j\omega\tau}d\tau$$

$$= \frac{A}{2}\int_{-\infty}^{\infty}(e^{j\omega_0\tau}\cdot e^{-j\omega\tau} + e^{-j\omega_0\tau}\cdot e^{-j\omega\tau})d\tau$$

Recall that the Fourier transform is also defined as

$$F[e^{j\omega_0\tau}] = 2\pi\delta(\omega - \omega_0).$$

Therefore,
$$G_n(f) = \frac{A}{2}[2\pi\delta(\omega - \omega_0) + 2\pi\delta(\omega + \omega_0)]$$

$$= \frac{A}{2}[\delta(f - f_0) + \delta(f + f_0)]$$

(b) The sketches for $R_n(\tau)$ and $G_n(f)$ are shown in Fig. (1).

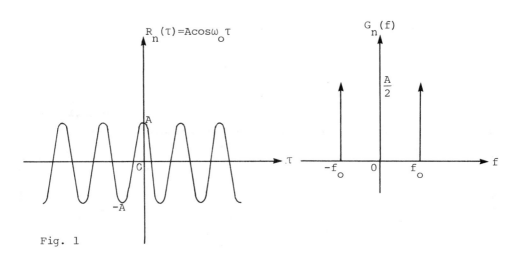

Fig. 1

(c) Similarly, if $R_n(\tau) = Ne^{-\omega_0|\tau|}$

Then $G_n(f) = \int_{-\infty}^{\infty} N e^{-\omega_0 |\tau|} \cdot e^{-j\omega\tau} d\tau$

$= N \left[\int_{-\infty}^{0} e^{(\omega_0 - j\omega)\tau} d\tau + \int_{0}^{\infty} e^{-(\omega_0 + j\omega)\tau} d\tau \right]$

$= N \left\{ \frac{1}{(\omega_0 - j\omega)} \left[e^{(\omega_0 - j\omega)\tau} \right]_{-\infty}^{0} \right.$

$\left. + \frac{1}{-(\omega_0 + j\omega)} \left[e^{-(\omega_0 + j\omega)\tau} \right]_{0}^{\infty} \right\}$

$= N \left[\frac{1}{(\omega_0 - j\omega)} + \frac{1}{(\omega_0 + j\omega)} \right]$

$= \frac{N(2\omega_0)}{\omega_0^2 + \omega^2}$

Since $\omega_0 = 2\pi f_0$,

$G_n(f) = N \left[\frac{2(2\pi f_0)}{(2\pi f_0)^2 + (2\pi f)^2} \right] = \frac{N/\pi f_0}{1 + (f/f_0)^2}$

The sketches of $R_n(\tau)$ and $G_n(f)$ are shown in Fig. (2) for

$R_n(\tau) = N e^{-\omega_0 |\tau|}$

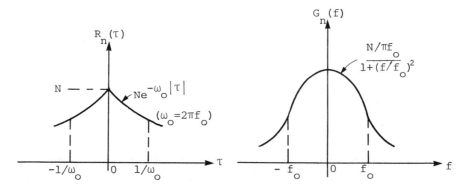

Fig. 2

(Note that f_0 is the 3-dB bandwidth.)

• **PROBLEM 13-4**

Given the correlation function and spectral-density pairs of two noise sources $n_1(t)$ and $n_2(t)$ as shown in Fig. 1, compare the inverse correlation time/bandwidth relation in each case.

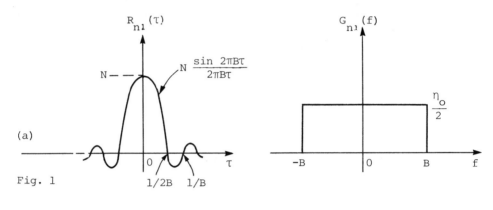

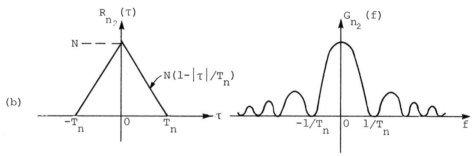

Fig. 1

Solution: Observe in Fig. (1a), the noise $n_1(t)$ is strictly bandlimited to B hertz. At $\tau = \pm 1/2B$ and integral multiples thereof, it is readily seen that $n_1(t)$ and $n_1(t+\tau)$ are always uncorrelated. Note that this is also the case for $|\tau| \geq T_n$ of $n_2(t)$. Now, observe in Fig. (1b), the measure of the bandwidth of $n_2(t)$ is the first zero crossing of $G_{n2}(f)$ at $f = 1/T_n$.

In both cases, if the bandwidth is chosen as 1 MHz, then samples spaced more than 1μs apart are essentially uncorrelated.

Note that in the case of $n_2(t)$, most of the noise power is concentrated about $1/T_n$. This indicates that the noise $n_2(t)$ rarely fluctuates at these rates because, little noise power appears at high frequencies as shown. Therefore, 1/bandwidth in the case of $n_2(t)$ is a measure of time between significant changes in $n_2(t)$. On the other hand, in the case of $n_1(t)$, all the power is concentrated in the range 0 to B hz. This case is of course, the random signal equivalent of a band-limited deterministic signal.

• **PROBLEM** 13-5

The power spectral density of a Gaussian noise of zero mean is given as

$$G_n(f) = 3 \ \mu V^2/Hz \qquad |f| \leq 1 \text{ kHz}$$
$$= 0 \qquad \text{elsewhere.}$$

(a) Determine the normalized power and the probability density function of the noise.

(b) If the noise is passed through a filter which will decrease the noise power by 50% at the output, find the probability density function of noise at the output of the filter.

Solution:

(a) For the noise, the normalized power is defined as

$$P = \int_{-\infty}^{\infty} G_n(f) df$$

Thus, since $G_n(f)$ is given,

$$P = \int_{-10^3}^{10^3} 3 \times 10^{-6} df$$

$$= 2 \int_{0}^{10^3} 3 \times 10^{-6} df$$

$$= 2(3 \times 10^{-6})(10^3)$$

$$= 6 \times 10^{-3} \text{ watt.}$$

The probability density function of the Gaussian noise is defined as,

$$f(n) = \frac{1}{\sqrt{2\pi\sigma^2}} e^{-(n-m)^2/2\sigma^2}$$

where m is the mean and σ^2 is the variance of the Gaussian noise function respectively. For the Gaussian distribution under consideration, $m = 0$ and $\sigma^2 = P = 6 \times 10^{-3}$ (Note: $\sigma^2 = E(n^2) = P$)

Thus,

$$f(n) = \frac{1}{\sqrt{2\pi(6 \times 10^{-3})}} \times e^{-\frac{n^2}{2(6 \times 10^{-3})}}$$

$$= \frac{1}{\sqrt{12\pi \times 10^{-3}}} e^{-\frac{n^2}{12 \times 10^{-3}}}$$

(b) Let the response of the filter be an impulse function $h(t)$, Then, the expected value of the noise at the output of the filter is

$$= E\left[\int_{-\infty}^{\infty} h(t - \tau) \cdot n(\tau) d\tau\right]$$

$$= \int_{-\infty}^{\infty} h(t - \tau) \, E[\overline{n(\tau)}] \, d\tau$$

Where $n(\tau)$ is the noise function.

However, assuming the filter to be an ideal filter, the expected value of noise at the output will be zero.

Since at the output of the filter the power is $\frac{1}{2}$ the power of $n(t)$, the variance of the output noise is

$$\sigma^2 = \frac{P}{2} = 3 \times 10^{-3} \text{ watt.}$$

Hence, the probability density function at the output of the filter will be

$$f_0(n_0) = \frac{1}{\sqrt{6 \times 10^{-3}}} e^{\left(\frac{-n_0^2}{6 \times 10^{-3}}\right)}$$

● **PROBLEM 13-6**

Given the normalized power of a Gaussian noise, $n_k(t) = a_k \cos(2\pi k \Delta f t) + b_k \sin(2\pi k \Delta f t)$ for narrowband noise, is 0.02 µW, determine the probability density functions of the coefficients a_k and b_k.

Solution: For a Gaussian noise, the probability density function is expressed as,

$$f(n) = \frac{1}{\sqrt{2\pi\sigma^2}} e^{-(n-m)^2/2\sigma^2}$$

where m is the mean of the distribution and σ^2 is its variance.

Since the Gaussian noise under consideration is a superposition of noise components a_k and b_k, the probability density functions of these individual components can be expressed as

$$f(a_k) = \frac{1}{\sqrt{2\pi \overline{a_k^2}}} e^{\left[\frac{-(a_k - m)^2}{2\,\overline{a_k^2}}\right]} \qquad (1)$$

$$\text{and} \quad f(b_k) = \frac{1}{\sqrt{2\pi \overline{b_k^2}}} e^{\left[\frac{-(b_k - m)^2}{2\,\overline{b_k^2}}\right]} \qquad (2)$$

where a_k is the component of noise at $t = t_1$ for which $\cos 2\pi k \Delta f t_1 = 1$ and $\sin 2\pi k \Delta f t_1 = 0$, and b_k is the component of noise at $t = t_2$ for which $\cos 2\pi k \Delta f t_2 = 0$ and $\sin 2\pi k \Delta f t_2 = 1$. $\overline{a_k^2} = \overline{b_k^2}$ = the variance of the Gaussian distribution = normalized power = 0.02×10^{-6} W.

Since for Gaussian noise, the mean is zero,

$$f(a_k) = \frac{1}{\sqrt{2\pi(0.02 \times 10^{-6})}} e^{\left[-\frac{a_k^2}{2(0.02 \times 10^{-6})}\right]}$$

$$= \frac{1}{\sqrt{0.04\pi \times 10^{-6}}} e^{\left[-\frac{a_k^2}{0.04 \times 10^{-6}}\right]}$$

$$\text{and} \quad f(b_k) = \frac{1}{\sqrt{0.04\pi \times 10^{-6}}} e^{\left[-\frac{b_k^2}{0.04 \times 10^{-6}}\right]}$$

● **PROBLEM 13-7**

Consider a carrier $m(t) = n(t) \cdot \cos(\omega_0 t + \theta)$, amplitude modulated by the noise $n(t)$, where θ is a random variable with a probability density

$$f(\theta) = \begin{cases} \frac{1}{2\pi} & -\pi \leq \theta \leq \pi \\ 0 & \text{elsewhere.} \end{cases}$$

If $E[n^2(t)] = \sigma^2$ and θ and $n(t)$ are independent, determine the power dissipated by $m(t)$.

Solution: The power dissipated by $m(t)$ is equal to the expectation of $m^2(t)$. Since θ and $n(t)$ are independent random variables,

$$E[m^2] = E[n^2(t) \cos^2(\omega_0 t + \theta)]$$

$$= E\{n^2(t) \times [\tfrac{1}{2}\cos(\omega_0 t + \theta - \omega_0 t - \theta) + \tfrac{1}{2}\cos(2\omega_0 t + 2\theta)]\}$$

$$= E\{n^2(t) \times \tfrac{1}{2}[1 + \cos(2\omega_0 t + 2\theta)]\}$$

Since θ and $n(t)$ are independent, it follows that

$$E[m^2] = \tfrac{1}{2} E_n[n^2(t)] + \tfrac{1}{2} E_n[n^2(t)] E_\theta[\cos(2\omega_0 t + 2\theta)]$$

$$= \tfrac{1}{2} E_n[n^2(t)]$$

$$= \tfrac{1}{2}\sigma^2$$

Note: θ is uniformly distributed between $-\pi$ to $+\pi$, its expected value is equal to zero. Thus $E_\theta[\cos(2\omega_0 t + 2\theta)] = 0$.

● **PROBLEM 13-8**

Noise $n(t)$ is used to amplitude modulate a carrier $m(t) = n(t) \cos(\omega_0 t + \theta)$ where θ is a random variable, uniformly distributed between $-\pi$ to $+\pi$.

If $n(t) = n_1 \cos(\omega_1 t + \phi_1) + n_2 \cos(k\omega_1 t + \phi_2)$, where n_1, n_2, ϕ_1 and ϕ_2 are uncorrelated, show that the power dissipated by $m(t)$ is $E(m^2) = \tfrac{1}{4}[E(n_1^2) + E(n_2^2)]$.

Solution:
$$E(m^2) = E[n^2(t)\cos^2(\omega_0 t + \theta)]$$

$$= E\left[n^2(t) \times 0.5\{1 + \cos(2\omega_0 t + 2\theta)\}\right]$$
$$= 0.5\ E\left[n^2(t)\right] + 0.5\ E\left[n^2(t)\cos(2\omega_0 t + 2\theta)\right]$$

Assuming $n(t)$ and θ to be independent, the above expression evaluates to

$$E(m^2) = 0.5\ E\left[n^2(t)\right] + 0.5\ E_n\left[n^2(t)\right] \times E_\theta\left[\cos(2\omega_0 t + 2\theta)\right]$$

Since θ is a random variable uniformly distributed between $-\pi$ and $+\pi$, its expected value is equal to zero.

Thus, $E_\theta\left[\cos(2\omega_0 t + 2\theta)\right] = 0$.

Therefore, $E(m^2) = 0.5\ E\left[n^2(t)\right]$ \hfill (1)

Substituting the expression given for $n(t)$ into equation (1) yields

$$E(m^2) = 0.5\ E\left[\{n_1 \cos(\omega_1 t + \phi_1) + n_2 \cos(k\omega_1 t + \phi_2)\}^2\right]$$
$$= 0.5\ E\left[n_1^2 \cdot \cos^2(\omega_1 t + \phi_1) + 2n_1 n_2 \cos(\omega_1 t + \phi_1) \cdot \cos(k\omega_1 t + \phi_2) + n_2^2 \cos^2(k\omega_1 t + \phi_2)\right]$$

Using the identity $\cos^2\theta = \tfrac{1}{2}(1 + \cos 2\theta)$,

$$n_1^2 \cos^2(\omega_1 t + \phi_1) = 0.5\left[1 + \cos(2\omega_1 t + 2\phi_1)\right]$$

$$E(m^2) = 0.5\ E\left[n_1^2 \times 0.5\{1 + \cos(2\omega_1 t + 2\phi_1)\}\right]$$
$$+ E(n_1 n_2)\ E\left[\cos(\omega_1 t + \phi_1)\cos(k\omega_1 t + \phi_2)\right]$$
$$+ 0.5\ E\left[n_2^2 \times 0.5\{1 + \cos(2k\omega_1 t + 2\phi_2)\}\right]$$
$$= 0.5\ E\left[0.5 n_1^2 + 0.5 n_1^2 \cos(2\omega_1 t + 2\phi_1)\right]$$
$$+ E(n_1 n_2)\ E\left[\cos(\omega_1 t + \phi_1)\cos(k\omega_1 t + \phi_2)\right]$$
$$+ 0.5\ E\left[0.5 n_2^2 + 0.5 n_2^2 \cos(2k\omega_1 t + 2\phi_2)\right] \quad (2)$$

Since n_1, n_2, ϕ_1 and ϕ_2 are uncorrelated,

$$0.5 n_1^2 \cdot \cos(2\omega_1 t + 2\phi_1) = 0$$
$$E(n_1 n_2)\ E\left[\cos(\omega_1 t + \phi_1)\cos(k\omega_1 t + \phi_2)\right] = 0$$

and $0.5 n_2^2 \cos(2k\omega_1 t + 2\phi_2) = 0$

Thus, equation (2) reduces to

$$E(m^2) = 0.25\left[E(n_1^2) + E(n_2^2)\right]$$

• **PROBLEM** 13-9

Figure (1) represents the two-sided power spectral density of noise n(t).

(a) Graphically represent the power spectral density of the product $n(t) \cos 2\pi f_1 t$.

(b) If the frequency range is from $-(f_2 - f_1)$ to $(f_2 - f_1)$ find the normalized power of the product.

(c) For the product $n(t) \cdot \cos 2\pi \left[(f_2 + f_1)/2\right] t$, repeat parts (a) and (b).

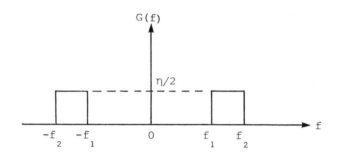

Fig. 1

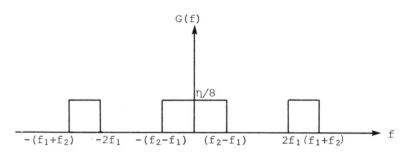

Fig. 2

Solution:

(a) The graphical representation of the power spectral density for the product $n(t) \cdot \cos(2\pi f_1 t)$ is shown in figure (2) below.

(b) The area under the curve in the frequency range $-(f_2 - f_1)$ to $(f_2 - f_1)$ is the normalized power of the product.

Therefore, $P(|f| \leq f_2 - f_1) = \frac{\eta}{8} \times 2(f_2 - f_1)$

$$= \frac{\eta}{4} (f_2 - f_1)$$

(c) For the product $n(t) \cdot \cos 2\pi \left[(f_2 + f_1) t/2\right]$, the power spectral density is plotted as shown in the figure (3).

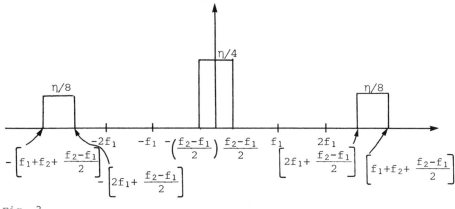

Fig. 3

Now, the normalized power is calculated as in part (b). Hence,

$$P(|f| \leq f_2 - f_1) = \frac{\eta}{4} \times 2\frac{(f_2 - f_1)}{2}$$

$$= \frac{\eta}{4}(f_2 - f_1)$$

• **PROBLEM 13-10**

In Fig. (1), the power spectral density $G_n(f)$ of a bandlimited noise waveform $n(t)$ is shown. Sketch the power spectral density of $n(t) \cdot \sin(2\pi \times 10^6 t)$ and calculate its normalized power in the range $-\infty$ to $+\infty$.

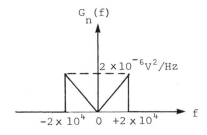

Fig. 1

Solution: The power spectral density of the product $\overline{n(t) \cdot \sin}(2\pi \times 10^6 t)$ is plotted in Fig. (2).

Note that f_c is the carrier frequency $= \frac{2\pi \times 10^6}{2\pi} = 10^6$ Hz

$$f_c + f_m = (10^6 + 2 \times 10^4)\text{Hz} = 1020 \text{ kHz}$$

$$f_c - f_m = (10^6 - 2 \times 10^4)\text{Hz} = 980 \text{ kHz}$$

and the new amplitude is $\dfrac{2 \times 10^{-6} V^2/Hz}{4} = 5.0 \times 10^{-7} V^2/Hz$.

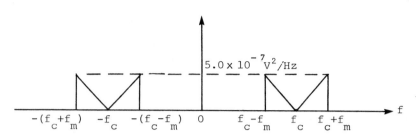

Fig. 2

Now, the normalized power $P(|f| < \infty)$ is the total area under the waveforms in Fig. (2).

Hence, $P(|f| < \infty) = \dfrac{[(f_c + f_m) - f_c] \times 5.0 \times 10^{-7}}{2} \times 4$

$= [(1.02 \times 10^6) - 10^6] \times 5.0 \times 10^{-7} \times 2$

$= 20$ m watt

● **PROBLEM** 13-11

The power spectral density of a noise signal $n(t)$ is shown in figure (1) and is given by the equation $n(t) = n_1(t) \cdot \cos(2\pi f_0 t) - n_2(t) \cdot \sin(2\pi f_0 t)$. Sketch the power spectral densities of $n_1(t)$ and $n_2(t)$, when

(i) $f_0 = f_1$

(ii) $f_0 = f_2$

(iii) $f_0 = \tfrac{1}{2}(f_1 + f_2)$

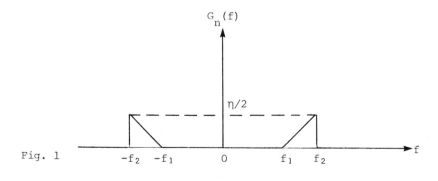

Fig. 1

Solution:

(i) For $f_0 = f_1$:

(a)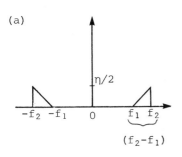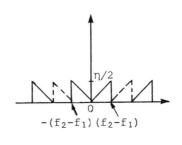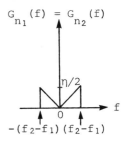

(ii) For $f_0 = f_2$:

(b)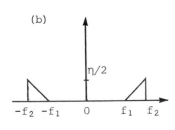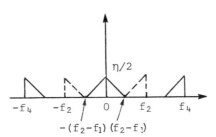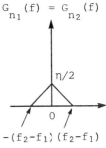

(iii) For $f_0 = \tfrac{1}{2}(f_2 + f_1)$:

(c)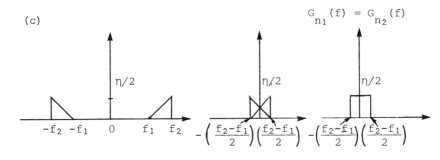

● **PROBLEM 13-12**

The superposition of spectral components of noise $n(t)$ can be expressed as

$$n(t) = n_1(t) \cos 2\pi f_0 t - n_2(t) \sin 2\pi f_0 t$$

where $n_1(t) = \lim_{\Delta f \to 0} \sum_{i=1}^{\infty} \left[a_i \cos 2\pi(i-I)\Delta f t + b_i \sin 2\pi(i-I)\Delta f t \right]$

and $n_2(t) = \lim_{\Delta f \to 0} \sum_{i=1}^{\infty} \left[a_i \sin 2\pi(i-I)\Delta ft - b_i \cos 2\pi(i-I)\Delta ft \right].$

Prove that $E(n_1^2) = E(n_2^2)$.

Solution:

$E(n_1^2) = E \left[\lim_{\Delta f \to 0} \sum_{j,i=1}^{\infty} \{a_j \cos 2\pi(j-I)\Delta ft + b_j \sin 2\pi(j-I)\Delta ft\} \right.$

$\left. \times \left(a_i \cos 2\pi(i-I)\Delta ft + b_i \sin 2\pi(i-I)\Delta ft \right) \right]$

$= \lim_{\Delta f \to 0} \sum_{j,i=1}^{\infty} \left[E(a_j a_i) \cos 2\pi(j-I)\Delta ft \cdot \cos 2\pi(i-I)\Delta ft \right.$

$+ E(a_j a_i) \cos 2\pi(j-I)\Delta ft \cdot \sin 2\pi(i-I)\Delta ft$

$+ E(b_j a_i) \sin 2\pi(j-I)\Delta ft \cdot \cos 2\pi(i-I)\Delta ft$

$\left. + E(b_j b_i) \sin 2\pi(j-I)\Delta ft \cdot \sin 2\pi(i-I)\Delta ft \right]$

$= \lim_{\Delta f \to 0} \sum_{j=1}^{\infty} \left[E(a_j^2) \{\cos 2\pi(j-I)\Delta ft\}^2 \right.$

$\left. + E(b_j^2) \{\sin 2\pi(j-I)\Delta ft\}^2 \right]$

$= \sum_{j=1}^{\infty} E(a_j^2)$

Similarly, $E(n_2^2)$ can be evaluated as

$E(n_2^2) = E \left[\lim_{\Delta f \to 0} \sum_{j,i=1}^{\infty} \{a_j \sin 2\pi(j-I)\Delta ft - b_j \sin 2\pi(j-I)\Delta ft\} \times \right.$

$\left. \{a_i \sin 2\pi(i-I)\Delta ft - b_i \cos 2\pi(i-I)\Delta ft\} \right]$

$= \lim_{\Delta f \to 0} \sum_{j,i=1}^{\infty} \left[E(a_j a_i) \sin 2\pi(j-I)\Delta ft \sin 2\pi(j-I)\Delta ft \right.$

$- E(a_j b_i) \sin 2\pi(j-I)\Delta ft \cdot \cos 2\pi(j-I)\Delta ft$

$- E(b_j a_i) \cos 2\pi(j-I)\Delta ft \cdot \sin 2\pi(i-I)\Delta ft$

$\left. + E(b_j b_i) \cos 2\pi(j-I)\Delta ft \cdot \cos 2\pi(i-I)\Delta ft \right]$

$= \lim_{\Delta f \to 0} \sum_{j=1}^{\infty} \left[E(a_j^2) \{\sin 2\pi(j-I)\Delta ft\}^2 + E(b_j^2) \cdot \right.$

$\left. \{\cos 2\pi(j-I)\Delta ft\}^2 \right]$

$$= \sum_{j=1}^{\infty} E(a_j^2)$$

Hence, $E(n_1^2) = E(n_2^2) = \sum_{j=1}^{\infty} E(a_j^2)$

• **PROBLEM 13-13**

The autocorrelation function of a random process $X(t) = A \cos(\omega_0 t + \theta)$ is given by $R_X(\tau) = (A^2/2)\cos(\omega_0 \tau)$ as shown in figure (1a), where A and ω_0 are constants. Determine the power spectrum $S_X(\omega)$ of $X(t)$.

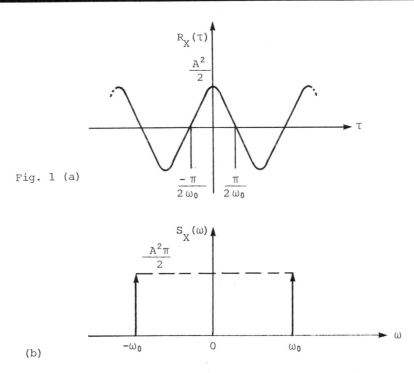

Fig. 1 (a)

(b)

Solution: The autocorrelation function $R_X(\tau) = (A^2/2)\cos(\omega_0 \tau)$ can be expressed in exponential form as follows:

$$R_X(\tau) = (A^2/4)\left(e^{j\omega_0 \tau} + e^{-j\omega_0 \tau}\right)$$

since $\dfrac{e^{j\omega_0 \tau} + e^{-j\omega_0 \tau}}{2} = \cos\omega_0 \tau$

Now, note that the inverse transform of a frequency domain impulse function is given by

$$\frac{1}{2\pi} \int_{-\infty}^{+\infty} \delta(\omega) e^{j\omega\tau} d\omega = \frac{1}{2\pi}$$

Thus, 1 and $2\pi\delta(\omega)$ form a Fourier transform pair, i.e., $1 \leftrightarrow 2\pi\delta(\omega)$. From the frequency-shifting property of Fourier transform, it follows that

$$e^{j\omega_0\tau} \leftrightarrow 2\pi\delta(\omega - \omega_0)$$

and

$$e^{-j\omega_0\tau} \leftrightarrow 2\pi\delta(\omega + \omega_0)$$

Using the above two results, the Fourier transform of $R_X(\tau)$ is

$$S_X(\omega) = \frac{A^2}{4}(2\pi)\left[\delta(\omega - \omega_0) + \delta(\omega + \omega_0)\right]$$

$$= \frac{A^2\pi}{2}\left[\delta(\omega - \omega_0) + \delta(\omega + \omega_0)\right]$$

The plot of $S_X(\omega)$ is shown in figure (1b).

• **PROBLEM 13-14**

Determine the power spectrum $S_W(\omega)$ of a wide-sense stationary noise process $W(t)$, which has an autocorrelation function given by $R_W(\tau) = Ae^{-3|\tau|}$ with $A = $ constant as shown in Fig. (1a).

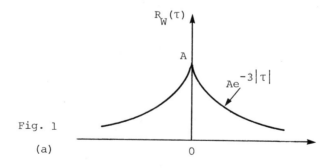

Fig. 1
(a)

Solution: The relationship between the autocorrelation $R_W(\tau)$ and the power spectral density $S_W(\omega)$ is given by

$$S_W(\omega) = \int_{-\infty}^{+\infty} R_W(\tau) e^{-j\omega\tau} d\tau$$

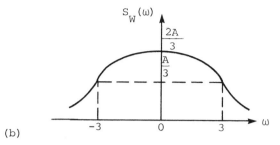

(b)

i.e., the power spectral density is equal to the Fourier transform of the autocorrelation function.

Hence, $S_W(\omega) = \int_{-\infty}^{+\infty} Ae^{-3|\tau|}e^{-j\omega\tau}d\tau$

$= A\int_0^\infty e^{-(3+j\omega)\tau}d\tau + A\int_{-\infty}^0 e^{(3-j\omega)\tau}d\tau$

$= A\left[\dfrac{e^{-(3+j\omega)\tau}}{-(3+j\omega)}\right]_0^\infty + A\left[\dfrac{e^{(3-j\omega)\tau}}{3-j\omega}\right]_{-\infty}^0$

$= \dfrac{-A}{3+j\omega}[0-1] + \dfrac{A}{3-j\omega}[1-0]$

$= \dfrac{A}{(3+j\omega)} + \dfrac{A}{(3-j\omega)}$

By rationalizing, we get

$S_W(\omega) = \dfrac{6A}{9+\omega^2}$ which is sketched in Fig. (1b).

• **PROBLEM 13-15**

Figure below shows a linear time-invariant network. If the input x(t) is a white noise of spectral density $S_X(\omega) = \eta/2$, where η is a positive constant, determine the output power spectrum and the average power.

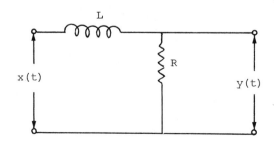

Fig. 1

Solution: From the network shown; the transfer function $H(\omega)$ is realized to be

$$H(\omega) = \frac{R}{R + j\omega L}$$

$$= \frac{1}{1 + \frac{j\omega L}{R}}$$

$$= \left(1 + \frac{j\omega L}{R}\right)^{-1}$$

The magnitude of $H(\omega)$ is

$$|H(\omega)| = \frac{1}{\sqrt{(1)^2 + \left(\frac{\omega L}{R}\right)^2}}$$

Therefore, $|H(\omega)|^2 = \dfrac{1}{1 + \left(\frac{\omega L}{R}\right)^2}$

The output power density spectrum for a linear time invariant network is defined as

$$S_Y(\omega) = S_X(\omega)|H(\omega)|^2 = \frac{\eta/2}{1 + \left(\frac{\omega L}{R}\right)^2}$$

To calculate the average power in $y(t)$, it is known that:

$$P_Y = \frac{1}{2\pi} \int_{-\infty}^{\infty} S_Y(\omega) d\omega$$

$$= \frac{\eta}{4\pi} \int_{-\infty}^{\infty} \frac{d\omega}{1 + \left(\frac{\omega L}{R}\right)^2}$$

$$= \frac{\eta}{4\pi} \left[\tan^{-1}\left(\frac{\omega L}{R}\right) \cdot \frac{R}{L}\right]_{-\infty}^{\infty}$$

$$= \frac{\eta R}{4\pi L} \left[\tan^{-1}(\infty) - \tan^{-1}(-\infty)\right]$$

$$= \frac{\eta R}{4\pi L} \left(\frac{\pi}{2} + \frac{\pi}{2}\right)$$

$$= \frac{\eta R}{4\pi L} (\pi)$$

$$= \frac{\eta R}{4L}$$

FILTERING

• **PROBLEM** 13-16

$f(t) = \int_{Kt-\alpha}^{Kt+\alpha} \frac{\sin x}{x} dx$ is the response of an ideal filter to a particular rectangular pulse input.

Suppose $\alpha \ll 1$. Find an approximate expression for $f(t)$.

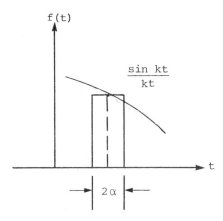

Fig. 1

Solution: $\frac{\sin x}{x}$ remains approximately constant at about $\frac{\sin Kt}{Kt}$ over the range 2α. The result is thus approximately equal to the area of the rectangle shown in Fig. (1).

Therefore,

$$f(t) \cong 2\alpha \frac{\sin Kt}{Kt}$$

• **PROBLEM** 13-17

In Fig. (1), the system function of an ideal filter is shown.

a) What is the response to $\delta(t) - \delta(t - 10^{-3})$?

b) What is the response to sinc $10^3 t$? (Hint: sinc $x = \frac{\sin \pi x}{\pi x}$)

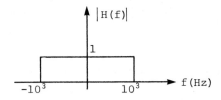

Fig.1

Solution:

a) The Fourier transform of $\delta(t)$ is $\int_{-\infty}^{\infty} \delta(t)e^{-\omega t}dt = 1$. Hence, the Fourier transform of the response to $\delta(t)$ is

$$1 \cdot H(f) = 1e^{-j2\pi f t_0}, \quad |f| \leq 10^3$$

$$= 0, \quad |f| > 10^3$$

From the curve of the angle of H(f) in Fig. (2),

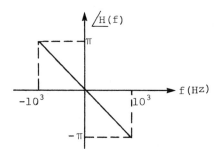

Fig. 2

$$\angle H(f) = -\frac{\pi}{10^3} f = -2\pi f t_0 \text{ where } t_0 = \tfrac{1}{2} \times 10^{-3} \text{ seconds.}$$

Now, the response to $\delta(t)$ is the inverse transform of $e^{-j2\pi f t_0}$, $|f| < 10^3$

Hence, $\int_{-10^3}^{10^3} e^{-j2\pi f t_0} e^{j2\pi f t} df = \left[\frac{e^{j2\pi(t-t_0)f}}{j2\pi(t-t_0)}\right]_{10^{-3}}^{10^3}$

$$= 2 \times 10^3 \frac{\sin 2\pi(t-t_0) \times 10^3}{2\pi(t-t_0) \times 10^3}.$$

and the response to $\delta(t - 10^{-3})$ is delayed by 10^{-3} seconds.

Let $\text{sinc } x = \frac{\sin \pi x}{\pi x}$. Then the response to $\delta(t) - \delta(t - 10^{-3})$ is

$2 \times 10^3 \text{ sinc } \left[2 \times 10^3 (t - \tfrac{1}{2} \times 10^{-3})\right] -$

$2 \times 10^3 \text{ sinc } \left[2 \times 10^3 (t - \tfrac{3}{2} \times 10^{-3})\right]$

b) Since H(f) is the Fourier transform of the unit impulse response, a by-product of part (a) is that the Fourier transform of $\text{sinc}(2 \times 10^3)t$ is rectangular, extending to 10^3 hz. sinc $10^3 t$ is spread out in time by a factor of two compared to sinc $2 \times 10^3 t$, and by the scale change property,

$$F\left[f(at)\right] = \frac{1}{|a|} F(j\tfrac{\omega}{a}) \quad \text{if} \quad \left[F\ f(t)\right] = F(j\omega),$$

and the Fourier transform of $\text{sinc}(10^3 t)$ is rectangular extending to $\tfrac{1}{2} \times 10^3$ hz.

This is entirely within the ideal filter range, so output is undistorted but delayed by t_0.

Hence, response = $\text{sinc}\left[10^3(t - \tfrac{1}{2} \times 10^{-3})\right]$.

• **PROBLEM** 13-18

The transfer function of an RC low-pass filter is given as

$$H(f) = \frac{1}{1 + jf/f_c} \quad \text{where } f_c \text{ is the 3-dB frequency.}$$

If the input to this filter is a white noise, find the output noise power N_0 of the filter.

Solution: For the RC filter, the output noise power N_0 is given as

$$N_0 = \int_{-\infty}^{\infty} G_{no}(f) df \tag{1}$$

where $G_{no}(f)$ is the output power spectral density of the noise.

Now, recall that the relation between the input and output power spectral density is

$$G_{no}(f) = G_{ni}(f) |H(f)|^2. \tag{2}$$

Since the input noise is white, the power spectral density $G_{ni}(f) = \frac{\eta}{2}$.

Substituting for $G_{ni}(f)$ and $H(f)$ into Eq. (2) yields,

$$G_{no}(f) = \frac{\eta}{2} \frac{1}{1 + (f/f_c)^2}$$

Hence, $N_0 = \int_{-\infty}^{\infty} G_{no}(f) df$

$= \int_{-\infty}^{\infty} \frac{\eta}{2} \frac{1}{1 + (f/f_c)^2} df$

letting $y = f/f_c$ then $dy f_c = df$

Hence, $N_0 = \int_{-\infty}^{\infty} \frac{\eta}{2} \frac{1}{1 + (y)^2} f_c \, dy$

Now, noting that $\int_{-\infty}^{\infty} \frac{dx}{(1 + x^2)} = \pi$

Therefore, $N_0 = \frac{\eta f_c \pi}{2}$

• **PROBLEM 13-19**

Fig. (1) represents the transfer function of a rectangular low-pass filter.

(a) If the input to the filter is a white noise, determine its output noise power N_0.

(b) Repeat part (a) if $H(f)$ is given as shown in Fig. (2).

(c) Repeat part (a) if,

(i) $H(f) = j2\pi\tau f$

(ii) $H(f) = \frac{1}{j\omega\tau} - \frac{e^{-j\omega T}}{j\omega\tau}$

and name the filter being used for both cases.

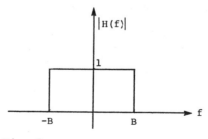

Fig. 1

Solution:

(a) Since the input noise is white, the input power spectral density is $G_{ni}(f) = \frac{\eta}{2}$. Therefore, the output power spectral density will be $G_{no}(f) = G_{ni}(f)|H(f)|^2$

Since
$$H(f) = \begin{cases} 1 & -B \leq f \leq B \\ 0 & \text{elsewhere} \end{cases}$$

$$G_{no}(f) = \begin{cases} \frac{\eta}{2} & -B \leq f \leq B \\ 0 & \text{elsewhere} \end{cases}$$

The output noise power N_0 is expressed as

$$N_0 = \int_{-\infty}^{\infty} G_{no}(f) df$$

$$= \int_{-B}^{B} \frac{\eta}{2} df$$

$$= 2 \int_{0}^{B} \frac{\eta}{2} df$$

$$= \eta B$$

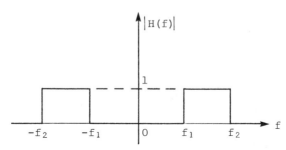

Fig. 2

(b) Similarly, the output noise power of a rectangular bandpass filter with transfer function as shown in Fig. (2) is

$N_0 = 2 \frac{\eta}{2}(f_2 - f_1) = \eta(f_2 - f_1)$, which is also the area under the $G_{no}(f)$ curve.

(c) i) The function $H(f) = j2\pi\tau f$ is the transfer function of a differentiating filter network, where τ is a time constant factor. For such a network, the output waveform is proportional to the time derivative of the input waveform. The output power spectral density in this case is,

$$G_{no}(f) = G_{ni}(f)|H(f)|^2$$

$$= \frac{\eta}{2} |j2\pi\tau f|^2$$

$$= \frac{\eta}{2} (4\pi^2\tau^2 f^2)$$

$$= 2\eta\pi^2\tau^2 f^2$$

Then
$$N_0 = \int_{-B}^{B} 2\eta\pi^2\tau^2 f^2 df$$

$$= 2\eta\pi^2\tau^2 \cdot 2 \int_{0}^{B} f^2 df$$

$$= 4\eta\pi^2\tau^2 \frac{B^3}{3}$$

Note that in this case, we assume that following the differentiator is a filter as in part (a).

ii) The transfer function $H(f) = \frac{1}{j\omega\tau} - \frac{e^{-j\omega T}}{j\omega\tau}$ represents a network which performs an integration over an interval T. Hence, an integrator.

The noise power at the filter output is

$$N_0 = \int_{-\infty}^{\infty} G_{no}(f) df$$

$$= \int_{-\infty}^{\infty} G_{ni}(f) |H(f)|^2 df$$

Note that $H(f) = \frac{1 - e^{-j\omega T}}{j\omega\tau}$

hence, $|H(f)| = \frac{|1 - e^{-j\omega T}|}{|j\omega\tau|} = \frac{|1 - (\cos\omega T - j\sin\omega T)|}{\omega\tau}$

$$= \frac{|(1 - \cos\omega T) + j\sin\omega T|}{\omega\tau}$$

$$= \frac{\sqrt{1 - 2\cos\omega T + 1}}{\omega\tau}$$

$$= \frac{\sqrt{2 - 2\cos\omega T}}{\omega\tau}$$

Thus, $|H(f)|^2 = \frac{2(1 - \cos\omega T)}{\omega^2\tau^2} = \frac{2(1 - \cos\omega T)}{4\pi^2 f^2 \tau^2}$

$$= \frac{1}{\pi^2 f^2 \tau^2} \cdot \frac{1}{2} \cdot (1 - \cos\omega T)$$

$$= \frac{1}{\pi^2 f^2 \tau^2} \cdot \frac{1}{2} \cdot (1 - \cos 2\pi f T)$$

Now, recall that $\cos(2\theta) = \cos^2\theta - \sin^2\theta$

and $1 = \cos^2\theta + \sin^2\theta$

$= \cos^2(\pi f T) + \sin^2(\pi f T)$

Hence, $|H(f)|^2 = \frac{1}{\pi^2 f^2 \tau^2} \cdot \frac{1}{2} \cdot (\cos^2\pi f T + \sin^2\pi f T - \cos^2\pi f T + \sin^2\pi f T)$

$$= \left(\frac{T}{\tau}\right)^2 \left(\frac{\sin \pi T f}{\pi T f}\right)^2$$

Thus, $N_0 = \int_{-\infty}^{\infty} \frac{\eta}{2} |H(f)|^2 df$

$$= \frac{\eta}{2}\left(\frac{T}{\tau}\right)^2 \int_{-\infty}^{\infty} \left(\frac{\sin \pi f T}{\pi f T}\right)^2 df$$

Note that $\int_{-\infty}^{\infty} \left(\frac{\sin x}{x}\right)^2 dx = \pi$

with $\pi f T = x$.

$\pi T df = dx$ or $df = \frac{dx}{\pi T}$

Therefore, $N_0 = \frac{\eta}{2}\left(\frac{T}{\tau}\right)^2 \int_{-\infty}^{\infty} \frac{1}{\pi T}\left(\frac{\sin x}{x}\right) dx$

$$= \frac{\eta T}{2\pi \tau^2} \times (\pi)$$

$$= \frac{\eta T}{2\tau^2}$$

● **PROBLEM** 13-20

The input to the cascade filter i(t) shown in Fig. (1) is an impulse function.

If given $|I(j\omega)|^2 = 4$, $H_2(j\omega) = 10 (\sin\omega/\omega)$

and the autocorrelation function

$$R_{ci}(\tau) = 80 \quad 3 \leq \tau \leq 5$$
$$= 0 \quad \text{elsewhere}$$

Plot the signal waveform $i(t)$, $b(t)$, $c(t)$ and the transfer functions $h_1(t)$ and $h_2(t)$ of the cascade filter.

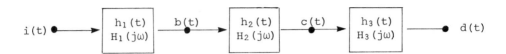

Fig. 1

Solution: Since $i(t)$ is an impulse function and $|I(j\omega)|^2 = 4$, then $A(j\omega)$ can be written as

$$A(j\omega) = 2e^{-j\omega x}$$

Now, realize that $c\delta(t) = \frac{1}{2\pi} \int_{-\infty}^{\infty} ce^{j\omega t} d\omega = F^{-1}[c]$ (1)

and $F[f(t - t_0)] = e^{-j\omega t_0} F(j\omega)$ (2)

Therefore, $i(t) = 2\delta(t - x)$, where x is unknown. This is shown in Fig. (2a).

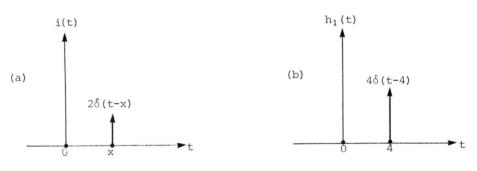

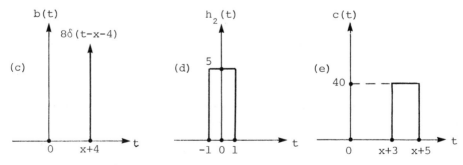

Fig. 2

Next, since $R_{ci}(\tau)$ is given and $i(t)$ is known, then $c(t)$ is obtained in Fig. (2e) by using

$$f(t - t_1) \circledast g(t - t_2) = R_{fg}\left[\tau - (t_1 - t_2)\right] \quad (3)$$

where $R_{fg}(\tau) = f(t) \circledast g(t)$

(i.e., The correlation of two delayed pulse signals is delayed by the difference of the two individual delays.)

$$f(t) \circledast \delta(t) = f(\tau) \quad (4)$$

Now, with $H_2(j\omega)$ given as $H_2(j\omega) = 10\left(\frac{\sin\omega}{\omega}\right)$, using the Fourier transform relationship for rectangular pulse signal,

i.e.,
$$F(j\omega) = \int_{-\tau/2}^{\tau/2} A e^{-j\omega t} dt = \tau A \left[\frac{\sin(\omega\tau/2)}{\omega\tau/2}\right] \quad (5)$$

Thus, $h_2(t)$ is obtained as shown in Fig. (2d).

Finally, with the convolution relationship,

$$c(t) = b(t) \circledast h_2(t)$$

and $c(t) = i(t) \circledast h_1(t) \circledast h_2(t)$

(note: $\circledast$ - correlation, $\circledast$ - convolution)

Then, using Eq. $f(t - t_1) \circledast g(t - t_2) = \psi_{fg}\left[t - (t_1 + t_2)\right]$

(i.e., The convolution of two delayed signals is delayed by the sum of the two individual delays) and $f(t) \circledast \delta(t) = f(t)$ (i.e., Any impulse signal convolved with an impulse function is just the original pulse function), $h_1(t)$ and $b(t)$ can be obtained as shown in Fig. (2b and 2c).

Note: There is insufficient information given here to determine the value of x.

● **PROBLEM** 13-21

Construct a Butterworth filter of order three (i = 3) with $\omega_m = 1$.

Solution: For an ideal low-pass filter, the amplitude characteristics can be approximated from its transfer function,

$$|H_i(\omega)| = \frac{1}{\sqrt{1 + \omega^{2i}}}$$

Since i = 3

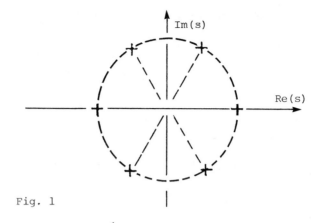

Fig. 1

$$|H(\omega)| = \frac{1}{\sqrt{1 + \omega^6}}$$

or $$|H(j\omega)| = \frac{1}{\sqrt{1 + \omega^6}}$$

$$|H(j\omega)|^2 = \frac{1}{1 + \omega^6}$$

But, $|H(j\omega)|^2 = H(j\omega) \cdot H^*(j\omega)$

where $H^*(j\omega)$ is the conjugate of $H(j\omega)$.

i.e., $|H(j\omega)|^2 = H(j\omega) \cdot H(-j\omega)$ \hfill (1)

Now, represent Eq. (1) in the s-plane

i.e., let $s = j\omega$; $\omega = \frac{s}{j} = -js$

$$\omega^6 = -s^6$$

Thus, $|H(s)|^2 = H(s) \cdot H(-s)$

$$= \frac{1}{1 - s^6}$$

which indicates that there are six poles of $|H(s)|^2$ as shown in figure (1).

Note that three of these poles are associated with $H(s)$ and the other three poles with $H(-s)$.

$H(s)$ is found from its poles as below:

$$H(s) = \frac{1}{(s-p_1)(s-p_2)(s-p_3)} = \frac{1}{s^3 + 2s^2 + 2s + 1}$$

where p_1, p_2 and p_3 are poles associated with $H(s)$. The above system transfer function can be synthesized to yield the network of figure (2), where $v(t)$ and $i(t)$ are the response and the source respectively.

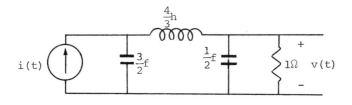

Fig. 2

Note that as i increases, the filter responses are more like an ideal filter. As i approaches infinity, the delay between input and output also approaches infinity.

• **PROBLEM 13-22**

The transfer function of a non-recursive digital filter is given as

$$h(n) = 1, 1, 2, 4 \quad n = 0, 1, 2, 3 \text{ respectively}$$

$$= 0 \quad n > 3$$

Derive and implement a block diagram for this filter.

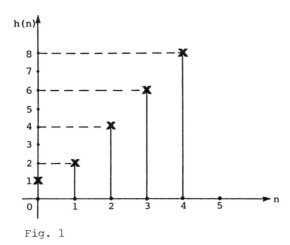

Fig. 1

Solution: A system which transforms one sequence of numbers to another sequence of numbers (i.e. from input to output) is called a digital filter. In general, the linear combination of previous input numbers is the current output number. Hence, convolution relationship can be written as

$$y(n) = \sum_{m=0}^{n} x(m) \cdot h(n - m) \qquad (1)$$

i.e. $x(n) \longrightarrow$ Digital Filter $\xrightarrow{y(n)}$

Digital filter outputs may also depend upon past output values in addition to past input values. Thus,

$$y(n) = \sum_{m=0}^{n} x(m) \cdot h(n-m) + \sum_{m=0}^{n=1} y(m) \cdot k(n-m) \qquad (2)$$

Now, expanding the summation of equation (1) yields,

$$y(n) = 8x(n-4) + 6x(n-3) + 4x(n-2) + 2x(n-1) + x(n)$$

The block diagram of a system to yield the above summation (output) corresponding to an input $x(n)$ is shown in figure (2).

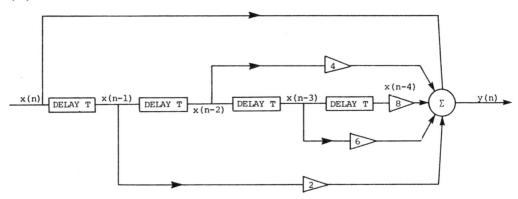

Fig. 2

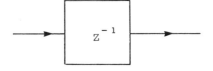

Fig. 3

The delay-T block in the figure above can be realized in practice by evaluating the z-transform of a system which essentially delays the input by one period. A z-transform block is shown in Fig. (3).

NOISE BANDWIDTH

● **PROBLEM** 13-23

A Gaussian filter with $|H(\omega)|^2 = e^{-\omega^2}$ is used to filter noise at the output of an amplifier stage. Determine the bandwidth of the filter.

Solution: If the input noise to this Gaussian filter has a power spectral density $G_{ni}(f)$, then the power spectral density of the output noise can be expressed as

$$G_{no}(f) = G_{ni}(f) \cdot |H(f)|^2$$

Assuming the input noise to be white, $G_{ni}(f) = \eta/2$ for all frequencies.

Therefore $G_{no}(f) = \frac{\eta}{2} e^{-\omega^2}$, where η is a constant. The output noise power of a filter is given by

$$N_0(\text{Gaussian}) = \int_{-\infty}^{\infty} G_{no}(f) df.$$

$$= \frac{\eta}{2} \int_{-\infty}^{\infty} e^{-(2\pi f)^2} df$$

$$= \frac{\eta}{2} \frac{1}{2\sqrt{\pi}}$$

$$= \frac{\eta}{4\sqrt{\pi}}$$

The bandwidth of the Gaussian filter can be evaluated by setting $N_0(\text{Gaussian}) = N_0(\text{Rectangular})$. In the presence of white noise, a rectangular (low-pass) filter with $H(f) = 1$ over its bandpass B_N would yield an output-noise power

$$N_0(\text{Rectangular}) = \frac{\eta}{2} \cdot 2B_N = \eta B_N$$

Now, by setting $N_0(\text{Rectangular}) = N_0(\text{Gaussian})$,

$$\eta B_N = \frac{\eta}{4\sqrt{\pi}}$$

Therefore, B_N (the noise bandwidth) $= \frac{1}{4\sqrt{\pi}}$

● **PROBLEM 13-24**

If the power transfer function of a system is given as $|H(\omega)|^2 = \dfrac{1}{1 + \left(\dfrac{\omega}{B}\right)^2}$, determine its noise bandwidth, B_N.

Note: B is the 3-dB bandwidth in radians/second and ω is the angular frequency.

Solution: To calculate the noise bandwidth B_N of a system, it is known that

$$B_N = \frac{\int_0^\infty |H(\omega)|^2 d\omega}{|H(0)|^2} \qquad (1)$$

Since $|H(\omega)|^2 = \dfrac{1}{1 + \left(\dfrac{\omega}{B}\right)^2}$,

then at $\omega = 0$, $|H(0)|^2 = \dfrac{1}{1 + 0} = 1$.

Substituting the values of $|H(0)|^2$ and $|H(\omega)|^2$ into equation (1)

$$B_N = \int_0^\infty \frac{B^2}{B^2 + \omega^2} d\omega$$

$$= B^2 \int_0^\infty \frac{1}{B^2 + \omega^2} d\omega$$

$$= B^2 \left(\frac{1}{B}\right) \left[\tan^{-1} \frac{\omega}{B}\right]_0^\infty$$

Noting that $\int \dfrac{dx}{a^2 + x^2} = \dfrac{1}{a} \tan^{-1} \dfrac{x}{a}$

Thus, $B_N = B \left[\tan^{-1} \infty - \tan^{-1} 0\right]$

$= B \left(\dfrac{\pi}{2} - 0\right)$

$= \dfrac{B\pi}{2}$

From the result, it is clear that B_N is 1.57 times larger than the system 3-dB bandwidth.

● **PROBLEM 13-25**

Given the transfer function of 4 different networks below,

$$H(\omega) = Ne^{-\alpha|\omega|} \qquad (1)$$
$$= Ne^{-\alpha\omega^2} \qquad (2)$$
$$= Sa(\omega t_0) \qquad (3)$$
and $$= G_{2w}(\omega), \qquad (4)$$

(a) Determine their equivalent noise bandwidth, B_N with respect to the zero frequency.

(b) If $H(\omega) = S_a\left[(\omega - \omega_0)t_0\right] + S_a\left[(\omega + \omega_0)t_0\right]$, determine the equivalent noise bandwidth with respect to ω_0.

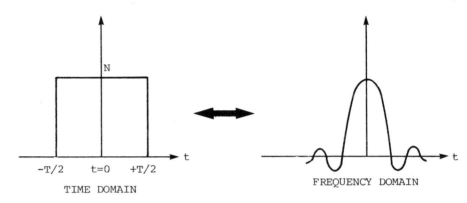

Fig. 1

Solution:

(a) The equivalent noise bandwidth, B_N with respect to some frequency f_0 is defined as

$$B_N = \frac{1}{|H(j2\pi f_0)|^2} \int_0^\infty |H(j2\pi f)|^2 \, df.$$

Now, B_N for the 4 different cases are calculated as follows:

(i) $H(\omega) = Ne^{-\alpha|\omega|}$

$$B_N = \frac{1}{N^2} \int_0^\infty N^2 e^{-4\pi\alpha f} \, df$$

$$= \int_0^\infty e^{-4\pi\alpha f} \, df$$

$$= \left[\frac{e^{-4\pi\alpha f}}{-4\pi\alpha}\right]_0^\infty$$

$$= \frac{1}{4\pi\alpha} \text{ cps}$$

(ii) $H(\omega) = Ne^{-\alpha\omega^2}$

699

$$B_N = \frac{1}{N^2} \int_0^\infty N^2 e^{-8\pi^2 \alpha f^2} \, df$$

$$= \int_0^\infty e^{-8\pi^2 \alpha f^2} \, df$$

$$= \frac{1}{2}\sqrt{\frac{\pi}{8\pi^2 \alpha}}$$

$$= \frac{1}{4\sqrt{2\pi\alpha}}$$

Note: $\int_{-\infty}^{+\infty} \underbrace{e^{-mx^2}}_{\text{even function}} dx = 2 \int_0^\infty e^{-mx^2} dx = \sqrt{\frac{\pi}{m}}$

(iii) $H(\omega) = Sa(\omega t_0)$

$$B_N = \int_0^\infty \frac{1}{2\pi t_0} \frac{\sin^2(2\pi f t_0)}{f^2} \, df$$

This can be solved, when we transform the function from the frequency domain to the time domain.

i.e., $\frac{1}{2\pi t_0} \frac{\sin^2(2\pi f t_0)}{f^2} = \frac{1}{2\pi t_0} \frac{\sin^2(2\pi f t_0)}{(2\pi t_0)^2 f^2} (2\pi t_0)^2$

$$= (2\pi t_0) \left[\frac{\sin(2\pi f t_0)}{2\pi f t_0}\right]^2$$

Now, let $N \cdot P_T(t) \leftrightarrow NT \frac{\sin(\omega T/2)}{(\omega T/2)}$ (See Fig. (1))

$$P_T(t) \leftrightarrow T \frac{\sin(\omega T/2)}{(\omega T/2)}$$

$$\frac{\sin(\omega T/2)}{(\omega T/2)} \leftrightarrow \frac{1}{T} P_T(t)$$

But since $t_0 = \frac{T}{2}$,

Hence, $\frac{\sin(\omega t_0)}{\omega t_0} \leftrightarrow \frac{1}{2t_0} P_{2t_0}(t)$

Now, since the width, $T = 2t_0$

Thus, $\dfrac{1}{2\pi t_0} \displaystyle\int_0^\infty \dfrac{\sin^2(2\pi f t_0)}{f^2} df = 2\pi t_0 \displaystyle\int_0^\infty \left[\dfrac{\sin(2\pi f t_0)}{2\pi f t_0}\right]^2 df$

$$= t_0 \int_0^\infty \left[\dfrac{\sin(\omega t_0)}{(\omega t_0)}\right]^2 d\omega$$

with $\omega = 2\pi f$

and let $F(\omega) = \sin(\omega t_0)/(\omega t_0)$

$$B_N = t_0 \int_0^\infty [F(\omega)]^2 d\omega$$

$$= \dfrac{t_0}{2} \int_{-\infty}^{+\infty} [F(\omega)]^2 d\omega$$

$$= \dfrac{t_0}{2} \cdot 2\pi \left[\dfrac{1}{2\pi} \int_{-\infty}^{+\infty} |F(\omega)|^2 d\omega\right]$$

$$= t_0 \pi \left[\int_{-\infty}^{+\infty} \left(\dfrac{1}{2t_0} P_{2t_0}(t)\right)^2 dt\right]$$

$$= \dfrac{t_0 \pi}{4 t_0^2} \left[\int_{-t_0}^{+t_0} dt\right]$$

Since the width is $2t_0$, the limits are from $(-t_0)$ to $+(t_0)$ because in the other regions it is zero. The amplitude is one as shown in Fig. (2).

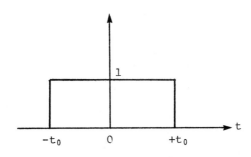

Fig. 2

Thus, $B_N = \dfrac{\pi}{4 t_0} \displaystyle\int_{-t_0}^{+t_0} 1 dt$

$$= \frac{\pi}{4t_0} \cdot 2 \int_0^{t_0} 1 dt$$

$$= \frac{\pi}{2t_0} \int_0^{t_0} dt$$

$$= \frac{\pi}{2t_0} \left[t_0 - 0 \right]$$

$$= \frac{\pi}{2} \text{ cps.}$$

(iv) $H(\omega) = G_{2w}(\omega)$

Here 2w represents the width as shown in Fig. (3).

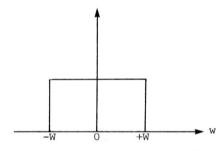

Fig. 3

Since $\omega = w$

$2\pi f = w$

Therefore, $f = \frac{w}{2\pi}$

$$B_N = 1 \int_0^{w/2\pi} 1 df$$

$$= \left[f \right]_0^{w/2\pi}$$

$$= \frac{w}{2\pi} - 0$$

Therefore, $B_N = \frac{w}{2\pi}$ cps

(b) The function here is the same as the function of case (iii), but it is shifted by $\pm\omega_0$. And the bandwidth in this case is double that of the bandwidth in (iii).

Also the area is twice as much as in (iii) because the contribution from both the positive and negative frequencies is equal.

From case (iii), $B_N = \frac{\pi}{2}$

Therefore, in this case, the bandwidth B_N is

$$B_N = 2(\frac{\pi}{2})$$

$$= \pi \text{ cps.}$$

SHOT NOISE

• **PROBLEM** 13-26

(a) A noise waveform is represented by a train of rectangular pulses $r_p(t)$. If the pulse amplitude A_p is a random variable having values 1, 2, 3, 4 ... 10 with equal probability, and the amplitudes are statistically independent of each other,

(a) Determine the power spectral density $G_n(f)$ of the noise. Assume the mean time between pulses to be T_s second. (Hint: $G_n(f) = \frac{1}{T_s} \overline{|R_p(f)|^2}$, $-\infty < f < \infty$)

(b) Repeat part (a) if $A_p = 1$ volt and consider only two pulses with width τ_1 or τ_2 with equal probability.

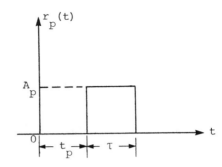

Fig. 1

Solution:

(a) Given $G_n(f) = \frac{1}{T_s} \overline{|r_p(f)|^2}$

where $R_p(f)$ is the Fourier transform of a single pulse $r_p(t)$

703

Thus, $G_n(f) = \frac{1}{T_s} E\left[\left|\int_{-\infty}^{\infty} r_p(t)e^{-j2\pi ft} dt\right|^2\right]$

$= \frac{1}{T_s} E\left[\left|\int_{t_p}^{t_p+\tau} A_p e^{-j2\pi ft} dt\right|^2\right]$

$= \frac{1}{T_s} E\left[\left|A_p \left(\frac{e^{-j2\pi f(t_p+\tau)} - e^{-j2\pi ft_p}}{-j2\pi f}\right)\right|^2\right]$

$= \frac{1}{T_s} E\left[A_p^2 \tau^2 \left(\frac{\sin\pi f\tau}{\pi f\tau}\right)^2\right]$

(Note: $\sin\theta = \frac{e^{j\theta} - e^{-j\theta}}{2j}$)

Since A_p is a random variable which can have values 1, 2, 3 ..., 10 with equal probability,

$G_n(f) = \frac{\tau^2}{T_s}\left(\frac{\sin\pi f\tau}{\pi f\tau}\right)^2 \sum_{p=1}^{10} A_p^2 \cdot P(A_p)$

$= \frac{\tau^2}{T_s}\left(\frac{\sin\pi f\tau}{\pi f\tau}\right)^2 \sum_{p=1}^{10} \frac{p^2}{10}$

$= \frac{(38.5)\tau^2}{T_s}\left(\frac{\sin\pi f\tau}{\pi f\tau}\right)^2 \quad -\infty < f < \infty$

(b) Similar calculation as part (a), but with pulse amplitude of 1 volt and p = 1 and 2.

Thus,

$G_n(f) = \frac{1}{T_s} E\left[\left|\int_{t_p}^{t_p+\tau_p} 1 \cdot e^{-j2\pi ft} dt\right|^2\right]$

$= \frac{1}{T_s} E\left[\left(\frac{\sin\pi f\tau_p}{\pi f}\right)^2\right]$

$= \frac{1}{T_s(\pi f)^2} \cdot \sum_{p=1}^{2} (\sin\pi f\tau_p)^2 \cdot P(\tau_p)$

$= \frac{1}{2T_s(\pi f)^2}\left[(\sin\pi f\tau_1)^2 + (\sin\pi f\tau_2)^2\right] \quad -\infty < f < \infty$

• **PROBLEM 13-27**

The dc plate current of a diode with a 10 k resistor load and operating temperature-limited is 0.01 m·amps as shown in the figure below. If the room temperature is at 22 °C and the bandwidth of operation is 5 KHz, determine the resultant rms noise voltage across the 10 k resistor.

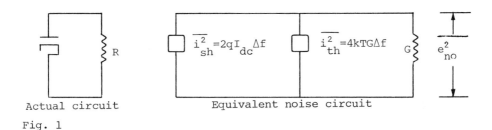

Actual circuit   Equivalent noise circuit

Fig. 1

Solution: Assume that the thermal noise from the resistor and shot noise from the tube are statistically independent, then, the moments of the thermal noise voltage $e_n(t)$ across the resistor can be expressed as

$$m_{10} = \overline{e_n} = 0$$

$$m_{20} = \mu_{20} = \overline{e_{th}^2} = \frac{4kT \Delta f}{G}$$ (Note: $e(t)$ is a normally distributed random variable)

and $\quad \overline{i_{th}^2} = 4kTG \Delta f$

There also exists a component of noise, the shot noise current (i_{sh}) due to the non uniform emission of charged electrons from the cathode of the diode. Under temperature limited operation, the moments for $i_{sh}(t)$ are expressed as

$$m_{10} = \overline{i_{sh}} = 0$$

$$m_{20} = \mu_{20} = \overline{i_{sh}^2} = 2q \cdot I_{dc} \cdot \Delta f.$$

where k is the Boltzmann's constant = 1.38×10^{-23} joule/°K,

q is the electron charge = 1.6×10^{-19} coulombs,

Δf is the bandwidth in Hz,

T is the temperature in degrees Kelvin,

and G is the conductance in mhos.

The total noise voltage $\overline{e_{no}^2}$ across the resistor can be expressed as

$$\overline{e_{no}^2} = \frac{\overline{i_{sh}^2}}{G^2} + \frac{\overline{i_{th}^2}}{G^2}$$

$$= \frac{(2q \cdot I_{dc} \cdot \Delta f)}{G^2} + \frac{4kTG\Delta f}{G^2}$$

$$= (2q \cdot I_{dc} \cdot \Delta f)R^2 + (4kT\Delta f)R$$

$$= (2 \times 1.6 \times 10^{-19} \times 0.01 \times 10^{-3} \times 5 \times 10^3)$$

$$(10 \times 10^3)^2 + (4 \times 1.38 \times 10^{-23} \times 295 \times 5 \times 10^3)$$

$$(10 \times 10^3)$$

$$\overline{e_{no}^2} = 2.4142 \times 10^{-12} \text{ volts}^2$$

Thus the total rms noise voltage is

$$\sqrt{\overline{e_{no}^2}} = 1.55376 \times 10^{-6} \text{ volts rms.}$$

● **PROBLEM 13-28**

What is the magnitude of R_{eq} in a pentode if $I_b = 10$ ma, $I_{c2} = 4$ ma, and $g_m = 6 \times 10^{-3}$ mhos. Consider also the effects of shot and partition noise.

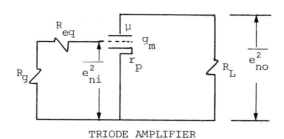

TRIODE AMPLIFIER

Fig. 1

Solution: When there is a random fluctuation in the division of current between the screen grid and the plate in a pentode, it produces an additional noise called partition noise and its effect is accounted for by the addition of another term to the expression for R_{eq} as in case of a triode shown in the figure.

Therefore, for a pentode:

$$R_{eq} = \frac{I_b}{I_b + I_{c2}} \left(\frac{2.5}{g_m} + \frac{20 I_{c2}}{g_m^2} \right) \quad (1)$$

where I_b and I_{c2} are the plate and screen grid currents respectively in amps, g_m is the transconductance in mhos, and R_{eq} is the resistance in ohms.

From Eq. (1), $(2.5/g_m)$ represents the shot noise and $(20\, I_{c_2}/g_m^2)$ represents the partition noise. Assuming shot noise and partition noise to be statistically independent, and substituting the given values into Equation (1):

$$R_{eq} = \frac{10}{14}\left(\frac{2.5}{6 \times 10^{-3}} + \frac{20 \times 4 \times 10^{-3}}{36 \times 10^{-6}}\right)$$

$$= \frac{10}{14}\left(\frac{2.5 \times 10^3}{6} + \frac{80 \times 10^3}{36}\right)$$

$$= 1884 \cdot 9205 \text{ mohms}$$

$$= 1.885 \text{ ohms}$$

• **PROBLEM 13-29**

Determine the mean-square noise voltage across the load resistor R_L of a common emitter transistor amplifier shown in Fig. (1a) below. The hybrid-pi small-signal equivalent model and an equivalent noise model are shown in Figs. (1b) and (1c) respectively. In the hybrid-pi small signal equivalent circuit, it is assumed that the losses across the bias resistors R_1 and R_2 are negligible, the capacitors c_1 and c_2 act like circuits in the frequency region of interest and that, the dynamic resistance of the collector is small compared to R_L.

In the equivalent noise circuit, $v_s^2 = 4kTR_sB$, $v_b^2 = 4kTr_bB$ and $v_L^2 = 4kTR_LB$. The main noise sources for the transistor are:

A) thermal noise, v_b^2, generated by the base-spreading resistance, r_b;

B) shot noises due to current fluctuations, i_b^2, in the base and i_c^2 in the collector.

Assume that the thermal noise generated by R_s and R_L are considerable and that these three noise sources are uncorrelated so that their mean-squared values add up.

Solution: The components of shot noise i_b^2 and i_c^2 in the base and collector circuits are given by

$$i_b^2 = 2\,e\,I_b B = \frac{2kTB}{r_\pi}$$

and $\quad i_c^2 = 2\,eI_c B$

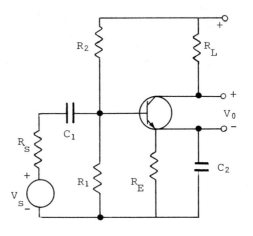

(a)

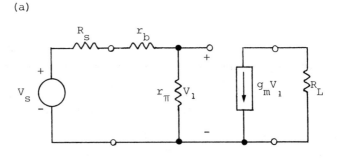

(b)

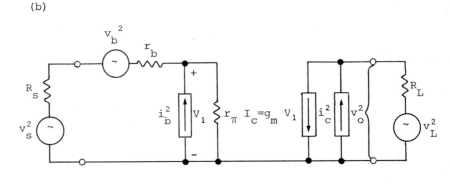

(c)

where $e = 1.6 \times 10^{-19}$ coulomb, and I_b and I_c are the quiescent base and collector currents.

Now, to determine these noise voltages and currents that are present in the collector circuit, first calculate the rms part of v_1 due to thermal noise and rms part of v_1 due to shot noise in the base circuit.

Then,

$$v_1^2 = \left(\frac{r_\pi}{r_\pi + r_b + R_s}\right)^2 (v_s^2 + v_b^2) + \left[\frac{r_\pi(r_b + R_s)}{r_\pi + r_b + R_s}\right]^2 i_b^2$$

In the above equation, the first term is due to the source

and base-spreading resistances, and the second term is due to shot noise of the base current.

The total mean-square output noise voltage, v_0^2 can be expressed as the sum of the mean-square output voltage due to v_1, $g_m^2 v_1^2 R_L^2$, the collector-circuit shot noise, $i_c^2 R_L^2$, and the thermal noise of R_L, which is v_L^2.

Therefore, $v_0^2 = g_m^2 v_1^2 R_L^2 + i_c^2 R_L^2 + v_L^2$

Substituting the value of v_1 in the above equation,

$$v_0^2 = g_m^2 R_L^2 \frac{r_\pi^2}{(r_\pi + r_b + R_s)^2} \left[4kTR_s B + 4kTr_b B + \frac{(r_b + R_s)^2 (2kTB)}{r_\pi} \right] + 2eI_c B R_L^2 + 4kTR_L B.$$

Further simplification yields

$$\frac{v_0^2}{B} = 4kTR_L \left[1 + \frac{eI_c R_L}{2kT} + \frac{g_m^2 R_L r_\pi^2 (R_s + r_b)}{(r_\pi + r_b + R_s)^2} \left(1 + \frac{r_b + R_s}{2r_\pi} \right) \right] V^2/Hz.$$

Assume $R_s = 1000\ \Omega$, $r_b = 100\ \Omega$, $R_L = 10{,}000\ \Omega$, $I_c = 1$ mA, and $I_c/I_b = 100$. Also, $g_m = eI_c/kT$ and $r_\pi = kT/eI_b$. At room temperature $e/kT = 40\ V^{-1}$, so that

$$g_m = 40 \times 10^{-3} = 0.04\ \text{mho}$$

and $\quad r_\pi = \dfrac{10^5}{40} = 2500\ \text{ohms}$

Hence, $\dfrac{v_0^2}{B} = 4(1.38 \times 10^{-23})(290)(10^4) \left[1 + \dfrac{(40)(10^{-3})(10^4)}{2} \right.$

$\left. + \dfrac{(0.04)^2(10^4)(2500)^2(1100)}{(3600)^2} \left(1 + \dfrac{1100}{5000} \right) \right]$

$= 1.6 \times 10^{-16}(1 + 200 + 10{,}355)\ V^2/Hz.$

It is clear from the above result that the first two terms in the parenthesis are negligible when compared to the last term. The first two terms are due to the load resistance noise and collector shot noise respectively, and the last term is due to shot noise of the base region and also due to the thermal noise of R_s and r_b.

Hence, the mean-square output noise voltage per hertz is

$$\frac{v_0^2}{B} = 1.6889 \times 10^{-12}\ V^2/Hz$$

If the bandwidth, B = 10 KHz, the rms noise voltage, v_0 will be

$$v_0^2 = 1.6889 \times 10^{-12} \times 10 \times 10^3$$
$$= 1.68896 \times 10^{-8} \text{ (volts)}^2$$

Therefore, $v_0 = 1.29959 \times 10^{-4}$ volts

$$v_0 = 0.13 \text{ mV}$$

CHAPTER 14

NOISE IN MODULATION SYSTEMS

NOISE IN AM

• PROBLEM 14-1

Consider at the output of a SSB modulator, a signal $p_1(t) = m(t)\cos(2\pi f_c t) + \hat{m}(t)\sin(2\pi f_c t)$ is being received. If the modulating signal, $m(t)$, and $\hat{m}(t)$ are given as follows:

$$m(t) = \sum_{k=1}^{m} A_k \cos(\omega_k t + \theta_k),$$

$$\hat{m}(t) = \sum_{k=1}^{m} A_k \sin(\omega_k t + \theta_k).$$

And the power spectral densities $\overline{m^2(t)} = \overline{\hat{m}^2(t)} = \sum_{k=1}^{m} \frac{A_k^2}{2}$.

Show the following:

(a) The input normalized power P_i of $p(t)$ is $\overline{m^2(t)} = \overline{\hat{m}^2(t)}$.

(b) If $m(t)$ is bandlimited to f_m, then the spectra of $m^2(t)$ and $\hat{m}^2(t)$ will extend from 0 to $2f_m$.

(c) Calculate the ratio P_o/P_i, where P_o is the output normalized power of a SSB demodulated signal.
(i.e., $p_i(t) \longrightarrow \boxed{\times} \longrightarrow$ baseband filter $\longrightarrow p_o(t)$
  $\cos 2\pi f_c t$)

711

Solution:

(a) Since the normalized power $P_i = \overline{p_i^2(t)}$ and $p_i(t)$ is given,

thus, $P_i = \overline{\left[m(t)\cos(2\pi f_c t) + \hat{m}(t)\sin(2\pi f_c t)\right]^2}$

$= \overline{m^2(t)\cos^2 2\pi f_c t} + \overline{2m(t)\hat{m}(t)\cos 2\pi f_c t \cdot \sin 2\pi f_c t}$

$+ \overline{\hat{m}^2(t)\sin^2 2\pi f_c t}$

Now, notice that $\overline{m^2(t)\cos^2 2\pi f_c t} = \frac{1}{2}\overline{m^2(t)} + \frac{1}{2}\overline{m^2(t)\cos(4\pi f_c t)}$

$\overline{\hat{m}^2(t)\sin^2 2\pi f_c t} = \frac{1}{2}\overline{\hat{m}^2(t)} + \frac{1}{2}\overline{\hat{m}^2(t)\cos(4\pi f_c t)}$

and $\sin 2\alpha = 2\sin\alpha\cos\alpha$

Hence, $P_i = \frac{1}{2}\overline{m^2(t)} + \frac{1}{2}\overline{m^2(t)\cos(4\pi f_c t)}$

$+ \overline{m(t)\hat{m}(t)\sin(4\pi f_c t)}$

$+ \frac{1}{2}\overline{\hat{m}^2(t)} + \frac{1}{2}\overline{\hat{m}^2(t)\cos(4\pi f_c t)}$

Since $m^2(t)\cos(4\pi f_c t)$ and $\hat{m}^2(t)\cos(4\pi f_c t)$ consist of a sum of sinusoidal waveforms in the range $2f_c \pm 2f_m$, the average of such a sum is zero. (Also for the term $m(t)\hat{m}(t)\sin(4\pi f_c t)$.)

Therefore, $P_i = \frac{1}{2}\overline{m^2(t)} + \frac{1}{2}\overline{\hat{m}^2(t)} = \overline{m^2(t)} = \overline{\hat{m}^2(t)}$

(b) $m^2(t) = \sum_{k=1}^{m} A_k \cos(\omega_k t + \theta_k) \sum_{j=1}^{m} A_j \cos(\omega_j t + \theta_j)$

$= \sum_{k=1}^{m} \sum_{j=1}^{m} A_k A_j \cos(\omega_k t + \theta_k)\cos(\omega_j t + \theta_j)$

$= \sum_{k=1}^{m} \sum_{j=1}^{m} \frac{A_k A_j}{2}\left\{\cos\left[(\omega_k + \omega_j)t + (\theta_k + \theta_j)\right] + \cos\left[(\omega_k - \omega_j)t + (\theta_k - \theta_j)\right]\right\}$

and similarly,

$\hat{m}^2(t) = \sum_{k=1}^{m} A_k \sin(\omega_k t + \theta_k) \sum_{j=1}^{m} A_j \sin(\omega_j t + \theta_j)$

$= \sum_{k=1}^{m} \sum_{j=1}^{m} \frac{A_k A_j}{2}\left\{-\cos\left[(\omega_k + \omega_j)t + (\theta_k + \theta_j)\right] + \cos\left[(\omega_k - \omega_j)t + (\theta_k - \theta_j)\right]\right\}$

From the above equations, all spectra extend from 0 to $2f_m$ is shown.

(c) The output signal $p_o(t)$ is

$$p_o(t) = p_i(t) \times \cos(2\pi f_c t)$$

$$= \left[m(t) \cos(2\pi f_c t) + \hat{m}(t) \sin(2\pi f_c t) \right] \cos(2\pi f_c t)$$

$$= m(t) \left[\frac{1}{2} + \frac{1}{2} \cos(4\pi f_c t) \right] + \frac{1}{2} \left[\hat{m}(t) \sin(4\pi f_c t) \right].$$

Note: $\sin \alpha \sin \beta = \frac{1}{2} \cos(\alpha - \beta) - \frac{1}{2} \cos(\alpha + \beta)$

$\cos \alpha \cos \beta = \frac{1}{2} \cos(\alpha - \beta) + \frac{1}{2} \cos(\alpha + \beta)$

$$p_o(t) = \frac{1}{2} m(t) + \frac{1}{2} \left[m(t) \cos(4\pi f_c t) + \hat{m}(t) \sin(4\pi f_c t) \right] \quad (1)$$

Since the second term of Eq. (1) will be filtered out,

hence, $P_o = \overline{\left[\frac{1}{2} m(t) \right]^2} = \frac{1}{4} \cdot \overline{m^2(t)}$

Therefore, $P_o/P_i = \frac{1}{4} \overline{m^2(t)} \Big/ \overline{m^2(t)} = \frac{1}{4}$

● **PROBLEM 14-2**

In a transmission system, a noise process $N(t)$ having power density $S_N(\omega) = \frac{P^2}{P^2 + \omega^2}$, where $P > 0$ is a constant, is mixed with a signal $A \cos(\omega_o t)$.

Determine:

(a) the average signal-to-noise ratio.

(b) the value of P such that SNR is maximum.

What is the effect of choosing this value of P?

Solution:
(a) The average power in a sinusoidal signal of peak amplitude A is $\frac{A^2}{2}$, which is also the power dissipated in a 1Ω resistor when driven by this signal.
The noise power is given by

$$P_N = \frac{1}{2\pi} \int_{-\infty}^{\infty} S_N(\omega) \, d\omega$$

713

Substituting in given $S_N(\omega)$,

$$P_N = \frac{1}{2\pi} \int_{-\infty}^{\infty} \frac{P^2}{P^2 + \omega^2} d\omega$$

$$= \frac{P^2}{2\pi} \left\{ \left[\frac{1}{P} \tan^{-1}\left(\frac{\omega}{P}\right) \right]_{-\infty}^{\infty} \right\}$$

Hence, $P_N = P/2$

Now define (S/N) as the signal-to-noise ratio. Then

$$(S/N) = \frac{A^2/2}{P/2} = \frac{A^2}{P}$$

(b) The signal-to-noise ratio (S/N) is maximum when the noise power tends to zero. i.e., $P \to 0$. The consequence of having $P \to 0$ is that the noise disappears because the noise power density spectrum $S_N(\omega)$ becomes a vanishingly narrow function around $\omega = 0$.

• **PROBLEM 14-3**

A baseband waveform v(t) with power spectral density

$$G_v(f) = \begin{cases} \frac{\eta_m |f|}{2f_m} & |f| < f_m \\ 0 & \text{elsewhere} \end{cases}$$

is transmitted using single-sideband modulation technique. Assuming cut-off frequency of the baseband filter to be f_m.

(a) Calculate the input and the output signal power.

(b) If the SSB signal is accompanied by a white noise, determine the signal-to-noise ratio at the output port.

(c) Repeat part (a) and (b) if double-sideband modulation is used for transmission.

Solution:

(a) The input signal power V_i is given as

$$V_i = \overline{v^2(t)} = \int_{-f_m}^{f_m} G_v(f) df$$

Hence, $V_i = 2 \int_0^{f_m} \frac{\eta_m}{2f_m} f \, df$

$= \frac{\eta_m}{f_m} \left(\frac{f^2}{2}\right)_0^{f_m}$

$= \frac{\eta_m f_m}{2}$

or graphically,

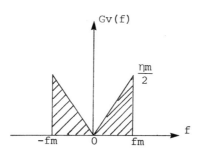

Fig. 1

V_i = the total area under the hatched portion of the curve as shown in Fig. (1).

Now, the output signal power is calculated similarly.

Observe in Fig. (2),

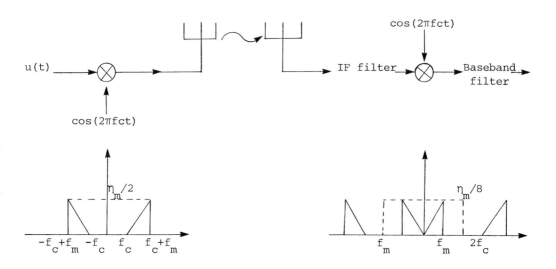

Fig. 2

Hence, output signal power = $\overline{\dfrac{v^2(t)}{4}} = \dfrac{\eta_m f_m}{8}$

• **PROBLEM 14-4**

Consider a sinusoidal wave at 900 Hz is to amplitude modulate up to 50% a carrier wave of frequency f_c and amplitude 10 mv. Also present is a thermal noise of two-sided power spectral density $\eta/2 = 10^{-3}$ watt/Hz as shown in Fig. (1). The received signal together with the noise is demodulated by multiplying with a local carrier of amplitude 1 volt. Find the output power after the filter shown in Fig. (2), corresponding to (a) signal, and (b) noise.

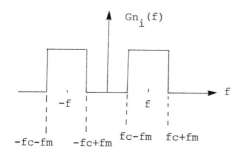

Fig. 1   Fig. 2

Solution:

(a) Consider the received signal to be of the form

$$v_i(t) = A m(t) \cos 2\pi f_c t$$

where $m(t) = m \cos 2\pi f_m t$ which represents an arbitrary modulating signal,

$$A = 10 \times 10^{-3} \text{ volt}$$

$$m = 50/100 = 0.5$$

and $f_m = 900$ Hz

The signal power before demodulation is then

$$V_i = \overline{v_i^2(t)} = \overline{A^2 m^2(t) \cos^2 2\pi f_c t}$$

$$= \tfrac{1}{2} \overline{A^2 m^2(t)} + \tfrac{1}{2} \overline{A^2 m^2(t) \cos(4\pi f_c t)}$$

(Note: $\cos^2 x = \tfrac{1}{2}(\cos 2x + 1)$)

Since $m(t)$ can be represented as a sum of sinusoidal spectral components, the term $m^2(t) \cos 4\pi f_c t$ consists of a sum of sinusoidal waveforms in the frequency range $2f_c \pm 2f_m$ whose average value is zero. Thus,

$$V_i = \overline{v_i^2(t)} = \tfrac{1}{2} \overline{A^2 m^2(t)}$$

716

Now, after demodulation, by multiplication with a local carrier ($\cos 2\pi f_c t$) and passing through the baseband filter, the signal at the output is $v_o(t) = m(t)/2$ with power

$$V_o = \overline{\frac{A^2 m^2(t)}{4}} = \frac{\overline{V_i}}{2}$$

Therefore, $V_o = \dfrac{A^2 \times \overline{m^2 \cos^2 2\pi f_m t}}{4}$

$$= \frac{10^{-4} \times 0.5^2 \times 0.5}{4}$$

$$= 3.125 \times 10^{-6} \text{ watt}$$

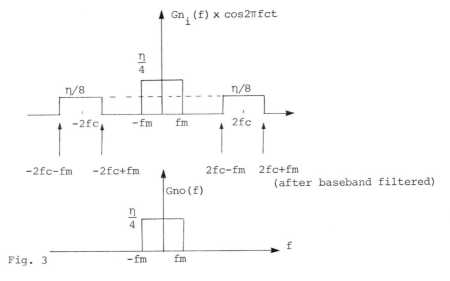

Fig. 3

(b) The noise in Fig. (2) is multiplied by $\cos 2\pi f_c t$ and then baseband filtered. See Fig. (3).

Hence,

$$N_o = \frac{\eta f_m}{2}$$

$$= 10^{-3} \times 900$$

$$= 0.9 \text{ watt}$$

● **PROBLEM 14-5**

The figure below shows the circuitry of a superheterodyne receiver. Using this receiver as a model,

(a) Derive expressions for signals at points L, M, N, and P.

(b) Derive an expression for the bandwidth of the radio-frequency amplifier.

(c) Sketch the spectra of AM signals in the superheterodyne receiver.

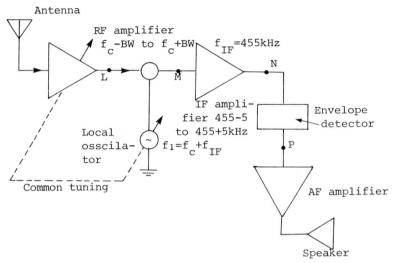

Solution:

(a) The signal furnished by an antenna to a receiver is usually very low in power while the required power may be of the order of tenth of a watt. In addition to the transmitted signal, the signal at the receiver also contains noise which must be filtered to aid in the detection of the information signal. The superheterodyne receiver carries out these two basic functions as an integrated unit. In general, superheterodyne operation involves the conversion of frequency from the tuned radio-frequency to the fixed intermediate frequency. It refers to the use of a frequency converter and fixed, tuned IF amplifier, then after which an envelope detector will be used for signal detection.

Now consider at point L, let the RF amplifier is tuned to the frequency f_c and an AM signal of the form as follows is received.

$$v_L(t) = g(t) \cos 2\pi f_c t, \qquad (1)$$

where $g(t) = k[1 + ms(t)]$, k = constant and

$s(t)$ = information signal

At point M, the above signal is mixed with the local oscillator frequency $f_\ell = f_c + f_{IF}$, generated by the local oscillator. Thus, the resulting signal at M is

$$v_M(t) = k_1 g(t) \left[\cos 2\pi f_{IF} t + \cos 2\pi(2f_c + 2f_{IF})t\right] \qquad (2)$$

The above equation consists of an IF frequency component at 455 kHz (first term in the brackets) and a sum frequency component (second term in the brackets) at a frequency which is definitely external to the range of frequencies between 455 - 5 kHz and 455 + 5 kHz of the IF amplifier response.(the range is fixed at 10 kHz which represents the channel bandwidth per station.) (See figure 1.)

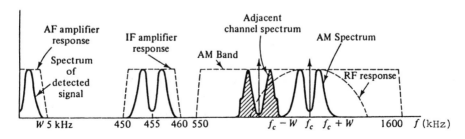

Fig. 1

Hence, when the signal at M passes through the IF amplifier, the sum frequency component will be rejected and the signal at point N has the form

$$v_N(t) = k_2 g(t) \cos 2\pi f_{IF} t. \qquad (3)$$

The output of the envelope detector follows the envelope of the modulated signal. Hence, rejecting all noise power and generates the signal at P.

$$v_P(t) = k_3 g(t) \qquad (4)$$

This signal is finally amplified by an audio frequency (AF) amplifier which is used to drive an audio output device such as a speaker.

(b) An important property of the RF amplifier stage is to provide the receiver system with the image-channel rejection facility. In the absence of this facility the image channels will be detected along with the desired signal. Hence, due to this property, the RF amplifier bandwidth. 2BW should be no less than 2BW (where BW is the channel bandwidth). (Note: The IF stage furnishes adjacent-channel selectivity.) To explain the phenomenon of image channel rejection, consider the following example. If f_ℓ is set at

$$f_\ell = f_c + f_{IF}$$

then

$$f_{RF} = f_c$$

(where f_c is the carrier frequency)

and its image frequency will be

$$f_{RF(i)} = f_c + 2f_{IF}$$

This image frequency is within the frequency band of the IF

stage. It is thus necessary for the RF stage to remove this image frequency and hence the name image-channel rejection.

The spectra of AM signals is shown in Fig. (1). Note that for standard AM broadcasting, the frequency band is 550-1600 kHz and the IF frequency is fixed at 455 kHz with a 10 kHz channel bandwidth per station.

• **PROBLEM** 14-6

The desired broadcasting station's frequency is given as 640 kHz, calculate the required image frequency if the station used a receiver of 470 kHz IF (intermediate-frequency).

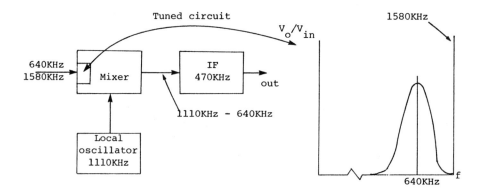

Solution: In order to determine the image frequency, it is first necessary to know the frequency of the local oscillator (LO). The relation between the LO frequency, the desired station's frequency and the intermediate-frequency is given by

$$LO = IF + \text{station's frequency}$$

Therefore, $\quad LO = 470 \text{ kHz} + 640 \text{ kHz}$

$$= 1110 \text{ kHz}$$

It is now required to determine the frequency component such that the sum of this frequency and the local oscillator frequency will deliver an output at the IF of 470 kHz.

Hence, $\quad A - 1110 \text{ kHz} = 470 \text{ kHz}$

$$A = 1110 \text{ kHz} + 470 \text{ kHz}$$

$$= 1580 \text{ kHz}.$$

Thus, the image frequency for the given standard broadcast band receiver is 1580 kHz.

Note: From this problem, it can be inferred that neglecting the image frequency on the standard broadcast band does not affect the receiver performance in a serious way. To illustrate this point further, consider the situation as illustrated in the figure below.

It can be seen that, since the image frequency of 1580 kHz is far away from the center frequency of 640 kHz of the tuned circuits, most of the image frequency will be attenuated at the tuned circuit prior to the signals entry into the mixer.

It is of importance to note however that the same situation will not be true at higher frequencies used by many communication receivers since the image frequency will be close to the center frequency of the tuned circuit.

• **PROBLEM 14-7**

A sinusoidal baseband signal of frequency f_m = 5 kHz is transmitted using DSB amplitude modulation. The carrier has a frequency f_c = 1 MHz and of amplitude 2 volts. If the signal to 10% amplitude modulated and during transmission, white noise of two-sided power spectral density 10^{-6} watt/Hz is added. Determine the following:

Given the demodulator has input-output characteristic; $v_o = 3 v_i^2$, and the IF filter, before the demodulator, with center frequency at 1.002 MHz and 10 kHz bandwidth, has a rectangular characteristic of unity gain.

(a) Find $v_o(t)$ output signal waveform at the demodulator and determine its normalized power.

(b) Find output noise power at the demodulator and the signal-to-noise-ratio.

Solution: (a) Let x(t) be the amplitude modulated signal of the form:

$$x(t) = A\left[1 + m(t)\right] \cos 2\pi f_c t$$
$$= 2\left[1 + 0.1 \cos 2\pi(5 \times 10^3 t)\right] \cos 2\pi \times 10^6 t$$
$$= 2\left[\cos(2\pi \times 10^6 t) + 0.05 \cos(2\pi \times 1005 \times 10^3 t)\right.$$
$$\left. + 0.05 \cos(2\pi \times 995 \times 10^3 t)\right]$$

(Note: modulation index = m = 0.1 for 10% amplitude modulated)

The output of the IF filter is

$$v_i(t) = (2 \cos 2\pi \times 10^6 t) + 0.1 \cos(2\pi \times 1005 \times 10^3 t)$$
$$v_o(t) = 3v_i^2(t) = 3\left[2 \cos(2\pi \times 10^6 t) + 0.1 \cos(2\pi \times 1005 \times 10^3 t)\right]$$
$$\approx 3\left[4 \cos^2(2\pi \times 10^6 t) + 0.4 \cos(2\pi \times 10^6 t)\cos(2\pi \times 1005 \times 10^3 t)\right]$$

$$= 6 + 6 \cos(2\pi \times 10^6 t) + 0.6 \cos(2\pi \times 2005 \times 10^3 t)$$
$$+ 0.6 \cos(2\pi \times 5000t)$$

Hence, $S_o(t) \simeq 0.6 \cos(2\pi \times 5000t)$

and $S_o = \overline{S_o^2(t)} = 0.6^2 \overline{\cos^2(2\pi \times 5000t)} = 0.18$ watt

(b) Now, let $n_2(t) = 6 \times 2n(t)\left[1 + m(t)\right] \cos \omega_c t + 3n^2(t)$
$$\simeq 12 n(t) \cos \omega_c t + 3n^2(t)$$

where $n(t) = n_c(t) \cos \omega_o t - n_s(t) \sin \omega_o t$
$$= n_c(t) \cos(2\pi \times 1002 \times 10^3 t) -$$
$$n_s(t) \sin(2\pi \times 1002 \times 10^3 t)$$

Since $n(t)$ has the power spectral density η/s for 997 kHz $\leq f <$ 1007 kHz, the power spectral density of the first term of $n_2(t)$, $12n(t)\cos \omega_c t$, is

$$Gn'(f) = \begin{cases} 36\eta & |f| \leq 2 \text{ kHz} \\ 18\eta & 2 \text{ kHz} < |f| \leq 7 \text{ kHz} \\ 0 & 7 \text{ kHz} < |f| \end{cases}$$

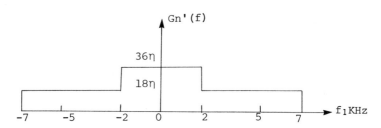

$N_o' = 36\eta \times 4000 + 18\eta \times 6000 = \frac{\eta}{2} \times 5.04 \times 10^5 =$
$10^{-6} \times 5.04 \times 10^{-5} \simeq 0.5$ watt

N_o'' is obtained by $N_o'' = 3\lambda^2 \eta^2 f_m^2$

Hence, $N_o'' = 3\lambda^2 \eta^2 f_m^2 = 3 \times 9 \times (2 \times 10^{-6})^2 \times (5 \times 10^3)^2$
$$= 0.0027 \text{ watt.}$$

The total noise power at the demodulator output is

$$N_o = N_o' + N_o'' = 0.5 + 0.0027 = 0.5027$$
$$\simeq 0.5 \text{ watt}$$

$$\frac{S_o}{N_o} = \frac{0.18}{0.5} = 0.36 \simeq -4.4 \text{ dB}$$

• **PROBLEM 14-8**

A demodulator has two input signals namely (a) a received signal at a resonant frequency of $\omega_o/2\pi$, and (b) a signal $A_o \cos(\omega_o t)$ generated by the local oscillator at the receiver.

The received signal has a noise power spectrum defined as follows:

$$S_X(\omega) = \begin{cases} P/2, & -\omega_o - (\alpha/2) < \omega < -\omega_o + (\alpha/2) \\ P/2, & \omega_o - (\alpha/2) < \omega < \omega_o + (\alpha/2) \\ 0, & \text{elsewhere} \end{cases}$$

where ω_o is the center frequency, α is the bandwidth and $P/2$ is the noise power density. Determine the noise power at the product demodulator output and at the output of the low-pass filter following the demodulator.

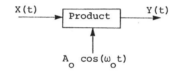

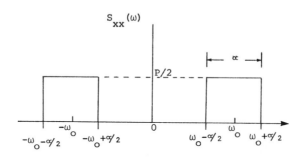

Solution: $Y(t)$ has an autocorrelation function given by;

$$R_Y(t, t+\tau) = E\left[Y(t)\, Y(t+\tau)\right]$$
$$= E\left[X(t)\, A_o \cos\omega_o t\, X(t+\tau) A_o \cos(\omega_o t + \omega_o \tau)\right]$$
$$= E\left[A_o^2\, X(t)\, X(t+\tau)\, \cos(\omega_o t)\cos(\omega_o t + \omega_o \tau)\right]$$

Thus, $R_Y(t, t+\tau) = \dfrac{A_o^2}{2} R_X(t, t+\tau)\left[\cos\omega_o\tau + \cos(2\omega_o t + \omega_o \tau)\right]$

The power density spectrum and the average of the autocorrelation function over a period of time form a Fourier transform pair. The inverse transform of which will yield the power density spectrum at the output of the demodulator.

Assuming X(t) to be time invariant and of constant mean (wide-sense stationary), the time average of $R_Y(t, t + \tau)$ is equal to

$$\frac{A_o^2}{2} R_X(\tau) \cos(\omega_o \tau).$$

The Fourier transform of the above equation yields

$$S_Y(\omega) = \int_{-\infty}^{\infty} R_Y(t, t + \tau) e^{-j\omega\tau} d\tau$$

$$= \frac{A_o^2}{2} \left[S_X(\omega - \omega_o) + S_X(\omega + \omega_o) \right]$$

Hence, the power density spectrum of the noise at the output of the demodulator is,

$$S_Y(\omega) = \begin{cases} P A_o^2/8 & -2\omega_o - (\alpha/2) < \omega < -2\omega_o + \alpha/2 \\ P A_o^2/4 & -\alpha/2 < \omega < \alpha/2 \\ P A_o^2/8 & -2\omega_o - (\alpha/2) < \omega < 2\omega_o + \alpha/2 \\ 0 & \text{otherwise} \end{cases}$$

Since the desired signal is in the frequency band $-\alpha/2 < \omega < \alpha/2$, the low-pass filter following the demodulator will not suppress the noise components in this frequency band, and hence will be present at the filter output as noise power given by

$$N_o = \frac{1}{2\pi} \int_{-\alpha/2}^{\alpha/2} \frac{P A_o^2}{4} d\omega$$

$$= \frac{P A_o^2 \alpha}{8\pi}$$

• **PROBLEM 14-9**

Compare AM and FM noise and show by using diagram.

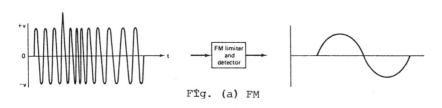

Fig. (a) FM

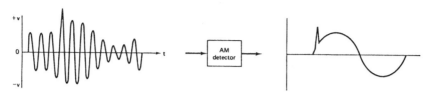

Fig. (b) AM

Solution: One of the most important advantages of FM over AM is FM's superior noise characteristics. Static noise is very common in AM but not on FM. Due to the reason that the amplitude changes in AM contain the modulating signal, any attempt to filter out the noise adversely affects the signal received. However, in FM, the modulating signal is not carried by amplitude changes but instead by frequency changes. Addition of external noise of some source during transmission can be clipped off by a limiter circuit and or through the use of detector circuits that are insensitive to amplitude changes.

Fig. (a) shows the noise removal action of an FM limiter circuit, while Fig. (b) shows that the noise spike feeds right through to the speaker in AM system. The advantage of using frequency modulation is clearly observed.

NOISE IN FM

• **PROBLEM 14-10**

The signal $g(t) = 4(1 + 0.2 \cos 2\pi \times 10^4 t) \cos 2\pi \times 10^5 t$ is applied to a filter which rejects lower sideband frequency components.

(a) For the filter output $g_o(t)$, write an expression and show that $g_o(t)$ is an angle modulated waveform of the form $g_o(t) = A(t) \cos \left[2\pi \times 10^5 t + \theta(t)\right]$

(b) Now, the filter input is applied to a converter having transfer function

$$H(f) = \begin{cases} j\sigma\omega & f_c = 10^5 \pm 10^4 \text{ Hz} \\ 0 & \text{elsewhere} \end{cases}$$

(where $o = \frac{1}{2}$ sec.) and then applied to an envelope demodulator. Write an expression for the output of the envelope demodulator.

Solution:

(a) First, simplify $g(t)$ as follows:

$g(t) = 4(1 + 0.2 \cos 2\pi \times 10^4 t) \cos 2\pi \times 10^5 t$

$$= 4 \cos 2\pi \times 10^5 t + 0.8 \left[\tfrac{1}{2} \cos 2\pi (10^5 - 10^4) + \tfrac{1}{2} \cos 2\pi (10^5 + 10^4) t \right]$$

$$= 4 \cos 2\pi \times 10^5 t + 0.4 \left[\cos 2\pi (10^5 - 10^4) t + \cos 2\pi (10^5 + 10^4) t \right]$$

Hence, after passing g(t) through the filter, the filter output $g_o(t)$ becomes

$$g_o(t) = 4 \cos 2\pi \times 10^5 t + 0.4 \cos 2\pi (10^5 + 10^4) t$$

$$= 4 \cos 2\pi \times 10^5 t + 0.4 \cos 2\pi \times 10^5 t \cos 2\pi \times 10^4 t$$

$$- 0.4 \sin 2\pi \times 10^5 t \sin 2\pi \times 10^4 t$$

$$= (4 + 0.4 \cos 2\pi \times 10^4 t) \cos 2\pi \times 10^5 t -$$

$$0.4 \sin 2\pi \times 10^5 t \sin 2\pi \times 10^4 t$$

$$= \left[(4 + 0.4 \cos 2\pi \times 10^4 t) + (0.4 \sin 2\pi \times 10^4 t)^2 \right]^{\tfrac{1}{2}}$$

$$\times \cos \left[2\pi \times 10^5 t + \tan^{-1} \left(\frac{0.4 \sin 2\pi \times 10^4 t}{4 + 0.4 \cos 2\pi \times 10^4 t} \right) \right]$$

$$= A(t) \cos \left[2\pi \times 10^5 t + \theta(t) \right]$$

Therefore, $A(t) = \left[(4 + 0.4 \cos 2\pi \times 10^4 t)^2 + (0.4 \sin 2\pi \times 10^4 t)^2 \right]^{\tfrac{1}{2}}$,

which when maximized is approximately equal to 4.

and $\theta(t) = \tan^{-1} \left(\dfrac{0.4 \sin 2\pi \times 10^4 t}{4 + 0.4 \cos 2\pi \times 10^4 t} \right) \simeq 0.1 \sin 2\pi \times 10^4 t$

Hence, $g_o(t)$ is of the form of an angle modulated waveform.

(b) From part (a), the voltage applied to the convertor

$$g_o(t) \simeq 4 \cos \left[2\pi \times 10^5 t + 0.1 \sin 2\pi \times 10^4 t \right]$$

Now, notice that the output of the converter is related to the input by the equation

$$\text{output} = \sigma \frac{d}{dt} \text{input}$$

due to the transfer function of the converter is conveniently chosen as $H(j\omega) = j\sigma\omega$.

Hence, the output is the derivative of $g_o(t)$ in the frequency range $10^5 \pm 10^4$ Hz with $\sigma = 1/2\pi$.

Thus, equal to $-\left[4\sigma \ 2\pi \times 10^5 + 0.1 \times 2\pi \times 10^4 \cos 2\pi \times 10^4 t \right]$

$$\times \sin \left[2\pi \times 10^5 + 0.1 \sin 2\pi \times 10^4 t \right]$$

$$= - \left[4 \ 10^5 + 0.1 \times 10^4 \cos 2\pi \times 10^4 t \right] \times$$

$$\sin[2\pi \times 10^5 + 0.1 \sin 2\pi \times 10^4 t]$$

Therefore, the output waveform of the envelope demodulator is, neglecting the dc-component, $-0.2 \times 10^4 \cos 2\pi \times 10^4 t$.

• **PROBLEM 14-11**

Consider the FM receiver system shown in Fig. (1a). If the input to the IF carrier filter is a signal of the form $v_i(t) = A \cos(\omega_c t) + k \int_{-\infty}^{t} m(\lambda) d\lambda$ plus a white noise of power spectral density $\eta/2$, determine the output signal to noise ratio of the receiver.

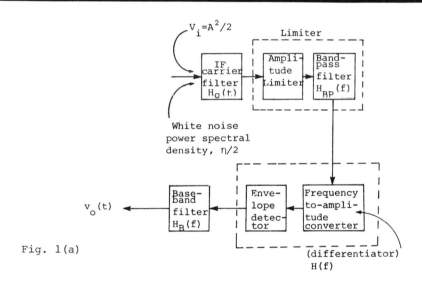

Fig. 1(a)

Solution: Let $m(t)$ be the frequency-modulating baseband waveform.

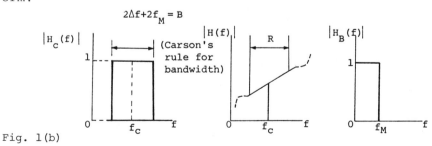

Fig. 1(b)

Now, observe from Fig. (1b), the IF carrier filter has a bandwidth of $BW = 2\Delta f + 2f_m$, which passes the signal with negligible distortion and eliminating all noise outside the bandwidth BW. This signal with its accompanying noise is passed through a limiter-discriminator and a baseband filter and appears at the output as a signal $v_o(t)$ and a noise waveform $n_o(t)$.

Now, let's proceed to the calculation of the output S/N ratio. The limiter circuit following the filter is used to suppress any variation in amplitude due to noise. The output of the limiter will thus be

$$v_L(t) = A_L \cos\left[\omega_c t + k \int_{-\infty}^{t} m(\lambda)\, d\lambda\right]$$

The limiter circuit is followed by a discriminator circuit consisting of an amplitude converter whose input and output are related as:

$$v_d(t) = \delta \frac{d}{dt} v_L(t) \tag{3}$$

which implies the fact that a multiplication by $j\omega$ in the frequency domain is equivalent to a differentiation in the time domain. i.e., $\delta \frac{d}{dt} \iff j\delta\omega$

Setting $\phi(t) = k \int_{-\infty}^{t} m(\lambda)\, d\lambda$, the signal at the output of the discriminator will be

$$v_d(t) = \delta A_L \left[\omega_c + \frac{d}{dt}\phi(t)\right]$$

If $\alpha = \delta A_L$, then

$$v_d(t) = \alpha\omega_c + \alpha \frac{d}{dt}\phi(t)$$
$$= \alpha\omega_c + \alpha k m(t)$$

The baseband filter following the discriminator rejects the dc component and passes the signal without any distortion resulting in an output signal.

$$v_o(t) = \alpha k m(t)$$

Therefore, the output signal power is

$$V_o = \alpha^2 k^2 \overline{m^2(t)}$$

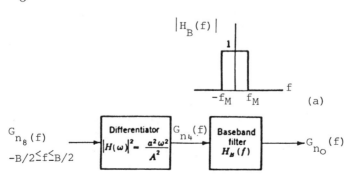

(a)

Fig. (2) shows: (a) the operations performed by the discriminator and baseband filter on the noise output of the limiter and (b) the variation of the power spectral density with frequency at the output of the FM discriminator.

Hence, the output noise power N_o can be calculated by computing the area of the shaded region in Fig. (2b).

Therefore,
$$N_o = \int_{-f_m}^{+f_m} G_{n_4}(f) df$$

$$= \frac{\alpha^2 \eta}{A} \int_{-f_m}^{f_m} 4\pi^2 f^2 df$$

$$N_o = \frac{8\pi^2}{3} \frac{\alpha^2 \eta}{A^2} f_m^3$$

where $G_{n_4}(f) = \frac{\alpha^2 \omega^2 \eta}{A^2} \quad |f| \leq \beta/2$

Now, the signal-to-noise ratio of the output can be evaluated as

$$\frac{S_o}{N_o} = \frac{\alpha^2 k^2 \overline{m^2(t)}}{(8\pi^2/3)(\alpha^2 \eta/A^2) f_m^3}$$

$$= \frac{3}{4\pi^2} \cdot \frac{k^2 \overline{m^2(t)}}{f_m^2} \frac{A^2/2}{\eta f_m}$$

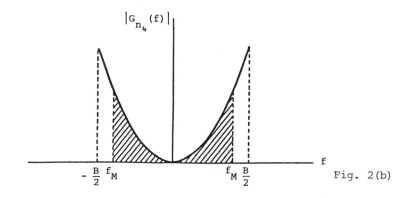

Fig. 2(b)

● **PROBLEM 14-12**

The output signal-to-noise ratio of two FM receivers is identical and is given as

$$\frac{S_o}{N_o} = \frac{3k^2 \cdot \overline{m^2(t)} \cdot A^2}{4\pi^2 \cdot f_m^2 \cdot 2\eta_m f_m}$$

Find the ratio of the IF bandwidths of these two receivers in order to pass 98% of the signal energy. Given two FM signals

1) $km(t) = \beta\omega_m \cos(\omega_m t)$ and

2) $km(t) =$ Gaussian signal with power spectral density

$$= \begin{cases} \eta_m/2 & |f| \leq f_m \\ 0 & \text{elsewhere} \end{cases}$$

Solution: The IF bandwidth for the signal $km(t) = \beta\omega_m \cos(\omega_n t)$ is defined as $BW_{IF_1} = 2(\beta + 1)f_m$

Assuming $\beta \gg 1$, $BW_{IF_1} \cong 2\beta f_m$.

Now, since $\left(\dfrac{S_o}{N_o}\right)_1 = \dfrac{k^2 \cdot \overline{m^2(t)} \cdot A^2}{4\pi^2 \cdot f_m^2 \cdot 2\eta_m f_m}$ is given and notice that

$$k^2 \cdot \overline{m^2(t)} = \dfrac{\beta^2 \omega_m^2}{2},$$

$$\left(\dfrac{S_o}{N_o}\right)_1 = \dfrac{3}{4\pi^2} \cdot \dfrac{\beta^2 \omega_m^2}{2f_m^2} \cdot \dfrac{A^2}{2\eta_m f_m}$$

In terms of BW_{IF_1},

$$\left(\dfrac{S_o}{N_o}\right)_1 = \dfrac{3}{4\pi^2} \cdot \dfrac{\beta^2 \cdot 4\pi^2 \cdot f_m^2}{2f_m^2} \cdot \dfrac{A^2}{2\eta_m f_m}$$

$$= \dfrac{3\beta^2 A^2}{4\eta_m f_m} \cdot \dfrac{4f_m^2}{4f_m^2}$$

$$= \dfrac{3}{2} \cdot \dfrac{4\beta^2 \cdot f_m^2}{4f_m^2} \cdot \dfrac{A^2}{2\eta_m f_m}$$

$$= \dfrac{3}{2} \cdot \left(\dfrac{2\beta f_m}{2f_m}\right)^2 \cdot \dfrac{A^2}{2\eta_m f_m}$$

$$= \dfrac{3}{2} \cdot \left(\dfrac{BW_{IF_1}}{2f_m}\right)^2 \cdot \dfrac{A^2}{2\eta_m f_m}$$

Similarly, for $km(t) =$ Gaussian, the IF bandwidth is given as $BW_{IF_2} = 4 \cdot 6(\Delta f_{rms}) = 4 \cdot 6 \times \dfrac{\sqrt{k^2 \overline{m^2(t)}}}{2\pi} = \dfrac{4 \cdot 6\sqrt{\eta_m \cdot f_m}}{2\pi}$

Hence, $\left(\dfrac{S_o}{N_o}\right)_2 = \dfrac{3 \; k^2 \cdot \overline{m^2(t)} \cdot A^2}{4\pi^2 \cdot f_m^2 \cdot 2\eta_m f_m}$

$$= \frac{3}{4\pi^2} \cdot \frac{1}{f_m^2} \cdot \eta_m f_m \cdot \frac{A^2}{2\eta_m f_m}$$

$$= \frac{3}{4\pi^2} \cdot \frac{1}{f_m^2} \cdot \left(\frac{\sqrt{\eta_m f_m}\, 2\pi(4\cdot 6)}{2\pi(4\cdot 6)}\right)^2 \cdot \frac{A^2}{2\eta_m f_m}$$

$$= \frac{3}{4\pi^2 f_m^2} \cdot \left(\frac{2\pi \cdot BW_{IF_2}}{4\cdot 6}\right)^2 \cdot \frac{A^2}{2\eta_m f_m}$$

Since $\left(\frac{S_o}{N_o}\right)_1 = \left(\frac{S_o}{N_o}\right)_2$,

therefore, $\dfrac{3}{2} \cdot \left(\dfrac{BW_{IF_1}}{2f_m}\right)^2 \cdot \dfrac{A^2}{2\eta_m f_m} = \dfrac{3}{4\pi^2 f_m^2} \cdot \left(\dfrac{2\pi BW_{IF_2}}{4\cdot 6}\right)^2 \cdot \dfrac{A^2}{2\eta_m f_m}$

$$\frac{3}{2} \cdot \left(\frac{BW_{IF_1}}{2f_m}\right)^2 = \frac{3}{4\pi^2 f_m^2} \cdot \left(\frac{2\pi BW_{IF_2}}{4\cdot 6}\right)^2$$

$$\frac{1}{2} \cdot \frac{BW_{IF_1}^2}{4f_m^2} = \frac{1}{4\pi^2 f_m^2} \cdot \frac{4\pi^2 BW_{IF_2}^2}{(4\cdot 6)^2}$$

$$\frac{BW_{IF_2}^2}{BW_{IF_1}^2} = \frac{4\cdot 6^2}{8}$$

Hence, $\dfrac{BW_{IF_2}}{BW_{IF_1}} \cong 1.63$

• **PROBLEM 14-13**

A frequency modulation (FM) receiver with input signal-to-noise ratio $S_i/\eta f_m$ = 40 dB is used to pick-up signals from different FM stations. If the transmitted FM signal is represented by $km(t) = \beta \cdot \omega_m \cdot \cos(\omega_m t)$, calculate output signal-to-noise ratio for different values of β such as 5, 10 and 105.

Solution: When an input signal frequency modulates a high frequency carrier at the transmitter end and the resulting FM signal is being corrupted by a white Gaussian noise of spectral density $\eta/2$, the output signal-to-noise ratio of a FM receiver is given by

$$\frac{S_o}{N_o} = \frac{3}{4\pi^2} \cdot \frac{k^2 \cdot \overline{m^2(t)}}{f_m^2} \cdot \frac{A^2/2}{\eta \cdot f_m} \tag{1}$$

where $\dfrac{A^2}{2}$ = input signal S_i

and $\overline{m^2(t)}$ = the normalized power of the signal.

Letting $\eta f_m = N_M$ and modifying Eq. (1) results in

$$\frac{S_o}{N_o} = \frac{3}{4\pi^2} \cdot \frac{\beta^2 \cdot \omega_m^2}{2f_m^2} \cdot \frac{S_i}{N_M} \quad \text{because} \quad k^2 \cdot \overline{m^2(t)} = \frac{\beta^2 \omega_m^2}{2}$$

$$= \frac{3}{4\pi^2} \cdot \beta^2 \cdot \frac{4\pi^2 \cdot f_m^2}{2f_m^2} \cdot \frac{S_i}{N_M}$$

$$= \frac{3}{2} \beta^2 \cdot \frac{S_i}{N_M}$$

Hence, expressing this ratio in dB, we get

$$\frac{S_o}{N_o} = 10\log\left(\frac{3}{2}\beta^2\right) + 40 \quad (dB)$$

Note: $\frac{S_o}{N_o} = \frac{3}{2}\beta^2 \frac{S_i}{N_m}$ where $\frac{S_i}{N_m}$ is given as 40 dB.

(a) Now, for $\beta = 5$, the ratio is

$$\frac{S_o}{N_o} = 10\log\left(\frac{3}{2} \times 25\right) + 40$$
$$= 55.7 \text{ dB}$$

(b) $\beta = 10$

$$\frac{S_o}{N_o} = 10\log\left(\frac{3}{2} \times 100\right) + 40$$
$$= 61.7 \text{ dB}$$

(c) $\beta = 105$

$$\frac{S_o}{N_o} = 10\log\left(\frac{3}{2} \times 105^2\right) + 40 = 82.185 \text{ dB}$$

● **PROBLEM 14-14**

Radio FM signals with maximum modulating frequency $f_{m(max)}$ = 7 kHz are transmitted. Knowing that the input signal-to-noise ratio is 3, calculate the output signal-to-noise ratio for the worst case.

Solution: The relation between the phase shift caused by the noise signal and the corresponding frequency deviation can be expressed as

$$\Delta f = \phi \times f_m \tag{1}$$

where Δf = frequency deviation,

ϕ = phase shift in radians,

and f_m = modulating signal frequency.

The worst-case output signal-to-noise ratio occurs when the phase shift ϕ of the resultant is maximum. Considering the figures below, it is obvious that the maximum phase shift of the resultant occurs when the noise and the desired signal are at right angles to each other.

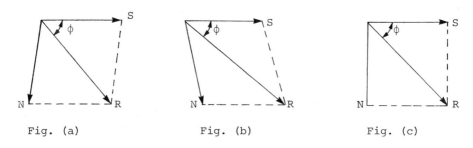

Fig. (a)   Fig. (b)   Fig. (c)

Thus, from Fig. (c) $\tan\phi = \frac{N}{S} = \frac{1}{S/N}$.

Given that S/N for the input signal = 3

$$\tan\phi = \frac{1}{3} = 0.333$$

$$\phi = \tan^{-1} 0.333 = 18.435° = 0.322 \text{ radian}$$

Therefore, from equation (1) the worst-case frequency deviation is $\Delta f_{min} = \phi \times f_m$

$$= 0.322 \times 7 \text{ kHz}$$

$$= 2.254 \text{ kHz}$$

For standard FM broadcasting, the deviation corresponding to full volume (maximum deviation) is 75 kHz. Thus, a 2.254 kHz worst-case noise deviation will result in an output signal-to-noise ratio of

$$\frac{75 \text{ kHz}}{2.254 \text{ kHz}} \cong 33$$

Now, realize that the inherent noise reduction capability of FM can be improved by reducing the maximum modulating frequency or by increasing the maximum allowed frequency deviation from the standard 75 kHz value. However, there is a trade off in increasing the allowable frequency deviation, since an increased bandwidth is required. In practice, many FM systems, utilized as communication links, operate with decreased bandwidths - called the narrow-band FM systems, and is usually operated with a 10 kHz maximum deviation.

• PROBLEM 14-15

A narrow band receiver with frequency modulation is operating at a maximum modulating frequency of 4 kHz. If the maximum frequency deviation is Δf_{max} = 12 kHz and the input signal-to-noise ratio is 5:1, calculate the worst case output signal-to-noise ratio.

Solution: In order to determine the worst case output signal-to-noise ratio, first, we have to recall the relationship between signal-to-noise ratio and frequency deviation, which is, for the worst case,

$$\frac{S}{N} = \frac{\Delta f_{max}}{\Delta f_{min}}$$

Now, consider the illustrations in Fig. (1), it is obvious that the maximum phase shift of the resultant occurs when the noise and the desired signal are at right angles to each other. Hence, the worst case frequency deviation due to noise can be calculated when the phase shift ϕ of the resultant is maximum.

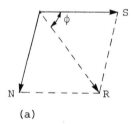
(a)
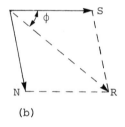
(b)
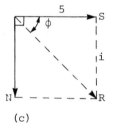
(c)

Thus, $\tan \phi = N/S = \frac{1}{S/N} = 1/5$

$= 0.2$

and $\phi = \tan^{-1} 0.2$

$= 11.31°$

$= 0.197$ radian

The relation between the frequency deviation, and the modulating signal frequency (f_m) is given by

$$\Delta f = \phi \times f_m$$

Therefore, $\Delta f_{min} = 0.197 \times 4 \times 10^3$ Hz $= 0.788$ kHz

The worst-case output signal-to-noise ratio can now be evaluated.

Thus, $\frac{S}{N}$ ratio $= \frac{\Delta f_{max}}{\Delta f_{min}} = \frac{12 \text{ kHz}}{0.788 \text{ kHz}}$

≈ 15

• **PROBLEM 14-16**

A preemphasis circuit of transfer function $|H_p(f)|^2 = k^2 f^2$ is used in an FM system, with the modulating signal having spectral density of $G(f) = \dfrac{G_o}{1 + (f/f_1)^2}$

where $f_1 \ll f_m$. Determine a suitable value of k^2 such that preemphasis does not affect the bandwidth of the channel.

Solution: The normalized power of the modulating signal is given by

$$P = \int_{-f_m}^{f_m} G(f)\, df$$

$$= \int_{-f_m}^{f_m} \dfrac{G_o}{1 + (f/f_1)^2}\, df$$

Now changing variables to $y = f/f_1$ and realizing that

$$\int_{-\infty}^{\infty} dy/(1 + y^2) = \pi$$

Hence,

$$P \cong \int_{-\infty}^{\infty} \dfrac{G_o}{1 + (f/f_1)^2}\, df$$

$$\cong G_o \int_{-\infty}^{\infty} \dfrac{1}{1 + (y)^2}\, f_1\, dy$$

$$\cong G_o \cdot f_1 \cdot \pi \qquad (1)$$

The necessary condition for the K^2 such that preemphasis does not increase the bandwidth is: The normalized power of the modulating signal $m(t)$ must be greater than or equal to the normalized power of the preemphasized signal.

Since the power spectral density of $m(t)$ is $G(f)$, the density of the preemphasized signal will be

$$\int_{-f_m}^{f_m} |H_p(f)|^2\, G(f)\, df$$

Thus, from the above condition we have

$$\int_{-f_m}^{f_m} G(f)\,df \geq \int_{-f_m}^{f_m} |H_p(f)|^2 G(f)\,df$$

$$G_o f_1 \pi \geq \int_{-f_m}^{f_m} \frac{G_o k^2 f^2}{1 + (f/f_1)^2}\,df$$

$$\geq G_o k^2 \int_{-f_m}^{f_m} \frac{f^2}{1 + (f/f_1)^2}\,df$$

Let $u = f/f_1$ then $du = 1/f_1\,df$.

$$G_o f_1 \pi \geq G_o k^2 f_1^3 \int_{-f_m/f_1}^{f_m/f_1} \frac{u^2}{1 + u^2}\,du$$

$$\geq G_o k^2 f_1^3 \int_{-f_m/f_1}^{f_m/f_1} \frac{u^2 + 1 - 1}{1 + u^2}\,du$$

$$\geq G_o k^2 f_1^3 \int_{-f_m/f_1}^{f_m/f_1} \left(1 - \frac{1}{1 + u^2}\right) du$$

$$\geq G_o k^2 f_1^2 \left[\int_{-f_m}^{f_m} df - \int_{-f_m}^{f_m} \frac{1}{1 + (f/f_1)^2}\,df\right]$$

Therefore,

$$G_o f_1 \pi \geq G_o k^2 f_1^2 (2f_m - \pi f_1)$$

Hence, $$k^2 \leq \frac{\pi}{f_1(2f_m - \pi f_1)}$$

• **PROBLEM 14-17**

In an FM communication system preemphasis and deemphasis are used in the transmitter and the receiver systems, respectively. See Fig. (1).

(a) Given the power spectral density $G_m(f)$ of the baseband signal m(t) as follows:

$$G_m(f) = \eta_m/2 \qquad -f_m < f < f_m$$

$$= 0 \qquad \text{elsewhere}$$

Determine the normalized power of the signal $m(t)$.

(b) If the preemphasis filter has transfer characteristic $|H_p(f)|^2 = A_1^2 + A_2^2 f^2$, determine the relationship between A_1 and A_2.

(c) Calculate the output noise (N_o) without deemphasis, and the ratio R of N_o to the noise output with deemphasis (N_{od}), given that

$$G_{n'}(f) = \frac{\alpha^2 \omega^2}{A^2}\eta \quad \text{for } |f| \leq f_{n'},$$

where $f_{n'} > f_m$.

(Note $n'(t)$ is the noise output obtained after the FM demodulator as indicated in Fig. (1)).

Solution:

(a) The normalized power P_m is defined as

$$P_m = \int_{-f_m}^{f_m} G_m(f) df$$

Hence, $$P_m = \int_{-f_m}^{f_m} \eta_m/2 \, df = \eta_m f_m$$

(b) Recall that if $G_m(f)$ is the power spectral density of $m(t)$, the density of $m_p(t)$ after the preemphasis filter is $|H_p(f)|^2 G_m(f)$, and it is required that the normalized power of $m(t)$ must be the same as the normalized power of the preemphasized signal $m_p(t)$.

Hence, $$P_m = \int_{-f_m}^{f_m} G_m(f) df = \int_{-f_m}^{f_m} |H_p(f)|^2 G_m(f) df$$

$$\eta_m f_m = \frac{\eta_m}{2} \int_{-f_m}^{f_m} |H_p(f)|^2 df$$

Thus,
$$2f_m = \int_{-f_m}^{f_m} |H_p(f)|^2 df$$

$$= \int_{-f_m}^{f_m} (A_1^2 + A_2^2 f^2) df$$

$$= 2A_1^2 f_m + \frac{2A_2^2 f_m^3}{3}$$

Therefore, the relationship between A_1 and A_2 is as follows:

$$2f_m = 2A_1^2 f_m + \frac{2A_2^2 f_m^3}{3}$$

$$1 = A_1^2 + \frac{A_2^2 f_m^2}{3}$$

$$A_1^2 = 1 - \frac{A_2^2 f_m^2}{3}$$

(c) The output noise without deemphasis is calculated as follows:

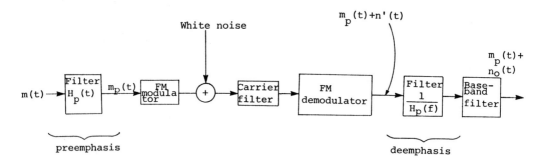

Observe in Fig. (1), since $G_{n'}(f)$ is given and the baseband filter passes frequencies up to f_m Hz,

$$N_o = \int_{-f_m}^{f_m} G_{n'}(f) df = \int_{-f_m}^{f_m} \frac{\alpha^2 \omega^2}{A^2} \eta df$$

$$= \int_{-f_m}^{f_m} \frac{\alpha^2 \eta}{A^2} (2\pi f)^2 df$$

$$= \frac{\alpha^2 \eta}{A^2} \int_{-f_m}^{f_m} 4\pi^2 f^2 \, df$$

$$= \frac{8\pi^2}{3} \frac{\alpha^2 \eta}{A^2} f_m^3$$

Hence,
$$N_{od} = \int_{-f_m}^{f_m} G_{n'}(f) \left|\frac{1}{H_p(f)}\right|^2 df$$

$$= \frac{\alpha^2 \eta 4\pi^2}{A^2} \int_{-f_m}^{f_m} f^2 \left|\frac{1}{H_p(f)}\right|^2 df$$

$$= \frac{\alpha^2 \eta 4\pi^2}{A^2} \, 2 \int_0^{f_m} f^2 \frac{1}{A_1^2 + A_2^2 f^2} df$$

As a result,

$$R = \frac{N_o}{N_{od}} = \frac{\frac{8\pi^2}{3} \frac{\alpha^2 \eta}{A^2} f_m^3}{\frac{\alpha^2 \eta 4\pi^2}{A^2} \, 2 \int_0^{f_m} f^2 \frac{1}{A_1^2 + A_2^2 f^2} df}$$

$$= \frac{f_m^3 / 3}{\int_0^{f_m} \frac{f^2}{A_1^2 + A_2^2 f^2} df}$$

● **PROBLEM 14-18**

In commercial FM broadcasting, preemphasis and deemphasis is commonly employed. In Fig. (1), a DSB-SC system is represented by a block diagram and the corresponding network of a preemphasis filter is shown in Fig. (3). If the baseband signal has spectral density as follows:

$$M(f) = \begin{cases} M_o \dfrac{1}{1 + (f/f_1)^2} & |f| \leq f_m \text{(the spectral range)} \\ 0 & \text{elsewhere} \end{cases} \quad (1)$$

where M_o = low frequencies spectral density

and f_1 = frequency at which M(f) has dropped from its

low-frequency value by 3 dB.

And the power spectral density of the output noise is shown in Fig. (2).

Determine the preemphasis improvement

$$R = \frac{N_o \text{ (preemphasis output noise power)}}{N_{od} \text{ (deemphasis output noise power)}}$$

if the preemphasis network is adjusted such that

f_2 (3 dB frequency of the network) $= \frac{1}{2\pi RC} = f_1$ and

that preemphasis does not affect the transmitted power.

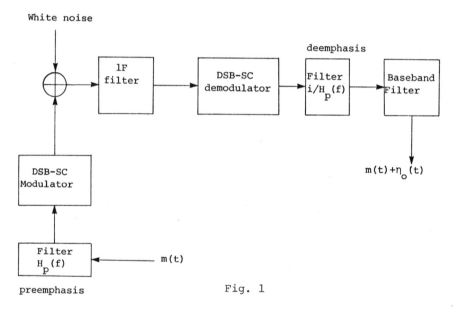

Fig. 1

Solution: Since the power spectral density of the output noise is given, the total output noise power N_o is the area under the plot in Fig. (2).

Therefore, $N_o = \frac{\eta}{4} \times 2f_m = \frac{\eta f_m}{2}$

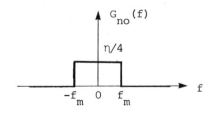

Fig. 2

The preemphasis improvement R is defined as

$$R = \frac{N_o}{N_{od}}$$

Notice that the output voltage $V_o(f) = rI(f)$ for the network in Fig.(3).

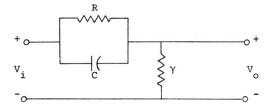

Fig. 3

Hence, the transfer function,

$$H_p(f) = \frac{V_o(f)}{I(f)} = \frac{r}{R}(1 + j\omega CR) = \frac{r}{R}\left(1 + j\frac{f}{f_2}\right) \quad (2)$$

where $\frac{r}{R} = k =$ product of the amplifier gain.

Now, in the present case, the noise output with deemphasis (N_{od}) can be expressed as

$$N_{od} = \int_{-f_m}^{f_m} \frac{\eta}{4} \frac{1}{|H_p(f)|^2} df$$

Thus,
$$R = \frac{\eta f_m/2}{\int_{-f_m}^{f_m} \eta/4 \cdot \frac{1}{|H_p(f)|^2} df}$$

$$= 2f_m \left[\int_{-f_m}^{f_m} \frac{df}{k^2 \left[1 + (f/f_2)^2\right]}\right]^{-1}$$

$$= 2f_m \left[\frac{2f_1}{k^2} \tan^{-1}\left(\frac{f_m}{f_2}\right)\right]^{-1}$$

$$= \frac{f_m k^2}{f_2 \tan^{-1}(f_m/f_2)} \quad (2)$$

Since $f_2 = f_1$, k can be determined as follows:

Using Eq. (1) with $f_1 = f_2$ and Eq. (2),

P = the normalized signal power

$$= \int_{-f_m}^{f_m} G(f)df = \int_{-f_m}^{f_m} |H_p(f)|^2 G(f)df$$

$$= \int_{-f_m}^{f_m} \frac{M_o \, df}{1 + (f/f_2)^2} = \int_{-f_m}^{f_m} k^2 M_o \, df$$

Therefore, $\quad 2M_o \int_0^{f_m} \frac{df}{1 + (f/f_2)^2} = 2k^2 M_o \int_0^{f_m} df$

$$2f_2 \tan^{-1}(f_m/f_2) = 2k^2 f_m$$

Thus, $\quad k^2 = \frac{f_2}{f_m} \tan^{-1}(f_m/f_2)$

Substituting the value for k^2 into Eq. (2),

$$R = \frac{f_2 \tan^{-1}(f_m/f_2)}{f_2 \tan^{-1}(f_m/f_2)} = 1$$

Thus, the advantage of using preemphasis is less pronounced in an AM system.

NOISE IN PCM AND DM

• **PROBLEM 14-19**

One method for one thousand signals multiplexing prior to long distance transmission is shown in Fig. (1). Each baseband signal occupies a bandwidth of 6 kHz and has lowest SSB carrier frequency = 10 kHz. Assume that the angle modulator is a frequency modulator and the normalized power for each signal is equal, the channel with the least noise turns out to have a signal-to-noise ratio of 80 dB. (a) If the signal-to-noise ratio lower than 20 dB were not acceptable, how many channels of the system would be usable.

(b) If instead, the angle modulator is a phase modulator (assuming that the total power transmitted being the same as (a)), calculate the signal-to-noise ratio in a channel.

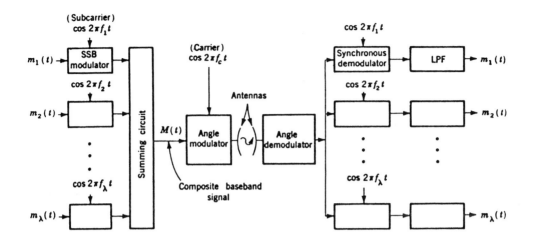

Solution:

(a) For the given system, the output noise power can be expressed as

$$N_o = \int_{-f_m}^{f_m} G_m(f)df = \frac{\alpha^2 \eta}{A^2} \int_{-f_m}^{f_m} 4\pi^2 f^2 df$$

$$= \frac{8\pi^2}{3} \frac{\alpha^2 \eta}{A^2} f_m^3$$

where $G_m(f) = \frac{\alpha^2 \omega^2 \eta}{A^2}$, $|f| \le \beta/2$

Thus, the output noise power of the nth channel is

$$N_{o,n} = \frac{2\alpha^2 \eta}{A^2} \int_{f_o + (n-1)f_m}^{f_o + nf_m} 4\pi^2 f^2 df$$

$$= \frac{8\pi^2 \alpha^2 \eta}{3A^2} \left[(f_o + nf_m)^3 - (f_o + (n-1)f_m)^3 \right]$$

where $f_o = 10 \times 10^3$ Hz and $f_m = 6 \times 10^3$ Hz

The output signal power $S_o = \alpha^2 \cdot k^2 \cdot \overline{m^2(t)}$

Thus, the output signal-to-noise ratio will be

$$\left(\frac{S_o}{N_o}\right)_{\text{channel } n} = \frac{3A^2 \cdot k^2 \cdot \overline{m^2(t)}}{8\pi^2 \eta} \cdot \frac{1}{(f_o + nf_m)^3 - (f_o + (n-1)f_m)^3}$$

$$= \left(\frac{S_o}{N_o}\right)_{\text{channel } 1} \times \frac{(f_o + f_m)^3 - f_o^3}{(f_o + nf_m)^3 - (f_o + (n-1)f_m)^3}$$

$$= (80 \text{ dB}) \times \frac{(16 \times 10^3 \text{ Hz})^3 - (10 \times 10^3 \text{ Hz})^3}{[(10 + 6n) \times 10^3]^3 - [(4+6n)10^3]^3}$$

$$= 10^8 \times \frac{(16)^3 - (10)^3}{(10 + 6n)^3 - (4 + 6n)^3}$$

$$\left(\frac{S_o}{N_o}\right)_{n=1000} = 476$$

$$= 26.7 \text{ dB} \quad (\because 10 \log 476 = P \text{ dB}$$
$$\therefore P = 26.7)$$

Now, notice that the signal-to-noise ratio of all one thousand signals combined together is greater than the prescribed minimum value of 20 dB. Hence, all one thousand channels of the system are usable.

(b) The integrator network in the phase-modulation system shown in Fig. (2b) is essentially a deemphasis circuit which includes a deemphasis filter with a transfer function $H_d(f)$.

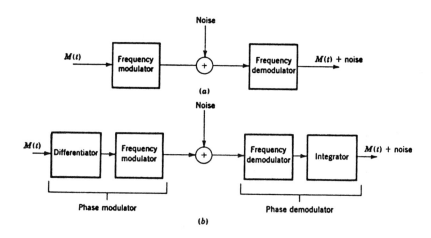

Fig. 2 Comparison of an FM system (a) with a phase-modulation system in (b).

The noise at the output of the demodulator is given by

$$N_{o,\text{top}} = \frac{2\alpha^2 \eta}{A^2} \int_{(n-1)f_m}^{nf_m} 4\pi^2 f^2 df$$

$$\simeq \frac{8\pi^2 \alpha^2 \eta n^2 f_m^3}{A^2}$$

where η = spectral density of the noise power

f_m = frequency range of an individual component signal

In the presence of a deemphasis filter, the output noise power of the top channel is

$$N_{od,\,top} = \frac{2\alpha^2\eta}{A^2} \int_{(n-1)f_m}^{nf_m} \frac{4\pi^2 f^2}{|H_p(f)|^2}\, df$$

Thus, the signal-to-noise ratio will be

$$\frac{S_o}{N_{od,\,top}} = \frac{\alpha^2 k^2 \overline{m^2(t)}}{\frac{2\alpha^2\eta}{A^2} \int_{(n-1)f_m}^{nf_m} \frac{4\pi^2 f^2 df}{|H_p(f)|^2}}$$

where $|H_d(f)|^2 = \dfrac{1}{|H_p(f)|^2} = \dfrac{1}{4\pi^2\tau^2 f^2}$

Therefore, $\dfrac{S_o}{N_o} = \dfrac{A^2 \cdot k^2 \cdot \overline{m^2(t)} \cdot \tau^2}{2\eta \cdot f_m}$

$= \dfrac{3A^2 k^2 \overline{m^2(t)}}{8\pi^2 \eta} \cdot \dfrac{1}{f_m^3 n^2}$ where n = number of signals being multiplexed.

$= \left(\dfrac{S_o}{N_o}\right)_{\text{channel 1 of (a)}} \times \dfrac{(f_o + f_m)^3 - f_o^3}{f_m^3 n^2}$

$= 10^8 \times \dfrac{(10 \times 10^3 + 6 \times 10^3)^3 - (10 \times 10^3)^3}{(6 \times 10^3)^3 (1000)^2}$

$= 10^8 \times \dfrac{(16000)^3 - (10000)^3}{(6000)^3 (1000)^2}$

≈ 1433

≈ 31.6 dB

● **PROBLEM 14-20**

If in pulse-code modulation, the sampling of the quantized waveform is flat-topped instead of instantaneous and the flat-topped sampling pulses have amplitude I/τ where τ = pulse duration, determine the power spectral density $G_{e_s}(f)$ of the sampled quantization error given that

$$G_{e_s}(f) = \frac{I^2}{T_s}\, \overline{e^2(kT_s)}.$$

Solution: The sampled quantization error waveform is given by

$$e_s(t) = e(t) \, I \sum_{k=-\infty}^{\infty} \delta(t - kT_s).$$

The Fourier transform of the pulse of an amplitude I/τ and a duration τ is

$$H(f) = I \frac{\sin \pi f \tau}{\pi f \tau}$$

Since the input of the flat-top system is $\sum_{k=-\infty}^{\infty} e(kT_s)\delta(t - kT_s)$, the power spectral density of the sampled quantization error is given by

$$G_{e_s}(f) = |H(f)|^2 \frac{\overline{e^2(kT_s)}}{T_s} = \frac{T^2 \, \overline{e^2(kT_s)}}{T_s} \left(\frac{\sin \pi f \tau}{\pi f \tau}\right)^2$$

● PROBLEM 14-21

A modulating signal m(t) is being bandlimited and then sampled. Due to this process, distortion occurs. Given that $G_m(f)$ = the power spectral density of m(t) and f_m = cut off frequency of the bandlimiting filter.

(a) Write an expression for the noise N_D caused by distortion in terms of $G_m(f)$.

(b) Given that $G_m(f) = G_o \, e^{-|f/f_1|}$, determine N_D.

If the signal m(t) is undergone the process of sampling and quantization at Nyquist rate, determine the output signal-to-noise ratio, i.e., $SNR_{output} = S_o/(N_D + N_q)$, where N_q is the quantization noise and S_o is the output signal power.

$\sum_{k=-\infty}^{\infty} I\delta(t-kT_s) \quad T_s = \frac{1}{2/M}$

Sampling pulses

$m_q u(t) = m_z(t) + e_z(t)$ White noise, n(t)

Encoder: Sampler (multiplier) → Analog-to-digital converter → ⊕ → Matched filter detector → Digital-to-analog converter

$m_q(t) = m(t) + e(t)$

Quantizer

m(t)

(rectangular filter with cutoff f_M) → Low-pass filter → $\tilde{m}_{qs}(t)$: Decoder

$\tilde{m}(t) = m_o(t) + n_q(t) + n_o(t)$

Solution:

(a) The noise N_D due to distortion can be evaluated as

$$N_D = \int_{-\infty}^{\infty} G_m(f)df - \int_{|f| \leq f_m} G_m(f)df$$

$$= \int_{|f| > f_m} G_m(f)df$$

$$= \int_{-\infty}^{-f_m} G_m(f)df + \int_{f_m}^{\infty} G_m(f)df$$

$$= 2 \int_{f_m}^{\infty} G_m(f)df \qquad (1)$$

(b) If, however, $G_m(f) = G_o e^{-|f/f_1|}$, then substituting for $G_m(f)$ into equation (1). That is

$$N_D = 2 \int_{f_m}^{\infty} G_o e^{-|f/f_1|} df$$

$$= 2 G_o f_1 e^{-f_m/f_1}$$

In order to calculate the output SNR, first we have to determine S_o and N_q as follows:

Consider the binary PCM encoder-decoder in Fig. (1), the signal at the output of the sampler is

$$m_{qs}(t) = m_s(t) + e_s(t) \qquad (2)$$

where $m_s(t) = m(t) I \sum_{k=-\infty}^{\infty} \delta(t - kT_s) \qquad (3)$

represents the signal component, and

$$e_s(t) = e(t) I \sum_{k=-\infty}^{\infty} \delta(t - kT_s) \qquad (4)$$

represents the sampled error function.

Considering the sampling pulses to be separated by a time T_s and having a strength I, the d·c component of the impulse train of equation (3) is I/T_s and the output of the baseband filter is (considering only the signal component)

$$m_o(t) = I/T_s \cdot m(t)$$

and the normalized signal output power is

$$\overline{m_o^2(t)} = \frac{I^2}{T_s^2} \overline{m^2(t)} \tag{5}$$

Now express $\overline{m^2(t)}$ in terms of the number M of quantization levels and the step size S. Assuming that the instantaneous value of $m(t)$ may fall anywhere in its allowable range with equal likelihood, the probability density of the instantaneous value of m is
$$f(m) = \frac{1}{MS}$$

Note that MS is the peak-to-peak range of the signal.

The variance $\overline{m^2(t)}$ of $m(t)$ is thus,

$$\overline{m^2(t)} = \int_{-MS/2}^{MS/2} m^2 f(m) \, dm$$

$$= \int_{-MS/2}^{MS/2} m^2 \cdot \frac{1}{MS} \, dm$$

$$= \frac{M^2 S^2}{12}$$

Hence, the output signal power

$$S_o = \overline{m_o^2(t)} = \frac{I^2}{T_s^2} \left(\frac{M^2 S^2}{12} \right) \tag{6}$$

Now, for the quantization noise N_q,

$$N_q = \int_{-f_m}^{f_m} G_{e_s}(f) \, df$$

where $G_{e_s}(f)$ is the power spectral density of the sampled quantization error, which is

$$G_{e_s}(f) = \frac{I^2}{T_s} \overline{e^2(t)}$$

$$= \frac{I^2}{T_s} \cdot \frac{S^2}{12}$$

Note that $\overline{e^2(t)} = \frac{S^2}{12}$ is the quantization error and generally for a periodic waveform

748

$$G_n(f) = \frac{1}{T_s} \left| \int_{-\infty}^{\infty} p(t) e^{-j2\pi ft} dt \right|^2$$

where p(t) consists of a single pulse.

Hence, if p(t) is an impulse of strength I, $G_n(f) = \frac{I^2}{T_s}$ $-\infty < f < +\infty$.

Therefore, $N_q = \frac{I^2}{T_s} \cdot \frac{S^2}{12} \cdot 2f_m = \frac{I^2}{T_s^2} \left(\frac{S^2}{12}\right)$ (7)

(Note at Nyquist rate $2f_m = 1/T_s$)

Finally, substituting S_o, N_D and N_q into the equation for output SNR,

$$\text{SNR}_{\text{output}} = \frac{I^2 M^2 S^2 / 12 \, T_s^2}{(I^2/T_s^2) \, 2 \, G_o f_1 \, e^{-f_m/f_1} + (I^2/T_s^2)(S^2/12)}$$

$$= \frac{M^2}{1 + (24 \, G_o f_1 / 3^2) e^{-f_m/f_1}}$$

● **PROBLEM 14-22**

The error due to thermal noise in a communication system is represented by a pulse train of amplitude Δm_s. Let $p(t) = I\Delta m_s/\tau$ be the pulse slope, $t \in [0, \tau]$ where $\tau = 1/2f_m$, calculate the power spectral density of the thermal noise, $G_{th}(f)$ and the output noise power due to thermal noise error, N_{th}. Assuming that the duration of each error pulse is $\frac{1}{2f_m}$ when the transmitted signal is sampled at the Nyquist rate.

Given: $\overline{(\Delta m_s)^2} = \left(\frac{2^{2N-1}}{3N}\right) S^2$ and $T = T_s/NP_e$ is the threshold equation $\approx 1/[(16)2^{2N}]$.

Solution: The power spectral density defined in terms of pulse slope as:

$$G(f) = \frac{1}{T_s} \left| \int_{-\infty}^{\infty} p(t) e^{(-j\pi ft)} dt \right|^2$$

$$= \frac{1}{T_s} \overline{|P(f)|^2} \qquad -\infty < f < \infty$$

Hence,
$$G_{th}(f) = \frac{1}{T} \cdot \overline{\left[\int_0^\tau \frac{I \Delta m_s}{\tau} e^{(-j\pi ft)} dt \right]^2}$$

$$= \frac{I^2 \overline{(\Delta m_s)^2}}{T} \left(\frac{\sin \pi f \tau}{\pi f \tau} \right)^2$$

Now, note that $\overline{(\Delta m_s)^2} = (2^{(2N-1)}/3N) \cdot S^2 \simeq (2^{(2N)}/3N)S^2$ and $T = T_s/NP_e$ are given. Substituting into Eq. (1),

$$G_{th}(f) = \frac{I^2 \left[2^{2N}/(3N) \right] S^2}{T_s/(N\, P_e)} \cdot \left[\frac{\sin (\pi f \tau)}{\pi f \tau} \right]^2$$

$$= \frac{2^{(2N)} \cdot S^2 \cdot P_e \cdot I^2}{3\, T_s} \cdot \left[\frac{\sin (\pi f \tau)}{\pi f \tau} \right]^2$$

Now, notice that $N_{th} = \int_{-f_m}^{f_m} G_{th}(f) \cdot df$.

Therefore,
$$N_{th} = \int_{-f_m}^{f_m} \frac{2^{(2N)} \cdot S^2 \cdot P_e \cdot I^2}{3\, T_s} \left(\frac{\sin \pi f \tau}{\pi f \tau} \right)^2 df$$

$$= \frac{2^{(2N)} \cdot S^2 \cdot P_e \cdot I^2}{3\, T_s} \int_{-f_m}^{f_m} \left(\frac{\sin \pi f \tau}{\pi f \tau} \right)^2 df$$

$$\simeq 0.756 \left[\frac{2^{(2N)} S^2 P_e I^2}{3\, T_s} \right]$$

• **PROBLEM 14-23**

A 4 kHz audio signal is transmitted using PCM technique. If the system operation is to be just above the threshold and the output signal-to-noise ratio is equal to 47 dB, find N, the number of binary digits needed to assign individual binary code designation to the M quantization level. i.e., $M = 2^N$.

Given that S_o = output signal power = $\dfrac{I^2}{T_s^2} \cdot \dfrac{M^2 S^2}{12}$

N_q = quantization noise power = $\frac{I^2}{T_s} \cdot \frac{S^2}{12}(2f_m)$

N_{th} = thermal noise power = $\left(\frac{I^2}{T_s^2}\right)\left(\frac{P_e \, 2^{2N} S^2}{3}\right)$

where P_e = error probability and

$$\left(2^{2N+2} P_e\right) = 0.26 \qquad (1)$$

(Note: Signal is sampled at $2f_s$ where f_s is the Nyquist rate).

Solution: The signal-to-noise ratio for phase-shift keying can be expressed as

$$\left(\frac{S_o}{N_o}\right)_{psk} = \frac{S_o}{N_q + N_{th}}$$

$$= \frac{\dfrac{I^2}{T_s} \cdot \dfrac{M^2 S^2}{12}}{\dfrac{I^2}{T_s} \dfrac{S^2}{(1)^2}(2f_m) + \left(\dfrac{I^2}{T_s^2}\right)\left(\dfrac{P_e \cdot 2^{2N} \cdot S^2}{3}\right)}$$

$$= \frac{M^2/12}{T_s \cdot 2f_m/12 + P_e\left(2^{2N}/3\right)}$$

$$= \frac{(2^N)^2}{2 T_s f_m + 4 P_e \, 2^{(2N)}}$$

$$= \frac{2^{(2N)}}{2 T_s f_m + 4 P_e \, 2^{(2N)}} \qquad (2)$$

Now, notice that $T_s = \dfrac{1}{f_s}$ since in this case, the signal is sampled at twice the Nyquist rate, $T_s = \dfrac{1}{2(2f_m)} = \dfrac{1}{4f_m}$.

Therefore, substituting $T_s = \dfrac{1}{4f_m}$ into Eq. (2),

$$\left(\frac{S_o}{N_o}\right)_{psk} = \frac{2^{(2N)}}{2f_m(1/4f_m) + 4 P_e \, 2^{(2N)}}$$

$$= \frac{2^{(2N)}}{(1/2) + 4 P_e \, 2^{(2N)}}$$

$$= \frac{2^{(2N+1)}}{1 + 2^{(2N+2)} \cdot P_e} \qquad (3)$$

Now, equating Eq. (3) to obtain N and knowing that

$$\left(\frac{S_o}{N_o}\right)_{psk} = 47 \text{ dB} = 5 \times 10^4.$$

Thus,

$$\frac{2^{2N+1}}{1 + 0.26} = 5 \times 10^4$$

$$2^{2N} = (5 \times 10^4)(1.26)$$

$$2^{2N} = 63000$$

$$N \simeq 8$$

• **PROBLEM** 14-24

Two audio signals, each occupying 3.9 kHz bandwidth, are transmitted using a pulse-code modulation technique over a time-division multiplexed channel. If the signals are sampled at the Nyquist rate and the output SNR $\geq$ 30 dB, with thermal-noise effects included, calculate:

(a) N, the number of binary digits required to keep the system operating above the threshold if

$$2^{(2N+1)} \cdot P_e = 0.26 \text{ where } P_e = \text{erfc}\sqrt{\left(\frac{1}{4N}\right) \frac{S_i}{nf_m}}$$

(b) Sketch the system and find the speed of the commutator.

Solution:

a) Recall that the SNR for phase-shift-keying is

$$\left(\frac{S_o}{N_o}\right)_{psk} = \frac{S_o}{N_q + N_{th}} = \frac{2^{(2N)}}{2 T_s f_m + 4 P_e \cdot 2^{(2N)}}$$

In this case, $T_s = \frac{1}{f_s} = \frac{1}{2f_m}$, hence,

$$\left(\frac{S_o}{N_o}\right)_{psk} = \frac{2^{(2N)}}{2f_m (1/2f_m) + 4 P_e \cdot 2^{(2N)}}$$

$$= \frac{2^{(2N)}}{1 + 4 P_e \cdot 2^{(2N)}}$$

$$= \frac{2^{(2N)}}{1 + 2^{(2N + 1)} \cdot P_e} \qquad (1)$$

Now, let $(S_o/N_o)_{psk} = 30$ dB $= 10^3$

Hence, $\left(\dfrac{S_o}{N_o}\right)_{psk} = \dfrac{2^{(2N)}}{1 + 0.26} = 10^3 \qquad (2)$

Solving this equation, the approximate value of N is found to be equal to 5.

Now, using Eqs. (1) and (2), with $P_e = \operatorname{erfc} \sqrt{(1/4N)(S_i/\eta f_m)}$, and $f_m = 3.9$ kHz.

$S_i/\eta \simeq 5.8 \times 10^5$ Hz

b) The commutator speed is equal to $2f_m = 2 \times 3.9$ kHz
$= 7.8$ krps.

and the transmitter-receiver system is sketched below.

• **PROBLEM 14-25**

For the delta-modulation system shown in figure (1),

(a) Determine the power of the quantization noise at the output of the baseband filter having a bandwidth f_m as shown. Assume that S = the step size.

(b) Given $s(t) = At$, where $A = s/\tau$, sketch $\Delta(t)$, the difference signal.

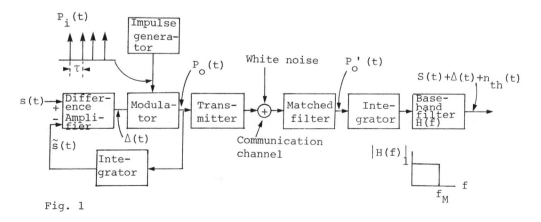

Fig. 1

Solution:

(a) Observe, in Fig. (1), $\Delta(t) = s(t) - \tilde{s}(t)$

where $s(t)$ is the input signal and $\tilde{s}(t)$ is the integrator output.

The output of the modulator $P_o(t)$, a pulse train, produced by applying $p_i(t)$ to the modulator, which polarity is depended on $\Delta(t)$. The waveform $p_o(t)$ is then fed back to an integrator, the output of which is designated $\tilde{s}(t)$. In general $\tilde{s}(t)$ is an approximation of the input signal $s(t)$. The signal $s(t)$ and $\tilde{s}(t)$ are then compared in a difference amplifier. The amplifier output is thus,

$$\Delta(t) = m(t) - \tilde{m}(t), \quad \text{as plotted in Fig. (2).}$$

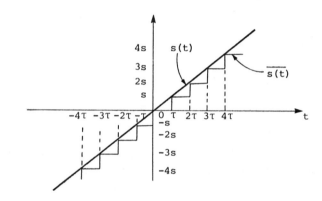

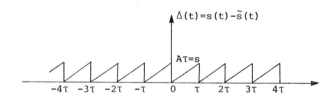

(b) Since the fundamental frequency of $\Delta(t)$ is $1/\tau$, which is much larger than the baseband filter bandwidth f_m, the baseband filter output is zero.

• **PROBLEM 14-26**

A decision has to be made whether a baseband signal should be modulated using delta modulation or pulse code modulation technique. If the signal-to-noise ratio, S_o/N_q where N_q = quantization noise and S_o = output signal power, during transmission for both modulations should be at least 30 dB, determine the bandwidths ratio for PCM to DM (Hint: $BW_{PCM}/BW_{DM} = \frac{f_m N}{(1/2\tau)}$ where N is the number of binary digits needed to assign individual binary code designation to the M quantization levels.

Solution: Recall for delta modulation, the output signal power $S_o = \frac{S^2}{2\omega_m^2 \tau^2}$ and the quantization noise $N_q = \frac{S^2 \tau f_m}{3}$, where S = the step size and $\omega_m = 2\pi f_m$ = angular modulating frequency.

Hence, $\left(\frac{S_o}{N_q}\right)_{DM} = \frac{S^2 / \left[2(2\pi f_m)^2 \tau^2\right]}{S^2 \tau f_m / 3}$

$$= \frac{S^2}{8\pi \cdot f_m^2 \tau^2} \cdot \frac{3}{S^2 \cdot \tau \cdot f_m}$$

$$= \frac{3}{8\pi^2 \cdot f_m^3 \cdot \tau^3}$$

Now, since $\left(\frac{S_o}{N_q}\right)_{DM}$ — 30 dB is given

Thus, setting $\frac{3}{8\pi^2 \cdot (f_m \tau)^3} = 10^3$

and solve for $f_m \tau$

Hence, $f_m \cdot \tau = \sqrt[3]{\frac{3}{8\pi^2 \cdot (10)^3}}$

Now, for pulse code modulation, recall: $\left(\frac{S_o}{N_q}\right)_{PCM} = 2^{(2N)} = 10^3$

Hence, solving for 2N, $2N = \frac{3}{\log_{10} 2}$

Therefore, the bandwidths ratio of $\frac{BW_{PCM}}{BW_{DM}} = f_m \tau 2N$

$$= \sqrt[3]{\frac{3}{8\pi^2 10^3}} \cdot \frac{3}{\log_{10} 2}$$

(Note: For a fix channel bandwidth ratio, the performance of DM is always less than PCM.)

● **PROBLEM 14-27**

Figure below shows the schematic of a delta-pulse-code-modulation (DPCM) system.

(a) Given m(t) = At, what would be the maximum value of A in order to avoid slope overloading?

(b) In the transmitted PCM waveform, τ', the bit duration is equal to $\frac{1}{2f_m N}$ where N is the number of bits/code word and $\frac{1}{2f_m}$ is the Nyquist rate, determine τ'/τ.

Assume $S/\tau = A_{max}$ where S is the step size and τ is the interval between samples held by a voltage level corresponding to a single bit.

(c) If $S/\tau = 2A_{max}$, sketch the waveforms $\tilde{m}(t)$ and $\Delta(t)$, where $\Delta(t) = m(t) - \tilde{m}(t)$.

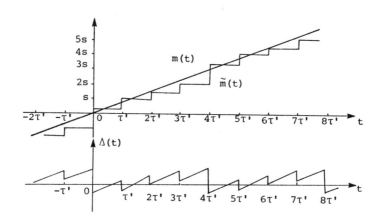

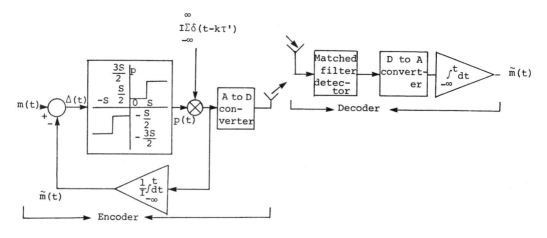

Solution:

In order to avoid slope overloading,

$$\frac{d\, m(t)}{dt} \cdot \tau' \leq \frac{3S}{2}$$

Hence, $\quad A \cdot \tau' \leq \dfrac{3S}{2}$

$$A_{max} = \dfrac{3S}{2\tau'}$$

(Note τ' is defined in question (b).)

(b) Since τ is equal to S/A_{max}, then

$$\tau'/\tau = \tau'/\left(S/A_{max}\right) = \dfrac{\tau'}{S} \cdot \dfrac{3S}{2\tau'} = 1.5$$

(c) Notice that if we have a bandlimited communication channel, the occurrence of intersymbol interference in DM and PCM systems will depend on the values $1/\tau$ and $1/\tau'$ to the channel bandwidth, respectively. Hence, for a fixed channel, $\tau'/\tau = 1$ is required if intersymbol interference in the cases of DM or PCM is to be the same.

Now, $\quad m(t) = A_{max} \cdot t = \left(\dfrac{S}{2\tau}\right)t = \left(\dfrac{0.5S}{\tau}\right)t$

The sketches for $\tilde{m}(t)$ and $\Delta(t)$ are in the figures below.

CHAPTER 15

COMMUNICATION SYSTEMS AND NOISE CALCULATIONS

TUNED CIRCUITS, AMPLIFIER, AND OSCILLATOR IN COMMUNICATION SYSTEMS

● **PROBLEM 15-1**

Assuming no resistive component, calculate the following for the circuit shown below.

(a) circuit impedance Z_0 at resonant frequency f_0.

(b) inductive reactance X_L and capacitive reactance X_C at resonance.

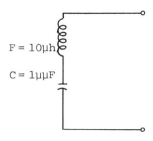

Solution: (a) To calculate the impedance Z_0 at the resonant frequency f_0, it is first necessary to find the resonant frequency.

For a series L-C circuit, the resonant frequency is

$$f_0 = \frac{1}{2\pi\sqrt{LC}} = \frac{1}{2\pi\sqrt{(10 \times 10^{-6})(1 \times 10^{-12})}}$$

$$= 50.33 \text{ MHz}$$

Since the impedance of a pure L-C series circuit is zero at the resonant frequency f_0, it acts like a short circuit,

i.e., $Z_0 = 0\,\Omega$

(b) Inductive and capacitive reactances are calculated as follows:

$$X_L = \omega L = 2\pi f L$$

and

$$X_C = \frac{1}{2\pi f C}$$

Hence, substituting values for $f = f_0$, $L = 10 \times 10^{-6}$ H and $C = 1 \times 10^{-12}$ F, at resonance,

$$X_L = 2\pi(50.33 \times 10^6)(10 \times 10^{-6})\,\Omega$$

$$= 3162.32\,\Omega$$

$$X_C = \frac{1}{2\pi(50.33 \times 10^6)(1 \times 10^{-12})}\,\Omega$$

$$= 3162.32\,\Omega$$

● **PROBLEM** 15-2

A series R-L-C circuit with $R = 100\,\Omega$, $C = 0.005\,\mu F$ is resonating at a frequency of 0.356 MHz when powered by an A.C. source of voltage V_0. Compute (a) inductive reactance and capacitive reactance at the resonant frequency (b) voltages across different components in the circuit at resonance.

Solution: Since the value of inductance is not specified, it can be computed, using the standard formula for the R-L-C circuit resonant frequency. The resonant frequency is given by

$$f_0 = \frac{1}{2\pi\sqrt{LC}}$$

Hence,

$$L = \frac{1}{(2\pi \cdot f_0)^2 C}$$

$$= \frac{1}{[2\pi \times (0.356 \times 10^6)]^2 (0.005 \times 10^{-6})}$$

$$= 40\,\mu H$$

(a) The inductive and capacitive reactances at resonance are equal i.e., $$X_{LO} = X_{CO}$$

Now, X_L at resonant frequency is

$$X_{LO} = 2\pi \cdot (356 \times 10^3)(40 \times 10^{-6})$$
$$= 89.43 \Omega$$

Hence, $$X_{CO} = 89.43 \Omega$$

(b) The total impedance at the resonant frequency is equal to the resistance.

i.e., $$Z_0 = R$$

Hence $$Z_0 = 100 \Omega$$

Using Ohm's law, the current flowing in the circuit can be determined as follows,

$$I_0 = \frac{V_0}{Z_0}$$
$$= \frac{V_0}{100} \text{ amps}$$

Assuming voltage $V_0 = 100V$, then $I_0 = 1$ amp. Hence,

V_{RO} = voltage across resistor at the resonant frequency f_0

$$= I_0 \cdot R$$
$$= 1 \cdot (100)$$
$$= 100V$$

$$V_{LO} = I_0 \cdot X_{LO} = V_{CO}$$
$$= 1 \cdot (89.43)$$
$$= 89.43V$$

● **PROBLEM 15-3**

Calculate the following for an R-L-C circuit shown below.

(a) Resonant frequency f_0 of the circuit,

(b) Figure of merit Q of the circuit before and after installing the external resistor,

(c) Bandwidth of the circuit before installing the external resistor, and

(d) Impedance of the circuit at resonance before an external resistor is added.

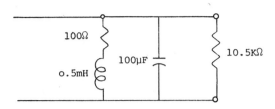

Solution: (a) Resonant frequency is calculated as follows.

$$f_0 = \frac{1}{2\pi\sqrt{LC}}$$

$$= \frac{1}{2\pi\sqrt{(0.5\times 10^{-3})(100\times 10^{-6})}}$$

$$= 711.77 \text{ Hz}$$

(b) The figure of merit before and after installing the external load can be calculated from the standard formula:

$$Q_1 = Q_{before} = \frac{X_L}{R}$$

$$= \frac{2\pi \cdot f_0 \cdot L}{R}$$

$$= \frac{2\pi(711.77)(0.5\times 10^{-3})}{100}$$

$$= 0.022.$$

$$Q_{after} = \frac{Z_T}{X_L}$$

$$= \frac{Z_T}{2\pi \cdot f_0 \cdot L} \quad \frac{Z_T}{2\pi(711.77)(0.5\times 10^{-3})}$$

Now, Z_T is the total impedance presented by the circuit with the load, and it is given by

$$Z_T = \frac{Z_0 \cdot R_X}{Z_0 + R_X} = \text{impedance of circuit after the external resistance is installed.}$$

$$= \frac{Q_1 X_L \cdot R_X}{Q_1 X_L + R_X}$$

$$= \frac{0.022(2\pi \times 711.77 \times 0.5\times 10^{-3})(10.5\times 10^3)}{0.049 + (10.5\times 10^3)}$$

$$= \frac{516.53}{10500.05}$$

$$= 0.05\Omega$$

Hence,
$$Q_{after} = \frac{0.05}{2\pi(711.77)(0.5 \times 10^{-3})}$$

$$= 0.022$$

(c) The BW of the circuit before installing the external resistance is,

$$BW = \frac{f_0}{Q_1}$$

$$= \frac{711.77}{0.022}$$

$$= 32.35 \text{ KHz}$$

(d) Impedance of the circuit at resonance is

$$Z_0 = Q_1 \cdot X_L$$

$$= 0.022 \times (2\pi \times 711.77 \times 0.5 \times 10^{-3})$$

$$= 0.049 \Omega$$

● **PROBLEM 15-4**

An amplifier in a common-emitter configuration as shown below is driving an external load (R_L) of value 1.7KΩ. This radio-frequency amplifier is tuned to a frequency $f = 1.42$ MHz by a tank circuit. Calculate (a) Q of the coil, (b) total impedance of output circuit, and (c) gain at 1.42 MHz, assuming following data

$$\beta = 60, \quad R_{in} = 450 \Omega$$

$$C = 200 \text{ pf} \quad R \text{ of coil} = 20 \Omega$$

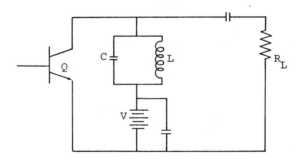

Solution: (a) Q of coil $= \frac{X_L}{R}$

$$= \frac{2\pi \cdot f \cdot L}{R_{(coil)}}$$

The value of L can be evaluated from the relation:

$$f = \frac{1}{2\pi\sqrt{LC}}$$

(Note $f = 1.42$ MHz is the resonant frequency)
Thus,

$$L = \frac{1}{4\pi^2 \cdot f_0^2 \cdot C} = \frac{1}{4\cdot\pi^2 \cdot (1.42 \times 10^6)^2 \cdot 200 \times 10^{-12}}$$

$$= 6.28 \times 10^{-5} \text{ H.}$$

Hence,
$$Q_C = \frac{2\pi \times (1.42 \times 10^6) \times (6.28 \times 10^{-5})}{20}$$

$$= 28.01$$

(b) Total impedance of the output circuit is

$$Z = Z_0 \parallel R_L \parallel R_{out}$$

Now assuming the output impedance of the transistor to be very high, we can approximate Z as:

$$Z = Z_0 \parallel R_L$$

$$= Q_1 \cdot X_L \parallel R_L$$

Hence, $\frac{1}{Z} = \frac{1}{(28.01)(2\pi \times 1.42 \times 10^6 \times 6.28 \times 10^{-5})} + \frac{1}{1700}$

$$= (6.37175 \times 10^{-5}) + (5.88 \times 10^{-4})$$

Therefore, $Z = 1.53$ KΩ

(c) The gain at resonance $= A_V = -\beta \cdot \frac{Z}{R_{in}}$

$$A_V = -60 \frac{1.53 \times 10^3}{450}$$

$$= -204$$

• **PROBLEM 15-5**

A radio frequency amplifier in common-emitter configuration with internal capacitance $C_{int} = 100$ pf is shown below. Modify the circuit arrangement to improve the overall gain BW product and calculate the value of C_N.

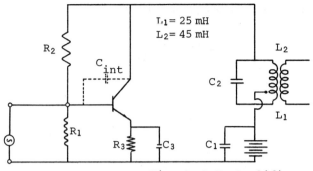

Fig. A R.F. Amplifier

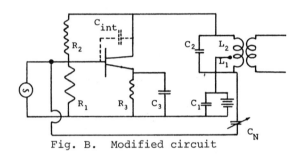

Fig. B. Modified circuit

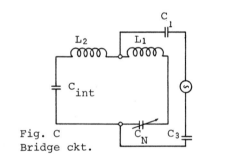

Fig. C
Bridge ckt.

<u>Solution</u>: The internal capacitance C_{int} contribute to the Miller capacitance, which reduces the gain bandwidth product of the amplifier. One way to improve GBW is to neutralize this additional capacitance by an adjustable capacitance. To neutralize this capacitance C_{int}, an external capacitance C_N is added as shown in Fig. (b). A resulting bridge is shown in Fig. (c). At radio frequency, all the by-pass capacitors are short circuited, resulting in a bridge, whose impedance equation at balance is

$$\frac{C_N}{C_{int}} = \frac{L_2}{L_1}$$

Therefore,

$$C_N = \frac{L_2}{L_1} C_{int}$$

$$= \frac{45 \times 10^{-3}}{25 \times 10^{-3}} \times 100 \times 10^{-12}$$

$$= 180 \text{ pF}.$$

• **PROBLEM 15-6**

A single-tuned IF amplifier shown in figure (a) operates at a standard intermediate frequency of 455 kHz and has a bandwidth of 9.7 kHz. Find the required turns ratio 'a' of the input autotransformer and the mid-frequency current gain with the following data,

$L' = 6.9 \mu H$ $R_p = 2k\Omega$

$r_{b'e} = R_{b'} = 1k\Omega$ $C_{b'e} = 1000 \text{ pF} = C_\pi$

$r_i = 5k\Omega$ $g_m = 0.15$

$R_L = 500\Omega$ $C_{b'c} = 4pF = C_\mu$

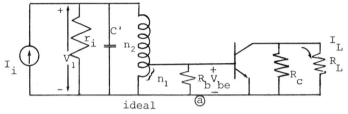

FIG. a   Tuned amplifier using an autotransformer.

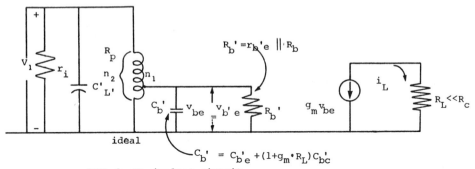

FIG. b   Equivalent circuit

Solution: The turns ratio of the ideal autotransformer is

$$a = \frac{n_1}{n_2} = \frac{V_{b'e}}{V_1} < 1$$

now, the total input capacitance will be

$C = C' + a^2 C_1$

$ = C' + a^2(C_\pi + C_\mu + g_m \cdot R_L \cdot C_\mu)$

$ = C' + a^2 [1000 + 4 + (0.1)(500)(4)] \times 10^{-12}$

$ = C' + (1204 \times 10^{-12} a^2)$

If R is the total input resistance, then

$$\frac{1}{R} = \frac{1}{r_i} + \frac{1}{R_p} + \frac{a^2}{R_b'}$$

where R_p is the parallel resistance of the transformer coil.

Thus,

$$\frac{1}{R} = \frac{1}{5000} + \frac{1}{2000} + \frac{a^2}{1000}$$

$$= 10^{-4}(7 + 10\,a^2) \tag{1}$$

With $\omega_0^2 = \frac{1}{L'C}$, where ω_0 = the center frequency, we have,

$$4.\pi^2[(455)^2 \times 10^6] = \frac{1}{(6.9 \times 10^{-6})(C' + 1204 \times 10^{-12}a^2)}$$

Thus,
$$C' + (1204)(10^{-12})a^2 = \frac{10^{-6}}{57.2}$$

$$\cong 0.017 \times 10^{-6}$$

Since $a \leq 1$, $C' \cong 0.017\ \mu F$

now, bandwidth is

$$BW = 9.7 \times 10^3 = \frac{1}{2\pi \cdot R \cdot C'} \cong \frac{1}{2\pi R(0.017 \times 10^{-6})}$$

Thus, $R \cong 938\,\Omega$

Substituting for R in equation (1) and solving,

$$a^2 \cong 0.4$$

or
$$a \cong 0.63$$

The mid-frequency current gain of the amplifier is,

$$A_m = -a \cdot g_m \cdot R$$

$$= -(0.63)(0.1)(930)$$

$$\cong -59.$$

● **PROBLEM 15-7**

A tuned cascode amplifier with two identical transistors is functioning as a tuned IF amplifier in a communication system whose intermediate frequency is 150 MHz. Calculate the current gain and the bandwidth of the amplifier given the following data.

$\beta = 112$ $g_m = 0.1$ mho
$C_\pi = 13.5$ pF $L_1 = 0.20$ μH
$C_\mu = 1$ pF Q for $L_1 = 150$
$R_S = 8.5$ kΩ $r_\pi = 200$ Ω

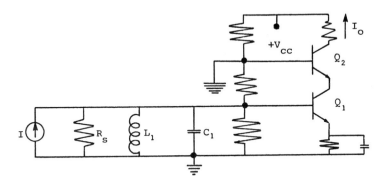

Solution: The total input capacitance is given as follows

$$C_{in} = C_1 + C_\pi + 2C_\mu$$
$$= (C_1 + 13.5 + 2) \text{ pF} \quad (1)$$

because Miller capacitance for a common-emitter amplifier is given by

$$C_M = C_\mu(1 + g_m \cdot R_L) = 2C_\mu$$

It should be noted that transistor Q_1 is acting like a C-E amplifier, while transistor Q_2 is cascoding. It can easily be shown that the input capacitance which is reflected on the input side is $2C_\mu$.

Now, this total input capacitance C_{in} forms a tank circuit with L_1. The resonant frequency of this tank circuit is the intermediate frequency f_0.

Hence,

$$f_0 = 150 \text{ MHz} = \frac{1}{2\pi\sqrt{L_1 C_{in}}}$$

Therefore,

$$C_{in} = \frac{1}{4\pi^2 \cdot f_0^2 \cdot L_1}$$

$$= 17.76 \text{ pF}.$$

From equation (1) it follows that

$$C_1 = 2.26 \text{ pF}$$

The total resistance across C_{in} is R' such that

$$\frac{1}{R'} = \frac{1}{R_S} + \frac{1}{Q_1 \cdot \omega_0 \cdot L_1} + \frac{1}{r_\pi}$$

$$= \frac{1}{8.5k\Omega} + \frac{1}{(150)(2\pi \times 150 \times 10^6)(0.20 \times 10^{-6})} + \frac{1}{200}$$

$$= 5.153 \times 10^{-3} \mho$$

R' = 194Ω

The BW of the amplifier is found to be

$$BW = \frac{1}{2\pi \cdot R' \cdot C_{in}}$$

$$= 46.178 \text{ MHz}$$

The current gain of the amplifier is

$$A = \frac{I_0}{I_S} = -g_m \cdot R' = -19.40$$

● **PROBLEM** 15-8

Design, using a suitable transistor, an intermediate frequency amplifier with f_0 = 455 kHz, bandwidth = 10.7 kHz, and voltage gain = $|A_v|$ = 50, to drive an external load of 600Ω.

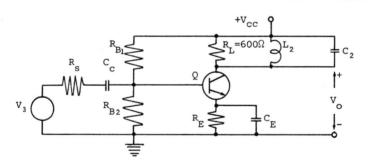

FIG. (a) Circuit

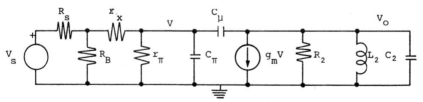

FIG. (b) Small-signal model

Solution: Since all the design calculations of the amplifier is based on the small signal model of the transistor, it is first necessary to choose a transistor with suitable parameters.

In order to do this, the small signal hybrid-π model should be equivalent to the circuit representation of the amplifier at 455 kHz. This model is valid if f_T of the transistor is very large as compared to $f_0 = 455$ kHz.

Now,
$$f_T = \frac{g_m}{2\pi(C_\pi + C_\mu)} \quad (1)$$

Choosing a transistor with the following parameters,

$g_m = 0.3$ mho, $\quad r_\pi = 500\Omega$

$r_x = 26\Omega \quad\quad C_\pi = 70$ pF and $C_\mu = 2.7$ pF

equation (1) gives

$$f_T = 6.57 \times 10^8 \text{ Hz}$$

$$\cong 6.6 \times 10^8 \text{ Hz}$$

$$\gg f_0$$

Hence, the hybrid-π model is valid at $f_0 = 455$ kHz. Assuming both biasing resistors to be very large, the voltage gain of the amplifier is

$$A_V = \frac{-g_m \cdot r_\pi \cdot R_L}{R_S + r_\pi + r_x}$$

$$-50 = \frac{-(0.3)(500)(600)}{R_S + 500 + 26}$$

Thus, $\quad R_{source} = 1274\Omega$

Now, the output circuit of the amplifier is a band-pass circuit with $R_L \| L_2 \| C_2$. The frequency response of this is shown in fig. (c).

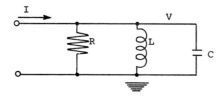

Band-pass circuit

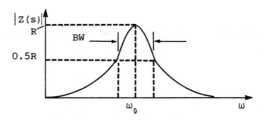

Band-pass response

The resonant frequency of this band-pass circuit is given by

$$\omega_0 = \frac{1}{\sqrt{L_2 C_2}} \qquad (2)$$

at frequency, $\omega = \omega_0 + \omega'$,

where ω' is a small frequency deviation from the resonance,

$$y(s) = \frac{1}{Z(s)} = G + j\omega_0 \cdot C_2 + j\omega' C_2 + \frac{1}{j\omega_0 L_2} + \frac{j\omega'}{\omega_0^2 \cdot L_2}$$

Since $\omega' << \omega_0$, $Z(s)$ can be approximated as,

$$Z(s) = \frac{1}{G + 2j\omega' C_2}$$

At $\omega' = \frac{G}{2C_2}$, $Z(s)$ is 3dB below its peak value, and the bandwidth of the resonant circuit in radians/second is thus twice ω'.

i.e., $\qquad$ B.W. $= 2\omega'$ rad/sec.

$$= \frac{G}{C_2}$$

$$= \frac{1}{RC_2}$$

where R = total output impedance

$$= R_L \parallel R_0$$

$$= R_L \text{ (assuming } R_0 \text{ of Q is very large)}$$

Hence,
$$C_2 = \frac{1}{600 \times (10.7) \times 10^3}$$

$$= 1.56 \times 10^{-7} F.$$

Using equation (2)

$$L_2 = \frac{1}{(2\pi \cdot f_0)^2 C_2} = 7.84 \times 10^{-7} H$$

The design of the circuit is shown in Fig. (1).

• **PROBLEM 15-9**

For a single tone-modulated signal

$$m(t) = \cos \omega_m t$$

Find the demodulated single sideband output g(t).

Solution: For $m(t) = \cos \omega_m t$ the Hilbert transform $\overset{v}{m}(t)$ is $m(t)$ retarded by 90°. The Hilbert transform is a linear operator. Thus
$$\overset{v}{m}(t) = \sin \omega_m t$$

In general the output is:
$$y(t) = \frac{A_r}{4}\left[m(t) \cos \Delta\phi \pm \overset{v}{m}(t) \sin \Delta\phi\right]$$

where A_r represents the effects of range attenuation

$\Delta\phi = \phi_c - \phi_0$

ϕ_c = overall channel phase shift

ϕ_0 = channel output phase shift.

This yields for the output
$$y(t) = \frac{Ar}{4} \cos \Delta\phi \cos \omega_m t \pm \frac{Ar}{f} \sin \Delta\phi \sin \omega_m t$$

$$= \frac{Ar}{2} \cos(\omega_m t \pm \Delta\phi)$$

Thus, when the modulation is a single tone or a series of harmonics, the effects due to a constant phase error between the local oscillator frequency and the radio frequency, produce a constant phase shift in the harmonics of the information signal. Thus, as $\Delta\phi$ drifts from 0 to $\pi/2$, the output undergoes increasing delay distortion, $\Delta\phi$, approaching an amount $\Delta\phi/\omega_m$ for each harmonic in $m(t)$.

When $m(t)$ is voice or music, the delay distortion produces no significant degradation. Since the human ear is insensitive to considerable phase or delay distortion of the above form. However, for pulse data such as in television video, facsimile, etc., the effect is critical.

● **PROBLEM 15-10**

A 400 kHz quartz crystal with parameters $L = 3.2H$, $C = 0.05$ pF and $R = 4k\Omega$ is used in an oscillator together with a standard differential amplifier RCA CA3000IC, as shown below. The crystal is mounted in a holder whose capacitance C_h is 6 pF. Determine:

(a) Bandwidth of the amplifier and show that the necessary condition for sustained oscillations is $R_s = \frac{R_x}{A-1}$ where $R_s = R_{S1} = R_{S2}$.

(b) The signal at pin no. 10, if a very low frequency signal is applied to pin no. 2 of CA3000IC, in fig. (2).

(c) An RC filter is placed as shown in fig. (3), what is the output signal if a very low frequency signal is applied to pin no. 2.

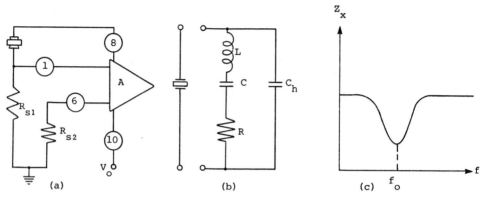

FIG. 1 (a) Crystal oscillator using RCA CA 3000 IC
(b) Electrical equivalent circuit of the crystal
(c) Impedance v/s frequency plot

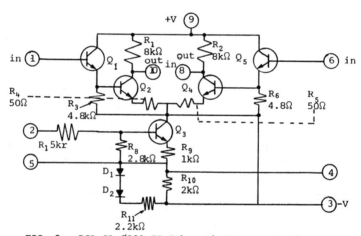

FIG. 2 RCA CA 3000 IC Schematic Representation

Solution: (a) Fig. 1(b) shows the equivalent circuit of a quartz crystal. The resonant frequency of the oscillator is the resonant frequency of the tank circuit shown in fig. 1(b). Usually, the capacitance offered by the crystal holder is very large as compared to the capacitance of the crystal itself. When two capacitors are in parallel, their equivalent capacitance is approximately equal to the smallest among them, (if other one is very large), hence C_h can be neglected.

The resonant frequency is thus given by,

$$f_0 = \frac{\omega_0}{2\pi} = \frac{1}{2\pi\sqrt{LC}}$$

and the bandwidth is equal to $\frac{1}{2\pi \cdot R \cdot C}$. Substituting all the values, the bandwidth of the amplifier is

$$= \frac{1}{2\pi(4 \times 10^3) \times (0.05 \times 10^{-12})}$$

$$= 7.95 \times 10^8 \text{ Hz} \qquad (1)$$

Fig 1(c) shows the variation of Z_X i.e. the impedance of the crystal as a function of frequency f.

From the schematic representation of the RCA CA3000IC, it can be seen that the noninverting output of the IC is used to provide proper polarity for positive feedback.

At f_0, $Z_X = R_X$ and if $R_{S1} = R_{S2} = R_S$, then the voltage at pin 1 is given by

$$\frac{V_1}{V_8} = \frac{R_S \| R_i}{R_X + R_S \| R_i} \qquad (2)$$

where V_8 is the voltage at pin no. 8 and R_i is the input impedance of the IC.

For sustained oscillations, the loop gain must be unity or $AV_1 = V_8$ where A is the gain of the difference amplifier.

Substituting for V_1 interms of V_8 from equation (2), the necessary condition for sustained oscillations is,

$$\frac{A \times (R_S \| R_i)}{R_X + (R_S \| R_i)} V_8 = V_8$$

$$\frac{A}{R_X + \left[\frac{R_S \cdot R_i}{R_S + R_i}\right]} \times \frac{R_S \cdot R_i}{R_S + R_i} = 1$$

$$\frac{A \times R_S \times R_i}{R_S \cdot R_i + (R_S + R_i)R_X} = 1$$

$$\frac{A \cdot R_S \cdot R_i}{R_S \cdot R_i + R_S \cdot R_X + R_i \cdot R_X} = 1$$

$$\frac{A \cdot R_S}{R_S + \frac{R_S}{R_i} \cdot R_X + R_X} = 1$$

Since $R_i \gg R_S$,

$$\frac{A \cdot R_S}{R_S + R_X} = 1$$

$$R_S = \frac{R_S + R_X}{A}$$

$$R_S - \frac{R_S}{A} = \frac{R_X}{A}$$

$$R_S\left(1 - \frac{1}{A}\right) = \frac{R_X}{A}$$

$$R_S = \frac{R_X}{A-1},$$

(b) When a sinusoidal signal of frequency lower than f_0 is applied to pin 2, the base-bias voltage to transistor Q_5 (in figure (2)), and also I_0 the sink current for the difference amplifier vary with the instantaneous value of the signal. The collector voltage at pin 10 is given by

$$v_0 = v_1 \cdot g_m \cdot I_0$$

where v_1 = voltage applied to pin 1

If the input signal is $V_m \cdot \sin\omega t$, then I_0 is proportional to V_m and if V_1 is also sinusoidal and of the form $V \cdot \sin\omega_0 t$, then the output is an AM signal with an ω_m envelope and an ω_0 carrier.

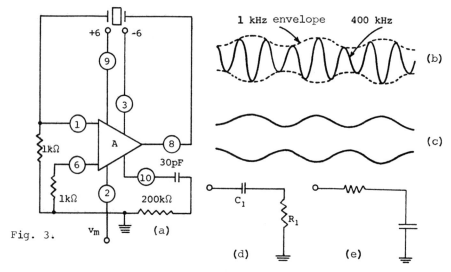

Fig. 3.

(c) The RC filter shown in fig. 3(a) is a high-pass filter. When a sinusoidal signal of freq. ω_m is applied at the input pin 2, an AM signal appears at the output, which is then passed through the R-C filter. Now if the frequency of the incoming signal is 5 kHz, C = 30 pF and R = 50 kΩ, then the outer envelope of the AM signal which is present at pin 10, may or may not be detected depending on the filter configuration.

Thus two cases arise.

Case (1):

If the filter configuration is as shown in figure 3 (d), then the lower cut-off frequency is

$$f_u = \frac{1}{2\pi \cdot C_1 \cdot R_1}$$

$$= \frac{1}{2\pi \cdot (30 \times 10^{-12})(200 \times 10^3)}$$

$$= 26.52 \text{ kHz}.$$

Since this cut-off frequency is much higher than the signal frequency of 5 kHz, the filter will block this frequency and will allow only the carrier frequency to pass, resulting in a waveform as shown in fig. 3(b).

Case (2):

If the resistor and capacitor are interchanged, then it will act like a low-pass filter and its higher cut-off frequency will be same as the lower cut-off frequency of the above filter. In this case, carrier frequency will be blocked and input signal will pass through, so only envelope of frequency 5 kHz will be detected and the waveform will be as shown in fig. 3(c).

• **PROBLEM 15-11**

Describe the operation of a ratio detector circuit shown in fig. (1) and prove that $\frac{V_1}{V_2}$ is independent of the input amplitude. Draw a circuit of the practical version of a ratio detector.

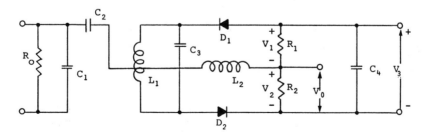

Fig. 1(a). Ratio detector circuit

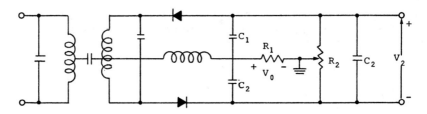

Fig. 1(b). Practical version of a ration detector

Solution: Consider the input tank circuit of Fig. (a) with inductor L_1 and C_3. Since the inductor L_1 is being center-tapped, at resonance, the voltage across resistors R_1 and R_2 become equal i.e., $V_1 = V_2$ and obviously the output voltage is $V_3 = 2V_1 = 2V_2$.

At a frequency away from the center frequency of the tank circuit, voltage across R_1 is slightly higher than that across R_2. From the circuit configuration and voltage versus frequency plot of the tank circuit, it can be shown that

$$\Delta V_1 = -\Delta V_2$$

Hence it can be seen that voltage V_3 always remains the same i.e., equal to $2V_1$.

Let I be the current in the input tuned circuit, then

$$V_1 = K_1 \cdot I \qquad (1)$$

and
$$V_2 = -K_2 \cdot I \qquad (2)$$

but $V_1 + V_2$ = constant

Hence from equation (1) and (2), $K_1 - K_2$ = constant = K.

Adding both sides of equation (1) and (2),

$$V_1 + V_2 = K \cdot I. \qquad (3)$$

dividing equation (3) by (2), we get

$$\frac{V_1}{V_2} + 1 = \frac{K}{-K_2} = \text{constant}$$

Hence $\frac{V_1}{V_2}$ = constant

Thus, it can be concluded that $\frac{V_1}{V_2}$ is independent of the amplitude of the input signal, and hence, a limiter stage is not necessary.

Fig. (b) shows a practical circuit in which resistors R_1 and R_2 are replaced by C_1 and C_2 together with a slight modification at the output side. In this case, at resonance, the output voltage V_0 becomes zero, which can be illustrated by applying KVL to circuit No. (1) and (2) of figure (b). Since, the rest of the circuitary is the same ratio $\frac{V_1}{V_2}$ is independent of the signal amplitude. This circuit is noise free because, by choosing the time constant R_2C_2 to be large, V_3 cannot follow rapid fluctuations in the amplitude of the input signal which results because of noise and hence V_2 is independant of the amplitude of the input signal and depends only on its frequency deviation.

THERMAL NOISE

• **PROBLEM 15-12**

For the circuit in Fig. (1), find the power density spectrum of the thermal noise voltage across the terminals xy, given noise voltage power density spectrum $P_v(\omega) = 2kTR_{xy}(\omega)$ and noise current power density spectrum $P_i(\omega) = 2kT\text{Re}[Y_{xy}(\omega)]$

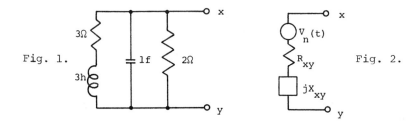

Fig. 1. Fig. 2.

Solution: Looking into the terminals xy in Fig. (1), the expression for the equivalent admittance can be written as

$$Y_{xy}(\omega) = 2 + j\omega + \frac{1}{3 + 3j\omega}$$

$$= \frac{(3 + 3j\omega)2 + (3 + 3j\omega)j\omega + 1}{3 + 3j\omega}$$

$$= \frac{6 + 6j\omega + 3j\omega - 3\omega^2 + 1}{3 + 3j\omega}$$

$$= \frac{7 + 9j\omega - 3\omega^2}{3 + 3j\omega}$$

Hence, the expression for the equivalent impedance between terminals xy is

$$Z_{xy}(\omega) = \frac{1}{Y_{xy}(\omega)}$$

$$= \frac{3 + 3j\omega}{7 + 9y\omega - 3\omega^2}$$

and
$$R_{xy}(\omega) = R_e[Z_{xy}(\omega)]$$

$$= R_e\left[\frac{3 + 3j\omega}{7 + 9j\omega - 3\omega^2}\right]$$

$$= \frac{3 + 3j\omega}{[(7-3\omega^2) + j9\omega]} \cdot \frac{(7-3\omega^2)-j9\omega}{(7-3\omega^2)-j9\omega}$$

$$= \frac{(3 + 3j\omega)[(7-3\omega^2) - j9\omega]}{(7-3\omega^2) - (j9\omega)^2}$$

$$= \frac{21 - 6j\omega + 18\omega^2 - 9j\omega^3}{49 + 9\omega^4 + 39\omega^2}$$

$$= \frac{21 + 18\omega^2 - j(6\omega + 9\omega^3)}{49 + 9\omega^4 + 39\omega^2}$$

Considering only the real part

$$R_{xy}(\omega) = \frac{21 + 18\omega^2}{49 + 9\omega^4 + 39\omega^2}$$

Fig. (1) can be represented by its equivalent circuit as shown in Fig. (2).

The power density spectrum of the noise voltage across xy is obtained as:

$$P_v(\omega) = 2kTR_{xy}(\omega)$$

$$= \frac{2kT(21 + 18\omega^2)}{49 + 9\omega^4 + 39\omega^2}$$

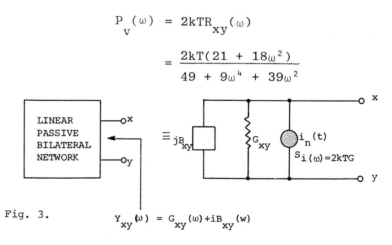

Fig. 3.

$$Y_{xy}(\omega) = G_{xy}(\omega) + iB_{xy}(\omega)$$

The Norton's equivalent is shown in Fig. (3) and the corresponding power density spectrum of the noise current $P_i(\omega)$ is given by

$$P_i(\omega) = 2kTR_e[Y_{xy}(\omega)]$$

$$R_e[Y_{xy}(\omega)] = \frac{7 + 9j\omega - 3\omega^2}{3 + 3j\omega} \cdot \left(\frac{3-3j\omega}{3-3j\omega}\right)$$

$$= \frac{(7 + 9j\omega - 3\omega^2)(3-3j\omega)}{(3 + 3j\omega)(3-3j\omega)}$$

$$= \frac{21-21j\omega + 27j\omega + 27\omega^2 - 9\omega^2 + 9j\omega^3}{9 + 9\omega^2}$$

Considering only the real part

$$R_e[Y_{xy}(\omega)] = \frac{21 + 18\omega^2}{9 + 9\omega^2}$$

$$= \frac{3(7 + 6\omega^2)}{3(3 + 3\omega^2)}$$

$$= \frac{7 + 6\omega^2}{3 + 3\omega^2}$$

Thus,
$$S_i(\omega) = 2kT \frac{7 + 6\omega^2}{3 + 3\omega^2}$$

• **PROBLEM 15-13**

Estimate the root mean square value of the noise voltage across the capacitor terminals XX' for the given R-C network.

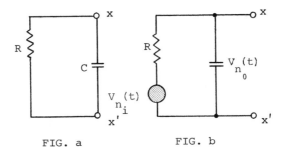

FIG. a   FIG. b

Solution: From figures (a) and (b) it is clear that the resistor R in (a) is replaced by a noiseless resistor R and a series noise voltage source $v_{ni}(t)$ in (b), where $v_{ni}(t) = R\, i_n(t)$.

$S_i(\omega) = 2kTR$ is the power density spectrum due to thermal noise in the resistor.

From the R-C network given; the transfer function $H(\omega)$ is calculated as:

$$H(\omega) = \frac{\frac{1}{j\omega C}}{R + \frac{1}{j\omega C}}$$

$$= \frac{1}{j\omega CR + 1}$$

The magnitude of $H(\omega)$ is

$$|H(\omega)| = \frac{1}{\sqrt{\omega^2 C^2 R^2 + 1}}$$

Therefore, $\quad |H(\omega)|^2 = \dfrac{1}{(\omega^2 C^2 R^2 + 1)}$

Hence, the power density spectrum is calculated as

$$S_0(\omega) = S_i(\omega)|H(\omega)|^2$$

$$S_0(\omega) = \frac{2kTR}{1 + \omega^2 C^2 R^2}$$

Note: $\tfrac{1}{2}\pi$ times the area under the power density spectrum gives the mean square value of the signal.

i.e.,

$$\overline{v_{no}^2} = \frac{1}{2\pi} \int_{-\infty}^{+\infty} S_0(\omega) d\omega$$

But

$$S_0(\omega) = S_i(\omega)|H(\omega)|^2$$

Therefore,

$$\overline{v_{no}^2} = \frac{1}{2\pi} \int_{-\infty}^{\infty} S_i(\omega)|H(\omega)|^2 d\omega$$

Since the power density spectrum is always an even function of ω,

$$\overline{v_{no}^2} = \frac{2}{2\pi} \int_{0}^{\infty} S_i(\omega)|H(\omega)|^2 d\omega$$

Hence, the root mean square value is calculated as

$$\sqrt{\overline{v_{no}^2}} = \left[\frac{1}{\pi} \int_{0}^{\infty} S_i(\omega)|H(\omega)|^2 d\omega\right]^{1/2}$$

$$= \left[\frac{1}{\pi} \int_0^\infty \frac{2kTR}{1 + \omega^2 C^2 R^2} \, d\omega \right]^{1/2}$$

$$= \left[\frac{2kTR}{\pi CR} \tan^{-1}(\omega CR) \Big|_0^\infty \right]^{1/2}$$

$$= \left[\frac{2kTR}{\pi CR} \tan^{-1}\infty - \tan^{-1}0 \right]^{1/2}$$

$$= \left[\frac{2kT}{\pi C} \left(\frac{\pi}{2} - 0\right)\right]^{1/2}$$

$$= \left[\frac{2kT\pi}{2\pi C}\right]^{1/2}$$

$$= \left[\frac{kT}{C}\right]^{1/2}$$

$$= \sqrt{\frac{kT}{C}}$$

● **PROBLEM 15-14**

For a 40k resistor at a room temperature of 27°C, find the rms thermal noise voltage for the following bandwidth range

(a) 0 to 10 kHz and

(b) 10 kHz to 25 kHz

Solution: For a given resistor, the thermal noise voltage $v_n(t)$ across the output terminals is a normal distribution and has a mean square voltage expressed as

$$\overline{v_n^2} = 4kTR \, df$$

The mean square value is also known as the second moment denoted as μ_{20}

Thus,
$$\mu_{20} = \overline{v_n^2} = 4kTR\,df \tag{1}$$

where k = Boltzmann's constant = 1.38×10^{-23} joules/°K

T = Absolute temperature in °K = $273 + 27 = 300°K$

R = Resistance in ohms

and df = bandwidth in Hz

Thus, from equation (1)

(a) $\overline{v_n^2} = 4(2.38 \times 10^{-23})(300)(40 \times 10^3)(10 \times 10^3)$

   $= 6.62 \times 10^{-12}$ volts^2

$\sqrt{\overline{v_n^2}} = 2.57 \times 10^{-6}$ volts rms

(b) $\overline{v_n^2} = 4(1.38 \times 10^{-23})(300)(40 \times 10^3)(15 \times 10^3)$

   $= 9.93 \times 10^{-12}$ volts^2

$\overline{v_n^2} = 3.15 \times 10^{-6}$ volts rms

Note: $\overline{v_n^2}$ in equation (1) is the normalized average power associated with the random noise voltage $v_n(t)$ within the frequency range df.

• **PROBLEM 15-15**

Figure (1) shows a series noise circuit consisting of two resistors R_1 and R_2 at absolute temperatures T_1 and T_2, respectively. If the two noise sources $v_{n_1}(t)$ and $v_{n_2}(t)$ are independent, develop its equivalent noise circuit shown in figure (2).

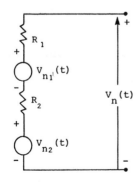

Fig. 1. Actual circuit

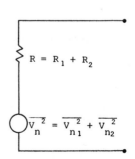

Fig. 2. Equivalent noise circuit

Solution: Figure (1) shows two series resistors R_1 and R_2 together with their associated noise voltages $v_{n_1}(t)$ and $v_{n_2}(t)$, respectively. $v_n(t)$ represents the total noise voltage across the output terminals for the above circuit.

It can be deduce from the above circuit (by considering the polarities) that

$$v_n(t) = v_{n_1}(t) + v_{n_2}(t)$$

It has been determined experimentally that the noise voltage $v_n(t)$ that appears across the terminals of a resistor network has a mean square voltage (Gaussian distribution) given by the relation

$$\overline{v_n^2} = 4kTR \, df$$

where k = Boltzmann's constant, T = Temperature in degrees Kelvin and df = the bandwidth.

Applying the above principle to the problem, the equivalent noise voltage across the output port of the two resistor circuits can be determined by evaluating the mean square value of $v_n(t)$.

Therefore, $\quad \overline{v_n^2} = \overline{(v_{n_1} + v_{n_2})^2}$

(where the bar at the top denotes the mean).

The mean square value is also known as the 2nd moment and is represented by μ_{20}.

Hence, $\quad \mu_{20} = \overline{v_n^2} = \overline{(v_{n_1} + v_{n_2})^2}$

$$= \overline{v_{n_1}^2 + 2v_{n_1} \cdot v_{n_2} + v_{n_2}^2}$$

$$= \overline{v_{n_1}^2} + \overline{v_{n_2}^2} + \overline{2v_{n_1} v_{n_2}}$$

Since $v_{n_1}(t)$ and $v_{n_2}(t)$ are statistically independent, the term $2v_{n_1} v_{n_2}$ in the above expression evaluates to zero.

Thus,

$$\mu_{20} = \overline{v_n^2} = \overline{v_{n_1}^2} + \overline{v_{n_2}^2}$$

● **PROBLEM 15-16**

A line amplifier of bandwidth 22 kHz has an input source resistance of 750Ω at a standard noise temperature of 290°K. If the input and output impedences of the amplifier are 750Ω each, determine the effective input source noise power, expressed in dbm.

Solution: To calculate the source noise power, we should first calculate the mean-square input noise voltage.

i.e.,
$$\overline{v^2} = 4kTR\,\Delta f$$

$$= 4(1.38 \times 10^{-23})(290)(750)(22000)$$

$$= 2.64 \times 10^{-13} \text{ volts.}$$

Therefore, average rms value is $\sqrt{\overline{v^2}} = 5.14 \times 10^{-7}$ volts rms and the source noise power = $\overline{v^2}/R$

$$= 3.52 \times 10^{-16} \text{ watts}$$

Therefore, the source noise power in dbm =

$$10 \log [3.52 \times 10^{-16}/10^{-3}]$$

$$= -124.53 \text{ dbm.}$$

Note: k is the Boltzmann's constant:

$$k = 1.38 \times 10^{-23} \text{ Joule per degree kelvin}$$

Δf is the bandwidth in cycles/second.

● **PROBLEM 15-17**

Consider two resistors R_x and R_y connected in parallel as shown in figure (1). If the resistors are in thermal equilibrium at absolute temperatures T_x and T_y, respectively, estimate the average rms noise voltage across terminals 'a' and 'b'.

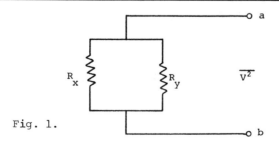

Fig. 1.

Solution: The rms noise voltage due to the combination of resistors R_x and R_y can be expressed in terms of their admittances G_x and G_y, and the associated noise current generators as shown in the equivalent circuit in figure (2)

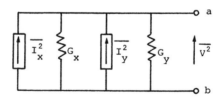

Fig. 2. Two resistors in parallel

The associated noise current generators can be evaluated as
$$\overline{I_x^2} = 4kT_x G_x \Delta f$$
and
$$\overline{I_y^2} = 4kT_y G_y \Delta f,$$

where k = Boltzmann's Constant = 1.38×10^{-23} joule/°K, and Δf = Bandwidth in hertz.

Thus, the total noise power is
$$\frac{\overline{I_x^2} + \overline{I_y^2}}{G_x + G_y} = \frac{4k\Delta f(T_x G_x + T_y G_y)}{G_x + G_y}$$

or
$$\overline{v^2}(G_x + G_y) = \frac{4k\Delta f(T_x G_x + T_y G_y)}{G_x + G_y}$$

Therefore, the average rms noise voltage will be
$$[\overline{v^2}]^{1/2} = \left[\frac{4k\Delta f(T_x G_x + T_y G_y)}{[G_x + G_y]^2}\right]^{1/2}$$

● **PROBLEM 15-18**

Given the resistor network in figure (1), calculate the rms noise voltage appearing across the output ports in a 100-kHz bandwidth at a room temperature of T = 290°K.

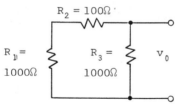

FIG. 1
Resistor network

$R_2 = 100\Omega$
$R_1 = 1000\Omega$
$R_3 = 1000\Omega$
v_0

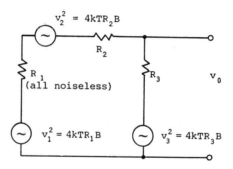

Fig. 2. Equivalent circuit for noise voltage calculation

<u>Solution</u>: The noise voltage across the output ports can be evaluated as a sum of the noise voltages due to each resistor across the output terminal. Thus, to calculate the rms noise voltage, it is first necessary to calculate the noise voltage due to individual resistors, sum the squares of the resistances of these individual resistors and finally obtain the square root of the resulting sum.

The noise voltage due to each resistor in the circuit can be calculated by the voltage division method. Considering the noise equivalent circuit in Fig. (2), we have

$$V_0^2 = V_{01}^2 + V_{02}^2 + V_{03}^2$$

where

$$V_{01} = (4kTR_1B)^{1/2} \cdot \frac{R_3}{R_1 + R_2 + R_3}$$

$$V_{02} = (4kTR_2B)^{1/2} \cdot \frac{R_3}{R_1 + R_2 + R_3}$$

$$V_{03} = (4kTR_3B)^{1/2} \cdot \frac{R_1 + R_2}{R_1 + R_2 + R_3}$$

k = Boltzmann's constant = 1.38×10^{-23} joules/°K

T = 290°K (Room temperature)

B = Bandwidth = 100×10^3 Hz

   = 10^5 Hz

Thus $$V_0^2 = (4kTB) \left[\frac{(R_1 + R_2) R_3^2}{(R_1 + R_2 + R_3)^2} + \frac{(R_1 + R_2)^2 R_3}{(R_1 + R_2 + R_3)^2} \right]$$

$$= (4 \times 1.38 \times 10^{-23} \times 290 \times 10^5) \left[\frac{(1100)(1000)^2}{(2100)^2} + \frac{(1100)^2(1000)}{(2100)^2} \right]$$

$$V_0^2 = 8.385 \times 10^{-13} \ V^2$$

∴ $V_0 = 9.16 \times 10^{-7}$ V(rms)

● **PROBLEM 15-19**

Determine the mean square value of the output noise $n_i(t)$ when a transducer with an internal resistance R_a is connected to an RC low-pass filter shown in figure (1b). The Thevenin's equivalent of the cascade is shown in figure (1c).

(Note: R_a in Fig. (1) is the internal resistance of the transducer.)

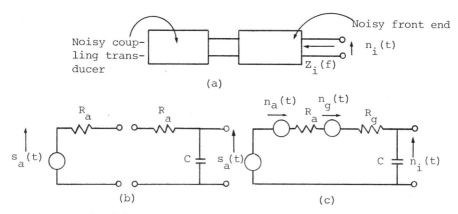

FIG. 1 (a) Noisy coupling transducer cascaded with noisy front end stages;
(b) Resistive transducer to be cascaded with RC low-pass filter;
(c) Thevenin equivalent of cascade.

Solution: The relation between the autocorrelation function, $R_{XX}(\omega)$, the power spectral density, $S_{XX}(\omega)$ and the mean square value of the output noise is given by the total average power.

i.e., Total average power =

$$E[X^2(t)] = \int_{-\infty}^{+\infty} x^2 f(x) dx$$

$$= R_{XX}(0)$$

$$= \int_{-\infty}^{\infty} S_{XX}(\omega) d\omega$$

$$= \int_{-\infty}^{+\infty} S_{XX}(f) \frac{d\omega}{2\pi}$$

Further simplification will result in

$$E[n_i(t)n_i(t+0)] = R_{n_i n_i}(\tau)\big|_{\tau=0}$$

$$= \int_{-\infty}^{+\infty} S_{n_i n_i}(f) \, e^{j\tau\omega} \frac{d\omega}{2\pi} \bigg|_{\tau=0}$$

where $R_{n_i n_i}$ and $S_{n_i n_i}$ are the autocorrelation and the spectral density of the predetection noise respectively.

The relationship between the input and output spectral densities is given by

$$S_{n_i n_i}(f) = |H(f)|^2 \, S_{n_a n_a}(f)$$

This can also be written as

$$S_{n_i n_i}(f) = |H(f)|^2 \, S_{nn}(f)$$

where $S_{nn}(f)$ is the spectral density of the driving side, and $H(f)$ is the system function acting on the noise spectra S_{nn}.

Now
$$S_{nn}(f) = 2kTR = \frac{n_0}{2} \quad \frac{\text{volt}^2}{\text{Hz}}$$

where k is the Boltzmann's constant

$$= 1.38 \times 10^{-23} \text{ Joule/}°K,$$

R is the resistance in ohms,

T is the absolute temperature in degrees Kelvin, and $n_0/2$ is the spectral density of white noise at the output of the bandpass filter.

Since the two sources are independent, it implies

$$S_{nn}(t) = \left[S_{n_a n_a}(f) + S_{n_g n_g}(f) \right]$$

$$= (2kTR_a + 2kTR_g)$$

$$= n_0 \tag{1}$$

The system function should include the effects of the components of the transducer in cascade with the front-end network. From figure (1c), it follows that

$$H(f) = \frac{1}{1 + \left(\frac{j\omega}{\alpha}\right)}$$

$$|H(f)|^2 = \frac{1}{1 + \left(\frac{\omega}{\alpha}\right)^2}$$

where
$$\alpha = \frac{1}{(R_a + R_g)C} \tag{2}$$

From equations (1) and (2),

$$S_{n_i n_i}(f) = |H(f)|^2 S_{nn}(f)$$ can be

written as

$$S_{n_i n_i}(f) = \frac{2kT(R_a + R_g)}{1 + \left(\frac{\omega}{\alpha}\right)^2}$$

The autocorrelation function of the pre-detection noise can be obtained by evaluating the inverse Fourier transform of $S_{n_i n_i}(f)$ and solving the integral using the Fourier transform pair.

i.e., $\dfrac{2ab}{b^2 + \omega^2} \longleftrightarrow a\, e^{-b|\tau|}, \quad b > 0$

Therefore, $R_{n_i n_i}(\tau) = \dfrac{kT}{C} e^{-\alpha|\tau|}$

Thus, the mean square value of the predetection noise is

$$E[n_i^2(t)] = R_{n_i n_i}(0) = \frac{kT}{C}$$

It is thus clear that noise increases as temperature increases and it is independent of R_a and R_g because it is a low-pass R-C circuit.

The increase of noise spectral intensity at the driving side is directly proportional to the resistance.

i.e., $$S_{nn}(f)_v = 2kTR$$

Since the Bandwidth (α), is inversely proportional to the resistance, the above equation does not exactly hold.

i.e., $$\alpha \sim 1/R$$

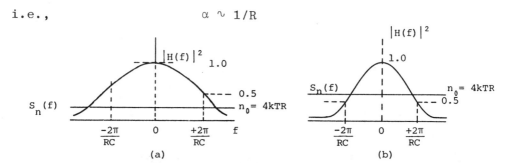

(a) (b)

In Equation (2), the area of the system amplitude function and the resistance R are inversely proportional to each other. Since the output spectral density is the product of equations (1) and (2), the effects of R are cancelled, which is clear from figure (2). As R_a or R_g increases, the bandwidth α of Equation (2) decreases, which in turn, decreases the area under $|H(f)|^2$, cancelling the effects of the increase in R. Since the noise power is not dependent on C,

$$E[n_i^2(t)] \text{ decreases as C increases}$$

because C is inversely proportional to the bandwidth.

Hence, it is impossible to isolate the contribution to the output noise due to the transducer resistance from the thermal noisiness of the front end.

NOISE POWER AND BANDWIDTH CALCULATIONS

● **PROBLEM** 15-20

Consider the circuit shown below. If the resonant frequency f_0 of the circuit is 4GHz and the 3dB bandwidth is 10^7 Hz, determine,

(a) the resistive component $R(f)$ of the impedance $Z(f)$ where $Z(f) = R(f) + jx(f) = $ impedance looking back into terminals p and g.

(b) the power spectral density $G(f)$ assuming noise occurs in the circuit is approximately white.

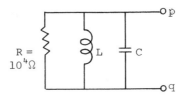

Solution: (a) Since this is a parallel R-L-C circuit, it is much easy to evaluate the admittance. Hence, looking back into terminals p and q, the admittance

$$Y(f) = \frac{1}{R} + \frac{1}{j2\pi fL} + j2\pi fC$$

Now, realize that $Z(f) = \frac{1}{Y(f)}$.

thus, $$R(f) = \text{Re}\left[Z(f)\right] = \text{Re}\left[\frac{1}{Y(f)}\right]$$

$$= \text{Re}\left[\frac{1}{\frac{1}{R} + \frac{1}{j2\pi fL} + j2\pi fC}\right] \qquad (1)$$

Since $f_0 = \dfrac{1}{2\pi\sqrt{LC}} = 4 \times 10^9$ Hz and 3dB bandwidth:

$B = \dfrac{1}{2\pi RC} = 10^7$ Hz

Eq. (1) simplify to

$$R(f) = \operatorname{Re}\left[\dfrac{1}{\dfrac{1}{R} + jf(2\pi C - 2\pi L)}\right]$$

$$= \operatorname{Re}\left[\dfrac{R}{1 + j\dfrac{f}{B}\left[1 - \left(\dfrac{f_0}{f}\right)^2\right]}\right]$$

$$= \dfrac{R B^2 f^2}{f^4 + (B^2 - 2f_0^2)f^2 + f_0^4},$$

$$= \dfrac{10^4 \times (10^7)^2 \times f^2}{f^4 + [(10^7)^2 - 2(4 \times 10^9)^2]f^2 + (4 \times 10^9)^4}$$

$$= \dfrac{10^{18} f^2}{f^4 - 3.1999 \times 10^{19} f^2 + 2.56 \times 10^{38}} \qquad (2)$$

(b) Since noise is white, the power spectral density $G(f) = \dfrac{\overline{v_n^2}}{2dF} = 2kTR(f)$ (3), where $\overline{v_n^2}$ = mean square voltage in narrow frequency band $dF = 4kTRdF$. Substituting Eq. (2) into Eq. (3),

$$G(f) = 2kTR(f) = 2kT \dfrac{10^{18} f^2}{f^4 - 3.1999 \times 10^{19} f^2 + 2.56 \times 10^{38}}$$

• **PROBLEM** 15-21

A quantum mechanical expression for the noise power spectral density $G(f)$ is given as

$$G_q(f) = \dfrac{hf/2}{e^{hf/kT} - 1}.$$

Find $G_q(f)$ if $\lambda \cong 1$ mm and 1 μm gives $T = 4°K$. Is the noise white in these regions?

(Note: $h = 6.62 \times 10^{-34}$ joule sec = Plank's constant

$k = 1.37 \times 10^{-23}$ joules/°K = Boltzmann's constant)

Solution: Since $f = \frac{C}{\lambda}$

$$G_q(f) = \frac{hC/\lambda 2}{e^{hC/\lambda kT}-1}$$

Hence,

$$G_q(f)\bigg|_{\lambda=1mm} = \frac{6.62 \times 10^{-34} \times 3 \times 10^8/2 \times 10^{-3}}{\exp(6.62 \times 10^{-34} \times 3 \times 10^8/10^{-3} \times 1.37 \times 10^{-23} \times 4)-1}$$

$$= 2.74 \times 10^{-32}$$

$$G_q(f)\bigg|_{\lambda=1\mu m} = \frac{6.62 \times 10^{-34} \times 3 \times 10^8/2 \times 10^{-6}}{\exp(6.62 \times 10^{-34} \times 3 \times 10^8/10^{-6} \times 1.37 \times 10^{-23} \times 4)-1}$$

$$= \frac{9.93 \times 10^{-28}}{e^{3624}-1}$$

$$\cong 9.93 \times 10^{-1598}$$

Now, notice that since $\frac{kT}{2} = \frac{1.37 \times 10^{-23} \times 4}{2} = 2.74 \times 10^{-23}$ which is a lot greater than $G_q(f)\big|_{\lambda=1mm}$ and $G_q(f)\big|_{\lambda=1\mu m}$, hence noise in these regions is not white.

(Note: hF << kT for white noise.)

● **PROBLEM** 15-22

(a) Determine the noise bandwidth BW_N of a parallel R-L-C filter with 3-dB bandwidth = B Hz.

(b) Repeat part (a) if the filter having 3dB bandwidth $f_c = \frac{1}{2\pi RC}$ has circuit shown below.

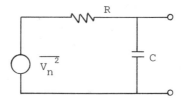

Solution: (a) The noise bandwidth is defined as

$$BW_N = \frac{1}{2|H(f_0)|^2}\int_{-\infty}^{\infty}|H(f)|^2 df:$$

Now, knowing $f_0 = \dfrac{1}{2\pi\sqrt{LC}}$, $B = \dfrac{1}{2\pi RC}$

and

$$|H(F)|^2 = \left|\dfrac{1}{\dfrac{1}{R} + j\omega C + \dfrac{1}{j\omega C}}\right|^2$$

$$= \dfrac{B^2 R^2 f^2}{f^4 - (2f_0^2 - B^2)f^2 + f_0^4}$$

Hence,

$$BW_N = \dfrac{1}{2|H(f_0)|^2} \int_{-\infty}^{\infty} |H(f)|^2 df$$

$$= \dfrac{1}{2R^2} \int_{-\infty}^{\infty} \dfrac{B^2 R^2 f^2}{f^4 - (2f_0^2 - B^2)f^2 + f_0^4} df$$

$$= \dfrac{1}{2R^2}(\pi B R^2)$$

$$= \left(\dfrac{\pi}{2}\right) B$$

(b) For the circuit shown, $H(f) = \dfrac{1}{1 + j\left(\dfrac{f}{f_c}\right)}$

Hence

$$BW_N = \dfrac{1}{2H^2(0)} \int_{-\infty}^{\infty} |H(f)|^2 df$$

$$= \dfrac{1}{2} \int_{-\infty}^{\infty} \dfrac{df}{1 + \left(\dfrac{f}{f_c}\right)^2}$$

$$= \dfrac{\pi f_c}{2}$$

$$= \dfrac{1}{4RC}$$

NOISE FIGURE AND NOISE TEMPERATURE CALCULATIONS

● **PROBLEM** 15-23

A transistor amplifier operating at a center frequency of 1 MHz at 27°C and bandwidth of 12 kHz is driven by a 750 Ω source. If the voltage gain and the input impedence of the amplifier are 50 and 750 Ω respectively, determine its noise figure.

793

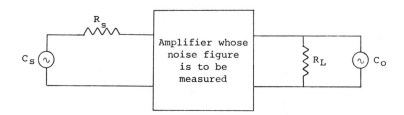

Solution: Figure (1) illustrates the simplest configuration for the measurement of noise figure of a given device. The input source e_s, is a signal generator with a 750 Ω source resistance and operating at 1 MHz.

In general, the noise figure (NF) of a device is expressed as

$$NF = 10 \log \frac{S_i/N_i}{S_o/N_o} \qquad (1)$$

where

S_i = input signal power

N_i = input noise power

S_o = output signal power

N_o = output noise power

The input noise power can be determined in terms of the rms voltage generated across the parallel combination of the source resistance R_s and the input impedance Z_{in} of the amplifier as,

$$N_i = \frac{e_N^2}{R} = \frac{\sqrt{4KT\Delta fR}}{R}$$

where e_N = rms voltage across R

R = parallel combination of Z_{in} and R_s

i.e. $\quad R = \dfrac{(R_s)(Z_{in})}{R_s + Z_{in}} = \dfrac{(750)(750)}{750 + 750} = 375 \; \Omega$

Hence $\quad e_N = \sqrt{4KT\Delta f}$

$\qquad = \sqrt{4 \times 1.38 \times 10^{-23} \times (273 + 27) \times 12 \times 10^3 \times 375}$

$\qquad = 0.2729 \; \mu V$

and $\quad N_i = (0.2729 \times 10^{-6})^2/R$

Under normal operating conditions, the signal voltage at the input is considerably at a higher level than the noise voltage generated by the $R_s//Z_{in}$ resistance.

Therefore setting $e_s = 1$ mV

the total input power S_i to the device is

$$S_i = e_s^2/R = (1 \times 10^{-3})^2/R$$

For an amplifier gain of 50, the output signal voltage will be

$$e_o = e_i \times 50 = 1 \times 10^{-3} \times 50 \text{ V}$$

With a load resistance R_L, the output signal power can be expressed as

$$S_o = e_o^2/R_L = (50 \times 10^{-3})^2/R_L$$

In the absence of any input signal, and with present day standards, the noise voltage present at the output of the amplifier can be estimated to be 25 μV.

Therefore, output noise power $N_o = (25 \times 10^{-6})^2/R_L$

Thus, from equation (1)

$$N \cdot F = 10 \log \frac{\dfrac{(1 \times 10^{-3})^2/R}{(0.2729 \times 10^{-6})^2/R}}{\dfrac{(50 \times 10^{-3})^2/R_L}{(25 \times 10^{-6})^2/R_L}}$$

$$= 10 \log 3.35$$

$$= 5.25 \text{ dB}.$$

• **PROBLEM 15-24**

For the network in the figure below let R_s = 1k, R_g = 2k and Δf (frequency range bandwidth being considered for positive frequencies only) = 5kHz   Find N_{ab} N_0', N_{cd} and N_0.

Where N_{ab} = available noise power from signal source at terminal ab.

N_{cd} = available noise power at the output terminals cd which arises within the network.

and

N_0' and N_0 = the available noise powers from the output terminals cd.

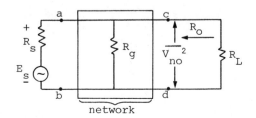

network

Solution: For a network such as given in figure, the incremental noise figure F_i is defined over a small frequency range Δf as
$$F_i = \frac{N_0}{N_0'} \tag{1}$$

Here N_0 and N_0' are the available noise powers from the output terminals cd, over the incremental frequency range Δf, when the network is noisy and noise-free respectively. It is assumed that the various noise sources contributing to N_0 and N_0' are statistically independent.

Eq. (1) may be expressed as

$$F_i = \frac{\overline{v_{no}^2}/4R_0}{\overline{v_{no'}^2}/4R_0} = \frac{\overline{v_{no}^2}}{\overline{v_{no'}^2}} = \frac{4kT\left(\frac{R_s R_g}{R_s + R_g}\right)\Delta f}{4kTR_s\left(\frac{R_g}{R_s+R_g}\right)^2 \Delta f} = 1 + \frac{R_s}{R_g}$$

where $\overline{v_{no}^2}$ and $\overline{v_{no'}^2}$ are the mean-square, open-circuit, output noise voltages when the network is noisy and noise-free, respectively. Also R_0 is the network internal resistance looking back into terminals cd, which in this case is $R_s R_g/(R_s + R_g)$

The available power gain G_a of the network in figure can be determined by

$$G_a = \frac{S_0}{S_{ab}} = \frac{N_0'}{N_{ab}} = \frac{\left(E_s \frac{R_g}{R_s + R_g}\right)^2 / \left(4 \frac{R_s R_g}{R_s + R_g}\right)}{E_s^2/4R_s} = \frac{R_g}{R_s + R_g}$$

(3)

where S_0 is the available signal power at the output terminals of a network and N_0' is the available noise power at the output terminals of a noise-free network. Also, S_{ab} and N_{ab} are the available signal and noise powers, respectively, at the output terminals of the signal source.

Now using Eq. (2) and (3),

$$F_i = 1 + \frac{R_s}{R_g} = 1 + \frac{1}{2} = 1.5$$

and
$$G_a = \frac{R_g}{R_s + R_g} = \frac{2}{1+2} = \frac{2}{3}$$

Then the available noise power from the signal source at terminal ab is
$$N_{ab} = kT\Delta f$$
$$= (4.07 \times 10^{-21})(5 \times 10^3)$$

$$= 2.04 \times 10^{-17} \text{ watts}$$

For a noise-free network, the available output noise power from terminals cd is

$$N'_0 = kTG_a \Delta f$$
$$= (4.07 \times 10^{-21})(\tfrac{2}{3})(5 \times 10^3)$$
$$= 1.36 \times 10^{-17} \text{ watts}$$

Then the available noise power at the output terminals cd which arises within the network is

$$N_{cd} = kTG_a(F_i - 1)\Delta f$$
$$= (4.07 \times 10^{-21})(\tfrac{2}{3})(0.5)(5 \times 10^3)$$
$$= 6.78 \times 10^{-18} \text{ watts}$$

Finally, the total available noise power at the output terminals cd is expressed as

$$N_0 = kTG_a F_i \Delta f$$
$$= (4.07 \times 10^{-21})(\tfrac{2}{3})(1.5)(5 \times 10^3)$$
$$= 2.04 \times 10^{-17} \text{ watts}$$

Note: (1) At the output terminals cd of the network, the total available noise power N_0 is the sum of the available noise power N_0' which originates at the signal source and the available noise N_n which arises within the network

(2) k = Boltzmann's constant = 1.38×10^{-23} joules/°K and T = absolute temperature in °K = 273 + °C.

● **PROBLEM 15-25**

A triode amplifier with the following parameters is shown below.

$R_p = R_L = 12 \text{ k}\Omega$, $R_s = 1 \text{ k}\Omega$, $C = 2 \times 10^{-9} \text{F}$,

$g_m = 2.5 \times 10^{-3}$ mho and $R_g = 80 \text{ k}\Omega$.

Determine (a) The noise figure of the amplifier,

(b) The rms noise voltage at the output,

and (c) Optimum value of the source resistance.

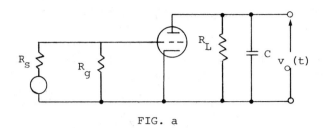

FIG. a

Solution: Figure (a) can be represented by its equivalent noise circuit as in figure (b). Here, all noise sources, (R_s, R_g and the triode) are replaced by their noiseless equivalent components along with an equivalent noise voltage source. Note that R_L in this case is assumed to be noiseless since its contribution to noise is negligible as compared to other components. The power density spectrum due to R_s, R_g and R_{eq} is $2kTR_s$, $2kTR_g$ and $2kTR_{eq}$ respectively.

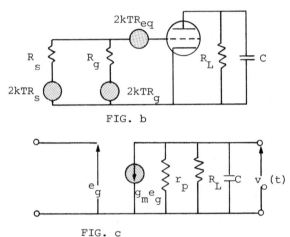

FIG. b

FIG. c

Let the relation between the output voltage v_0 and the grid voltage e_g be represented by the transfer function $H(\omega)$. In other words, $H(\omega)$ is the transfer function relating v_0 to the equivalent noise voltage source due to R_{eq}. The relation between v_0 and the noise voltage due to R_g can be expressed in terms of $H(\omega)$ as follows:

$$\frac{R_s}{R_s + R_g} H(\omega)$$

Similarly, the correspondance between v_0 and the noise voltage source due to R_s is

$$\frac{R_g}{R_s + R_g} H(\omega)$$

Now, the power spectral density, at the output, of the noise voltages due to R_{eq}, R_g and R_s is

$$2kTR_{eq}|H(\omega)|^2 \qquad (1)$$

$$2kTR_g\left(\frac{R_s}{R_s+R_g}\right)^2|H(\omega)|^2 \qquad (2)$$

and
$$2kTR_s\left(\frac{R_g}{R_s+R_g}\right)^2|H(\omega)|^2 \text{ respectively.} \qquad (3)$$

(a) For an amplifier, the noise figure is given by

$$F = 1 + \frac{S_{nao}}{S_{nso}}$$

where S_{nao} and S_{nso} are the power density spectrum of the noise in the load due to the amplifier and the source respectively. These can be expressed as

$$S_{nao}(\omega) = 2kTR_{eq}|H(\omega)|^2 + 2kTR_g\left(\frac{R_s}{R_s+R_g}\right)^2|H(\omega)|^2 \qquad (4)$$

and
$$S_{nso}(\omega) = 2kTR_s\left(\frac{R_g}{R_s+R_g}\right)^2|H(\omega)|^2 \qquad (5)$$

Thus
$$F = 1 + \frac{R_{eq} + R_g\left(\frac{R_s}{R_s+R_g}\right)^2}{R_s\left(\frac{R_g}{R_s+R_g}\right)^2}$$

$$= 1 + \frac{R_{eq}(R_s + R_g)^2 + R_g(R_s)^2}{R_s(R_g)^2}$$

Now $R_s = 1K\Omega$, $R_g = 80 \times 10^3 \Omega$ and

$$R_{eq} = \frac{2.5}{g_m} = \frac{2.5}{2.5 \times 10^{-3}} = 10^3 \Omega$$

Therefore

$$F = 1 + \frac{10^3(1 \times 10^3 + 80 \times 10^3)^2 + 80 \times 10^3(1 \times 10^3)^2}{1 \times 10^3(80 \times 10^3)^2}$$

$$= 1 + 0.01$$

$$= 1.01$$

(b) To determine the rms value of the output noise voltage, it is required to find the transfer function $H(\omega)$. From the equivalent circuit of figure 1(c), the transfer function relating the grid voltage e_g and the output voltage v_0 is given by,

$$H(\omega) = -g_m Z$$

where $\dfrac{1}{Z} = \dfrac{1}{r_p} + \dfrac{1}{R_L} + j\omega C$

$$= \dfrac{1}{12 \times 10^3} + \dfrac{1}{12 \times 10^3} + (2 \times 10^{-9})(j\omega)$$

$$= 0.083 \times 10^{-3} + 0.083 \times 10^{-3} + (2 \times 10^{-9})(j\omega)$$

$$= 0.166 \times 10^{-3} + (2 \times 10^{-9})(j\omega)$$

$$= 2 \times 10^{-9} (0.083 \times 10^6 + j\omega)$$

Thus

$$H(\omega) = -\dfrac{2.5 \times 10^{-3}}{2 \times 10^{-9}(0.083 \times 10^6 + j\omega)}$$

$$= -\dfrac{1.25 \times 10^6}{(0.083 \times 10^6 + j\omega)}$$

and

$$|H(\omega)|^2 = \dfrac{1.56 \times 10^{12}}{(6.88 \times 10^9 + \omega^2)} \qquad (7)$$

The mean square noise voltage at the output can be expressed as

$$v_0^2 = \dfrac{1}{\pi} \int_0^\infty S_{nto}(\omega)\, d\omega$$

where $S_{nto}(\omega) = S_{nao}(\omega) + S_{nso}(\omega)$

$\qquad\qquad$ = Power spectral density of the noise at the output.

From equations (4), (5) and (7),

$$S_{nto} = 2kT|H(\omega)|^2 \left[R_{eq} + R_g \left(\dfrac{R_s}{R_s + R_g}\right)^2 + R_s \left(\dfrac{R_g}{R_s + R_g}\right)^2 \right]$$

$$= 2 \times 1.38 \times 10^{-23} \times 290 \times \dfrac{1.56 \times 10^{12}}{(6.88 \times 10^9 + \omega^2)} \times 1987.16$$

$$= \dfrac{2.48 \times 10^{-5}}{(6.88 \times 10^9 + \omega^2)}$$

Therefore
$$v_0^2 = \frac{1}{\pi}\int_0^\infty \frac{2.48 \times 10^{-5}}{6.88 \times 10^9 + \omega^2}\,d\omega$$

$$= \frac{2.48 \times 10^{-5}}{\pi} \times \frac{1}{0.82 \times 10^5}\tan^{-1}\frac{\omega}{0.82 \times 10^5}\bigg|_0^\infty$$

$$= \frac{1.24 \times 10^{-10}}{0.82}$$

$$= 1.512 \times 10^{-10}$$

The rms noise voltage will be
$$\sqrt{v_0^2} = 12.2\ \mu V$$

(c) From equation (6) the noise figure is

$$F = 1 + \frac{R_{eq}(R_s + R_g)^2 + R_g R_s^2}{R_s R_g^2}$$

General, $R_s \ll R_g$. Under this condition, the above equation reduces to

$$F \cong 1 + \frac{R_{eq}}{R_s} + \frac{R_s}{R_g} \tag{8}$$

The maximum or optimum value R_s can be evaluated by differentiating equation (8) w.r.t. R_s and equating it to zero.

Thus
$$\frac{dF}{dR_s} = \frac{-R_{eq}}{R_s^2} + \frac{1}{R_g} = 0$$

or
$$R_s = \sqrt{R_{eq} R_g}$$

$$= \sqrt{10^3 \times 80 \times 10^3}\ \text{ohm}$$

$$\cong 9 \times 10^3\ \text{ohm}$$

$$= 9\ \text{K ohm}$$

The corresponding noise figure is

$$F = 1 + \frac{10^2}{9 \times 10^3} + \frac{9 \times 10^3}{80 \times 10^3}$$

$$= 1.22$$

● **PROBLEM 15-26**

A two-stage amplifier shown in figure (1) is constructed using two identical tubes with $g_m = 4.9 \times 10^{-3}$ mho and $r_p = 15000\Omega$. Calculate:

(a) noise figure of individual stages

(b) gain and the maximum output power across terminals a,b.

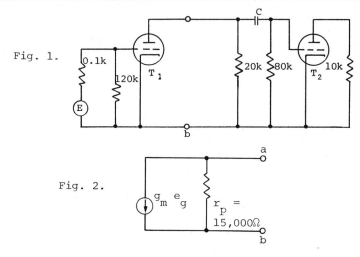

Fig. 1.

Fig. 2.

Solution: (a) To determine the noise figure of a two-stage amplifier, it is first necessary to find the noise figure of an individual stage.

The noise figure of a single-stage vacuum-tube amplifier is expressed as,

$$F = 1 + \frac{S_{na}}{S_{ns}}$$

$$= 1 + \frac{R_{eq}(R_s + R_g)^2 + R_g \cdot R_s^2}{R_s \cdot R_g^2} \quad (1)$$

where
$$R_{eq} = \frac{2.5}{g_m} \quad (2)$$

now for the first stage T_1:

$$R_{eq} = \frac{2.5}{4.9} \times 10^3 \Omega$$

$$= 510.204 \Omega$$

$$R_s = 100\Omega, \quad R_g = 120 \times 10^3 \Omega$$

Substituting in equation (1) we get

$$F_1 = 1 + \frac{510.204(100 + 120000)^2 + 1200000000}{100 \times (120000)^2}$$

$$= 6.11 \qquad (3)$$

Now in the case of the second stage T_2, R_g is a parallel combination of 20K and 80K resisters, thus

$$R_g = \frac{20K \times 80K}{100K}$$

$$= 16000 \Omega.$$

R_s of the second stage is obviously the output impedance of the first stage which is r_p of tube T_1.

$$\therefore R_s = r_p = 15000 \Omega.$$

Hence noise-figure is

$$F_2 = 1 + \frac{(510.204)(15000 + 16000)^2 + (16000)(15000)^2}{(15000) \cdot (16000)^2}$$

$$= 2.065$$

(b) At a particular frequency, the gain is:

$$G_a = \frac{\text{available signal power at the load}}{\text{available signal power at the source}}$$

For the first stage, $R_s = 100 \Omega$. If the strength of the incoming signal is E volts, then available power is obtained by connecting 100Ω across the source terminals.

Thus, the maximum available power at the source is,

$$\left(\frac{E}{200}\right)^2 \times 100 = \frac{E^2}{400} \text{ Watts} \qquad (4)$$

The equivalent circuit of the first stage when looked into from the output terminals is shown in Fig. (2).

Let e_g = grid voltage for the first stage

g_m = transconductance of the first stage.

Then $g_m e_g$ will represent an equivalent current source as shown in figure (2).

Referring to Fig. (1), the grid voltage e_g is the drop across the 120KΩ resistor, hence

$$e_g = \frac{120 \times 10^3}{(120 \times 10^3) + (0.1 \times 10^3)} E$$

$$= 0.99E$$

hence,

$$g_m \cdot e_g = 2.5 \times 10^{-3} \times 0.99E$$

$$= 2.498 \times 10^{-3} E.$$

Maximum power is the power delivered by the first stage to a matched impedance across a,b.

Max. power is the power delivered to a resistor of value $r_p = 15000\Omega$.

Therefore maximum output power across ab is

$$= \left(\frac{g_m \cdot e_g}{2}\right)^2 \times 15000$$

$$= \left(\frac{2.498 \times 10^{-3} \times E}{2}\right)^2 \times 15000$$

$$= 0.0234 E^2 \qquad (5)$$

● **PROBLEM 15-27**

The following figure shows the block schematic of a communication receiver consisting of a mixer stage followed by an IF amplifier. Assuming both the gain G_{a_1} and the noise figure F_{i_1} of the mixer stage to be constant over the bandpass of the IF amplifier, calculate:

(a) Noise figure F of the mixer intermediate frequency amplifier combination.

(b) Total noise at the output of the IF amplifier in watts.

(c) Noise power at point C from the antenna, i.e., N_{ac}

(d) Noise N_{n_1c} at the IF amplifier output, generated by the noisy mixer stage.

(e) Noise at point C which is generated by the amplifier stage.

(f) Strength of the signal at the input of mixer so as to have unity $\frac{S}{N}$ ratio of the amplifier output.

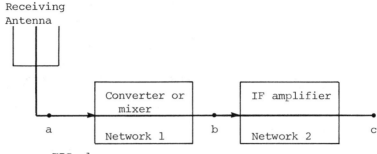

FIG. 1

$F_{i1} = 11.20\text{db} = 13.18$ $F_{e2} = 5.26\text{db} = 3.357$

$G_{m1} = 4.9\text{db} = .32$ $G_{m2} = 70\text{dB} = 10^7$

$B_{e1} = 17\text{mc}$ $B_{e2} = 6\text{mc}$

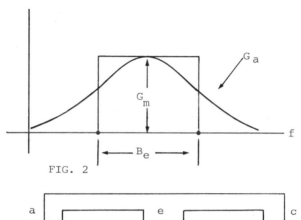

FIG. 2

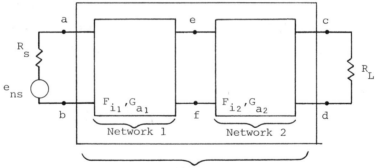

Cascade networks

FIG. 3

Solution: (a) The overall or effective noise figure is

$$F_e = \frac{N_0}{N_0'}$$

$$= \frac{\int_0^\infty k.T.F_i.G_a \cdot dF}{\int_0^\infty k.T.G_a \cdot df} \tag{1}$$

where

$$F_i = \frac{S_s/N_s}{S_0/N_0} = \frac{N_0}{N_0'} = \frac{N_0}{G_a \cdot N_s}$$

$$= \text{incremental noise figure.}$$

Since $k \cdot T$ is a constant, equation (1) simplifies to

$$F_e = \frac{N_0}{N_0'} = \frac{k \cdot T \int_0^\infty F_i \cdot G_a \cdot df}{k \cdot T \cdot B_e \cdot G_m}$$

$$= \int_0^\infty F_i \cdot G_a \cdot df \Big/ (B_e \cdot G_m) \qquad (2)$$

Fig. (2) shows variation of G_a with frequency f where

$$B_e \cdot G_m = \int_0^\infty G_a \cdot df \qquad (2a)$$

It is obvious that G_m is the maxima of the curve and B_e is called equivalent noise bandwidth.

Let us assume that the mixer and If amplifier stage is replaced by two blocks as shown in fig. (3), their respective gains and noise figures are shown in respective blocks. In this fig., the antenna is replaced by a voltage source e_{ns} with source resistance R_s. This network is driving R_L, say a loudspeaker in the case of a radio set.

The effective noise figure is $F_{e_{12}}$ which is

$$F_{e_{12}} = \frac{N_0}{N_0'} = \frac{k \cdot T \int_0^\infty F_{i_{12}} \cdot G_{a_{12}} \cdot df}{k \cdot T \, B_{e_{12}} \cdot G_{m_{12}}}$$

$$= \frac{\int_0^\infty F_{i_{12}} \cdot G_{a_{12}} \cdot df}{B_{e_{12}} \cdot G_{m_{12}}} \qquad (3)$$

where $B_{e_{12}} \cdot G_{m_{12}} = \int_0^\infty G_{a_{12}} \, df$

In order to calculate the effective noise figure, incremental noise figure must be known and it is given by

$$F_{i_{1n}} = F_{i_1} + \frac{F_{i_2}-1}{G_{a_1}} + \frac{F_{i_3}-1}{G_{a_1} \cdot G_{a_2}} + \cdots + \frac{F_{i_n}-1}{G_{a_1} \cdot G_{a_2} \cdots G_{a(n-1)}}$$

$$(4)$$

Hence
$$F_{e_{12}} = \frac{\int_0^\infty F_{i_1} G_{a_1} G_{a_2} df + \int_0^\infty F_{i_2} G_{a_2} df - \int_0^\infty G_{a_2} \cdot df}{B_{e_{12}} \cdot G_{m_{12}}}$$

where $G_{a_{12}} = G_{a_1} \cdot G_{a_2}$

Now assuming $G_{a_1} = G_{m_1}$ and $B_{e_{12}} = B_{e_{21}}$

$$F_{e_{12}} = \frac{F_{i_1} G_{m_1} \int_0^\infty G_{a_2} df}{G_{m_1} \cdot G_{m_2} \cdot B_{e_2}} + \frac{\int_0^\infty F_{i_2} G_{a_2} df}{G_{m_1} \cdot G_{m_2} \cdot B_{e_2}} - \frac{\int_0^\infty G_{a_2} df}{G_{m_1} \cdot G_{m_2} \cdot B_{e_2}}$$

$$= F_{i_1} + \frac{F_{e_2}}{G_{m_1}} - \frac{1}{G_{m_1}}$$

$$= F_{i_1} + \frac{F_{e_2} - 1}{G_{m_1}} \quad \text{where } G_{m_{12}} = G_{m_1} \cdot G_{m_2} \quad (5)$$

Using these equations, the quantities can be calculated as follows:

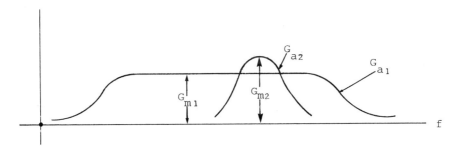

FIG. 4   Plot of G_{a1} and G_{a2} versus frequency

Using Eq. (5) and the data given in Fig. (1), we have

$$F_{e_{12}} = F_{i_1} + \frac{F_{e_2} - 1}{G_{m_1}}$$

$$= 13.18 + \frac{3.357 - 1}{0.32}$$

$$= 20.54$$

It can be seen that the low value of G_{m_1} makes the IF amplifier noise very significant

(b)
$$N_c = k \cdot T \int_0^\infty F_{i_{12}} \cdot G_{a_{12}} \cdot df \quad \text{using Eq. (3)}$$

$$= k \cdot T \cdot F_{e_{12}} \cdot B_{e_{12}} \cdot G_{m_{12}}$$

$$= (407 \times 10^{-20})(20.54)(6 \times 10^6)(10^7 \times 0.32)$$

$$= 1.60 \times 10^{-6} \text{ W}.$$

where $kT = (1.38 \times 10^{-23}) \times 290$

$$= 0.407 \times 10^{-20}$$

(c) The noise power at point "a" goes through the mixer and the amplifier and is given by

$$N_a = k \cdot T \cdot B_{e12}$$

Then

$$N_{ac} = N_a \cdot G_{m12} = k \cdot T \cdot B_{e12} \cdot G_{m12}$$

$$= (0.407 \times 10^{-20})(6 \times 10^6)(\frac{10^7}{.32})$$

$$= 7.63 \times 10^{-7} \text{ watts}.$$

(d) The noise generated in the mixer, mainly shot noise and stray noise picked up by the antenna, appears at the output of the IF amplifier. This can be expressed as,

$$N_{n_1 b} = k \cdot T \cdot B_{e2} \cdot G_{m1} \cdot (F_{i_1} - 1)$$

Then $N_{n_1 c} = N_{n_1 b} \cdot G_{m2}$

$$= (0.407 \times 10^{-20})(6 \times 10^6)(0.32)(12.18)(10^7)$$

$$= 9.518 \times 10^{-7} \text{ watts}.$$

(e) $N_{n_2 c} = k \cdot T \cdot B_{e2} \cdot G_{m2} \cdot (F_{e2} - 1)$

$$= (0.407 \times 10^{-20})(6 \times 10^6)(10^7)(2.357)$$

$$= 5.755 \times 10^{-7} \text{ watts}.$$

It can be seen that

$$N_c = N_{ac} + N_{n_1 c} + N_{n_2 c}$$

(f) Now assuming that the signal power spectral density at point "a" is a constant,

$$S_a = \frac{S_c}{G_{m12}}$$

$$= \frac{N_c}{10^7 \times 0.32}$$

$$= 5 \times 10^{-13} \text{ W}.$$

The incremental noise temperature is defined as the ratio of available noise power from the output terminals to the available noise from a resistor. Hence

$$t_i = \frac{N_0}{k \cdot T \cdot \Delta f} = \frac{k \cdot T \cdot F_i \cdot G_a \, \Delta f}{k \cdot T \cdot \Delta f}$$

$$= F_i \cdot G_a$$

Hence

$$t_{i_1} = F_{i_1} \cdot G_{a_1} = F_{i_1} \cdot G_{m_1}$$

$$= \frac{13.18}{3.125}$$

$$= 4.21$$

$$= 6.24 \text{ db}.$$

• **PROBLEM** 15-28

An amplifier of power gain 20 db has an input consisting of 100 µW signal power and 1 µW noise power. If the amplifier contributes an additional 100 µW of noise, determine

(a) the output signal to noise ratio,

(b) the noise factor and

(c) the noise figure.

Solution: Power gain G = 20dB

i.e., $\quad 10 \log G = 20$ dB

therefore, $\quad G = 100$

(a) The output signal S_{output} from the amplifier is

$$S_{output} = S_{input} \times \text{Gain}$$

$$= 100 \times 10^{-6} \times 100$$

$$= 10^{-2} \text{ W}$$

The output noise is calculated as follows:

$$N_{output} = N_{input} \, G + \text{Noise generated by amplifier}$$

$$= (1 \times 10^{-6} \times 100) + (100 \times 10^{-6})$$

$$= 200 \ \mu W$$

$$\therefore \quad \frac{S_{output}}{N_{output}} = \text{output } \frac{S}{N} \text{ ratio} = \frac{10^{-2}}{2 \times 10^{-4}} = 50$$

(b) The noise factor F of a system is the ratio of input S/N to output S/N.

Thus,
$$F = \frac{S_{input}/N_{input}}{S_{output}/N_{output}}$$

$$= \frac{100 \times 10^{-6}/1 \times 10^{-6}}{50}$$

$$= 2.$$

(c) The noise figure NF = 10 log F

$$= 10 \log 2$$
$$= 3 dB.$$

• **PROBLEM 15-29**

Calculate the noise temperature T_s of the series combination of two noisy resistors shown in Fig. 1(a). Assumed that the individual resistors are at different physical temperatures.

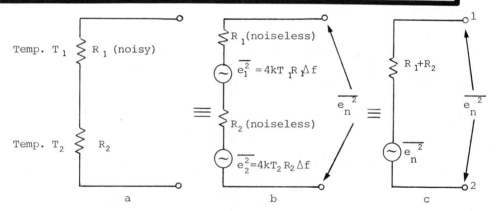

FIG. 1 (a) Series combination of two noisy resistors R_1 and R_2 at temperatures T_1 and T_2
(b) Equivalent ckt. with noiseless resistors and noise voltage generators
(c) Thevenin equivalent with a single noise voltage generator

Solution: The noise voltage $e_n(t)$ that appears across the terminals of a resistor R is found to have a Gaussian distribution. Assuming a very narrow band of frequency df, the mean square voltage is equal to,

$$\overline{e_n^2} = 4k \cdot T \cdot R \cdot df \qquad (1)$$

where T is the temperature in Kelvin, and k is the Boltzmann's constant. Further, this voltage is independent of the center frequency of the filter with bandwidth df. Fig. (b) depicts the equivalent circuit of Fig. (a). In this circuit, the noisy resistors are replaced by a combination of noiseless resistors and noise voltage generators. Fig. (c) is the equivalent of Fig. (b) after applying Therenin's theorem.

Since each resistor acts independently as far as its contribution to total noise is concerned, the total noise voltage, across terminals 1 and 2, is

$$\overline{e_1^2(t)} + \overline{e_2^2(t)} = \overline{e_n^2(t)} \qquad (2)$$

Substituting for $\overline{e_1^2(t)}$ and $\overline{e_2^2(t)}$ as in Eq. (1),

$$4k \cdot T_1 \cdot R_1 \cdot df + 4k \cdot T_2 \cdot R_2 \cdot df = 4k \cdot T_S \cdot (R_1 + R_2) df$$

or

$$T_S = \frac{T_1 R_1 + T_2 R_2}{R_1 + R_2}$$

This shows that the noise temperature of a source is not necessary equal to its physical temperature.

If, however, $T_1 = T_2 = T$, then $T_S = T$.

● **PROBLEM** 15-30

Find the mean squared terminal noise voltage in a network consisting of a parallel combination of a resistor R and a capacitor C. R and C are in thermal equilibrium at the standard noise temperature 290°K. Assume that the effective frequency band is infinite.

Solution: Since the capacitor C and the resistor R are connected in parallel, the effective impedance Z_C is given by

$$Z_C = \frac{1}{\frac{1}{R} + j\omega C} = \frac{R}{1 + j\omega RC} = \frac{R(1 - j\omega RC)}{1 + \omega^2 R^2 C^2}$$

$$= \frac{R}{1 + \omega^2 R^2 C^2} - j\frac{\omega CR^2}{1 + \omega^2 R^2 C^2}$$

In a system, the contribution to noise, by the imaginary part of the effective impedance is negligible when compared to that of the real part.

The real part of an impedance is in general a function of frequency.

Hence, the voltage noise source equation may be written as

$$\overline{V}^2 = 4kT \int_{f_1}^{f_2} R_e(Z) df$$

Since the effective frequency band is infinite,

$$\overline{V}^2 = 4kT \int_0^\infty \frac{R}{1 + \omega^2 C^2 R^2} df \qquad (1)$$

Let $u = \omega RC = 2\pi f RC = 2\pi RCf$

Therefore, $du = 2\pi RC\, df$

Now, substituting the above results in equation (1),

$$\overline{V}^2 = \frac{4kTR}{2\pi RC} \int_0^\infty \frac{du}{1 + u^2}$$

$$= \frac{2kT}{\pi C} \int_0^\infty \frac{du}{1 + u^2}$$

$$= \frac{2kT}{\pi C} \left[\tan^{-1} u\right]_0^\infty$$

$$= \frac{2kT}{\pi C} \left[\frac{\pi}{2} - 0\right]$$

$$\overline{V}^2 = \frac{kT}{C}$$

where T is the absolute temperature in degrees Kelvin.
k is the Boltzmann's constant

$$= 1.38 \times 10^{-23} \text{ joule/degree Kelvin}$$

Δf is the bandwidth in hertz.

• **PROBLEM 15-31**

Two resistors R_1 and R_2 at temperatures T_1 and T_2 respectively are connected in series as shown in the figure. If a capacitor C_1 is placed across R_1, compute the effective noise temperature of the combination.

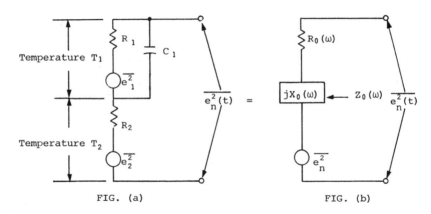

FIG. (a) FIG. (b)

Solution: Since both the resistors are subjected to different temperatures, they can be shown by pure resistors R_1 and R_1, each in series with a voltage source of value $\overline{e_1^2}$ and $\overline{e_2^2}$. Let $e_n(t)$ be the total voltage across the output terminals.

The mean-squared voltage $\overline{e_{n_1}^2(t)}$, due to R_1 is given by

$$\overline{e_{n_1}^2(t)} = \overline{e_1^2(t)} \left| \frac{1}{1+j\omega R_1 C_1} \right|^2$$

$$= \frac{\overline{e_1^2(t)}}{1 + \omega^2 R_1^2 C_1^2} \tag{1}$$

and that due to R_1 is

$$\overline{e_{n_2}^2(t)} = \overline{e_2^2(t)} \tag{2}$$

The total voltage $\overline{e_n^2(t)} = \frac{2k \cdot T \cdot R \cdot d\omega}{\pi}$

$$= \overline{e_{n_1}^2(t)} + \overline{e_{n_2}^2(t)}$$

$$= 2k \left[\frac{T_1 \cdot R_1}{1 + \omega^2 \cdot R_1^2 \cdot C_1^2} + T_2 \cdot R_2 \right] \frac{d\omega}{\pi} \tag{3}$$

The total output impedance of the series combination is

$$Z(\omega) = R_1 \| \frac{1}{j\omega C_1} + R_2$$

$$= R_2 + \frac{R_1 \left(\frac{1}{j\omega C_1}\right)}{R_1 + \frac{1}{j\omega C_1}}$$

$$= R_2 + \frac{R_1(1 - j\omega R_1 C_1)}{1 + \omega^2 \cdot R_1^2 \cdot C_1^2}$$

$$= R_2 + \left(\frac{R_1}{1+\omega^2 \cdot R_1^2 \cdot C_1^2}\right) - j\left(\frac{\omega \cdot R_1^2 \cdot C_1}{1+\omega^2 \cdot R_1^2 \cdot C_1^2}\right)$$

Hence, resistive part is $= R_2 + \dfrac{R_1}{1+\omega^2 \cdot R_1^2 \cdot C_1^2}$

$$= R.$$

From the definition of T_S we have

$$\overline{e_n^2(t)} = 2k \cdot T_S \cdot R \frac{d\omega}{\pi}$$

$$= 2k \cdot T_S \left[R_2 + \frac{R_1}{1+\omega^2 \cdot R_1^2 \cdot C_1^2}\right] \frac{d\omega}{\pi} \quad (4)$$

Comparing equations (3) and (4)

$$T_S \left[R_2 + \frac{R_1}{1+\omega^2 R_1^2 C_1^2}\right] = \frac{T_1 R_1}{1+\omega^2 R_1^2 C_1^2} + T_2 R_2$$

$$\therefore T_S = \frac{T_1 R_1 + T_2 R_2 (1+\omega^2 R_1^2 C_1^2)}{R_1 + R_2 (1+\omega^2 R_1^2 C_1^2)}$$

● **PROBLEM** 15-32

The three stages of a cascaded amplifier, have power gains $G_1 = 15$, $G_2 = 10$ and $G_3 = 5$ respectively. If the effective input noise temperatures are $T_{e1} = 1350K$, $T_{e2} = 1700K$ and $T_{e3} = 2500K$, determine,

(a) overall effective noise temperature and

(b) ordering of the cascaded stages to achieve optimum noise performance.

Solution: (a) For an M-stage cascade amplifier, the effective noise temperature is given as,

$$T_e = T_{e1} + \frac{T_{e2}}{G_1} + \frac{T_{e3}}{G_1 G_2} + \cdots + \frac{T_{eM}}{G_1 G_2 \cdots G_{M-1}}$$

where T_{eM} and G_M are the effective noise temperature and power gain of the Mth stage.

Since it is a 3-stage cascade amplifier, the overall noise temperature is

$$T_e = 1350 + \frac{1700}{15} + \frac{2500}{150}$$

$$= 1350 + 113.33 + 16.66$$

$$= 1479.99K \cong 1480K \text{ (approximately)}$$

(b) From the previous stage, it is observed that the total contribution to T_e by the second and third stages are much smaller than that of the first stage, since G_1 and G_1G_2 are in the denominator.

To reduce noise and to improve noise performance, the stage with the highest gain should be first, followed by other stages with decreasing gains.

• **PROBLEM 15-33**

Compute the noise temperature of the combination of two resistors R_1 and R_2, subjected to temperatures $T_1 = 280°K$ and $T_2 = 400°K$ respectively.

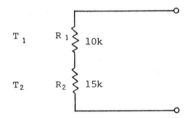

Solution: The noise voltage across a resistor has a mean square form expressed as

$$v^2 = 4kTBR$$

where   k = Boltzmann's constant

T = Temperature in degree Kelvin

and   B = Bandwidth of operation.

Thus, for a series circuit with resistors R_1 and R_2 at temperatures T_1 and T_2, the total noise voltage will be

$$v_n^2 = 4kT_1R_1B + 4kT_2R_2B \qquad (1)$$

The equivalent resistance of the combination is $R_1 + R_2$ and the total noise power is

$$P = \frac{v_n^2}{4(R_1 + R_2)} \qquad (2)$$

The general expression for the equivalent noise temperature is

$$T = \frac{P}{K \cdot B}$$

Substituting for P from equation (2)

$$T = \frac{R_1 \cdot T_1 + R_2 \cdot T_2}{R_1 + R_2} \qquad (3)$$

$$= \frac{(10 \times 10^3)(280) + (15 \times 10^3)(400)}{(25 \times 10^3)}$$

$$= 352°K.$$

• **PROBLEM 15-34**

A block diagram of a receiver system with an antenna is shown below. The loss factor of the antenna lead-in cable is $L = 1.5$ dB $= F_1$. The noise figure and gain of different stages are shown in the respective blocks.

Calculate, (a) the overall noise figure and noise temperature of the system,

(b) the noise figure and noise temperature of the system with cable and preamplifier stage interchanged.

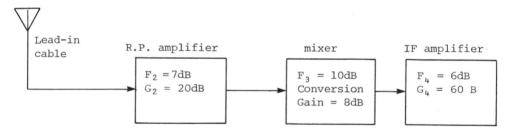

Solution: (a) The overall noise figure of an M-stage cascaded system is given by

$$F_{t_1} = F_1 + \frac{F_2 - 1}{g_{a_1}} + \frac{F_3 - 1}{g_{a_1} \cdot g_{a_2}} + \ldots$$

where $g_a = \dfrac{\text{available power spectral density at the two port output}}{\text{available power spectral density at the source output}}$

Substituting values of F_1, F_2 etc. we obtain,

$$F_{t_1} = 1.41 + \frac{5.01 - 1}{1/1.41} + \frac{10 - 1}{100/1.41} + \frac{3.98 - 1}{(100)(6.3)/1.41}$$

$$= 1.41 + 5.65 + 0.127 + (6.67 \times 10^{-3})$$

$$= 7.19$$

$$= 8.57 \text{ dB}.$$

The overall effective noise temperature is

$$T_e = T_0(F_{t_1} - 1)$$

$$= 290(7.19 - 1)$$

$$= 1795.1°K.$$

(b) After interchanging the cable and the preamplifier stage, the new noise figure is

$$F_{t_2} = 5.01 + \frac{1.41-1}{100} + \frac{10-1}{100/1.41} + \frac{3.98-1}{100(6.3)/1.41}$$

$$= 5.14777$$

$$= 7.116 \text{ dB}.$$

The new temperature is

$$T_e = 290(5.14777 - 1)$$

$$= 1202.9°K.$$

These two are essentially determined by the noise level of the RF preamplifier.

● **PROBLEM 15-35**

Assuming a bandwidth B of 16 MHz and T_s = 290°K, calculate the following, for a radio receiver whose gain G = 50 dB and noise figure F = 20 dB.

(a) Effective front-end temperature T_e.

(b) Noise N_i due to the receiver as well as due to the source.

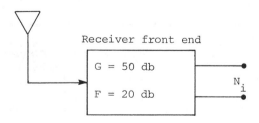

Solution: (a) The temperature T_e of the receiver is calculated using the equation for noise figure, stated below.

$$F = \text{noise figure} = 1 + \frac{T_e}{T_s}$$

Hence

$$T_e = (F-1)T_s$$
$$= (20-1)290$$
$$= 5510°K$$

This is not the physical temperature of the front-end, but the equivalent temperature of a resistor at the input that would produce the same amount of predetection noise as the receiver with noise figure of 20 dB.

(b) The noise at the output end is

$$N_i = G \cdot N_a \cdot F$$

where $N_a = k \cdot T_0 \cdot B$ = Average antenna noise.

T_0 = source input temperature = 290°K.

Hence
$$N_i = G \cdot k \cdot T_0 \cdot B \cdot F$$
$$= 10^5 (1.38 \times 10^{-23})(290)(16 \times 10^6)100$$

because G and F must be expressed as ratios

$$10 \log_{10} G = 50$$

Therefore $G = 10^5$

and
$$10 \log_{10} F = 20$$

Therefore $F = 100$

and
$$N_i = 0.64 \; \mu W.$$

● **PROBLEM 15-36**

A communication system with a preamplifier of noise figure 2 dB and gain 20 dB is shown below. A connecting cable with noise figure $F = 4$ and gain $G = 0.33$ is used for interconnecting the amplifier and the receiver of gain $G = 80$ dB and noise figure $F = 13$ dB. Assuming a suitable bandwidth and a loss less antenna lead in cable, calculate the system's noise figure.

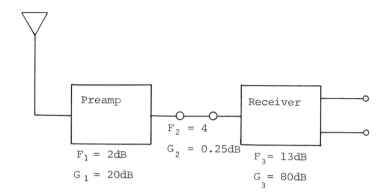

Solution: Assuming equal temperatures for all the three stages i.e., preamplifier, connecting cable and receiver, the noise figure of the system is

$$F = F_1 + \frac{(F_2-1)}{G_1} + \frac{(F_3-1)}{G_1 G_2} \tag{1}$$

This is obtained from the general formula for the noise figure of a system with M stages.

Substituting values for all quantities in equation (1) yields,

$$F = 2 + \frac{(4-1)}{20} + \frac{(13-1)}{20 \times 0.25} = 4.55$$

$$= 6.58 \text{ dB}.$$

Note that, the addition of a preamplifier results in the reduction of the overall noise figure of the system.

The overall temperature of the system can be expressed as

$$T_e = T_{e_1} + \sum_{K=2}^{M} \frac{T_{e_K}}{\prod_{\ell=1}^{K-1} G_e}$$

$$= T_{e_1} + \frac{T_{e_2}}{G_1} + \frac{T_{e_3}}{G_1 G_2}$$

The effective temperatures T_{e_2} and T_{e_3} can be determined from the identity

$$F = 1 + \frac{T_e}{T_s} \tag{2}$$

or

$$T_e = (F-1) T_s$$

where T_s = Source temperature in °K = 290°K (assumed)

Therefore $T_{e_1} = (F_1-1)T_s = (2-1) \times 290 = 290$,

$$T_{e_2} = (F_2-1)T_s = (4-1) \times 290 = 870$$

and $T_{e_3} = (F_3-1)T_s = (13-1) \times 290 = 3480$

Thus
$$T_e = 290 + \frac{870}{20} + \frac{3480}{20 \times 0.25}$$

$$= 1029.5°K$$

Expressing the noise figure in terms of the overall temperature and the source temperature as in equation (2),

$$F = 1 + \frac{T_e}{T_s}$$

$$= 1 + \frac{1029.5}{290}$$

$$= 4.55$$

$$= 6.58 \text{ db.}$$

● **PROBLEM** 15-37

An antenna lead-in cable with attenuation loss L=4 and gain G = 0.25 is used for feeding signal to a receiver with F = 10 db and gain G = 80 db. Assuming that cable is being subjected to physical temperature of 290°K, calculate

(a) noise figure of the system.

(b) overall noise temperature.

Solution: This is basically a two-stage system in which antenna lead-in cable forms the first stage and the receiver, the second stage.

The noise figure of the cable is F_1 = L = 4 because the cable is at a physical temperature $T = T_s = 290°K$ and at this temperature,
$$F_1 = L = 1/G = 4.$$

The noise figure of the system is

$$F = F_1 + \frac{(F_2-1)}{G_1}$$

where F_2 is the noise figure of the receiver = 10 db.

820

Thus,
$$F = 4 + \frac{(10-1)}{1/4}$$
$$= 40$$
$$= 16.02 \text{ db}.$$

Let T_e be the effective receiver system temperature, hence T_e of the cable, which forms the first stage is given by

$$F = 1 + \frac{T_{e_1}}{T_s}$$

hence
$$T_{e_1} = (4-1) T_s$$
$$= 870°K.$$

Similarly
$$T_{e_2} = (10-1) 290$$
$$= 2610°K$$

The overall noise temperature can be determined from the equality.

$$T_e = T_{e_1} + \frac{T_{e_2}}{G_1}$$
$$= 11,310°K.$$

CHAPTER 16

DATA TRANSMISSION

● **PROBLEM** 16-1

A noise signal n(t) having an autocorrelation function $R_n(\tau) = \sigma^2 \cdot e^{-|\tau/\tau_0|}$ is applied to an integrator along with a binary voltage signal.

(a) Consider only the noise signal, determine its power spectral density and the mean square value $\left[n_0(T)\right]^2$ of the noise at the output of the integrator at time t = T. Assume noise is input to the integrator at time t = 0.

(b) Find the ratio of the signal to the rms noise voltage at the integrator output at time t = T, where T is the time for which voltage signal is maintained either at +v or -v volts.

<u>Solution</u>: (a) The power spectral density, $G_n(f)$, of the noise is related to the autocorrelation function as follows:

$$G_n(f) = \int_{-\infty}^{\infty} R_n(\tau) \cdot e^{(-j2\pi \cdot f \cdot \tau)} \cdot d\tau$$

Hence, substitute in given $R_n(\tau)$,

$$G_n(f) = \int_{-\infty}^{\infty} \sigma^2 e^{-\left(\left|\frac{\tau}{\tau_0}\right| - j2\pi \cdot f \cdot \tau\right)} d\tau$$

$$= 2\sigma^2 \int_0^\infty e^{\left(-\frac{\tau}{\tau_0}\right)} \cos(2\pi f\tau) \cdot d\tau$$

$$= \frac{2\sigma^2 \cdot (1/\tau_0)}{\left(\frac{1}{\tau_0}\right)^2 + (2\pi \cdot f)^2}$$

$$= \frac{2\sigma^2 \cdot \tau_0}{1 + (2\pi \cdot \tau_0 \cdot f)^2}$$

(b) The mean square value of the noise is evaluated as follows:

$$\overline{[n_0(T)]^2} = \overline{\left[\frac{1}{T}\int_0^T n(t) \cdot dt\right]^2}$$

$$= \frac{1}{T^2}\int_0^T dt_1 \int_0^T \overline{n(t_1)n(t_2)} dt_2$$

$$= \frac{1}{T^2}\int_0^T dt_1 \int_0^T R(t_1 - t_2) dt_2$$

$$= \frac{1}{T^2}\int_0^T dt_1 \int_0^T \sigma^2 \cdot e^{\left(\frac{-|t_1-t_2|}{\tau_0}\right)} \cdot dt_2$$

$$= \frac{\sigma^2}{T^2}\int_0^T dt_1 \left[\int_0^{t_1} e^{\frac{-(t_1-t_2)}{\tau_0}} dt_2 + \int_{t_1}^T e^{\frac{(t_1-t_2)}{\tau_0}} dt_2\right]$$

$$= \frac{\tau_0 \cdot \sigma^2}{T^2}\int_0^T \left[2 - e^{(t_1/\tau_0)} \cdot e^{(-T/\tau_0)} - e^{(-t_1/\tau_0)}\right] dt_1$$

$$= \frac{2\tau_0 \cdot \sigma^2}{T^2}\left[T - \tau_0(1 - e^{-T/\tau_0})\right]$$

(c) The integrator output of the signal at time T is

$$|S_0(T)| = \frac{VT}{T}$$

and the rms noise voltage output is

$$[n_0(T)]_{rms} = \{\overline{[n_0(T)]^2}\}^{\frac{1}{2}}$$

823

Hence, the ratio

$$\frac{|S_0(T)|}{[n_0(\tau)]_{rms}} = \frac{VT/\tau}{\{[n_0(T)]^2\}^{\frac{1}{2}}}$$

$$= \frac{VT/\tau}{\frac{\sqrt{2\tau_0} \cdot \sigma}{\tau} \sqrt{T-\tau_0(1-e^{-T/\tau_0})}}$$

$$= \frac{VT}{\sqrt{2\tau_0} \cdot \sigma \cdot \sqrt{T-\tau_0(1-e^{-T/\tau_0})}}$$

● **PROBLEM** 16-2

A receiver system shown in Fig. 1 is used for receiving a binary coded baseband PCM signal of transmitting voltage +v or -v volt. Assuming that the threshold voltage used for distinguishing the transmitted signal is v_t and white Gaussian noise is added during transmission.

(a) Compute the probability of making an error in decision.

(b) Find v_t so that the probability of error is minimum. Given the probability of transmitting +v or -v is 0.75 and 0.25 respectively.

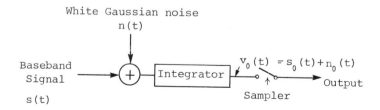

Fig. 1

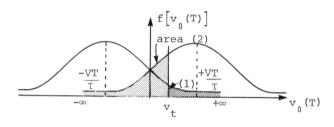

Fig. 2.

<u>Solution</u>: (a) Since the probability of transmitting +v or -v is equal, hence,

$$P[+v] = P[-v] = 1/2$$

The output of the integrator before the sampler is $v_0(t) = s_0(t) + n_0(t)$. The signal output for the given input will consist of a train of the ramp of duration T. At the end, the ramp voltage will be $+\frac{vT}{\tau}$ or $-\frac{vT}{\tau}$ depending on whether $+v$ or $-v$ is transmitted. The Gaussian white noise $n_0(t)$ starts with $n_0(0) = 0$ and has random value $n_0(T)$ at the end of each interval. Hence, the output is

$$v_0(T) = s_0(T) + n_0(T) \tag{1}$$

The plot of $f[v_0(T)]$, i.e., the probability density of output against $v_0(T)$ is shown in Fig. 2. Its distribution is also Gaussian, but peak occurs at $\frac{vT}{\tau}$ or $\frac{-vT}{\tau}$ depending on input $+v$ or $-v$. This shifting of peak is a result of $s_0(T) = \frac{\pm vT}{\tau}$. If $-v$ has been transmitted and if the noise is excessive, the output can be more than v_t, resulting in an error. The probability of such error i.e., $v_0(T) > v_t$ when $-v$ is transmitted is

$$P[v_t < v_0(T), -v] = \text{area of shaded portion (1). (see Fig.2)}$$

$$= \int_{v_t}^{\infty} \frac{e^{\frac{-(n_0^2(T))}{2\sigma_0^2}}}{\sqrt{2\pi \cdot \sigma_0^2}} \, dn_0(T) \tag{2}$$

Hence, putting the value of $n_0(T)$ in terms of $v_0(T)$, Eq.(2) results

$$P\left[v_t < v_0(t), -v\right] = \int_{v_t}^{\infty} \frac{e^{\frac{-(v_0 - vT/\tau)^2}{2\sigma_0^2}}}{\sqrt{2\pi\sigma_0^2}} \, dv_0$$

Similarly,

$$P\left[v_0(T) < v_t, +v\right] = \int_{-\infty}^{v_t} \frac{e^{\frac{-(v_0 - vT/\tau)^2}{2\sigma_0^2}}}{\sqrt{2\pi\sigma_0^2}} \, dv_0$$

Hence, total probability of making an error is

$$P_e = P[+v]P[v_0(T) < v_t, +v] + P[-v]P[v_t < v_0(\tau), -v]$$

$$= \frac{1}{2}\left[\int_{-\infty}^{v_t} \frac{e^{\frac{-(v_0 - vT/\tau)^2}{2\sigma_0^2}}}{\sqrt{2\pi \cdot \sigma_0^2}} \, dv_0 + \int_{v_t}^{\infty} \frac{e^{\frac{-(v_0 - vT/\tau)^2}{2\sigma_0^2}}}{\sqrt{2\pi \cdot \sigma_0^2}} \, dv_0\right] \tag{3}$$

(b) In order to find the value of v_t which will result in the minimum value of P_e, it is necessary to differentiate equation (3). Hence, in equation (3), dv_0 is replaced by dv_t. Now, the output voltage v_0 is either less than or greater than v_t. Hence, $dv_0 = dv_t$ and limits of integration will remain the same.

The first derivative of P_e is

$$\frac{dP_e}{dv_t} = \frac{3}{4} \frac{e^{-\frac{(v_t - vT/\tau)^2}{2\sigma_0^2}}}{\sqrt{2\pi\sigma_0^2}} - \frac{1}{4} \frac{e^{-\frac{(v_t - vT/\tau)^2}{2\sigma_0^2}}}{\sqrt{2\pi\sigma_0^2}}$$

and

$$\frac{d^2 P_e}{dv_t^2} = -ve$$

Now, to find the value of a minima, the first derivative is equated to zero. Hence,

$$\frac{3}{4} \frac{e^{-\frac{(v_t - vT/\tau)^2}{2\sigma_0^2}}}{\sqrt{2\pi \cdot \sigma_0^2}} - \frac{1}{4} \frac{e^{-\frac{(v_t - vT/\tau)^2}{2\sigma_0^2}}}{\sqrt{2\pi\sigma_0^2}} = 0$$

$$3 e^{\frac{v_t vT}{\sigma_0^2 \tau}} - e^{\frac{-v_t vT}{\sigma_0^2 \tau}} = 0$$

$$e^{\frac{v_t vT}{\sigma_0^2 \tau}} = \frac{1}{3} e^{\frac{-v_t vT}{\sigma_0^2 \tau}}$$

$$e^{\frac{-2 v_t vT}{\sigma_0^2 \tau}} = \frac{1}{3}$$

$$\frac{-2 v_t vT}{\sigma_0^2 \tau} = \ln 1/3$$

$$v_t = - \frac{\sigma_0^2 \tau}{2vT} \ln 1/3$$

Note: (P_e) min can be calculated as follows:

$$(P_e)\min = \frac{3}{4} \int_{-\infty}^{-\frac{\sigma_0^2 \tau}{2vT} \ln 3} \frac{e^{-\frac{(v_0 - vT/\tau)^2}{2\sigma_0^2}}}{\sqrt{2\pi\sigma_0^2}} dv_0$$

$$+ \frac{1}{4} \int_{-\frac{\sigma_0^2 \tau}{2vT} \ln 3}^{\infty} \frac{e^{-\frac{(v_0 - vT/\tau)^2}{2\sigma_0^2}}}{\sqrt{2\pi\sigma_0^2}} dv_0$$

● **PROBLEM 16-3**

An integrate and dump receiver is used for receiving a noise corrupted signal. The signal is represented by $\pm v$ and 0 with equal probability. Find an expression for the threshold voltage v_t so that the probability of error is independent of signal.

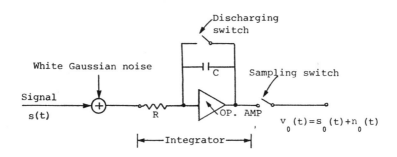

Figure 1.

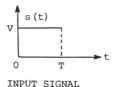

INPUT SIGNAL

Figure 2.

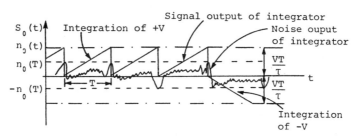

OUTPUT OF THE SIGNAL RECEIVER

Figure 3.

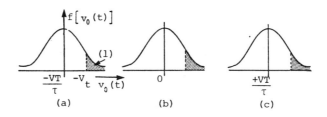

Figure 4.

Solution: A binary-encoded PCM baseband signal consists of a time sequence of voltage levels +v, 0 and -v volts. When the noise is mixed with the input signal, the sampled value is generally different from 0 or ±v. If the sample value is positive, the transmitted voltage was +v and if the sample value is negative, the level was -v. When the voltage level is zero, the sample value can be positive or negative due to noise. Hence, in this case probability of error is high.

Now, since there is equal likelihood of transmitting +v, 0, -v volts, their probabilities are equal i.e.,

$$P[+v] = P[0] = P[-v] = 1/3 \qquad (1)$$

Now the output of the receiver is

$$v_0(t) = s_0(t) + n_0(t)$$

Since the probability density of the noise sample $n_0(T)$ is Gaussian and symmetrical around zero, the probability density of the integrator output $v_0(t)$ is also Gaussian and it is symmetrical about $\frac{+vT}{T}$, 0, or $\frac{-vT}{T}$ depending upon input signal. This is shown in Fig. 4. In the figure, the density of the output is plotted against output voltage.

Now, let v_t be the threshold voltage. When -v volts is transmitted at the output of the integrator, output voltage $v_0(t)$ results, which is the sum of $s_0(t)$ and $n_0(t)$. Since

$$v_0(T) = s_0(T) + n_0(T)$$

$$\therefore n_0(T) = v_0(T) - s_0(T)$$

$$= v_0 + \frac{vT}{T} \quad \text{(dropping T)}$$

Now, at the output of the receiver, -'ve voltage should be generated so that it can be interpreted that -v was transmitted.

Letting $\frac{vT}{T} = v_t$ = threshold voltage, the probability that $n_0(T) > v_t$ is the area of shaded portion (1) and is given by

$$P_e[-v] = \int_{-v_t}^{\infty} \frac{e^{-\frac{(v_0 + \frac{vT}{T})^2}{2\sigma_0^2}}}{\sqrt{2\pi\sigma_0^2}} dv_0 \qquad (3)$$

Similarly, referring to Fig. 4(b) and (c) results

$$P_e[0] = 2 \int_{v_t}^{\infty} \frac{e^{-\frac{v_0^2}{2\sigma_0^2}}}{\sqrt{2\pi\sigma_0^2}} dv_0 \qquad (4)$$

and

$$P_e[+v] = \int_{-\infty}^{v_t} \frac{e^{-\frac{(v_0 - vT/\tau)^2}{2\sigma_0^2}}}{\sqrt{2\pi\sigma_0^2}} \, dv_0 \tag{5}$$

The probability of error will be independent of the signal if and only if

$$P_e[-v] = P_e[0] = P_e[+v] \tag{6}$$

Hence,

$$\int_{-v_t}^{\infty} \frac{e^{-\left(\frac{v_0 + vT/\tau}{\sqrt{2\sigma_0^2}}\right)^2}}{\sqrt{2\pi\sigma_0^2}} \, dv_0 = 2 \int_{v_t}^{\infty} \frac{e^{-\frac{v_0^2}{2\sigma_0^2}}}{\sqrt{2\pi\sigma_0^2}} \, dv_0 \tag{7}$$

Letting

$$x_1 = \frac{v_0 + \frac{vT}{\tau}}{\sqrt{2}\sigma_0} \quad \text{and} \quad x_2 = \frac{v_0}{\sqrt{2}\sigma_0}$$

$$dx_1 = \frac{1}{\sqrt{2}\sigma_0} \, dv_0, \quad dx_2 = \frac{1}{\sqrt{2}\sigma_0} \, dv_0$$

Hence,

$$\sqrt{2}\sigma_0 \int_{-v_t}^{\infty} \frac{e^{-x_1^2} dx_1}{\sqrt{2\pi}\,\sigma_0} = 2 \int_{v_t}^{\infty} \sqrt{2}\,\sigma_0 \frac{e^{-x_2^2} dx_2}{\sqrt{2\pi}\,\sigma_0}$$

$$x_1 = \left(-v_t + \frac{vT}{\tau}\right)/\sqrt{2}\sigma_0 \quad x_2 = \frac{v_t}{\sqrt{2}\,\sigma_0}$$

$$\frac{1}{\sqrt{\pi}} \int_{-v_t + \frac{vT}{\tau}}^{\infty} e^{-x_1^2} dx_1 = 2 \frac{1}{\sqrt{\pi}} \int_{v_t}^{\infty} e^{-x_2^2} dx_2 \tag{8}$$

$$x_1 = \left(-v_t + \frac{vT}{\tau}\right)/\sqrt{2}\sigma_0 \quad x_2 = \frac{v_t}{\sqrt{2}\,\sigma_0}$$

$$\text{erfc}\left[\frac{\frac{vT}{\tau} - v_t}{\sqrt{2}\sigma_0}\right] = 2\,\text{erfc}\left(\frac{v_t}{\sqrt{2}\,\sigma_0}\right)$$

● **PROBLEM 16-4**

An integrate and dump receiver as shown in Fig. 1 is used to process a binary signal of value ±2v with time period T. This signal is computed during the process of transmission, by a white Gaussian noise of density 10^{-4} volt^2/Hz. Write an expression to find T_{min} such that probability of error is less than 10^{-4}.

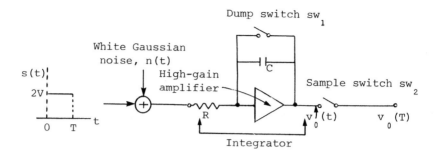

AN INTEGRATE AND DUMP RECEIVER

Figure 1.

<u>Solution</u>: Consider a binary-encoded PCM baseband signal consisting of a time sequence of voltage levels +2v or -2v. When this signal is corrupted by the white Gaussian noise, the sampled output of the integrator is different from ±2v.

The probability of such error is

$$P_e = \frac{1}{2} \text{erfc} \left(\frac{E_s}{\eta} \right)^{\frac{1}{2}}$$

where $E_s = (2v)^2 T$ = signal energy of a bit.

Thus,

$$P_e = \frac{1}{2} \text{erfc} \left(\frac{4v^2 T}{\eta} \right)^{\frac{1}{2}}$$

where $\eta = 2 \times$ power spectral density of Gaussian noise.

Substituting all known quantities, the probability of error is

$$P_e = \frac{1}{2} \text{erfc} \left(\frac{4v^2 T}{2 \times 10^{-4}} \right)^{\frac{1}{2}}$$

Since the probability of error must be less than 10^{-4}, so

$$\frac{1}{2} \text{erfc} \left(\frac{4v^2 T}{2 \times 10^{-4}} \right)^{\frac{1}{2}} < 10^{-4}$$

From the above expression, the value of T_{min} can be determined.

• **PROBLEM 16-5**

A binary signal is transmitted in the form of a voltage waveform, having constant amplitude of either +v or -v volts over a noisy communication channel. Assuming equal probability of transmitting ±v volts and presence of a noise whose power spectral density is $G_n(f) =$

$G_0 / \left[1 + \left(\frac{f}{f_1} \right)^2 \right]$, calculate:

(a) The transfer function H(f) of an optimum filter to be used for this signal.

(b) The probability of error, P_e, when an optimum filter is used.

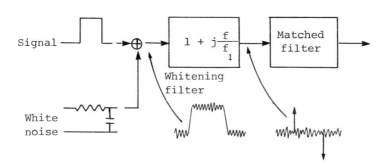

Solution: (a) The Schwary inequality states that given arbitrary complex functions X(f) and Y(f), of a common variable f, then

$$\left| \int_{-\infty}^{\infty} X(f) Y(f) df \right|^2 \leq \left[\int_{-\infty}^{\infty} |X(f)|^2 df \int_{-\infty}^{\infty} |Y(f)|^2 df \right]$$

The equal sign applies for $X(f) \cdot KY^*(f)$, where K is a constant and $Y^*(f)$ is the complex conjugate of Y(f).

Now apply the Schwary inequality to

$$\gamma^2 = \frac{P_0^2(T)}{\sigma_0^2} = \frac{\left| \int_{-\infty}^{\infty} H(f) P(f) e^{(j2\pi Tf)} df \right|^2}{\int_{-\infty}^{\infty} |H(f)|^2 G_n(f) df} \qquad (1)$$

(Note: The optimum filter is defined as the filter which maximizes the ratio γ.)

By making the identification

$$X(f) = \sqrt{G_n(f)}\, H(f) \qquad (2)$$

and

$$Y(f) = \frac{1}{\sqrt{G_n(f)}}\, P(f) e^{(j2\pi Tf)} \qquad (3)$$

Using eqs.(2), (3) and the Schwary inequality, we rewrite eq.(1) as

$$\frac{P_0{}^2(T)}{\sigma_0{}^2} = \frac{\left|\int_{-\infty}^{\infty} X(f)Y(f)df\right|^2}{\int_{-\infty}^{\infty} |X(f)|^2 df} \leq \int_{-\infty}^{\infty} |Y(f)|^2 df \qquad (4)$$

or

$$\frac{P_0{}^2(T)}{\sigma_0{}^2} \leq \int_{-\infty}^{\infty} |Y(f)|^2 df = \int_{-\infty}^{\infty} \frac{|P(f)|^2}{G_n(f)} df \qquad (5)$$

The ratio $\frac{P_0{}^2(T)}{\sigma_0{}^2}$ will attain its maximum value when the equal sign in eq.(5) may be employed as in the case when $X(f) = KY^*(f)$.

Then from eqs.(4) and (5), the optimum filter which yields a maximum ratio $\frac{P_0{}^2(T)}{\sigma_0{}^2}$ has a transfer function

$$H(f) = k \frac{P^*(f)}{G_n(f)} e^{j2\pi fT} \qquad (6)$$

Using eq.(6)

$$H(f) = K \frac{H(f/f_1)^2}{G_0} e^{(j2\pi fT)} \left[\int_0^T V e^{(-j2\pi ft)} dt\right]^*$$

$$= \frac{KVT}{G_0} \left[1 + \left(\frac{f}{f_1}\right)^2\right] \left(\frac{\sin \pi fT}{\pi fT}\right) e^{(-j\pi fT)}$$

(b) Since from eq.(5), the maximum ratio is

$$\int_{-\infty}^{\infty} \frac{|P(f)|^2}{G_n(f)} df = \int_{-\infty}^{\infty} \frac{1 + (f/f_1)^2}{G_0} \left(\frac{\sin \pi ft}{\pi f}\right)^2 df = \infty$$

Hence,

$$P_e = \frac{1}{2} \operatorname{erfc}\left[\frac{1}{2\sqrt{2}} \cdot \sqrt{\int_{-\infty}^{\infty} \frac{|P(f)|^2}{G_n(f)} df}\right] = 0$$

Note:

The optimum filter consists of the whitening filter and the matched filter. The whitening filter is given by $1 + j(f/f_1)$ which contains a differentiator. Therefore, if the input is pulse signals, then the output of the whitening filter is an impulse sequence. This gives infinite signal energy.

• **PROBLEM 16-6**

Digital data is to be transmitted through a baseband system with $N_0 = 10^{-7}$ W/Hz and the received signal amplitude $A = 20$mV. (a) If 10^3 bits per second (bps) are transmitted, what is P_E (the average probability error)? (b) If 10^4 bps are transmitted, to what value must A be adjusted in order to attain the same P_E as in (a)?

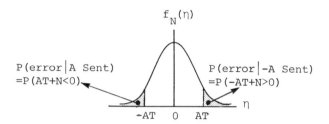

ILLUSTRATION OF ERROR PROBABILITIES FOR BINARY SIGNALING.

Figure 1.

Solution: (a) To solve part (a), note that in Figure 1 shown, an error can occur if -A is transmitted and $-AT + N > 0$. The probability of this event is the same as the probability that $N > AT$ which can be written as

$$P(E|-A) = \int_{AT}^{\infty} \frac{e^{-\eta^2/N_0 T}}{\sqrt{\pi N_0 T}} \, d\eta$$

$$= \frac{1}{2} \, \mathrm{erf}\left(A\sqrt{\frac{T}{N_0}}\right)$$

for

$$f_N(\eta) = \frac{e^{-\eta^2/N_0 T}}{\sqrt{\pi N_0 T}}$$

which is the area to the right of $\eta = AT$ in Fig. 1. The average probability of error is

$$P_E = P(E/+A)P(+A) + P(E/-A)P(-A)$$

Since P(E/A) is also equal to $\frac{1}{2} \, \mathrm{erfc}\left(A\sqrt{\frac{T}{N_0}}\right)$, by substituting and noting that $P(+A)+P(-A) = 1$, $P_E = \frac{1}{2} \, \mathrm{erfc}\left(A\sqrt{\frac{T}{N_0}}\right)$ is obtained.

Thus, in order to avoid the square root, let $S \triangleq A^2 T/N_0$. (1) Eq.(1) can be integrated in two ways. First, since in each signal pulse, energy E_s is

$$E_s = \int_{t_0}^{t_0+T} A^2 \, dt = A^2 T$$

Hence,

$$S = \frac{A^2 T}{N_0} = \frac{E_s}{N_0} = \frac{\text{signal energy/pulse}}{\text{noise power spectral density}} \quad (2)$$

Second, recall that a rectangular pulse of duration T seconds has amplitude spectrum AT sincTf, and a rough measure of its bandwidth is BW = $\frac{1}{T}$

Hence,
$$S = \frac{A^2}{N_0 (1/T)} = \frac{A^2}{N_0 \, BW}$$

which is the ratio of signal power to noise power in the signal bandwidth. Note: S is referred to as the signal-to-noise ratio (SNR).

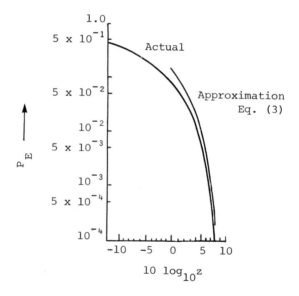

Figure 2.

Now, using eq.(2) and information given

$$S = \frac{A^2 T}{N_0} = \frac{(0.02)^2 (10^{-3})}{10^{-7}} = 4$$

Using approximation $P_E \approx \dfrac{e^z}{2\sqrt{\pi z}}$, $z \gg 1$ \quad (3)

and note that $\text{erfc}(u) \approx \dfrac{e^{-u^2}}{u\sqrt{\pi}}$, $u \gg 1$ (asymptotic expansion for the complementary error function) (see Fig. 2)

therefore,
$$P_E \approx \frac{e^{-4}}{2\sqrt{4\pi}} = 2.58 \times 10^{-3}$$

(b) Part (b) is solved by finding A such that
$$A^2(10^{-4})/(10^{-7}) = 4,$$
which gives
$$A = 63.2 \text{mV}.$$

• **PROBLEM 16-7**

A suitable matched filter is to be designed for a signal $s(t)$, which is either 0 or $A\cos(2\pi \cdot f_0 \cdot t)$ during a time period $T = \frac{n}{f_0}$. Find the transfer function of the matched filter for processing noise corrupted signals $s(t)$. What is the probability of error, P_e, if the noise has power spectral density $G_n(f) = \eta/2$. What will be the transfer function and P_e if the signal is $\pm A(1 - \cos 2\pi \cdot f_0 \cdot t)$?

Solution: An optimum filter which yields a maximum ratio $P_0^2(T)/B_0^{-2}$ is called a matched filter. Since the input noise is white in this case, $G_n(f) = \eta/2$ and $H(f)$ is given as

$$H(f) = \frac{KP^*(f)e^{-j2\pi fT}}{\eta/2}$$

Hence,

$$H(f) = \frac{2K}{\eta} e^{-j2\pi fT} \left[\int_{-\infty}^{\infty} \{s_1(t) - s_2(t)\} e^{-j2\pi ft} dt \right]^*$$

$$= \frac{2K}{\eta} e^{-j2\pi fT} \int_{0}^{T} A\cos 2\pi f_0 t \, e^{j2\pi fT} dt$$

$$= \frac{2K}{\eta} e^{-j2\pi fT} \int_{0}^{n/f_0} \frac{A}{2} \left[e^{j2\pi(f+f_0)t} + e^{j\pi(f-f_0)t} \right] df$$

$$= \frac{AK}{\eta} e^{-j2\pi nf/f_0} \left[\frac{e^{j2\pi(f+f_0)t}}{j2\pi(f+f_0)} + \frac{e^{j2\pi(f-f_0)t}}{j2\pi(f-f_0)} \right]_{0}^{n/f_0}$$

$$= \frac{AK}{\eta} e^{j2\pi n f/f_0} \left[\frac{e^{j2\pi n f/f_0} - 1}{j2\pi(f+f_0)} + \frac{e^{j2\pi n f/f_0} - 1}{j2\pi(f-f_0)} \right]$$

$$= \frac{AK}{\eta} e^{-j2\pi nf/f_0} \left[\frac{[e^{j2\pi n f/f_0} - 1] 2f}{j2\pi(f^2 - f_0^2)} \right]$$

835

$$= \frac{AK}{\eta} \frac{f[1 - e^{-j2\pi n f/f_0}]}{j\pi(f^2 - f_0^2)}$$

$$= \frac{2AKf}{\pi\eta} \left[\frac{\sin(\pi n f/f_0)}{f^2 - f_0^2}\right] e^{-j\pi n f/f_0}$$

and

$$P_e = \frac{1}{2} \text{erfc} \left[\frac{P_0^2(T)}{8\sigma_0^2}\right]_{MAX}^{1/2} = \frac{1}{2} \text{erfc} \left[\frac{1}{4\eta} \int_0^T S_1^2(t) dt\right]^{\frac{1}{2}}$$

$$= \frac{1}{2} \text{erfc} \left[\frac{1}{4\eta} \int_0^{n/f_0} (A\cos 2\pi f_0 t)^2 dt\right]^{\frac{1}{2}}$$

$$= \frac{1}{2} \text{erfc} \left[\frac{A^2}{4\eta} \int_0^{n/f_0} \frac{1+\cos 4\pi f_0 t}{2} dt\right]^{\frac{1}{2}}$$

$$= \frac{1}{2} \text{erfc} \left[\frac{A^2}{8\eta} \left[t + \frac{\sin 4\pi f_0 t}{4\pi f_0}\right]_0^{n/f_0}\right]^{\frac{1}{2}}$$

$$= \frac{1}{2} \text{erfc} \left[\frac{A^2 n}{8\eta f_0}\right]^{\frac{1}{2}}$$

(b) If $s(t) = \pm A(1 - \cos 2\pi f_0 t)$,

then
$$H(f) = \frac{2K}{\eta} e^{-j2\pi fT} \left[\int_{-\infty}^{\infty} \{S_1(t) - S_2(t)\} e^{-j2\pi ft} dt\right]^*$$

$$= \frac{2K}{\eta} e^{j2\pi fT} \int_0^T 2A(1-\cos 2\pi f_0 t) e^{j2\pi ft} dt$$

$$= \frac{4AK}{\eta} e^{-j2\pi n f/f_0} \int_0^{n/f_0} \left[1 - \frac{e^{j2\pi(f+f_0)t} + e^{j2\pi(f-f_0)t}}{2}\right] dt$$

$$= \frac{4AK}{\eta} e^{j2\pi n f/f_0} \left[t - \frac{e^{j2\pi(f+f_0)t}}{j4\pi(f+f_0)} - \frac{e^{j2\pi(f-f_0)t}}{j4\pi(f-f_0)}\right]_0^{n/f_0}$$

$$= \frac{4AK}{\eta} e^{-j2\pi n f/f_0} \left[\frac{n}{f_0} - 2f \frac{(e^{j2\pi n f/f_0}-1)}{j4\pi(f^2-f_0^2)}\right]$$

$$= \frac{4AK}{\eta} e^{-j2\pi n f/f_0} \left[\frac{n}{f_0} - \frac{f}{\pi} \cdot \frac{\sin(\pi n f/f_0)}{f^2-f_0^2} e^{j\pi n f/f_0}\right]$$

$$= \frac{4AKn}{\eta f_0} e^{j2\pi n f/f_0} - \frac{4AKf}{\pi\eta} \frac{\sin(\pi n f/f_0)}{f^2-f_0^2} e^{j2\pi n f/f_0}$$

and $Pe = \frac{1}{2} \text{erfc} \left(\frac{E_s}{\eta} \right)^{\frac{1}{2}} = \frac{1}{2} \text{erfc} \left[\frac{1}{\eta} \int_0^{n/f_0} A^2(1-\cos 2\pi f_0 t)^2 df \right]^{\frac{1}{2}}$

$= \frac{1}{2} \text{erfc} \left[\frac{A^2}{\eta} \int_0^{n/f_0} \{1 - 2\cos 2\pi f_0 t + \cos^2 2\pi f_0 t\} dt \right]^{\frac{1}{2}}$

$= \frac{1}{2} \text{erfc} \left[\frac{A^2}{\eta} \left(\frac{3}{2} \frac{n}{f_0} \right) \right]^{\frac{1}{2}} = \frac{1}{2} \text{erfc} \left[\frac{-3A^2 n}{2\eta f_0} \right]^{\frac{1}{2}}$

● **PROBLEM** 16-8

The impulse response of a filter matched to a pulse signal $s(t)$ is $h_0(t) = s(t_0 - t)$.

If $s(t) = \begin{cases} A, & 0 \le t \le T \\ 0, & \text{elsewhere} \end{cases}$

determine the response of the filter to $s(t)$.

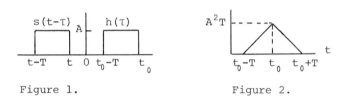

Figure 1. Figure 2.

Solution: The necessary condition for a filter to operate linearly is $t_0 > T$, since it will have nonzero impulse response for $t < 0$.

The response of the filter to $s(t)$ is

$$y(t) = h_0(t) * s(t) = \int_{-\infty}^{\infty} h_0(\tau) s(t-\tau) d\tau$$

The factors within the integral are shown in Fig. 1 and the response of the filter to $s(t)$ is shown in Fig. 2.

From the Fig. 2, it is obvious that the output signal has its peak at $t = t_0$. The signal-to-noise ratio is also at its peak at $t = t_0$, since at this point the noise is minimum. It is for this reason that in digital signaling, $t_0 = T$.

• **PROBLEM 16-9**

Sketch the impulse response $h_0(t)$ for an optimum filter which is used for processing a signal $m(t) = f(t)$, so that $k(j\omega) = 1$. If the signal and noise power spectral densities are $s_f(\omega) = \dfrac{1}{\omega^2+b^2}$ and $s_x(\omega) = c$ respectively, find $H_0(j\omega)$ and the mean-square error $\overline{|e(t)|_m^2}$ using the following relations:

$$H_0(j\omega) = \text{optimum transfer function for the actual filter}$$

$$= \frac{s_f(\omega) \cdot k(j\omega)}{s_f(\omega) + s_x(\omega)} \quad (1)$$

and

$$\overline{|e(t)|_m^2} = \text{a minimum value for the mean-square output error.}$$

$$= \frac{1}{2\pi} \int_{-\infty}^{\infty} \frac{s_f(\omega)\, s_x(\omega)}{s_f(\omega)+s_x(\omega)} \cdot |k(j\omega)|^2 d\omega \quad (2)$$

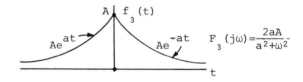

Figure 1.

Solution: From eqs. (1) and (2) we obtain

$$H_0(j\omega) = \frac{1/c}{\omega^2+k^2}$$

and

$$\overline{|e_n(t)|_m^2} = \frac{1}{2\pi} \int_{-\infty}^{\infty} \frac{1}{\omega^2+k^2}\, d\omega$$

where $k^2 = b^2 + 1/c$ and n is an integer. Comparing Fig. 1 and the expression for $H_0(j\omega)$ gives the impulse response $h_0(t)$ shown in Fig. 2a.

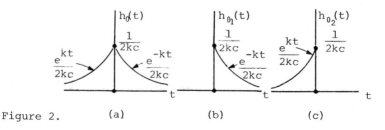

Figure 2. (a) (b) (c)

Now, to evaluate the mean-square error integral, write the integrand as

$$\frac{1}{\omega^2+k^2} = \frac{1}{-(j\omega)^2+k^2}$$

$$= \frac{1}{(j\omega+k)(-j\omega+k)}$$

then $E(j\omega) = j\omega + k = a_0(j\omega) + a_1$

where $n = 1$, $a_0 = 1$, and $a_1 = k$ is obtained.

Now, since $N(j\omega)N^*(j\omega) = 1 = b_0$, the minimum mean-square error is

$$\overline{|e_1(t)|^2}_m = \frac{b_0}{2a_0 a_1} = \frac{1}{2k}$$

● **PROBLEM** 16-10

A signal receiver with matched filter is used for the reception of a signal $s(t) = \pm 2\left(\frac{t}{T}\right)$, with time period T. During the transmission to this signal a white noise of power spectral density 10^{-6} volt^2/Hz is added. Sketch the signal appearing at the output of the receiver and calculate the minimum interval T, so that P_e will not exceed 10^{-4}.

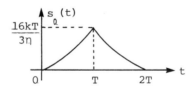

Solution: (a) The impulse response of the filter to a unit strength impulse applied at $t = 0$, is

$$h(t) = F^{-1}\left[H(f)\right] = \frac{2k}{\eta} \int_{-\infty}^{\infty} P^*(f) e^{-j2\pi fT} e^{j2\pi ft} \, df$$

$$= \frac{2K}{\eta} \int_{-\infty}^{\infty} P^*(f) e^{j2\pi f(t-T)} \, df \qquad (1)$$

Since the filter is physically realizable, it will have an impulse response which is real (not complex). Therefore, $h(t) = h^*(t)$.

Hence, from eq.(1)

$$h(t) = \frac{2K}{\eta} \int_{-\infty}^{\infty} P(f) e^{j2\pi f(T-t)} \, df$$

$$= \frac{2K}{\eta} P(T-t)$$

since the error probability $p(t) = s_1(t) - s_2(t)$ (the difference signal).

Thus,

$$h(t) = \frac{2K}{\eta} \left[s_1(T-t) - s_2(T-t) \right]$$

$$= \frac{8K}{\eta T} (T-t) u(T-t), \quad t \geq 0$$

Now, the signal output at the receiver $s_0(t)$ is evaluated as

$$s_0(t) \Big|_{\substack{s(t) = \frac{2t}{T} \\ n(t) = 0}} = \int_0^t h(\lambda) s(t-\lambda) \, d\lambda$$

$$= \int_0^t \left[\frac{8K}{\eta T} (T-\lambda) \right] \left[\frac{2}{T} (t-\lambda) \right] d\lambda$$

$$= \frac{16K}{\eta T^2} \int_0^t \left[Tt - (T+t)\lambda + \lambda^2 \right] d\lambda$$

$$= \frac{16K}{\eta T^2} \left[Tt\lambda - (T+t)\frac{\lambda^2}{2} + \frac{\lambda^3}{3} \right]_0^t$$

$$= \frac{16K}{\eta T^2} \left[Tt^2 - \frac{Tt^2}{2} - \frac{t^3}{2} + \frac{t^3}{3} \right]$$

$$= \frac{8K}{\eta T} t^2 \left(1 - \frac{t}{3T} \right)$$

(b) Since P_e is defined as

$$P_e = \frac{1}{2} \text{erfc} \left(\frac{E_s}{\eta} \right)^{\frac{1}{2}} \leq 10^{-4} \quad \text{where } E_s = V^2/T \text{ is the signal energy of a bit.}$$

or

$$\left(\frac{E_s}{\eta} \right)^{\frac{1}{2}} = \left(\frac{4}{3} \frac{T}{\eta} \right)^{\frac{1}{2}} \geq 2.63$$

Therefore,

$$T_{min} = (2.63)^2 \times \frac{3}{4} \times 10^{-6} \approx 5.2 \times 10^{-6} \text{ sec}$$

● **PROBLEM 16-11**

A phase-shift keying system suffers from imperfect synchronization. Determine ϕ so that the

$$P_e(PSK) = P_e(DPSK) = 10^{-5}$$

Solution: The probability error P_e for imperfect synchronization PSK is

$$P_e = \tfrac{1}{2} \, \text{erfc} \sqrt{\frac{E_s}{\eta} \cos^2 \phi}$$

where $E_s = V^2/T$ is the signal energy of a bit

and for DPSK,

$$P_e = \tfrac{1}{2} \, e^{-E_s/\eta}$$

Now, set

$$\tfrac{1}{2} \, e^{-E_s/\eta} = \tfrac{1}{2} \, \text{erfc} \sqrt{\frac{E_s}{\eta} \cos^2 \phi} = 10^{-5}$$

Solving for E_s/η and $(E_s/\eta)\cos^2\phi$

$$\frac{E_s}{\eta} = -\ln 2 + 5 \ln 10 \approx 10.82$$

and

$$\sqrt{\frac{E_s}{\eta} \cos^2 \phi} \approx 3$$

Hence,

$$\cos \phi = \frac{3}{\sqrt{10.82}} = 0.91$$

and

$$\phi \approx 24°$$

● **PROBLEM 16-12**

A sinusoidal signal within the frequency range of 20-60 Hz modulates a carrier at 2KHz. Derive all the necessary formulas and determine the value of the capacitor C of the detector circuit as shown in Fig. 1 so that envelope distortion is minimum. Assume modulation index m = 0.7.

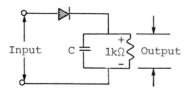

Figure 1.

Solution: For a detector circuit shown in Fig. 1, the output voltage is equivalent to the envelope of the amplitude modulated wave and can be expressed as

$$v = V(1 + m \cos\omega t)$$

where $\omega = 2\pi \times$ modulating frequency and

$m =$ modulation index.

The rate of change of the envelope of the modulation is therefore

$$\frac{dv}{dt} = -m\omega V \sin\omega t. \tag{1}$$

At any instant of time, the voltage across a capacitor, which is discharging through a resistor R, is given by the relation

$$v_c = V_0 \, e^{-t/RC}$$

Thus, the rate of discharge of the capacitor is

$$\frac{dv_c}{dt} = -\frac{1}{RC} V_0 \, e^{\frac{-t}{RC}} = -\frac{v_c}{RC} \tag{2}$$

In the above circuit, the voltage across the capacitor is equal to the output voltage. Therefore,

$$v_c = V(1 + m \cos\omega t),$$

and from equation (2)

$$\frac{dv_c}{dt} = \frac{-V(1 + m \cos\omega t)}{RC} \tag{3}$$

The necessary condition for the minimum envelope distortion is that at any instant of time, the rate of discharge of the capacitor must be greater than the rate of change of the envelope.

i.e., $\dfrac{V(1 + m \cos\omega t)}{RC} > m\omega V \sin\omega t$

Therefore,

$$\frac{1}{R\omega C} > \frac{m \sin\omega t}{1 + m \cos\omega t} \tag{4}$$

In order that equation (4) is valid, the left-hand size should always be greater than the right-hand size. The maximum value of the R·H·S corresponding to a value of t can be determined by differentiating the R·H·S with respect to 't' and equating it to zero.

Thus, $\dfrac{d}{dt}\left(\dfrac{m \sin\omega t}{1 + m \cos\omega t}\right) = 0$

$$0 = \frac{(1+m\cos\omega t)m\omega\cos\omega t + (m\sin\omega t)m\omega\sin\omega t}{(1+m\cos\omega t)^2}$$

$$0 = \frac{m\omega\cos\omega t + m^2\omega\cos^2\omega t + m^2\omega\sin^2\omega t}{(1+m\cos\omega t)^2}$$

$$0 = \frac{m\omega\cos\omega t + m^2\omega}{(1+m\cos\omega t)^2}$$

Therefore,

$$\cos\omega t = -m$$

and

$$\sin\omega t = \sqrt{1-m^2}$$

Thus, the maximum value of the $\frac{m\sin\omega t}{1+m\cos\omega t}$ in equation (4) evaluated to $m/\sqrt{1-m^2}$ and from equation (4)

$$\frac{1}{R\omega C} > \frac{m}{\sqrt{1-m^2}} \qquad (5)$$

where $\frac{1}{\omega C}$ in equation (5) corresponds to the reactance of C at the modulating frequency f. This reactance is inversely proportional to the frequency and hence, in order for equation (5) to be valid, the highest modulation frequency should be at 60 Hz with R = 1KΩ and m = 0.7. The value of C corresponding to the minimum envelope distortion is evaluated as follows:

$$\frac{1}{R\omega C} > \frac{m}{\sqrt{1-m^2}},$$

$$\frac{1}{1 \times 10^3 \times 2\pi \times 60 \times C} > \frac{0.7}{\sqrt{1-(0.7)^2}}$$

Thus, $\qquad C < 2.7 \ \mu F$

Therefore, for the given detector circuit the ideal value of capacitance is 2.7 µF.

● **PROBLEM 16-13**

If the frequency offset Ω in FSK satisfies $\Omega T = n\pi$, the signals $s_1(t)$ and $s_2(t)$ are orthogonal.

(a) Prove this statement.

(b) Calculate P_e.

Solution: (a) Proof:

For an orthogonal function,

$$\int_0^T s_1(t) \cdot s_2(t) \cdot dt = \int_0^T A \cdot \cos(\omega_0+\Omega)t \times A \cdot \cos(\omega_0-\Omega)t \, dt$$

$$= \frac{A^2}{2} \int_0^T \left[\cos 2\omega_0 t + \cos 2\Omega t\right] dt$$

$$= \frac{A^2}{2} \int_0^T \cos 2\Omega t \cdot dt$$

$$= \frac{A^2}{2} \left[\frac{\sin 2\Omega t}{2\Omega}\right]_0^T$$

$$= \frac{A^2}{4\Omega} \sin(2\Omega T)$$

$$= \frac{A^2}{4\Omega} \sin(2n\pi) = 0$$

(b) $P_0(t)\big|_{s(t)=s_1(t)} = K \int_0^T s_1(\lambda)\left[s_1(\lambda) - s_2(\lambda)\right] d\lambda$

$$= K \int_0^T s_1^2(\lambda) d\lambda$$

$$= K \cdot A^2 \int_0^T \cos^2(\omega_0 + \Omega)\lambda \cdot d\lambda$$
(Because $s_1(t)$ and $s_2(t)$ are orthogonal)

$$= \frac{K \cdot A^2 \cdot T}{2}\left[1 + \frac{\sin 2\omega_0 T}{2(\omega_0 + \Omega)T}\right]$$

$$= \frac{K \cdot A^2 \cdot T}{2}$$

$P_0(t)\big|_{s(t)=s_2(t)} = K \int_0^T s_2(\lambda)\left[s_1(\lambda) - s_2(\lambda)\right] d\lambda$

$$= -K \int_0^T s_2^2(\lambda) d\lambda = -KA^2 \int_0^T \cos^2(\omega_0 - \Omega)\lambda \cdot d\lambda$$

$$= \frac{-K \cdot A^2 \cdot T}{2}\left[1 + \frac{\sin 2\omega_0 T}{2(\omega_0 - \Omega)T}\right]$$

$$\approx \frac{-K \cdot A^2 \cdot T}{2}$$

and
$$\sigma_0^2 = KA \int_0^T n(\lambda)\left[\cos(\omega_0+\Omega)\lambda - \cos(\omega_0-\Omega)\lambda\right]^2 d\lambda$$

$$= \frac{K^2 \cdot A^2 \cdot \eta}{2} \int_0^T \left[\cos(\omega_0+\Omega)\lambda - \cos(\omega_0-\Omega)\lambda\right]^2 d\lambda$$

$$= \frac{K^2 \cdot A^2 \cdot T \cdot \eta}{2}\left[1 + \frac{\omega_0 \sin 2\omega_0 T}{2(\omega_0^2-\Omega^2)T} - \frac{\sin 2\omega_0 T}{2\omega_0 T}\right]$$

$$\approx \frac{K^2 \cdot A^2 \cdot T \cdot \eta}{2}$$

Hence,
$$P_e = \int_{\frac{K \cdot A^2 \cdot T}{2}}^{\infty} \frac{e^{-\frac{n_0^2(T)}{2\sigma_0^2}}}{\sqrt{2\pi\sigma_0^2}} dn_0(T)$$

$$\approx \frac{1}{2} \text{erfc}\sqrt{\frac{A^2 T}{4\eta}}$$

$$= \frac{1}{2} \text{erfc}\sqrt{0.5 \frac{E_s}{\eta}}$$

where $E_s = A^2 T/2$.

• **PROBLEM 16-14**

In a M-ary PSK transmission system, M signals of the form $A \cdot \cos(\omega_0 t + \theta_i)$ for M values of i, are generated locally for correlation with the noisy received signal. Compute the different values of θ_i such that the probability of error for each θ_i is the same. Draw the block diagram of a correlator detector.

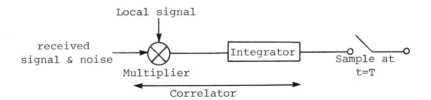

A COHERENT SYSTEM OF SIGNAL RECEPTION

Figure 1.

Solution: Fig. 1 shows a coherent system for signal reception. The input is a waveform $s_1(t)$, which is corrupted by the white noise $n(t)$. Interval T, as stated in the problem is the bit length. The received signal plus noise is multiplied by a locally generated waveform and passed through an integrator, the output of which is sampled at t = T. The operation of mulitplying the received signal plus noise by a locally generated waveform is known as correlation, and the receiver is called a correlator.

In the case of PSK, the input signal is of the form $s_1(t) = A \cdot \cos\omega_0 t$ or $s_2(t) = -A \cdot \cos\omega_0 t$. Hence, the local signal must be $s_1(t) - s_2(t) = 2A \cdot \cos\omega_0 t$.

When the locally generated signals are of the form $2A \cdot \cos(\omega_0 t + \phi)$, the probability of error is

$$P_e = \frac{1}{2} \operatorname{erfc} \sqrt{\frac{E_s^2}{\eta} \cos^2\phi}$$

$$= \frac{1}{2} \operatorname{erfc} \sqrt{\frac{A^2 T}{2\eta} \cos^2\phi}$$

where E_s is the signal energy and 2η is the power spectral density of noise.

When M locally generated signals of the form $\cos(\omega_0 t + \theta_i)$ are used, the probability of error in each case depends on the value of $\cos\phi$. Hence, for equal probability of error, the following relation must be satisfied:

$$\cos\theta_1 = \cos\theta_2 = \ldots = \cos\theta_M.$$

Using a standard solution for the equations, it follows that

$$\theta_1 = \frac{2\pi}{M}, \quad \theta_2 = \frac{4\pi}{M} \ldots$$

i.e., for equal probability of error P_e, the phase offset θ_i must be an even multiple of $\frac{\pi}{M}$, including zero.

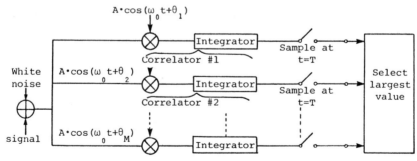

Figure 2. CORRELATOR DETECTOR

The detector circuit to be used is shown in Fig. 2. The operation of this circuit is similar to that described earlier except that only the largest value of the sampled output is selected.

● **PROBLEM 16-15**

A channel is a path for the transmission of information between two or more points. Other names that are used for it are circuit, facility, line, path and wire. A channel is further classified as (a) simplex, (b) half-duplex and (c) full-duplex. Define each term and use examples to illustrate.

Solution: A channel is classified as:

a) Simplex if it carries information in only one direction. For example, a line can be used to link a level measuring instrument to a computer.

b) Half-duplex if it is carrying information in either direction but not simultaneously.

c) Full-duplex if it can carry information in either direction at the same time. For example, the telephone line.

Figures below show simple electrical circuits illustrating the above classification.

T_x = Transmitter

R_x = Receiver

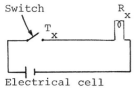
(a) Simplex

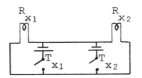
(b) Half Duplex

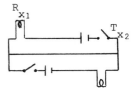
(c) Full Duplex

● **PROBLEM 16-16**

Explain what a parallel transmission is. Give examples to illustrate and state what the advantages and disadvantages of using such a transmission.

Solution: When using parallel transmission, each element of a code is transmitted along its own channel, and hence, the total code is transmitted at the same instant. For example, an 8-bit code needs 8 channels. It is expensive because a number of channels are used.

Often the cost of running a terminal equipment turns out to be cheaper using parallel transmission. It is commonly used for short line lengths where the user has control over both the transmitter and receiver. Some machines designed for parallel operations use separate channels by multiplexing all the channels on one line. Frequency division multiplexing is also used.

• PROBLEM 16-17

Using block diagrams show how multiplexing techniques are used.

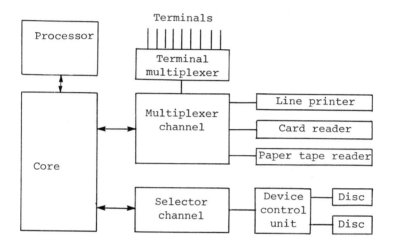

Solution: Multiplexing techniques are used to attach a number of slow peripherals to the system, as shown in the figure. In this case, the multiplexer channel (input/output bus) controls the transfer of data to and from the peripherals attached to it, all of which may be operating simultaneously.

• PROBLEM 16-18

Use block diagrams to illustrate three approaches to interface a communication multiplexer to a system.

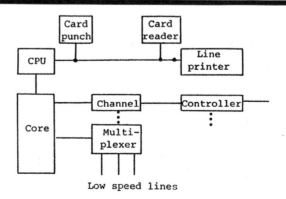

Figure a.

848

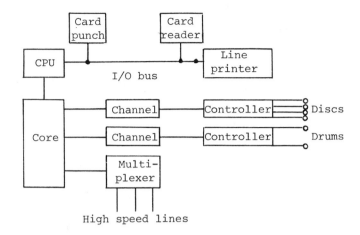

Figure b.

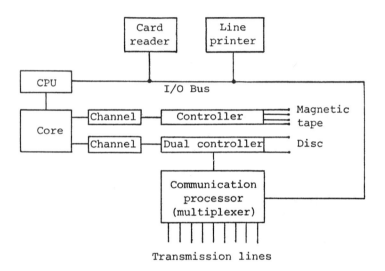

Figure c.

Solution: (1) Figure (a) shows a number of low-speed lines multiplexed on to the input/output bus of a system; the devices might typically be teletypes or similar devices.

(2) A number of high-speed lines can be multiplexed directly into core. The lines might be to remote batch terminals, or to other computers.

Figure (c) shows a dual controller of a disc with the communication processor using this channel to pass data to and from the disc; control information is passed via the input/output bus of the main processor. This type of configuration might be suitable for a data-collection system.

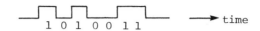
● PROBLEM 16-19

There are two types of serial transmissions, (a) asynchronous and (b) synchronous. Explain each one and discuss their advantages and disadvantages.

```
    _   _ _   _ _
 __| |_| | |_| | |_____  → time
  1 0 1 0 0 1 1
```
ASCII Code for the letter s

Figure 1.

```
Space (0) ┌Start┐                                       Stop
Mark  (1) │element│ I₁   I₂   I₃   I₄   I₅  │element│
          0    20   40   60   80  100  120   150
                                               time
                                               m sec
```

Figure 2.

<u>Solution</u>: (a) Asynchronous: The information in the form of characters is transmitted one bit at a time (i.e., serially) along a channel. For example, the ASCII code for the letter S is 01010011 and it would be transmitted as shown in Fig. 1.

Asynchronous transmission is the form used by most electromechanical serial devices such as teleprinters, teletypewriters, etc. Every character consists of three parts, a start element, the information, and a stop element, as shown in Fig. 2. Normally the information is sent as a fixed number of bits, usually 5 or 8.

Note: The start element = 1 information unit.

The stop element is either 1, 1½ or 2 times the length of the information unit.

The <u>advantages</u> of asynchronous serial transmission are that the <u>characters</u> can be generated easily by electromechanical devices and can be easily used to drive such devices, as teleprinters; that the character is itself complete, it contains its own synchronising information, namely the start and stop elements; that it does not matter how regular or irregular in time the codes are transmitted or received.

The <u>disadvantages</u> are: that a considerable section of the character is not bearing any message information, it is thus not very efficient; that synchronization is dependent upon recognition of the stop/start transition, which is fairly easily missed or falsely discovered in the presence of noise; an electrical path always distorts the signal, asynchronous serial transmission is sensitive to distortion; that the speed must be limited because of the sensitivity to distortion.

(b) Synchronous: A serial signal stream is transmitted along one channel (as before) except that in this case there are no stop and start elements. Synchronization is carried out by gathering the data into blocks, for example 100 characters, and prefixing every block with a unique code which will be recognized by the receiver. Detection of the unique code causes the receiver to synchronize with the signal for the whole of that block.

The <u>advantages</u> of the synchronous system are: (1) efficiency: since the proportion of message to synchronising information is greater; (2) the synchronising code can be made complicated (e.g. the characters SYN SYN SYN), (3) not that much sensitive to distortion and can operate at high speed; (4) a common timing clock is used for transmitter and receiver.

The <u>disadvantages</u> are: (1) if a synchronising error occurs then a whole block is lost; (2) the characters must be sent in blocks and not as they become available.

● **PROBLEM** 16-20

A typical interface device known as the British Standard Interface is used for parallel data transmission as shown in the figure below. Explain the operation of this system taking into consideration the fact that it imposes no timing restriction.

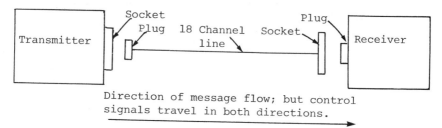

DATA TRANSMISSION

Figure 1.

<u>Solution</u>: The parallel transmission of a data in the form of eight bit characters is possible on full-duplex basis using two interfaces, each one at both ends. The data from the transmitter is transmitted using this interface along an 18-channels line. The function performed by each one of these lines is shown in Fig. 2. Line 2 serves as a reference, lines 3 to 8 are the controlling lines. The information in the form of an 8-bit character is transmitted on lines 11 to 18.

Line number	Direction of signal flow		Line description
	TRANSMITTER (SOURCE)	RECEIVER (ACCEPTOR)	
1	←———————→		Screen or protective earth
2	←———————→		Zero voltage reference
3	←———————		Acceptor Operable
4	———————→		Source Operable
5	←———————		Acceptor Control
6	———————→		Source Control
7	←———————		Acceptor Error
8	———————→		Source Terminate
9	———————→		Parity Valid
10	———————→		Parity digit
11	———————→		Bit 1 ⎫
12	———————→		Bit 2 ⎪
13	———————→		Bit 3 ⎪
14	———————→		Bit 4 ⎬ DATA
15	———————→		Bit 5 ⎪
16	———————→		Bit 6 ⎪
17	———————→		Bit 7 ⎪
18	———————→		Bit 8 ⎭

Figure 2.

Now the data is transmitted using the following handshaking procedure. When the receiver is ready to accept the data, it sets logic 1 on line 5. As soon as the transmitter detects logic 1 on line 5, it sets line 6 to logic 1 from previous logic 0 state, and puts the data on lines 11 to 18 in the form of 8 bits. By setting logic 1 on line 6, the transmitter asks the receiver to read the data. Then the receiver reads the data, with no time limitation. All the logic levels on data lines are stable as long as logic 1 persist on the line. After reading, it sets the line 5 to logic 0. After detecting 0, then the transmitter puts new data character on the data lines and sets line 6 to logic 0. If something is wrong in the data transmission, it is detected by the receiver, using parity bit. So line 7 is used to indicate an error and it asks for retransmission. Once line 5 is logic 0, transmitter can set new data whenever it is required.

When line 7 is used, the transmitter must wait until line 5 goes to 0, so status of line 7 can be used to decide whether previous transmission was correct or wrong.

Since no time restriction is involved, fast and slow devices can operate together at a speed which in turn depends on the sum of their operating times and channel delays.

● **PROBLEM** 16-21

Discuss the salient points of a typical main processor software that deals and responds to an "interupt" from a communication system.

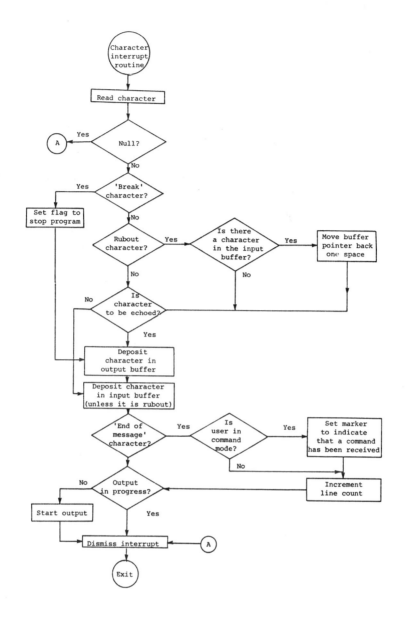

Figure 1.

Solution: The piece of software package that deals with the interrupt from a communication system is always incorporated within the operating system software of the computer. The software package depends mainly on the application rather than on what sort of transmission method is used. The manner in which the transmitted character will be handled depends mainly on the application.

As an example of a main processor software for communication system, a typical situation is a line-sharing system used to handle a terminal.

In this case, the communication processor will hand over either a single character at a time or a group of characters, to the main processor, i.e., C.P.U. when the single characters are handed over to the main processor, it is its duty to process all these received characters and form units which will convey information. Obviously, in this situation, valuable CPU time will be diverted in doing this composing. If on the other hand, the communication processor is speedy so that it can transmit blocks of characters at a time, then the CPU will not be disturbed. In some time-sharing systems, transmission is on character-by-character basis, in others, it is the other way around.

A typical interrupt routine followed by a main processor as soon as it receives a character from a communication processor is shown in Fig. 1. This system is operating on a full-duplex mode, and main processor will handle character echoing.

CHAPTER 17

INFORMATION THEORY AND CODING

INFORMATION THEORY AND ENTROPY

● **PROBLEM** 17-1

A man is informed that, when a pair of dice were rolled, the resulting sum was a "seven". How much information is there in this message?

Solution: By definition, the amount of information received in a message is given as

$$\text{Information received} = \log_2 \left[\frac{\text{Probability at the receiver of the event after the message is received.}}{\text{Probability at the receiver of the event before the message is received.}} \right]$$

Assuming the "noiseless" case (where the man is certain that the received information is correct), the probability of the event at the receiver after the message is received is unity. Next, calculate the probability of occurrence of a seven when two dice are tossed. A seven can be obtained in the following mutually exclusive ways:

Die #1	Die #2	Total of 2 dice
1	6	7
2	5	7
3	4	7
4	3	7
5	2	7
6	1	7

Each of the individual possibilities has a probability of occurrence of

$$\frac{1}{6} \times \frac{1}{6} = \frac{1}{36}$$

Since six mutually exclusive possibilities, each having a probability of 1/36, will give a "seven", the probability of occurrence of a seven is

$$6 \times \frac{1}{36} = \frac{1}{6}$$

Therefore,

$$\text{information received} = \log_2\left(\frac{1}{(1/6)}\right) = \log_2(6) = 2.585 \text{ bits}$$

● **PROBLEM** 17-2

The statistical data for a certain community indicate that 25% of all girls are blondes, 50% of all girls and 75% of all blondes have blue eyes. If it is known that a girl has blue eyes, how much additional information is available if one is told that she is also a blonde?

Solution: From the given data,

P_1 = P(Blonde) = probability that a girl is blonde irrespective of the color of her eyes = 0.25.

$P_2 = P_{\text{Blonde}}$ (Blue eyes) = probability that a blonde girl has blue eyes = 0.75.

P_3 = P (Blue eyes) = probability that a girl is blue eyed irrespective of the color of her hair.

It is required to determine the probability P_X so that a blue-eyed girl is blonde.

i.e., $\qquad P_X = P_{\text{blue eyes}}$ (Blonde)

Now, let P_4 = P (Blonde, Blue eyes) = probability that a blonde girl is blue eyed.

then $\qquad P_4 = P_1 \cdot P_2 = P_3 P_X$

Thus, $\qquad P_X = \frac{P_1 \cdot P_2}{P_3}$

Knowing that a girl has blue eyes, the additional information available in knowing that she is a blonde is

$$\log_2\left(\frac{1}{P_x}\right) = \log_2\left(\frac{P_3}{P_1 \cdot P_2}\right)$$

$$= (-\log_2 P_1 - \log_2 P_2 + \log_2 P_3)$$

$$= (-\log_2 0.25 - \log_2 0.75 + \log_2 0.5)$$

$$= (2 + 0.415 - 1)$$

$$= 1.415 \text{ bits.}$$

● **PROBLEM 17-3**

A television picture is composed of approximately 300,000 individual picture elements, each of which can attain ten different and distinguishable brightness levels (such as black and shades of gray) for proper contrast. If, for any picture element, the probability of occurrence of the ten brightness levels are equal, and 30 picture frames are being transmitted per second, determine the bandwidth of the T.V. video signal. Assume that the signal-to-noise ratio of 1000 (30 db) is required for proper reproduction of any picture.

Solution: For any information channel, the bandwidth of the signal will depend on the channel capacity which in turn is dependent on the amount of information being transmitted. With the given data, the information per second can be evaluated as follows:

Since, for each picture element, the probability of occurrence of the ten different levels is equal, the information per picture element = $\log_2 10$

$$= 3.32 \text{ bits/element}$$

With 300,000 picture elements per picture frame, the information per picture frame = $300{,}000 \times 3.32$

$$= 996{,}000 \text{ bits/picture frame.}$$

Since 30 picture frames are being transmitted per second, the total information per second is

$$= 996{,}000 \times 30$$

$$= 29.88 \times 10^6 \text{ bits/second.}$$

In other words, the video signal under consideration is an information signal of 29.88×10^6 bits/second. The transmission of which requires a channel of capacity $C = 29.88 \times 10^6$ bits/second.

The relation between the channel capacity C, and the channel bandwidth BW is

$$C = BW \log_2\left(1 + \frac{S}{N}\right) \text{ bits/second}$$

where $\frac{S}{N}$ = signal to noise ratio = 1000

Therefore, $C = BW \log_2 (1001) = 29.88 \times 10^6$

and $BW \cong 3.0 \text{ MHz}$

Therefore, the bandwidth of the video signal = 3 MHz.

● **PROBLEM 17-4**

Find the entropy of the source $S = \{S_1, S_2, S_3\}$ with probabilities $P(S_1) = \frac{1}{2}$ and $P(S_2) = P(S_3) = \frac{1}{4}$.

Fig. 1

An information source

<u>Solution</u>: An information source illustrated in the figure above, is known as a zero-memory source and is defined by the source alphabet S together with the probabilities with which the symbols occur: $P(S_1), P(S_2), \ldots, P(S_n)$.

The entropy, or average amount of information, for such a source is defined as

$$H(S) \overset{\Delta}{=} \sum_S P(S_i) \log_2 \frac{1}{P(S_i)} \text{ bits/symbol}$$

With the given probabilities, the entropy is calculated as follows:

$$H(S) = \frac{1}{2} \log_2 2 + \frac{1}{4} \log_2 4 + \frac{1}{4} \log_2 4$$

$$= \frac{3}{2} \text{ bits/symbol}.$$

Note that this average amount of information per symbol generated by the source can also be interpreted as the average amount of uncertainty one obtained without first considering the source output.

● **PROBLEM** 17-5

If six messages having probabilities $\frac{1}{4}, \frac{1}{4}, \frac{1}{8}, \frac{1}{8}, \frac{1}{8}$ and $\frac{1}{8}$ can be accommodated in a single communication system, determine the entropy.

Solution: Entropy or, the average information per message interval is defined as,

$$H = \sum_{i=1}^{n} P_{x_i} \log_2 \left(\frac{1}{P_{x_i}}\right),$$ where n is the number of messages.

Since there are six messages and the corresponding probabilities are given, the entropy is

$$H = \frac{1}{4} \log_2 4 + \frac{1}{4} \log_2 4 + \frac{1}{8} \log_2 8 + \frac{1}{8} \log_2 8$$

$$+ \frac{1}{8} \log_2 8 + \frac{1}{8} \log_2 8$$

$$= \frac{1}{2} + \frac{1}{2} + \frac{3}{8} + \frac{3}{8} + \frac{3}{8} + \frac{3}{8}$$

$$= 2.50 \text{ bits/message.}$$

● **PROBLEM** 17-6

Figure (A) and (B) illustrate a general information channel and the corresponding channel diagram, respectively. For the set of input signals shown, determine the priori and posteriori entropy.

$$A \begin{Bmatrix} a_1 \\ a_2 \\ \vdots \\ a_r \end{Bmatrix} \longrightarrow \boxed{P(b_j/a_i)} \longrightarrow \begin{Bmatrix} b_1 \\ b_2 \\ \vdots \\ b_s \end{Bmatrix} B$$

Figure A

An information channel

Solution: For an information channel, if H(A), a priori entropy, is the average number of bits required to represent a symbol from a source having priori probabilities $P(a_i)$, $i = 1, 2, \ldots n$; then H(A) can be expressed as

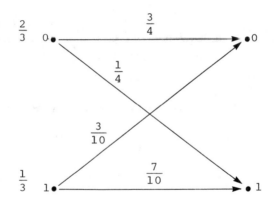

Fig. B

A noisy information channel

$$H(A) = \sum_A P(a) \log_2 \frac{1}{P(a)} \qquad (1)$$

Also, if b_j is the received symbol and $H(A/b_j)$, a posteriori entropy, is the average number of bits required to represent a symbol from a source having probabilities $P(a_i/b_j)$, $i = 1, 2 \ldots n$; then the posteriori entropy of A is

$$H(A/b_j) = \sum_A P(a/b_j) \log_2 \frac{1}{P(a/b_j)} \qquad (2)$$

Thus, from equation 1, the a priori entropy of the set of input signals for the given channel is

$$H(A) = \frac{2}{3} \log_2 \frac{3}{2} + \frac{1}{3} \log_2 3$$

$$= 0.918 \text{ bit.}$$

The posteriori probabilities corresponding to the symbol 0 and symbol 1 received at the output respectively are

$$P_r\{a=0/b=0\} = \frac{\left(\frac{2}{3}\right)\left(\frac{3}{4}\right)}{\left(\frac{12}{20}\right)} = \frac{5}{6}$$

and $$P_r\{a=1/b=1\} = \frac{\left(\frac{1}{3}\right)\left(\frac{7}{10}\right)}{\left(\frac{12}{30}\right)} = \frac{7}{12}$$

Thus, if the received symbol is 0, the posteriori entropy is

$$H(A/0) = \frac{5}{6} \log_2 \frac{6}{5} + \frac{1}{5} \log_2 5$$

$$= \tfrac{5}{6}(.26) + \tfrac{1}{5}(2.322)$$

$$= 0.68 \text{ bit.}$$

If, however, the received symbol is 1, the posteriori entropy is

$$H(A/1) = \tfrac{7}{12}\log_2 \tfrac{12}{7} + \tfrac{5}{12}\log_2 \tfrac{12}{5}$$

$$= 0.980 \text{ bit.}$$

The above results show that the uncertainty regarding the validity of the received signal decreases if a '0' is received, and increases if a '1' is received.

• **PROBLEM 17-7**

In a coin tossing experiment C, let H and T refer to the events of obtaining a head or a tail, respectively. Determine the average information associated with such an experiment, when

(a) $P(H) = P(T) = 0.5$ (honest coin)

(b) $P(H) = 3/8$ and $P(T) = 5/8$ (biased coin).

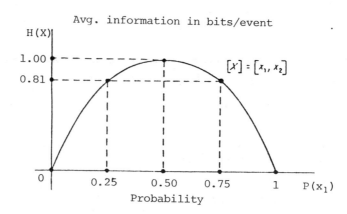

Solution: The average information per message interval (or for each tossing of a coin in the present case) can be expressed as

$$H(C) = E[I(X_i)]$$

$$= -\sum_{i=1}^{m} P(X_i) \log \frac{1}{P(X_i)} \qquad (1)$$

where $I(X_i)$ is the total information in one complete experiment, and $P(X_i)$ is the probability of occurrence of each individual toss.

(a) From equation (1) the average information associated with the tossing of an honest coin can be evaluated as

$$H(C) = P(H) \log_2 \frac{1}{P(H)} + P(T) \log_2 \frac{1}{P(T)}$$

$$= \frac{1}{2} \log_2 2 + \frac{1}{2} \log_2 2.$$

$$= 1 \text{ bit/toss.}$$

(b) The average information associated with the tossing of a biased coin is

$$H(C) = \frac{3}{8} \log_2 \frac{8}{3} + \frac{5}{8} \log_2 \frac{8}{5}$$

$$= 0.954 \text{ bits/toss.}$$

Note: This problem illustrates that the average information is maximum when the different events have equal probabilities and that a deviation from this condition will lead to a reduction in the average information.

On further evaluation, it can be shown that, if $P(T) = 1$ and $P(H) = 0$ or vice-versa, the average information $H(C)$ is zero. These properties are illustrated by the plot of $H(X)$ versus $P(X_1)$ in figure below, where X_1, X_2 are events which are mutually exclusive and exhaustive.

• **PROBLEM** 17-8

Messages $Q_1, \ldots Q_m$ have probabilities $p_1, \ldots p_m$ of occurring

(a) Find an expression for H. (the average information per message)

(b) If we are to transmit three messages, i.e., M=3, determine H in terms of p_1 and p_2 by using the result that $p_1 + p_2 + p_3 = 1$.

(c) Calculate p_1 and p_2 for $H = H_{(max)}$ by letting $\partial H / \partial p_1 = 0$ and $\partial H / \partial p_2 = 0$.

(d) Repeat part (c) for the case of M messages.

Solution: (a) It is supposed that messages $Q_1, \ldots, Q_m$ have probabilities $p_1, \ldots, p_m$ of occurring. Now suppose further that during a long period of transmission a sequence of L messages have been generated. Then, if L is very large, expect that p_1L messages of Q_1, p_2L messages of Q_2, etc., will have occurred in the sequence. The total information in such a sequence will be

$$I_{total} = p_1 L \log_2 \frac{1}{p_1} + p_2 L \log_2 \frac{1}{p_2} + \ldots$$

The average information per message interval, represented by the symbol H, will then be

$$H = \frac{I_{total}}{L} = p_1 \log_2 \frac{1}{p_1} + p_2 \log_2 \frac{1}{p_2} + \ldots = \sum_{K=1}^{M} P_K \log_2 \frac{1}{P_K}$$

$$= -\sum_{K=1}^{M} P_K \log_2 P_K \tag{1}$$

(b) $H = -P_1 \log_2 P_1 - P_2 \log_2 P_2 - P_3 \log_2 P_3$

$\quad\quad = -P_1 \log_2 P_1 - P_2 \log_2 P_2 - (1-P_1-P_2) \log_2(1-P_1-P_2).$

(c) First, obtain $\frac{\partial H}{\partial P_1}$ and $\frac{\partial H}{\partial P_2}$

$$\frac{\partial H}{\partial P_1} = -1 - \log_2 P_1 + 1 + \log_2 (1-P_1-P_2) = \log_2 \left(\frac{1-P_1-P_2}{P_1}\right)$$

and

$$\frac{\partial H}{\partial P_2} = -1 - \log_2 P_2 + 1 + \log_2 (1-P_1-P_2) = \log_2 \left(\frac{1-P_1-P_2}{P_2}\right)$$

Then by setting $\frac{\partial H}{\partial P_1} = C$ and $\frac{\partial H}{\partial P_2} = 0$, obtain

$$1 - P_1 - P_2 = P_1 = P_2$$

Hence, $\quad P_1 = P_2 = P_3 = 1/3$ for $H = H_{(max)}$.

(d) Using the relation $P_M = 1 - \sum_{K=1}^{M-1} P_K$, Eq. (1) is rewritten as follows:

$$H = -\sum_{K=1}^{M-1} P_K \log_2 P_K - \left(1 - \sum_{K=1}^{M-1} P_K\right) \log_2 \left(1 - \sum_{K=1}^{M-1} P_K\right)$$

$$\frac{\partial H}{\partial P_\ell} = -1 - \log_2 P_1 + 1 + \log_2 \left(1 - \sum_{K=1}^{M-1} P_K\right) = \log_2 \left(\frac{1 - \sum_{k=1}^{M-1} P_k}{P_\ell}\right)$$

From $\partial H / \partial P_1 = 0$, $l = 1, 2, 3, \ldots, (M-1)$,

$$1 - \sum_{K=1}^{M-1} P_K = P_1, \quad l = 1, 2, 3, \ldots, (M-1),$$

or $\quad P_1 = P_2 = \ldots = P_{M-1} = 1 - \sum_{K=1}^{M-1} P_K$

Hence, H is maximum if

$$P_1 = P_2 = \ldots = P_{M-1} = \frac{1}{M}$$

and $P_M = 1 - \sum_{K=1}^{M-1} P_K = \frac{1}{M}.$

• **PROBLEM 17-9**

For a discrete random variable X, the average information or entropy associated with n possible outcomes is expressed as

$$H(X) = E\{I(X_j)\} = -\sum_{j=1}^{n} p(X_j) \log_2 p(X_j). \qquad (1)$$

Where $P(X_j)$ is the probability associated with the event X_j.

Considering a binary source with probabilities $p(1) = \alpha$ and $p(0) = (1-\alpha) = \beta$, evaluate the entropy of the source as a function of 'α' and plot $H(\alpha)$ for $0 \leq \alpha \leq 1$.

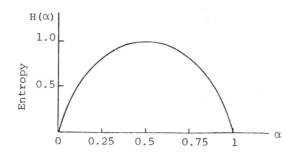

Fig. 1

Entropy for a binary source

Solution: For a binary source with the given probabilities, the entropy $H(\alpha)$ can be expressed [from equation (1)] as

$$H(\alpha) = -\alpha \log_2 \alpha - (1-\alpha) \log_2 (1-\alpha).$$

It is known that the entropy, or average value, will be a maximum when all outcomes or events have equal probabilities. Thus, in the case of a binary source, the entropy will be a maximum when $\alpha = \beta = 1/2$, or in other words the uncertainty of the output is a maximum when $\alpha = \beta = 1/2$. If $\alpha \neq 1/2$, one of the symbols is more likely to occur than the other. Hence, the uncertainty as to which symbol appears at the source output is reduced.

If α or β is equal to zero, then the uncertainty is also zero since it is obvious as to which symbol must occur at the output. The plot of H(α) corresponding to different values of α is shown in figure (1).

● **PROBLEM** 17-10

The probabilities associated with the inputs X_1, X_2, X_3 and the outputs Y_1, Y_2 of the channel shown in figure (1) are

$$P(X_1) = \frac{1}{18}, \quad P(X_2) = \frac{1}{3}, \quad P(X_3) = \frac{5}{11},$$

$$P(Y_1) = \frac{1}{4}, \quad \text{and} \quad P(Y_2) = \frac{3}{4} \quad \text{respectively.}$$

If the joint probabilities are defined as

$$[P(XY)] = \begin{bmatrix} \frac{1}{18} & 0 \\ \frac{2}{5} & \frac{1}{4} \\ \frac{1}{20} & \frac{1}{3} \end{bmatrix}$$

determine the following:

(a) Joint entropy H(XY),

(b) Conditional entropy H(Y/X), and

(c) Conditional entropy H(X/Y).

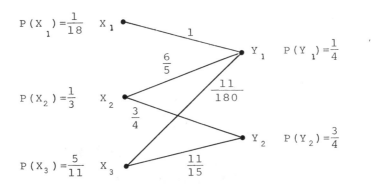

Fig. 1.

Solution: (a): The joint entropy or the average information H(XY) is defined as:

$$H(XY) = E[I(x_i y_i)]$$

$$= -\sum_{i=1}^{3}\sum_{j=1}^{2} P(x_i y_j) \log_2 P(x_i y_j)$$

$$= [-\tfrac{1}{18} \log_2 \left(\tfrac{1}{18}\right) - \tfrac{2}{5} \times \log_2 \left(\tfrac{2}{5}\right) - \tfrac{1}{20} \log_2 \left(\tfrac{1}{20}\right)$$

$$- \tfrac{1}{4} \log_2 \left(\tfrac{1}{4}\right) - \tfrac{1}{3} \log_2 \left(\tfrac{1}{3}\right)]$$

$$= (0.230 + 0.528 + 0.216 + 0.5 + 0.527)$$

$$= 2.005 \text{ bits/character.}$$

This joint entropy can be represented as an area as shown in figure (2)

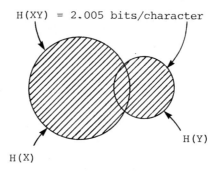

Fig. 2.

(b): For a transmission channel, the conditional entropy $H(Y/X)$ is expressed as,

$$H(Y/X) = -\sum_{i=1}^{3}\sum_{j=1}^{2} P(x_i y_j) \log_2 \frac{P(x_i y_j)}{P(x_i)}$$

$$= -\tfrac{1}{18} \log_2 \left(\tfrac{\tfrac{1}{18}}{\tfrac{1}{18}}\right) - \tfrac{2}{5} \times \log_2 \left(\tfrac{\tfrac{2}{5}}{\tfrac{1}{3}}\right) - \tfrac{1}{20} \log_2 \left(\tfrac{\tfrac{1}{20}}{\tfrac{5}{11}}\right)$$

$$- \tfrac{1}{4} \log_2 \left(\tfrac{\tfrac{1}{4}}{\tfrac{1}{3}}\right) - \tfrac{1}{3} \log_2 \left(\tfrac{\tfrac{1}{3}}{\tfrac{5}{11}}\right)$$

$$= 0 + 0.105 + 0.159 + 0.103 + 0.152$$

$$= 0.519 = \text{bits/character.}$$

(c) With $H(X/Y) = -\sum_{i=1}^{3}\sum_{j=1}^{2} P(x_i y_j) \log_2 \frac{P(x_i y_j)}{P(y_j)}$,

the conditional entropy $H(X/Y)$ for the given channel will be,

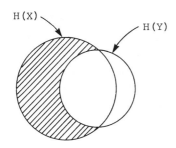

H(X/Y) = 0.752 bits/character

Fig. 3.

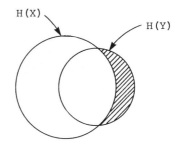

H(Y/X) = 0.519 bits/character

Fig. 4.

$$H(X/Y) = -\frac{1}{18}\log_2\left(\frac{\frac{1}{18}}{\frac{1}{4}}\right) - \frac{2}{5}\log_2\left(\frac{\frac{2}{5}}{\frac{1}{4}}\right)$$

$$- \frac{1}{20}\log_2\left(\frac{\frac{1}{20}}{\frac{1}{4}}\right) - \frac{1}{4}\log_2\left(\frac{\frac{1}{4}}{\frac{3}{4}}\right) - \frac{1}{3}\log_2\left(\frac{\frac{1}{3}}{\frac{3}{4}}\right)$$

$$= 0.121 - 0.271 + 0.116 + 0.396 + 0.390$$

$$= 0.752 \text{ bits/character.}$$

The representation of the conditional entropies H(X/Y) and H(Y/X) in terms of their areas is shown in figures (3) and (4) respectively.

● **PROBLEM 17-11**

A message is transmitted in the form of dots and dashes, with the probability of occurrence of a dash at 0.33.

(a) Calculate the information in bits conveyed by a single dash and a single dot.

(b) Determine the average information in a dot-dash code, and

(c) Compute the rate of information transmission if a dot lasts for 10 msec and intersymbol interval is also 10 msec.

It can be assumed that a dash is three times as long as a dot.

Solution: (a) For noiseless transmission of an entity, k, with the probability of occurrence P_k, the amount of information conveyed can be expressed as,

$$I_k = \log_2 \frac{1}{P_k}$$

867

Therefore, the information in a dot over a noiseless channel is:

$$I_{dot} = \log_2 \frac{1}{P_{dot}} \quad \left(P_{dot} = 1 - P_{dash}\right)$$

$$= \log_2 \frac{1}{0.66}$$

$$\cong 0.585 \text{ bits.}$$

Similarly,

$$I_{dash} = \log_2 \frac{1}{0.33}$$

$$\cong 1.585$$

(b) Average information in a dot-dash code is:

$$H = P[dot]I_{dot} + P[dash]I_{dash}$$

$$= \tfrac{2}{3} \log_2 1.5 + \tfrac{1}{3} \log_2 0.33$$

$$= 0.918 \text{ bits}$$

(c) If r is the rate of generation of a message, and H is the average information in a message, then the average number of bits of information per second is

$$R = r \cdot H$$

Hence

$$R = r \cdot (0.918) \text{ bits/sec.}$$

where

$$\tfrac{1}{r} = \text{average time/message}$$

$$= (\tfrac{2}{3} \times 20) + (\tfrac{1}{3} \times 40) \text{ msec.}$$

$$= \frac{1}{37.5}$$

Hence, r = 37.5

and   R = 34.5 bits/sec.

• **PROBLEM 17-12**

An alphabet, independent of intersymbol influence and with equal probability of occurrence of all letters, consists of 8 consonants and 8 vowels. If the interpretation of the consonants is always correct and that of the vowels is only 50% correct, being mistaken for other vowels otherwise, determine the average rate of information transmission per symbol. Assume that all vowels involved in errors have equal probabilities of error.

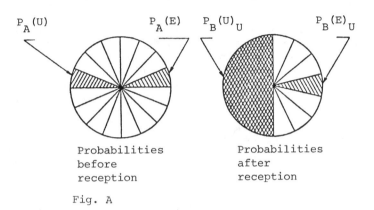

Fig. A

Solution: Since there are equal numbers of consonants and vowels in the alphabet, it is evident that 50% of the letters received will be consonants, and 50% will be vowels, of which 25% are incorrect vowels and 25% are correct vowels. The amount of information in the received signal corresponding to the three components can be evaluated as follows.

(1) Correct consonants: The information content per letter can be expressed as,

$$I = \log_2\left(\frac{P_B}{P_A}\right) \text{ bits/letter,} \qquad (1)$$

where P_A = probability before reception.

and P_B = probability after reception.

Since there are 16 letters in the alphabet (8 consonants and 8 vowels) and all letters are equally probable, the probability before reception of each letter = 1/16 = P_A.

Also, since all the consonants are received without error, the probability after reception of each consonant letter is 1.

Therefore, from equation (1),

$$\text{Information/letter} = \log_2\left(\frac{1}{\frac{1}{16}}\right) = 4 \text{ bits/letter}$$

(2) Correct vowels: In this case, $p_A = \frac{1}{16}$ and $p_B = \frac{1}{2}$ (since only half the vowels are correct).

Hence, Information/letter = $\log_2\left(\frac{\frac{1}{2}}{\frac{1}{16}}\right)$ = 3 bits/letter

(3) Incorrect vowels: Here $p_A = \frac{1}{16}$

The probability after reception can be determined from figure A as follows:

Suppose that E and U are two vowels, and a U is received corresponding to a transmission of E. Having received U, it is required to determine the probability that the sent vowel was an E. Knowing that a vowel was sent, the probability that it was a U, the incorrect vowel is $\frac{1}{2}$. However, since all letters are equally probable, the other half of the probability is equally divided among the remaining 7 vowels including E. Thus, the probability that the transmitted vowel was

$$E = \frac{1}{7} \times \frac{1}{2} = \frac{1}{14} = p_B$$

Therefore, Information/letter = $\log_2\left(\frac{\frac{1}{14}}{\frac{1}{16}}\right)$

$= (\log_2 16 - \log_2 14)$

$= (4 - 3.807)$

$= 0.193$ bits/letter.

It is important to note that the reception of an incorrect vowel will increase the probability of the following received vowels being correct. Thus, average information/symbol will be

$$= \left(\frac{50}{100} \times 4\right) + \left(\frac{25}{100} \times 3\right) + \left(\frac{25}{100} \times 0.2\right)$$

$= 2.8$ bits/symbol.

SIGNAL TRANSMISSION: CHANNEL REPRESENTATION AND CHANNEL CAPACITY

● PROBLEM 17-13

A sinusoidal signal in the frequency range 0.4 to 4.4 KHz is being sampled at a rate equal to four times the Nyquist rate. If this signal is received by a receiver of signal-to-noise ratio $\frac{S_i}{\eta} = 10^6$ after being quantized by a 256 level quantizer at the transmission end, calculate

(a) The time T between samples.

(b) The number of bits N and the probability of error if the quantized signal is encoded into a binary PCM waveform.

(c) The probability of error, P_e, if the signal is encoded into 1 to 256 orthogonal signals.

Solution: (a) $\quad T = \dfrac{1}{\text{sampling frequency}}$

$= \dfrac{1}{4 \times \text{Nyquist rate}}$

$= \dfrac{1}{4 \times (2 f_M)}$

$= \dfrac{1}{4 \times 2 \times 4000}$

$= 3.125 \ \mu\text{sec}.$

(b) Number of bits $N = \log_2 256$

$= \dfrac{\log_{10} 256}{\log_{10} 2}$

$= 8$

The probability of error is given by

$$P_e = \tfrac{1}{2} \operatorname{erfc} \sqrt{\dfrac{V^2 T}{\eta}}$$

$$= \tfrac{1}{2} \operatorname{erfc} \sqrt{\dfrac{S_i T}{N} \dfrac{1}{\eta}}$$

$$= \tfrac{1}{2} \operatorname{erfc} \sqrt{10^6 \times 3.125 \times 10^{-6} \times 8}$$

$$= \tfrac{1}{2} \operatorname{erfc} (5).$$

(c) $P_e = 1 - \left(\dfrac{1}{\sqrt{\pi}}\right)^{256} \displaystyle\int_{-\infty}^{\infty} e^{-y^2} \left(\int_{-\infty}^{\sqrt{\frac{S_i}{\eta} \cdot T}} e^{-x^2} dx\right)^{256} dy$

$$= 1-\left(\frac{1}{\sqrt{\pi}}\right)^{256} \int_{-\infty}^{\infty} e^{-y^2}\left(\int_{-\infty}^{\sqrt{3.125}+y} e^{-x^2} dx\right)^{256} dy$$

$$= 1-\left(\frac{1}{\sqrt{\pi}}\right)^{256} \int_{-\infty}^{\infty} e^{-y^2}\left(\int_{-\infty}^{\sqrt{3.125}+y} e^{-x^2} dx\right)^{256} dy$$

$$\cong 10^{-5}.$$

● **PROBLEM 17-14**

The input probabilities of a binary input-output channel are $P(X_1) = P(X_2) = 0.5$, and the transition probabilities are defined by the following matrix

$$P[Y/X] = \begin{bmatrix} 0.7 & 0.3 \\ 0.4 & 0.6 \end{bmatrix}$$

Determine the output probabilities and the joint probability matrix.

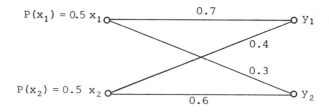

Binary symmetric channel

Solution: For a binary input-output channel, with known input probabilities, the output probabilities can be directly derived from the channel transition matrix as follows.

Let $\qquad [P(X)] = [P(X_1) \; P(X_2)]$ and

$\qquad\qquad [P(Y)] = [P(Y_1) \; P(Y_2)]$

be the row matrices corresponding to the input and output probabilities respectively.

If $P(Y/X)$ is the matrix defining the transition probabilities, then

$$[P(Y)] = [P(X)][P(Y/X)]$$

Therefore $P(Y) = [0.5 \quad 0.5] \begin{bmatrix} 0.7 & 0.3 \\ 0.4 & 0.6 \end{bmatrix}$

$$= [0.55 \quad 0.45].$$

The joint probability matrix is

$$P[X,Y] = \begin{bmatrix} 0.5 & 0 \\ 0 & 0.5 \end{bmatrix} \begin{bmatrix} 0.7 & 0.3 \\ 0.4 & 0.6 \end{bmatrix}$$

$$= \begin{bmatrix} 0.35 & 0.15 \\ 0.2 & 0.3 \end{bmatrix}$$

● **PROBLEM** 17-15

For the binary symmetric channel shown in figure 1, determine the channel capacity.

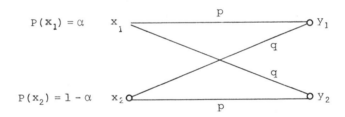

Fig. 1

Binary Symmetric Channel

Solution: For a transmission channel, the channel capacity can be determined by evaluating the maximum information being transmitted through the channel.

The total information in the channel can be expressed as

$$I(X;Y) = H(Y) - H(Y/X) \tag{1}$$

where by definition

$$H(Y/X) = - \sum_{i=1}^{2} \sum_{j=1}^{2} P(x_i, y_j) \log_2 P(y_j, x_i)$$

and $H(Y)$ is the average information per message with the given probabilities (see figure).

Hence,

$$H(Y/X) = -\alpha p \cdot \log_2 p - (1-\alpha) \cdot p \cdot \log_2 p$$

$$-\alpha q \cdot \log_2 q - (1-\alpha) \cdot q \cdot \log_2 q$$

$$= -p \cdot \log_2 p - q \cdot \log_2 q$$

From equation (1),

$$I(X;Y) = H(Y) + p \log_2 p + q \log_2 q \qquad (2)$$

which will be maximum when $H(Y)$ is maximum. In general, the average information in any channel will be a maximum when all the events have equal probabilities. Since in the present case, the output is binary, $H(Y)$ is maximum when both the outputs have equal probability i.e. $\frac{1}{2}$, and is achieved for equally likely inputs.

Thus,
$$H(2) = P(x_1) \log_2 \frac{1}{P(x_1)} + P(x_2) \log_2 \frac{1}{P(x_2)}$$

$$= \tfrac{1}{2} \log_2 2 + \tfrac{1}{2} \log_2 2$$

$$= 1 \text{ bit/symbol}.$$

Therefore, equation (2) yields,

$$I(X;Y) = 1 + p \log_2 p + q \log_2 q$$

$$= 1 - H(p)$$

where $H(p) = -p \log_2 p - (1-p) \log_2 (1-p)$

The figure below illustrates the channel capacity for a binary symmetric channel. It is important to note that if $p=0$ or 1, the channel capacity is one bit per symbol i.e. the output is completely determined by the channel input.

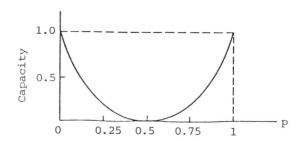

Fig. 2.

Capacity of binary symmetric channel

● **PROBLEM 17-16**

Assume that the discrete channel shown in the figure is noiseless. Determine its channel capacity.

Solution: For a transmission channel, the channel capacity

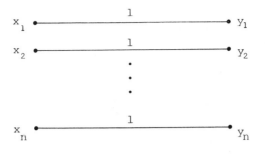

Fig. 1.

Noiseless Channel

C can be defined as the maximum information per symbol which can be transmitted through the channel.

i.e., $$C = \max[I(X;Y)]$$

It is of importance to note that the maximization process eliminates the effect of the source probabilities or the channel capacity and hence, the channel capacity is only a function of the channel transition probabilities.

Now, the correspondence between the total signal in the channel [$I(X;Y)$], and the average signal can be expressed as

$$I(X;Y) = H(X) - H(X/Y)$$

where $H(X/Y) = - \sum_{i=1}^{n} \sum_{j=1}^{n} P(X_i, Y_j) \log_2 P(X_i, Y_j).$

Since the channel under consideration is a noiseless channel all $P(X_i, Y_j)$ and $P(X_i/Y_j)$ are zero except when $i = j$, in which case $P(X_i/Y_j) = 1$. Thus, for a noiseless channel $H(X/Y) = 0$ and $I(X;Y) = H(X)$.

Now $$H(X) = \sum_{j=1}^{n} P(X_i) \log_2 \frac{1}{P(X_i)} \qquad (1)$$

It is know that the average information $H(X)$ is a maximum when all the events or source symbols have equal probabilities, in which case equation (1) reduces to

$$\max[H(X)] = \sum_{j=1}^{n} \frac{1}{n} \log_2 n = \log_2 n = \max[I(X;Y)].$$

Thus,

$$C = \max[I(X;Y)] = \log_2 n.$$

• **PROBLEM** 17-17

A three input information channel with all three input symbols equally probable, is defined by the channel matrix,

$$\begin{bmatrix} 0.5 & 0.3 & 0.2 \\ 0.2 & 0.3 & 0.5 \\ 0.3 & 0.3 & 0.4 \end{bmatrix}$$

Assuming that a maximum-likelihood decision rule is used, determine the error probability, P_E, of the channel.

Solution: In any information channel, it is necessary that the error probability always be a minimum. This can be achieved in practice by the use of the maximum-likelihood decision rule which chooses for each output symbol, the input symbol with the highest probability and in effect, minimizes the error probability.

From the channel matrix, for the outputs b_1, b_2 and b_3, the highest probability corresponds to the inputs $a_1(0.5)$, $a_3(0.3)$ and $a_2(0.5)$, respectively. With all the inputs equally probable (i.e., $\frac{1}{3}$), the probability of error can be expressed as

$$P_E = \frac{1}{3}[(0.2 + 0.3) + (0.3 + 0.3) + (0.2 + 0.4)]$$

$$= \frac{17}{30}$$

Note that P_E is the sum of the probability of error for each individual output.

• **PROBLEM** 17-18

Since the symbol error for each transmitted symbol in a coherent binary Frequency Shift Keying (FSK) system is equally probable, a binary symmetric channel model is chosen for FSK transmission. If the transmitter power is 1000 Watts and 10,000 symbols/sec are transmitted through a channel having an attenuation of 30db between the transmitter and detector,

(a) determine the channel matrix.

(b) If the symbol rate at the source is reduced by 25%, evaluate the corresponding channel matrix.

Assume, the noise power spectral density

$$N_o = 2 \times 10^{-5} \text{ W/Hz} \quad \text{and} \quad P_E = \tfrac{1}{2} \text{ erf} \left(\sqrt{\frac{E_S}{2N_o}} \right)$$

Solution: With a 30db attenuation in the channel, the signal power P_R received at the detector corresponding to a transmitted signal of 1000 Watts is

$$P_R = (1000)(10^{-3}) = 1 \text{ Watt}.$$

The received signal consists of 10,000 symbols each with energy

$$E_S = P_R T = \frac{1}{10,000} = 10^{-4} \text{ Joule}.$$

The error probability can now be estimated from the identity

$$P_E = \tfrac{1}{2} \text{ erfc}\left(\sqrt{\frac{E_S}{2N_o}}\right)$$

$$P_E = \tfrac{1}{2} \text{ erfc}\left(\sqrt{\frac{10^{-4}}{2 \times 2 \times 10^{-5}}}\right)$$

$$= \tfrac{1}{2} \text{ erfc } [1.581]$$

$$= 0.0127 \qquad (1)$$

The channel matrix will thus be

$$P[Y/X] = \begin{bmatrix} (1-0.0127) & 0.0127 \\ 0.0127 & (1-0.0127) \end{bmatrix}$$

$$= \begin{bmatrix} 0.9873 & 0.0127 \\ 0.0127 & 0.9873 \end{bmatrix}$$

(b) Assuming all other parameters held constant, a reduction in the source symbol rate of 25% will result in a rate of 7500 symbols/second. Hence, the resulting received energy per symbol is,

$$E_S = \frac{1}{7500} = 1.333 \times 10^{-4} \text{ Joule}$$

and the probability of error in this case will be

$$P_E = \tfrac{1}{2} \text{ erfc}\left(\sqrt{\frac{1.333 \times 10^{-4}}{2 \times 2 \times 10^{-5}}}\right)$$

$$= 0.0049. \qquad (2)$$

The channel matrix will thus be

$$P[Y/X] = \begin{bmatrix} (1-0.0049) & 0.0049 \\ 0.0049 & (1-0.0049) \end{bmatrix}$$

$$\begin{bmatrix} 0.9951 & 0.0049 \\ 0.0049 & 0.9951 \end{bmatrix}$$

Comparing the error probabilities as given by (1) and (2) it is observed that a 25% reduction in the source symbol rate results in an improvement of the error probability by almost a factor of 3.

• **PROBLEM 17-19**

The cost of making the correct decisions for the given binary channel is zero, i.e., $c(d_1/x_1) = c(d_2/x_2) = 0$. The cost of making the incorrect decision d_1 is twice as much as that of making the incorrect decision d_2. That is to say, $c(d_1/x_2) = 2c(d_2/x_1)$ which decision should be made, first,

(a) When a y_1 is received, and

(b) When a y_2 is received?

Given $\quad P(y_1/x_1) = 0.8, \ P(y_2/x_2) = 0.6$,

and $\quad P(x_1) = 0.8$

Solution: For a binary channel if, $P(x_1/y_j)$ and $P(x_2/y_j)$ are the probabilities that y_j was received corresponding to transmissions X_1 and X_2 respectively, then the average conditional cost of making decision d_1 when y_j is received can be expressed as:

$$C(d_1/y_j) = C(d_1/x_1)\,P(x_1/y_j) + C(d_1/x_2)\,P(x_2/y_j) \quad (1)$$

And the average conditional cost of making decision d_2 when y_j is received is

$$C(d_2/y_j) = C(d_2/x_1)\,P(x_1/y_j) + C(d_2/x_2)\,P(x_2/y_j) \quad (2)$$

The choice between decisions d_1 and d_2 is dependent on their conditional costs such that the average cost involved should be a minimum. In other words, decision d_1 is chosen whenever $C(d_1/y_j) < C(d_2/y_j)$ and vice-versa. If, however, the conditional costs are equal, then either of the decisions may be chosen.

From equations (1) and (2), the likelihood ratio $L(y_j)$ can be defined as

$$L(y_j) = \frac{P(y_j/x_1)}{P(y_j/x_2)} > \frac{P(x_2)}{P(x_1)} \cdot \frac{C(d_1/x_2)-C(d_2/x_2)}{C(d_2/x_1)-C(d_1/x_1)} = L_t \quad (3)$$

where L_t is defined as the threshold value of the likelihood ratio. From equation (3) when $L(y_j) > L_t$, the decision chosen is d_1 and vice-versa.

(a) For the given binary channel, considering the reception of y_1, equation (3) yields:

$$L_t = \frac{P(x_2)}{P(x_1)} \cdot \frac{C(d_1/x_2)}{C(d_2/x_1)} = \frac{0.200}{0.800} \times \frac{2}{1} = 0.500$$

and

$$L(y_1) = \frac{P(y_1/x_1)}{P(y_1/x_2)} = \frac{0.800}{0.400} = 2$$

Since $L(y_1) = 2 > L_t = 0.500$, it follows that an X_1 has been transmitted or, the decision made is d_1. If $L(y_1)$ had been smaller than L_t, the decision would be d_2.

(b) Considering the reception of a y_2,

$$L(y_2) = \frac{P(y_2/x_1)}{P(y_2/x_2)} = \frac{0.200}{0.600} = 0.333$$

In this case $L(y_2) = 0.333 < L_t = 0.500$ and hence, it follows that decision d_2 (that an x_2 has been transmitted) would be made. If $L(y_2)$ had been greater than L_t, then decision d_1 (that an x_1 has been transmitted) would be made.

CODING AND ERROR-DETECTION (SYNDROME)

● **PROBLEM** 17-20

Explain the method of algebraic coding for a message source generating M equally likely messages and compute the $\overline{\overline{H}}$ matrix if 11111 and 00000 are transmitted repeatedly as coded symbols.

Solution: Let a message source be generating M messages with equal probability of occurrence. These are encoded in M words of length k such that $M = 2^k$, with no redundancy. The encoding process is carried out in such a way as to deliver one bit of information per coded word. To each of these coded words of length k bits, r redundant bits are added, resulting in k + r bits.

Due to the addition of redundant bits to each coded word, there results 2^{k+r} words while the number of possible words conveying messages are only 2^k. The format of a transmitted code word is:

$$a_1 a_2 \text{----} a_k c_1 c_2 \text{----} c_r$$

where a_i is the ith bit of the message code word and c_j is the jth redundant bit. These redundant bits are the parity-check bits.

In the problem, the repeatedly transmitted coded symbols are 11111 and 00000. So k+r=5. The number of redundant bits in each word is r=4, hence k=1.

Thus the corresponding $\bar{H}$ matrix will be.

$$\bar{H} = \begin{bmatrix} h_{11} & 1 & 0 & 0 & 0 \\ h_{21} & 0 & 1 & 0 & 0 \\ h_{31} & 0 & 0 & 1 & 0 \\ h_{41} & 0 & 0 & 0 & 1 \end{bmatrix}$$

This matrix must satisfy the equality $\bar{H}\bar{T} = 0$.

Hence,

$$\bar{H}\bar{T}_1 = \begin{bmatrix} h_{11} & 1 & 0 & 0 & 0 \\ h_{21} & 0 & 1 & 0 & 0 \\ h_{31} & 0 & 0 & 1 & 0 \\ h_{41} & 0 & 0 & 0 & 1 \end{bmatrix} \begin{bmatrix} 1 \\ 1 \\ 1 \\ 1 \\ 1 \end{bmatrix} = \begin{bmatrix} h_{11} & 1 \\ h_{21} & 1 \\ h_{31} & 1 \\ h_{41} & 1 \end{bmatrix} = \begin{bmatrix} 0 \\ 0 \\ 0 \\ 0 \end{bmatrix}$$

Similarly,

$$\bar{H}\bar{T}_2 = \begin{bmatrix} h_{11} & 1 & 0 & 0 & 0 \\ h_{21} & 0 & 1 & 0 & 0 \\ h_{31} & 0 & 0 & 1 & 0 \\ h_{41} & 0 & 0 & 0 & 1 \end{bmatrix} \begin{bmatrix} 0 \\ 0 \\ 0 \\ 0 \\ 0 \end{bmatrix} = \begin{bmatrix} 0 \\ 0 \\ 0 \\ 0 \end{bmatrix}$$

Therefore, $h_{11} = h_{21} = h_{31} = h_{41} = 1$, and hence

$$\bar{H} = \begin{bmatrix} 1 & 1 & 0 & 0 & 0 \\ 1 & 0 & 1 & 0 & 0 \\ 1 & 0 & 0 & 1 & 0 \\ 1 & 0 & 0 & 0 & 1 \end{bmatrix}$$

● **PROBLEM 17-21**

Eight decimal numbers from 0 to 7 are binary-encoded.

(a) For each decimal number, express it in 3 digit binary form.
(b) Add a single parity-check bit to the end of each code word.
(c) Each 4-bit code word forms a $\bar{T}$ matrix. Prove the following:

If $\bar{H} = [1111]$, $\bar{H}\bar{T} = 0$ for each $\bar{T}$ and if a single error occurs, $\bar{H}\bar{T} = 1$.

Solution: (a)

decimal number	binary representation
0	000
1	001
2	010
3	011
4	100
5	101
6	110
7	111

(b) The single parity-check bit code is an example of an algebraic code. Keep in mind the definition of addition when a single parity-check bit is added to each code word.

Note: Addition
0+0=0
0+1=1
1+0=1
1+1=0

Hence,

decimal number	parity-check bit code
0	0000
1	0011
2	0101
3	0110
4	1001
5	1010
6	1100
7	1111

(c) $\bar{T} = [a_1 a_2 a_3 c_1]^t$

$\bar{H}\bar{T} = [1111][a_1 a_2 a_3 c_1]^t$

$= a_1 + a_2 + a_3 + c_1 = 0$ for every $\bar{T}$

If a single error, say a_2, occurs, a message $\bar{T}' = [a_1 \bar{a}_2 a_3 c_1]^t$ will be received, where $a_1 + a_2 + a_3 + c_1 = 0$.

Hence, $\bar{H}\bar{T}' = [1111][a_1 \bar{a}_2 a_3 c_1]^t$

$= a_1 + \bar{a}_2 + a_3 + c_1$

$= a_1 + a_2 + a_3 + c_1 + a_2 + \bar{a}_2$

$= a_2 + \bar{a}_2 = 1$

Note: $a_2 + a_2 = 0$

for $a_2 = 0$ or 1.

• **PROBLEM** 17-22

Consider five messages having probabilities $\frac{1}{2}$, $\frac{1}{4}$, $\frac{1}{8}$, $\frac{1}{16}$ and $\frac{1}{16}$ respectively.

(a) Find the entropy, H.

(b) One technique used in constructing an optimum code is to list the messages in order of decreasing probability and divide the list into two equally or almost equally probable groups. The messages in the top group are given the value 0, and the messages in the bottom group are given the value 1, The same procedure is now employed for each group, separately. The procedure is continued until no further division is possible. Find the code for each message.

(c) Find the average number of bits/message, then compare with H.

<u>Solution</u>:   (a) $H = P_1 \log_2 \frac{1}{P_1} + P_2 \log_2 \frac{1}{P_2} + \ldots$

Hence,

$H = \left(\frac{1}{2} \log_2 2\right) + \left(\frac{1}{4} \log_2 4\right) + \left(\frac{1}{8} \log_2 8\right) + \left(\frac{1}{16} \log_2 (16) \times 2\right)$

$= \frac{1}{2} + \frac{1}{2} + \frac{3}{8} + \left(\frac{1}{4} \times 2\right)$

$= 1.875$ bits/message

(b)

	Message	Code
$Q_1 (P_1 = \frac{1}{2})$	Q_1	0
$Q_2 (P_2 = \frac{1}{4})$	Q_2	10
$Q_3 (P_3 = \frac{1}{8})$	Q_3	110
$Q_4 (P_4 = \frac{1}{16})$	Q_4	1110
$Q_5 (P_5 = \frac{1}{16})$	Q_5	1111

(c) The average number of bits per message is

$$\frac{1}{2} \times 1 + \frac{1}{4} \times 2 + \frac{1}{8} \times 3 + \frac{1}{16} \times 4 + \frac{1}{16} \times 4 = 1.875 \text{ bits},$$

which is equal to the average information per message.

• **PROBLEM** 17-23

Determine the code words corresponding to all possible 4-bit message words given the algebraic code consists of 4-bit message words and 3 parity check bits.

The matrix $\bar{H}$ is defined as,

$$\bar{H} = \begin{bmatrix} 1 & 1 & 0 & 1 & 1 & 0 & 0 \\ 1 & 0 & 1 & 1 & 0 & 1 & 0 \\ 0 & 1 & 1 & 1 & 0 & 0 & 1 \end{bmatrix}$$

Solution: Let a_1, a_2, a_3, and a_4 refer to the message bits and c_1, c_2, and c_3 refer to the parity check bits. With a column matrix

$$\bar{T} = \begin{bmatrix} a_1 \\ a_2 \\ a_3 \\ a_4 \\ c_1 \\ c_2 \\ c_3 \end{bmatrix}$$ representing the transmitted code word,

the parity bits are selected to satisfy the linear equations corresponding to the matrix equation $\bar{H}\bar{T} = 0$.

Carrying out the multiplication, the matrix equation $\bar{H}\bar{T} = 0$ gives three linear simultaneous equations. (Also note that since $\bar{H}$ is a 3×7 matrix and $\bar{T}$ is a 7×1 matrix, the product $\bar{H}\bar{T}$ is a 3×1 matrix.)

$$a_1 + a_2 + a_4 + c_1 = 0$$

$$a_1 + a_3 + a_4 + c_2 = 0$$

$$a_2 + a_3 + a_4 + c_3 = 0$$

Thus, c_1, c_2, and c_3 are selected such that the combination of c_1 with the first, second and fourth bits, the combination of c_2 with the first, third and fourth bits and the combination of c_3 with the second, third and fourth bits, respectively, will result in even parity. With these three properties under consideration the 16-message words will be

Message Word	Code Word
0000	0000000
0001	0001111
0010	0010011
0011	0011100
0100	0100101
0101	0101010
0110	0110111
0111	0111001
1000	1000110
1001	1001001
1010	1010101
1011	1011010
1100	1100011
1101	1101100
1110	1110000
1111	1111111

When these code words are transmitted, the reciever forms the product of the recieved message $\bar{R}$ and the $\bar{H}$ matrix. If the product $\bar{H}\bar{R} \neq 0$, it indicates that $\bar{R}$ is an incorrect message and thus the existence of errors introduced during transmission.

● **PROBLEM 17-24**

(a) Obtain a (15, 11) block code and find all the codewords.

(b) Obtain the $\bar{H}$ matrix.

(c) Illustrate with an example that a single error can be corrected.

Solution: (a) The matrix $\bar{H}$ as a rectangular matrix with k rows and n columns is given by

$$\overline{H} = \begin{bmatrix} \overbrace{h_{11} \quad h_{12} \quad \cdots \quad h_{1k}}^{a_1 \quad \cdots \quad a_k} & \overbrace{1 \quad 0 \quad 0 \quad \cdots \quad 0}^{c_1 \quad \cdots \quad c_t} \\ h_{21} \quad h_{22} \quad \cdots \quad h_{2k} & 0 \quad 1 \quad 0 \quad \cdots \quad 0 \\ \cdots\cdots\cdots\cdots\cdots\cdots & \cdots\cdots\cdots\cdots \\ h_{r_1} \quad h_{r_2} \quad\quad h_{rk} & 0 \quad 0 \quad 0 \quad \cdots \quad 1 \end{bmatrix}$$

Since $k = 11$ and $r = 4$, and notice that

(1) k is the length of each M coded message words, where $M = 2^k$

(2) r is the number of redundant bits added to each message code word, and

(3) the transmitted code word has $(k+r) = n$ bits.

With the above information, the $\overline{H}$ matrix is obtained as follows:

	a_1	a_2	a_3	a_4	a_5	a_6	a_7	a_8	a_9	a_{10}	a_{11}	c_1	c_2	c_3	c_4
$\overline{H} =$	1	1	1	1	1	1	1	0	0	0	0	1	0	0	0
	1	1	1	1	0	0	0	1	1	1	0	0	1	0	0
	1	1	0	0	1	1	0	1	1	0	1	0	0	1	0
	1	0	1	0	1	0	1	1	0	1	1	0	0	0	1

(b) There are $2^k = 2^{11} = 2048$ code words. The code words are as follows:

	a_1	a_2	a_3	a_4	a_5	a_6	a_7	a_8	a_9	a_{10}	a_{11}	c_1	c_2	c_3	c_4
m_1	0	0	0	0	0	0	0	0	0	0	0	0	0	0	0
m_2	0	0	0	0	0	0	0	0	0	0	1	1	0	1	1
m_3	0	0	0	0	0	0	0	0	1	0	0	1	0	1
⋮															
m_{2046}	1	1	1	1	1	1	1	1	0	1	1	0	1	0
m_{2047}	1	1	1	1	1	1	1	1	1	0	1	1	1	0
m_{2048}	1	1	1	1	1	1	1	1	1	1	1	1	1	1

(c) Let's assume that a single error has been made in the code word m_{2046} so that the received sequence is 111101111011010.

Then $\overline{S} = \overline{H}\overline{R} = \overline{H}[111101111011010]^t$

$$= [1011]^t$$

Since the syndrome is identical with the fifth column of the matrix $\bar{H}$, the fifth digit, a_5 is in error, and hence the correct word is 111111111011010.

• **PROBLEM 17-25**

If a binary code is to represent 110 different codewords, calculate the number of bits required and evaluate its efficiency as compared to a decimal system to accomplish the same task.

Solution: In a binary system, the number of bits B, required to represent N different code words is given as

$$B = \log_2 N$$

Thus, to represent 110 different codewords, the number of bits required are

$$B = \log_2 110$$

$$= 6.781.$$

In other words, 7 bits are required to represent 110 codewords.

Thus, the efficiency will be

$$\eta_1 = \frac{6.78}{7} \times 100\% = 97 \text{ percent.}$$

Similarly, the number of digits required to represent 110 different codewords in the decimal system is

$$D = \log_{10} 110$$

$$= 2.041$$

i.e., a total of 3 digits are required to represent 110 codewords.

The efficiency will thus be

$$\eta_2 = \frac{2.04}{3} \times 100\% = 68 \text{ percent.}$$

• **PROBLEM 17-26**

An encoded message consisting of three-bits message words and one parity check bit are sent via Frequency Shift Keying (FSK). Determine the probability of an undetected error at the receiver, given that the signal to noise ratio at the receiver is 8 dB.

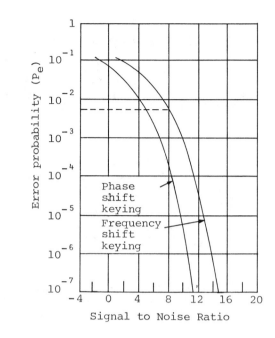

Table for Bit error rate

<u>Solution</u>: A single parity check bit in the coded message can detect only odd bit errors and even bit errors, if they exist, will pass through undetected at the receiver. Thus, in a 4-bit codeword, an error will pass undetected if the codeword undergoes either two or four bit errors, whereas a single or three bit error will be detected at the receiver. For a receiver with a signal to noise ratio of 8 dB, the bit error rate can be approximated from the table as 5×10^{-3}.

Thus, the probability of occurrence of a two bit error

$$= 5 \times 10^{-3} \times 5 \times 10^{-3} \text{ (assuming independence)}$$

$$= 25 \times 10^{-6}.$$

In a 4-bit codeword, the number of possible combinations of 2 bits are 6 (i.e., first and second bit, first and third bit, first and fourth bit, second and third bit, second and fourth bit and third and fourth bit, respectively). Hence, the probability of two bit erros for a 4-bit word is

$$6 \times (5 \times 10^{-3})^2 \times (1-5 \times 10^{-3})^2$$

$$= 1.49 \times 10^{-4}.$$

The probability of occurrence of 4-bit errors in a 4-bit word

$$= (25 \times 10^{-6})^2$$

$$= 625 \times 10^{-12}.$$

Thus, the joint probability or the probability of either two or four bit erros per message can be approximated to 1.49×10^{-4}. Since 625×10^{-12} is small compared to 1.49×10^{-4}, it is neglected. In other words, for the given system, there will exist about 3 wrong decoded or undetected errors for every 20,000 transmitted messages.

● **PROBLEM 17-27**

For the zero-memory information source described by the table, find the average length (L_{avg}) of binary code needed to encode the symbols.

Source symbol	Symbol probability P_i
s_1	$\frac{1}{4}$
s_2	$\frac{1}{4}$
s_3	$\frac{1}{4}$
s_4	$\frac{1}{4}$

<u>Solution</u>: In the case of an instantaneous binary code, the number of bits (or in general, the average code length (L_{avg}) must be greater than or equal to the entropy $H(S)$. In order to satisfy this constraint, it is required that

$$L = \log_2\left(\frac{1}{P_i}\right) \text{ for all i.}$$

i.e., $\log_2\left(\frac{1}{P_i}\right)$ must always be an integer for each i and the probabilities P_i associated with the symbols must all be of the form $(1/r)^{\alpha_i}$ where α_i is also an integer.

The entropy (the average information per message interval) is defined as.

$$H = \sum_{i=1}^{n} P_i \log_2\left(\frac{1}{P_i}\right).$$

For the zero-memory source at hand, the entropy

$H = \frac{1}{4} \log_2 4 + \frac{1}{4} \log_2 4 + \frac{1}{4} \log_2 4 + \frac{1}{4} \log_2 4$

$= 2 \text{ bits/symbol}.$

Thus, in order to encode the given source symbols into a binary code, the minimum code length or the number of bits required is 2. Since the probability of each source symbol is $\frac{1}{4} = (\frac{1}{2})^2$, a compact binary code must have four words, each of length 2 bits. One possible binary code corresponding to the given symbols is

$$s_1 \to 00$$
$$s_2 \to 01$$
$$s_3 \to 10$$
$$s_4 \to 11.$$

Another equally useful code that is possible is:

$$s_1 \to 01$$
$$s_2 \to 10$$
$$s_3 \to 11$$
$$s_4 \to 00.$$

● **PROBLEM** 17-28

Construct an instantaneous trinary (three-state) code for a zero-memory source with corresponding source symbols and probabilities as shown in the table (1) below.

Source symbol	Symbol probability P_i
s_1	$\frac{1}{3}$
s_2	$\frac{1}{3}$
s_3	$\frac{1}{9}$
s_4	$\frac{1}{9}$
s_5	$\frac{1}{27}$
s_6	$\frac{1}{27}$
s_7	$\frac{1}{27}$

Solution: For any instantaneous code, the average code

length (L) must be greater than or equal to the corresponding entropy. In general entropy in r-ary units (e.g. binary, trinary etc.) can be expressed as

$$H_r(s) = \sum_i P_i \log_r \frac{1}{P_i} \quad \text{in r-ary units.}$$

where i is the number of symbols and P_i is the probability of occurrence of the ith symbol. Thus, for the given zero-memory source, the entropy associated with a trinary code is

$$H_3(s) = \sum_{i=1}^{7} P_i \log_3 \frac{1}{P_i}$$

$$= (\tfrac{1}{3} \log_3 3 + \tfrac{1}{3} \log_3 3 + \tfrac{1}{9} \log_3 9 + \tfrac{1}{9} \log_3 9$$

$$+ \tfrac{1}{27} \log_3 27 + \tfrac{1}{27} \log_3 27 + \tfrac{1}{27} \log_3 27)$$

$$= (\tfrac{1}{3} + \tfrac{1}{3} + \tfrac{2}{9} + \tfrac{2}{9} + \tfrac{3}{27} + \tfrac{3}{27} + \tfrac{3}{27})$$

$$= \tfrac{13}{9} \text{ trinary units/symbol.}$$

Thus, in order to construct an instantaneous trinary code for the given zero-memory source, the minimum number of trinary symbols per source symbol should be at least 13/9. This lower bound can be achieved if and only if the word length, L, is chosen such that

$$L = \log_3 \left(\frac{1}{P_i} \right) \quad \text{for all i.} \tag{1}$$

In order that equation (1) is valid, $\log_3 \left(\frac{1}{P_i} \right)$ must always be an integer and the probability P_i associated with the source symbol must be of the form $(1/r)^{\alpha_i}$ where α_i is also an integer.

Using equation (1) and table (1), the trinary code length for each given source symbol is evaluated as shown in the table (1) below.

source symbol	s_1	s_2	s_3	s_4	s_5	s_6	s_7
code length	1	1	2	2	3	3	3

The trinary code words corresponding to these code lengths are:

$$s_1 \to 0 \text{ (code length = 1)}$$
$$s_2 \to 1$$
$$s_3 \to 20 \text{ (code length = 2)}$$
$$s_4 \to 21$$
$$s_5 \to 220 \text{ (code length = 3)}$$
$$s_6 \to 221$$
$$s_7 \to 222$$

(Note: For trinary code, the numbers 0, 1 and 2 are used.)

The average code length will thus be:

$$L_{(avg)} = \sum_{i=1}^{7} P_i L_i \quad \text{where } L_i \text{ is the word length of the } i^{th} \text{ trinary code.}$$

i.e., $L_{(avg)} = \frac{1}{3}(1) + \frac{1}{3}(1) + \frac{1}{9}(2) + \frac{1}{9}(2) + \frac{1}{27}(3)$

$$+ \frac{1}{27}(3) + \frac{1}{27}(3)$$

$$= \frac{2}{3} + \frac{4}{9} + \frac{9}{27}$$

$$= \frac{13}{9} \text{ trinary symbols/source symbol.}$$

This is equal to the entropy.

• **PROBLEM 17-29**

For a codeword s_j, the Hamming weight $w(s_j)$ is defined as the number of ones in that code word, and $d(s_i, s_j)$ or d_{ij}, the Hamming distance is defined as the number of positions in which s_i and s_j differ. If $d_{ij} = w(s_i \oplus s_j)$, where $\oplus$ denotes modulo two addition (which is binary addition without a carry),

(a) Evaluate the Hamming distance between $s_1 = 101101$ and $s_2 = 001100$.

(b) If 1101011 is received corresponding to a code which consists of codewords [0001011, 1110000, 1000110, 1111011, 0110110, 1001101, 0111101, 000000], determine the decoded codeword.

Solution: (a) $d_{12} = w(s_1 \oplus s_2)$

Now $s_1 \oplus s_2 = 101101 \oplus 001100$

$$= 100001.$$

Geometrical representation of codewords

```
        x x x x x | x x x x x
        x x ⊗ x x | x x x x x
        x ⊗ c'⊗ x | x x C x x
        x x ⊗ x x | x x x x x
        x x x x x | x x x x x
        ----------+----------
        x x x x x | x x x x x
        x x x x x | x x x x x
        x x C x x | x x C x x
        x x x x x | x x x x x
        x x x x x | x x x x x
```

Thus, $\quad d_{12} = w(100001)$

$\qquad\qquad = 2$

which states that s_1 and s_2 differ in 2 positions. The figure below shows a geometric representation of codewords where the 4 "c's" and the 96 "x's" are the possible received words. The code words (c's) are each distance five apart and the circled x's have a Hamming distance of one from c'. In general, the figure illustrates the concept of minimum-distance-decoding, wherein the decoded codeword is the codeword closest in Hamming distance to the received word.

(b) From the principle of minimum-distance-decoding, the decoded codeword is the codeword with the minimum Hamming distance to the received word 1101011. The Hamming distances of the code words to 1101011 can be evaluated as follows:

$w(0001011 \oplus 1101011) = 2 \qquad w(0110110 \oplus 1101011) = 5$

$w(1110000 \oplus 1101011) = 4 \qquad w(1001101 \oplus 1101011) = 3$

$w(1000110 \oplus 1101011) = 4 \qquad w(0111101 \oplus 1101011) = 4$

$w(1111011 \oplus 1101011) = 1 \qquad w(0000000 \oplus 1101011) = 5$

It thus follows that 1111011, the codeword with a Hamming distance of one to 1101011 is the decoded codeword.

• **PROBLEM 17-30**

A 7 bit algebraic code consists of 4 message bits and 3 parity check bits. If while transmitting a message word 1010, an error occurs in the fourth bit, define and find the syndrome, given that

$$\overline{H} = \begin{bmatrix} 1 & 1 & 0 & 1 & 1 & 0 & 0 \\ 1 & 0 & 1 & 1 & 0 & 1 & 0 \\ 0 & 1 & 1 & 1 & 0 & 0 & 1 \end{bmatrix}$$

Solution: If $\vec{E}$ represents an error vector which contains a '1' in every bit and $\bar{T}$ represents a transmission vector, then the recieved vector $\bar{R}$ is defined as

$$\bar{R} = \bar{T} + \vec{E}$$

where + represents a modulo-2 addition. At the receiver, the transmitted signal can be regained by the error correction of the received signal, or in other words, by isolating the vector $\vec{E}$. Multiplying the received vector $\bar{R}$ by the matrix $\bar{\bar{H}}$ yields

$$\bar{\bar{H}}\bar{R} = \bar{\bar{H}}(\bar{T} + \vec{E}) = \bar{\bar{H}}\bar{T} + \bar{\bar{H}}\vec{E}. \qquad (1)$$

In any transmission system, the parity bits in the algebraic code are selected to satisfy the linear equations corresponding to the matrix equation $\bar{\bar{H}}\bar{T} = 0$

Thus, equation (1) reduces to

$$\bar{\bar{H}}\bar{R} = \bar{\bar{H}}\vec{E}$$

The product $\bar{\bar{H}}\vec{E}$ is defined as the error syndrome, denoted as $\bar{S}$.

To determine the characteristics of the error syndrome, it is assumed that during transmission, error occurs only in a single bit and hence the error vector $\vec{E}$ would have a single '1' and zeros in the remaining positions. Thus, the syndrome or the product $\bar{\bar{H}}\vec{E}$ would be a vector of one column, the column of H corresponding to the 1 bit of the error vector.

For a message word 1010, the corresponding transmitted code word is 1010101, and with a one bit error the received word is 1011101.

The product vector $\bar{\bar{H}}\bar{R}$ would thus be

$$\bar{\bar{H}}\bar{R} = \begin{bmatrix} 1 & 1 & 0 & 1 & 1 & 0 & 0 \\ 1 & 0 & 1 & 1 & 0 & 1 & 0 \\ 0 & 1 & 1 & 1 & 0 & 0 & 1 \end{bmatrix} \begin{bmatrix} 1 \\ 0 \\ 1 \\ 1 \\ 1 \\ 0 \\ 1 \end{bmatrix}$$

$$= \begin{bmatrix} 1 \\ 1 \\ 1 \end{bmatrix}$$

The resultant vector is identical to the fourth column of

the $\bar{\bar{H}}$ matrix and hence it follows that an error has occurred in the fourth bit position.

Note: An important factor to be considered is that, to correct a single error using the above technique, it is essential that $\bar{\bar{H}}$ does not contain any duplicate columns and also that $\bar{\bar{H}}$ should not contain any column with all zeros. This is due to the fact that the product $\bar{\bar{H}}\bar{R}$ will yield a column vector consisting of all zeros if no error exists in the received signal, and hence the presence of a column consisting of all zeros in $\bar{\bar{H}}$ would result in an ambiguity.

CHAPTER 18

ANTENNAS

FUNDAMENTAL LAWS OF RADIATION

• **PROBLEM 18-1**

An electromagnetic wave of 40KHz is radiated by an antenna. Find the time taken by this wave to reach a point 60 km away.

Solution: The time taken by the wave is

$$T = d/c$$

where

d = distance, in meters

$c = 3 \times 10^8$ meters/sec

which is the velocity of light (and of the wave) in free space. Hence,

$$T = \frac{60 \times 1000}{3 \times 10^8} = 2 \times 10^{-4} \text{ seconds}$$

$$= 0.2 \text{ msec.}$$

Note that radiation travels at the speed of light.

• **PROBLEM 18-2**

A satellite in a synchronous orbit is beaming microwave power to earth. If the beamwidth of the transmitting antenna on the satellite is 0.1°, and the distance from the earth's surface is 36,000 (km), what is the size of the spot illuminated by the antenna on the earth's surface? Assume a circular spot or beam area as shown in the figure.

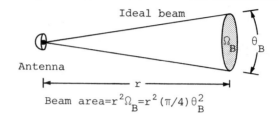

Ideal beam

Antenna

Beam area=$r^2\Omega_B = r^2(\pi/4)\theta_B^2$

ILLUSTRATION OF THE CONCEPT OF A BEAM SOLID ANGLE FOR AN IDEAL ANTENNA.

Solution: The beamwidth in radians is:

$$\theta_B = \frac{\pi}{180°} \times 0.1° = 1.745 \times 10^{-3} \text{ (radian)}$$

and from

$$\Omega_B = \frac{\pi}{4} \theta_B^2$$

$$\Omega_B = \pi/4 \times (1.745 \times 10^{-3})^2$$

$$= 2.39 \times 10^{-6} \text{ (steradian)}$$

Then the area of the spot is $r^2\Omega_B$, or

$$A_{spot} = 2.39 \times 10^{-6}(36,000 \times 10^3)^2 = 3.10 \times 10^9 \text{ square meters.}$$

• **PROBLEM 18-3**

What should be the length of a half-wave dipole antenna when it is receiving a radio signal of frequency 6 MHz. Assume that the velocity of electromagnetic waves on the antenna is equal to the speed of light.

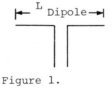

Figure 1.

Solution: In Fig. (1), a dipole antenna is shown. It is given that the signal frequency f = 6 MHz and the velocity C = 3 x 10^8 m/sec speed of light. It is known that, the length of a half-wave dipole antenna, L_{dipole} is equal to half the wave length of its radiation.

Then, from the equation, $f\lambda = C$,

λ can be expressed as $\lambda = \frac{C}{f}$.

Therefore, $\lambda = \frac{3 \times 10^8}{6 \times 10^6}$

$$\lambda = 50 \text{ meters}$$

Hence, the required length of the dipole, $L_{dipole} = \frac{\lambda}{2} = \frac{50}{2} = 25$ meters.

● **PROBLEM 18-4**

Assuming that a dipole antenna is placed 2 meters above the earth's surface, determine its operating frequency if the antenna is placed one-quarter of a wavelength above the earth's surface.

Solution: Since the antenna is placed one-quarter of a wave length above the earth's surface,

Let $\quad h = \frac{\lambda}{4} \quad$ (1)

where h is the distance of the antenna from the earth's surface.

Therefore, $\lambda = 4h$.

Now, the frequency f is given by

$$f = c/\lambda \quad (2)$$

Substituting for $\lambda = 4h$ in Eq. (2)

$$f = \frac{c}{4h}$$

But $\quad h = 2m$

Therefore, $f = \frac{c}{8} = \frac{3 \times 10^8}{8}$

$$= 37.5 \times 10^6 \text{ Hz}$$

$$= 37.5 \text{ MHz}.$$

● **PROBLEM 18-5**

Calculate the length of a half-wave dipole antenna so as to optimally radiate a signal of 300 MHz when the velocity factor for the antenna element is

    (A)   0.70

    (B)   1.0

Solution:

(A) Use the relation $f\lambda = kc$ to calculate the wave length of the radiated signal.

where λ = wavelength of the radiation.

         c = velocity of light in vacuum = 3×10^8 m/s.

         k = velocity factor for the antenna.

Thus,

$$\lambda = \frac{kc}{f} = \frac{(0.70)(3 \times 10^8) \text{ msec}^{-1}}{300 \times 10^6 \text{ sec}^{-1}} = 0.7 \text{ meters}.$$

Since a half-wave dipole antenna is to be used, the length of the antenna is $\lambda/2 = 0.35$ meters.

(B) Use the same relation as in part (a) except that, now, $k = 1.0$.

Thus, $\quad = \dfrac{(1.0)(3 \times 10^8) \text{ msec}^{-1}}{300 \times 10^6 \text{ sec}^{-1}}$

$= 1$ meter.

Therefore, the length of the antenna must be $\lambda/2$ or 0.5 meters.

● **PROBLEM 18-6**

A Marconi antenna shown in figure (1) below is transmitting at a frequency of 120 megahertz. Calculate its optimum length by assuming that the velocity factor of the antenna as 0.9.

Figure 1.

Solution: For a Marconi vertical antenna operating at a given frequency, the optimum length is equivalent to a quarter of its wavelength.

Now, to find the wavelength, λ, it is known that:

$$f\lambda = kC,$$

where f is the frequency in Hertz,

λ is the wavelength in meters,

C is the velocity of light = 3×10^8 m/sec

and k is the velocity factor for the antenna.

Therefore, $\lambda = \dfrac{kC}{f}$

$= \dfrac{(0.9)(3 \times 10^8) \text{ msec}^{-1}}{120 \times 10^6 \text{ sec}^{-1}}$

$= 2.25$ meters.

Hence, for a quarter wavelength,

$$\dfrac{\lambda}{4} = \dfrac{2.25 \text{ m}}{4} = 0.5625 \text{ m}.$$

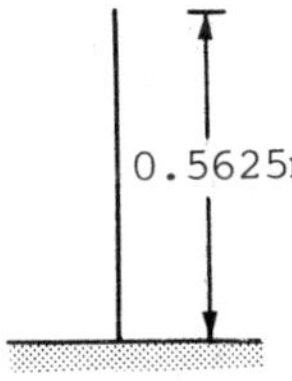

Figure 2.

Thus, the optimum length would be 0.5625 meters as shown in figure 2.

ANTENNA CHARACTERISTICS

● **PROBLEM 18-7**

For the two-antenna system shown in the figure, one antenna is transmitting a 1 GHz signal 30 meters above moist earth. The receiving antenna is 16 km away and 10 meters above moist earth. If the relative permitivity of moist earth is $\varepsilon_r = 12$ and the conductivity $\sigma = 0.02$ mho/meter, show that the signal is almost perfectly reflected by the earth to the receiver.

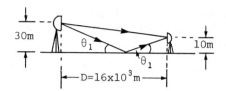

Solution: When the distance between the transmitter and receiver is small enough for the earth to be considered flat, the energy transmitted may reach the receiver by a direct path between the antennas or by a route involving reflection from the earth's surface. From electromagnetic theory, the reflection coefficient Γ is given as

$$\Gamma = \frac{\eta_2 \sin \theta_1 - \eta_1 \sin \theta_2}{\eta_2 \sin \theta_1 + \eta_1 \sin \theta_2}$$

and

$$\eta = \sqrt{\frac{\mu}{\left(\varepsilon + \frac{\sigma}{j\omega}\right)}}$$

where η = characteristic impedance of the medium. (consider air medium #1, moist earth medium #2)

θ_1 = angle of incidence of the energy

θ_2 = angle of refraction

σ = conductivity of the medium

$\varepsilon = \varepsilon_o \varepsilon_r$ = electric permitivity

$\varepsilon_o = \frac{1}{36\pi} \times 10^{-9}$ F/m

ε_r = relative permitivity of the medium

$\mu = \mu_o \mu_r$ = magnetic permeability

$\mu_o = 4\pi \times 10^{-7}$ H/m

$\mu_r \simeq 1$ for non-magnetic materials.

Thus, for air (or free space) $\varepsilon = 1$, $\sigma = 0$

$$\eta_1 = \sqrt{\frac{4\pi \times 10^{-7}}{\varepsilon_o + \frac{(0)}{j\omega}}} = \sqrt{\frac{4\pi \times 10^{-7}}{\frac{1}{36\pi} \times 10^{-9}}} = 120\pi \simeq 377 \text{ ohms}$$

for moist earth, $\varepsilon_r = 12$, $\sigma = 0.02$

$$\eta_2 = \sqrt{\frac{4\pi \times 10^{-7}}{\frac{12}{36\pi} \times 10^{-19} + \frac{0.02}{j(2\pi \times 10^9)}}} = \sqrt{\frac{4\pi \times 10^{-7}}{1.06 \times 10^{-10} - j2.812 \times 10^{-23}}}$$

Because the imaginary part is over ten orders of magnitude less than the real part, we can approximate to get

$$\eta_2 \approx \sqrt{\frac{4\pi \times 10^{-7}}{1.06 \times 10^{-10}}} = 108.881 \text{ ohms}$$

$$\therefore \eta_2 \approx 109 \text{ ohms}$$

From the figure, $\theta_1 = \tan^{-1}\frac{(30 + 10)m}{16 \times 10^3 \text{ m}} = 0.0025$ radians

at the reflection point, the angle of incidence is so near grazing that:

$$\tan \theta_1 \approx \sin \theta_1 \approx \theta_1 = 0.0025 \text{ radians.}$$

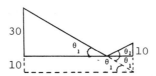

Thus, $\cos \theta_1 \approx 1$.

Now, to find θ_2, use Snell's law:

$$\frac{\cos \theta_1}{\cos \theta_2} = \sqrt{\frac{\varepsilon_2}{\varepsilon_1}} = \sqrt{\frac{\varepsilon_o \varepsilon_{r_2}}{\varepsilon_o \varepsilon_{r_1}}} = \sqrt{12} = 2\sqrt{3} = 3.464$$

Hence,

$$\cos \theta_2 = \frac{1}{3.464}$$

or $\theta_2 = 73.22°$

Thus, $\sin \theta_2 = 0.957$

This gives

$$\Gamma = \frac{\text{Reflected field}}{\text{Incident field}} = \frac{109(0.0025) - 120\pi(0.957)}{109(0.0025) + 120\pi(0.957)}$$

$$= -0.9985$$

$$\approx -1$$

It is thus obvious that moist earth behaves as a perfect reflector.

When the transmission distance is great compared with the antenna heights, reflection is practically complete and takes place with a reversal of phase as indicated by the negative sign. The two waves do not cancel out in spite of the phase reversal because the direct and indirect paths have different lengths.

• **PROBLEM 18-8**

For a short dipole antenna, evaluate the maximum effective aperture A_{em}.

Solution:
$$A_{em} = \frac{V^2}{4SR_r} \quad (1)$$

where V = Voltage induced in the short dipole.

R_r = Radiation Resistance.

S = Poynting vector.

The voltage induced in the short dipole is a maximum when the dipole is parallel to the incident electric field E. Hence,

$V = E\ell$ volts.

The Poynting vector

$$S = \frac{E^2}{Z_0} \text{ watts/meter}^2$$

where Z_0 = intrinsic impedance of the medium through which the radiation passes (for air or vacuum $Z_0 = \sqrt{\mu_0/\epsilon_0}$).

The radiation resistance is $R_r = \sqrt{\frac{\mu_0}{\epsilon_0}} \left[\frac{(\beta\ell)^2}{6\pi} \right]$ ohms.

Substituting these values for V, S, and R_r into equation (1), the maximum effective aperture of a short dipole is

$$A_{em} = \frac{3}{8\pi} \lambda^2 = 0.119\lambda^2.$$

Thus, regardless of how small the dipole is, it can collect power over an aperture of 0.119 wavelength² and deliver it to a receiver. It is assumed here that the dipole is lossless. However, in practice, losses are present due to the finite conductivity of the dipole conductor so that the actual effective aperture is less than A_{em}.

• **PROBLEM 18-9**

A short dipole has a radiation resistance $R_r = \sqrt{\mu_0/\epsilon_0} \frac{(\beta\ell)^2}{6\pi}$ ohms. Find A_{em} the maximum effective aperture of this dipole.

Solution: The maximum effective aperture is given by

$$A_{em} = V^2/4SR_r \tag{1}$$

where

V = rms voltage induced by the passing wave
   = $E\ell$ volts

S = Poynting vector of the incident wave
   = E^2/Z_0 watts/meter^2

where $Z_0 = \sqrt{\mu_0/\epsilon_0}$ ohms.

Substituting these values and the given radiation resistance in (1)

$$A_{em} = \frac{V^2}{4SR_r} = \frac{E^2\ell^2}{4\frac{E^2}{Z_0}\left[\frac{\sqrt{\mu_0/\epsilon_0}\,\beta^2\ell^2}{6\pi}\right]} = \frac{E^2\ell^2 Z_0 6\pi}{E^2\ell^2 Z_0 4\beta^2}$$

$$= \frac{6\pi}{4\beta^2} = \frac{6\pi}{4\left(\frac{2\pi}{\lambda}\right)^2} = \frac{6\lambda^2}{16\pi}$$

$$= \frac{3}{8\pi}\lambda^2 = 0.119\,\lambda^2 \text{ square meters.}$$

Hence the required maximum effective aperture is $0.119\,\lambda^2$ square meters.

Note that the maximum effective aperture is solely wavelength dependent. Hence, since λ is proportional to frequency, the maximum effective aperture can also be said to be strictly frequency dependent.

• PROBLEM 18-10

The directivity of an antenna is 50 and the antenna operates at a wavelength of 4 meters. What is its maximum effective aperture?

Solution: The maximum effective aperture of an antenna, A_{em}, is given by

$$A_{em} = \frac{D\lambda^2}{4\pi} \qquad \text{where } D = \text{directivity}$$
$$\lambda = \text{radiation wavelength}$$

$$A_{em} = \frac{50 \times 16}{4\pi} = 63.6 \text{ meter}^2$$

● PROBLEM 18-11

The radiated electromagnetic field associated with a propagating wavefront moves towards the XZ plane and towards a dipole antenna. The electric field vector of the incident radiation is on the Y axis and is oriented at 45° to the XY plane. Find the effective area of the antenna if the wavelength of the radiation is 0.5 meters and the antenna is perpendicular to the ZY plane and on the X axis as shown in the figure.

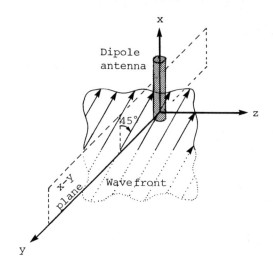

Solution: The amount of cross sectional area presented by an antenna to the received radiation is called the effective cross sectional area of the antenna and is a function of the antenna gain, and the orientation of the antenna to the wavefront of the radiation.

If the voltage received by the antenna is given as

$$V_R = \frac{\lambda E_i RG_R}{2\pi(120)} \cos \psi \qquad (1)$$

where V_R = Received voltage,

λ = Wavelength of the radiation,

E_i = Magnitude of incident electric field on antenna,

ψ = Angle between wavefront of the incident radiation, and the plane of the antenna,

R = Antenna radiation resistance,

G_R = Gain of the receiving antenna,

then the power absorbed by the antenna is

$$P = \tfrac{1}{2}I_R^2 R = \tfrac{1}{2}\frac{V_R^2}{R}.$$ (2)

Substituting equation (1) into equation (2) gives

$$P = \frac{1}{8}\frac{\lambda^2}{\pi^2}\frac{E_i^2}{120} G_R \cos\psi$$

$$= A_E \frac{E_i^2}{240\pi}$$

where $A_E = \dfrac{G_R \lambda^2}{4\pi} \cos\psi$ = effective area of the antenna.

The angle ψ in this case is zero because the wavefront is in the plane of the antenna. The gain of the dipole is given by:

$$G_R = 1.67 \left[\frac{\cos(\pi/2 \cos\theta)}{\sin\theta}\right]$$

where θ is the angle between the antenna and the electric field. In this case $\theta = 45°$, thus,

$$G_R = 1.67 \left[\frac{\cos(\pi/2 \cos 45°)}{\sin 45°}\right]$$

$$= 1.67 \left[\frac{\cos\left(\pi/2 \left(\frac{1}{\sqrt{2}}\right)\right)}{\left(\frac{1}{\sqrt{2}}\right)}\right]$$

$$= 1.67 \left[\sqrt{2} \cos(63.64°)\right]$$

$$= 1.05.$$

Hence, the effective area is

$$A_E = \frac{(1.05)(0.5)^2}{4\pi}$$

$$= 0.021 \text{ m}^2.$$

● **PROBLEM 18-12**

Construct a parasitic beam antenna as shown in Fig. (1), with a dipole, a reflector and a director. Assume the velocity coefficient of the antenna is 0.80 and that the dipole is cut for a 200 MHz signal.

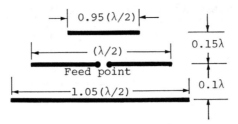

Figure 1.

Solution: The wavelength of the frequency radiated is the determining characteristic of any antenna. For this parasitic beam antenna, the dipole must be one half wavelength long. The wavelength for an antenna is calculated using

$$KC = f\lambda$$

where K = velocity coefficient,

C = speed of light in vacuum = 3×10^8 m/s,

f = frequency of radiation,

and λ = wavelength of radiation.

Therefore, for this antenna

$$\lambda = \frac{(0.80)(3 \times 10^8) \text{m/sec}}{(200 \times 10^6) 1/\text{sec}} = 1.2 \text{ meters}$$

Thus, the dipole must be $\lambda/2$ or 0.6 meters long.

In a beam antenna, the reflector must be longer and the director must be shorter than the dipole. The actual length used for these elements will determine the exact shape of the radiation pattern for the beam. Typically, the reflector is in the range of 5% longer and the director is in the range of 5% shorter than the dipole (half-wavelength). i.e.,

for f = 200 MHz

reflector length = $(1.05)(\frac{\lambda}{2})$ = $(1.05)(0.6)$ = 0.63 meter

director length = $(0.95)(\frac{\lambda}{2})$ = $(0.95)(0.6)$ = 0.57 meter.

The reflector and director are placed anywhere from 0.1λ to 0.25λ away from the dipole (again depending on the radiation pattern sought).

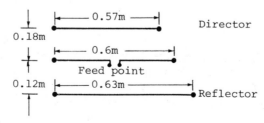

Figure 2.

The beam antenna has maximum radiation emanating from the direction perpendicular to the dipole element in the plane of the array. Similar antennas are used for home television reception.

director spacing = $(0.15)\lambda$ = $(0.15)(1.2)$ = 0.18 meter.

refelctor spacing= $(0.1)\lambda$ = $(0.1)(1.2)$ = 0.12 meter.

See Fig. (2)

● **PROBLEM 18-13**

An antenna radiates a power of 8 kw and draws a current of 20 A. Calculate its radiation resistance.

Solution: The radiation resistance is the theoretical resistance that would dissipate the same amount of power, in heat, as is radiated from the antenna in the form of electromagnetic energy.

It is given that;

Power, P = 8 kw = 8,000 watts

and current, I = 20 amperes.

Now, to determine the value of radiation resistance, R_{rad}, use the equation for power, which is given by

$$P = VI.$$

But $V = I \cdot R$ from Ohm's Law

Therefore, $P = (I \times R)(I) = I^2 R$

Hence, R_{rad} can be expressed as

$$R_{rad} = \frac{P}{I^2}$$

$$= \frac{8000}{400}$$

$$= 20 \text{ ohms}.$$

● **PROBLEM 18-14**

Find the radiation resistance of a dipole antenna $\frac{1}{10}$ wavelength long.

Solution: An antenna's total radiated power can be expressed in terms of the amount of power absorbed by an equivalent resistance called the radiation resistance. The radiation resistance of the antenna up to a quarter wavelength long (other than very short dipoles) can be calculated from:

$$R = 20(\beta \ell)^2 = 80\pi^2 \left(\frac{\ell}{\lambda}\right)^2 \text{ ohms}$$

where λ = radiated wavelength,

β = phase constant = $2\pi/\lambda$,

and ℓ = antenna length (in wave lengths).

Hence, $R = 80\pi^2 \left(\frac{\lambda/10}{\lambda}\right)^2 = 7.896$ ohms.

Note: Provided the far field is known as a function of angle, the radiation resistance can be found as follows:

From $P = \frac{1}{2} I_0^2 R$ watts

where P = radiated Power,

I_0 = amplitude of terminal current (amps),

and

$$P = \frac{1}{2} \int_S \text{Re} \left[H_\phi H_\phi^* Z\right] ds = \frac{1}{2} \int_S |H_\phi|^2 \text{Re}[Z] ds \text{ watts}$$

where Z = impedance of transmission medium (for air Z = Z_0 = 120π ohms).

The radiation resistance at the terminals of an antenna is given by

$$R = \frac{120\pi}{I_0^2} \int_S |H|^2 ds \text{ ohms.}$$

where $|H|$ = amplitude of far H field (amp/meter).

● **PROBLEM 18-15**

Suppose an antenna has a power input of 40π W and an efficiency of 98 percent. If the radiation intensity has been found to have a maximum value of 200 W/unit solid angle, find the directivity and gain of the antenna.

Solution: The average power per unit solid angle, U_{avg} is given by

$$P_{avg} = \frac{P_{rad}}{4\pi} = \frac{40\pi(0.98)}{4\pi} = 9.8 \text{ watts/steradian}$$

where P_{rad} is the power of the antenna's radiation.
Since the directivity, D, is defined as

$$\text{Directivity} = \frac{\text{Radiation intensity per unit solid angle}}{\text{Average power per unit solid angle}},$$

$$D = \frac{200 \text{ watt/sr}}{9.8 \text{ watt/sr}} = 20.41$$

or in decibels,

$$D = 10 \log_{10}(20.41) = 13.10 \text{ dB}$$

To find the gain G, recall that for an antenna

$$G = \frac{\text{radiation intensity in a given direction}}{\text{power density of an isotropic radiator}}.$$

For the isotropic radiator, the radiation intensity U is:

$$U = \frac{P_{in}}{4\pi} = 10 \text{ watt/steradian}.$$

Hence,

$$G = \frac{200 \text{ W/sr}}{10 \text{ W/sr}} = 20$$

or in decibels,

$$G = 10 \log_{10}(20) = 13.01 \text{ dB}$$

• **PROBLEM 18-16**

Ten half-wave dipoles are used in the structure of a log-periodic antenna array. The design factor is 0.7 and the smallest dipole has a length $\ell_1 = 0.03$ meter. Find:

(A) the cut-off frequencies.

(B) the passband.

Solution: For a log-periodic array antenna, both the spacing and the length of each dipole has a special fixed relationship, resulting in a constant ratio between each of the adjacent elements. That is

$$\frac{D_2}{D_1} = \frac{D_3}{D_2} = \frac{D_4}{D_3} = \ldots = \frac{\ell_2}{\ell_1} = \frac{\ell_3}{\ell_2} = \frac{\ell_4}{\ell_3} = \ldots \text{ etc.} \qquad (1)$$

where: D_n = distance of dipole n from the vertex formed near the feed point (see figure), in meters,

and ℓ_n = length of dipole n in meters.

The inverse of this ratio is the design factor ζ.

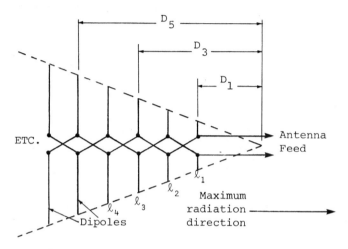

The figure shows a simple log-periodic array constructed from a group of dipoles of various lengths. Maximum radiation is in a direction measured from the largest dipole to the smallest.

PART (A): Since the antenna length and its frequency are inversely related, the highest cut-off frequency is found as follows:

the smallest dipole has a length

$$\ell_1 = \frac{\lambda}{2} = 0.03 \text{ meter.}$$

Thus, $\lambda = 2\ell_1 = 0.06$ meter.

Hence, from $C = f\lambda$, the upper cut-off frequency is

$$f = \frac{3 \times 10^8 \text{ m/sec}}{0.06 \text{ m}} = 5 \text{ GHz.}$$

To find the lower cut-off frequency, make use of the inverse of equation (1) to find the size of the largest dipole.

Thus,

$$\frac{\ell_1}{\ell_2} = \frac{\ell_2}{\ell_3} = \ldots = \frac{\ell_9}{\ell_{10}} = 0.7 \text{ (design factor)}$$

Hence, $\ell_2 = \ell_1/0.7$

$\ell_3 = \ell_2/0.7 = \ell_1/(0.7)^2$

$\ell_4 = \ell_3/0.7 = \ell_1/(0.7)^3$

etc.

$\therefore \quad \ell_{10} = \ell_9/(0.7) = \ell_1/(0.7)^9$

$$= 0.06/(0.7)^9 = 1.487 \text{ meters.}$$

Then the lower cut-off frequency will be,

$$f = \frac{3 \times 10^8 \text{ m/sec}}{1.487 \text{ m}}$$

$$= 201.768 \times 10^6 \text{ Hz}$$

$$= 201.768 \text{ MHz}$$

PART (B): The resulting passband is the difference between the upper and lower cut-off frequencies:

Hence, $5 \text{ GHz} - 201.768 \text{ MHz} = 5 \times 10^9 - 201.768 \times 10^6 \text{ Hz}$

$$= 4.798 \times 10^9 \text{ Hz}$$

$$= 4.798 \text{ GHz.}$$

● **PROBLEM 18-17**

The bandwidth and cut-off frequency of an antenna are 0.8 MHz and 44 MHz respectively. Calculate the Q-factor of the antenna.

Solution: It is given that,

Bandwidth, BW = 0.8 MHz.

Cut-off frequency, f_o = 44 MHz.

Now, to determine the Q-factor of an antenna, use the equation for bandwidth, which is given by

$$BW = \frac{f_o}{Q}$$

Therefore, Q can be expressed as;

$$Q = \frac{f_o}{BW} = \frac{44 \times 10^6}{0.8 \times 10^6}$$

$$Q = 55$$

DIRECTIONAL CHARACTERISTICS OF SIMPLE ANTENNAS

● **PROBLEM** 18-18

Figure 1 shows an antenna which applies 4 kW of power in the forward direction and 700 watts of power in the reverse direction. Calculate the front to back ratio of this antenna in dB.

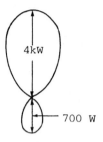

Figure 1.

Solution: The power applied in the forward direction, P_F = 4 kW = 4000 watts and, the power applied in the reverse direction, P_B = 700 watts.

The front to back ratio of an antenna (A_{FB}) is the ratio of powers in the front direction to the back direction expressed in decibels.

Therefore, $A_{FB} = 10 \log_{10} \dfrac{P_F}{P_B}$

$= 10 \log_{10} \dfrac{4000}{700}$

$= 10 \log_{10} 5.71428$

$= 10(0.75696)$

$= 7.5696$ dB

● **PROBLEM** 18-19

Determine the reflection factor of an horizontal antenna for the reflection angles, 30°, 60° and 90° respectively, when the antenna is placed one quarter wavelength above the earth's surface. It is given that the signal strength for this antenna is unity.

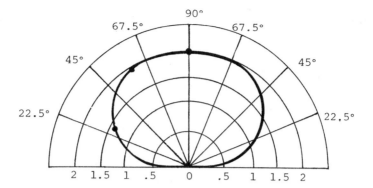

Solution: The reflection factor is the increase in signal strength at the evaluation point due to reflection from the earth's surface. When the reflection angle, α = 30°, the angle of lag between the directed and reflected ray is calculated as:

Lag angle = 180° + 2h sin α

where h is the height of the antenna above the earth's surface. Since the antenna is placed a quarter wavelength above the earth's surface,

h = λ/4 = 90°.

Therefore, lag angle = 180° + 2 × 90° × sin 30°

= 180° + (180°)(0.5)

= 270°

This angle of lag 270° is equal to a lead angle of (360 − 270) = 90°

Therefore, the reflection factor for a signal strength of unity is:

$$(1)\cos\left(\frac{90}{2}\right) + (1)\cos\left(\frac{90}{2}\right) = 1.414.$$

Now, when the angle of reflection is 60°

angle of lag = 180° + 2h sin 60°

But h = 90°

Therefore, angle of lag = 180° + 180° sin 60°

= 336°

Therefore, the lead angle = 360° − 336° = 24°.

Then, the reflection factor is calculated as:

$$(1)\cos\left(\frac{24}{2}\right) + (1)\cos\left(\frac{24}{2}\right)$$

$$= \cos 12° + \cos 12°$$

$$= 1.96.$$

Similarly, for $\alpha = 90°$,

angle of lag $= 180° + 2(h)\sin \alpha$

$$= 180° + 2(90°)\sin 90°$$

$$= 360°$$

Therefore, lead angle $= 360° - 360°$

$$= 0°$$

Hence, the reflection factor

$$= (1)\cos 0° + (1)\cos 0°$$

$$= 1 + 1 = 2.$$

Hence, the signal strength is doubled by the reflection. This means that both signals are in phase at the equivalent point.

The figure below shows a plot of the reflection factor for the horizontal dipole antenna.

Note: This is not the same as a plot of the radiation pattern. The radiation pattern is obtained by superposing the plot and the antenna's free-space radiation pattern.

● **PROBLEM 18-20**

For point-to-point communication at higher frequencies, the desired radiation pattern is a single narrow lobe or beam. To obtain such a characteristic (at least approximately) a multi-element linear array is usually used. An array is linear when the elements of the array are spaced equally along a straight line. In a uniform linear array the elements are fed with currents of equal magnitude and having a uniform progressive phase shift along the line. The pattern of such an array can be obtained by adding vectorially the field strengths due to each of the elements. For a uniform array of non-directional elements the field strength would be

$$E_T = E_0 \left| 1 + e^{j\psi} + e^{j2\psi} + e^{j3\psi} + \ldots + e^{j(n-1)\psi} \right| \quad (1)$$

where

$$\psi = \beta d \cos \phi + \alpha$$

and α is the progressive phase shift between elements. (α is the angle by which the current in any element leads the current in the preceding element.)

Compute the pattern of such a linear array.

Solution: Eq. (1) is viewed as a geometric progression and written in the form

$$\frac{E_T}{E_0} = \sum_{k=0}^{n-1} \left| e^{jk\psi} \right| = \left| \frac{1 - e^{jn\psi}}{1 - e^{j\psi}} \right|.$$

where $\left|\frac{E_T}{E_0}\right| = |A| =$ the array factor.

This result will be simplified using Euler's form for sin x:

$$\sin x = \left(\frac{e^{jx} - e^{-jx}}{2j} \right)$$

Rearranging,

$$2j(\sin x) = (e^{jx} - e^{-jx})$$
$$2j(\sin x)e^{jx} = e^{jx}(e^{jx} - e^{-jx})$$
$$= (e^{2jx} - 1)$$

∴ $(\sin x) = -j0.5\, e^{-jx}(e^{2jx} - 1)$

$$\sin x = j0.5\, e^{-jx}(1 - e^{2jx}) \quad (A)$$

Similarly,

$$\sin nx = j0.5\, e^{-jnx}(1 - e^{2jnx}) \quad (B)$$

Dividing the magnitude of equation (B) by the magnitude of equation (A) gives

$$\left| \frac{\sin nx}{\sin x} \right| = \left| \frac{1 - e^{2jnx}}{1 - e^{2jx}} \right|$$

where $\left| \frac{e^{-jnx}}{e^{-jx}} \right| = 1$

Hence,

$$\frac{E_T}{E_0} = \left| \frac{1 - e^{jn\psi}}{1 - e^{j\psi}} \right|$$

$$= \left| \frac{\sin \frac{n\psi}{2}}{\sin \frac{\psi}{2}} \right| \quad (2)$$

The maximum value of this expression is n and occurs when $\psi = 0$. This is the principal maximum of the array. Since $\psi = \beta d \cos\phi + \alpha$ the principal maximum occurs when

$$\cos\phi = -\frac{\alpha}{\beta d}$$

For a broadside array the maximum radiation occurs perpendicular to the line of the array at $\phi = 90$ degrees, so $\alpha = 0$ degrees. For an end-fire array the maximum radiation is along the line of the array at $\phi = 0$, so $\alpha = -\beta d$ for this case.

The expression (2) is zero when

$$\frac{n\psi}{2} = \pm k\pi, \quad k = 1,2,3,\ldots$$

These are the nulls of the pattern. Secondary maximas occur approximately midway between the nulls, when the numerator of expression (2) is a maximum, that is when

$$\frac{n\psi}{2} = \pm(2m+1)\frac{\pi}{2} \quad m = 1,2,3,\ldots$$

The first secondary maximum occurs when

$$\frac{\psi}{2} = \frac{+3\pi}{2n}$$

(note that $\psi/2 = \pi/2n$ does not give a maximum). The amplitude of the first secondary lobe is

$$\left|\frac{1}{\sin(\psi/2)}\right| = \left|\frac{1}{\sin(3\pi/2n)}\right|$$

$$\simeq \frac{2n}{3\pi} \quad \text{for large n}$$

The amplitude of the principal maximum was n, so the amplitude ratio of first secondary maximum to principal maximum is $2/3\pi = 0.212$. This means that the first secondary maximum is about 13.5 dB below the principal maximum, and this ratio is independent of the number of elements in the uniform array, as long as the number is large.

● **PROBLEM 18-21**

(A) Extend the concept of the array factor for the two element linear array shown, to any number of elements (N), derive the array factor for an array with N elements.

(B) Generalize this result so that the array factor is explicitly expressed as a function of the number of antennas in the array, the relative amplitudes and phases of the currents in them and their relative spacings.

(C) Find the array factor and sketch the normalized radiation pattern for horizontally stacked dipoles with separation $\lambda_0/4$ and $3\lambda_0/4$ respectively, and $\alpha_n = -\pi/4$.

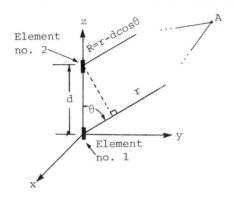

TWO ELEMENT ARRAY. Each antenna is radiating energy out to point A where the distance d<<A.

Solution:

(A) The array factor for any given array is the resultant pattern produced by a similar array of isotropic antennas. The radiation pattern of any array can be simulated from the array factor.

In the figure, the two elements are assumed to be radiating energy out to point A. The distance from either antenna to point A is much greater than the distance, d, separating the two antennas. The electric field at A, due to element No.1 is

$$\vec{E}_1 = f(\theta,\phi) \frac{e^{-j\beta_0 r}}{r} \qquad (1)$$

where $f(\theta,\phi)$ is some function dependent on the spherical angles θ and ϕ, and β_0 is the propagation constant = $2\pi/\lambda_0$.

The electric field at A due to element No. 2 is

$$\vec{E}_2 = f(\theta,\phi) \frac{e^{-j\beta_0 r}}{(r-d\cos\theta)} e^{+j\beta_0 d\cos\theta} \qquad (2)$$

The distances of each element to point A are related and can be expressed as

$$R = r - d\cos\theta.$$

Since it is assumed that $d \ll r$ (the distance from either antenna to point A is much greater than the separation between antennas)

$$R \simeq r.$$

This is called a far-zone approximation. (Actually, the distance from each antenna to point A differs by $-d\cos\theta$. Hence, equation (2) can be written as

$$\vec{E}_2 = f(\theta,\phi) \frac{e^{-j\beta_0 r}}{r} e^{+j\beta_0 d\cos\theta}$$

At point A, the total electric field is the superposition of the electric fields due to element No. 1 and element No. 2, i.e.,

$$\vec{E}_A = (\vec{E}_1 + \vec{E}_2) = f(\theta,\phi) \frac{e^{-j\beta_0 r}}{r}(1 + e^{+j\beta_0 d\cos\theta}) \quad (3)$$

where the magnitude of the term in parenthesis in equation (3) is the array factor for a two element array.

Thus, for any number (N) of antennas in the array that are equally spaced, the resulting field at A will be the superposition of each separate electric field using the far-zone approximation.

i.e., $$\vec{E} = f(\theta,\phi) \frac{e^{-j\beta_0 r}}{r} \left[1 + \sum_{n=1}^{N-1} e^{+j\beta_0 nd\cos\theta} \right]$$

where the magnitude of the bracketed term is the array factor.

(B) The generalized form, to indicate that the array factor is a function of the number of elements, their current amplitudes and phases and relative spacings is

$$\vec{E} = f(\theta,\phi) \frac{e^{-j\beta_0 r}}{r} \left[1 + \sum_{n=1}^{N-1} c_n e^{+j(\beta_0 nd\cos\theta + \alpha_n)} \right] \quad (4)$$

where c_n = relative current amplitude.

α_n = relative current phase (i.e., the phase shift in the excitation current of each element).

$= -\beta_0 d\cos\theta_0$ (where θ_0 is a constant).

Note: When $\alpha_n = 0$, $\theta_0 = 90°$. In this case the maximum radiation is directed normal to the array axis (broadside array).

For $\alpha_n \neq 0$, $\theta_0 < 90°$ (end-fire array).

(C) For horizontally stacked dipole arrays:

Separation	Array Factor	Normalized Radiation Patterns
$\dfrac{\lambda_0}{4}$	$\cos\dfrac{\pi}{4}(\cos\theta-1)$	End fire array $\alpha_n = -\dfrac{\pi}{4}$
$\dfrac{3\lambda_0}{4}$	$\cos\dfrac{3\pi}{4}(\cos\theta-1)$	End fire array $\alpha_n = -\dfrac{3\pi}{4}$

• **PROBLEM 18-22**

In the figure below, each four elements linear array has separation distance $d = \dfrac{\lambda_0}{4}$. Find the array factor for this array if the phase shift in the excitation current of each element is $\alpha = -\dfrac{\pi}{4}$.

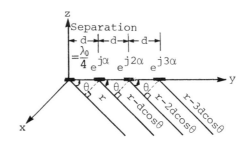

Solution: The phase shift in the excitation currents of each element is $-\pi/4$. The phase shift of excitation between the first and second antennas is $e^{j\alpha_n}$, between the second and the third is $e^{j2\alpha_n}$, and between the third and the fourth is $e^{j3\alpha_n}$. The array factor $|A|$ is the magnitude of the superposition of these factors. Thus,

$$\begin{aligned}
E_{TOTAL} &= e^{-j\beta_0 \gamma} + e^{-j\beta_0(r-d\cos\theta)+j\alpha_n} + e^{-j\beta_0(r-2d\cos\theta)+2j\alpha_n} \\
&\quad + e^{-j\beta_0(r-3d\cos\theta)+3j\alpha_n} \\
&= e^{-j\beta_0 r}\left[1 + e^{j(\beta_0 d\cos\theta+\alpha_n)} + e^{j(2\beta_0 d\cos\theta+2\alpha_n)} \right. \\
&\quad \left. + e^{j(\beta_0 3d\cos\theta+3\alpha_n)}\right]
\end{aligned}$$

$$\begin{aligned}
|A| = |1| &\left[1 + e^{j(\beta_0 d\cos\theta+\alpha_n)} + e^{j(2\beta_0 d\cos\theta+2\alpha_n)} \right. \\
&\left. + e^{j(\beta_0 3d\cos\theta+3\alpha_n)}\right] \\
&= \left|\sum_{x=0}^{N-1} e^{jx(\beta_0 d\cos\theta+\alpha_n)}\right| \\
&= \left|\frac{1 - e^{Nj(\beta_0 d\cos\theta+\alpha_n)}}{1 - e^{j(\beta_0 d\cos\theta+\alpha_n)}}\right| \qquad (1)
\end{aligned}$$

This result can be simplified using Euler's form for sin x:

$$\text{if } \sin x = \frac{e^{jx} - e^{-jx}}{2j}$$

then,

$$2j(\sin x) = (e^{jx} - e^{-jx})$$

$$2j(\sin x)e^{jx} = e^{jx}(e^{jx} - e^{-jx})$$

$$= (e^{2jx} - 1)$$

Therefore, $\sin x = -j0.5\, e^{-jx}(e^{2jx} - 1)$

$$= j0.5\, e^{-jx}(1 - e^{2jx}) \qquad (2)$$

Similarly,

$$\sin Nx = j0.5\, e^{-jNx}(1 - e^{2jNx}) \qquad (3)$$

Dividing the magnitude of equation (3) by the magnitude of equation (2) gives equation (1), where $x = \left[(\tfrac{1}{2})(\beta_0 d\cos\theta+\alpha_n)\right]$.

i.e., $\dfrac{\sin Nx}{\sin x} = \left|\dfrac{1 - e^{2jNx}}{1 - e^{2jx}}\right|$ where $\left|\dfrac{e^{-jNx}}{e^{-jx}}\right| = 1$

Thus,

$$|A| = \left| \frac{1 - e^{Nj(\beta_0 d \cos\theta + \alpha_n)}}{1 - e^{j(\beta_0 d \cos\theta + \alpha_n)}} \right| = \frac{\sin(\frac{N}{2})(\beta_0 d \cos\theta + \alpha_n)}{\sin(\frac{1}{2})(\beta_0 d \cos\theta + \alpha_n)}$$

Substituting the given values $d = \lambda_0/4$

$$\alpha = -\pi/4$$

$$\beta_0 = \frac{2\pi}{\lambda_0} \text{ and } N=4 \text{ gives,}$$

$$|A| = \frac{\sin \frac{4}{2}\left(\frac{\pi}{2}\cos\theta - \frac{\pi}{4}\right)}{\sin \frac{1}{2}\left(\frac{\pi}{2}\cos\theta - \frac{\pi}{4}\right)} = \frac{\sin\pi(\cos\theta - \frac{1}{2})}{\sin\frac{\pi}{2}(\cos\theta - \frac{1}{2})}$$

Note: Since the phase shift term α_n is not equal to zero, this antenna forms an end-fire array.

POWER RELATIONS

• **PROBLEM** 18-23

For a simple communications system, the transmitting and receiving antennas used are both half-wave dipoles of 75 Ω impedance. The receiver circuit has an internal impedance of 150Ω. Given a radiated signal of 100 volts at the transmitter and a path-loss of 10 volts due to attenuation, find the power delivered to the receiver.

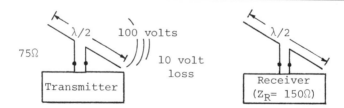

Solution: The power delivered to the receiver is the power actually dissipated by the impedance of the receiver (where the standard power relation $P = V^2/z$ is used). Knowing the receiver's impedance, the voltage of the signal at the receiver can be evaluated. At the receiving antenna, the effective source voltage E_R is:

E_R = (Radiated voltage) - (Transmission losses).

   = 100v - 10v.

   = 90 volts

The radiated signal induces voltages throughout the transmission path. Hence, the magnitude of the received voltage is the voltage induced in the receiver (V_R) which is

$$V_R = \frac{Z_R}{Z_R + Z_T}(E_R) = \frac{150}{150 + 75}(90v) = 60 \text{ volts}.$$

Therefore, the power (P_R) delivered to the receiver is

$$P_R = \frac{(V_R)^2}{Z_R} = \frac{(60)^2}{150} = 24 \text{ watts}.$$

Note: The transmitter power was $P_T = \frac{(V_T)^2}{Z_T} = \frac{(100)^2}{75} = 133\frac{1}{3}$ watts. Thus, $109\frac{1}{3}$ watts (or 82%) is lost before reception.

• PROBLEM 18-24

Two antennas, 48 km apart, have transmitting and receiving gains of 40 dBw (i.e., with respect to 1 watt) each. For a transmitter frequency of 4 GHz, find the required transmitter power so that the received power is 10^{-6} watts (-60 dBw).

Solution: The power received or transmitted by an antenna can be expressed as

$$S_R = \frac{S_T}{4\pi d^2} G_T G_R \frac{\lambda^2}{4\pi} = \frac{S_T G_T G_R}{\left(4\pi \frac{d}{\lambda}\right)^2}$$

where S_R = received power at the antenna,

S_T = transmitted power at the antenna,

G_T = transmitter antenna gain,

G_R = receiver antenna gain,

λ = radiated wavelength,

and d = distance between sending and receiving points.

Before calculating the result, let us justify the relation used. An isotropic radiator is an antenna that radiates an equal amount of power in all directions. Thus, the power density received at a given distance from the antenna (in watts/meter^2) is the transmitter power divided by the surface area of the sphere, tangent to the measurement point with the antenna at its center.

i.e., $$S_R = \frac{S_T}{4\pi d^2}$$

Antennas can be made to radiate in specific directions. Thus, concentrating the radiated power towards specific receiving areas. This focusing effect is denoted by assigning to the antenna a gain G_T over the isotropic radiation. The effective power density is now

$$S_R = \frac{S_T}{4\pi d^2} G_T$$

For aperture antennas (i.e., parabolic, horn, lens, etc.), which are most commonly used at higher frequencies, the antenna gain is given by:

$$G_T = \frac{4\pi A_T \eta_T}{\lambda^2}$$

Where A_T is the transmitting antenna area,

η_T is an antenna efficiency parameter,

and λ is, as before, the radiated wavelength,

The receiver will also be of the same type as the transmitter so that the receiver area A_R and receiver efficiency η_R diminish the receiver power by the factor $A_R \eta_R$. Thus,

$$\frac{S_R}{A_R \eta_R} = \frac{S_T}{4\pi d^2} G_T$$

or

$$S_R = \frac{S_T G_T A_R \eta_R}{4\pi d^2},$$

and since the receiver gain is $G_R = \dfrac{4\pi A_R \eta_R}{\lambda^2}$

Substituting for $A_R \eta_R$ gives:

$$S_R = \frac{S_T}{4\pi d^2} G_T G_R \frac{\lambda^2}{4\pi} = \frac{S_T G_T G_R}{(4\pi d/\lambda)^2}$$

Expressing the above in decibels gives

$$10\log_{10} S_R = 10\log_{10} S_T + 10\log_{10} G_T + 10\log_{10} G_R -$$

$$20\log_{10}(4\pi d/\lambda)$$

Now, to find S_T, from the given information

-60 dBw $= 10\log_{10} S_T + 40$ dBw $+ 40$ dBw $-$

$$20\log_{10}\left(4\pi\frac{48 \times 10^3 m}{(3 \times 10^8/4 \times 10^9)}\right)$$

-60 dBw $= 10\log_{10} S_T + 80$ dBw $- 138.1$ dBw

-1.892 dBw $= 10\log_{10} S_T$

(This means S_T is 1.892 dB below 1 watt)

Thus, 0.647 watts $= S_T$

Therefore, 0.647 watts must be transmitted in order to receive 10^{-6} watt at the receiver.

• **PROBLEM** 18-25

If S_R is the power radiated by a transmitting antenna and S_T is the power (maximum) in the load of a receiving antenna, show that

$$\frac{S_R}{S_T} = \frac{A_{Tem} A_{Rem}}{(d\lambda)^2}$$

where A_{Tem} and A_{Rem} are the maximum effective apertures of transmitting and receiving antennas, d is the distance between the two antennas and λ is the wave length. Both the antennas are situated in free space.

Solution: The useful power picked up by the receiving antenna is given by the product of effective aperture and average Poynting vector in the oncoming wave

$$S_R = A_{Rem} P_{AVERAGE} \tag{1}$$

The power density at the receiver is the power density of an isotropic radiator ($S_T/4\pi d^2$) times the transmitting antenna gain.

$$S_R = \frac{S_T A_{Rem} G_T}{4\pi d^2} \tag{2}$$

where S_T = power transmitted,

$$G_T = 4\pi A_{Tem}/\lambda^2 = \text{antenna gain}, \tag{3}$$

and

A_{Rem} = effective aperture of receiver.

Substituting (3) in (2) gives,

$$\frac{S_R}{S_T} = \frac{A_{Rem}}{x^2} \cdot \frac{4\pi A_{Tem}}{4\pi d^2} = \frac{A_{Tem} A_{Rem}}{(d\lambda)^2}$$

Hence the proof.

• **PROBLEM** 18-26

A communication link between two $\lambda/2$ dipole antennas is established in a free space environment. If the transmitter delivers 1 kW of power to the transmitting antenna and the transmitter gain G_T is 1.64, how much power will be received by a receiver connected to the receiving dipole 500 km from the transmitter if the frequency is 200 MHz? Assume that the path between dipoles is normal to each dipole, and the dipoles are perfectly aligned. See Fig. (1)

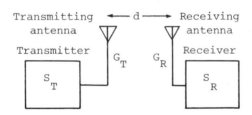

Figure 1.

Solution: For a $\lambda/2$ dipole antenna with transmitter gain $G_T = 1.64$, the effective aperture of the transmitting antenna $\left(\text{from } G_T = \frac{4\pi A_{TE}}{\lambda^2}\right)$ is given as

$$A_{TE} = \frac{G_T \lambda^2}{4\pi}$$

$$= \frac{1.64 \left(\frac{3 \times 10^8}{2 \times 10^8}\right)^2}{4\pi}$$

$$= 0.294 \text{ m}^2$$

$$= A_{RE} = \text{effective receiver area}$$

This must be true since the antennas are identical in type and differ only in their use. Thus,

$$G_T = G_R = 1.64$$

$$A_{TE} = A_{RE} = 0.294 \text{ m}^2$$

The power received, S_R, is given by:

$$S_R = \frac{S_T G_T A_{RE}}{4\pi d^2} = \frac{S_T G_T G_R}{(4\pi d/\lambda)^2}$$

$$= \frac{(10^3)(1.64)(0.294)}{4\pi(5 \times 10^5)^2}$$

Thus,

$$S_R = 1.5329 \times 10^{-10} \text{ watts}$$

• **PROBLEM** 18-27

For free-space transmission, derive the basic transmission path loss, L. Also, determine the basic transmission loss between a ground based antenna and an antenna on an aircraft at distances of 1, 10, 100 and 200 m, and

(a) at a frequency of 300 MHz,

(b) at a frequency of 3 GHz.

Solution: For a power S_T radiated from an isotropic antenna, the power density at a distance d is $S_T/4\pi d^2$. If the transmitting antenna has a gain G_T in the desired direction, the power density is increased to $G_T S_T/4\pi d^2$. If the effective area of the receiving antenna is $A_{RE} = \lambda^2 G_R/4\pi$, the power received will be

$$S_R = \frac{G_T S_T}{4\pi d^2} \frac{\lambda^2 G_R}{4\pi} = \frac{\lambda^2 G_T G_R}{(4\pi d)^2} S_T \text{ in watts.}$$

Hence, the ratio of received to transmitted power is

$$\frac{S_R}{S_T} = \frac{\lambda^2 G_R G_T}{(4\pi d)^2}.$$

The basic transmission path loss, L, is defined as the reciprocal of this ratio, expressed in decibels, for transmission between isotropic antennas ($G_T = G_R = 1$). That is:

$$L = 10 \log_{10} \frac{(4\pi d)^2}{\lambda^2} = 10 \log_{10} \left(4\pi \frac{d}{\lambda}\right)^2.$$

(a) At a frequency of f = 300 MHz:

For distance = 1 meter:

$$L = 10 \log_{10} \left[4\pi \frac{1 \text{ meter}}{\left(\frac{3 \times 10^8 \text{ msec}^{-1}}{3 \times 10^8 \text{ Hz}} \right)} \right]^2$$

$= 21.984$ dB

For distance = 10 meters:

$$L = 10 \log_{10} \left[4\pi \frac{10 \text{ m}}{\left(\frac{3 \times 10^8 \text{ msec}^{-1}}{3 \times 10^8 \text{ Hz}} \right)} \right]^2$$

$= 41.984$ dB

For distance = 100 meters:

$$L = 10 \log_{10} \left[4\pi \frac{100 \text{ m}}{\left(\frac{3 \times 10^8 \text{ msec}^{-1}}{3 \times 10^8 \text{ Hz}} \right)} \right]^2$$

$= 61.984$ dB

For distance = 200 meters:

$$L = 10 \log_{10} \left[4\pi \frac{200 \text{ m}}{\left(\frac{3 \times 10^8 \text{ msec}^{-1}}{3 \times 10^8 \text{ Hz}} \right)} \right]^2$$

$= 68.005$ dB

(b) At a frequency of 3 GHz

For distance = 1 meter:

$$L = 10 \log_{10} \left[4\pi \frac{1 \text{ m}}{\left(\frac{3 \times 10^8 \text{ msec}^{-1}}{3 \times 10^9 \text{ Hz}} \right)} \right]^2$$

$L = + 41.984$ dB

For distance = 10 meters:

$$L = 10 \log_{10} \left[4\pi \frac{100 \text{ m}}{\left(\frac{3 \times 10^8 \text{ msec}^{-1}}{3 \times 10^9 \text{ Hz}} \right)} \right]^2$$

$= 61.984$ dB (see fig. 1)

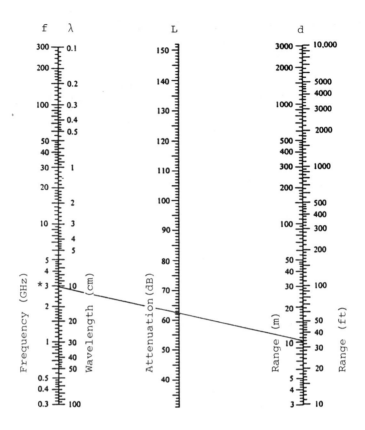

NOMOGRAM OF TRANSMISSION PATH LOSS FOR FREE SPACE.

Figure 1.

For distance = 100 meters:

$$L = 10 \log_{10} \left[4\pi \frac{100 \text{ m}}{\left(\frac{3 \times 10^8 \text{ msec}^{-1}}{3 \times 10^9 \text{ Hz}} \right)} \right]^2$$

= 81.984 dB

For distance = 200 meters:

$$L = 10 \log_{10} \left[4\pi \frac{200 \text{ m}}{\left(\frac{3 \times 10^8 \text{ msec}^{-1}}{3 \times 10^9 \text{ Hz}} \right)} \right]^2$$

= 88.005 dB

The nomogram in Fig. (1) can also be used to calculate L. In Fig. (1), '*' indicated the case where d = 10 meters and f = 3 GHz so that L = 61.984 ≃ 62 dB.

CHAPTER 19

TRANSMISSION LINES

A REFLECTED PULSE

• PROBLEM 19-1

A pulse of voltage is applied to the line of Fig. 1 by throwing the switch to position 2 at t = 0, and returning the switch to position 1 at $t = t_1$, where t_1 is less than the time T it takes a wave to travel the length of the line. (The short circuit is provided so the sending end impedance is zero regardless of the voltage.) Analyze the waves on the line.

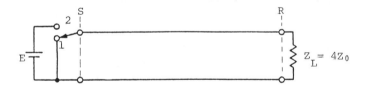

Terminated line arranged for transmission of a pulse of voltage.

Fig. 1.

Solution: The voltage wave is shown in Fig. 2a, for several values of t, and the current wave is shown in Fig. 2b. The current is shown in units of E/Z_0.

When the wave reaches the receiving end, it is reflected with a reflection coefficient Γ_L:

$$\Gamma_L = \frac{Z_L - Z_0}{Z_L + Z_0} = \frac{4Z_0 - Z_0}{4Z_0 + Z_0} = \frac{3}{5}$$

The current wave is reflected in the ratio

$$\frac{i^-}{i^+} = -\frac{3}{5}$$

The reflected waves of voltage and current, as well as the total voltage and current, are also shown in Fig. 2. It can be noted that v_L/i_L is always equal to $4Z_0$.

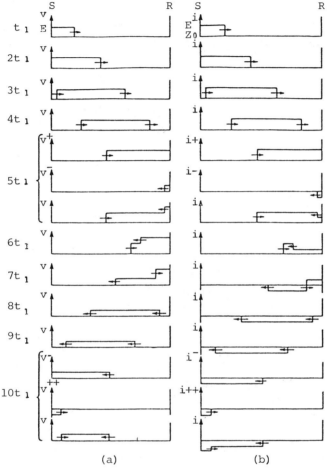

(a) Voltage waves and (b) current waves, on line of Fig. 1 at various instants of time. Arrows indicate directions of motion of wavefronts.

Fig. 2.

When the reflected waves return to the sending end, they "see" a short circuit, so they are again reflected. The reflection coefficient at the sending end is -1 for voltage and -1 for current. Observe carefully that the incident wave is now the backward traveling wave, and the reflected wave travels in the forward direction. Thus

$$\Gamma_S = \frac{v^{++}}{v^-} = -\frac{i^{++}}{i^-}$$

where v^{++}, i^{++} now represent the (second) reflected wave traveling in the forward direction: This wave is also shown in Fig. 2.

In the foregoing example the reflection at the sending end gave rise to no difficulty because the battery had been disconnected before the wave returned. Even if the battery is not disconnected, however, it has no effect on reflections, except for its internal resistance Z_S, because the reflected and rereflected waves may be considered separately from the original incident wave and are merely superimposed upon it.

• **PROBLEM** 19-2

The circuit of Fig. 1 is in steady-state condition just before the switch is closed, shorting out part of the load resistance. Draw the distance-time plots for voltage and current, and plot V_L, I_L versus time after $t = 0$.

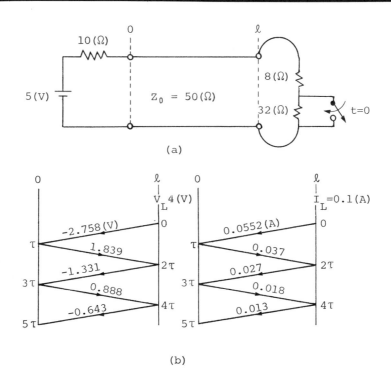

Fig. 1.

Solution: At $t = 0-$,

$$V_L = V^+ + V^- = V^+(1 + \rho_L) = 5\left(\frac{40}{50}\right) = 4 \text{ Volts}$$

The load reflection coefficient Γ_L is:

$$\Gamma_L = (Z_L - Z_0)/(Z_L + Z_0)$$

$$\Gamma_L = \frac{40 - 50}{40 + 50} = -\frac{1}{9} = -0.1111.$$

Then
$$V^+ = \frac{4V}{\frac{8}{9}} = 4.5 \text{ Volts}$$

and
$$V^- = -\frac{1}{9}(4.5V) = -0.5 \text{ Volts}.$$

Prior to the closing of the switch, the load voltage can be considered as composed of an incident voltage of 4.5 Volts and a reflected voltage of -0.5 Volts. Closing the switch cannot instantaneously change the incident voltage, but the reflected voltage does change instantly to maintain boundary conditions. Then, at t = 0+,

$$V^+ = \text{same} = 4.5 \text{ Volts}$$

$$\Gamma_L = \frac{8 - 50}{8 + 50} = -0.724$$

$$V^- = [-0.724(4.5)] = -3.258 \text{ Volts}.$$

Thus,
$$\Delta V^- = [-3.258 - (-0.5)] = -2.758 \text{ Volts}$$

and
$$V_L(t = 0+) = [4.5 - 3.258] = 1.242 = (4.0 - 2.758) \text{Volts}$$
$$= [V_L(t = 0-) + \Delta V^-] \text{ Volts}.$$

The change in reflected current can readily be calculated from ΔV^- and Z_0. Since voltage and current are reflected with opposite polarity,

$$\Delta I_L = \Delta I^- = \frac{\Delta V^-}{Z_0} = \frac{2.758}{50} = 0.0552 \text{ Amp}$$

and
$$I_L(t = 0+) = I_L(t = 0-) + \Delta I_L^-$$
$$= \frac{5}{10 + 40} + 0.0552 = 0.1552 \text{ Amp.}$$

To calculate the load voltage and current as a function of time, T_G must be known. Hence,

$$\Gamma_G = \frac{Z_G - Z_0}{Z_G + Z_0}$$

$$\Gamma_G = \frac{10 - 50}{10 + 50} = -0.667 = -\frac{2}{3}$$

Now write V_L and I_L as a function of time.

$$V_L = \{4.0 + 2.758[-u(t) + \tfrac{2}{3}u(t - 2\tau)$$

$$- 0.724\left[\tfrac{2}{3}\right]u(t - 2\tau)$$

$$+ 0.724\left(\tfrac{2}{3}\right)^2 u(t - 4\tau) - (0.724)^2\left(\tfrac{2}{3}\right)^2 u(t - 4\tau)$$

$$+ \ldots]\}\text{Volts}$$

$$= [4.0 - 2.758u(t) + 0.507u(t - 2\tau)$$

$$+ 0.245u(t - 4\tau) + \ldots]\text{Volts}$$

and

$$I_L = \{0.1 + 0.0552[u(t) + \tfrac{2}{3}u(t - 2\tau)$$

$$+ 0.724\left(\tfrac{2}{3}\right)u(t - 2\tau)$$

$$+ 0.724\left(\tfrac{2}{3}\right)^2 u(t - 4\tau) + (0.724)^2\left(\tfrac{2}{3}\right)^2 u(t - 4\tau)$$

$$+ \ldots]\} \quad (A)$$

$$= [0.1 + 0.0552u(t) + 0.063u(t - 2\tau)$$

$$+ 0.031u(t - 4\tau) + \ldots] \text{ Amp.}$$

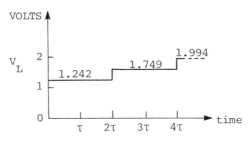

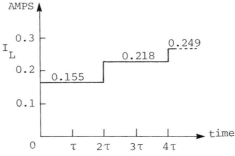

Fig. 2.

Both V_L and I_L are plotted in Fig. 2.

933

• **PROBLEM 19-3**

For the circuit shown in Figure (1), use a reflection diagram to prove that soon after the switch is closed at time $t = 0$, the voltage at the short circuit (load) will be zero and the current will stabilize at $\frac{V}{Z_0}$ Amps.

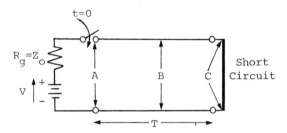

Fig. 1. (attenuation = 0)

Solution: From the circuit, it can be seen that after the switch is closed at $t = 0$, a step function of voltage is sent towards the short circuit from the matched generator.

With the load being a short circuit, the reflection coefficient at point C is:

$$\Gamma_L = \frac{0 - Z_0}{0 + Z_0} = -1$$

and so the total reflection of the voltage wave takes place at the load.

At the generator, the reflection coefficient of the input is similarly found to be

$$\Gamma_{IN} = \frac{Z_0 - Z_0}{Z_0 + Z_0} = 0$$

which follows from the fact that the generator is matched to the line.

Hence, the totally reflected voltage waves moving towards the generator will cancel any further voltage waves moving towards the load, leaving the voltage at the load equal to zero. However, since the forward and reflected voltage waves are of opposite sign, the corresponding current waves are additive. This can be explained by recalling that

$$V(l,t) = V^+ e^{-j\beta l} + V^- e^{+j\beta l}$$

$$I(l,t) = I^+ e^{-j\beta l} - I^- e^{+j\beta l}$$

where βl is the electrical length of the line so that the reflected current wave is always of opposite sign with respect to the reflected voltage wave. The initial current surge will be, from Ohm's law at the input, $\frac{V}{2Z_0}$. The battery current will thus be $\frac{V}{2Z_0}$ from t = 0 until the reflected wave arrives at 2T raising the current by a further $\frac{V}{2Z_0}$ to give $\frac{V}{Z_0}$ (this is the value determined by the static behavior of the circuit).

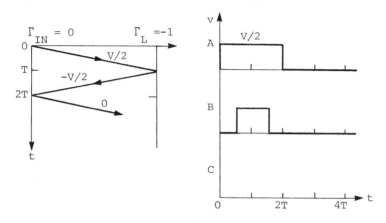

Fig. 2. Reflection diagram and voltage waveforms for a short-curcuit line.

At the load, the current will be zero until the first current wave arrives at t = T whereafter the currents for the incident and reflected waves add instantaneously to give a short-circuit current of $\frac{V}{Z_0}$.

● **PROBLEM 19-4**

The circuit shown in Fig. (1) contains no energy dissipating elements. Using a reflection diagram, show that if the constant battery voltage V is maintained, the short circuit current would build up to infinity.

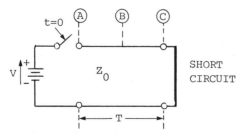

Fig. 1. (attenuation = 0)

<u>Solution</u>: At the short circuit, the current is zero until the arrival of the first voltage wave. At t = T, the currents

935

for the incident and reflected waves add instantaneously to give a short circuit current of $\frac{2V}{Z_0}$. The voltage reflection coefficient at the battery will be -1 (i.e., $\Gamma_{IN} = -1$) so that a new voltage wave of V travels towards the short circuit where is adds another increment of $\frac{2V}{Z_0}$ to the current at t=3T.

Thus, both of the current waveforms increases towards infinity in a step-like manner as is shown in figures 2(a) and 2(b).

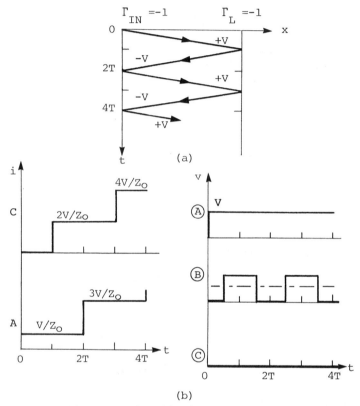

Fig. 2. (a) Reflection diagram for a short-curcuit line supplied by a battery with zero internal resistance; (b) corresponding voltage and current waveforms.

Note: In practice, there are no perfectly lossless systems so that the current would be limited by the series resistance of the circuit (internal source resistance plus the fact that even the best short circuit may have a very small resistance value).

• **PROBLEM 19-5**

An input step voltage of 12 volts is applied to a transmission line with characteristic impedance $Z_0 = 100\,\Omega$. The internal resistance of the source is $R_S = 11\,\Omega$ and the one-way time delay of the line is t_d.

Sketch the waveforms at the output when the line is terminated in a load $R_L = 900\Omega$.

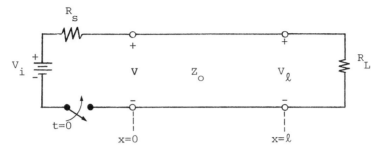

Fig. 1. Voltage source V_i with impedance R_s, is connected to load R_L by means of a transmission line with characteristic impedance Z_o.

Solution: The reflection coefficient due to the load at $x = \ell$ is Γ_R and is given by:

$$\Gamma_R = \frac{Z_L - Z_0}{Z_L + Z_0} = \frac{900 - 100}{900 + 100} = 0.80 \text{ where } Z_L = R_L = 900\Omega.$$

The reflection coefficient due to the mismatch at the input end, i.e., at $x = 0$, is Γ_S and is given by:

$$\Gamma_S = \frac{Z_S - Z_0}{Z_S + Z_0} = \frac{11 - 100}{11 + 100} = -0.80 \text{ where } Z_S = R_S = 11\Omega.$$

At time $t = 0$, the 12 volts step input will appear at the line at $x = 0$ as a step of amplitude

$$V = V_1 = \left(\frac{Z_0}{Z_0 + R_S}\right)V = \left(\frac{100}{100 + 11}\right)12 = 10.81 \text{ volts}.$$

The voltage front of V_1 will arrive at the load at time $t = t_d$ and immediately there will appear a reflected front V_2 where

$$V_2 = \Gamma_R V_1 = 0.8(10.81) = 8.65 \text{ volts}.$$

Thus, at time $t = t_d$, the load voltage V_ℓ will jump from zero volts to

$$V_\ell = V_1 + \Gamma_R V_1 = (10.81 + 8.65) = 19.46 \text{ volts}.$$

Now, V_2 goes back to the input end and is reflected as V_3 where

$$V_3 = \Gamma_S V_2 = (-0.80)(8.65) = -6.93 \text{ volts}.$$

V_3 gets back to the line end after t_d seconds, i.e., at time $t = (t_d + 2t_d) = 3t_d$. This gives rise to

$$V_4 = 0.8 V_3$$
$$= 0.8(-6.93) = -5.55 \text{ volts}.$$

Hence, at $t = 3t_d^+$ (a time just after $3t_d$) the total load voltage is:

$$(V_1 + V_2 + V_3 + V_4) = (10.81 + 8.65 - 6.93 - 5.55)$$

$$= 6.98 \text{ volts.}$$

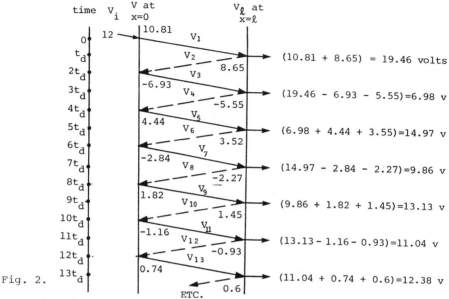

Fig. 2.

At $x = \ell$, each solid line and each dotted line add to get a total which is then summed with the previous (output) value of V_ℓ to give the current value of the output voltage.

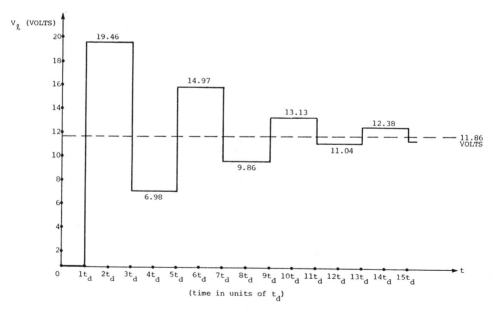

Fig. 3. Voltage wave form at $x=\ell$ due to reflection.

Continuing in this manner Fig. (2) shows the values while Fig. (3) gives the waveforms. The waveform oscillates back and forth about an asymptotic value. The limiting value is the same along the line and is found to be:

$$V_x \text{ (at all x and at } t=\infty) = V_i\left(\frac{R_L}{R_L + R_S}\right)$$

$$= 12\left(\frac{900}{900 + 11}\right)$$

$$= 11.86 \text{ volts.}$$

● **PROBLEM** 19-6

For the 50 ohm lossless transmission line system shown in Fig. (a), a transmission line is terminated in a capacitor and is fed from a 150 volt D.C. generator whose internal resistance is matched to the line. If the switch is closed at time t = 0, sketch the waves after t = 0.

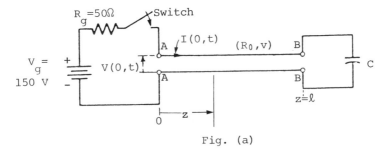

Fig. (a)

Solution: At the input end of the transmission line, i.e., at terminals A - A, the voltage wave is given by

$$V(0,t) = 75 \, u\left(t - \frac{0}{v}\right),$$

i.e., half of the voltage is energizing the transmission line and the other half is appearing across R_g.

After a delay of $\frac{z}{v}$, this wave travels a distance z, away from the input end A - A and is represented as

$$V(z,t) = V^+\left(t - \frac{z}{v}\right) = 75u\left(t - \frac{z}{v}\right)V \quad 0 < t < \frac{\ell}{v} \quad (1)$$

The load (capacitor) is at position $z = \ell$ and the voltage wave will consist of the forward and backward moving components $V^-\left(t + \frac{\ell}{v}\right)$ due to reflection at the load. Hence,

$$V(\ell,t) = V^+\left(t - \frac{\ell}{v}\right) + V^-\left(t + \frac{\ell}{v}\right). \quad (2)$$

If the current wave is written as

$$I(\ell,t) = \frac{V^+\left(t - \frac{\ell}{v}\right) - V^-\left(t + \frac{\ell}{v}\right)}{R_0}$$

Then
$$V(\ell,t) = 2V^+\left(t - \frac{\ell}{v}\right) - R_0 I(\ell,t). \quad (3)$$

The equivalent circuit can thus be represented as in Fig.(b), where the current-voltage relationship for the capacitor is
$$I(\ell,t) = C\frac{d}{dt}V(\ell,t) \quad (4)$$

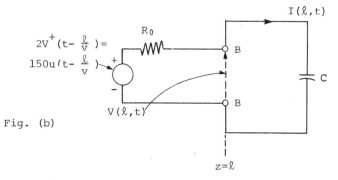

Fig. (b)

Using equations (1) and (4), and substituting these into equation (3), gives the following differential equation:
$$R_0 C \frac{d}{dt} V(\ell,t) + V(\ell,t) = 150\, u\left(t - \frac{\ell}{v}\right)$$

which simplifies to:
$$\frac{d}{dt} V(\ell,t) + \frac{1}{R_0 C} V(\ell,t) - \frac{150}{R_0 C} u\left(t - \frac{\ell}{v}\right) = 0$$

The solution of which is the voltage across the capacitor. Hence,
$$V(\ell,t) = \left[150\, u\left(t - \frac{\ell}{v}\right)\right]\left[1 - e^{-\left(t - \frac{\ell}{v}\right)/R_0 C}\right]$$

Using the above result in the original equation for the voltage wave, equation (2) gives the value of reflected voltage at $z = \ell$ as:
$$V^-\left(t + \frac{\ell}{v}\right) = V(\ell,t) - V^+\left(t - \frac{\ell}{v}\right)$$
$$= \left[150u\left(t - \frac{\ell}{v}\right)\right]\left[1 - e^{-\left(t - \frac{\ell}{v}\right)/R_0 C}\right] - 75u\left(t - \frac{\ell}{v}\right)$$
$$= 75u\left(t - \frac{\ell}{v}\right) - \left[150u\left(t - \frac{\ell}{v}\right)e^{-\left(t - \frac{\ell}{v}\right)/R_0 C}\right] \quad (5)$$

At any given location z, the reflected wave which is represented by equation (5), is delayed by $\left(\frac{\ell-z}{v}\right)$ so
$$V^-\left(t + \frac{z}{v}\right) = 75u\left(t + \frac{z}{v} - \frac{2\ell}{v}\right) - \left[150u\left(t + \frac{z}{v} - \frac{2\ell}{v}\right) \times e^{-\left[t + \left(\frac{z}{v}\right) - \left(\frac{2\ell}{v}\right)\right]/R_0 C}\right] \quad (6)$$

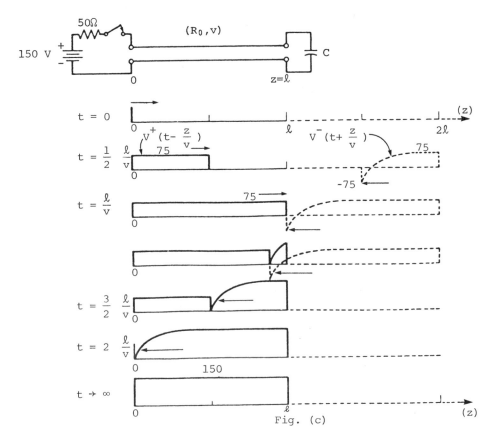

Hence, the total voltage on the line is the superposition of equations (1) and (7) as shown in Fig. (c).

● **PROBLEM 19-7**

A lossless 0.2km long, 50 ohm coaxial cable with $\varepsilon_r = 2.25$ is shown in the diagram below. Find the voltage and current waves on the line at time $t = 0$ (the time the source was switched on, so that $V_g(t) = 150u(t)$ volts).

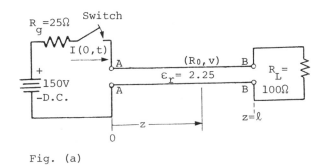

Fig. (a)

Solution: At the terminals A - A of the equivalent input circuit shown in Fig. (1),

$$V(0,t) = V^+\left(t - \frac{0}{v}\right) = 150u(t)\frac{50}{25+50} = 100u(t) \text{ Volts}$$

$$I(0,t) = I^+\left(t - \frac{0}{v}\right) = \frac{V^+\left(t - \frac{0}{v}\right)}{50} = 2u(t) \text{ Amps.}$$

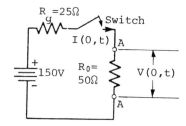

The velocity of propagation is given by

$$v = \frac{1}{\sqrt{\mu_0 \epsilon_0 \epsilon_r}} = \frac{c}{\sqrt{\epsilon_r}} = \frac{3 \times 10^8 \text{m/S}}{\sqrt{2.25}} = 2 \times 10^8 \text{m/s}$$

At any given point z along the line, the incident waves are delayed by $\frac{z}{v}$ seconds so that, for $t < \frac{\ell}{v}$,

$$V(z,t) = V_1^+\left(t - \frac{z}{v}\right) = 100u\left(t - \frac{z}{v}\right) \text{ Volts}$$

$$I(z,t) = I_1^+\left(t - \frac{z}{v}\right) = 2u\left(t - \frac{z}{v}\right) \text{ Amps}$$

Hence, the voltage and current waves will arrive at the load after a delay of

$$\frac{\ell}{v} = \frac{200\text{m}}{2 \times 10^8 \text{m/s}} = 1 \times 10^{-6} \text{sec(or 1 } \mu \text{ sec).}$$

At the load, the reflection coefficient is defined as

$$\Gamma(\ell) \equiv \frac{V^-\left(t + \frac{\ell}{v}\right)}{V^+\left(t - \frac{\ell}{v}\right)}$$

$$= \frac{R_L - R_0}{R_L + R_0}$$

$$= \frac{100 - 50}{100 + 50}$$

$$= \frac{50}{150}$$

$$= \frac{1}{3}$$

Thus, the reflected voltage at $z = \ell$ is found to be

$$V_1^-\left(t + \frac{\ell}{v}\right) = \Gamma(\ell)V_1^+\left(t - \frac{\ell}{v}\right) = 33.3u\left(t - \frac{\ell}{v}\right)$$

which is reflected towards the generator as

$$V_1^-\left(t + \frac{z}{v}\right) = 33.3u\left(t - \frac{\ell}{v} - \frac{\ell - z}{v}\right) = 33.3u\left(t + \frac{z}{v} - \frac{2\ell}{v}\right)$$

where $\frac{2\ell}{v} = 2\,\mu\text{ seconds}$.

Figure (3) shows that the reflected wave appears to originate from an "image position" $z = 2\ell$ and arrives at the load after 1 μ second and reaches the generator after 2 μ seconds.

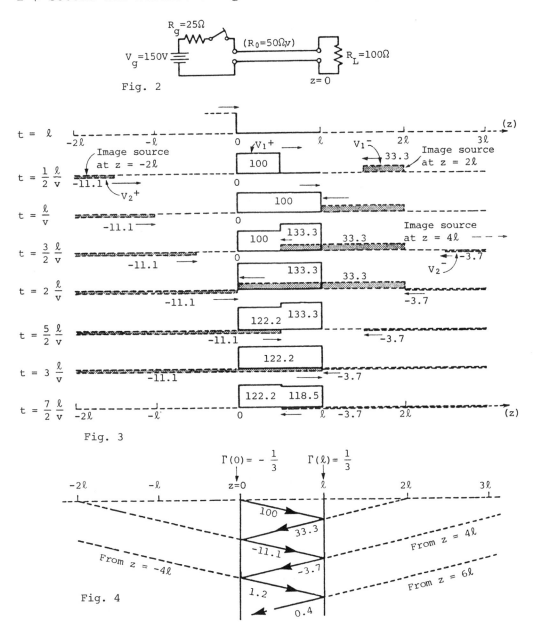

Figures. (2) Transmission-line system. (3) Equivalent input circuit. (4) Echo diagram.

The reflection coefficient at the generator is

$$\Gamma(0) \quad \frac{V_2^+\left(t - \frac{0}{v}\right)}{V^-\left(t + \frac{0}{v}\right)} = \frac{R_g - R_0}{R_g + R_0}$$

thus,

$$\Gamma(0) = \frac{25 - 50}{25 + 50} = -\frac{1}{3}$$

Therefore, the second reflection at the generator ($z = 0$) is given by

$$V_2^+\left(t - \frac{0}{v}\right) = \Gamma(0)V_1^-\left(t + \frac{0}{v}\right) = \left(-\frac{1}{3}\right)33.3u\left(t + \frac{0}{v} - \frac{2\ell}{v}\right)$$

$$= -11.1u\left(t - \frac{2\ell}{v}\right) \text{ Volts,}$$

which is delayed by $\frac{z}{v}$ seconds as it moves to the load so that

$$V_2^+\left(t - \frac{z}{v}\right) = -11.1u\left(t - \frac{z}{v} - \frac{2\ell}{v}\right) \text{ Volts.} \quad (1)$$

Repeating the above procedure, equation (1) will produce a third echo at the load (which appears to originate from $z = 4\ell$). This process will continue indefinitely approaching the steady state value of 120 volts. This is found by not considering the drop due to R_0,

i.e.,

$$V_{L(S.S.)} = \left(\frac{100}{100 + 25}\right)(150) = 120.$$

The total voltage wave at any time is given by

$$V(z,t) = V^+\left(t - \frac{z}{v}\right) + V^-\left(t + \frac{z}{v}\right) \text{ Volts.}$$

The sum of all positive and negative travelling waves can be seen in the diagram of Fig. (4), and is expressed as

$$V(z,t) = V^+\left(t - \frac{z}{v}\right) + V^-\left(t + \frac{z}{v}\right)$$

$$= \left[V_1^+\left(t - \frac{z}{v}\right) + V_2^+\left(t - \frac{z}{v}\right) + \ldots\right]$$

$$+ \left[V_1^-\left(t + \frac{z}{v}\right) + V_2^-\left(t + \frac{z}{v}\right) + \ldots\right]$$

$$= \left[100u\left(t - \frac{z}{v}\right) - 11.1u\left(t - \frac{z}{v} - \frac{2\ell}{v}\right) + 1.2u\left(t - \frac{z}{v} - \frac{4\ell}{v}\right) - \ldots\right]$$

$$+ \left[33.3u\left(t + \frac{z}{v} - \frac{2\ell}{v}\right) - 3.7u\left(t + \frac{z}{v} - \frac{4\ell}{v}\right)\right.$$

$$\left. + 0.4u\left(t + \frac{z}{v} - \frac{6\ell}{v}\right) - \ldots\right]$$

CHARACTERISTIC IMPEDANCE AND LINE INPUT IMPEDANCE

• **PROBLEM** 19-8

A screened telephone cable has the following parameters at 10 KHz:

$$L = 0.7 \times 10^{-3} \, H/Km$$

$$C = 0.05 \times 10^{-6} \, F/Km$$

$$R = 28 \, ohm/Km$$

$$G = 1 \times 10^{-6} \, mho/Km.$$

Determine the characteristic impedance, phase constant, and attenuation constant for this cable.

Solution: The characteristic impedance of the cable is given by:

$$Z_0 = \sqrt{\frac{(R + j\omega L)}{(G + j\omega C)}}$$

where R, L, G, and C are the distributed line parameters for resistance, inductance, conductance, and capacitance respectively.

At 10KHz, $\omega = 2\pi f = 6.28 \times 10^4$ rad/sec

Thus,

$$Z_0 = \sqrt{\frac{[28 + j(6.28 \times 10^4)(0.7 \times 10^{-3})]}{[1 \times 10^{-6} + j(6.28 \times 10^4)(0.05 \times 10^{-6})]}}$$

$$= \sqrt{\frac{(28 + j\, 43.98)}{(1 \times 10^{-6} + j\, 3.14 \times 10^{-3})}}$$

$$= \sqrt{\frac{52.14 \underline{/57.52^0}}{3.14 \times 10^{-3} \underline{/89.98^0}}}$$

$$= \sqrt{1.66 \times 10^4 \underline{/-32.46^0}}$$

$$= 128.83 \underline{/-16.23^0} \, \Omega$$

$$Z_0 = (123.69 - j36.01) \Omega$$

The propagation constant, γ, is given by

$$\gamma = \sqrt{(R + j\omega L)(G + j\omega C)}$$

$$= \sqrt{[28 + j(6.28 \times 10^4)(0.7 \times 10^{-3})][1 \times 10^{-6} + j(6.28 \times 10^4)(0.05 \times 10^{-6})]}$$

$$= \sqrt{(52.12\underline{/57.50^0})(3.14\times10^{-3}\underline{/89.98^0})}$$

$$= \sqrt{0.1637\underline{/147.48^0}}$$

$$= 0.4\underline{/73.74^0}$$

$$= 0.112 + j0.384$$

$$= \alpha + j\beta$$

where α is the attenuation constant and β is the phase constant.

Thus, $\quad \alpha = 0.112$ Neper/Km

$$= 0.112 \text{ Np/Km} \times 8.686 \text{ dB/Np} = 0.9728 \text{ dB/Km}$$

and $\quad \beta = 0.384$ rad/sec

● **PROBLEM 19-9**

Consider the following parameters of an air-spaced telephone line at 10 kHz:

$$L = 2.4 \text{ mH/km}$$

$$C = 0.005 \text{ μF/km}$$

$$R = 28 \Omega/\text{km}$$

$$G = 3 \text{ μmho/km}$$

(a) Find the line characteristic impedance, the attenuation for a 10kHz signal in dB/km and the wavelength

(b) If the line is to be loaded with inductors at 2km intervals so that it provides distortionless transmission, calculate the value for the inductors and the new value for the characteristic impedance, assuming that the inductors have negligible resistance.

Solution: (a) From the distributed line parameters given, the characteristic impedance is

$$Z_0 = \sqrt{\frac{(R+j\omega L)}{(G+j\omega C)}} \quad \begin{pmatrix} \text{where at 10 kHz} \\ \omega = 2\pi f = 6.28 \times 10^4 \end{pmatrix}$$

$$= \sqrt{\frac{28 + j150.80}{3\times10^{-6}+j(3.14\times10^{-4})}}$$

$$= \sqrt{\frac{153.37\underline{/79.48^0}}{3.14\times10^{-4}\underline{/89.45^0}}}$$

$$= \sqrt{4.88\times10^5\underline{/-9.97^0}}$$

$$= 698.57\underline{/-4.99^0}$$

$$= (695.9 - j60.7)\Omega$$

Similarly, the propagation constant in terms of distributed parameters is given as follows:

$$\gamma = (\alpha + j\beta) = \sqrt{(R+j\omega L)(G+j\omega C)}$$

$$= \sqrt{[28+j150.80][3\times10^{-6}+j(3.14\times10^{-4})]}$$

$$= \sqrt{[153.37\underline{/79.48^0}][3.14\times10^{-4}\underline{/89.45^0}]}$$

$$= \sqrt{4.82\times10^{-2}\underline{/168.93}}$$

$$= 0.22\underline{/84.47}$$

$$= 0.212 + j0.218$$

Therefore, separating and equating real and imaginary parts from both sides results in

$$\alpha = 0.212 \text{ Neper/km} = 0.212 \times 8.686\text{dB} = 0.184 \text{ dB/km}$$

$$\beta = 0.218 \text{ rad/km} = \frac{2\pi}{\lambda}$$

Thus, the wavelength $\lambda = \frac{2\pi}{0.218} = 28.82$ km

(b) The necessary condition for distortionless transmission is that the ratios (L/R) and (C/G) must be equal. Therefore,

$$L = \frac{RC}{G} = \frac{(28)(0.005\times10^{-6})}{3\times10^{-6}} = 46.67 \times 10^{-3} \text{H/km}$$

Hence, the extra inductance needed is $(46.67 - 2.4) \times 10^{-3}$ = 44.27 mH/km.

For the loaded line, the characteristic impedance becomes

$$Z_0 = \sqrt{\frac{(R+j\omega L)}{(G+j\omega C)}} = \sqrt{\frac{R(1+j\omega L/R)}{G(1+j\omega C/G)}} = \sqrt{\frac{R}{G}} = \sqrt{\frac{L}{C}} = \sqrt{\left(\frac{28}{3\times10^{-6}}\right)}$$

$$= 3055 \, \Omega$$

• PROBLEM 19-10

(A) Derive the expression for the characteristic impedance of the symmetric T-section attenuator shown in the figure below. If $R_1 = 50\Omega$ and $R_2 = 100\Omega$, find the characteristic impedance R_0.

(B) Derive the attenuation per section in decibels and calculate its value for R_1 and R_2 of part (A).

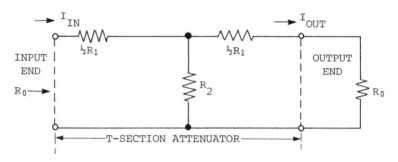

Solution: (A) Assume that the T-section attenuator is terminated in its characteristic impedance R_0 (which is a pure resistance). This is done so that the input impedance due to the unloaded T-section can be calculated. The input impedance will be equal to the overall resistance of the network in the diagram to the right of the input end. Combining these resistors results in

$$R_0 = \frac{1}{2}R_1 + \frac{(\frac{1}{2}R_1 + R_0)R_2}{\frac{1}{2}R_1 + R_0 + R_2} \; \Omega$$

Hence, multiplying both sides by $(\frac{1}{2}R_1 + R_0 + R_2)$ and simplifying:

$$\tfrac{1}{2}R_1 R_0 + R_0^2 + R_2 R_0 = \tfrac{1}{4}R_1^2 + \tfrac{1}{2}R_1 R_0 + R_1 R_2 + R_0 R_2$$

Thus,

$$R_0^2 = \tfrac{1}{4}R_1^2 + R_1 R_2$$

$$= R_1 R_2 \left(1 + \frac{R_1}{4R_2}\right)$$

Therefore,

$$R_0 = \sqrt{\left[R_1 R_2 \left(1 + \frac{R_1}{4R_2}\right)\right]}.$$

Now, $R_1 = 50\Omega$ and $R_2 = 100\Omega$. Hence, the characteristic impedance is

$$R_0 = \sqrt{\left[50(100)\left(1 + \frac{50}{4(100)}\right)\right]}$$

$$= \sqrt{\left[5000\left(1 + \tfrac{1}{8}\right)\right]}$$

$$= \sqrt{5625}$$

$$= 75 \text{ ohms.}$$

(B) The attenuation in decibels is defined in terms of a power ratio as follows:

$$\text{attenuation in dB} = 10 \log_{10}\left(\frac{P_{in}}{P_{out}}\right) = 10 \log_{10}\left(\frac{I_{in}^2 R_{in}}{I_{out}^2 R_{out}}\right)$$

$$= 20 \log_{10}\left(\frac{I_{in}}{I_{out}}\right) + 10 \log_{10}\left(\frac{R_{in}}{R_{out}}\right)$$

where P_{in} = power delivered to the input end

P_{out} = power measured at the output end

I_{in} = current into the T-section attenuator

I_{out} = current out of the T-section attenuator

R_{in} = resistance looking into the input end of the T-section

R_{out} = resistance looking into the output end of the T-section

and $R_{in} = R_{out} = R_0$ because the junction is symmetric.

Hence,

$$10 \log_{10}\left(\frac{R_{in}}{R_{out}}\right) = 0$$

From the figure:

$$I_{out} = \left(\frac{R_2}{R_2 + \frac{1}{2}R_1 + R_0}\right)I_{in}$$

Thus,

$$\frac{I_{in}}{I_{out}} = \frac{R_2 + \frac{1}{2}R_1 + R_0}{R_2}$$

$$= 1 + \frac{1}{2}\left(\frac{R_1}{R_2}\right) + \frac{R_0}{R_2}$$

Substituting the value of R_0 as found in part (A):

$$\frac{I_{in}}{I_{out}} = 1 + \frac{1}{2}\left(\frac{R_1}{R_2}\right) + \frac{1}{R_2}\sqrt{[R_1 R_2 + \tfrac{1}{4}R_1^2]}$$

$$= 1 + \frac{1}{2}\left(\frac{R_1}{R_2}\right) + \sqrt{\left[\frac{R_1}{R_2} + \frac{1}{4}\left(\frac{R_1}{R_2}\right)^2\right]}.$$

Therefore, the attenuation per section is equal to

$$20 \log_{10}\left\{1 + \frac{1}{2}\left(\frac{R_1}{R_2}\right) + \sqrt{\left[\frac{R_1}{R_2} + \frac{1}{4}\left(\frac{R_1}{R_2}\right)^2\right]}\right\} \text{ dB.}$$

For $R_1 = 50\Omega$ and $R_2 = 100\Omega$, the attenuation is:

$$\text{Attenuation} = 20 \log_{10} \left\{ 1 + \frac{1}{2}\left(\frac{50}{100}\right) + \sqrt{\left[\frac{50}{100} + \frac{1}{4}\left(\frac{50}{100}\right)^2\right]} \right\} \text{dB}$$

$$= 20 \log_{10} \{1 + \frac{1}{4} + \sqrt{9/16}\} \text{ dB}$$

$$= 20 \log_{10} \{1 + \frac{1}{4} + \frac{3}{4}\} \text{ dB}$$

$$= 20 \log_{10} \{2\} \text{dB}$$

$$= 10 \log_{10} \{4\} \text{dB}$$

$$= 6.021 \text{ dB}.$$

Thus, this T-section attenuator will perfectly match a 75Ω load to another 75Ω load without affecting the overall loading of the power source while the power at the output end is reduced by 6.021 dB (or by a factor of 4).

● **PROBLEM 19-11**

For the coaxial cable shown in Fig. (1), of the attenuation is zero and $\varepsilon = \varepsilon_0 \varepsilon_r$ where $\varepsilon_r = 2$ at a frequency of 29MHz. Find:

(A) The characteristic impedance $\hat{Z}_0$, the phase constant β, and the phase-velocity v_p of this transmission line.

(B) If it is found that at this frequency the dielectric exhibits a loss tangent of

$$\frac{\varepsilon''}{\varepsilon'} = 200 \times 10^{-6},$$

find the new values of Z_0, β, and v_p. Is the attenuation still zero?

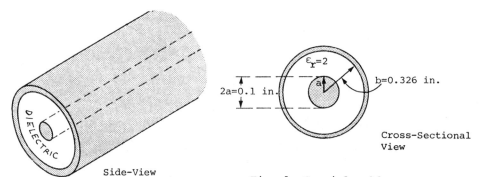

Fig. 1 Coaxial cable

Solution: (A) In a coaxial cable, the transverse electromagnetic mode propagates at the speed of light for the dielectric with zero cut-off frequency. The characteristic impedance for the coaxial cable can be found from the electromagnetic theory to be

$$\hat{Z}_0 = \frac{\hat{\eta}_{TEM}}{2\pi} \cdot \ln\left[\frac{b}{a}\right]$$

where

$$\hat{\eta}_{TEM} \equiv \frac{\gamma}{j\omega\varepsilon} = \frac{j\omega\sqrt{\mu\varepsilon}}{j\omega\varepsilon} = \sqrt{\frac{\mu}{\varepsilon}}$$

γ is the propagation constant $= \alpha + j\beta$

$\mu = \mu_0\mu_r = 4\pi \times 10^{-7}$ H/m with $\mu_r = 1$ for non-magnetic materials

$= $ magnetic permeability

$\varepsilon = \varepsilon_0\varepsilon_r$ is the electric permitivity

where
$$\varepsilon_0 = \frac{1}{36\pi} \times 10^{-9} \text{ F/m}.$$

If the attenuation, α, is negligible, then the propagation constant is:

$$\gamma = j\beta = j\omega\sqrt{\mu\varepsilon}.$$

Hence,

$$\hat{\eta} = \sqrt{\frac{\mu_0}{2\varepsilon_0}} = \frac{120\pi}{\sqrt{2}} = 266.57\Omega.$$

Thus,

$$\hat{Z}_0 = \frac{\hat{\eta}}{2\pi}\ln\left(\frac{b}{a}\right)$$

$$= \frac{266.57}{2\pi}\ln\left(\frac{0.326}{0.1}\right)$$

$$= 50.136\Omega$$

$$\cong 50\Omega$$

The phase constant β is found to be

$$\beta = \omega\sqrt{\mu_0(2\varepsilon_0)} = \frac{2\pi(20\times 10^6)\sqrt{2}}{3\times 10^8}$$

$$= 0.592 \text{ rad/meter}$$

(where $C = \frac{1}{\sqrt{\mu\varepsilon_0}} = 3\times 10^8$ m/s).

Finally, the phase velocity can be found using relation

$$v_p = \frac{\omega}{\beta} = \frac{\omega}{\omega\sqrt{\mu 2\varepsilon_0}}$$

$$= \frac{1}{\sqrt{\mu\varepsilon_0(2)}} = \frac{c}{\sqrt{2}}$$

$$= 2.121 \times 10^8 \text{ m/sec}.$$

which is 70.71% of the speed of light in free space. Note that β and v_p are independent of the cable's physical dimensions.

(B) With the loss tangent as given, $\hat{\eta}$ is now

$$\hat{\eta} = \frac{\sqrt{\frac{\mu}{\varepsilon}}}{\left[1 + \left(\frac{\varepsilon''}{\varepsilon'}\right)^2\right]^{\frac{1}{4}}} e^{j(\frac{1}{2})\arctan(\varepsilon''/\varepsilon')}$$

$$= \eta e^{j\theta} \Omega$$

Hence,

$$\hat{\eta} = \frac{120\pi/\sqrt{2}}{\left[1 + (200\times 10^{-6})^2\right]^{\frac{1}{4}}} e^{j(\frac{1}{2})\arctan(200\times 10^{-6})}$$

$$= 266.57 \, e^{j(0.0001)} \, \Omega$$

$$\cong 266.57 \quad \text{(essentially unchanged from part (a)).}$$

Therefore,

$$\hat{Z}_0 \cong 50 e^{j(0.001)} \Omega \cong 50\Omega.$$

The constants α and β are evaluated using the small loss approximation formula as follows:

$$\beta \cong \omega\sqrt{\mu\varepsilon} \left[1 + \frac{1}{8}\left(\frac{\varepsilon''}{\varepsilon'}\right)^2\right]$$

$$\alpha \cong \frac{\omega\sqrt{\mu\varepsilon}}{2}\left(\frac{\varepsilon''}{\varepsilon'}\right)$$

Therefore,

$$\beta = 0.592\left[1 + \frac{(0.0002)^2}{8}\right]$$

$$\cong 0.592 \text{ rad/meter}.$$

$$\alpha = \frac{0.592}{2}(200\times 10^{-6}) = 59.2 \times 10^{-6} \text{ Np/m}.$$

This value of α implies that the wave decays to e^{-1} of its original value at a distance $d = \alpha^{-1} = \left(\frac{1}{59.2\times 10^{-6}}\right)$ or 16.8 km, which is 10.4 miles. Since β is essentially unchanged Np is again 2.121×10^8 m/sec in the lossy dielectric. Although α is extremely small in certain applications such as TV signal transmission, it is not negligible always.

• **PROBLEM 19-12**

Starting from the voltage and current equations for a long transmission line

$$V = V_{IN} \cosh\gamma\ell - I_{IN}Z_0 \sinh\gamma\ell \qquad (1)$$

$$I = I_{IN} \cosh\gamma\ell - \frac{V_{IN}}{Z_0} \sinh\gamma\ell \qquad (2)$$

where γ is the propagation constant

   ℓ is the length of the line

   Z_0 is the characteristic impedance of the line

   the subscript 'IN' denotes source, or input values

(A) Determine an expression for the input impedance in terms of the characteristic impedance Z_0, the load impedance Z_L, and Z_{oc} (the input impedance when Z_L is replaced by an open circuit).

(B) If $Z_0 = 70\Omega$, $Z_L = 100\underline{/45°}\,\Omega$, $Z_{oc} = 100\underline{/-45°}\,\Omega$ and $V_{IN} = 100$ volts rms find the power input to the transmission line.

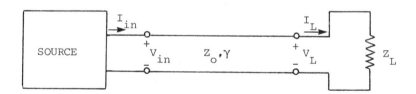

Solution: (A) From the given information, the voltage and current values at the load end can be expressed as:

$$V_L = V_{IN} \cosh\gamma\ell - I_{IN}Z_0 \sinh\gamma\ell \qquad (3)$$

$$I_L = I_{IN} \cosh\gamma\ell - \frac{V_{IN}}{Z_0} \sinh\gamma\ell \qquad (4)$$

where $V_L = I_L Z_L$ at the load.

Substituting the value of V_L and equation (4) into equation (3) gives the input impedance of the line as:

$$Z_{IN} = \frac{V_{IN}}{I_{IN}} = Z_0 \left[\frac{Z_L \cosh\gamma\ell + Z_0 \sinh\gamma\ell}{Z_0 \cosh\gamma\ell + Z_L \sinh\gamma\ell} \right] \qquad (5)$$

If Z_L is replaced by an open circuit (i.e., $Z_L = \infty$), equation (5) reduces to

$$Z_{IN} = Z_{oc} = Z_0\left(\frac{\cosh\gamma\ell}{\sinh\gamma\ell}\right) = Z_0 \coth\gamma\ell$$

Solving for $\cosh\gamma\ell$,

$$\frac{Z_{oc}}{Z_0} = \sinh\gamma\ell = \cosh\gamma\ell \tag{6}$$

Finally, if equation (6) is substituted into equation (5), the desired expression for Z_{IN} results:

$$Z_{IN} = Z_0 \left[\frac{\frac{Z_{oc}}{Z_0}\sinh\gamma\ell + \frac{Z_0}{Z_L}\sinh\gamma\ell}{\frac{Z_{oc}Z_0}{Z_L Z_0}\sinh\gamma\ell + \sinh\gamma\ell} \right]$$

$$= \left[\frac{Z_{oc} + \frac{Z_0^2}{Z_L}}{\frac{Z_{oc}}{Z_L} + 1} \right] \left[\frac{\sinh\gamma\ell}{\sinh\gamma\ell}\right]$$

Hence,

$$Z_{IN} = \left[\frac{Z_{oc}Z_L + Z_{oc}^2}{Z_{oc} + Z_L} \right]. \tag{7}$$

(B) Using the values given in the problem with expression (7) from part (a) gives

$$Z_{IN} = \left[\frac{(100\underline{/45°})(100\underline{/-45°}) + (70)^2}{(100\underline{/-45°}) + (100\underline{/45°})} \right]$$

$$= \left[\frac{(10\times 10^3 \underline{/0°}) + (70)^2}{\left[\frac{100}{\sqrt{2}}(1-j)\right] + \left[\frac{100}{\sqrt{2}}(1+j)\right]} \right]$$

$$= \frac{10 \times 10^3 + 4.9 \times 10^3}{\left(\frac{200}{\sqrt{2}}\right)}$$

$$= \frac{(\sqrt{2})14.9 \times 10^3}{200}$$

Thus,
$$Z_{IN} = 105 \text{ ohms}.$$

The input power is given by

$$P_{IN} = \frac{V_{IN(rms)}^2}{Z_{IN}} = \frac{(100)^2}{105} = 95.24 \text{ watts}.$$

• **PROBLEM 19-13**

A 100-mile telephone line has a series resistance of 4 ohms/mile, an inductance of 3 mh/mile, a leakage conductance of 1 μmho/mile, and a shunt capacitance of 0.015 μf/mile, at an angular frequency ω = 5000. At the sending end there is a generator supplying 100 volts peak, at 5000 radians per second, in series with a resistance of 300 ohms. The load at the receiving end consists of 200-ohm resistor. Find the voltage and current as functions of z, and calculate their values at the midpoint of the line.

Solution: $Z = (R + j\omega L) = 4 + j15 = 15.53 \ \underline{/75.1^0}$

$Y = (G + j\omega C) = (1 + j75)10^{-6} = 75 \times 10^{-6} \ \underline{/82.15^0}$

$\gamma = \sqrt{ZY} = 0.0342 \ \underline{/82.15^0} = 0.00466 + j0.0338$

$Z_0 = \sqrt{Z/Y} = 455 \ \underline{/-7.1^0} = 452 - j56.1$

The load impedance $Z_L = 200$. The reflection coefficient

$\Gamma = \dfrac{Z_L - Z_0}{Z_L + Z_0}$ is found to be $-0.396 \ \underline{/-7.6^0}$.

Calculation of the input impedance Z_i requires evaluation of $e^{-2\gamma \ell}$. As $\ell = 100$, this exponential becomes

$e^{-2\gamma \ell} = e^{-0.932} e^{-j6.76} = 0.394 \ \underline{/-6.76^0} = 0.394 \ \underline{/-27.3^0}$

The product $\Gamma e^{-2\gamma \ell}$ becomes $-0.128 + j0.0890$. Calculation of Z_i yields $Z_i = 353 \ \underline{/3.3^0} = 353 + j20.1$.

The sending-end current $\bar{I}_S$ is $\bar{V}_g / (Z_g + Z_i)$. The impedance Z_g is 300 ohms resistive, and $\bar{V}_g = 100 \ \underline{/0^0}$. Therefore, $\bar{I}_S = 100/(300 + 353 + j20.1) = 0.153 \ \underline{/-1.8^0}$. The sending-end voltage $\bar{V}_S$ is $\bar{I}_S Z_i$, or $54 \ \underline{/1.5^0}$, or $54 + j1.41$.

The coefficients $\bar{V}_1$ and $\bar{V}_2$ are

$\bar{V}_1 = \tfrac{1}{2}(\bar{V}_S + \bar{I}_S Z_0) = 61.4 \ \underline{/-4.35^0}$

$\bar{V}_2 = \tfrac{1}{2}(\bar{V}_S - \bar{I}_S Z_0) = -9.58 \ \underline{/-39.35^0})$

Consequently, the line voltage $\bar{V}$ and the line current $\bar{I}$, as functions of z, are

$\bar{V} = \left[(61.4 \ \underline{/-4.35^0})e^{-\gamma z} - (9.58 \ \underline{/-39.35^0})e^{\gamma z}\right]$ volts

$\bar{I} = \left[(0.135 \ \underline{/2.75^0})e^{-\gamma z} + (0.021 \ \underline{/-32.3^0})e^{\gamma z}\right]$ Amps with $\gamma = 0.00466 + j0.0338 = \alpha + j\beta$.

At the midpoint of the line, z is 50, and γz is $0.233 + j1.69$. The exponential of $-\gamma z$ becomes $e^{-0.233} e^{-j1.69}$, or $0.791 \underline{/-96.8°}$. The exponential of γz becomes $1.262 \underline{/96.8°}$. Therefore,

$$\overline{V} = \left[(61.4 \underline{/-4.35°})(0.791 \underline{/-96.8°}) - (9.58 \underline{/-39.35°})(1.262 \underline{/96.8°})\right] \text{volts.}$$

Evaluation gives $\overline{V} = 60.0 \underline{/-105.4°}$. This is the phasor voltage, maximum value, at $z = 50$ miles. Similarly, the current is found to be $\overline{I} = 0.0829 \underline{/-87.4°}$. The instantaneous voltage and current at the midway point are

$$v = 60 \sin(5000t - 105.4°) \text{ volts}$$

$$i = 0.0829 \sin(5000t - 87.4°) \text{ Amps}$$

The complex impedance where each phasor is multiplied by $e^{j\omega t}$, then the real component is used at the midway point, looking toward the load, is the ratio of $\overline{V}$ to $\overline{I}$. This gives an impedance of $725 \underline{/-18°}$ ohms.

The voltage and current are readily determined at any point z, although the calculations are tedious. The values of $\overline{V}$ and $\overline{I}$ at the sending end, at the 50-mile point, and at the receiving end are given below.

$\overline{V}_S = 54 \underline{/1.5°}$ volts $\overline{I}_S = 0.153 \underline{/-1.8°}$ Amps

$\overline{V}_{50} = 60 \underline{/-105.4°}$ volts $\overline{I}_{50} = 0.0829 \underline{/-87.4°}$ Amps

$\overline{V}_L = 23.4 \underline{/-193°}$ volts $\overline{I}_L = 0.117 \underline{/-193°}$ Amps

If the line were terminated in its characteristic impedance, there would be no reflected wave, and the voltage and current would decrease exponentially with respect to the axial coordinate z. However, in this problem there is reflection. The voltage rises and then falls as z increases, and the current falls and then rises. The sum of the incident and reflected waves results in this standing-wave effect.

● **PROBLEM 19-14**

A section of a low-loss transmission line is 0.40 wavelength long and is terminated in a short circuit. Its characteristic impedance is 73 ohms. The frequency of operation is 200 MHz. Determine the input reactance of the line section, and the value of inductance or capacitance to which it is equivalent.

Solution: In general, for a transmission line, the input impedance is given by:

$$Z_{in} = Z_0 \left[\frac{Z_L + Z_0 \tanh(\gamma \ell)}{Z_0 + Z_L \tanh(\gamma \ell)} \right]$$

where
$$\gamma = (\alpha + j\beta) = \text{the propagation constant}$$

For a short circuit, $Z_L = 0$ and since the line is low-loss the attenuation $[\alpha = 0]$.

The equation reduces to
$$Z_{in} = Z_0 \tanh(\beta \ell).$$

Making use of the relation $\tanh(\beta \ell) = j\tan(\beta \ell)$ gives
$$Z_{in} = jZ_0 \tan(\beta \ell) = jZ_0 \tan\left(\frac{2\pi}{\lambda}\right)\ell$$

$$Z_{in} = j(73)\tan\left[\frac{2\pi}{\lambda}\right](0.4\lambda)$$

$$= j73 \tan(2.51 \text{ rad})$$

$$= -j53.4 \text{ ohms}$$

This is a capacitive reactance, and the equivalent value of capacitance, C_{in}, is obtained when solving the equation
$$Z_{in} = \frac{1}{j\omega C_{in}} = -j\left(\frac{1}{\omega C_{in}}\right)$$

Therefore,
$$53.4 = \frac{1}{\omega C_{in}}$$

or
$$C_{in} = \frac{1}{2\pi \times 200 \times 10^6 \times 53.4}$$

$$= 14.9 \times 10^{-12} \text{ Farads}$$

● **PROBLEM** 19-15

A ten kilometer long telephone line has the following constants per loop kilometer:

R = 196 ohms

L = 7.1 mH

C = 0.09 μF

G = 0.

For an input of 10 volts at frequency $f = \frac{5 \times 10^3}{2\pi}$ Hz,

(a) Derive an expression for the current at the far end of the transmission line when the line is terminated by a short circuit. Assume the general equations:

$$V_x = V_S \cosh\gamma x - (I_S Z_0)\sinh\gamma x$$

$$I_x = I_S \cosh\gamma x - (V_S/Z_0)\sinh\gamma x$$

(b) Find the value of the characteristic impedance Z_0 of the line.

(c) Find the value of the propagation constant γ (show the attenuation and phase constant explicitly).

(d) Find the magnitude and phase of the current at the short circuit.

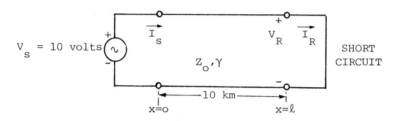

Solution: (A) As seen in the diagram, let ℓ be the line length. At the far end of the line we have

$$x = \ell = 10 \text{ km}$$

$$V = 0 \text{ volts}$$

and let $I_x = I_R$.

The general equations now become:

$$0 = V_S \cosh\gamma\ell - I_S Z_0 \sinh\gamma\ell \quad (1)$$

$$I_R = I_S \cosh\gamma\ell - (V_S/Z_0)\sinh\gamma\ell \quad (2)$$

Rearranging equation (1) gives

$$I_S = \frac{V_S \cosh\gamma\ell}{Z_0 \sinh\gamma\ell} \quad (3)$$

Substituting equation (3) into equation (2)

$$I_R = \frac{V_S}{Z_0}\left(\frac{\cosh\gamma\ell}{\sinh\gamma\ell}\right)\cosh\gamma\ell - \frac{V_S}{Z_0}\sinh\gamma\ell\left(\frac{\sinh\gamma\ell}{\sinh\gamma\ell}\right)$$

$$= \frac{V_S}{Z_0}\left[\frac{\cosh^2\gamma\ell - \sinh^2\ell\omega}{\sinh\gamma\ell}\right]$$

Thus, the expression for the current at the receiving end is

$$I_R = \frac{V_S}{Z \sinh \gamma \ell}, \text{ since } \cosh^2 \gamma \ell - \sinh^2 \gamma \ell = 1$$

(B) To calculate Z_0, the characteristic impedance:

$$Z_0 = \sqrt{\frac{R + j\omega L}{G + j\omega C}}$$

From the given information,

$R = 196$ ohms

$f = \frac{5 \times 10^3}{2\pi}$ Hz, hence $\omega = 2\pi f = 5 \times 10^3$ rad

$L = 7.1$ mH, hence, $\omega L = (5 \times 10^3)(7.1 \times 10^{-3}) = 35.5$

$C = 0.09 \mu F$, hence, $\omega C = (5 \times 10^3)(0.09 \times 10^{-6}) = 4.5 \times 10^{-4}$

Thus,

$$Z_0 = \sqrt{\frac{196 + j35.5}{j4.5 \times 10^{-4}}}$$

$$= 10^2 \sqrt{7.9 - j43.5}$$

$$= 10^2 \sqrt{44.2 \underline{/80^0}}$$

$$= 10^2 \sqrt{44.2} \underline{/\tfrac{1}{2} \times 80^0}$$

$$\cong 10^2 (6.65) \underline{/40^0}$$

$$\cong 665 \underline{/40^0} \; \Omega$$

(C) To calculate γ, the propagation constant:

$$\gamma = \sqrt{(R + j\omega L)(G + j\omega C)}$$

$$= \sqrt{(196 + j35.5)(j4.5 \times 10^{-4})}$$

$$= 10^{-2} \sqrt{(-160 + j882)}$$

$$= 10^{-2} \sqrt{897 \underline{/180 - 80^0}}$$

$$= 10^{-2} \times 29.95 \underline{/50^0}$$

$$\cong 0.3 \underline{/50^0}$$

Since $\gamma = \alpha + j\beta$

Hence, $\alpha \cong 0.193$ Np/km = attenuation factor

and $\beta \cong 0.23$ radian/km = phase constant

(D) First, $\sinh \gamma \ell$ must be found. Since $\ell = 10$ km, therefore $\gamma \ell = 10(0.3 \underline{/50^0}) \cong 3 \underline{/50^0}$, hence, $\gamma \ell \cong 1.93 + j2.3$

Now, $\gamma\ell = \alpha\ell + j\beta\ell$, therefore

$$\sinh\gamma\ell = \sinh(\alpha\ell + j\beta\ell)$$
$$= (\sinh \alpha\ell)\cos\beta\ell + j(\cosh \alpha\ell)\sin \beta\ell$$
$$= (\sinh 1.93)\cos 2.3 + j(\cosh 1.93)\sin 2.3$$

RADIANS TO DEGREES AND DECIMALS

Radians	Degrees	Radians	Degrees	Radians	Degrees	Radians	Degrees
1	57.2958	0.1	5.7296	0.01	0.5370	0.001	0.0573
2	114.5916	.2	11.4592	.02	1.1459	.002	.1146
3	171.8873	.3	17.4592	.03	1.7189	.003	.1719
4	229.1831	.4	22.9183	.04	2.2918	.004	.2292
5	286.4789	.5	28.6479	.05	2.8648	.005	.2865
6	343.7747	.6	34.3775	.06	3.4377	.006	.3438
7	401.0705	.7	40.1070	.07	4.0107	.007	.4011
8	458.3662	.8	45.8366	.08	4.5837	.008	.4584
9	515.6620	.9	51.5662	.09	5.1566	.009	.5157
10	572.9578	1.0	57.2958	.10	5.7296	.010	.5730

TABLE 1

The value of $\beta\ell$ is in radians and from table 1

$$2.3 \text{ rad} = 131.8°$$

Since $e^{1.93} = 6.89$.

Therefore $e^{-1.93} = \dfrac{1}{e^{1.93}} = \dfrac{1}{6.89} = 0.145$.

Now, $\sinh(1.93) = \dfrac{1}{2}\left[e^{1.93} - e^{-1.93}\right] = 3.372$

and $\cosh(1.93) = \dfrac{1}{2}\left[e^{1.93} + e^{-1.93}\right] = 3.517$

Therefore,

$$\sinh\gamma\ell = 3.372(\cos 131.8°) + j3.517(\sin 131.8°)$$
$$= 3.372(-0.665) + j3.517(0.7455)$$
$$= -2.245 + j2.625$$
$$= 3.45\underline{/130.6°}$$

Finally, taking V_S as the reference vector,

$$I_R = \dfrac{10\underline{/0°}}{(665\underline{/40°})(3.45\underline{/130.6°})} \times 1000\text{mA}$$
$$\cong 4.36\underline{/90.6°}\text{mA}$$

• **PROBLEM 19-16**

A $\frac{5\lambda}{4}$ long transmission line with input impedance of 50Ω is connected to another lossless transmission line of the same length but of input impedance Z_0 = 75Ω. This combination is terminated in a pure resistance as shown in the figure. Calculate the input impedance of first line due to such combination.

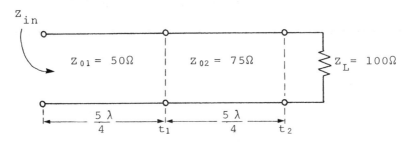

Fig. 1.

Solution: The normalized input impedance of any low-loss section of transmission line of an odd-number of quarter-wavelength long is the reciprocal of the normalized terminal load impedance connected to that section.

The impedance looking in at plane t_2 is 100Ω (as given). The impedance looking in at plane t_1 is 56.25Ω, where the procedure used is shown explicitly:

$\frac{100\Omega}{75\Omega}$ = 1.33   - normalized

$\frac{1}{1.33}$ = 0.75   - reciprocal

(0.75 × 75Ω) = 56.25Ω   - un-normalized to find the impedance at plane t_1.

The input impedance of the first line with characteristic impedance Z_{01} is then

$\frac{56.25\Omega}{50\Omega}$ = 1.125   - normalized

$\frac{1}{1.125}$ = 0.8889   - reciprocal

(0.8889 × 50Ω) = 44.44Ω   - un-normalized to find the impedance at the input plane.

Therefore, Z_{in} = 44.44Ω for the transmission line system shown.

TRANSMISSION LINE DISTRIBUTED PARAMETERS

• **PROBLEM 19-17**

A transmission line used for carrier telephony is 10 miles long. At 795.77 Hz the cable's open circuit input impedance (Z_{in} for a load of $Z_L = \infty \, \Omega$) is $Z_{oc} = 2.93 \times 10^3 \, \underline{/26^0} \, \Omega$ and the short circuit input impedance (Z_{in} for a load of $Z_L = 0 \, \Omega$) is $Z_{sc} = 260 \, \underline{/-32^0} \, \Omega$. Find the distributed line parameters R, L, G, and C for this cable.

Solution: The characteristic impedance Z_0 and the propagation constant γ can be expressed in terms of the distributed line parameters as follows:

$$Z_0 = \sqrt{\frac{(R + j\omega L)}{(G + j\omega C)}}$$

$$\gamma = \sqrt{(R + j\omega L)(G + j\omega C)}$$

Therefore,

$$\gamma Z_0 = R + j\omega L$$

and

$$\frac{\gamma}{Z_0} = G + j\omega C$$

so that if both γ and Z_0 are known, all the distributed line parameters can be found.

The input impedance of a transmission line in terms of the characteristic impedance and the load impedance is:

$$Z_{in} = Z_0 \left[\frac{Z_L \cosh \gamma \ell + Z_0 \sinh \gamma \ell}{Z_0 \cosh \gamma \ell + Z_L \sinh \gamma \ell} \right]$$

where γ is the propagation constant
and ℓ is the line length.

When $Z_L = \infty$; $Z_{in} = Z_{oc} = Z_0 \coth \gamma \ell$ \hfill (1)

when $Z_L = 0$; $Z_{in} = Z_{sc} = Z_0 \tanh \gamma \ell$ \hfill (2)

To find Z_0, multiply equations (1) and (2) to get:

$$Z_{oc} Z_{sc} = Z_0^2$$

or

$$Z_0 = \sqrt{Z_{oc} Z_{sc}}$$

$$= \sqrt{(2.93 \times 10^3 \, \underline{/26^0})(260 \, \underline{/-32^0})}$$

$$= \sqrt{761.8 \times 10^3 \angle -6°}$$

Thus,
$$Z_0 = 872.812 \angle -3° \text{ ohms.}$$

Solving equation (2) for $\tanh \gamma \ell$ gives

$$\tanh \gamma \ell = \frac{Z_{sc}}{Z_0} = \frac{260 \angle -32°}{872.812 \angle -3°}$$

$$= 0.298 \angle -29°$$

$$= 0.2605 - j0.1444$$

To find $\gamma \ell$ recall that if $\tanh \gamma \ell = x$, then $\gamma \ell = \frac{1}{2} \ln \left(\frac{1+x}{1-x} \right)$ for any value of x, real or complex.

Therefore,
$$\gamma \ell = \frac{1}{2} \ln \left[\frac{1 + (0.2605 - j0.1444)}{1 - (0.2605 - j0.1444)} \right]$$

$$= \frac{1}{2} \ln \left[\frac{1 + 0.2605 - j0.1444}{1 - 0.2605 + j0.1444} \right]$$

$$= \frac{1}{2} \ln \left[\frac{1.2605 - j0.1444}{0.7395 + j0.1444} \right]$$

$$= \frac{1}{2} \ln \left[\frac{1.2687 \angle -6.535}{0.7535 \angle 11.049°} \right]$$

$$= \frac{1}{2} \ln (1.684 \angle -17.584°)$$

$$= \frac{1}{2} \ln (1.684 \angle -0.3069 \text{ rad})$$

Now,
$$\ln (r \angle \theta) = \ln r \cdot e^{j(\theta + 2n\pi)}$$

$$= \ln r + j(\theta + 2n\pi)$$

Note: The $2n\pi$ term must be introduced since the vector $re^{j(\theta + 2n\pi)}$ is the same for all values of n, but the logarithm of the vector depends on the value of n. This ambiguity arises because the line may be longer than one wavelength but this would not be observable from the calculations.

Hence,
$$\gamma \ell = \frac{1}{2} \left[(\ln 1.684) + j(-0.3069 + 2n\pi) \right]$$

$$= \frac{1}{2} \left[(0.5212) - j0.3069 + j2n\pi \right]$$

$$= 0.2606 - j0.1535 + jn\pi$$

with $\ell = 10$ miles

$$\gamma = 0.02606 - j0.01535 + j\left(\frac{n\pi}{10} \right)$$

$$= \alpha + j\beta.$$

Since α and β must both be positive, the smallest physically possible value of β is obtained when n = 1. This gives

$$\gamma = 0.0261 - j0.01535 + j(\pi/10)$$

$$= 0.0261 + j0.2988 = 0.2999 \,\underline{/85.016°}$$

Hence,
$$\alpha = 0.0261 \text{ Np/mile}$$

$$\beta = +0.2988 \text{ rad/mile}$$

Therefore,
$$\gamma Z_0 = R + j\omega L$$

$$= (0.2999 \,\underline{/85.016°})(872.812 \,\underline{/-3°})$$

$$= 261.79 \,\underline{/82.016°}$$

$$= 36.36 + j259.26$$

Equating real and imaginary parts gives

$$R = 36.36 \, \Omega/\text{mile}$$

$$\omega L = 259.26 \, \Omega/\text{mile} \text{ and since } \omega = 2\pi(795.77)$$

$$\cong 5000 \text{ rad/sec}$$

$$L = \frac{259.26}{5 \times 10^3} = 51.85 \times 10^{-3} \text{ H/mile}$$

Thus,
$$L = 51.85 \text{ mH/mile}$$

Similarly,
$$\gamma/Z_0 = G + j\omega C$$

$$= \frac{0.2999 \,\underline{/85.016°}}{872.82 \,\underline{/-3°}} = 343.6 \times 10^{-6} \,\underline{/88.016°}$$

$$= 11.90 \times 10^{-6} + j343.40 \times 10^{-6}$$

Therefore,
$$G = 11.90 \times 10^{-6} \text{ mho/mile}$$

$$= 11.9 \text{ μmho/mile}$$

$$\omega C = 343.40 \times 10^{-6}$$

or
$$C = \frac{343.4 \times 10^{-6}}{5000} = 68.679 \times 10^{-9} \text{ F/mile}$$

$$= 0.0687 \text{ μF/mile}$$

● PROBLEM 19-18

The coaxial transmission line for carrier telephony between Chicago and New York City is 1450 km long. The distributed parameters of this line are:

$$R = \text{negligible}$$
$$L = 264.083 \ \mu H/km$$
$$G = \text{negligible}$$
$$C = 44.428 \ pF/km$$

(A) How long is a telephone signal delayed in transmission? Is this delay noticeable?

(B) At a particular point on the line, the RMS value of the signal current is 65 mA. What is the RMS signal voltage at this point and how much signal power is passing through this point?

Solution: (A) Since the resistance and conductance per km are negligible, the propagation constant for the line is

$$\gamma = \sqrt{(R + j\omega L)(G + j\omega C)}$$
$$= \sqrt{(j\omega L)(j\omega C)}$$
$$= j\omega\sqrt{LC}$$
$$= \alpha + j\beta.$$

Therefore,
$$\beta = \omega\sqrt{LC}$$

This result does not necessarily imply that $\alpha = 0$, only that $\alpha \ll \beta$ and, as such, it can be neglected.

The phase velocity, v_p, of the signal on the line is

$$v_p = \frac{\omega}{\beta} = \frac{\omega}{\omega\sqrt{LC}} = \frac{1}{\sqrt{(264.083 \times 10^{-6})(44.428 \times 10^{-9})}}$$

$$= \frac{1}{\sqrt{1.173 \times 10^{-11}}}$$

$$= 2.9198 \times 10^5 \ km/sec$$

$$= 2.9198 \times 10^8 \ m/sec = 97.326\% \text{ of the speed of light in the free space.}$$

The time delay, t, for a signal sent over this line is

$$t = \frac{\text{distance}}{v_p} = \frac{1450 \times 10^3 \ m}{2.9198 \times 10^8 \ m/s}$$

$$= 4.966 \times 10^{-3} \text{ seconds}$$

$$= 4.966 \text{ milliseconds.}$$

Although transmission delay is both subjective and dependent on signal characteristics, evidence shows that for signals of reasonable strength and clarity, delays up to 50 milliseconds cause little inconvenience. Thus, a 4.966 msec delay would not be noticeable.

(B) To find the RMS signal voltage and RMS signal power at a particular point, the power relation is to be used:

$$P = IV = I^2 Z_0$$

where the characteristic impedance of the line, Z_0, is

$$Z_0 = \sqrt{\frac{R + j\omega L}{G + j\omega C}} = \sqrt{\frac{j\omega L}{j\omega C}}$$

$$= \sqrt{\frac{L}{C}}$$

$$= \sqrt{\frac{264.083 \times 10^{-6}}{44.428 \times 10^{-9}}}$$

$$\cong 77.1 \Omega$$

Hence, the RMS signal power P is

$$P = (I)^2 Z_0$$

$$= (0.065)^2 (77.1)$$

$$= 325.75 \text{ milliwatts}$$

and therefore, the RMS signal voltage is

$$P/I = V$$

or

$$V = 5.012 \text{ volts RMS.}$$

REFLECTION AND TRANSMISSION COEFFICIENTS AND VOLTAGE STANDING WAVE RATIO

• **PROBLEM 19-19**

A radio-frequency circuit operating at f = 20MHz is shown below in Fig. (a). Ignoring losses for the lines

(A) express the line lengths in terms of the wavelength on each line;

(B) find the input impedance looking into line 1 (at A - A) using the Smith chart and the average power delivered to the load;

(C) find the location of V_{min} and V_{max} on the Smith chart. Also find the VSWR for lossless line 2.

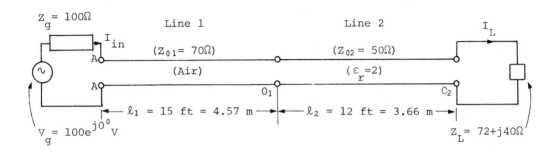

Fig. (a)

Solution: (A) Since line 1 has lossless dielectric (i.e., air), the attenuation (α) is zero and the propagation constant β_0 is

$$\beta_0 = \omega\sqrt{\mu\varepsilon} = \frac{2\pi}{\lambda_0} \quad \text{where } \lambda_0 = \lambda_1 = \text{free space wavelength}$$

Hence,

$$\lambda_1 = \frac{2\pi}{\beta_0} = \frac{c}{f} = \frac{3 \times 10^8 \text{ m/sec}}{20 \times 10^6 \text{ Hz}} = 15 \text{ meters}$$

Hence, length 1 = 15 ft = 4.57m

$$= \frac{4.57 \text{ m}}{15 \text{ m}/\lambda_1}$$

$$= 0.305\, \lambda_1$$

For line 2, with dielectric of relative permitivity $\varepsilon_r = 2$

$$\beta_2 = \omega\sqrt{\mu\varepsilon_0\varepsilon_r} = \beta_0\sqrt{\varepsilon_r} = \frac{2\pi\sqrt{2}}{\lambda} = 0.592$$

Thus,

$$\lambda_2 = \frac{2\pi}{\beta_2} = \frac{2\pi}{0.592}$$

$$= 10.61 \text{ meters}$$

This gives length 2 = 12 ft = 3.66m

$$= \frac{3.66 \text{ m}}{10.61 \text{ m}/\lambda_2}$$

$$= 0.346\lambda_2$$

(B) To find the input impedance using the Smith chart, first normalize Z_L with respect to Z_{02}. The normalized load impedance at the origin O_2 is

$$\bar{Z}_L(0_2) = \frac{Z_L(0_2)}{Z_{02}} = \frac{72 + j40}{50} = 1.44 + j0.8$$

This is plotted on the Smith chart in Fig. (b) as point A.

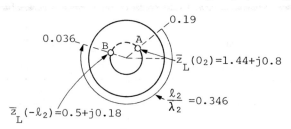

Fig. (b)

Now moving along the 50Ω line, rotate point A by $0.346\lambda_2$ towards the generator along the circle of constant reflection coefficient of magnitude $|\Gamma|$, where Γ is given as

$$\Gamma = \frac{Z_L - Z_{02}}{Z_L + Z_{02}} = \frac{22 + j40}{122 + j40} = 0.260 + j0.243$$

$$= 0.356 \underline{/43^0}$$

Doing this, gives point B which has the value

$$\bar{Z}_L(-\ell_2) = 0.5 + j0.18.$$

The actual line impedance (un-normalized) must be continuous at the junction so that un-normalizing $Z_L(-\ell_2)$ gives $Z_L(-\ell_2)$ which is

$$Z_L(-\ell_2) = \bar{Z}_L(-\ell_2)Z_{02}$$

$$= (0.5 + j0.18)(50)$$

$$= 25 + j9 \ \Omega$$

$$= Z_1(0_1)$$

Normalizing $Z_1(0_1)$ with respect to line 1 gives

$$\bar{Z}_1(0_1) \equiv \frac{Z_1(0_1)}{Z_{01}} = \frac{25 + j9}{70} = 0.358 + j0.128$$

which is plotted as point C in Fig. c.

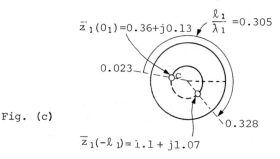

Fig. (c)

Rotating point C by $0.305\lambda_1$ towards the generator gives the normalized input impedance as

$$\bar{Z}_{in} = \bar{Z}_1(-\ell_1) = 1.1 - j1.07$$

Thus,
$$Z_{in} = Z_1(-\ell_1) = \bar{Z}_1(-\ell_1)Z_{01} = (1.1 - j1.07)70$$
$$= 78 - j73.5 \; \Omega$$
$$= 107.2 \; \underline{/-43.3^0} \; \Omega$$

To find the power delivered to the load, from the equivalent input circuit as shown in Fig. (d), proceed as follows.

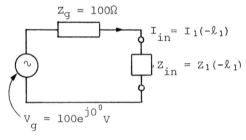

Fig. (d)

$$I_{in} = \frac{100}{(178 - j73.5)} = 0.518 \; \underline{/22.4^0} \; \text{Amps.}$$

Thus,
$$P_{IN} = \frac{1}{2}|I_{in}|^2 Z_{IN} \cos\theta_Z = \text{the power into the line}$$

$$= \frac{1}{2}\left[(0.518)^2 \; 107.2(\cos 43.3)^0\right]$$

$$= 10.467 \; \text{watts}$$

Since both lines are lossless, the power into the lines is the power delivered to the load. Hence,

$$P_{LOAD} = P_{in} = 10.467 \; \text{watts.}$$

(C) As found earlier in part (B) the magnitude of the reflection coefficient is

$$|\Gamma| = 0.356.$$

Hence, the VSWR is:

$$\text{VSWR} = \frac{1 + |\Gamma|}{1 - |\Gamma|} = \frac{1 + 0.356}{1 - 0.356} = \frac{1.356}{0.644} = 2.11$$

This could also have been found directly from the Smith chart from the rightmost intersection of the $|\Gamma|$ circle and the real axis at $\bar{r} = 2.11$. This point also coincides with V_{max}.

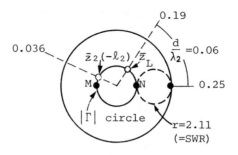

Fig. (e)

This is plotted as point N on Fig. (e). This point is $0.06\lambda_2$ towards the generator from the load, or in meters, the distance d is equal to $0.06(10.61) = 0.637$ meters from the load as shown in Fig. (f).

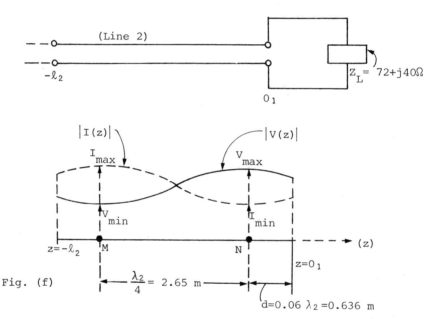

Fig. (f)

V_{min} is $\frac{\lambda}{4}$ towards the generator from V_{max} and is shown as point M and is 2.65 m from the load.

• **PROBLEM 19-20**

A 50Ω slotted air-line is connected to a short length of lossless 50Ω cable which in turn is connected to a load Z_L. With Z_L in place at a frequency of 500MHz, the VSWR is measured to be 3.2 and a voltage minimum occurs at a scale position of 19.4 cm along the slotted line. When Z_L is replaced by a short circuit, a null is observed at the 11.2 cm position. Find the value of Z_L.

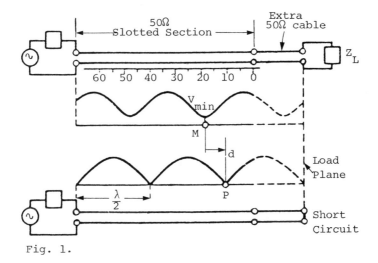

Fig. 1.

Solution: The addition of an extra 50Ω cable line does not change the impedance measurement procedure because slotted line measurements are not concerned with the physical distance to the load. Only the locations of relative voltage minima and nulls are used.

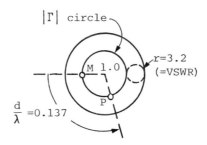

Fig. 2.

First, draw the VSWR circle on the Smith chart. The VSWR circle (i.e., with r = 3.2) is indicated in Fig.(2). The distance from the center of the chart to the intersection of the real axis and the VSWR circle gives the magnitude of the reflection coefficient $|\Gamma|$ (see Fig. 2). Now draw the circle of constant magnitude of reflection coefficient. The voltage minimum corresponds to the minimum impedance found from the leftmost intersection of the $|\Gamma|$ circle and the real axis.

The distance from the minimum to the null is

$$(19.4 \text{cm} - 11.2 \text{ cm}) = 8.2 \text{ cm} = 8.2 \times 10^{-2} \text{ meter.}$$

For a frequency of 500 MHz the wavelength is $\lambda = \frac{c}{f}$

$$\lambda = \frac{3 \times 10^3 \text{ m/s}}{500 \times 10^6 \text{Hz}} = 0.6 \text{ meters.}$$

When the short circuit is connected, the null is 8.2 cm closer to the load. Therefore, on the Smith chart, rotate the null on the constant Γ circle towards the load by $\frac{d}{\lambda}$ where

$$\frac{d}{\lambda} = \frac{8.2 \times 10^{-2} m}{0.6 \ m} = 0.1367$$

resulting in point P (see Fig. 2).

The point P on the Smith chart represents the normalized impedance

$$\overline{Z}_L = 0.65 - j0.93$$

When $\overline{Z}_L$ is un-normalized the result is the load impedance calculated as follows:

$$Z_L = \overline{Z}_L(Z_0) = (0.65 - j0.93)50$$

$$= 32 - j46 \ \Omega$$

• **PROBLEM** 19-21

A 50 V pulse is incident on a 30 ohm load ($Z_L = R_L$) in a line with characteristic resistance of $Z_0 = R_0 = 50$ ohms. Find the following:

a) Incident Current Pulse, b) Reflection Coefficient, c) Reflected Voltage Pulse, d) Reflected Current Pulse, e) Net voltage and current at the load.

Solution: a) The incidend current pulse is

$$\frac{V^+(0,t)}{R_0} = \frac{50}{50} = 1 \text{ Ampere.}$$

b) The reflection coefficient

$$\Gamma_L = \frac{R_L - R_0}{R_L + R_0} = \frac{30 - 50}{30 + 50} = -0.25.$$

c) Reflected voltage pulse =

$$V^+(0,t) \times \Gamma_L = 50(-0.25) = -12.5V.$$

d) Reflected current pulse =

$$I^+(0.t) \times \Gamma_L = 1(-(-0.25)) = 0.25 \text{ Amperes}$$

e) Net voltage at the load = (50 - 1.25) = 37.5V.
Net current at the load = (1 + 0.25) = 1.25 Amperes.

• **PROBLEM** 19-22

Find the reflection coefficient for voltage waves at the load end of a 50 ohm transmission line terminated with a load impedance of 25 - j75 ohms.

Solution: The reflection coefficient Γ can be found when the characteristic impedance of the line Z_0 and the load impedance Z_L are known, using

$$\Gamma = \frac{Z_L - Z_0}{Z_L + Z_0}$$

Thus,

$$\Gamma = \frac{(25 - j75) - (50 + j0)}{(25 - j75) + (50 + j0)} = \frac{-(25 + j75)}{75 - j75}$$

$$= \frac{-79.057 \underline{/71.5651^0}}{106.066 \underline{/-45^0}} = -0.745 \underline{/116.5651^0}$$

● **PROBLEM 19-23**

Find the input impedance of a 50 ohm line terminated in +j50 ohms, for a line length such that $\beta \ell = \pi$ radian.

Solution: The reflection coefficient (Γ_L) at the load is:

$$\Gamma_L = \frac{Z_L - Z_0}{Z_L + Z_0} = \frac{j50 - 50}{j50 + 50} = \frac{50(j - 1)}{50(1 + j)} = \frac{-(1 - j)}{(1 + j)}$$

$$= -\frac{(1 - j)(1 - j)}{(1 + j)(1 - j)} = -\frac{1}{2}(1 - j)^2 = \frac{+2j}{2} = +j = 1\underline{/+90^0}$$

Hence, the voltage reflection coefficient at the input end of the line is

$$\Gamma_{in} = \Gamma_L e^{-j2\beta\ell}$$

$$= (+j)(e^{-j2\pi})$$

$$= (1\underline{/+90})(1\underline{/0^0})$$

$$= 1\underline{/+90}$$

$$= +j$$

Since,
$$Z = Z_0 \left[\frac{1 + \Gamma}{1 - \Gamma} \right]$$
then
$$Z_{in} = Z_0 \left[\frac{1 + \Gamma_{in}}{1 - \Gamma_{in}} \right]$$

$$= 50 \frac{[1 + (+j)]}{[1 - (+j)]}$$

$$= \frac{50(1 + j)}{(1 - j)}$$

$$= \frac{50(1+j)^2}{2}$$

$$= \frac{50(+2j)}{2}$$

Thus,
$$Z_{in} = +50j \text{ ohms.}$$

This illustrates the property of a half-wave length transformer (i.e., unchanged impedances).

• **PROBLEM** 19-24

In the transmission line circuit of Fig. (A), the source voltage is 10 volts rms and the internal source impedance Z_S equals the line's characteristic impedance of 50Ω. If the line is 100λ long with total attenuation of 6 dB and is terminated with a load Z_L, find:

the reflection loss in dB,
the power supplied to the line,
the power dissipated in the line,
and the power delivered to the load for

$Z_L = 50\Omega$ and $Z_L = 150\Omega$.

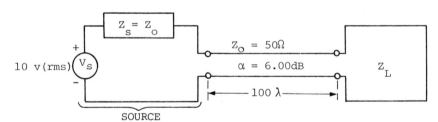

Fig. A.

Solution: A) For $Z_L = 50\Omega$, the load's reflection coefficient is

$$\Gamma = \frac{Z_L - Z_0}{Z_L + Z_0} = \frac{50 - 50}{50 + 50} = 0$$

hence, $|\Gamma| = 0$.

The reflection loss is, by definition, the ratio of the incident power to the transmitted power in decibels. If $|\Gamma|$ is known, we can use the formula.

Reflection loss in dB = $10 \log_{10} \left(\frac{1}{1 - |\Gamma|^2} \right)$.

Thus,

Reflection loss in dB = $10 \log_{10} \left(\frac{1}{1} \right) = 0 \text{dB}$.

In this case, there is no reflected power. Hence, incident power equals reflected power.

To find the power reaching the load, first, find the maximum available power that could ideally reach the load if there was no attenuation present on the line.

Note: For maximum power transfer from the source to the load, the load impedance must equal the characteristic impedance of the line, i.e.,

$$Z_0 = Z_L = 50 \ \Omega$$

The maximum power delivered to the load is equal to the power delivered to the line if there is no attenuation due to the line. The equivalent circuit for the system is shown in Fig.(B).

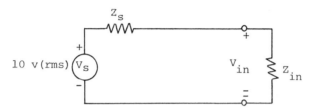

Fig. B.

Note: $P_{in} = Re\left[\dfrac{V_{in}^2}{Z_{in}}\right]$

where P_{in} = power into the line

V_{in} = input line voltage

Z_{in} = input line impedance

Now, $V_{in} = \left(\dfrac{Z_{in}}{Z_S + Z_{in}}\right) V_S = \left(\dfrac{50}{100}\right)(V_S)$

$Z_{in} = Z_0 = 50\Omega.$

Hence, $P_{in} = \dfrac{(\frac{1}{2}V_S)^2}{50} = \dfrac{\frac{1}{4}(100)}{50} = 0.5$ watts.

The power delivered to the load is the power input to the line times the loss due to the 6dB of attenuation from the line. Thus,

$$P_L = (0.5)(-6dB) = \text{power to the load}$$

where $-6dB = 6dB \text{ loss} \cong 0.25$

Therefore,

$$P_L = (\text{power to the load}) = 0.5(0.25) = 0.125 \text{ watts}$$

where $Z_L = 50\Omega$.

The difference between power supplied to the input and power delivered to the load is the power lost in the line attenuation.

Hence,
$$P_{LOST} = P_{in} - P_L = (0.5\omega - 0.125\omega) = 0.375 \text{ watts}.$$

Hence, the load receives only one-third of the power that is delivered to the transmission line.

(B) For $Z_L = 150\Omega$

the reflection-coefficient is

$$\Gamma = \frac{150 - 50}{150 + 50} = \frac{1}{2} = (0.5 + j0) = 0.5\underline{/0°}$$

or $|\Gamma| = 0.5$

The Reflection loss is thus,

$$\text{Refl. loss} = 10 \log_{10}\left[\frac{1}{1 - (0.5)^2}\right]$$

$$= 1.2494 \text{ dB}$$

$$\cong 1.25 \text{ dB}$$

The power delivered to the line input is given by the equivalent circuit of Fig.(c).

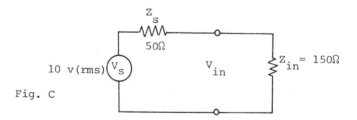

Fig. c

Now,

$$P_{in} = R_e\left[\frac{(V_{in})^2}{Z_{in}}\right]$$

$$= \frac{(7.5)^2}{150} = 0.375 \text{ watts}$$

(Note: $Z_{in} = 150\Omega$ because the net impedance transformation at a distance of 100λ away from Z_L still leaves $Z_L = 150\Omega$ on the Smith chart and each 0.5λ rotation returns the impedance to the same position.)

The power delivered to the load will now be

$$P_L = P_{in}(\alpha) = 0.375\omega(0.25) = 0.09375 \text{ watt}$$

Hence, the power dissipated by the line is

$$P_{LOST} = P_{in} - P_L = 0.375\omega - 0.09375 \text{ watt} = 0.28125 \text{ watt}.$$

● **PROBLEM 19-25**

Find the voltage reflection and transmission coefficients measured at the load, the voltage standing wave ratio (VSWR) on the line and the effective impedance measured at a distance of $\lambda/8$ from the load, for a transmission line with characteristic impedance $Z_0 = 50\Omega$ and load impedance $Z_L = (100 + j100)\Omega$.

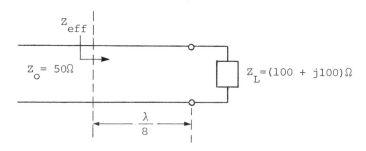

Solution: The reflection coefficient (Γ) is given by:

$$\Gamma = \frac{Z_L - Z_0}{Z_L + Z_0}$$

$$= \frac{(100 + j100) - 50}{(100 + j100) + 50}$$

$$= \frac{(50 + j100)}{(150 + j100)}$$

$$= \frac{111.80 \underline{/63.43^0}}{180.28 \underline{/33.69^0}}$$

$$= 0.62 \underline{/29.74^0}$$

$$= 0.538 + j0.308$$

(Note: Γ can also be found from the Smith chart.)

The transmission coefficient (T) is found by using relation

$$T = \frac{2Z_L}{Z_L + Z_0} = 1 + \Gamma$$

Hence,
$$T = \frac{2(100 + j100)}{(100 + j100) + 50}$$

$$= \frac{200 + j200}{150 + j100}$$

$$= \frac{282.84 \underline{/45^0}}{180.28 \underline{/33.69^0}}$$

$$= 1.57 \underline{/11.3^0}$$

$$= 1.539 + j0.3076$$

The voltage Standing Wave Ratio is found as follows:

$$\text{VSWR} = \frac{1 + |\Gamma|}{1 - |\Gamma|} = \frac{1 + (0.62)}{1 - (0.62)} = \frac{1.62}{0.38} = 4.26$$

(This also could have been found from the Smith Chart.)

The transformed or effective impedance at a distance $\frac{1}{8}\lambda$ from the load can be found by using the Smith Chart. First, normalize Z_{LOAD} by dividing it by Z_0 and plotting this on the chart. This point corresponds to 0.209λ on the scale marked towards the generator. Now, rotate the point along the circle of constant Γ towards the generator by the amount $\lambda/8$ (i.e., by 0.125λ). Plot the point and read the value of normalized impedance $\lambda/8$ away from the load as

$$\overline{z_{eff}} = (0.8 - j1.4).$$

Thus, the effective impedance is

$$Z_{eff} = (\overline{z_{eff}})Z_0$$

$$= (0.8 - j1.4)50\Omega$$

$$= (40 - j70)\Omega$$

• **PROBLEM 19-26**

A 60 volt source produces a 100 MHz signal and has internal impedance of 300 ohms. The source is connected to a 2 meter long parallel wire line with dielectric constant such that the velocity on the line, v_p = 2.5×10^8 m/sec, and characteristic impedance = 300 ohms. If the line is terminated by an antenna with an input impedance of 300 ohms:

(A) Is the line matched?

(B) What is the reflection coefficient and voltage standing wave ratio?

(C) Calculate the wavelength, the phase constant and the electrical length of the line.

(D) Find V_{in}, V_L, I_{in}, I_L, and P_{in}.

(E) Repeat parts (a), (b), and (d), if an identical antenna is connected in parallel with the first one. Also, find the voltage at the load.

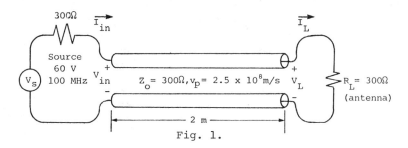

Fig. 1.

Solution: (A) Since the load is equal to the characteristic impedance, by definition the line is matched.

(B) For a matched line, the reflection coefficient must be zero. This can be seen from

$$\Gamma = \frac{Z_L - Z_0}{Z_L + Z_0} = \frac{300 - 300}{300 + 300} = 0$$

When the reflection coefficient is zero, the voltage standing wave ratio (VSWR) must be unity:

$$\text{VSWR} = \frac{1 + |\Gamma|}{1 - |\Gamma|} = \frac{1 + |0|}{1 - |0|} = 1$$

(C) For this two-wire line, the wavelength can be found from the relation,

$$v_p = f\lambda$$

where: v_p = phase velocity on the transmission line in m/s

f = frequency of signal (Hz)

λ = wavelength (m)

Thus,

$$2.5 \times 10^8 \text{ m/s} = 100\text{MHz}(\lambda)$$

$$\lambda = 2.5\text{m}$$

The phase constant (β) is given as:

$$\beta = \frac{2\pi}{\lambda} = \frac{\omega}{v_p}$$

Hence,

$$\beta = \frac{2\pi}{2.5\text{m}} = 0.8\pi \text{ rad/m}$$

or

$$\beta = \frac{2\pi \times (100 \times 10^6 \text{Hz})}{2.5 \times 10^8 \text{ m/s}} = 0.8\pi \text{ rad/m}$$

The electrical length of the line is equal to $\beta\ell$ where ℓ is the length of the line in terms of wavelengths at the frequency of the signal.

Hence,
$$\beta\ell = \frac{2\pi\ell}{\lambda} = \frac{2\pi}{2.5\text{m}}(2\text{m})$$

$$= 0.8\pi(2)$$

$$= 1.6\pi \text{ radians}$$

$$= 288°$$

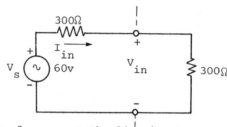

Fig. 2.　　at the line input

The input impedance offered to the voltage source is 300Ω and since the internal impedance of the source is also 300Ω, the voltage at the input to the line is half of 60v, or 30 volts. The equivalent circuit is shown in the diagram. Therefore, $V_{in} = 30\cos 2\pi \times 10^8 t$ and the voltage at the load is V_L. Since there is no reflection or attenuation on the line, V_L has the same amplitude as V_{in} but is delayrd by 1.6π rad. Thus, $V_L = 300\cos(2\pi \times 10^8 t - 1.6\pi)$. The input current is

$$I_{in} = \frac{V_{in}}{Z_{in}} = \frac{30\cos 2\pi \times 10^8 t}{300} = 0.1\cos 2\pi \times 10^8 t \text{ Amp}$$

and so the load current is

$$I_L = 0.1\cos(2\pi \times 10^8 t - 1.6\pi) \text{ Amp}.$$

The average power delivered to the input of the line by the source is equal to that delivered to the load by the line so

$$P_{in} = P_L = \frac{1}{2} \text{Re}\left[I_{in} V_{in}\right]$$

$$= \frac{1}{2}(0.1)(30)$$

$$= 1.5 \text{ watt}.$$

(E) Now, the line is no longer matched to the load. The new load impedance is thus 300Ω||300Ω = 150 ohms. The reflection coefficient is

$$\Gamma = \frac{Z_L - Z_0}{Z_L + Z_0} = \frac{150 - 300}{150 + 300} = \frac{150}{450} = \frac{1}{3}.$$

The new standing wave ratio is

$$\text{VSWR} = \frac{1 + |\Gamma|}{1 - |\Gamma|} = \frac{1 + \left|\frac{1}{3}\right|}{1 - \left|\frac{1}{3}\right|} = \frac{4/3}{2/3} = 2.$$

The input impedance presented to the source must now be found using

$$Z_{in} = Z_0 \left[\frac{Z_L + jZ_0 \tan\beta\ell}{Z_0 + jZ_L \tan\beta\ell}\right]$$

$$= 300 \left[\frac{150 + j300\tan(1.6\pi)}{300 + j150\tan(1.6\pi)}\right]$$

$$= \frac{4.5 \times 10^4 + j9 \times 10^4(-3.078)}{300 + j150(-3.078)}$$

$$= \frac{2.806 \times 10^5 /\underline{-80.772^0}}{550.566 /\underline{-56.983^0}}$$

$$= 509.699 /\underline{-23.790^0}$$

$$= 466.391 - j205.603 \text{ ohms}.$$

This is a capacitive impedance. Physically, this means that this 2 meter length of line stores more energy in its electric field than in its magnetic field. The input current is

$$\vec{I}_{in} = \frac{\vec{V}_{in}}{Z_{in}}$$

$$= \frac{\vec{V}_S \left(\frac{Z_{in}}{Z_{in} + Z_S}\right)}{Z_{in}}$$

$$= \frac{\vec{V}_S}{Z_{in} + Z_S}$$

$$= \frac{60}{(766.391 - j205.603)}$$

$$= 0.0756 /\underline{15.017^0} \text{ Amps}.$$

Thus,

$$I_{in} = 0.0756 \cos(2\pi \times 10^8 t + 0.262) \text{ Amp}$$

Since the line is still lossless, the power supplied to the input is delivered to the load. Thus,

$$P_{in} = P_L = \frac{1}{2} \text{Re}\left[I_{in}^2 Z_{in}\right] = \frac{1}{2}(0.0756)^2(466.391)$$

$$= 1.333 \text{ watts}$$

This power must divide equally between each antenna, so each antenna now receives 0.667 watts.

The voltage across the antenna input terminals V_L is found from the power relation $P_L = \frac{1}{2}(V_L^2/R_L)$.

Hence,

$$1.333 = \frac{1}{2}\left(\frac{|V_L|^2}{150}\right)$$

or

$$|V_L| = 20 \text{ volts.}$$

Resulting

$$V_L = 20(\cos 2\pi \times 10^8 t - 1.6\pi) \text{ volts.}$$

• PROBLEM 19-27

A low loss transmission line, operating at a frequency of 400MHz, with air as dielectric media, is used for radio transmission. Assuming negligible attenuation and 50Ω characteristic impedance, calculate:

(a) Reflection coefficient Γ

(b) VSWR and

(c) Wavelength of the transmitted signal.

Solution: The voltage reflection coefficient produced by the terminated load impedance is:

$$\Gamma = \frac{Z_L - Z_0}{Z_L + Z_0} = \frac{Z_L/Z_0 - 1}{Z_L/Z_0 + 1}$$

where Z_L is the load impedance and Z_0 is the characteristic impedance of the transmission line. Thus,

$$\Gamma = \frac{(0.4 - j1.6) - 1}{(0.4 - j1.6) + 1}$$

$$= \frac{-0.6 - j1.6}{1.4 - j1.6}$$

$$= \frac{1.71 /\!\!-\!110.56°}{2.13 /\!\!-\!48.81°}$$

$$= 0.80 /\!\!-\!61.75°$$

The voltage standing wave ratio is

$$\text{VSWR} = \frac{1 + |\Gamma|}{1 - |\Gamma|} = \frac{1 + 0.8}{1 - 0.8} = \frac{1.80}{0.2} = 9.0$$

The wavelength on the air dielectric transmission line is

$$\lambda = c/f = (3 \times 10^8)/(400 \times 10^6) = 0.75 \text{ meters}.$$

● **PROBLEM 19-28**

A low loss transmission line is delivering high frequency power to an antenna of impedance Z_L. Find the necessary condition so that the lowest VSWR will occur on the transmission line.

<u>Solution</u>: The problem is to show that for arbitrarily fixed Z_L and variable real $Z_0 = R_0$, the magnitude of the reflection coefficient (Γ) is minimized by $Z_0 = |Z_L| + j0 = R_L$. This is because the VSWR is a function of $|\Gamma|$. So if VSWR $= \left[\dfrac{1 + |\Gamma|}{1 - |\Gamma|}\right]$ is minimized, then $|\Gamma|$ will be minimized. The reflection coefficient (Γ) can be written as

$$\Gamma = \frac{Z_L - Z_0}{Z_L + Z_0}$$

so that the magnitude of this expression will be

$$|\Gamma| = \left|\frac{Z_L - Z_0}{Z_L + Z_0}\right| \tag{1}$$

Letting the value of Z_L as $R_L + jX_L$ results:

$$|\Gamma| = \left|\frac{(R_L + jX_L) - Z_0}{(R_L + jX_L) + Z_0}\right| = \sqrt{\left|\frac{R_L + jX_L - Z_0}{R_L + jX_L + Z_0}\right|^2}$$

the magnitude squared is obtained by multiplying the numerator and denominator by its respective complex conjugate. Thus,

$$|\Gamma| = \sqrt{\frac{(R_L - Z_0)^2 + X_L^2}{(R_L + Z_0)^2 + X_L^2}} = \sqrt{\frac{R_L^2 - 2R_L Z_0 + Z_0^2 + X_L^2}{R_L^2 + 2R_L Z_0 + Z_0^2 + X_L^2}}$$

or

$$|\Gamma|^2 = \frac{R_L^2 - 2R_L Z_0 + Z_0^2 + X_L^2}{(R_L^2 + 2R_L Z_0 + Z_0^2 + X_L^2)} \tag{2}$$

Where it is to be noted that if $|\Gamma|$ is a minimum, then $|\Gamma|^2$ will also be a minimum for the magnitude squared of the reflection coefficient.

To find the minimum of equation (2), set it equal to zero and take its derivative with respect to Z_0. Hence,

$$0 = \frac{(R_L^2 + 2R_L Z_0 + Z_0^2 + X_L^2)(-2R_L + 2Z_0) - (R_L^2 - 2R_L Z_0 + Z_0^2 + X_L^2)(2R_L + 2Z_0)}{(R_L^2 + 2R_L Z_0 + Z_0^2 + X_L^2)(R_L^2 + 2R_L Z_0 + Z_0^2 + X_L^2)}$$

After evaluating the above expression, we have

$Z_0^2 = R_L^2 + X_L^2$, and since $Z_0 = R_L + jX_L$, to minimize $|\Gamma|$, the term X_L must be zero. A practical solution to this situation would be to make use of a single stub matching or a quarter wavelength transformer.

● **PROBLEM** 19-29

Evaluate the voltage reflection and transmission coefficient for each of the three cases where a wave is traveling towards the junction on one of the three lines shown in the figure. In each case, calculate the voltage standing wave ratio that results.

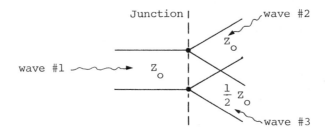

Fig. 1.

Solution: Case #1: In the case where wave #1 travels on the line of characteristic impedance Z_0, such system can be represented by an equivalent diagram as below:

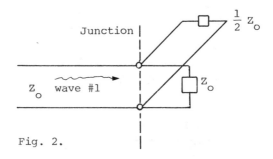

Fig. 2.

To find the voltage reflection coefficient (Γ), use

$$\Gamma = \frac{Z_L - Z_0}{Z_L + Z_0}$$

where Z_L, the load impedance, is the combination of Z_0 and

$\frac{1}{2} Z_0$. Thus,

$$Z_L = Z_0 \parallel \tfrac{1}{2} Z_0$$

$$= \frac{(Z_0)(\tfrac{1}{2} Z_0)}{(Z_0 + \tfrac{1}{2} Z_0)}$$

$$= \frac{\tfrac{1}{2} Z_0^{\,2}}{\tfrac{3}{2} Z_0}$$

$$= \frac{\tfrac{1}{2} Z_0}{\tfrac{3}{2}}$$

$$= \tfrac{1}{3} Z_0$$

Therefore,

$$\Gamma = \frac{\tfrac{1}{3} Z_0 - Z_0}{\tfrac{1}{3} Z_0 + Z_0}$$

$$= \frac{-\tfrac{2}{3} Z_0}{\tfrac{4}{3} Z_0}$$

$$= -\tfrac{1}{2}$$

or

$$= \tfrac{1}{2} \,/180°$$

The transmission coefficient is given by T as

$$T = \frac{2 Z_L}{Z_L + Z_0} = (1 + \Gamma)$$

$$= \frac{2(\tfrac{1}{3} Z_0)}{\tfrac{1}{3} Z_0 + Z_0}$$

$$= \frac{\tfrac{2}{3} Z_0}{\tfrac{4}{3} Z_0}$$

Thus,

$$T = \tfrac{1}{2}$$

The voltage standing wave ratio is:

$$\text{VSWR} = \frac{1 + |\Gamma|}{1 - |\Gamma|} = \frac{1 + |\tfrac{1}{2}|}{1 - |\tfrac{1}{2}|} = \frac{3/2}{1/2} = 3$$

Case #2: The case where wave #2 travels on the line of characteristic impedance Z_0 is equal to case #1 because the equivalent system will be the same as before. Hence,

$$\Gamma = -\frac{1}{2}$$

$$T = \frac{1}{2}$$

$$VSWR = 3$$

Case #3: In the case where wave #3 travels on the line of characteristic impedance $\frac{1}{2}Z_0$, an equivalent diagram representing the system is shown in Figure (3):

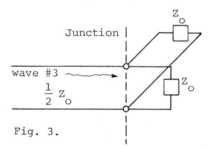

Fig. 3.

Now, notice that $Z_L = Z_0 \| Z_0 = \dfrac{Z_0^2}{2Z_0} = \dfrac{1}{2} Z_0$

Thus,

$$\Gamma = \frac{\frac{1}{2}Z_0 - \frac{1}{2}Z_0}{\frac{1}{2}Z_0 + \frac{1}{2}Z_0} = 0$$

and

$$T = \frac{2(\frac{1}{2}Z_0)}{\frac{1}{2}Z_0 + \frac{1}{2}Z_0} = \frac{Z_0}{Z_0} = 1$$

and

$$VSWR = \frac{1 + |0|}{1 - |0|} = 1$$

In this case, there is no reflection i.e., total transmission results.

• **PROBLEM 19-30**

Three transmission lines each of characteristic impedance Z_0 are connected in a star configuration as shown below. Determine (a) resistor R, to make SWR at the junction equal to zero, (b) the attenuation of a wave, when it crosses the junction with zero reflection coefficient, and the power loss.

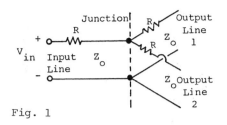

Fig. 1

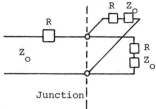

Fig. 2.

Solution: (a) Viewed from each line, an equivalent circuit can be drawn as in Figure (2):

The circuit to the right of the junction can then be combined to form an equivalent load as shown in Fig. (3).

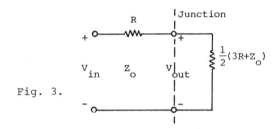

Fig. 3.

For the standing wave ratio to be zero, the line's characteristic impedance Z_0 must be matched with the above total load impedance.

The total load impedance seen by the line is

$$Z_L = R + \frac{1}{2} R + \frac{1}{2} Z_0 = \frac{1}{2} (3R + Z_0)$$

For the match,

$$Z_0 = Z_L$$
$$= \frac{3}{2} R + \frac{1}{2} Z_0$$

Hence, $R = \frac{1}{3} Z_0$. This is the value, R must have for the standing wave ratio to be zero.

(b) The attenuation will be the ratio of line voltages between the input and output lines. Thus,

$$\text{attenuation} = \frac{V_0}{V_{in}} = \frac{\frac{1}{2} Z_0}{\frac{1}{2} Z_0 + \frac{3}{2} R} = \frac{\frac{1}{2} Z_0}{\frac{1}{2} Z_0 + \frac{1}{2} Z_0} = \frac{1}{2}$$

Therefore, the output power to each line is one-quarter of the incident power, and half of the incident power is dissipated in the resistors.

SMITH CHART

• **PROBLEM 19-31**

If a transmission line with characteristic impedance Z_0 = 50 ohms is connected to the following load impedance Z_L, find the normalized load impedance $\bar{z}_L$ and plot these values on a Smith chart.

(a) $Z_L = 10\Omega$

(b) $Z_L = 50\Omega$

(c) $Z_L = 100\Omega$

(d) $Z_L = 50 + j50\Omega$

(e) $Z_L = 75 - j325\Omega$

(f) $Z_L = 20 - j72\Omega$

(g) $Z_L = 0\Omega$ (short circuit)

(h) $Z_L = \infty\Omega$ (open circuit)

IMPEDANCE OR ADMITTANCE COORDINATES

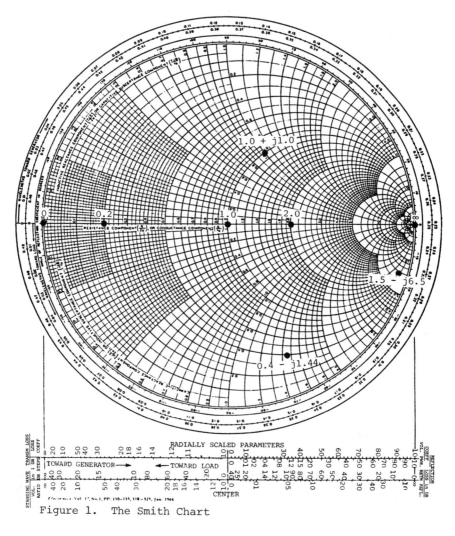

Figure 1. The Smith Chart

Solution: The process of normalization is accomplished by dividing the load impedance value by the characteristic impedance of the transmission line. Thus,

Normalized impedance = $\bar{z}_L = \dfrac{Z_L}{Z_0}$ which is dimensionless.

(A) $\bar{z}_L = 10\Omega/50\Omega = 0.2$

(B) $\bar{z}_L = 50\Omega/50\Omega = 1.0$

(C) $\bar{z}_L = 100\Omega/50\Omega = 2.0$

(D) $\bar{z}_L = \dfrac{50 + j50\Omega}{50\Omega} = 1.0 + j1.0$

(E) $\bar{z}_L = (75 - j325\Omega)/50\Omega = 1.5 - j6.5$

(F) $\bar{z}_L = (20 - j72\Omega)/50\Omega = 0.4 - j1.44$

(G) $\bar{z}_L = \dfrac{0\Omega}{50\Omega} = 0$

(H) $\bar{z}_L = \dfrac{\infty\Omega}{50\Omega} = \infty$

Using the impedance coordinates on the Smith chart to plot these points gives the values shown in the figure.

• **PROBLEM 19-32**

Find, using the Smith chart, the reciprocal of $1.2 - j0.8$.

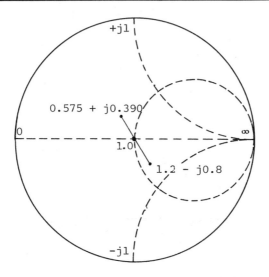

Solution: First enter $1.2 - j0.8$ on the Smith chart (figure shown); then swing the point, keeping the distance to the center constant, through $180°$ either clockwise or counterclockwise, the direction will not matter. The result, read directly, is

$$0.575 + j0.390$$

This is a useful step in many matching problems.

• **PROBLEM** 19-33

The reflection coefficient at the load of an ideal (lossless) transmission line is found to be
$-0.30 + j0.55$.

(A) Using the Smith chart, find the corresponding VSWR for this transmission line.

(B) Find the VSWR analytically.

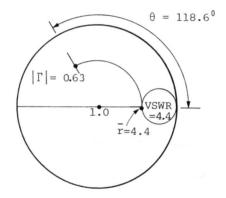

Fig. A. The numerical value for a VSWR coordinate circle is equal to the numerical value of the normalized resistance coordiante circle ($\bar{r} > 1$) to which it is tangent.

Solution: (A) To use the Smith chart, the reflection coefficient must be converted from the rectangular coordinate to the polar form. Hence, resulting:

$$\Gamma = -0.30 + j0.55 = M \underline{/\theta}$$

where
$$M = \sqrt{(-0.30)^2 + (0.55)^2} \cong 0.63$$

$$\theta = \tan^{-1} \frac{(+0.55)}{(-0.30)} \cong 118.6°$$

Thus,

$$\Gamma = 0.63 \; \underline{/118.6°}$$

and the magnitude of the reflection coefficient is 0.63.

Now, draw the reflection coefficient magnitude circle on the Smith chart. The voltage standing wave ratio (VSWR) value is always equal to the larger of the two values of the normalized resistance that intersects the reflection coefficient magnitude circle.

Hence, $\bar{r} = 4.4$,

and the VSWR = 4.4.

(B) The voltage standing wave ratio (VSWR) is given by

$$\text{VSWR} = \frac{1 + |\Gamma|}{1 - |\Gamma|}$$

where $|\Gamma|$ is the magnitude of the reflection coefficient expressed in polar form.

Thus,
$$\text{VSWR} = \frac{1 + |0.63|}{1 - |0.63|} = \frac{1.63}{0.37} = 4.40541.$$

The analytic approach is always more accurate but the Smith chart is quicker and easier to use and usually gives more than enough accuracy for practical situations.

● **PROBLEM 19-34**

The load impedance on a 50 ohm line is

$$Z_L = 50(1 + j)$$

What is the admittance of the load? First use direct computation, then check by use of the Smith chart.

Solution: By direct computation,

$$Y_L = \frac{1}{Z_L} = \frac{1}{50(1 + j)} = \frac{(1 - j)}{100} = 0.01(1 - j).$$

To use the Smith chart, find the normalized impedance at A in the figure:

$$\bar{z}_L = 1 + j$$

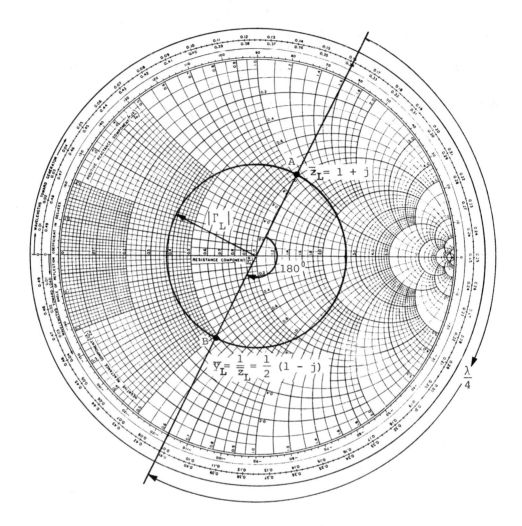

The Smith chart offers a convenient way to find the reciprocal of a complex number by using the property that the normalized impedance reflected back by a quarter wavelength inverts. Thus, the normalized admittance is found by locating the normalized impedance and rotating this point by $180°$ about the constant $|\Gamma_L|$ circle.

The normalized admittance that is the reciprocal of the normalized impedance is found by locating the impedance a distance $\lambda/4$ away from the load end at B:

$$\bar{y}_L = 0.5(1 - j) \rightarrow Y_L = \bar{y}_L Y_0 = (1 - j)/100$$

Note that the point B is just $180°$ away from A on the constant $|\Gamma_L|$ circle. For more complicated loads the Smith chart is a convenient way to find the reciprocal of a complex number.

● **PROBLEM** 19-35

A transmission line has a pure real characteristic impedance. At a certain point on this transmission line, measurements are made that show the line impedance at this point also to be pure and real. If the reflection coefficient magnitude is measured to be 0.60,

(A) Find all the possible values of normalized impedance at this point;

(B) If further measurements give that the VSWR = 4.00, what will be the normalized impedance value at this point?

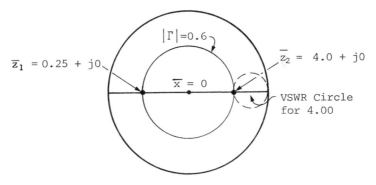

Two values of normalized pure resistance that produce a reflection coefficient of 0.60.

Solution: (A) To find all the possible values of the normalized impedance, first draw the circle of constant reflection coefficient magnitude on the Smith chart (see Fig.). This circle will be the locus of all normalized impedances which give rise to a reflection coefficient magnitude of 0.60. Only two of these points are pure and real numbers; these will be where the circle intersects the real axis. These points are:

$$\bar{z}_1 = 0.250 + j0$$

and $$\bar{z}_2 = 4.00 + j0.$$

Thus, there are two answers to this problem

(B) Once it is known that the VSWR is 4.00, drawing the VSWR circle will show the normalized impedance to be the point where the VSWR circle, the real axis and the reflection coefficient magnitude circle all coincide. Hence, the normalized impedance is found to be

$$\bar{z} = 4.00 + j0.$$

Note: The un-normalized value will depend on the characteristic impedance value of the line.

● **PROBLEM** 19-36

Find the length of an open circuited lossless stub that has a normalized input reactance of 0.75.

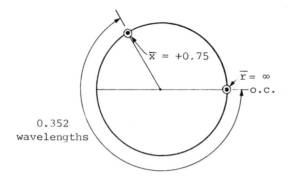

Fig. 1. The normalized input reactance of a section of lossless line with open circuit termination.

Solution:
If one physically measures the stub starting from the open circuit point to the point where its normalized impedance was 0.75, its length would be known. This idea can be extended to the Smith chart. Enter the Smith chart at the position corresponding to the open circuit or infinite normalized impedance point where the horizontal axis intersects the outer boundary as in Fig.(1). All normalized reactance values (zero imaginary component of normalized impedance) lie on the boundary of the Smith chart. Rotate in a clockwise direction (WAVELENGTHS TOWARDS GENERATOR; the generator being implicitly at the opposite end of the terminating open circuit load) until the point $\bar{x}$ = 0.75 is reached.

Adding the final value in wavelengths to the initial value gives the distance in wavelengths:

   distance = $(0.102\lambda + 0.25\lambda) = 0.352\lambda$ (see Fig.)

Note: Other answers are 0.852λ, 1.352λ, 1.852λ, etc., where each addition of 0.5 wavelength corresponds to a further $360°$ rotation around the Smith chart.

● **PROBLEM** 19-37

(A) If a 50 ohm lossless transmission line is short circuited, find the shortest length from the short circuit for which the input will have a capacitive susceptance of 0.025 mhos.

(B) Find the length if a 75 ohm lossless line is now used.

(C) Find the length if a 300 ohm lossless line were used.

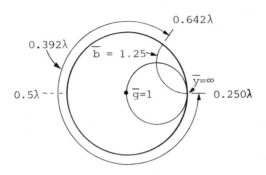

Normalized input susceptance of a lossless line section with short circuit termination.

Solution: (A) This problem will be solved using the Smith chart with admittance coordinates. All admittance values plotted on the Smith chart must first be normalized with respect to the transmission line admittance Y_0:

$$Y_0 = \frac{1}{Z_0} = \frac{1}{50} = 0.02 + j0 \text{ mhos}$$

Hence, the normalized capacitive susceptance ($\bar{b}$) is

$$\bar{b} = \frac{0.025 \text{ mho}}{0.02 + j0 \text{ mho}} = 1.25$$

i.e., the normalized input admittance must be

$$\bar{y}_{in} = 0 + j1.25$$

To find the normalized input admittance, start from the short circuit admittance point on the Smith chart; this is the point where the normalized admittance is infinite ($\bar{y} = \infty$) at the right of the chart. The required length is the angular distance in WAVELENGTHS TOWARDS GENERATOR over which a short circuit will transform to the desired value of normalized admittance. Move on the perimeter of the chart in the clockwise direction (where marked as WAVELENGTHS TOWARDS GENERATOR) until coming to the point $0 + j1.25$.

To find the angular distance, note that we started from the 0.250λ point and proceeded past the zero (or in this case 0.5λ) point to the 0.642λ point on the outer scale. The difference

$$(0.642\lambda - 0.250\lambda) = 0.392\lambda$$

gives the distance from the short circuit where the normalized input susceptance will be 1.25, and so, the actual input susceptance will be 0.025 mho.

Note: If the problem were not asking for the shortest length, then successive revolutions of 0.5λ would give answers of 0.892λ, 1.392λ, etc., which would be equally correct.

(B) For $Z_0 = 75$ ohms,

$$Y_0 = \frac{1}{Z_0} = \frac{1}{75} = 0.0133 + j0 \text{ mhos}.$$

$$\bar{b} = \frac{0.025 \text{ mho}}{0.01333 \text{ mho}} = 1.875$$

Following the same procedure as in part (a), gives the distance as

$$(0.672\lambda - 0.250\lambda) = 0.422\lambda.$$

(C) For $Z_0 = 300$ ohms,

$$Y_0 = \frac{1}{Z_0} = \frac{1}{300} = 0.00333 + j0 \text{ mhos}$$

and

$$\bar{b} = \frac{0.025 \text{ mho}}{0.00333 \text{ mho}} = 7.5$$

The distance could be found using the same procedure as in part (a).

Therefore, the distance is

$$(0.728\lambda - 0.250\lambda) = 0.478\lambda.$$

MATCHING

● **PROBLEM 19-38**

What are the required length and impedance of a quarter-wave transformer that will match a 100Ω load to a 50Ω line at $f = 10$ GHz for an air-filled line?

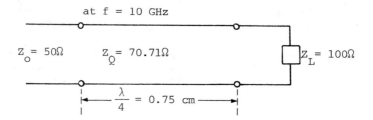

Solution:

A quarter-wave transformer's impedance is found from the formula

$$Z_Q = \sqrt{Z_0 Z_L}$$

so

$$Z_Q = \sqrt{(50)(100)}$$

$$= \sqrt{5000}$$

Thus, $Z_Q = 70.71 \Omega$ is the required impedance.

For an air-filled line it is known that the velocity of wave propagation is the same as the velocity of light c, where

$$c = 3 \times 10^8 \text{ m/sec}$$

Since $c = f\lambda$, the length of the quarter-wave transformer is one-quarter the wavelength corresponding to the frequency used.

Hence,

$$\lambda = \frac{3 \times 10^8 (m/s)}{100 \times 10^8 (1/s)}$$

$$= 0.03 \text{ meters}$$

Therefore,

$$\frac{\lambda}{4} = \frac{0.03}{4} = 0.75 \text{ cm}$$

is the required length for the quarter-wave matching section.

• **PROBLEM** 19-39

At 40MHz, an antenna array has an input impedance of 36Ω. The generator supplying power to the antenna has an output impedance of 500Ω and is about 30 meters away from the antenna terminals. If the generator is connected to a parallel wire transmission line with characteristic impedance of 500Ω, design a quarter-wavelength transformer using parallel-wire transmission line on which the phase velocity is 97% of the free-space velocity of light, to connect the antenna to the main transmission line and to provide an impedance matching.

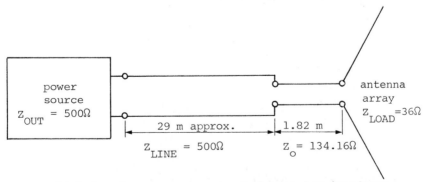

A high impedance source connected to a low impedance load by a quarter wavelength matching transformer.

Solution:
The characteristic impedance of a transmission line section used as a quarter-wavelength transformer is given by:
$$Z_0 = \sqrt{Z_{LINE} Z_{LOAD}}$$

where Z_{LINE} = Characteristic impedance of the line to be matched

Z_{LOAD} = Impedance of the load to be matched

Thus,
$$Z_0 = \sqrt{(500)(36)} = 134.16 \, \Omega$$

(This value can be obtained for a parallel wire line by spacing the conductors a few diameters apart.)

The length of the impedance transformer must be one-quarter of the operating wavelength. The free-space wavelength at the operating frequency of 40MHz is

$$\frac{3 \times 10^8 \text{ m/sec}}{40 \times 10^6 \text{ sec}^{-1}} = 7.5 \text{ meters}$$

If the phase velocity is 97% of the free-space value, the wavelength at the operating frequency will be

$$(0.97)(7.5m) = 7.275 \text{ meters.}$$

Hence, the quarter-wavelength transformer must be $\frac{7.275m}{4}$ = 1.82 meters long.

• **PROBLEM 19-40**

A signal source operating at 50MHz has an output impedance of 20 ohms. It is to supply power through a coaxial line to a load impedance of 150 + j40 ohms. Is it possible to design a quarter-wave transformer to match the source to the load?

Solution: For a quarter-wave transformer with characteristic impedance Z_0:
$$Z_0 = \sqrt{Z_{LINE} Z_{LOAD}}$$

where Z_{LINE} = characteristic impedance of line to be matched

Z_{LOAD} = impedance of load to be matched

This can be rewritten as
$$Z_0^2 = (Z_{LINE} Z_{LOAD}),$$

where Z_0 is purely resistive. Hence, the first requirement on the load and the line impedances to be matched is that they must have phase angles of equal magnitude and opposite sign. Secondly, their geometric mean which will be Z_0 must be a value which is physically achievable as a characteristic impedance for the type of transmission line being used. Therefore, in practice, the quarter-wave transformer is used to connect resistive loads to resistive output impedances.

In this problem the phase angles of the two impedances are not equal, therefore a quarter-wave transformer <u>cannot</u> provide perfect impedance matching. The load impedance can be made real only if a series capacitor of $-j40$ ohms is added. The matching transformer would then have a characteristic impedance $= \sqrt{(20)(150)} = 54.77$ ohms. For a coaxial line with air dielectric, the line wavelength equals the free-space wavelength. So at 50 MHz,

$$\lambda = \frac{3 \times 10^8 \text{ m/sec}}{50 \times 10^6 \text{ Hz}} = 6 \text{ meters}$$

Hence, the impedance transformer would be $\frac{\lambda}{4}$ or 1.5 meters long.

● **PROBLEM** 19-41

A dipole antenna (shown in the figure) having impedance of 72Ω at 150MHz is driven from a parallel-wire line having a 300Ω characteristic impedance Z_0. The feed-line conductors are spaced 2h = 0.75 in. apart. Using the accompanying graph, design a quarter-wave section of parallel-wire air-line that will match the 72Ω load to the 300Ω line at this frequency.

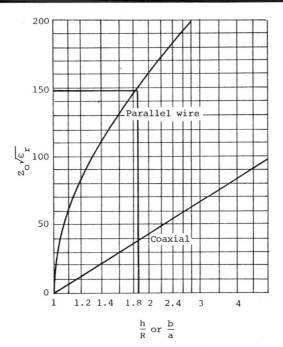

Fig. A Characteristic impedance of lossless coaxial and parallel-wire lines.

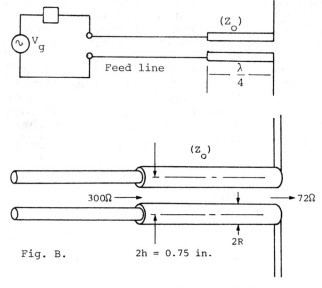

Fig. B. 2h = 0.75 in.

Solution: The characteristic impedance of the quarter-wave transformer is obtained from its load impedance and requires output impedance Z_{in} = 300Ω by use of

$$Z_0 = \sqrt{Z_{in} Z_L}$$

$$= \sqrt{(72)(300)}$$

$$\cong 147 \, \Omega.$$

At 150MHz, the wavelength (λ) on the line of an air dielectric lossless transformer section is obtained by using

$$\lambda = c/f$$

$$= \frac{3 \times 10^8 \text{ m/sec}}{150 \times 10^6 \text{ Hz}}$$

$$= 2 \text{ meters}$$

This gives the required length: $\frac{\lambda}{4}$ = 0.5m for the quarter-wave section.

The physical dimensions of the parallel-wire air-line are found by using the graph for the characteristic impedance of lossless parallel-wire lines. From this graph, when Z_0 = 147Ω then the ratio of line spacing to line radius is

$$\left(\frac{h}{R}\right) = 1.85.$$

given the spacing of 2h = 0.75 inches for the quarter-wave transformer. Thus, the conductor diameter 2R is found to be

$$\left(\frac{h}{R}\right) = \frac{2h}{2R} = 1.85.$$

Hence, 2R = 0.405 inches.

• **PROBLEM 19-42**

For the system shown in Figure (1), at the design frequency f_d, the VSWR on the 200Ω line is unity. Explain how to use the Smith chart to determine the percentage by which the frequency can be varied above and below the design value so that the 200Ω main line never has a VSWR greater than 1.30 on it. Assume that the load impedance Z_L does not vary with frequency.

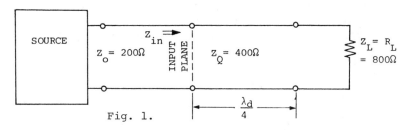

Fig. 1.

Solution: First, it is necessary to understand how to find the overall VSWR of the 200Ω main line at the design frequency f_d. The steps used are outlined below:

(1) Normalize the load impedance with respect to Z_Q and plot on the Smith chart.
$$\bar{z}_L = Z_L/Z_Q$$
$$= 200 + j0$$

(2) Rotate on the circle of constant reflection coefficient magnitude ($|\Gamma|$) by $0.25\lambda_d$ towards the generator. This corresponds to transformation of the load by the 400Ω line. Now, find (normalized) $\bar{z}_{in}$ at the input plane.

(3) Un-normalize $\bar{z}_{in}$ with respect to Z_Q to get Z_{in}
$$Z_{in} = \bar{z}_{in}(Z_Q) = \bar{z}_{in}(400) = 0.5 + j0$$

Finally,

(4) Normalize Z_{in} with respect to $Z_0 = 200Ω$ and plot on the Smith chart to get overall normalized impedance on the 200Ω main line:
$$\bar{z}_{main} = \frac{Z_{in}}{Z_0} = 1 \quad \text{at design frequency } f_d.$$

Away from the design frequency f_d, the length of the transformer is no longer $0.25\lambda_d$. Its electrical length ℓ is given by
$$\ell = \frac{\lambda_d}{4} = \left(\frac{\lambda}{4}\right)\left(\frac{\lambda_d}{\lambda}\right) = \frac{\lambda}{4}\left(\frac{f}{f_d}\right).$$

Thus, for example, if $f = 0.9(f_d)$ the effective length of the matching section is
$$0.25\lambda(0.9)$$
$$= 0.225\lambda$$

(not $0.25\lambda_d$) so the normalized load impedance $\bar{z}_L$ is transformed into
$$\bar{z}_{in} = 0.51 - j0.12.$$

The locus of all points after un-normalizing with respect to the 400Ω line and normalizing with respect to the 200Ω main line forms a circle of values for a range of frequencies above and below the design frequency. This circle passes through the center point of the chart at f_d because here the VSWR = 1.

The Smith chart has the useful property that when points are re-normalized by a constant factor, they are translated to a new circle whose center also lies on the real axis as described above. In this case, the impedance values lie on the circle passing through the points

$\bar{r} = 1$ and

$\bar{r} = 5$ (as indicated in Fig.2).

IMPEDANCE COORDINATES

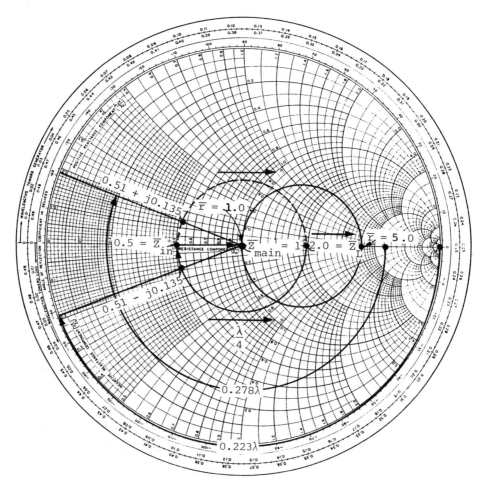

Fig. 2. Values at the input plane.

Therefore, as the frequency varied, the transformer provides a VSWR ranging from unity (where the electrical length of the transformer is $0.25\lambda_d = 0.25\lambda$) to 5.

In Fig.(3), the circle of constant $|\Gamma|$ corresponds to a VSWR of 1.30. This intersects the plotted locus line at normalized impedances of $1.02 \pm j0.27$. These values transform back to the values on the first Smith chart in Fig.(2) corresponding to normalized impedances

$$0.51 \pm j0.135$$

where we have multiplied $1.02 \pm j0.27$ by $\left(\dfrac{200}{400}\right)$ on the $|\Gamma| = 2$ circle of Fig.(2).

IMPEDANCE COORDINATES

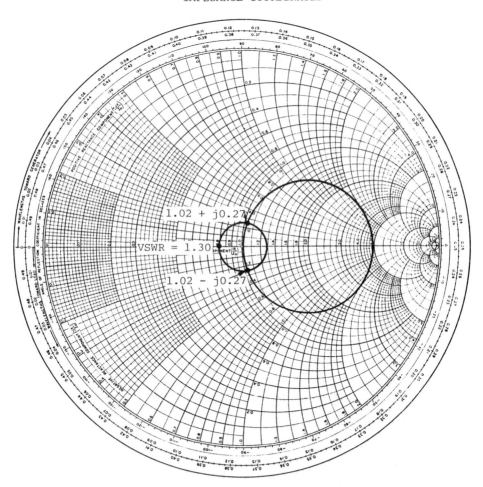

Fig. 3. Drawing radial lines through the points

Drawing radial lines through the points shows that the normalized load impedance of $\overline{z}_L = 2.0$ transforms to those values at the input plane if the transformer has electrical lengths of 0.223λ and 0.278λ. The transformer section which is 0.25 wavelengths at the design frequency is

$$0.278\lambda \text{ at } 1.11f_d \quad \text{and}$$

$$0.223\lambda \text{ at } 0.89f_d$$

The transmission line system therefore has a total bandwidth of ±11% (or 22%) within which the VSWR on the main 200Ω line is less than 1.30.

● **PROBLEM** 19-43

A transmission line has a voltage standing wave ratio (VSWR) of 2 and a voltage minimum at $\lambda/4$ from the load. What is the value of the normalized load $\bar{z}_L$ connected to this line, and also, find the position and length of a single stub tuner to match the load to this line.

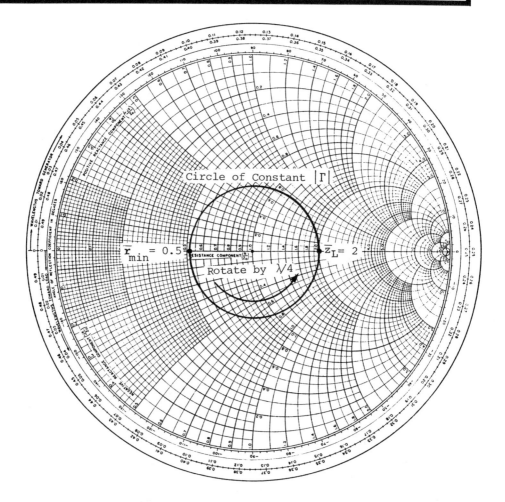

Solution:

Draw the circle of constant reflection coefficient Γ on the Smith chart. A voltage minimum is always associated with a minimum value of resistance for a given load because

$$R_{MAX} = \frac{V_{MAX}}{I_{MAX}}$$

$$R_{MIN} = \frac{V_{MIN}}{I_{MAX}}$$

From the Smith Chart it is seen that $\overline{r}_{min} = 0.5$. Thus, if $\overline{r}_{min}$ is $\lambda/4$ away from the load, rotate on the circle of constant Γ by $\lambda/4$ towards the load to read off the normalized value of load impedance $\overline{z}_L$. Therefore,

$$\overline{z}_L = 2$$

Using the Smith chart, the steps used to find a single shunt stub match are:

(1) Convert $\overline{z}_L$ to $\overline{y}_L$ and use only normalized admittance coordinates on the Smith chart. ($\overline{z}_L = 1/\overline{y}_L$).

(2) Start from $\overline{y}_L$ and rotate along the circle of constant reflection coefficient Γ until the $\overline{g} = 1$ circle is intersected. Call this point $\overline{y}_P$.

(3) Measure the difference in wavelengths between $\overline{y}_L$ and $\overline{y}_P$ on the "towards the generator" scale. This gives D_1 in wavelengths.

(4) Multiply the imaginary part of $\overline{y}_P$ by -1 and plot the point $\overline{y}_{STUB} = (0 + j\,\overline{b}_P)$ where $j\,\overline{b}_{P_1}$ is the value of the imaginary part of $\overline{y}_P$ after multiplied by -1.

(5) Measure the distance in wavelengths (on the "towards the generator" scale) between the points $\overline{y} = \infty$ and $\overline{y}_{STUB}$. This gives D_2 in wavelengths.

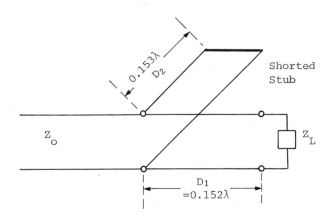

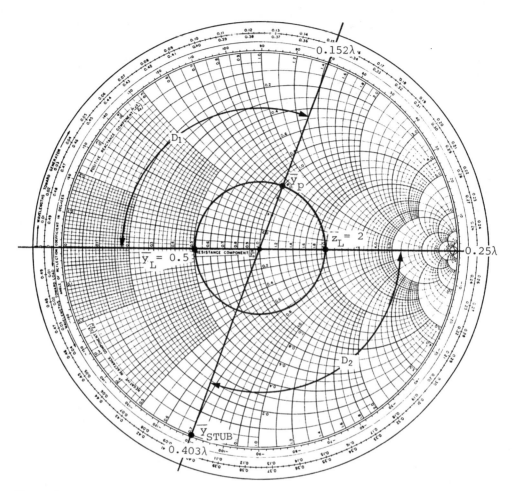

Therefore,

$D_1 = (0.152\lambda - 0\lambda) = 0.152\lambda$

$D_2 = (0.403\lambda - 0.25\lambda) = 0.153\lambda$

● **PROBLEM 19-44**

A transmission line is terminated by a normalized load $\bar{z}_L = 2$.

(A) Use a single stub tuner to match this load to the line at frequency f_1 and wavelength λ_1.

(B) If the wavelength is now increased by ten percent, what is the new value of stub susceptance and voltage standing wave ratio (VSWR) on the line?

(C) If at wavelength λ_1, the stub is placed $\frac{\lambda_1}{2}$ further toward the generator from its original position find the VSWR on the line when the wavelength is increased by ten percent.

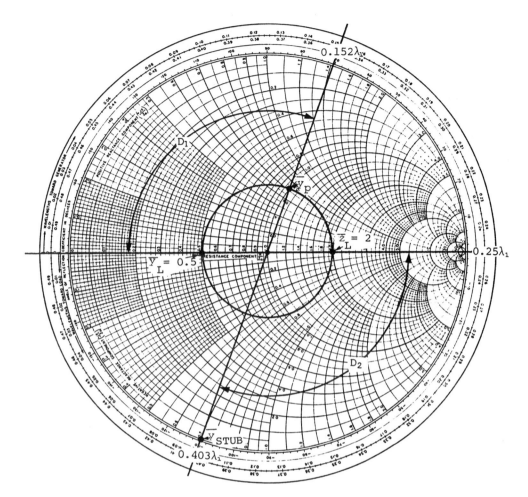

Solution:

Part (A)

For a single shunt stub match using a Smith chart:

(1) convert $\bar{Z}_L$ to $\bar{y}_L$ and use only normalized admittance coordinates
$$\bar{z}_L = 1/\bar{y}_L$$

(2) Start from $\bar{y}_L$ and rotate along the circle of constant Γ until the $\bar{g} = 1$ circle is intersected. Call this point $\bar{y}_P$.

(3) Measure the difference in wavelengths between $\bar{y}_L$ and $\bar{y}_P$ on the "towards the generator" scale. This gives D_1, the distance from the load to the stub, in wavelengths.

(4) Multiply the imaginary part of $\bar{y}_P$ by -1 and plot the

1007

point $\bar{y}_{STUB} = (0 + j\bar{b}_{P_1})$ where $j\bar{b}_{P_1}$ is the value of the imaginary part of $\bar{y}_P$ after multiplication by -1.

(5) Measure the distance in wavelengths (on the "towards the generator" scale) between the points $\bar{g} = \infty$ and $\bar{y}_{STUB}$.
This gives the length of the stub, D_2, in wavelengths. Thus, for a match

$$\bar{y}_{IN} = \bar{y}_P + \bar{y}_{STUB} = (1 + j\bar{b}_P) + (-j\bar{b}_{P_1}) = 1$$

$$\left[\text{where } j\bar{b}_P = -j\bar{b}_{P_1}\right]$$

For frequency f_1

$$\bar{z}_L = (2 + j0)$$

$$\bar{y}_L = (\tfrac{1}{2} + j0)$$

$$\bar{y}_P = (1 + j0.7)$$

$$D_1 = (0.152\lambda_1 - 0\lambda_1) = 0.152\lambda_1$$

and $$D_2 = (0.403\lambda_1 - 0.25\lambda_1) = 0.153\lambda_1$$

Now, because the load is matched to the line, the standing wave ratio (SWR) is one [i.e., $\Gamma = 0$].

Part (B)

If the physical distances D_1 and D_2 remain fixed and only the wavelength changes, D_1 and D_2 will now represent different fractions for the new wavelength $\lambda_2 = (1.1)\lambda_1$.

For λ_2, the distances in terms of λ_1 are

$$D_1 = (0.152)\lambda_2 = (0.152)(1.1\lambda_1)$$

$$= 0.1672\lambda_1$$

$$D_2 = (0.153)\lambda_2 = (0.153)(1.1\lambda)$$

$$= 0.1683\lambda_1$$

For this new wavelength, the match is not perfect and so the resulting standing wave ratio must be found.

The length D_1 rotates $\bar{y}_L$ to a new point $\bar{y}_{P_2} = (1.15 + j0.75)$

The length D_2 gives $\bar{y}_{STUB_2}$ a new value also. $\bar{y}_{STUB_2} = (0 - j0.56)$

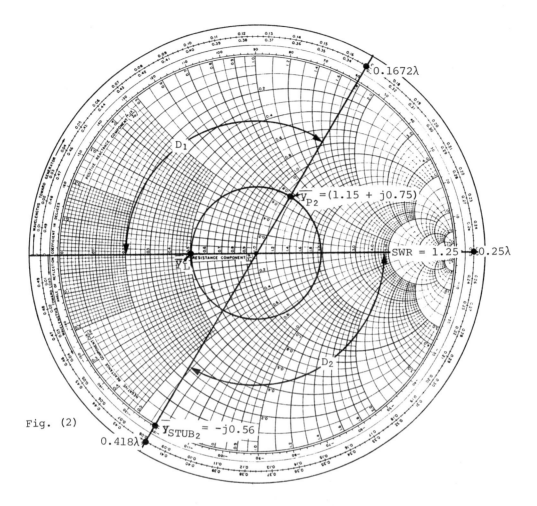

Fig. (2)

For the transmission line

$$\bar{y}_{in_2} = \bar{y}_{P_2} + \bar{y}_{STUB_2}$$

$$= (1.15 + j0.75) + (-j0.56)$$

$$= 1.15 + j0.19$$

Plot $\bar{y}_{in_2}$ and construct the circle of constant Γ from which the standing wave ratio can be found.

The standing wave ratio is now 1.25 for $\lambda_2 = (1.1)\lambda_1$

$$SWR = \frac{1 + |\Gamma|}{1 - |\Gamma|} = \frac{1 + |0.11|}{1 - |0.11|} = \frac{1.11}{0.89} = 1.25$$

Part (C)

If at the wavelength λ_1 the stub were placed $\lambda_1/2$ farther toward the generator from its original position, D_1 would have been $0.652\lambda_1$ and D_2 is unchanged. Now, when $\lambda_2 = (1.1)\lambda_1$ the resulting standing wave ratio is found using the same technique as before:

$$\bar{y}_{P_3} = 1.8 + j0.55 \quad \text{so,} \quad D_1 = 0.7172\lambda_1 = 0.2172\lambda_1$$

and $\bar{y}_{STUB_3} = 0 - j0.56$ giving, $D_2 = 0.1683\lambda_1$

Hence,

$$\bar{y}_{in_3} = \bar{y}_{P_3} + \bar{y}_{STUB_3}$$

$$+ 1.8 + j0.55 - j0.56$$

$$\bar{y}_{in_3} = 1.8 - j0.01$$

Plotting $\bar{y}_{in_3}$ and constructing the circle of constant Γ from which the standing wave ratio can be found gives,

$$SWR = \frac{1 + |\Gamma|}{1 - |\Gamma|} = \frac{1 + |0.282|}{1 - |0.282|} = \frac{1.282}{0.718} = 1.79$$

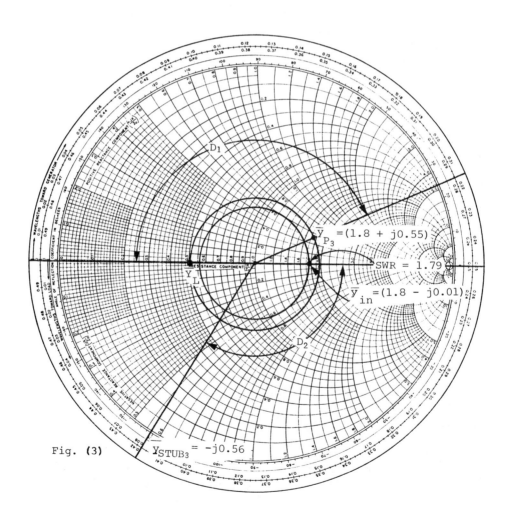

Fig. (3)

Conclusions

From part A, a perfect match can be achieved for only one wavelength (or frequency). Comparing the results of parts B and C of this problem, the solution corresponding to the first intersection of the g = 1 circle (the shortest length for D_1) gives the best performance when the wavelength is varied. This illustrated the greater frequency sensitivity of the match when the stub is placed farther from the load.

● **PROBLEM 19-45**

A load impedance $Z_1 = (43.1 - j17.3)\Omega$ terminates a 100Ω line. Using the Smith chart, find the VSWR on the line and then design a single-stub tuner to match this load to the line. (i.e., Find D_1 and D_2.)

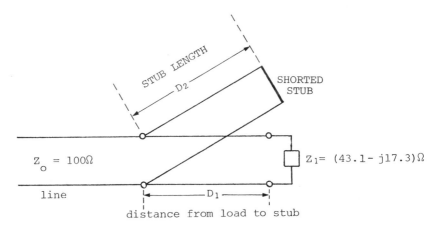

Fig. (1)

Solution: Any two admittances can be matched by a single stub tuner. The load admittance is first transformed along a section of line according to

$$\bar{y} = \frac{\bar{y}_1 + j\tan\beta\ell}{1 + \bar{y}_1 \tan\beta\ell}$$

(where the normalized admittance $\bar{y} = Y/Y_0 = YZ_0$) to provide the necessary conductance, then a shunt stub is added to alter the susceptance to the required value.

There are two unknowns to be determined. D_1 is the distance from the load to the location of the short circuited stub (measured in wavelengths) for the frequency of interest. D_2 is the distance from the short circuit to its connection with the line, also measured in wavelengths. To use the Smith chart for stub matching of a load to a line, all impedances must be transformed to correspond to admittances and then admittance coordinates should be used on the chart. The Smith chart utilizes values which are normalized with respect to the characteristic impedance of the line to be matched.

STEP (1) TO NORMALIZE AND PLOT $\bar{y}_1$:

Given the characteristic impedance $Z_0 = 100\Omega$ and the load impedance $Z_1 = (43.1 - j17.3)\Omega$, transform Z_1 into normalized admittance coordinates $\bar{y}_1 = (\bar{g}_1 + j\bar{b}_1)$. This can be done in two ways, analytically or graphically on the Smith chart.

For the analytical process of normalization, divide Z_1 by Z_0 which gives $\bar{Z}_1$, then take the reciprocal giving $\bar{y}_1$ and plot.

$$\frac{Z_1}{Z_0} = \frac{(43.1 - j17.3)\Omega}{100\Omega} = (0.431 - j0.173) = \bar{Z}_1$$

$$\frac{1}{\bar{Z}_1} = \frac{1}{(0.431 - j0.173)} = \frac{(0.43 + j0.173)}{(0.431)^2 + (0.173)^2}$$

$$= (2 + j0.8) = \bar{y}_1$$

Using the graphical method, obtain $\bar{Z}_1$ as before and plot. Now, moving on the circle of constant reflection coefficient Γ (having its center at $\bar{g} = 1$ and radius equal to $\bar{Z}_1$) proceed to the point that is $\pm 180°$ (diametrically opposite to) $\bar{Z}_1$ and on the perimeter of the circle.

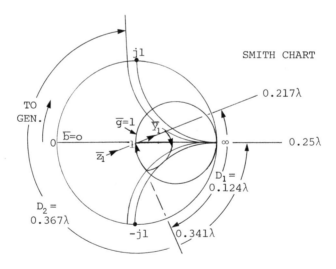

Fig. (2)

The VSWR, before matching, can be read directly from the chart as the value where the circle of constant Γ intersects the $\bar{b} = 0$ (horizontal) axis. Otherwise, (analytically)

$$\text{VSWR} = \frac{1 + |\Gamma|}{1 - |\Gamma|}$$

$$= \frac{1 + |0.41|}{1 - |0.41|}$$

Thus,
$$\text{VSWR} = 2.39$$

STEP (2) DETERMINATION OF LENGTH D_1:

To find D_1, move from the point $\bar{y}_1$ along the circle of constant Γ in the direction marked "towards the load" to the point of intersection between the constant Γ circle and the circle where $\bar{g} = 1$. Call this point $\bar{y}_P$.

$$\bar{y}_P = (1 - j0.9)$$

Using the scale of wavelengths towards the generator on the outer edge of the Smith chart, find the difference in wavelengths from the starting point at $\bar{y}_1$ to the new point $\bar{y}_P$, this difference is D_1 in wavelengths.

Hence, $D_1 = (0.341\lambda - 0.217\lambda) = 0.124\lambda$

STEP (3) DETERMINATION OF THE LENGTH D_2:

The point $\bar{y}_P$ has a real and imaginary part. To find D_2, take the imaginary part of $\bar{y}_P$ and change its sign (i.e., multiply by -1). Call this $j\bar{b}_{P_1}$.

$$j\bar{b}_{P_1} = (-j0.9)(-1) = (+j0.9)$$

Plot the point $(0 + j\bar{b}_{P_1}) = \bar{y}_{STUB}$ where $\bar{b}_{P_1}$ is the imaginary part of $\bar{y}_P$ after its sign has been changed.

To measure D_2 in wavelengths, use the admittance value for the short circuit at $\bar{g} = \infty$ (0.25λ) as the initial point. Next, find the value in wavelengths from the generator that corresponds to the final value at $\bar{y}_{STUB}$. This difference gives D_2 in wavelengths. Hence,

$$D_2 = (0.617\lambda - 0.25\lambda) = 0.367\lambda$$

The solution outlined above is only one of two basic solutions. The distance D_1 can be increased until the constant Γ circle intersects the $\bar{g} = 1$ circle for the second time at the point $\bar{y}_{P(NEW)} = (1 + j0.9)$. The solution then proceeds as before. The increased line lengths for this latter solution make the tuner performance more sensitive to frequency variation, and so, reduce the useful bandwidth.

Thus, in conclusion,

$$\bar{z}_1 = (0.431 - j0.173), \quad \bar{y}_1 = (2.0 + j0.8)$$

VSWR without matching = 2.41

Preferred solution (see diagram) $D_1 = 0.124\lambda$, $D_2 = 0.367\lambda$.

Alternative solution $D_1 = 0.442\lambda$, $D_2 = 0.133\lambda$.

• **PROBLEM 19-46**

A 72Ω antenna is connected to a 12 meter long feed line of characteristic impedance $Z_0 = 300Ω$ which is made of parallel-wire air dielectric cable. Match this antenna to the line so that a 150 MHz transmitter connected at the input end of the line will see no reflections.

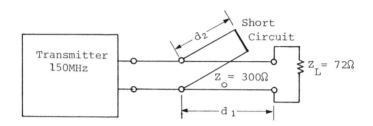

Solution: To match the antenna (load) to this feed line, use a length of shorted 300Ω transmission line as a single shorted stub match.

For a short circuited single stub match, plot all values on the Smith chart using admittance coordinates.

The load admittance is $Y_L = \frac{1}{Z_L} = \frac{1}{72}$ mho

The characteristic line admittance is

$$Y_0 = \frac{1}{Z_0} = \frac{1}{300} \text{ mho.}$$

Hence, the normalized load admittance to be plotted on the Smith chart is

$$\bar{y}_L = \frac{Y_L}{Y_0} = \frac{1/72}{1/300} = 4.1667$$

Let this point be $\bar{y}_L$.

Maintaining a constant distance from $\bar{y}_L$ to the origin (at the center) of the Smith chart, draw a circle of constant magnitude of reflection coefficient $|\Gamma|$, where this circle and the $\bar{g} = 1$ circle intersect. Note the value of the normalized susceptance $j\bar{b}$.

i.e., $j\bar{b} = -j1.6$

The amount of rotation of point $\bar{y}_L$ until it intersects $\bar{g} = 1$ circle is measured from the perimeter of the Smith chart to be $0.071\lambda = d_1$.

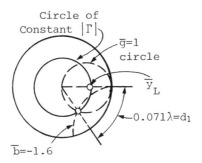

Fig. 1.

For 150 MHz, the wavelength λ is 2 meters. Therefore, the stub must be located at $0.071\lambda = 0.71(2)$ or 0.142 meters from the load. See figure 1.

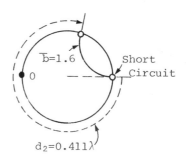

Fig. 2.

To cancel the inductive susceptance $(-j\bar{b} = -j1.6)$ at this point, the required stub length is found by moving on the perimeter of the chart from the point of short circuit $(\bar{y} = \infty)$ until arriving at a point $j\bar{b} = +j1.6$. Then, this rotation is measured to the stub length in terms of wavelengths. This length is:

$$d_2 = 0.411\lambda$$
$$= 0.411(2) = 0.822 \text{ meters}$$

(See Fig. 2)

● **PROBLEM 19-47**

A horn antenna is fed from a rectangular waveguide in which the TE_{10} mode propagates. At the junction, a reflection coefficient $\Gamma = 0.3\, e^{j\pi/4}$ is produced.
(a) What is the normalized input admittance to the horn?
(b) Find the required normalized susceptance and the spacing in guide wavelengths from the junction of an inductive diaphragm that will match the waveguide to the horn.

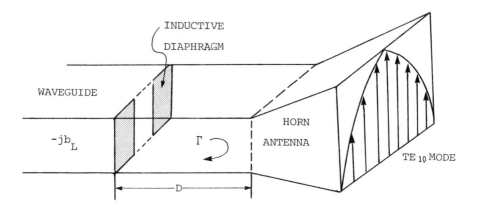

Solution: (a) The horn antenna acts as a load to the waveguide and the inductive diaphragm matches this load to the guide. Thus, this is essentially a single shunt stub matching problem where first the value of normalized admittance ($\bar{y}_L$) must be found from the given reflection coefficient.

The load impedance can be found using the relation

$$Z_L = Z_0 \left[\frac{1 + \Gamma}{1 - \Gamma}\right]$$

Hence,

$$\bar{z}_L = \frac{Z_L}{Z_0} = \left[\frac{1 + \Gamma}{1 - \Gamma}\right]$$

$$= \frac{1 + 0.3\underline{/45°}}{1 - 0.3\underline{/45°}} = \frac{1 + (0.212 + j0.212)}{1 - (0.212 + j0.212)}$$

$$= \frac{1.212 + j0.212}{0.788 - j0.212} = \frac{1.23\underline{/9.92°}}{0.82\underline{/-15.06°}}$$

therefore,

$$\bar{z}_L = 1.50 \underline{/+24.98°}$$

$$= 1.35 + j0.64$$

The input admittance $\bar{y}_L$ of the horn antenna is:

$$\bar{y}_L = 1/(\bar{z}_L)$$

$$= \frac{1}{1.36 + j0.64} = \frac{1.35 - j0.64}{(1.35)^2 + (0.64)^2}$$

Therefore,

$$\bar{y}_L = 0.6 - j0.275.$$

(b) To find the distance D using the Smith chart, start from $\bar{y}_L$ and rotate along the circle of constant reflection coefficient Γ until $\bar{g} = 1$ circle is intersected. (Let this point be $\bar{y}_P$.) Now, measure the difference in wavelengths between $\bar{y}_L$ and $\bar{y}_P$ towards the generator side. This gives D in guide wavelengths as shown.

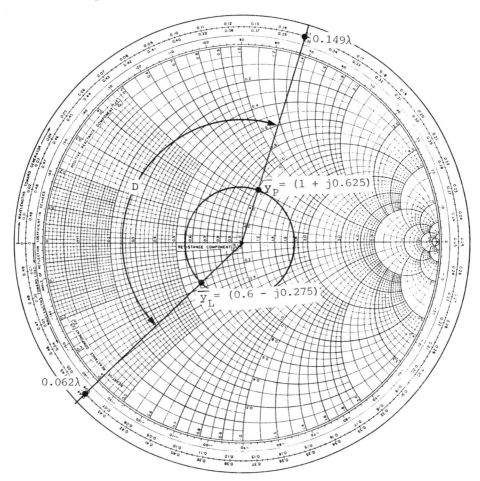

Thus, $D = (0.149\lambda + 0.062\lambda) = 0.211\lambda$ and therefore, the diaphragm is 0.211λ from the horn antenna.

The value of normalized susceptance $-j\bar{b}_L$ is the amount of susceptance needed to make $\bar{y}_{in} = 1$, where

$$\bar{y}_{in} = \left[\bar{y}_P + \bar{y}_L\right]$$

$$1 = (1 + j0.625) + (0 - j\bar{b}_L)$$

Thus,
$$-j\bar{b}_L = -j0.625$$

or
$$\bar{b}_L = 0.625.$$

Therefore, the value of normalized susceptance is

$$-j\bar{b}_L = -j0.625.$$

• **PROBLEM** 19-48

In the following circuit, what is the smallest value of d that will make the resistive part of Z_{in} equal to 50 ohms at the plane t_1? Find the required value of jX to make Z_{in} equal to 50 ohms at the plane t_2.

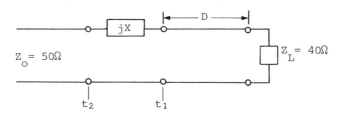

Solution:

Essentially the problem is to find a series single stub-match between Z_0 and the impedances to the right of plane t_2.

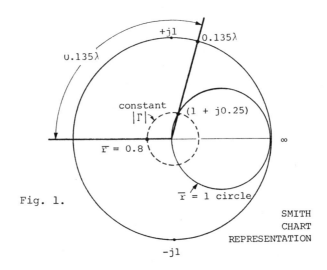

Fig. 1.

SMITH CHART REPRESENTATION

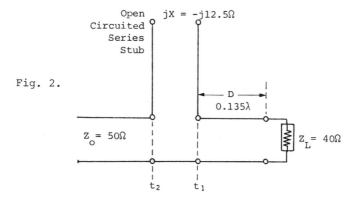

Fig. 2.

1018

First, normalize Z_L with respect to Z_0 and plot this on a Smith chart.
$$\bar{z}_L = Z_L/Z_0 = 40/50 = 0.8$$

Note: For a series single-stub match use impedance coordinates. For $\bar{z}_L = (0.8 + j0)$ draw the circle of constant reflection coefficient Γ and rotate along the perimeter of the circle until it intersects with the $r = 1$ circle at $(1 + j0.25) = \bar{z}_{in}$.

For this value of $\bar{z}_{in}$, when un-normalized, it is found that the real part of Z_{in} is 50Ω. Hence, the smallest value of D is found by noting the difference on the scale marked "towards the generator" of the final point $\bar{z}_{in}$ at 0.135λ and the initial point $\bar{z}_L$ at 0λ. Therefore,
$$D = (0.135\lambda - 0\lambda) = 0.135\lambda.$$

To make Z_{in} equal to 50Ω at plane t_2, Z_0 must be matched to the combined load of $Z_{in} + jX$.

In normalized form, this will be
$$\bar{z}_0 = \bar{z}_{in} + j\bar{x}$$
$$1 = (1 + j0.25) + j\bar{x}$$

Therefore,
$$j\bar{x} = -j0.25$$

Thus, when un-normalized, jX is
$$jX = Z_0(j\bar{x}) = 50(-j0.25)$$
$$jX = -j12.5\Omega$$

● **PROBLEM 19-49**

If the normalized load impedance is $\bar{z}_L = 2$, what is the position, D_1 away from the load impedance and the length D_2 (in wavelengths) of an open stub required for a series connected stub, match the specified load to the line?

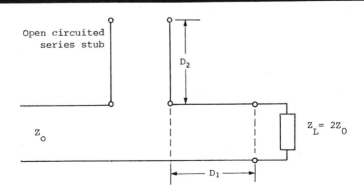

Solution: If $\bar{z}_L = 2$ then $Z_L = 2Z_0$ because

$$\bar{z}_L = \frac{Z_L}{Z_0}. \text{ Thus, } \bar{z}_L Z_0 = Z_L = 2Z_0$$

Using impedance coordinates on the Smith chart, plot $\bar{z}_L$ and rotate that point towards the generator on the circle of constant reflection coefficient until $\bar{r} = 1$ circle is intersected. Let this point be $\bar{z}_p$. The difference between $\bar{z}_p$ and $\bar{z}_L$ gives D_1 in wavelengths.

To find D_2, plot the imaginary part of $\bar{z}_p$ multiplied by (-1). Move from the point where $\bar{r} = \infty$ (the point of infinite normalized resistance on the open circuited stub) to the imaginary part of $\bar{z}_p$ multiplied by (-1). The difference between these two points gives the distance D_2 in wavelengths. Hence, from the Smith chart it can be seen that:

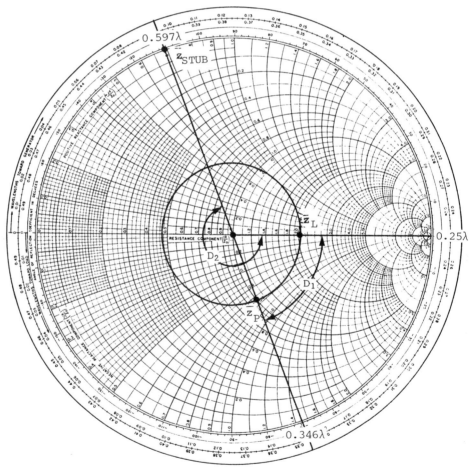

$\bar{z}_L = (2 + j0)$ and $\bar{z}_{STUB} = (+j0.7)$ and $\bar{z}_p = (1 - j0.7)$. Therefore,

$D_1 = (0.346\lambda - 0.25\lambda) = 0.096\lambda$ and
$D_2 = (0.597\lambda - 0.25\lambda) = 0.347\lambda$

Conclusion: The stub should be 0.096λ away from the load and 0.347λ away from the point of open circuit to the connection point on the line.

• **PROBLEM** 19-50

Consider the terminated line with two short-circuited stubs portrayed in Fig. 1. The position at which the stubs connect to the line is fixed, as shown, but the stub lengths, d_1 and d_2, are adjustable. This kind of arrangement is called a double-stub tuner. The load $Z_L = 50 + j100$ ohms. The line and stubs have a characteristic impedance $Z_0 = R_0 = 100$ ohms. Find the shortest values of d_1 and d_2 such that there is no reflected wave at A.

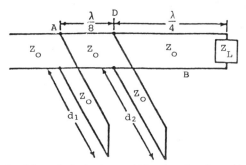

Fig. 1. Double-stub tuner with short-circuited stubs

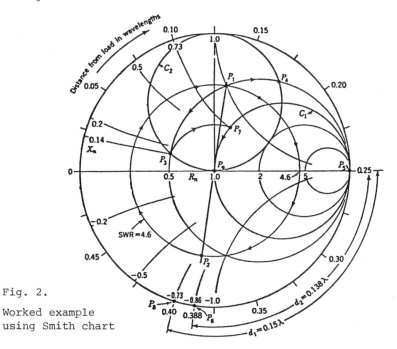

Fig. 2. Worked example using Smith chart

Solution: The normalized value of the load impedance is

$$\bar{z} = \frac{50 + j100}{100} = 0.5 + j1.0$$

The chart (Fig. 2) is entered at this normalized impedance as indicated by the point P_1. Constructing the circle of constant reflection coefficient magnitude, (Γ), through P_1, we note that the SWR at B (Fig. 1) is 4.6. Next, constructing the diametric line through P_1, locate P_2 halfway around the constant $|\Gamma|$ circle from P_1. Thus, the normalized load admittance is $0.4 - j0.8$. Now, moving clockwise along the constant $|\Gamma|$ circle from P_2 a distance of $\frac{1}{4}$ wavelength away from the load (toward the generator), arrive back at P_1. Thus at the point D the normalized admittance of the main line (looking toward the load) is $0.5 + j1.0$. Since the reflection at A must be zero, the admittance of the main line at A (without the stub of length d_1 connected) must fall on the circle marked C_1 (Fig 2.). Therefore, at the junction of the stub of length d_2 the admittance must fall on this circle rotated back (counterclockwise) $\frac{1}{4}$ wavelength to the position indicated by the circle marked C_2.

The admittance added by the stub of length d_2 will cause the total admittance to move from P_1 along a constant conductance line. In order to end up on the circle C_2, move either to the left, arriving at P_3, or to the right, arriving at P_4. Moving to P_3 results in shorter stubs; so make the stub of such length as to bring the total admittance to P_3. This requires a stub admittance (pure susceptance) of

$$\bar{y} = -j(1.0 - 0.14) = 0j0.86.$$

A short-circuited stub has an infinite SWR so that the admittance at points along the stub are on the circle at the periphery of the chart. At the short circuit the admittance is infinite (point P_5). Therefore, in order to present a value

$$\bar{y} = -j0.86 \text{ (point } P_6)$$

the stub length must be given by

$$d_2 = 0.388 - 0.25 = 0.138 \text{ wavelength.}$$

Next, moving along the constant $|\Gamma|$ curve from P_3 to P_7, the line admittance at A is $\bar{y} = 1.0 + j0.73$. Hence a stub admittance of $\bar{y} = 0j0.73$ is required in order to make the total normalized admittance at A equal to $1.0 + j0$, and therefore the actual impedance at A equal to $100 + j0$ ohms. A value

$$\bar{y} = -j0.73$$

falls at point P_8. Therefore the length of the stub is given by

$$d_1 = 0.40 - 0.25 = 0.15 \text{ wavelength.}$$

Connecting this stub brings the total admittance (or impedance) to the center of the chart (point P_9).

To summarize, the required stub lengths are

$$d_1 = 0.15 \text{ wavelength}$$

$$d_2 = 0.138 \text{ wavelength}.$$

If the movement had been to P_4 instead of to P_3, one would have ended up with longer stubs, namely,

$$d_1 = 0.443 \text{ wavelength}$$

$$d_2 = 0.364 \text{ wavelength}.$$

● **PROBLEM** 19-51

The load impedance $Z_L = 50(1 + j)$ on a 50-ohm line is to be matched by a double-stub tuner of $\frac{3}{8}\lambda$ spacing. What stub lengths ℓ_1 and ℓ_2 are necessary?

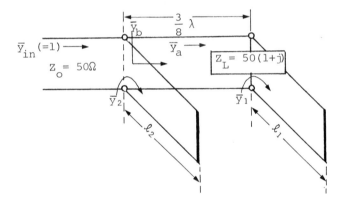

Solution: The normalized load impedance $\bar{z} = 1 + j$ corresponds to a normalized load admittance:

$$\bar{y} = 0.5(1 - j)$$

With the above in mind, rotate the $\bar{g} = 1$ circle by $\frac{3}{8}\lambda$ towards the load and then plot $\bar{y}$ on the Smith chart. There are two possible solutions for $\bar{y}_a$; each corresponds to the intersection of the rotated $\bar{g} = 1$ circle and the circle of constant $\bar{g} = 0.5$ (see Fig. 1a)

$$\bar{y}_{a_1} = 0.5 - j0.14$$

or $$\bar{y}_{a_2} = 0.5 - j1.85$$

1023

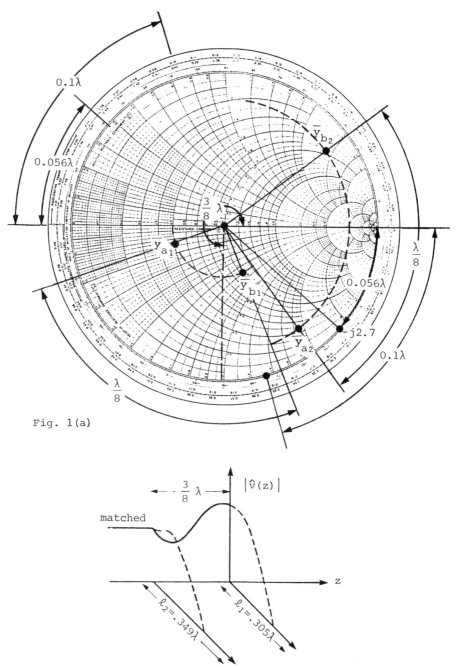

Fig. 1(a)

Fig. 1(b)
The voltage standing wave pattern.

Then find $\bar{y}_1$ by solving for the imaginary part of the equation $\bar{y}_a = \bar{y}_1 + \bar{y}_L$:

$$\bar{y}_1 = j\,\text{Im}(\bar{y}_a - \bar{y}_L) = \begin{cases} 0.36j \rightarrow \ell_1 = 0.305\lambda \\ \text{or} \\ -1.35j \rightarrow \ell_1 = 0.1\lambda \end{cases}$$

Note: Where large bandwidth is needed, it is better to choose the result which provides the shortest stub length.

By rotating the $\bar{y}_a$ solutions by $\frac{3}{8}\lambda$ back to the generator (270° clockwise, which is equivalent to 90° counterclockwise), their intersection with the $\bar{g} = 1$ circle gives the solutions for $\bar{y}_b$ as

$$\bar{y}_{b_1} = 1.0 - 0.72j$$
$$\bar{y}_{b_2} = 1.0 + 2.7j$$

This requires $\bar{y}_2$ to be

$$\bar{y}_2 = -j\ \text{Im}(\bar{y}_b) = \begin{cases} 0.72j \rightarrow \ell_2 = 0.349\lambda \\ \text{or} \\ -2.7j \rightarrow \ell_2 = 0.056\lambda \end{cases}$$

So that $\quad\bar{y}_{in} = (\bar{y}_2 + \bar{y}_b) = 1.$

The voltage standing wave pattern along the line and stubs is shown in Figure 1b. Note the continuity of voltage at the junctions. The actual stub lengths can be those listed plus any integer multiple of $\lambda/2$.

• **PROBLEM 19-52**

Given that

$$Y_L = (8 + j8) \times 10^{-3}\ \mho$$
$$Y_0 = 20 \times 10^{-3}\ \mho = \frac{1}{Z_0}$$

Use one short-circuited stub in parallel with the load and one short-circuited stub in parallel with the line at $d = 0.22\lambda$; match the line and find the length of the required stubs.

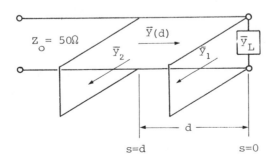

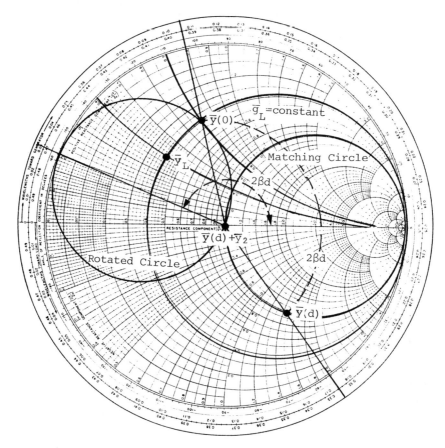

Solution: With reference to the given figure:

1. Draw the rotated $\bar{g} = 1$ circle (rotate 0.22λ counterclockwise).

2. Normalized $\bar{y}_L$ with respect to Y_0 and locate $\bar{y}_L = 0.4 + j0.4$.

3. Follow the $\bar{g}_L = 0.4$ curve to the point $\bar{y}(0) = 0.4 + j0.7$, where it intersects the rotated circle.

4. $\bar{b}_1$ will be the difference between the final and initial susceptances. Calculate $\bar{b}_1 = 0.7 - 0.4 = 0.3$.

5. From $\bar{y}(0) = 0.4 + j0.7$, follow the $|\Gamma_0|$ = constant circle a distance $2\beta d = 0.22\lambda$ "toward the generator" to the point $\bar{y}(d)$ on the $\bar{g} = 1$ matching circle.

6. Read $\bar{y}(d) = 1 - j1.45$.

7. The required normalized susceptance of the matching stub at $s = d$ is 1.45, the negative of the imaginary part of $\bar{y}(d)$.

The corresponding short-circuited stub lengths are found via the same process used for single stub matching.

For $\bar{b}_1 = 0.3$: $\ell_1 = 0.297\lambda$

For $\bar{b}_2 = 1.45$: $\ell_2 = 0.404\lambda$

Note:

There are two points of intersection. The entire procedure is valid for either point of intersection, but it leads to different matching-stub lengths.

• **PROBLEM** 19-53

For maximum power transfer from a source to a load, the load impedance must be conjugately matched to the internal impedance of the source (i.e. the impedances form a complex conjugate pair). It is desired to connect a load impedance Z_L = 50 + j55 Ω to a high frequency source with source impedance Z_S = 50 - j55 Ω operating at 150 MHz. If the load must be no less than 2.5 meters away from the source, find the length and characteristic impedance of the section of air dielectric transmission line that must be used to connect the load to the generator and, at the same time, present at the generator terminals an impedance equal to the load impedance.

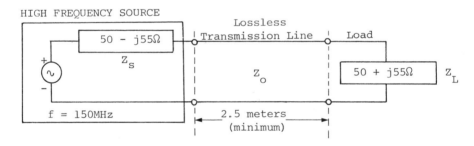

Solution: To present at the generator terminals an input impedance exactly equal to the imaginary load impedance, a half-wavelength transformer (i.e., a line section which is integer multiple of half-wavelength, must be employed).

For a lossless transmission line, the normalized input line impedance is given by

$$\bar{z}_{in} = \frac{Z_{in}}{Z_0} = \frac{\bar{z}_L + j\tan\beta\ell}{1 + j\bar{z}_L \tan\beta\ell}$$

where β = phase constant = $2\pi/\lambda$

ℓ = length of line in terms of wavelengths.

When the line wavelength is an integer multiple of $\frac{\lambda}{2}$ (i.e.,

$\ell = \frac{n\lambda}{2}$), the normalized input impedance is:

$$\bar{z}_{in} = \frac{\bar{z}_L + j\tan \frac{2\pi}{\lambda} \frac{n\lambda}{2}}{1 + j\bar{z}_L \tan \frac{2\pi}{\lambda} \frac{n\lambda}{2}}$$

$$= \frac{\bar{z}_L + j\tan n\pi}{1 + j\bar{z}_L \tan n\pi}$$

$$= \bar{z}_L \quad (\text{since } \tan n\pi = 0 \text{ for all values of } n)$$

Illustrating the fact that the impedance pattern is repeated after each half-wavelength along the line. For an air dielectric line, the wavelength at the line is equal to the free space value which is given by

$$\frac{c}{f} = \lambda,$$

where c = speed of light in vacuum = 3×10^8 m/sec

f = frequency in Hertz

λ = wavelength in meters

Thus,
$$\lambda = \frac{3 \times 10^8 \text{ m/s}}{150 \times 10^6 \text{ Hz}} = 2 \text{ meters}.$$

(so, $\lambda/2$ = 1 meter).

The shortest suitable length for a transmission line to connect the load to the source would be 3 meters or $\frac{3\lambda}{2}$.

Note: A section $\frac{\lambda}{2}$ or $\frac{2\lambda}{2}$ meters long would be shorter than the 2.5 meter constraint.

The peak values of current and voltage along the transmission line will be the lowest if the characteristic impedance of the lossless line is equal to the magnitude of the complex load impedance. In this case,

$$Z_0 = \sqrt{(50)^2 + (55)^2}$$

$$= 74.33 \text{ ohms}$$

$$\cong 75 \text{ ohms}$$

Note: At frequencies other than 150MHz, the transformer section of line will not be precisely an integral number of half-wavelength long, and so, its input impedance will not be identically equal to the load impedance.

INDEX

Numbers on this page refer to **PROBLEM NUMBERS**, not page numbers

Absolutely convergent, 4-9, 6-21
Absolutely integrable, 4-7, 4-10
Accumulate-tilt(Δ):
   due to initially flat top of a pulse, 11-22
Adaptive delta modulation, 12-20, 12-21
Adder-multiplier circuit, 9-8
Addition formula, 5-13, 5-23
Algebraic coding, 12-20
Amplitude modulated receiver, 9-20
Amplitude modulated suppressed carrier, 11-11
Amplitude modulation:
   double-sideband, 14-3, 14-7
   quadrature, 12-12
   single-sideband, 11-14, 14-3
Amplitude spectrum, 4-21
Analog signal, 12-8, 12-9
Analog to digital converter, 11-27, 12-9, 12-17
Analytical expression, 2-1, 2-2
Angle modulated waveform, 14-10
Angle modulation, 10-1
   two-tone 10-1

Angle of:
   lag, 18-19
   reflection, 18-19
Angular frequency offset, 9-23
Antenna array, 18-16, 19-39
   log-periodic, 18-16
Antenna gain, 18-11
Array:
   broad-side, 18-21
   end-fire, 18-21, 18-22
   horizontally stacked dipole, 18-21
   linear, 18-20
   uniform, 18-20
Array factor, 18-21, 18-22
Asynchronous transmission, 16-19
Attenuation, 12-1, 19-10, 19-19
   constant, 19-8
   loss, 15-35
   property, 4-8, 5-7
Audio-frequency amplifier, 14-5
Autocorrelation, 15-19
Autocorrelation function, 6-16, 6-30 to 6-34, 8-58, 13-2, 13-3, 13-13, 13-14, 13-20, 14-8, 16-1
   of a random process, 13-13
Average value of:

Numbers on this page refer to **PROBLEM NUMBERS**, not page numbers

a periodic waveform, 3-12
a random variable, 8-53

Balanced modulators, 9-13, 9-14, 12-22
Bandlimited signal, 11-19
Bandpass circuit, 15-8
Bandpass response, 15-8
Bandpass signal, 11-2
Bandwidth, 12-13, 15-3, 15-7, 15-10, 15-22
   baseband, 12-13
   three-dB, 13-24
Baseband signal, 9-3, 9-18, 11-24
   binary encoded, 16-3, 16-4
Baye's rule, 8-11, 8-12
Bessel coefficient, 10-16
Bessel's equation, 10-34, 10-36
   of order one, 10-35
   of order p, 10-37
   of order zero, 10-34
Bits:
   parity check, 17-20, 17-21, 17-26
   redundant, 17-20
British standard interface, 16-20
Butterworth filter, 13-21
   3rd order, 13-21

Carson's rule, 10-15, 10-18, 10-30
Cascade amplifier, 15-30
   tuned 15-7
Cascade filter, 13-20
Cauchy density function, 8-54
Central limit theorem, 8-49

Chain rule for integration, 6-31, 6-34, 8-36
Channel:
   binary. 17-19
   binary symmetric, 17-15, 17-18
   phase shift, 15-9
Channel capacity, 17-3
Channel characteristics:
   bandwidth, 12-8, 12-9, 12-10, 12-11, 12-15
   time constant, 11-22
Channel multiplexing, 11-21
Channel transmission, 11-22
Characteristic equation, 2-10, 2-21
Characteristic functions, 3-13
Characteristic impedance, 19-5, 19-8, 19-9, 19-18
Chopper-type modulator, 9-6
Clipping circuit, 11-26
Coaxial cable, 19-11
Coherent system, 16-14
   in DPSK, 12-22
Commutator, 11-7, 12-15, 14-24
Companding:
   logarithmic, 12-3
   Mu-law, 12-3
Comparator, 11-23, 11-24
Complementary error function, 12-4, 16-5
Complex Fourier coefficient, 6-12
Complex Fourier series, 3-17
Compressor, 12-3
   concept of:
   an orthogonal system, 3-7
   convergence, 6-20
   mean squared deviation, 3-20
Conditional probability, 8-8 to 8-11
Continuous random variable, 8-13, 8-20, 8-25, 8-29
Continuous spectrum, 4-1
Convergence of the integral, 4-1, 4-10

Numbers on this page refer to **PROBLEM NUMBERS**, not page numbers

Convergence properties, 5-4
Converter, 14-10
Convolution, 4-16, 4-25
    finite, 2-13
    graphical representation, 11-5, 11-14
    operation, 11-13
    relationship, 13-20, 13-21
    theorem, 5-47 to 5-50, 5-52
Correlation, 13-10
    coefficient, 8-50, 8-59
    function, 13-4
Correlator detector, 16-14
Cosine series, 13-15, 13-16
Counter, 11-27
Cramer's rule, 7-2
C.R. Cahn's sampling theorem, 11-10
Crosstalk:
    ratio, 11-22
    signal, 11-22
Cumulative distribution function, 8-14, 8-21, 8-24, 8-26, 8-27, 8-35
Cumulative normal distribution 8-27, 8-33
Current division principle, 1-2, 1-3, 1-11, 7-4, 7-5, 7-33

D.C. steady state:
    circuit, 1-2
    condition, 1-3
Decoder, 12-15
Decoding:
    minimum distance, 17-29
Decommutator, 11-7, 12-15
Deemphasis, 14-17
    filter, 14-19
Definition of:
    a limit, 5-2
    exponential order, 5-3
    Fourier transform, 4-2

Delay, 19-18
Delay distortion, 15-9
Delay module, 12-23
Delay T block, 13-21
Delta function, 4-16, 11-18
    property, 7-3
Delta modulation system, 14-25
Delta modulator, 12-16, 12-17, 12-18, 12-19
    adaptive, 12-20, 12-21
Demodulation scheme, 4-26
Demodulator, 14-8
    synchronous, 12-23
Derivative property, 4-8
    of Laplace transform, 5-7, 5-33
Design factor, 18-16
Detector circuit, 16-12
Difference amplifier, 12-18, 12-19
Differential equation, 1-9, 1-18 to 1-20, 7-21
Dipole antenna, 18-3, 18-4, 18-11, 18-12, 19-41
    half-wave, 18-3 to 18-5, 18-16, 18-23
    horizontal, 18-19
    short, 18-8, 18-9
    1/10 wavelength, 18-14
Dirac delta function, 4-11
Directivity, 18-10, 18-15
Dirichlet conditions, 6-1
Discrete random variable, 8-29
Discrete spectrum of Fourier coefficients, 3-1
Distortion:
    in amplitude modulation receiver, 9-20
    in pulse code modulation system, 12-9, 12-10
    in pulse modulation system, 11-1, 11-2, 11-4, 11-9, 11-22
    in single-sideband modulation, 9-27
Distortionless transmission,

Numbers on this page refer to **PROBLEM NUMBERS**, not page numbers

19-9
Distributed line parameters,
    19-8, 19-9, 19-17, 19-18
Double conversion, 9-18, 9-19
Double-stub tuner, 19-51
DPSK receiver, 12-23
Dynamic resistance, 13-29

Effective aperture, 18-8 to
    18-11, 18-25
Effective impedance, 19-25
Encoder, 12-15
    binary, 12-25
Encoding process, 17-20
Energy:
    dissipated in resistor, 1-5
    stored in inductors, 1-5
Energy spectral density,
    6-32, 6-34
Entropy, 17-4, 17-5, 17-9,
    17-27, 17-28
    joint, 17-10
    conditional, 17-10
Envelope demodulator, 14-10
Envelope detection, 9-27
Envelope detector, 9-22,
    9-25, 14-5
Envelope distortion, 16-22
Envelope modulated signal,
    9-27
Equivalent noise bandwidth,
    7-15, 13-25, 15-26
Error vector, 17-30
Euler's:
    equation, 3-11
    formula, 5-4, 18-20, 18-22
    identity, 4-4, 10-1, 10-3
Existence theorem, 5-1, 5-2
Expander, 12-3
Expansion theorem, 5-37
Expected value, 8-51
Exponential Fourier series,
    3-11, 3-12, 6-3 to 6-5,
    11-2, 11-15

Exponential tilt, 11-22
Even functions, 3-6

Far-zone approximation, 18-21
Federal communications, 10-8
    commission regulations,
    10-9
Figure of merit, 15-3
Filter:
    baseband, 14-4, 14-11
    Gaussian, 13-23
    high-pass, 11-22
    ideal, 13-17
    low-pass, 6-13, 7-13, 9-20,
        11-2, 11-11, 11-18, 12-10,
        13-18, 13-19, 14-8, 15-19
    non-recursive digital,
        13-22
    rectangular bandpass,
        4-17, 13-19
First order differential, 2-12
    equation, 2-21
FM discriminator, 14-11
FM receiver, 14-12, 14-13
Focussing effect, 18-23
Fourier:
    coefficients, 3-4, 3-6, 6-3,
        6-4, 6-6, 11-10
    cosine transform, 4-12
    expansion, 6-10
    integral formula, 4-10
    inversion integral, 4-9
    series, 3-1, 3-2, 3-5
    series expansion, 3-10
    sine series, 3-2, 3-3, 3-13,
        3-14
    sine transform, 4-12
    spectrum, 4-1
Frequency:
    convolution property, 11-15
    domain, 4-17
    domain integration, 4-19
    distribution, 4-23
    offset in FSK, 16-13

Numbers on this page refer to **PROBLEM NUMBERS**, not page numbers

    multiplier, 10-13
    response, 4-23, 15-8
    selective network, 10-33
    shifting property, 4-26, 6-17, 6-18, 13-13
    spectrum, 9-5
Frequency deviation, 10-2, 10-6 to 10-10, 10-12, 10-22, 10-23, 10-25, 14-14, 14-15
Frequency-shift keying(FSK), 12-25
Frequency modulated suppressed carrier, 10-4
Frobenius method, 10-35 to 10-37
Full-duplex, 16-15

Gain-bandwidth product, 15-5
Gamma function property, 10-37
Gated demodulator, 9-12
Gate function, 4-14, 10-19, 11-15
Gaussian:
    density, 8-53
    distribution, 15-15, 15-28
    probability density function, 8-27
    probability function, 4-2
    random variable, 8-40
    signal, 14-12
General uniform convergence theorem, 3-19
Guard band, 11-4, 11-21, 11-22, 12-10
Guard time, 11-22

Half-duplex, 16-15

Half-wave rectified sine function, 5-40
Half-wave symmetry, 3-27, 3-30
Half-wavelength transformer, 19-23, 19-53
Hamming weight distance, 17-29
Handshaking, 16-20
Harmonic oscillator equation, 1-9
Heisenberg's uncertainty principle, 4-2
Heterodyning, 10-10, 10-12
High-pass RC circuit, 6-28
Hilbert transform, 15-9
Holding circuit, 11-19
Homogenous:
    equation, 1-20
    differential equation, 2-12, 2-21
    solution, 1-9
Horn antenna, 19-47
Hybrid parameters, 7-5

IF amplifier, 14-5, 15-7, 15-8
IF bandwidth, 14-12
IF frequency, 14-5, 14-6
Image channel rejection, 14-5
Image frequency, 9-16 to 9-19, 14-6
    due to response rejection, 9-16
Image position, 19-7
Imperfect synchronization, 16-11
Impulse:
    function, 2-3, 2-22
    response, 2-13, 16-10
    sampler, 11-12
Incremental noise figure and temperature, 15-26
Indicial equation, 10-35
Inductive diaphragm, 19-47

Numbers on this page refer to **PROBLEM NUMBERS**, not page numbers

Infinite trigonometric series, 3-1
Initial value theorem, 5-45, 5-46
Inner product space, 3-31
Integrate and dump receiver, 16-3, 16-4
Integrating factor, 2-10
Integration by parts, 3-2, 3-6, 3-7
Integrator, 12-18
Intersymbol interference, 12-8, 12-10, 14-27, 17-12
Intrinsic impedance, 18-8
Inverse Fourier:
   integral, 4-2
   transform, 4-1
Ionospheric fading conditions, 9-14
Isotropic:
   antenna, 18-21, 18-27
   radiator, 18-15
Iteration, 4-8

Joint density function, 8-34, 8-35

Kirchhoff's:
   current law, 1-3, 1-16, 7-27, 7-28
   voltage law, 7-20, 7-22, 7-23, 7-31

LC tuned circuit, 9-22
Lead angle, 18-19
Legendre:
   functions, 3-32
   polynomials, 3-32
L'Hospital's rule, 3-27, 5-2, 5-5, 5-10 to 5-12
Limiter-discriminator, 14-11
Limiting process, 4-1
Linear delta modulation, 12-20
Linearity property, 5-7
Linear time invariant system, 6-7, 13-15
Linear transformation, 3-31
Local oscillator, 14-6, 14-8
Loss factor, 15-32
Loss tangent, 19-11
Low-pass RC circuit, 6-27

Main processor, 16-21
Marconi antenna, 18-6
Marginal densities, 8-29, 8-33
Marginal distribution, 8-47
Matched filter, 12-4, 16-7, 16-10
Matched line, 19-26
Matrix, 17-18, 17-21
   channel, 17-17
   channel transition, 17-14
   transformation, 3-31
Maxima:
   principal, 18-20
   secondary, 18-20
Maximum-likelyhood decision, 17-17
Mean convergence, 6-20
Mean-square:
   deviation, 6-20
   noise voltage, 13-29
   norm, 3-20
   value, 6-26, 15-13, 15-14, 15-19, 16-1
Method of undetermined coefficients, 1-9
Miller capacitance, 15-5, 15-7
Mixing, 13-7
Mixing involving noise, 13-8

Numbers on this page refer to **PROBLEM NUMBERS**, not page numbers

Modulation index, 9-1, 10-6, 10-7, 10-14, 10-23, 14-7, 16-12
Modulator:
   PSK, 12-13
   single-sideband, 14-1
Moment, 13-27
M-stage cascade amplifier, 15-32
Multiplexing:
   PCM, 12-15
   techniques, 16-17
   time-division, 11-7, 11-20, 11-22, 12-13, 12-14
Multiplexer of data transmission, 16-17, 16-18
Multiplier, 9-13, 9-17, 9-21, 10-10, 10-13
Mutually exclusive events, 8-3, 8-7
Mutually orthogonal functions, 3-33

Narrowband:
   frequency modulator, 10-32
   receiver, 14-15
Natural logarithm, 5-23
Natural response, 2-9
Noise:
   equivalent bandwidth, 13-24, 13-25
   factor, 15-27
   Gaussian, 13-5, 13-6
   in FM-AM, 14-9
   in PM, 14-19
   level, 15-32
   root mean square, 15-17, 15-18, 15-24
   white, 12-2, 12-4, 14-7, 16-14
Noise figure, 15-23 to 15-27, 15-33, 15-34
Noise temperature, 15-29
Nonhomogeneous equation, 2-10
Non-symmetric continuous function, 3-13
Normalized impedance, 19-16, 19-19, 19-35
Normalized load, 19-31
Normalized power, 6-9, 6-10, 6-13, 6-15, 6-24 to 6-26, 6-29, 13-1, 13-2, 13-5, 13-6, 13-9, 13-10, 14-1, 14-7, 14-13, 14-16, 14-18, 15-14
Nyquist interval, 11-6
Nyquist rate, 11-4, 11-18, 11-20, 14-23, 17-13

Odd function, 3-3
Odd symmetry, 3-30
Ohm's law, 1-21
One-port network, 7-17
On-off keying(OOK), 12-25
Open stub, 19-49
Optimum code, 17-22
Optimum filter, 16-5, 16-7, 16-9
Optimum functions, 3-33
Optimum length, 18-6
Optimum threshold, 12-4
Orthogonal:
   functions, 3-33
   set, 3-32
   signals, 17-13
   system, 3-4, 3-13
   transformation, 3-31
Orthogonality:
   relations, 3-21
   properties of sine and cosine, 3-1

Parallel data transmission,

Numbers on this page refer to **PROBLEM NUMBERS**, not page numbers

16-16
Parasitic beam antenna, 18-12
Parseval's:
   equality, 6-20
   theorem, 4-9, 6-7, 6-14, 6-21 to 6-26, 6-34
Partition noise, 13-28
Partial fractions:
   expression, 7-16
   technique, 5-25, 5-26, 5-27, 5-30
Particular solution, 1-9, 1-20
Passband, 18-16
Percentage modulation, 9-3, 9-4, 10-8, 10-9
Periodic:
   function, 3-5, 6-6
   waveform, 3-11, 6-14, 6-31
Periodicity:
   condition, 3-5
   factor, 3-10
   properties, 3-9
Phase:
   constant, 19-8, 19-11
   offset, 9-23
   spectrum, 4-29
   velocity, 19-11, 19-18, 19-39
Phase angle measurement, 10-31
Phase deviation, 10-2, 10-22
   instantaneous, 10-3, 10-11, 10-20
Phase shift, 2-13
Phase shifter, 11-14
Phase shift keying system, 12-22, 12-24, 16-11
   differential, 12-22, 12-23
   four phase, 12-24
   eight phase, 12-24
   M-ary, 16-14
Piecewise:
   continuous, 3-5, 4-10
   differentiability, 3-6, 3-13
   function, 3-5
Point of discontinuity, 3-4, 3-5, 3-9
Pole zero diagram, 7-6, 7-7
Power spectral density, 6-12, 6-13, 6-16, 6-29, 6-31 to 6-33, 9-25, 10-24, 13-1 to 13-4, 13-9, 13-10, 13-23, 14-1 to 14-4, 14-8, 14-12, 14-17, 15-21, 15-24 to 15-26, 16-1, 16-7, 16-9, 17-18
   due to white noise, 13-11, 13-15
   due to wide-sense stationary noise process, 13-14
   of preemphasis circuit, 14-16
   of sampled quantization error, 14-20, 14-21
   of thermal noise voltage, 15-12
Power transmission, 10-17
Poynting vector, 18-8, 18-9
   average, 18-25
Predetection noise, 15-19
Preemphasis, 14-16, 14-17
   filter, 14-17, 14-18
Preemphasis improvement(R), 14-18
Principle of inclusion and exclusion, 8-6
Probability:
   mass function, 8-21
   posteriori, 17-5
Probability density function, 8-15, 8-19, 8-24, 8-25, 8-27, 8-29, 12-2
   of coefficients, 13-6
   of noise, 13-5
Probability of error, 12-4, 12-6, 16-2, 16-4, 17-13, 17-17
Product demodulator, 14-8
Product modulator, 11-15, 11-18
Product vector, 17-30
Propagation constant, 19-9, 19-11, 19-15
Properties of expectation, 8-50

Numbers on this page refer to **PROBLEM NUMBERS**, not page numbers

Pulse amplitude modulation (PAM), 11-12, 11-21, 11-23, 11-24, 11-26
Pulse code modulation(PCM) system, 12-4, 12-7, 12-9, 12-11, 12-25
   binary, 12-4, 12-8, 12-10
   delta, 12-17, 14-27
   time-division multiplexing, 12-13, 12-14, 14-24
Pulse duration modulation (PDM), 11-23, 11-25, 11-26
Pulse generator, 12-18
Pulse time modulation(PTM), 11-23, 11-24

Quadrature multiplexing, 9-9
Quality factor, 9-19, 9-22, 18-17
Quantization error, 12-5, 14-21
   mean-square, 12-5, 12-6, 12-9
   sampled waveform, 14-20
Quantization level, 12-2, 12-3, 12-5, 12-7, 12-8, 12-13, 12-14, 14-21, 14-23
Quantization noise, 12-9, 14-26
Quantizer, 12-15, 12-18
Quantum mechanical expression, 15-21
Quarter-wavelength transformer, 19-38 to 19-41

Radiation:
   intensity, 18-15
   resistance, 18-8, 18-9, 18-13, 18-14

Radiation pattern, 18-19, 18-20
   in free space, 18-19
Radio frequency amplifier, 15-5
Raised cosine shaping, 12-24
Ramp:
   function, 2-1, 2-12
   generator, 11-27
   synchronized, 11-26
   voltage, 16-7
   waveform, 11-23, 11-24, 11-26
Random process, 8-58
Random variable, 8-13, 8-19, 8-27, 13-1
   discrete, 17-9
   independent, 13-7
   normally distributed, 13-27
   uniform, 13-1, 13-8
Ratio:
   signal to noise, 17-3
   front to back, 18-18
Ratio detector circuit, 15-11
Rayleigh function, 10-24
Rayleigh density function, 8-17, 8-53, 8-57
Reactance:
   inductive, 15-1 to 15-3
   capacitive, 15-1 to 15-3
Reactance tube modulator, 10-28
Real valved function's properties, 3-18
Reconstitution formula, 11-8
Reflected power, 19-24
Reflected waves, 19-1
Reflection:
   coefficient, 18-7, 19-1, 19-3, 19-7, 19-13
   diagram, 19-3
   factor, 18-19
   loss, 19-24
Relative permitivity, 19-19
Repeater, 12-1
Residual pilot carrier, 9-2
Resonant frequency, 7-15, 9-22, 10-13, 14-8, 15-1,

Numbers on this page refer to **PROBLEM NUMBERS**, not page numbers

15-2
Riemann integrable and
   differentiable, 3-19
Roll-off, 12-12, 12-13
   sinusoidal, 12-24
Root mean squared bandwidth,
   6-29
Rotation, 3-31

Sampling:
   flat-topped, 11-18, 11-22,
     11-23, 14-20
   formula, 11-10
   frequency, 11-2, 11-6
   natural, 11-15, 11-16
   property, 2-3
   rate, 11-3, 11-6, 11-7,
     11-16, 11-21, 11-27, 12-12
Sampling function, 4-4, 4-23,
   11-5, 11-10, 11-20
   spectrum, 11-5
Sampling theorem, 11-1, 11-3,
   11-5, 11-16, 11-20, 11-21,
   11-27, 12-9
Sawtooth waveform, 3-28,
   3-29, 5-44
Schwary inequality, 16-5
Second moment, 8-53, 15-4,
   15-15
Sensitivity, 10-33
Serial data transmission,
   16-19
   asynchronous, 16-19
   synchronous, 16-19
Scale change property, 13-17
Shifted waveforms, 2-4
Shifting property, 4-8, 4-11
Shifting theorem, 5-26, 6-4
Short-circuited stub, 19-50,
   19-52
Shot noise, 13-27 to 13-29
Sideband:
   double, 9-2
   frequencies, 9-5

   lower, 9-1
   power, 9-1, 9-2
   upper, 9-1
Signal:
   energy, 16-4
   power, 6-15, 19-18
   strength, 15-26
Signal distortion, 9-26
Signal-to-noise ratio:
   in AM, 14-2, 14-3
   in DM, 14-26
   in FM, 14-11 to 14-15,
     14-19
   in PLM, 14-26
   in PSK, 14-23
Signal transmission, 11-22
Simple loop methods, 1-3
Simplex, 16-15
Single-sideband modulated
   signal, 9-25
Single-sideband suppressed
   carrier signal, 9-24
Single-stub tuner, 19-45
Single-tuned IF amplifier,
   15-6
Slope-overload, 12-16, 12-18,
   12-19, 14-27
Sinusoidal steady state:
   amplitude response, 7-14
   analysis, 7-15, 7-22, 7-23,
     7-30
   transfer function, 7-19
Slotted line, 19-20
Small loss approximation
   formula, 19-11
Small signal model, 15-8
   hybrid pi, 13-29
Smith chart, 19-19, 19-31 to
   19-33, 19-37
Snell's law, 18-7
Solid angle, 18-15
Spectral distortion, 11-9
Spectrum envelope, 6-14
Spectrum pattern, 11-2
Square chopper, 9-16
Square law demodulator:
   characteristics, 14-7
Square law diode modulator,

Numbers on this page refer to **PROBLEM NUMBERS**, not page numbers

9-25, 9-26
Square law modulator, 9-11
Square wave function, 6-3
Standing wave effect, 19-13
Static behavior, 19-3
Statistically independent,
   13-27, 13-28, 15-15, 15-23
Steady state:
   condition, 1-2, 7-21, 19-2
   current, 7-23 to 7-25
   phasor representation,
     7-22
   response, 1-7, 1-8, 2-13,
     7-13, 7-20, 7-22
   transfer function, 7-13
   voltage, 7-23, 7-24, 7-27,
     7-32
Step size, 12-2, 12-3, 12-5,
   12-6, 12-9, 12-16, 12-18
   to 12-21, 14-21, 14-26
Step-size limiting, 12-16,
   12-18
Straightline variation, 3-29
Stretched impulses, 11-20
Stretched impulse width,
   11-17
Stub, 19-36
Sturm-Liouville problem, 3-13
Substitution property, 5-7
Superheterodyne receiver,
   14-5
Superposition theorem, 7-33
Sustained oscillations, 15-10
Swinging rate, 10-6
Switch:
   discharging, 16-3
   sampling, 16-3
Symmetric functions, 3-6
Symmetric limits, 4-5, 4-14,
   4-17, 4-22, 4-25
Symmetric T-section
   attenuator, 19-10
Synchronization demodulation,
   11-14
Synchronous detector, 9-16
Synchronous transmission,
   16-19
Syndrome, 17-24, 17-30

Taylor series expansion, 7-3
Telephone transmission line,
   12-12, 12-14
Theoretical variance, 8-42
Thermal equilibrium, 15-17
Thermal noise, 13-27, 13-29,
   14-22, 15-12, 15-14
Thevenin equivalent circuit,
   2-8, 2-17, 2-18
Thevenin's theorem, 7-28,
   7-31
Threshold, 14-23, 14-24
   voltage, 16-3
Time convolution theorem,
   6-23
Time domain integration, 4-19
Time domain response, 4-20
Time scaling property, 4-6,
   6-19, 6-23
Time shifting:
   property, 6-18, 6-19
   theorem, 4-18, 7-3
Tone modulation, 15-9
Transconductance, 13-28
Transfer parameter, 4-3
Transient response, 1-7,
   1-17, 1-19, 2-24, 4-24,
   7-22
Transistor reactance
   modulator, 10-27
Transmission bandwidth,
   12-12, 12-13, 12-25
Transmitter-receiver system,
   14-24
Transverse electromagnetic
   mode, 9-11
Trigonometric Fourier series,
   3-4 to 3-6, 6-2, 11-3
Trigonometric identities, 3-1
Trigonometric polynomial,
   3-21
Trinary code, 17-28
Triode amplifier, 15-25
   nosie figure, 15-25
T-section, 19-10
Tuned radio frequency
   receiver, 9-22
Two-port network, 4-29

Numbers on this page refer to **PROBLEM NUMBERS**, not page numbers

Two-stage amplifier, 15-26
   noise figure, 15-26

pattern, 19-51
ratio, 19-20, 19-25, 19-26,
   19-30, 19-33, 19-35, 19-42

Uncertainty, 17-4
Unit impulse function, 4-23
Uniformly convergent, 3-1
Unit step function, 2-1 to
   2-4, 4-14, 5-9, 5-16 to
   5-19, 5-29, 6-18
Upper single-sideband signal,
   6-17

Waveguide, 19-47
Wave number spectrum, 4-1
Weierstrass M-test, 5-1
Weight function, 3-7, 6-2
Whitening filter, 16-5
Wide-sense stationary, 8-58,
   8-59, 14-8
Wordlength, 12-8
Word time slot, 12-8

Variance, 3-20, 8-27, 12-9
   of quantization error, 12-2
Velocity coefficient, 18-12
Velocity factor, 18-5, 18-6
Velocity of propagation, 19-7
Voice message transmission,
   11-21, 12-14
Voltage division concept, 1-2,
   1-12, 1-15, 1-17, 7-1,
   7-12, 7-19, 7-27, 7-30
Voltage standing wave:

Y-parameters, 7-4

Zero-memory source, 17-4,
   17-27, 17-28
Z-transform, 13-21

THE PROBLEM SOLVERS

 Research and Education Association has published Problem Solvers in:

ADVANCED CALCULUS
ALGEBRA & TRIGONOMETRY
AUTOMATIC CONTROL SYSTEMS / ROBOTICS
BIOLOGY
BUSINESS, MANAGEMENT, & FINANCE
CALCULUS
CHEMISTRY
COMPUTER SCIENCE
DIFFERENTIAL EQUATIONS
ECONOMICS
ELECTRICAL MACHINES
ELECTRIC CIRCUITS
ELECTROMAGNETICS
ELECTRONIC COMMUNICATIONS
ELECTRONICS
FINITE MATHEMATICS
FLUID MECHANICS/DYNAMICS

GEOMETRY:
PLANE • SOLID • ANALYTIC
HEAT TRANSFER
LINEAR ALGEBRA
MECHANICS
NUMERICAL ANALYSIS
OPERATIONS RESEARCH
OPTICS
ORGANIC CHEMISTRY
PHYSICAL CHEMISTRY
PHYSICS
PRE-CALCULUS
PSYCHOLOGY
STATISTICS
STRENGTH OF MATERIALS & MECHANICS OF SOLIDS
TECHNICAL DESIGN GRAPHICS
THERMODYNAMICS
TRANSPORT PHENOMENA
MOMENTUM • ENERGY • MASS
VECTOR ANALYSIS

HANDBOOK OF MATHEMATICAL, SCIENTIFIC, AND ENGINEERING FORMULAS, TABLES, FUNCTIONS, GRAPHS, TRANSFORMS

If you would like more information about any of these books, complete the coupon below and return it to us.

RESEARCH and EDUCATION ASSOCIATION
505 Eighth Avenue • New York, N. Y. 10018
Phone: (212)695-9487

Please send me more information about your Problem Solver Books.

Name ....
Address ....
City .... State ....